# Oversampling Delta-Sigma Data Converters

*Theory, Design, and Simulation*

Edited by

**James C. Candy**
AT&T Bell Laboratories

**Gabor C. Temes**
Oregon State University

A Selected Reprint Volume
IEEE Circuits and Systems Society, *Sponsor*

The Institute of Electrical and Electronics Engineers, Inc., New York

© 1992 by
THE INSTITUTE OF ELECTRICAL AND ELECTRONICS ENGINEERS, INC.
345 East 47th Street, New York, NY 10017-2394

**ISBN 0-87942-285-8**

**IEEE Order Number : PC0274-1**

Printed in the United States of America
10  9  8  7  6  5  4  3  2

**Library of Congress Cataloging-in-Publication Data**

Oversampling delta-sigma data converters : theory, design, and
simulation / edited by James C. Candy, Gabor C. Temes.
   p.    cm.
   "IEEE Circuits and Systems Society, sponsor."
   Reprints of articles from various IEEE publications, 1962—1990.
   "A selected reprint volume."
   Includes bibliographical references and index.
   ISBN 0—87942—285—8 :
   1. Analog-to-digital converters.   2. Digital-to-analog converters.
   3. Pulse code modulation.   I. Candy, James C.   II. Temes, Gabor, C.,
TK7887.6.09    1992
621.39′814—dc20                91-19110
                                       CIP

# Contents

**Preface**        ix

**Introduction**        xi

Oversampling Methods for A/D and D/A Conversion
J. C. Candy and G. C. Temes

**Part 1: Basic Theory and Analysis**

An Analysis of Nonlinear Behavior in Delta-Sigma Modulators        33
S. H. Ardalan and J. J. Paulos
*IEEE Transactions on Circuits and Systems*, June 1987

A Use of Limit Cycle Oscillations to Obtain Robust Analog-to-Digital Converters        44
J. C. Candy
*IEEE Transactions on Communications*, March 1974

The Structure of Quantization Noise from Sigma-Delta Modulation        52
J. C. Candy and O. J. Benjamin
*IEEE Transactions on Communications*, September 1981

Multistage Sigma-Delta Modulation        60
W. Chou, P. W. Wong, and R. M. Gray
*IEEE Transactions on Information Theory*, July 1989

Oversampled Sigma-Delta Modulation        73
R. M. Gray
*IEEE Transactions on Communication*, May 1987

Quantization Noise Spectra        81
R. M. Gray
*Transactions on Information Theory*, November 1990

Double-Loop Sigma-Delta Modulation with dc Input        106
N. He, F. Kuhlmann, and A. Buzo
*IEEE Transactions on Communication*, April 1990

A Unity Bit Coding Method by Negative Feedback        115
H. Inose and Y. Yasuda
*Proceedings of the IEEE*, November 1963

Design of Stable High Order 1-Bit Sigma-Delta Modulators        127
T. Ritoniemi, T. Karema, and H. Tenhunen
*IEEE Proceedings of the International Symposium on Circuits and Systems'90*, May 1990

Reduction of Quantizing Noise by Use of Feedback        131
H. A. Spang III and P. M. Schultheiss
*IRE Transactions on Communications Systems*, December 1962

Oversampled, Linear Predictive and Noise-Shaping Coders of Order $N > 1$        139
S. K. Tewksbury and R. W. Hallock
*IEEE Transactions on Circuits and Systems*, July 1978

## Part 2: Design, Simulation Techniques, and Architectures for Oversampling Converters

Design Methodology for ΣΔM ..... 153
B. P. Agrawal and K. Shenoi
*IEEE Transactions on Communication*, March 1983

Table-Based Simulation of Delta-Sigma Modulators ..... 163
R. J. Bishop, J. J. Paulos, M. B. Steer, and S. H. Ardalan
*IEEE Transactions on Circuits and Systems*, March 1990

Simulating and Testing Oversampled Analog-to-Digital Converters ..... 168
B. E. Boser, K.-P. Karmann, H. Martin, and B. A. Wooley
*IEEE Transactions on Computer-Aided Design*, June 1988

A Use of Double Integration in Sigma Delta Modulation ..... 174
J. C. Candy
*IEEE Transactions on Communication*, March 1985

An Oversampling Analog-to-Digital Converter Topology for High-Resolution Signal
Acquisition Systems ..... 184
L. R. Carley
*IEEE Transactions on Circuits and Systems*, January 1987

Digitally Corrected Multi-Bit ΣΔ Data Converters ..... 192
T. Cataltepe, A. R. Kramer, L. E. Larson, G. C. Temes, and R. H. Walden
*IEEE Proceedings of the International Symposium on Circuits and Systems'89*, May 1989

A Higher Order Topology for Interpolative Modulators for Oversampling A/D Converters ..... 196
K. C.-H. Chao, S. Nadeem, W. L. Lee, and C. G. Sodini
*IEEE Transactions on Circuits and Systems*, March 1990

One Bit Higher Order Sigma-Delta A/D Converters ..... 206
P. F. Ferguson Jr., A. Ganesan, and R. W. Adams
*IEEE Proceedings of the International Symposium on Circuits and Systems'90*, May 1990

Optimization of a Sigma-Delta Modulator by the Use of a Slow ADC ..... 209
A. Gosslau and A. Gottwald
*IEEE Proceedings of the International Symposium on Circuits and Systems'88*, June 1988

Circuit and Technology Considerations for MOS Delta-Sigma A/D Converters ..... 213
M. W. Hauser and R. W. Brodersen
*IEEE Proceedings of the International Symposium on Circuits and Systems'86*, May 1986

Technology Scaling and Performance Limitations in Delta-Sigma Analog-Digital
Converters ..... 219
M. W. Hauser
*IEEE Proceedings of the International Symposium on Circuits and Systems'90*, May 1990

Delta-Sigma A/Ds with Reduced Sensitivity to Op Amp Noise and Gain ..... 223
P. J. Hurst and R. A. Levinson
*IEEE Proceedings of the International Symposium on Circuits and Systems'89*, May 1989

Multibit Oversampled Σ–Δ A/D Converter with Digital Error Correction ..... 227
L. E. Larson, T. Cataltepe, and G. C. Temes
*Electronics Letters*, August 1988

An Improved Sigma-Delta Modulator Architecture ..... 229
T. C. Leslie and B. Singh
*IEEE Proceedings of the International Symposium on Circuits and Systems'90*, May 1990

A 13 Bit ISDN-Band Oversampled ADC Using Two-Stage Third Order Noise Shaping ..... 233
L. Longo and M. Copeland
*IEEE Proceedings of the Custom Integrated Circuits Conference*, January 1988

A 16-Bit Oversampling A-to-D Conversion Technology Using Triple-Integration Noise Shaping — 237
Y. Matsuya, K. Uchimura, A. Iwata, T. Kobayashi, M. Ishikawa, and T. Yoshitome
*IEEE Journal of Solid-State Circuits*, December 1987

Improved Signal-to-Noise Ratio Using Tri-Level Delta-Sigma Modulation — 245
J. J. Paulos, G. T. Brauns, M. B. Steer, and S. H. Ardalan
*IEEE Proceedings of the International Symposium on Circuits and Systems'87*, May 1987

A Second-Order High-Resolution Incremental A/D Converter with Offset and Charge Injection Compensation — 249
J. Robert and P. Deval
*IEEE Journal of Solid-State Circuits*, June 1988

Improved Double Integration Delta-Sigma Modulations for A to D and D to A Conversion — 255
Y. Shoji and T. Suzuki
*IEEE Proceedings of the International Symposium on Circuits and Systems'87*, May 1987

Oversampling A-to-D and D-to-A Converters with Multistage Noise Shaping Modulators — 259
K. Uchimura, T. Hayashi, T. Kimura, and A. Iwata
*IEEE Transactions on Acoustics, Speech, and Signal Processing*, December 1988

Architectures for High-Order Multibit $\Sigma\Delta$ Modulators — 266
R. H. Walden, T. Cataltepe, and G. C. Temes
*IEEE Proceedings of the International Symposium on Circuits and Systems'90*, May 1990

Constraints Analysis for Oversampling A-to-D Converter Structures on VLSI Implementation — 270
A. Yukawa
*IEEE Proceedings of the International Symposium on Circuits and Systems'87*, May 1987

## Part 3: Implementations and Applications of Oversampling A/D Converters

Design and Implementation of an Audio 18-bit Analog-to-Digital Converter Using Oversampling Techniques — 279
R. W. Adams
*Journal of the Audio Engineering Society*, March 1986

The Design of Sigma-Delta Modulation Analog-to-Digital Converters — 293
B. E. Boser and B. A. Wooley
*IEEE Journal of Solid-State Circuits*, December 1988

A Noise-Shaping Coder Topology for 15 + Bit Converters — 304
L. R. Carley
*IEEE Journal of Solid-State Circuits*, April 1989

A Dual-Channel Voice-Band PCM Codec Using $\Sigma\Delta$ Modulation Technique — 311
V. Friedman, D. M. Brinthaupt, D.-P. Chen, T. W. Deppa, J. P. Elward, Jr., E. M. Fields,
J. W. Scott, and T. R. Viswanathan
*IEEE Journal of Solid-State Circuits*, April 1989

MOS ADC-Filter Combination That Does Not Require Precision Analog Components — 317
M. W. Hauser, P. J. Hurst, and R. W. Brodersen
*ISSCC Digital Technical Papers*, February 1985

A Multistage Delta-Sigma Modulator without Double Integration Loop — 320
T. Hayashi, Y. Inabe, K. Uchimura, and T. Kimura
*ISSCC Digital Technical Papers*, February 1986

An Oversampled Sigma-Delta A/D Converter Circuit Using Two-Stage Fourth Order Modulator — 322
T. Karema, T. Ritoniemi, and H. Tenhunen
*IEEE Proceedings of the International Symposium on Circuits and Systems'90*, May 1990

A 12-Bit Sigma-Delta Analog-to-Digital Converter with 15-MHz Clock Rate 326
R. Koch, B. Heise, F. Eckbauer, E. Engelhardt, J. A. Fisher, and F. Parzefall
*IEEE Journal of Solid-State Circuits*, December 1986

Area-Efficient Multichannel Oversampled PCM Voice-Band Coder 333
B. H. Leung, R. Neff, P. R. Gray, and R. W. Brodersen
*IEEE Journal of Solid-State Circuits*, December 1988

An 18b Oversampling A/D Converter for Digital Audio 340
K. Matsumoto, E. Ishii, K. Yoshitate, K. Amano, and R. W. Adams
*ISSCC Digital Technical Papers*, February 1988

A 14-Bit 80-kHz Sigma-Delta A/D Converter: Modeling, Design, and Performance
Evaluation 342
S. R. Norsworthy, I. G. Post, and H. S. Fetterman
*IEEE Journal of Solid-State Circuits*, April 1989

Fully Differential CMOS Sigma-Delta Modulator for High Performance Analog-to-Digital
Conversion with 5 V Operating Voltage 353
T. Ritoniemi, T. Karema, H. Tenhunen, and M. Lindell
*IEEE Proceedings of the International Symposium on Circuits and Systems'88*, June 1988

A High-Resolution CMOS Sigma-Delta A/D Converter with 320 kHz Output Rate 359
M. Rebeschini, N. van Bavel, P. Rakers, R. Greene, J. Caldwell, and J. Haug
*IEEE Proceedings of the International Symposium on Circuits and Systems'89*, May 1989

A CMOS Slope Adaptive Delta Modulator 363
J. W. Scott, W. Lee, C. Giancario, and C. G. Sodini
*ISSCC Digital Technical Papers*, February 1986

Stereo 16-Bit Delta-Sigma A/D Converter for Digital Audio 365
D. R. Welland, B. P. Del Signore, E. J. Swanson, T. Tanaka, K. Hamashita, S. Hara, and
K. Takasuka
*Journal of the Audio Engineering Society*, June 1989

## Part 4: Digital Filters for Oversampling A/D Converters

Using Triangularly Weighted Interpolation to Get 13-Bit PCM from a Sigma-Delta
Modulator 377
J. C. Candy, Y. C. Ching, and D. S. Alexander
*IEEE Transactions on Communication*, November 1976

A Voiceband Codec with Digital Filtering 385
J. C. Candy, B. A. Wooley, and O. J. Benjamin
*IEEE Transactions on Communication*, June 1981

Decimation for Sigma Delta Modulation 400
J. C. Candy
*IEEE Transactions on Communication*, January 1986

Multirate Filter Designs Using Comb Filters 405
S. Chu and C. S. Burrus
*IEEE Transactions on Circuits and Systems*, November 1984

Interpolation and Decimation of Digital Signals—A Tutorial Review 417
R. E. Crochiere and L. R. Rabiner
*Proceedings of the IEEE*, March 1981

Wave Digital Decimation Filters in Oversample A/D Converters 449
E. Dijkstra, L. Cardoletti, O. Nys, C. Piguet, and M. Degrauwe
*IEEE Proceedings of the International Symposium on Circuits and Systems'88*, June 1988

A Design Methodology for Decimation Filters in Sigma Delta A/D Converters     453
E. Dijkstra, M. Degrauwe, J. Rijmenants, and O. Nys
*IEEE Proceedings of the International Symposium on Circuits and Systems'87*, May 1987

On the Use of Modulo Arithmetic Comb Filters in Sigma Delta Modulators     457
E. Dijkstra, O. Nys, C. Piguet, and M. Degrauwe
*IEEE Proceedings on the International Conference on Acoustics, Speech, and Signal
Processing'88*, April 1988

Nine Digital Filters for Decimation and Interpolation     461
D. J. Goodman and M. J. Carey
*IEEE Transactions on Acoustics, Speech, and Signal Processing*, April 1977

A Novel Architecture Design for VLSI Implementation of an FIR Decimation Filter     467
H. Meleis and P. Le Fur
*IEEE Proceedings of the International Conference on Acoustics, Speech, and Signal
Processing'85*, March 1985

Efficient VLSI-realizable Decimators for Sigma-Delta Analog-to-Digital Converters     471
T. Saramäki and H. Tenhunen
*IEEE Proceedings of the International Symposium on Circuits and Systems'88*, June 1988

## Part 5: Theory and Implementations of Oversampling D/A Converters

Double Interpolation for Digital-to-Analog Conversion     477
J. C. Candy and A.-N. Huynh
*IEEE Transactions on Communication*, January 1986

A 16-Bit 4th Order Noise-Shaping D/A Converter     482
L. R. Carley and J. Kenney
*IEEE Proceedings of the Custom Integrated Circuits Conference*, 1988

A CMOS Stereo 16-Bit D/A Converter for Digital Audio     486
P. J. A. Naus, E. C. Dijkmans, E. F. Stikvoort, A. J. McKnight, D. J. Holland, and W. Bradinal
*IEEE Journal of Solid-State Circuits*, June 1987

**Author Index**     491

**Subject Index**     493

**Editor's Biographies**     499

# Preface

THE purpose of this book is to provide a convenient introductory and reference work on oversampling analog-to-digital and digital-to-analog converters. The introduction briefly traces the development of oversampling converters through the past 20 years, and tells why the technique has become popular in the last few years. The book includes a new tutorial paper on oversampling methods, and a number of previously published key papers on the theory and basic concepts, as well as papers describing architectures and their simulation, papers describing actual implementation of circuits and their performance, together with papers concerned primarily with decimation and ones concerned with oversampling digital-to-analog conversion.

In selecting papers for the book, particular care was taken to feature the large number of options that are available to designers of oversampling converters. Designers can select from an ever-increasing collection of architectures for modulators and demodulators each with its own distinct advantages and disadvantages. They can take advantage of trade-offs between oversampling ratio, resolution, and circuits complexities, and choose from numerous designs of digital decimation filters for the analog-to-digital converters and interpolation filters for the digital-to-analog converter.

The papers included in the book cover a very wide range of topics. Those concerned with modulators are dealing with the design of high speed analog filters contained in nonlinear feedback loops. In addition to the interesting practical problems involved in designing the circuits, there also are theoretical problems including some that have not yet been solved. To make up for these deficiencies the book includes papers describing computer simulation programs that are useful for designing and evaluating the performance of the circuits. The design of efficient and economic decimating and interpolating filters poses interesting problems in digital signal processing, as also does the design of the digital demodulator. Numerous novel circuits for these applications are described in many of the papers included in the volume.

# Introduction

THE notion of using artificially high sampling rates and single-bit code words to represent analog signals has been of interest ever since delta modulation [29]* was first proposed in 1946. Since that time many variants of delta modulation have been suggested, and the idea of converting its output signal into a PCM or DPCM form sampled at the Nyquist rate had arisen [24] by the late 1960s. One attempt simply counted the bits from the delta modulator, representing a high bit as $+1$ and a low bit as $-1$, and sampled the count at the Nyquist rate. But the resolution obtained by this technique was disappointing; oversampling ratios of the order 5,000 were needed in the delta modulation to get a good reproduction of speech signals. It was Goodman who pointed out [12] that counting was not enough, more effective digital filtering was needed to prevent the high-frequency modulation noise from folding into the signal band. This surely was the birth of the use of oversampling methods for A/D and D/A conversion. Unfortunately, at that time digital filters for this use were inordinately expensive.

Some time later, a related technique was developed [5], [111] for use in a videophone application. It was called *interpolation*. The idea was to digitize the signal with a very coarse quantizer, typically 16 levels, and cause the output to bounce between levels at high speed so that its average over the Nyquist interval was an accurate representation of the sample value. The average could be generated digitally at moderate cost. These circuits were found to be reliable and very tolerant of imperfections in the analog components. It was also found that their performance was much improved by taking a triangularly weighed average [89] of the fast samples instead of an ordinary average.

The quantizers used in these interpolating converters were based on a noise-shaping technique [9] that had been proposed by Cutler in 1954. His idea was to take a measure of the quantization error in one sample and subtract it from the next input sample. He also pointed out that this process could be iterated on itself or applied to DPCM. The most popular form of this noise-shaping quantization had been given the name *delta sigma modulation* [20] by Inose and Yasuda in 1962. It is performed by a delta modulator with the input signal added to the input of the integrator instead of to its output, thus eliminating the need for integration at the receiver. Unfortunately, due to a misunderstanding, the words *delta sigma* were interchanged and taken to be *sigma delta*. Both names are still used for describing this modulator. The survival of the second one is possibly due to an unwillingness to break up the familiar term *delta modulation* for designating feedback quantization with a single-bit code. We will henceforward use the original name, *delta sigma modulation*.

* Reference numbers in this introduction refer to the bibliography provided in the first paper of this book.

By the middle 1970s, integrated circuit technology had developed to a stage where it was realistic to build digital low-pass filters not only for suppressing the high-frequency noise of the modulator but also for contributing the anti-aliasing function for the input signal. Most designers prefer to use a delta-sigma modulator rather than a delta modulator, because it is more robust; delta modulation is used only when there is good reason for differentiating the signal. At that time, however, oversampling methods were not well accepted, despite their advantages of being predominantly digital and tolerating low-precision analog circuits. The interest of circuit designers lay with more conventional methods, particularly those employing switched-capacitor techniques for the analog filters and the converter networks.

In the late 1980s, when the line widths of integrated circuits fell below 2 $\mu$m, conditions changed. The use of powerful digital signal processors to perform functions that had previously been done in the analog domain demanded high resolution in the A/D and D/A converters. At the same time, the voltage range on chips had fallen to 5 V, and as the circuit densities grew the chip environment became more noisy. It was therefore difficult to get high resolution by conventional methods. Another stimulant for the use of oversampling methods has been the wide acceptance of digital recording for high fidelity audio, calling for inexpensive but high-resolution conversion between analog and digital forms. In recent years oversampling methods have become popular, at least for lower-frequency applications.

Now that a number of oversampling converters have been designed and manufactured, we have confirmation that the technique can indeed reliably provide high resolution with high yield, without trimming. In fact, 16-bit resolution is now commonplace. An important advantage that has been observed is the possibility of designing and scheduling a new converter with nearly as much assurance as a digital circuit. It is not unusual to hear that the first samples of circuits implemented in silicon have worked and met all specifications.

As confidence has grown in oversampling methods, there has been an unexpected change in design motivation. In the early days the purpose was to simplify the analog and digital circuits as much as possible, at the cost of having a high sampling rate. Now there is a willingness to accept more complex analog circuits and considerably more complex digital circuits in order to lower the oversampling ratio. We find modulators that include fifth-order filters [47], and cascaded modulators in [10], [75] in modern oversampled A/D converters. This trend is extending the application of oversampling methods to higher signal frequencies. Whatever the application or the technology, oversampling methods provide a wide range of trade-offs involving such factors as oversampling ratio, order of the filters used in the modulator, type of modulator, number of quantization levels in the modulated signal, number of stages and order of the filters in the decimator, in addition to equivalent choices for the D/A design. It sometimes appears that a designer's task would be easier if there were fewer choices to make!

An important difference between conventional converters and oversampled ones involves testing and specifying their performance. With conventional converters there is a one-to-one correspondence between input and output sample values, and hence one can describe their accuracy by comparing the values of corresponding input and output samples. In contrast, there is no similar correspondence in oversampling converters, because they inherently include digital low-pass filters, and hence each input sample value contributes to a whole train of output samples. Consequently, it has been useful to borrow techniques from communication technology to describe the performance of oversampling converters. Thus we measure their rms noise under various conditions, the distortion they introduce into sinusoidal signals, and their frequency responses. An important task in the design of an oversampled converter is therefore the calculation of rms values of modulation noise, and its spectral density. Examples of such calculations are given in the first paper, which is intended to serve as a tutorial introduction to the subject. That paper also contains a selected bibliography of papers related to oversampling methods. Another large bibliography is included in the review paper [19].

The other papers in this book are grouped into five sections: The first contains theoretical work and discussions of basic ideas. The second section contains papers concerned with the actual design of systems, their simulation, and description of architectures. The third section contains descriptions of implementation of oversampling converters and their application. The fourth section contains papers that are concerned primarily with decimation (descriptions of decimators can also be found in Sections 1, 2, and 3). Section 5 contains papers concerned primarily with oversampling D/A converters.

# Oversampling Methods
# for A/D and D/A Conversion

James C. Candy      Gabor C. Temes

*Abstract*—This paper is a review of oversampling methods used for A/D and D/A conversion. It brings together many of the results and ideas presented by the other papers in this book, and attempts to evaluate and compare the various techniques to show when they can be useful. Several mathematical methods for determining the performance of oversampled converters are presented. Although these methods depend, for the most part, on approximations, they are useful for obtaining initial design parameters. Final designs nearly always need to be confirmed by means of simulations.

Detailed design of circuits for these converters entail a number of important choices and trade-offs. These are discussed, along with the effects of having nonideal components. To assist in making a good choice of circuit configuration, a number of architectures for digital modulators are compared. A method for designing decimation filters suitable for use with these modulators is given, along with several designs of oversampling D/A converters.

## 1. INTRODUCTION

OVERSAMPLING methods have recently become popular because they avoid many of the difficulties encountered with conventional methods for A/D and D/A conversion. Conventional converters, illustrated in Fig. 1, have attributes that make it difficult to implement their circuits in fine-line VLSI technology. Chief among these is the use of analog filters, the need for high-precision analog circuits, and their vulnerability to noise and interference. The virtue of conventional methods is their use of a relatively low sampling frequency, usually the Nyquist rate of the signal (i.e., twice the signal bandwidth).

The low-pass filter at the input to the conventional encoder in Fig. 1 attenuates high-frequency noise and out-of-band components of the signal that alias into the signal when sampled at the Nyquist rate. The properties of this filter are usually specified for each application. The A/D circuit can take a number of different forms, such as flash converters for fast operation, successive-approximation converters for moderate rates, and ramp converters for slow applications. At the decoder a filter smooths the sampled output of the D/A circuit; the amount of smoothing required is usually part of the specification of the converter. The circuits of these conventional converters require high-accuracy analog components in order to achieve high overall resolution.

Oversampling converters, illustrated in Fig. 2, can use simple and relatively high-tolerance analog components, but require fast and complex digital signal processing stages. They modulate the analog input into a simple digital code, usually single-bit words, at a frequency much higher than the Nyquist rate. We shall show that the design of the modulator can trade resolution in time for resolution in amplitude in such a way that imprecise analog circuits may be used. The

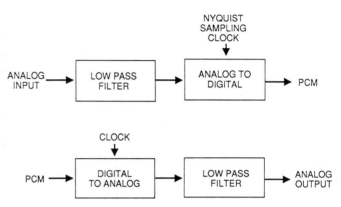

Fig. 1.  Conventional pulse code modulation, including analog filters for controlling aliasing noise and smoothing the output.

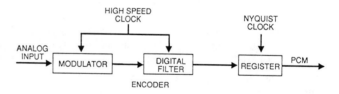

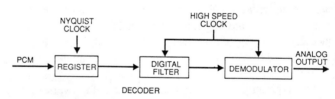

Fig. 2.  Oversampling pulse code modulation. The modulation and demodulation occurs at sufficiently high sampling rate that digital filters can provide most of the antialiasing and smoothing functions.

use of high-frequency modulation and demodulation can eliminate the need for abrupt cutoffs in the analog antialiasing filters at the input to the A/D converter, as well as in the filters that smooth the analog output of the D/A converter. Digital filters are used instead [12, 90, 102] as illustrated in Fig. 2. A digital filter smoothes the output of the digital modulator, attenuating noise, interference, and high-frequency components of the signal before they could alias into the signal band when the code is resampled at the Nyquist rate. Another digital filter interpolates the code in the decoder to a high-word-rate signal before it is demodulated to analog form.

Oversampling converters make extensive use of digital processing, taking advantage of the fact that fine-line VLSI is better suited for providing fast digital circuits than for provid-

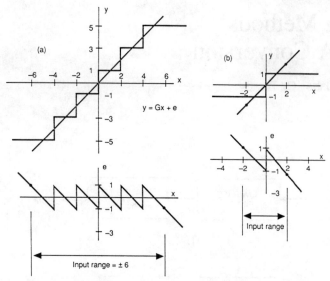

Fig. 3. (a) An example of a uniform multilevel quantization characteristic that is represented by linear gain $G$ and error $e$. (b) For two-level quantization the gain $G$ is arbitrary.

ing precise analog circuits. Because their sampling rate usually needs to be several orders of magnitude higher than the Nyquist rate, oversampling methods are best suited for relatively low frequency signals. They have found use in such applications as digital audio, digital telephony, and instrumentation. Future applications in video and radar system are imminent as faster technologies become available.

This paper is organized into five main sections. Following this introduction, the second section describes some basic properties of the quantization noise. It then introduces delta-sigma modulation as a technique for shaping the spectrum of quantization noise, moving most of the noise power to high frequencies, well outside the band of the signal, where it is removed by the digital filters. A number of other modulators are also described. The third section compares properties of the most commonly used circuit architectures for oversampling modulators and discusses their suitability for practical implementation. The fourth section discusses the design of digital filters that decimate the modulated signal, converting it from short digital words occurring at a high rate into long words at the Nyquist rate. The fifth section describes oversampling D/A converters.

## 2. DIGITAL MODULATION

### 2.1 Quantization

Quantization of amplitude and sampling in time are at the heart of all digital modulators. Periodic sampling at rates more than twice the signal bandwidth need not introduce distortion, but quantization does, and our primary objective in designing modulators is to limit this distortion. We begin our discussion with some properties of quantization that will be useful for specifying the noise of modulators. Figure 3a shows a uniform quantization that rounds off a continuous signal $x$ to odd integers in the range $\pm 5$. In this example the

level spacing $\Delta$ is 2. We will find it useful to represent the quantized signal $y$ by a linear function $Gx$ with an error $e$; that is,

$$y = Gx + e. \tag{1}$$

The gain $G$ is the slope of the straight line that passes through the center of the quantization characteristic so that, when the quantizer does not saturate (i.e., when $-6 \leq x \leq 6$), the error is bounded by $\pm \Delta / 2$. Notice that the above consideration remains applicable to a two-level (1-bit) quantizer, as illustrated in Fig. 3b, but in this case the choice of gain $G$ is arbitrary.

The error is completely defined by the input, but if the input changes randomly between samples by amounts comparable with the threshold spacing without causing saturation, then the error is largely uncorrelated from sample to sample [3] and has equal probability of lying anywhere in the range $\pm \Delta / 2$. If we further assume that the error has statistical properties that are independent of the signal, then we can represent it by a noise, and some important properties of modulators can be determined. In many cases, experiments have confirmed these properties, but here are two important instances where they may not apply: when the input is stationary, and when it changes regularly by a multiple or submultiples of $\Delta$ between sample times, as can happen in feedback circuits.

If we treat the quantization error $e$ as white noise having equal probability of lying anywhere in the range $\pm \Delta / 2$, its mean square value is given by

$$e_{\text{rms}}^2 = \frac{1}{\Delta} \int_{-\Delta/2}^{+\Delta/2} e^2 \, de = \frac{\Delta^2}{12}. \tag{2}$$

For the ensuing discussion of the spectral densities of the noise, we shall employ a one-sided representation of frequencies; that is, we assume that all the power is in the $0 \leq f < \infty$ range. When a quantized signal is sampled at frequency $f_s = 1/\tau$, all of its noise power folds into the frequency band $0 \leq f < f_s/2$. Then, if it is white, the spectral density of the sampled noise [3] is given by

$$E(f) = e_{\text{rms}} \left( \frac{2}{f_s} \right)^{1/2} = e_{\text{rms}} \sqrt{2\tau}. \tag{3}$$

Now we can examine an example of oversampling PCM. A signal in frequency band $0 \leq f < f_0$, to which a dither signal contained in the band $f_0 < f \leq f_s/2$ is added, is PCM-encoded at $f_s$. The oversampling ratio, defined as the ratio of the sampling frequency $f_s$ to the Nyquist rate $2f_0$, is given by the integer

$$\text{OSR} = \frac{f_s}{2f_0} = \frac{1}{2f_0 \tau}. \tag{4}$$

If the dither is sufficiently large to whiten and decorrelate the quantization error, the noise power that falls into the signal

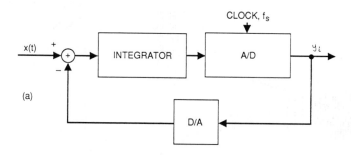

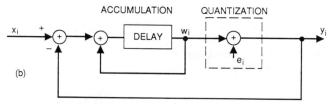

Fig. 4.  (a) A block diagram of a delta-sigma quantizer and (b) its sampled-data equivalent circuit.

band will be

$$n_0^2 = \int_0^{f_0} e^2(f)\, df = e_{rms}^2(2f_0\tau) = \frac{e_{rms}^2}{OSR}. \qquad (5)$$

Thus, we have the well-known result that oversampling reduces the in-band rms quantization noise $n_0$ by the square root of the oversampling ratio. Therefore, each doubling of the sampling frequency decreases the in-band noise by 3 dB, increasing the resolution by only half a bit.

### 2.2 Delta-Sigma Modulation

*2.2.1 A First-Order Feedback Circuit.* A more efficient oversampling quantizer is the delta-sigma modulator [20] shown in Fig. 4a. Although delta-sigma modulators usually employ two-level quantization, we start our discussion by assuming the modulator contains a multilevel uniform quantizer with unity gain $G = 1$. The input to the circuit is fed to the quantizer via an integrator, and the quantized output is fed back and subtracted from the input. This feedback forces the average value of the quantized signal to track the average input. Any difference between them accumulates in the integrator and eventually corrects itself. Figure 5 illustrates the response of the circuit to a ramp input; it shows how the quantized signal oscillates between two levels that are adjacent to the input in such a manner that its local average equals the average input [5].

*2.2.2 Modulation Noise in Busy Signals.* We analyze the modulator by means of the equivalent circuit shown in Fig. 4b. Here, the quantization is represented as an added error $e$ in accordance with (1), with the gain $G$ set to unity. Because this is a sampled-data circuit, we represent the integration by accumulation. Then it is easily seen that the output of the accumulator is

$$w_i = x_{i-1} - e_{i-1}, \qquad (6)$$

and the quantized signal is

$$y_i = x_{i-1} + (e_i - e_{i-1}). \qquad (7)$$

Thus, this circuit differentiates the quantization error, making the modulation noise the first difference of the quantization error while leaving the signal unchanged, except for a delay.

To calculate the effective resolution of the delta-sigma modulator, we now assume that the input signal is sufficiently busy so that the error can be treated as white noise which is uncorrelated with the signal. The spectral density of the modulation noise

$$n_i = e_i - e_{i-1} \qquad (8)$$

may then be expressed as

$$N(f) = E(f)\,|1 - e^{-j\omega\tau}| = 2e_{rms}\sqrt{2\tau}\,\sin\!\left(\frac{\omega\tau}{2}\right). \qquad (9)$$

Figure 6 compares this spectral density with that of quantization noise $E(f)$ when the oversampling ratio is 16. Clearly, feedback around the quantizer reduces the noise at low frequencies, but increases it at high frequencies.

The noise power in the signal band is

$$n_o^2 = \int_0^{f_0} |N(f)|^2\, df \approx e_{rms}^2 \frac{\pi^2}{3}(2f_0\tau)^3, \qquad f_s^2 \gg f_0^2, \qquad (10)$$

and its rms value is

$$n_0 = e_{rms}\frac{\pi}{\sqrt{3}}(2f_0\tau)^{3/2}. \qquad (11)$$

Each doubling of the oversampling ratio thus reduces this noise by 9 dB and provides 1.5 bits of extra resolution. This improvement in the resolution requires that the modulated signal be decimated to the Nyquist rate with a sharply selective digital filter. Otherwise, the high-frequency components of the noise will spoil the resolution when it is sampled at the Nyquist rate. Some early oversampling converters employed primitive decimation. One merely averaged the output samples of the modulator over each Nyquist interval $(2f_0)^{-1}$ to get a PCM signal. References [5], [15], and [89] show that the rms value of the noise in this PCM can be expressed as $\sqrt{2}\,e_{rms}(2f_0\tau)$. They also show that taking a triangularly weighted sum over the Nyquist interval gives an rms noise $4e_{rms}(2f_0\tau)^{3/2}$. An optimization of these techniques for attenuating the high-frequency noise is given in reference [25]. These decimators permit more noise to alias into the signal band than do the ones that use filters with impulse responses longer than a Nyquist interval, but the techniques may be useful because their circuit implementations can be very simple.

The above derivation of the average properties of modulation noise depends on representing the quantization error as white uncorrelated noise. But the analysis in references [14-17], which does not depend on this assumption, shows

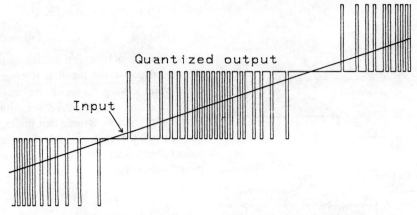

Fig. 5. The response of a multilevel delta-sigma quantizer to a ramp input. A two-level response is obtained by curtailing input amplitudes to a range of values that lie between two quantization levels.

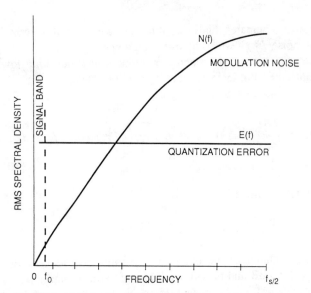

Fig. 6. The spectral density of the noise $N(f)$ from delta-sigma quantization compared with that of ordinary quantization $E(f)$.

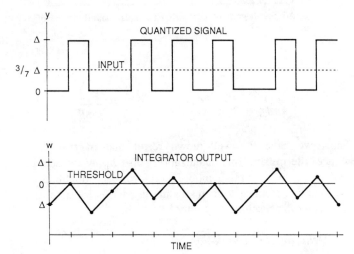

Fig. 7. Waveforms in a delta-sigma circuit for a constant input situated $3\Delta/7$ above a quantization level.

that Eq. (11) may apply even when the error is not white. Moreover, they also show that the error is rarely truly white.

**2.2.3 Pattern Noise from Delta-Sigma Modulation with DC Inputs.** When the input to the modulator is a dc signal, the quantized signal bounces between two levels, keeping its mean equal to the input. Figure 7 demonstrates that the oscillation may be repetitive; it returns to its staring condition after seven clock periods. The frequency of repetition depends on the input level: in this example the input is $3\Delta/7$ away from a level, and it results in a pattern that repeats every seven clock periods. When the repetition frequency lies in the signal band, the modulation is noisy, but when it does not, the modulation is quiet.

Figure 8 shows how the in-band rms noise depends on the value of the dc input, for a delta-sigma modulator having quantization levels at $\pm 1$ and an oversampling ratio of 16. The decimating filter that processes the output is the one described in Section 4. There are peaks of noise adjacent to

integer divisions of the space between levels; elsewhere the noise is small. This structure of the quantization error is called *pattern noise*. The larger peaks can far exceed the expected noise level calculated from Eq. (11), which is 41 dB in this example.

Surprisingly, it is also quite easy to get a mathematical expression [6, 15, 22] for the noise from delta-sigma modulation with dc input. Let $x$ be the input to a modulator, and let $Y'$ be an adjacent quantization level. The modulator output can then be expressed as

$$y(t) = Y' + \sum_l \sum_k \frac{\sin(\pi l x')}{\pi l} \exp\left(\pi j \frac{lx' + k}{\tau} t\right), \quad (12)$$

where $x' = (x - \bar{Y'})/\Delta$. Thus, $y(t)$ has a dc component equal to the input $x$, together with tones of frequency $lx' + k$. The tones that lie in the signal band represent the significant modulation noise. A sum of the powers of these tones, taking account of the response of the decimation filter, gives a good description of the pattern noise [6].

The following properties of pattern noise are noteworthy:

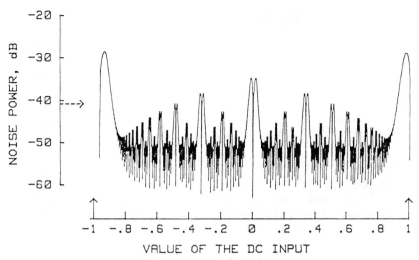

Fig. 8. The noise from delta-sigma modulation for dc inputs. Quantization levels are at ±1, and the noise is plotted for dc inputs between these levels. Peaks of noise occur adjacent to integer divisions of the level spacing.

- The *height* of each peak is inversely proportional to the oversampling ratio.
- The *width* of each peak is inversely proportional to the oversampling ratio.
- The *power* in each peak is inversely proportional to the oversampling ratio cubed.
- The *height* and *width* of each peak are inversely proportional to the denominator of the fraction that describes the position of the peak within the quantization interval relative to the level spacing. This fraction is $3/7$ in Fig. 7.
- The average noise represented by the graph is given by Eq. (11).
- About half of the total power is in the end peaks, one-sixteenth in the center peaks.

The noise pattern in Fig. 8 can be integrated against time as a function of a changing input to get a measure of the noise introduced into the signal. Figure 9 shows the signal-to-noise ratio for two sine waves, plotted against their amplitudes. Curve (a) corresponds to sine waves centered between levels, curve (b) for sine waves offset from center. Zero-dB input level corresponds to a peak amplitude equal to $\Delta/2$. For comparison, the dotted line shows the values predicted from Eq. (11). The resolution is better than this prediction when large peaks are not included, but is worse when they are. The dependence of noise on signal value and the fact that noise is composed of tones are reasons why this modulation is rarely used. When it is, dithering is often needed to randomize the noise and destroy tones that would be disturbing in audio applications [8, 78].

### 2.2.4 Dead Zones in Delta-Sigma Modulation.
The second graph in Fig. 7 show the output of the integrator for a steady input equal to $3/7$. Notice that this waveform can be raised by up to $\Delta/7$, with respect to the quantizer threshold level, without changing the sequence of decisions. A change of level at the output of the integrator corresponds to an

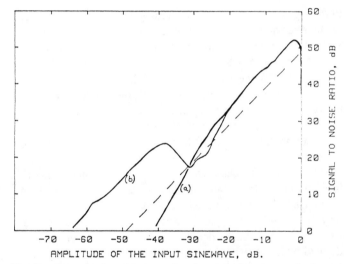

Fig. 9. A graph of modulation noise plotted against the amplitude of applied sine waves; 0 dB corresponds to an amplitude of $\Delta/2$. Curve (a) is for sine waves centered midway between levels, curve (b) is for sine waves biased $\Delta/64$ away from center. The dashed line is the calculated noise.

impulse at its input; consequently, this "dead zone" is noticeable only in rapidly changing inputs. The location and extent of the dead zones correspond in position and size to the peaks of noise in Fig. 8. For most applications, the pattern noise is more noticeable than are the dead zones. However, if the integrator has low dc gain, the dead zones can be significant.

We next describe some practical properties of the delta-sigma circuit because this simple modulator is useful for illustrating feedback quantization. Knowledge of its properties will help us explain improved modulators.

### 2.2.5 Influence of Circuit Parameters on Delta-Sigma Modulation [41]
*Net Gain in the Feedback Loop.* Our discussion has so far assumed unity sample gain in every component of the

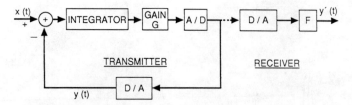

Fig. 10. A block diagram of a delta-sigma modulator including a gain $G$ in the feedforward path, and a nonlinearity $F$ in the representative decoder.

modulator. Figure 10 shows a modulator that includes a constant gain $G$ in the forward path of the feedback. Small deviation of this from unity will have little effect on the overall properties, provided the net gain in the feedback loop is large. The gain of the accumulator is

$$H(f) = \frac{z^{-1}}{1 - z^{-1}} \approx (j\omega\tau)^{-1} = \left[ j\pi(2f\tau) \right]^{-1},$$

$$f\tau \ll 1. \quad (13)$$

In the signal band this has a modulus greater than one-quarter of the oversampling ratio given in Eq. (4), which is usually sufficiently large. Measurements on real modulators and simulations [5, 36, 43] have demonstrated that with small gains (i.e., $G < 0.7$) the circuit responds sluggishly to changing inputs. With gains greater than 1.3, the circuit oscillates between more than two levels and eventually goes unstable when $G \geq 2$, as can be predicted from the linearized model. For most applications, $\pm 10\%$ changes of gain from unity are tolerable.

*Positioning the Quantization Thresholds.* The quantizer in a multilevel delta-sigma modulator usually takes the form of a flash A/D, which has short delay. Misplaced thresholds may be regarded as a nonlinearity of its gain, because the gain of the quantizer, defined in (1), is the level spacing divided by the threshold spacing. Such nonlinearity following the high gain of the integration in the forward path of the feedback loop has little effect on the baseband properties of the modulator [5]. Misplacing the thresholds by as much as a quarter of their spacing is usually tolerable.

*Positioning the D/A Quantization Levels.* Misplaced levels of the D/A are more serious than misplaced thresholds because they introduce nonlinearity directly into the signal [5, 38, 61]. The feedback action forces the average value of the quantized signal, $y$, to track the input, even when levels are misplaced. If the effective D/A converter at the receiver in Fig. 10 matches the one in the transmitter, the output $y'$ will track $y$. When it does not, the mismatch can be represented by a nonlinearity $F$ at the receiver. Such nonlinearity must usually be very small, and this calls for highly accurate D/A converters [38].

*Two-Level Quantization.* Use of two-level quantization avoids the need for matching level spacings. A misplaced level, as illustrated in Fig. 11, now introduces a change of quantization range and a dc offset, neither of which need be critical. Two-level quantization requires only one threshold,

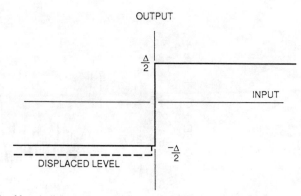

Fig. 11. A diagram illustrating a misplaced level in two-level quantization.

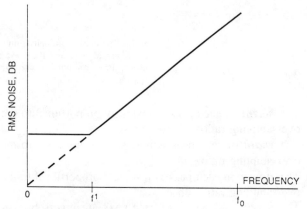

Fig. 12. An illustration of the effect of leakage in the integrator on the spectral density of modulation noise.

so the concept of gain $G$ in Eq. (1) is now arbitrary. Nevertheless, the results in Eqs. (9) and (11) apply [14–17].

Two-level modulators can have very robust circuits: the threshold need not be accurately positioned because it is preceded by the high dc gain of the integrator. The sample gain of the integrator is not critical because it drives only a single-threshold stage. The quantization levels need be positioned only to accommodate the range of the input signal.

*Leakage in the Integrator.* When the integrator in Fig. 4a includes leakage, its transfer function is given by

$$H(z) = \frac{z^{-1}}{1 - \alpha z^{-1}}, \quad (14)$$

and its dc gain is

$$H_0 = H(1) = \frac{1}{1 - \alpha}, \quad \alpha < 1. \quad (15)$$

The output of the modulator can be expressed as

$$Y = \frac{z^{-1}X}{1 + (1 - \alpha)z^{-1}} + \frac{(1 - \alpha z^{-1})E}{1 + (1 - \alpha)z^{-1}}. \quad (16)$$

There is increased noise at low frequency, as illustrated schematically in Fig. 12. If the dc gain of the integrator is at least equal to the oversampling ratio, the increase in baseband noise is less than 0.3 dB. This condition also ensures

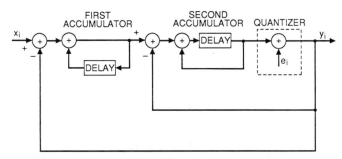

Fig. 13. A second-order delta-sigma quantizer.

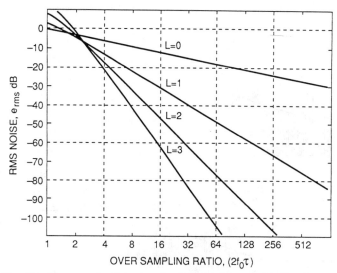

Fig. 14. The rms noise that enters the signal band for oversampling ratios in the range 1 through 512, assuming busy input signals. Graphs are plotted for ordinary quantization without feedback $L = 0$, and first-, second-, and third-order delta-sigma quantization. Zero dB of noise corresponds to that of PCM sampled at the Nyquist rate. A common level spacing is assumed.

that the dead zone described in Section 2.2.4 is not troublesome; it could be with smaller gains.

## 2.3 High-Order Modulators

### 2.3.1 Predicting In-Band Values of Quantization Error.

In a delta-sigma circuit, feedback via an integrator shapes the spectrum of the modulation noise, placing most of its energy outside the signal band. In general, the characteristics of the filter included in the feedback loop determine the shape of the noise spectrum [1, 27]. In this section we discuss a number of filters and circuit structures that are improvements on ordinary delta-sigma circuits.

The objective of using improved filters is to reduce the net noise in the signal band. To do this well, we need to subtract from the quantization error a quantity whose in-band component is a good prediction of the in-band error. Ordinary delta-sigma modulation subtracts the previous value of the quantization error from the present error (Eq. 7). Higher-order prediction should give better results than this first-order prediction.

### 2.3.2 Noise in High-Order Delta-Sigma Modulation.

There are several circuit arrangements that give second-order predictions of the quantization error [9, 26, 30]. The one shown in Fig. 13 is easy to build and is tolerant of circuit imperfection. It is an iteration [9] of delta-sigma feedback loops. The output of this modulator can be expressed as

$$y_i = x_{i-1} + (e_i - 2e_{i-1} + e_{i-2}),   (17)$$

so that the modulation noise is now the second difference of the quantization error. The spectral density of the noise is

$$N(f) = E(f)(1 - e^{-j\omega\tau})^2.   (18)$$

For busy signals

$$|N_2(f)| = 4e_{rms}\sqrt{2\tau} \sin^2\left(\frac{\omega\tau}{2}\right),   (19)$$

and the rms noise in the signal band is

$$n_0 = e_{rms}\frac{\pi^2}{\sqrt{5}}(2f_0\tau)^{5/2}, \qquad f_s^2 \gg f_0^2.   (20)$$

This noise falls by 15 dB for every doubling of the sampling frequency, providing 2.5 extra bits of resolution [26, 35, 36].

The technique can be extended to provide higher-order predictions by adding more feedback loops to the circuit [26].

In general, when there are $L$ loops and the system is stable, it can be shown that the spectral density of the modulation is

$$|N_L(f)| = e_{rms}\sqrt{2\tau}\left(2 \sin\left(\frac{\omega\tau}{2}\right)\right)^L,   (21)$$

and for oversampling ratios greater than 2, the rms noise in the signal band is given approximately by

$$n_0 = e_{rms}\frac{\pi^L}{\sqrt{2L+1}}(2f_0\tau)^{L+1/2}.   (22)$$

This noise falls $3(2L + 1)$ dB for every doubling of the sampling rate, providing $L + 0.5$ extra bits, but we shall see that there are difficulties in implementing their circuits when $L > 2$. Figure 14 plots the in-band noise against the oversampling ratio for examples of PCM and one, two, and three feedback loops. These graphs are derived from result (21), which assumes white uncorrelated noise, and this may not be valid unless the signal is sufficiently busy to randomize the errors or unless there is sufficient dithering. It is fortunate that the noise has been found to be nearly random in second- and higher-order modulators [36] even when the signal is constant. The randomness is provided by the retention of the noise in the prediction filter stages, and depends on these filters having long-term memory.

Figure 5 illustrates the output of a first-order modulator oscillating between two levels adjacent to the input value. The noise ranges in amplitude between $\pm\Delta$, which is consistent with Eq. (8). The output of a second-order modulator oscillates predominantly between three levels, but occasionally reaches a fourth, which is consistent with the expression for the noise in Eq. (17). Figure 15a shows the output of a second-order modulator having quantization levels at integer values, when it responds to an input ramp, and Table 1 lists

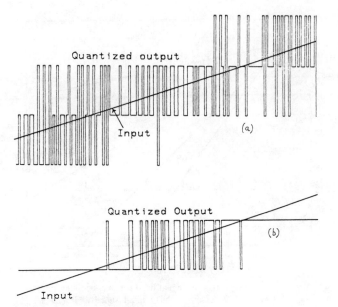

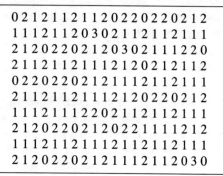

Fig. 15. Response of a second-order delta-sigma quantizer to ramp input for both multilevel and two-level quantization.

## TABLE 1
### TWO HUNDRED OUTPUT SAMPLES GENERATED BY A SECOND-ORDER DELTA-SIGMA MODULATOR RESPONDING TO A DC INPUT EQUAL TO 1.3. THE QUANTIZATION ROUNDS OFF REAL NUMBERS TO INTEGER VALUES.

```
0 2 1 2 1 1 2 1 1 2 0 2 2 0 2 2 0 2 1 2
1 1 1 2 1 1 2 0 3 0 2 1 1 2 1 1 2 1 1 1
2 1 2 0 2 2 0 2 1 2 0 3 0 2 1 1 1 2 2 0
2 1 1 2 1 1 2 1 1 1 2 1 2 0 2 1 2 1 1 2
0 2 2 0 2 2 0 2 1 2 1 1 1 2 1 1 2 1 1 1
2 1 1 2 1 1 2 1 1 1 2 1 2 0 2 2 0 2 1 2
1 1 1 2 1 1 1 2 2 0 2 1 1 2 1 1 2 1 1 1
2 1 2 0 2 2 0 2 1 2 0 2 2 1 1 1 1 2 1 2
1 1 1 2 1 1 2 1 1 1 2 1 1 2 1 1 2 1 1 1
2 1 2 0 2 2 0 2 1 2 1 1 1 2 1 1 2 1 2 0 3 0
```

its output for a steady input of 1.3. The output includes levels 0, 1, 2, and sometimes 3 in seemingly random order, but keeping its average equal to the input. These measurements commenced with arbitrary initial values in the integrators. Starting with integer values results in repetitive patterns in the output.

### 2.3.3 Dynamic Range of the Modulators.
The oscillation of the output signal uses up some of the dynamic range of the circuit, and if the quantizer is not to overload the input amplitudes to the modulator must be appropriately limited. Figure 16 shows the range of inputs that can be accommodated in several modulators, each employing six-level quantization. Ordinary PCM requires that its input be restricted to $\pm A$ in order that the quantization error lie in the range $\pm \Delta /2$. Inputs to a first-order delta-sigma modulator must be restricted to $\pm B$ for them to be interpolated by oscillation

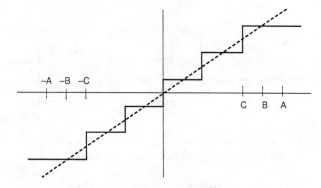

Fig. 16. The range of input amplitudes that can be accommodated by multilevel quantizers. Ordinary quantization accommodates inputs in the range $\pm A$, and first-order delta-sigma quantization accommodates $\pm B$ without overloading. Second-order delta-sigma accommodates $\pm C$ with small probability of overloading.

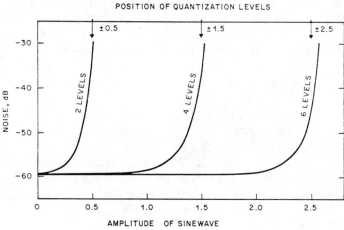

Fig. 17. Noise introduced into sine waves of various amplitudes by second-order delta-sigma quantization with either two, four, or six quantization levels.

between two levels. Inputs to a second-order modulator must be restricted to $\pm C$ to prevent frequent overloading of the quantizer. This permits the output to oscillate between three quantization levels, but overload can occur occasionally when the oscillations attempt to reach a fourth level that is not available.

When large inputs cause the quantizer to overload, the modulation noise increases, as illustrated in Fig. 17. This plots the noise introduced into sine waves of various amplitude, during second-order modulation [36]. Three cases are shown: two-level, four-level, and six-level modulation with quantization levels placed at $\pm 0.5$, $\pm 1.5$, and $\pm 2.5$. These results show that the "excess noise" due to overloading increases quite slowly. Even the two-level modulation has a useful range, despite the fact that theoretically it is overloaded for all conditions, except for zero input with zero initial conditions. The response of a second-order modulator with two-level quantization to a ramp input is illustrated in Fig. 15b. Examples of such modulators are given in references [73, 77, 81, 84].

For small input amplitudes, the noise in these modulators

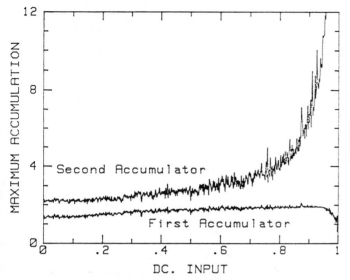

Fig. 18. Maximum amplitude of signals in the accumulators of a second-order delta-sigma modulator. The two quantization levels are at ±1. In practice the first accumulation is often clipped at ±2 and the second at ±4.

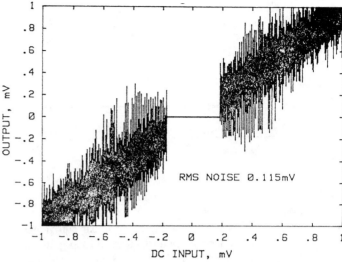

Fig. 19. An illustration of the dead zone caused by leakage in the accumulators of second-order delta-sigma quantization. The dc gain of the accumulators is 64, and the oversampling ratio is also 64. The range of input and output amplitudes that can be accommodated is ±1 V, and the noise is less than that of 12-bit PCM.

agrees with Eq. (20). Simulations show that the excess noise introduced into larger signals appears as odd-order harmonic distortion of sine waves and includes a minute increase in the gain of the fundamental. The excess noise decreases with increased oversampling ratio, but increases with the frequency of the applied sine wave. A complete characterization of the excess noise is not yet available to us, but attempts have been made to analyze it [2].

### 2.3.4 Influence of Circuit Parameters on Second-Order Modulators

*Circuit Tolerances.* The second-order modulator in Fig. 13, like the first-order one in Fig. 4, is very tolerant of circuit imperfections, especially when two-level quantization is employed. Compared with first-order systems, the second-order modulator has one more design parameter available: it is the ratio of the gains of the two feedback paths. The outer path dominates in determining the low-frequency properties of the circuit, while the inner path serves to stabilize the system, and it determines the high-frequency properties. Matching their relative gains to within ±5% is usually satisfactory. Sometimes, the gain of the inner loop is increased from unity in order to compensate for delay in the outer loop.

*Range of Integration.* An important parameter is the range of signal amplitudes that must be accommodated at the outputs of the integrators. Simple theory gives an adequate description of these signals for a multilevel quantizer that does not saturate. The output of the first integrator is given by $x_i - e_{i-1} + e_{i-2}$, the second by $x_i - 2e_{i-1} + e_{i-2}$, with $e$ and $x$ bounded by $\pm \Delta /2$. These results are inadequate for describing two-level quantization, and we have to resort to simulation. Figure 18 plots the maximum signal level at the output of the integrators as a function of a dc input level. The signal in the first integrator remains well

bounded, but the second one becomes very large for $x > 0.8$. Clipping this signal at about three times the range of the input signal has little effect on the overall performance of the modulator [36, 64].

*Leakage in the Integrators.* First-order modulators need integrators with dc gains $H_0$ which are at least equal to the oversampling ratio, in order to have low noise. Calculations of noise in second-order modulators indicate that somewhat lower gains could be tolerated, since now at dc the two integrator amplifiers are cascaded. But there is another consideration: leakage can permit the oscillations of the quantized signal to settle into regular patterns, because there is insufficient long-term memory to randomize it. This is most noticeable at the center of the range where the output can settle into a $+1, -1, +1, -1$ pattern. The effect is illustrated by Fig. 19, which shows the filtered output of a modulator responding to a very slowly changing ramp. The full range of the output is ±1 V; this is such an expanded scale that the noise is apparent. At the center of the range the output locks into the pattern and the input is ignored in the $-0.2$ mV $< 0 < 0.2$ mV range. It may be shown that the width of the dead zone is $1.5 \Delta H_0^{-2}$, and for this to be less than twice the rms quantization noise requires that the dc gain of the integrators satisfy

$$H_0 \ge (2 f_0 \tau)^{-5/4}. \qquad (23)$$

The dead zone is seldom noticeable because it is present only in very slowly changing signals: it takes time for the oscillations to settle into a pattern. Such a dead zone could actually be useful for audio applications, which need a very quiet idle state.

### 2.3.5 Limit Cycles in Third-Order Delta-Sigma Modulators.

Simple linear theory predicts that the third-order modulator shown in Fig. 20 has an rms noise given by Eq.

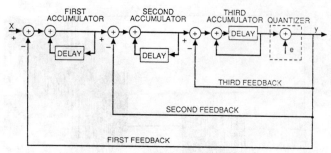

Fig. 20. A third-order delta-sigma quantizer.

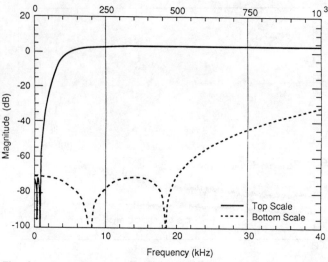

Fig. 21. Spectral densities of the noise from a generalized feedback quantizer of the type shown in Fig. 22.

(22) with $L = 3$. This can be realized in practice with a multilevel quantizer that does not overload [30, 36]; but the circuit is much more sensitive to circuit values than the first- and second-order circuits. For example, the linear equivalent circuit of this modulator becomes unstable with loop gains of only 1.15, compared with 2.0 and 1.33 for first- and second-order modulators.

If the quantizer saturates, the modulator can settle into a large-amplitude low-frequency limit cycle [30]. In this state the clipped signals fed back via the two inner loops are negligible compared with the signals in the integrators. The outer loop dominates; it contains three integrators and a delay, a basis for instability. Two-level third-order circuits cannot escape from this condition. These circuits can be made stable by clipping the outputs of the integrators or including other nonlinearities that make the inner feedback effective when the quantizer saturates. The noise performance of modulators with these arrangements is considerably worse than Eq. (21) predicts. Better performance is obtained by redesigning the filter used in the feedback loop.

**2.3.6 Noise Shaping Using Filters with Nonmonotonic Transfer Functions.** The delta-sigma modulators described so far contain filters in their feedback loops that have multiple poles at dc and zeros at high frequency which serve to stabilize the circuits. The frequency responses of these filters fall monotonically through the range 0 to $f_s/2$. The noise at the output of the modulator is shaped approximately as the inverse of this filter characteristic. Attempts have been made to replace this filter with a more rectangular high-pass filter [27, 39, 41, 47, 61, 62, 71]. The poles are distributed through the signal band in order to lower the in-band noise. The zeros are chosen to flatten the filter response at high frequency, in order to reduce the high-frequency noise and prevent it from using up dynamic range. A noise spectrum [39] obtained from such a modulator is given in Fig. 21.

Two-level, fourth- and fifth-order modulators have been successfully built with this technique. Their circuits, illustrated in Fig. 22, are based on cascaded integrators [39] with feedback branches used to position the poles and feedforward ones dimensioned to position the zeros. The danger of these circuits locking into high-amplitude limit cycles is avoided by allowing the integrators to clip at quite small amplitudes. The input amplitudes are limited to less than $\pm \Delta /4$ to avoid distorting the signal. Alternative structures have been reported in [41].

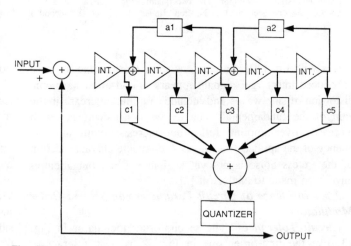

Fig. 22. A fifth-order feedback quantizer. The loop gain includes two complex poles and one real pole with zeros positioned to ensure stability. The quantization is two-level.

The noise performances of these modulators are poorer than that anticipated by Eq. (21), but better than that obtained from a second-order delta-sigma modulator. For example, 16-bit encoding of 20-kHz signals has been obtained by using a fourth-order modulation at 3 MHz. A major advantage of these modulators is the fact that their noise can be devoid of audible tones.

### 2.4 Some Alternative Modulator Structures

**2.4.1 Error Feedback.** Noise-shaping quantization was first described in reference [9], using the structure shown in Fig. 23. In this circuit, the difference between the input and the output signals of the quantizer is a measure of the quantization error which is fed back and subtracted from the next input sample. The circuit is algebraically equivalent to the delta-sigma modulator in Fig. 4, but it has the serious practical disadvantage that inaccuracies in the analog subtractors have a strong impact on the modulator's properties. We

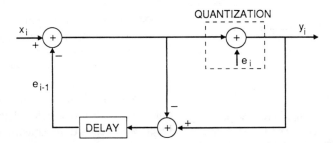

Fig. 23.   A quantizer with error feedback.

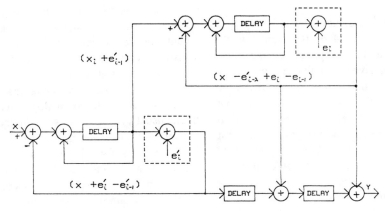

Fig. 24.   A cascade of two first-order delta-sigma modulators. The second one serves to digitally encode the quantization error $e'$ of the first so that it may be removed from the output.

shall see, however, that the circuit can be used as a demodulator, because there the processing is performed digitally. The circuit can be generalized by replacing the delay by a prediction filter, design methods for which are given in references [1] and [28].

*2.4.2 Cascaded Modulators.* The performance of a modulator can be improved by taking a measure of its noise, digitizing that measure in a second modulator, and combining the output of the two modulators in a way that cancels the noise of the first modulator. This technique was proposed for use with two delta modulators [10], and has since been widely applied to delta-sigma modulators [50, 75]. Figure 24 shows a method for cascading two first-order delta-sigma modulators. The output of the integrator in the first modulator is fed to the second modulator. Its output is digitally differenced and subtracted from the output of the first modulator to provide the net output of the circuit. We have used $e'$ to denote the quantization error in the first modulator and $e$ that of the second one. When scaling factors are ignored it can be shown that the net output of the circuit may be written as

$$y_i = x_{i-2} + (1 - g)(e'_{i-2} - e'_{i-3})$$
$$+ (e_i - 2e_{i-1} + e_{i-2}), \quad (24)$$

where $g$ is a measure of the accuracy of the circuit. It depends on a number of parameters, including the precision of component values and the dc gain of the first integrator. Ideally, $g$ is unity; then the noise of the first modulator does not contribute to the output. The remaining noise is the

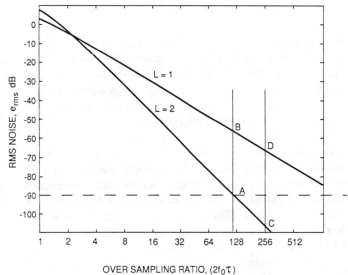

OVER SAMPLING RATIO, $(2f_0\tau)$

Fig. 25.   A graphical comparison of noise sources in the cascaded modulators of Fig. 24.

second difference of the quantization error from the second modulator: it is in the same form as the noise of a second-order delta-sigma modulator given in Eq. (17).

A first consideration in the design of cascaded modulators is: how close to unity must be the factor $g$? This can be determined from Fig. 25, which was derived from Fig. 14. As an example, suppose we are designing a modulator like the one in Fig. 24 to provide a resolution equivalent to a 15-bit PCM, which corresponds to a noise of $-90$ dB in

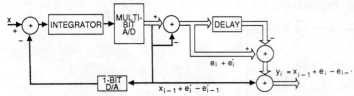

Fig. 26. A first-order delta-sigma modulator with two-level feedback. Its quantization error $e'$ is measured by a multibit A/D circuit and eliminated from the output. Only the first difference of the error $e$ from the multilevel quantization contaminates the output.

Fig. 25. An ideal second-order modulator oversampling by a factor of 120, represented by point $A$ on the graph, would just meet the requirement. At this sampling rate the noise from first-order modulation is given by point $B$ and is only $-57$ dB. We therefore need sufficient precision to make the term $1 - g$ much less than $(-90) - (-57) = -33$ dB; that is, $g$ must be well within $3\%$ of unity. The need for such precision is alleviated by raising the sampling rate. For example, at an oversampling ratio of 256, corresponding to points $C$ and $D$ on the graph, the second-order noise is $-108$ dB, well below the requirement. The first-order noise is $-68$ dB, and we need to reduce this by only 22 dB (i.e., keep $g$ within $10\%$ of unity) to achieve $-90$ dB noise. Scaling factors which are often included in these circuits, but are not included in Eq. (24), can amplify the noise terms $e'$ and $e$ in Eq. (24).

Because of the difficulty in obtaining adequate precision, the noise from this cascaded circuit is often dominated by the first-order term, which will include peaks of pattern noise. They can be avoided by using a second-order modulator as the first stage [34, 49, 52, 59]. When the performance is constrained by inaccuracy of circuit components, there is little advantage in using higher than first-order circuits as the second stage, or including a third stage.

When circuits are made with sufficient accuracy to eliminate the noise of the first stage, the cascaded modulator has several important advantages. For example, when $1 - g = 0$, the circuit in Fig. 23 provides a second-order modulation with a 2-bit output, yet the feedback loops are first-order with 1-bit digital signals. The output can oscillate between four levels, as shown in the first curve of Fig. 15, and there will be no excess noise introduced unless the inputs overload the first-order modulators. Adding another stage to the cascade can provide third-order modulation [50, 59] without incurring the dangers associated with third-order feedback loops.

An ingenious circuit that can be interpreted as a cascaded modulator is described in reference [48]. This technique is illustrated in Fig. 26, which shows a delta-sigma modulator that contains a multilevel A/D, only the sign-bit of which drives the feedback. The complete digital word from the A/D is reduced by an amount equal to the feedback bit (this is equivalent to inverting the sign bit of a two's complement code), and the resulting code is differentiated and added to the feedback bit to provide the output. This circuit can be regarded as a 1-bit delta-sigma modulator cascaded with an ordinary PCM encoder that serves to cancel the noise of the 1-bit modulation. The net output can resemble that of a first-order multilevel delta-sigma modulator, yet it contains only a 1-bit D/A in the feedback. The technique can obviously be extended for use in higher-order modulators.

### 2.4.3 Delta Modulation.
Most early work on oversampling was concerned with delta modulation [12, 13, 24]. Later work turned its attention to delta-sigma modulation because its circuits are more robust. The main difference between the techniques is that delta-sigma modulators, and other noise-shaping encoders, change the spectrum of the noise but leave the signal unchanged. By contrast, delta modulation and other signal-predictive encoders shape the spectrum of the modulated signal but leave the quantization noise unchanged at the receiver. It is the need for a filter at the receiver to restore the signal that makes signal-predicting encoders vulnerable to analog circuit inaccuracies. These filters usually have high gain in the signal band and thus magnify any distortion introduced by the D/As.

Figure 27 is a block diagram of a delta modulator and its demodulator. It transmits the first difference of the signal; consequently, an integrator is needed at the receiver. The output is contaminated with the quantization error directly, and it saturates by clipping the derivative of the signal. In contrast, the output of a delta-sigma modulator is contaminated by the difference of the quantization error and it saturates by clipping amplitudes. To compare these modulators, we now calculate their signal-to-noise ratios for sinusoidal inputs. The largest sine wave that the delta-sigma modulator will accommodate without saturating has peak values $\Delta/2$. Its rms noise is given by Eq. (11), and therefore the maximum rms signal-to-noise ratio can be expressed as

$$\text{SNR}_{\text{dsm}} = \frac{\Delta}{2\sqrt{2}} \frac{\sqrt{3}}{\pi e_{\text{rms}}} (2f_0\tau)^{-3/2} = \frac{\sqrt{4.5}}{\pi} (2f_0\tau)^{-3/2}.$$

$$(25)$$

The peak value of the largest sine wave that the delta modulator will accommodate is slope-limited and is given by $\Delta/\omega\tau$, where $\omega$ is the angular frequency of the sinusoidal signal. If $f'$ is the frequency at which the signal source delivers its steepest slope, the amplitudes of sine waves will be constrained to $\Delta/2\pi f'\tau$. The rms in-band noise introduced into the signal is given by Eq. (3). Therefore the signal-to-noise ratio is given by

$$\text{SNR}_{\text{dm}} = \frac{\sqrt{6}}{\pi} \left(\frac{f_0}{f'}\right)(2f_0\tau)^{-3/2}.$$

$$(26)$$

For general signals $f' = f_0$, and hence delta modulation has only 1 dB higher signal-to-noise ratio than the delta-sigma modulation. For some speech signals $f' = f_0/3$; then the delta modulation has a 10-dB advantage. Delta modulation can be useful for applications where $f'$ is much less than $f_0$ and when clipping slopes is more tolerable than clipping amplitudes, as in some audio and video applications.

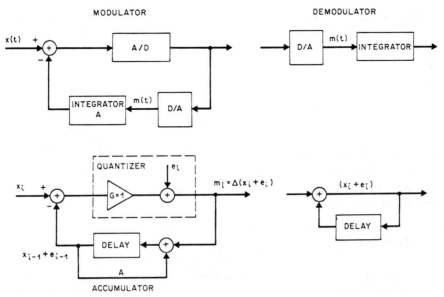

Fig. 27. A delta modulator and demodulator with its sampled-data equivalent circuit.

## 3. A COMPARISON OF ARCHITECTURES FOR MODULATORS

### 3.1 Criteria for Choosing the Architecture

In addition to the circuit configurations discussed in Sections 2.2 and 2.3 and illustrated in Figs. 4, 13, 20, and 22, various other architectures have been proposed for oversampling modulators. The motivation for developing a number of alternative systems was to match the configuration to the specific application. In comparing these architectures, we can use various criteria, including

- Achievable resolution, or signal-to-noise ratio, for a given oversampling ratio OSR
- Complexity of the modulator and its associated decimation filter
- Sensitivity of the performance to such nonideal effects as component-matching errors, as well as to amplifier imperfections such as finite gain, bandwidth, and slew rate; to noise; to nonlinearity; to supply voltage and temperature variations
- Stability properties
- External and/or additional trimmed components, if needed
- Dithering, if needed
- Linear signal range.

Clearly, these criteria may dictate different trade-offs in differing applications. For example, in a low-frequency system such as in voice telephony it is feasible to use a simple (say first-order single-bit) modulator and to achieve the required resolution by choosing a large oversampling ratio. By contrast, in a high-frequency application such as a radar receiver, a large oversampling ratio may not be possible and the designer may need to opt for a higher-order and/or multibit modulator, at the cost of increase complexity and sensitivity and reduced stability.

To make the choice of the optimal configuration easier, the next section lists and compares the more commonly used oversampling modulator configurations in terms of the criteria listed above. Although new architectures keep appearing, the list reflects the state of the art up to early 1990.

### 3.2 Architectures for Oversampling Modulators

***3.2.1 The First-Order Delta-Sigma Modulator.*** The simplest, most robust and stable modulator architecture is the one shown in Fig. 4 and discussed in Section 2.2. It can be argued that this is the only modulator deserving the name *delta-sigma* [20]. As shown, this modulator has the following properties:

- Its signal-to-noise ratio can be predicted from Eq. (11), which may be written in the form

$$\frac{n_0}{\Delta} = \frac{\pi}{6}(\text{OSR})^{-3/2}, \tag{27}$$

or from Fig. 14. Both indicate that the signal-to-noise ratio is relatively low and improves slowly with increased OSR.
- The complexity of the modulator is low.
- The performance is not sensitive to nonideal effects, such as component matching or amplifier imperfections. For single-bit quantization, the operation is also tolerant of A/D and D/A errors.
- Under usual circumstances, in the absence of excessive amounts of delay in the loop and/or excessive integrator gain or phase error, the modulator is inherently stable.
- First-order modulators may need trimming only if they use multibit internal quantizers.
- The noise-versus-signal-amplitude characteristics have large peaks for dc or slowly varying signals (cf. Fig. 8), and the noise includes single-frequency tones. Hence,

unless the signal is known to be very busy (i.e., rapidly varying and nearly random), a high-frequency dither signal uncorrelated with the input should be injected. This has the effect of redistributing the energy present in the noise peaks over the entire amplitude range and of eliminating the tones.

- The linear input signal range is large, as illustrated by points $\pm B$ in Fig. 16.

### 3.2.2 The Second-Order Delta Sigma Modulator. The
second-order modulator illustrated in Fig. 13 is often used in fully integrated oversampling converters. Its properties were discussed in Section 2.3, and are summarized here.

- Its signal-to-noise ratio can be predicted from Eq. (20), which may be written in the form

$$\frac{n_0}{\Delta} = \frac{\pi^2}{\sqrt{60}}(\text{OSR})^{-5/2}, \qquad (28)$$

or from Fig. 14. For practical values of the oversampling ratio, the resolution is much higher, and grows faster with oversampling ratio, than that of the first-order modulator. Thus, for OSR = 128, a quantization noise consistent with a 16-bit resolution is possible.
- The second-order modulator is somewhat more complex than the first-order one and usually requires a more elaborate digital decimation filter.
- The performance depends somewhat more critically on component matching and other nonidealities than that of the first-order modulator. However, the noise shaping at low frequencies is more insensitive to amplifier gain, since now there are two amplifiers in the loop.
- The stability of the second-order modulator is not as robust as that of the first-order one. A 30% increase of loop gain and/or additional delay in the loop will cause instability.
- If a single-bit quantizer is used, there is normally no need to use trimming or external components. However, to avoid saturation, the maximum signal levels at the amplifier outputs should be adjusted by scaling, and amplitude limiters may be needed in the loop [62, 64].
- For sufficiently rapid and randomly varying input signals, a second-order modulator may not need a dither signal, since the correlation between the quantization error and the input signal is reduced by the additional feedback loop.
- As points $\pm C$ in Fig. 16 illustrate, the linear quantization range of the second-order modulator is narrower than that of a first-order one. Optimum scaling of the signal amplitudes should be used to improve the dynamic range of the modulator.

### 3.2.3 The Second-Order Cascaded Modulator. As discussed in Section 2.3, the quantization noise may be reduced, and hence the resolution of the conversion improved, by using a higher-order loop filter. This, however, impairs or destroys the stability of the loop. An architecture which

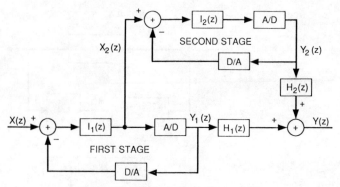

Fig. 28. A cascade of two first-order modulators.

provides higher-order noise shaping combined with the robust stability of a first- or second-order system, is the cascade or MASH configuration [50, 75]. The basic concept has been described in Section 2.4.2 and is illustrated again in Fig. 28. The $z$-transformed digital output signal $Y_1$ of the first loop can be written in the form

$$Y_1 = \frac{I_1 X}{1 + I_1} + \frac{E_1}{1 + I_1}, \qquad (29)$$

where $E_1$ is the $z$-transformed quantization error of the first-stage modulator and $I_1$ is the transfer function of the first-stage loop filter. The analog output signal of the loop filter is

$$X_2 = \frac{I_1(X - E_1)}{1 + I_1}. \qquad (30)$$

Since the input of the second-stage is $X_2$, its corresponding digital output signal is

$$Y_2 = \frac{I_2 X_2 + E_2}{1 + I_2}. \qquad (31)$$

The output of the complete modulator is, by Eqs. (29–31),

$$Y = H_1 Y_1 + H_2 Y_2 = \left[ H_1 + \frac{H_2 I_2}{1 + I_2} \right] \frac{I_1 X}{1 + I_1}$$
$$+ \left[ H_1 - \frac{I_1 I_2 H_2}{1 + I_2} \right] \frac{E_1}{1 + I_1} + \frac{H_2 E_2}{1 + I_2}. \quad (32)$$

To cancel the first stage quantization error $E_1$ in $Y$, let

$$\frac{H_1}{I_1} = \frac{H_2 I_2}{1 + I_2}. \qquad (33)$$

Then, the factor of $X$ in $Y$ becomes

$$\left[ H_1 + \frac{H_1}{I_1} \right] \frac{I_1}{1 + I_1} = H_1. \qquad (34)$$

To transmit $X$ undistorted, let $H_1 = z^{-k}$, where $k$ is the delay between the analog input signal and the digital output signal for each stage. Then, by Eq. (33), the factor of $E_2$ in $Y$ is given by

$$\frac{H_2}{1 + I_2} = \frac{H_1}{I_1 I_2} = \frac{z^{-k\tau}}{I_1 I_2}. \qquad (35)$$

In conclusion, neglecting the delay factor $z^{-k}$, the output signal is

$$Y = X + \frac{E_2}{I_1 I_2}. \qquad (36)$$

Comparing Eq. (36) with Eq. (29), it is clear that in the baseband frequency range where $|I_1|, |I_2| \gg 1$ the system of Fig. 28 is equivalent to a single-stage modulator containing a loop filter with transfer function $I_1 I_2$. Thus, for example, if $I_1 = I_2 = z^{-1}/(1 - z^{-1})$, then neglecting delays

$$Y = X + (1 - z^{-1})^2 E_2, \qquad (37)$$

and the error term will be noise-shaped by a second-order high-pass filter. Similarly, if $I_1$ and $I_2$ are second-order functions, $E_2$ will be filtered by a fourth-order characteristic, and so on. The stability properties of the system are, however, equivalent to those of a modulator containing only a single-loop filter $I_1$ or $I_2$.

A shortcoming of the scheme is that the once-filtered noise of the first stage will be perfectly canceled only if the condition given in Eq. (33) holds exactly. Because at baseband frequencies $|I_2| \gg 1$, Eq. (33) requires that $I_1 = H_1/H_2$. Since $I_1$ is the transfer function of a sampled-data analog circuit (e.g., a switched-capacitor integrator), it cannot exactly equal the digitally realized right-hand side of the relationship. Even a small error in $I_1$ will cause a leakage of the poorly filtered error signal of the first stage into the output $Y$. This results in a difficult design problem for the $I_1$ block. Moreover, the noise cancellation depends on the matching of the level spacings of the two D/A converters, because these spacings determine the gains of the two modulator stages.

In conclusion, the second-order modulator using the cascade configuration has the following properties:

- Its signal-to-noise ratio can be derived from Eqs. (20) or (28), or from Fig. 14. As is the case for other second-order modulators, the resolution is much better than that of a first-order one.
- The complexity of the modulator is more than twice that of a first order one and is higher than that of the single-stage second-order circuit.
- The stability of the second-order cascade modulator is as robust as that of a first-order one.
- As discussed, the performance is very sensitive to non-idealities of the first-stage analog integrator. It can be shown [83] that especially the phase error of $I_1$ can increase the noise dramatically. The resulting decrease in signal-to-noise ratio is particularly great when the oversampling ratio exceeds 100. The required op-amp gain is about 10 times higher than that for a single-stage configuration. However, the amplifier bandwidth requirement is more relaxed, because the two separate feedback loops have larger stability margin, and the amplifiers can settle simultaneously rather than sequentially. The slew rate required of the amplifiers is some-

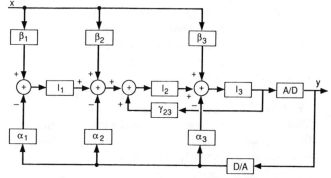

Fig. 29. A design for a single-stage third-order modulator. $I_1$ and $I_2$ represent delayed integration $z^{-1}/(1 - z^{-1})$, and $I_3$ simple integration $1/(1 - z^{-1})$. $\alpha$ and $\beta$ represent multiplication by $\alpha_i$ or $\beta_i$ together with delay $z^{-1}$, and $\gamma_{32}$ multiplication without delay.

what higher for the cascaded structure than for the single-stage one.

- The performance is sensitive to the matching of the level spacings of the two D/A circuits.
- The cascade modulator may be realized without external and/or trimmed components if it includes only single-bit quantizers.
- Since there is very little correlation between the input signal and the quantization errors, it is not essential to use dithering in cascaded modulators or any well-designed modulator of order 2 or greater.
- The linear input range of the cascaded modulator is the same as that of a first-order one. However, the digital output is a 2-bit data stream as a result of combining the 1-bit data $Y_1$ and $Y_2$. Hence, the cascaded modulator has larger output dynamic range than a single-stage second-order modulator.

**3.2.4 Higher-Order Single-Stage Modulators.** Higher-order modulators were briefly discussed in Section 2.3, and two possible implementations were shown in Figs. 20 and 22. A more recent architecture [41] for a third-order single-stage modulator is shown in Fig. 29. It contains a cascade of three integrators, each receiving inputs from the modulator's input signal, from the comparator output, and from the preceding integrator. An extra branch ($\gamma_{23}$) is included to generate a transmission pole at a nonzero frequency. The properties of this and similar high-order single-stage modulators are the following:

- The resolution can be very high; 20-bit accuracy has been demonstrated [63, 69] for audio-frequency converters. For a stable stage of order $L$, the signal-to-noise ratio increases by $L + 0.5$ bits for each doubling of the oversampling ratio, as can be seen from (22).
- The complexity of the modulator increases with $L$. In addition to the $L$ amplifiers needed, limiters are often used to restrict the signal amplitudes at the output of all integrators to prevent instability, and sometimes reset switches are used to discharge the feedback capacitors.
- Because higher-order ($L \geq 3$) modulators can only be conditionally stable, their sensitivities to variations in

circuit parameters may be high, especially if the $Q$'s of some of the closed-loop poles are high. However, since these circuits contain three or more cascaded amplifiers and the loop gain at or near dc equals the product of the gains, the modulator's performance is insensitive to finite op-amp gain effects.

- Also, due to the conditional stability, there are states which result in oscillations with increased amplitudes. Usually, such conditions are caused by large signals at the integrator outputs and can be eliminated by limiting these signals to relatively small values.

### 3.2.5 Higher-Order Cascade Modulators.

Cascaded modulators of order $L > 2$ can be realized by cascading three or more first-order stages [50, 59]. In such a structure, which is a direct extension of Fig. 28, each stage processes the quantization error of the previous stage. The outputs are then combined in such a manner that the quantization noise of all stages except the last one is canceled. The only remaining quantization noise will thus be the last-stage one, high-pass filtered by an $L$th-order transfer function $(1 - z^{-1})^L$.

In practice, second-order effects associated mostly with the analog integrators $I_1, I_2, \ldots, I_{L-1}$ limit the suppression of quantization noise originating from stages 1 through $L - 1$. Thus, it is impractical to use more than three stages. For a third-order system with an input-output signal delay of $3\tau$, choosing $H_1 = z^{-3}$, $H_2 = z^{-3}/I_1$, and $H_3 = 1/I_1 I_2$ gives, after a simple but lengthy derivation,

$$Y = X + \frac{E_3}{I_1 I_2 I_3}. \qquad (38)$$

Here, as before, $I_i$ is the transfer function of the analog filter in the $i$th stage, $E_i$ is the quantization error in the $i$th stage, and $H_i$ is the transfer function of the $i$th stage digital postfilter. This shows that the quantization noise in the output is the thrice-filtered quantization error of the third stage.

The demands that accurate noise cancellation puts on the performance (especially that of the first-stage integrator) are extremely hard to satisfy [49, 53]. Thus, while the ideal resolution of this circuit is superior to that of a lower-order modulator, and there are no stability problems, the practical realization of the analog circuitry requires great care.

An alternative technique for achieving third- or higher-order filtering with stable performance is to use second-order loop filters in the stages of the cascaded structure. Such a system [34, 49, 52, 59] is shown in Fig. 30, where the first stage performs as a second-order modulator. As before, the second stage is a first-order modulator which converts the quantization error of the first stage into a digital signal. If we choose $H_1$ to be a delay and $H_2$ to be proportional to $1/I_1 I_2$, the output will contain only the second-stage quantization error $E_2$ filtered by a third-order high-pass function.

Clearly the technique can be extended to higher-order ($N > 3$) modulators. In reference [76], a fourth-order modulator circuit constructed from two cascaded second-order

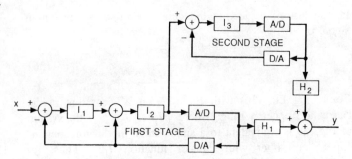

Fig. 30.   A cascade of a second-order modulator with a first-order one.

stages is described. This circuit also used interstage scaling to reduce requirements on the amplifier gain and capacitor accuracy.

These circuits have the same ideal performance as a single-stage high-order $L = 3$ or 4 modulator, but with the robust stability of a second-order loop. Also, they are not as sensitive to the imperfections as a cascade structure containing only a first-order modulator as the first stage. The noise in the output $Y_1$ of the first stage is now second-order-filtered, and even if some of this noise leaks to the modulator output $Y$, its effect on baseband noise is not as devastating as in a modulator using a first-order filter in the first stage.

Theory [17] as well as experience [49] shows that no dithering is required in these high-order cascade circuits.

### 3.2.6 Delta-Sigma Modulators Using Multibit Quantizers.

As discussions of Section 2.1 indicated, for a fixed input signal range the quantization noise decreases exponentially with the number $M$ of bits used in the quantizer. This is true since the rms noise $e_{rms}$ is proportional to the quantization step size $\Delta$, which in turn is proportional to $2^{-M}$. Thus every additional bit used in the quantizer reduces the noise level by 1 bit. In addition, the performance of the feedback loops using multibit quantizers follows the predictions of linearized theory more closely than that of single-bit systems. Hence, multibit modulators tend to be more stable than single-bit ones, particularly the high-order modulators. Finally, since the out-of-band noise from a multibit modulator is lower than that from single-bit ones, the complexity of the digital decimation filters needed to suppress it can be reduced.

These advantages are balanced by the added complexity required by the multibit internal A/D converter (which must be a flash-type circuit) and especially by the problems associated with realizing the D/A converter in the feedback path. The latter must be fast, and its conversion error, including nonlinearity, must be very small, less than half a least significant bit of the final (as opposed to the internal) output word. To see the need for such accuracy, we may note from Fig. 4a that any D/A conversion error added to the D/A output signal is directly subtracted from the input signal $x(t)$. Thus, it will appear in the digital output $y_i$ with unchanged amplitude. In addition, since in a multibit modulator the value of $y_i$ is quite close to that of the input $x(t)$, the correlation between $x(t)$ and the D/A error will be large.

Thus the D/A error will usually have a large baseband energy content and will generate harmonic distortion of sinusoidal input signals.

There are several strategies for dealing with this problem. The most obvious one is to use external components for the crucial elements of the D/A converter and/or to use trimming of these elements to the required accuracy. However, this raises the cost of the system considerably. An alternative method is to randomize the errors introduced by the imperfections (i.e., destroy the correlation between the signal and the D/A noise). This has been achieved by constructing the D/A converter from equal-valued components, resistors or capacitors, and choosing for each conversion a different set of elements according to some pseudorandom rule [66]. Thus, there will be very little correlation between the input signal and the conversion error, and the spectrum of the error power will be nearly white. Therefore, the in-band noise power will be reduced by the oversampling ratio. Alternatively, the elements of the D/A converter may be rotated in and out of the circuit by using a "barrel-shifting" scheme [85]. Let the frequency of the periodic element allocation be $f_{sw}$, which is chosen to be well above the baseband. Since the barrel-shifting process is approximately equivalent to an amplitude modulation, with $f_{sw}$ acting as carrier and the conversion error as the modulating signal, ideally the error spectrum is shifted by $f_{sw}$ out of baseband. In practice, variation of the signal $y_i$ with time interferes with the process, and the baseband error is merely reduced but not eliminated. Compared to the randomizing strategy, a 10- to 20-dB improvement in signal-to-noise ratio resulted [85].

A reduction in the quantization noise resulting from two-level modulation has been achieved by merely including a third level in the quantizer [51]. It was possible to preserve the D/A converter linearity because a single reference voltage was used to provide all three levels.

An alternative approach to the realization of a modulator with multibit internal quantizer is to use digital correction of the D/A conversion error [38]. The basic scheme is shown in Fig. 31. The D/A converter and the random-access memory (RAM) have the same inputs, and they are both sampled-data stages. Hence, if the RAM is programmed to provide the exact digital equivalent of the D/A analog output level for any digital input word, the D/A and RAM output signals will be equivalent to each other for any D/A output. In the baseband the D/A output signal spectrum is very nearly the same as the input signal spectrum, due to the high-gain feedback loop. Hence, the baseband spectrum of the digital RAM output data will also be very close to the analog input spectrum. The programming of the RAM can be performed on-chip by changing the configuration of the modulator stages and feeding the system with a multibit digital ramp. This calibration may be performed at power-up time or continuously, using duplicated components. The correction principle is applicable to both the single-stage and cascade modulator configurations.

A recently proposed technique [48] combines the simplic-

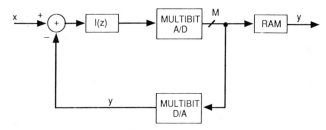

Fig. 31. Correction for imperfection in the multibit D/A using a random-access look-up table for compensating the A/D code. The digital output of the RAM corresponds to the analog output of the D/A.

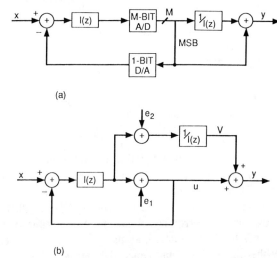

Fig. 32. (a) A technique for cascading a single-bit modulator with a PCM encoding of its error. (b) An equivalent circuit for this modulator. $e_1$ represents the single-bit quantization error, and $e_2$ the multibit quantization error.

ity and inherent linearity of the single-bit D/A converter with the reduced quantization error obtainable with a multibit A/D converter. The basic system implementing this modulator was discussed in Section 2.4.2 and is shown in Fig. 32a. This figure illustrates that only the most significant bit of the multibit word from the A/D is fed back to the D/A. An equivalent system in Fig. 32b indicates that the system consists of a single-bit delta-sigma modulator stage and a cascaded second stage, which here is not a delta-sigma loop but a multibit A/D converter combined with a digital postfilter $1/I$. Linearized analysis reveals that the $z$-transformed outputs of the two stages are

$$U = \frac{IX + E_1}{1 + I} \qquad (39)$$

and

$$V = \frac{X - E_1}{1 + I} + \frac{E_2}{I}, \qquad (40)$$

so that the overall output is

$$Y = U + V = X + \frac{E_2}{I}. \qquad (41)$$

Because $e_2$ is the quantization error of the $M$-bit A/D

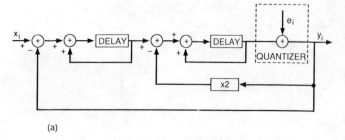

(a)

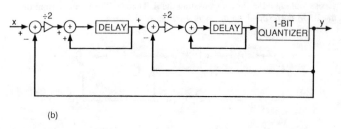

(b)

Fig. 33. (a) A second-order delta-sigma quantizer in which both integrators have delay in their forward path; a gain of 2 in the inner feedback compensates for the extra delay. (b) A scaling of signals in the integrators.

converter, the error in $y$ is $2^{M-1}$ times smaller than that in $u$, which is the output of the single-bit modulator. Notice that the entire A/D converter here is not in the feedback loop so that its nonlinearities are not suppressed by the feedback action. However, the error signal due to the multibit A/D nonidealities is high-pass-filtered, along with the quantization noise, by the digital filter, and hence they are suppressed in baseband.

Two modifications may make the system of Fig. 32a even more useful. The multibit A/D converter may be realized by a more economical pipelined system, since it is not in a feedback loop. This makes it necessary to delay $u$ to ensure proper timing. Also, the quantization noise may be further reduced if $x$ is subtracted from the second-stage input before A/D conversion. This allows interstage scaling, which reduces $e_2$.

### 3.3 Scaling Signal Amplitudes in Modulators

The amplitude of signals propagating in stages of a modulator can be scaled to make full use of their dynamic ranges and to simplify the design of the circuits. Such scaling may also arrange for the clipping which naturally occurs at the output of amplifiers to limit the range of signal amplitudes in accordance with the needs of the modulator design.

We illustrate these techniques by a few examples. Figure 33a shows a second-order delta-sigma modulator which includes a gain of 2 in its inner feedback path [23]. This compensates for the delay that has been moved into the forward path of the first integrator. It is easy to show that this circuit is functionally equivalent to the circuit in Fig. 13. It is, however, much easier to implement because both of its integrators introduce a one-sample delay into the signal path. A further improvement is illustrated in Fig. 33b, which assumes single-bit modulation. The 0.5 attenuation [64] at the input to both integrators ensures that they have sufficient

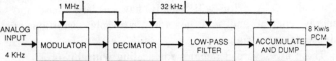

Fig. 34. Decimating the output of the modulator in two stages, from 1 MHz to 32 kHz and then to 8 kHz.

range to accommodate the signals, as explained in Section 2.3.4, *Range of Integration*.

The signal amplitudes in the cascade structure in Fig. 28 usually need to be scaled. For example, if the loop filters $I_i$ are simple integrators, then the signal $x_{i-1} - e_{i-1}$ fed to the second stage has twice the range of the input $x$ to the first stage. This causes the quantization error of the second stage to be effectively twice as large as that of the first one. This can be corrected by feeding the second stage with the difference between the input and output signals of the quantizer of the first stage; that is,

$$X_2 = (x_{i-1} - e_{i-1}) - (x_{i-1} + e_i - e_{i-1}) = -e_i, \quad (42)$$

which has the same range as $x$. The same result may be achieved by subtracting a delayed input $x_{i-1}$ from the output of the first-stage integrator in Fig. 28 before feeding it to the second stage [10].

### 4. DECIMATING THE MODULATED SIGNAL

#### 4.1 Multistage Decimation

The output of the modulator represents the input signal together with its out-of-band components, modulation noise, circuit noise, and interference. The digital filter shown in the encoder of Fig. 2 serves to attenuate all of the out-of-band energy of this signal so that it may be resampled at the Nyquist rate without incurring significant noise penalty because of aliasing.

A fairly simple filter would suffice to remove the modulation noise alone because its spectrum rises slowly; for example, the slope is 12 dB per octave for second-order delta-sigma modulation. But abrupt low-pass filters are often needed to remove the out-of-band components of the signal, and such filters are expensive to build at elevated sampling rates. In practice it nearly always pays to perform the decimation in more than one stage [97]. This is illustrated in Fig. 34, using the example of 4-kHz telephone signals that have been modulated at 1 MHz. The first stage of decimation lowers the word rate from 1 MHz to 32 kHz, an intermediate decimation frequency which is four times the Nyquist rate. The filter of this stage is designed predominantly to remove modulation noise, because this noise dominates at high frequencies. Out-of-band components of the signal dominate at the lower frequencies, and these are attenuated by the low-pass filter before the signal is resampled at its Nyquist rate by the accumulate-and-dump circuit. As the signal propagates through the filters and resampling stages, the word length increases from 1 to 16 bits in order to preserve the resolution. We describe the design of these decimating circuits

18

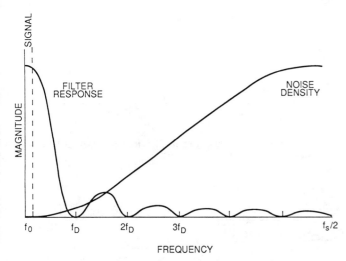

Fig. 35. Decimating with a filter having sinc² frequency response. $f_s$ is the modulation rate, $f_D$ the intermediate decimation frequency, and $0 \le f \le f_0$ the signal band.

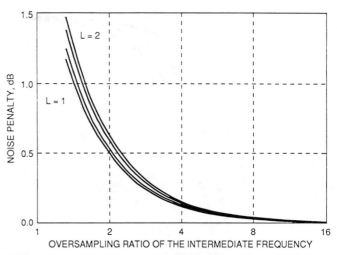

Fig. 36. The increase in noise caused by decimation with sinc$^{L+1}$ filters. $N = f_s/f_D$ is the decimation ratio. Results are plotted for first-order delta-sigma modulation $L = 1$, and second-order $L = 2$.

individually and explain why the intermediate frequency is chosen to be four times the Nyquist rate.

### 4.2 Design of the First-Stage Decimator

Figure 35 illustrates the action of the decimator. Frequencies below $f_0$ form the signal band, $f_D$ is the intermediate decimation frequency, and $f_s$ is the modulation frequency. The raised-cosine curve represents the spectral density of the quantization noise arising from second-order modulation. When this noise is sampled at $f_D$, its components in the vicinity of $f_D$ and harmonics of $f_D$ fold into the signal band. Consequently, it is sensible to place the zeros of the decimating filter at these frequencies. There is no need for an abrupt cutoff at $f_0$ because noise in the range $f_0$ to $f_D - f_0$ folds on itself without entering the signal band. A small droop in the response over the signal band can easily be compensated in the next stage of decimation.

A convenient filter for this decimation has a frequency response based on sampled $\mathrm{sinc}\,(\pi f/f_D)$ functions. An example is shown in Fig. 35. The simplest of these decimators is the accumulate-and-dump circuit. If its input samples are $x_i$ occurring at rate $f_s$ and its output samples are $y_k$ occurring at $f_D$, then

$$ y_k = \frac{1}{N} \sum_{i=N(k-1)}^{Nk-1} x_i, \tag{43} $$

where the decimation ratio $N$ is the integer ratio of the input frequency to the output frequency,

$$ N = \frac{f_s}{f_D}. \tag{44} $$

The transfer function of the filter is

$$ H(z) = \frac{Y(z)}{X(z)} = \frac{1}{N} \sum_{i=0}^{N-1} z^{-i} = \frac{1}{N} \frac{1 - z^{-N}}{1 - z^{-1}}, \tag{45} $$

and its frequency response is given for $z = e^{j\omega\tau}$ by

$$ H(e^{j\omega\tau}) = \frac{\mathrm{sinc}\,(\pi f N \tau)}{\mathrm{sinc}\,(\pi f \tau)}. \tag{46} $$

This has zeros at $f_D$ and all of its harmonics in the range $f_0 \le f < f_s$. This simple filter was used for decimation in oversampled A/D converters [5] at a time when it was important to have simple digital processing circuits. Much better performance can now be obtained by using a filter function that is represented by a product of sinc functions [89, 90, 91]. It has been shown [15, 89, 91] that a filter function $[\mathrm{sinc}\,(\pi f N \tau)/\mathrm{sinc}\,(\pi f \tau)]^{(L+1)}$ is close to being optimum for decimating the signal from delta-sigma modulators of order $L$ that have noise spectral density given by Eq. (21). The penalty for using this decimation is typically less than 0.5 dB increase in noise. Figure 36 plots the penalty against the oversampling ratio $(2f_0 N \tau)^{-1}$ of the intermediate frequency.

When the intermediate frequency is four times the Nyquist rate, the penalty is about 0.14 dB, but it increases as the intermediate frequency is lowered. Another factor that influences the choice of intermediate frequency is the droop in the frequency response of the filter at the edge of the signal band $f_0$. This is plotted in Fig. 37 against the intermediate oversampling ratio for filters of various orders and decimation ratios. With third-order decimation and an intermediate oversampling ratio of 4, the droop is about 2.75 dB, but it increases rapidly if the intermediate oversampling ratio is lowered. It usually is inconvenient to compensate for more than 3 dB of droop.

Besides attenuating the modulation noise, the filter must also provide sufficient attenuation of the high-frequency components of the signal that alias into the signal when resampled at the intermediate frequency. We can see in Fig. 35 that this attenuation is least at the frequency $f_D - f_0$. The attenuation at this frequency is plotted in Fig. 38 for various conditions;

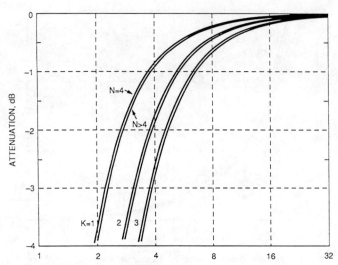

Fig. 37. The attenuation of $\text{sinc}^K$ decimation filters at the edge of the signal band $f_0$. This droop needs to be compensated.

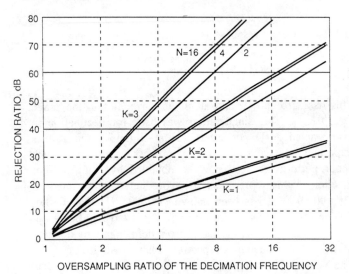

Fig. 38. A graph of $\text{sinc}^K\{\pi(f_D - f_0)N\tau\}/\text{sinc}^K\{\pi(f_D - f_0)N\}$. It is the attenuation of out-of-band components of the signal at frequency $f_D - f_0$ for $\text{sinc}^K$ decimation. $N$ is the decimation ratio $f_D/f_0$. This attenuation should meet the antialiasing requirements of the application. This graph may also be used to predict the attenuation of an interpolator, in which case the abscissa is the oversampling ratio at the input to the stage.

it is about 50 dB for an intermediate oversampling ratio of 4, third-order decimation $\text{sinc}^3$, and decimation ratios $N \geq 16$.

An intermediate oversampling ratio of about 4 and $\text{sinc}^K$ decimation has favorable characteristics for use with delta-sigma modulators in many applications. Lower intermediate oversampling ratios result in rapidly deteriorating characteristics, while higher ratios give less favorable design requirements for the low-pass filter in the next stage. The results in Fig. 36 do not apply to modulators that have sharply rising noise spectral densities, such as described in Section 2.3.6. One of the penalties of using these modulators is that they require more complex decimation filters [62, 69, 82] than do the ordinary delta-sigma modulators that have the noise spectral densities of (21) that rise more slowly with frequency.

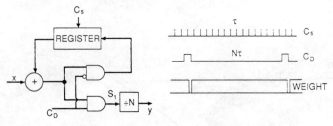

Fig. 39. An accumulate-and-dump circuit. $C_S$ is the input clock, and $C_D$ is the output clock.

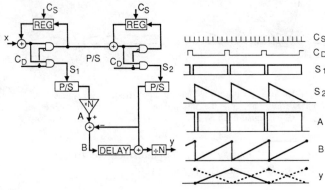

Fig. 40. A $\text{sinc}^2$ decimating circuit. A diagram of weighting factors of the input samples in the summation at various points in the circuit is given. P/S denotes a parallel to serial change in format.

We show in the next section that $\text{sinc}^K$ decimators have very simple implementations.

### 4.3 Implementing Sinc Decimators

Good designs of decimating circuits include resampling within the filter in such a way that unneeded values are not calculated [93]. The input section of the filter processes the bits of short words in parallel at the high input rate. After resampling, the later section of the filter can process the bits of the longer words serially at the lower output frequency.

Figure 39 shows an implementation of the accumulate-and-dump function given by Eq. (45). Input words are added to the contents of the register. The sum is placed back in the register, except at the time of every $N$th input word, when the sum $S_1$ is diverted to the output while the content of the register is reset.

Figure 40 shows a related implementation of the $\text{sinc}^2$ decimation [90]. It comprises a cascade of two accumulate-and-dump circuits. Their outputs, $S_1$ and $S_2$, are separately converted from parallel to serial words and combined to give the net output. The output of the first accumulator, $S_1$, is $NH(z)X(z)$, where $H(z)$ is given by (45), and the output of the second is given by

$$\frac{S_2}{X} = \sum_{i=0}^{N-1} \sum_{k=0}^{i} z^{-k} = \sum_{i=0}^{N-1} (i+1) z^{-i}$$

$$= \frac{1 - z^{-N}}{(1 - z^{-1})^2} - \frac{N z^{-N}}{1 - z^{-1}}. \qquad (47)$$

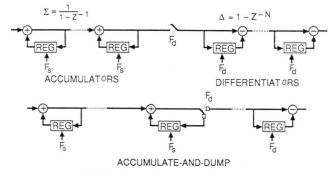

Fig. 41. A sinc$^K$ decimating circuit that comprises $K$ accumulators, followed by resampling and $K$ differentiators. All additions are modulo $2^n$, $n$ being the number of bits required in the output word. In the second circuit resampling is performed by an accumulate-and-dump.

An expression for the signal filtered by the second-order sinc function can then be constructed as

$$\frac{1}{N^2}\left(\frac{1-z^{-N}}{1-z^{-1}}\right)^2 X = \frac{1}{N^2}\{(NS_1 - S_2)z^{-N} + S_2\}, \quad (48)$$

and this can be implemented at the low output word rate because the expression does not involve a short delay $z^{-1}$ associated with the fast clock.

These decimating circuits are relatively simple to implement because the first accumulator needs to hold no more than $\log_2(N) + b$ bits and the second only $\log_2(N(N+1)/2) + b$ bits, $b$ being the number of bits in an input word. When $N$ is a power of 2, the multiplication and division by $N$ in Figs. 39 and 40 are merely changes in the significance of bits.

A third-order decimator can be designed by appending a third accumulator which generates a sum $S_3$. This is combined with the other accumulations according to the relationship

$$\left(\frac{1-z^{-N}}{1-z^{-1}}\right)^3 X = (1-z^{-N})^2 S_3 + Nz^{-N}(1-z^{-N})$$

$$\cdot\left(S_2 + \frac{S_1}{2}\right) + \frac{N^2 z^{-N}}{2}(1+z^{-N})S_1. \quad (49)$$

An alternative method for designing decimators that gives simpler circuits for third- and higher-order filters [92, 99] is illustrated by the first circuit in Fig. 41. The input signal is fed to a cascade of $K$ accumulators which, in normal operation, are not reset. This provides the filter function $(1-z^{-1})^{-K}$. The signal is then resampled at rate $f_D$ and fed to a cascade of $K$ differentiators to generate the decimated function $[(1-z^{-N})/(1-z^{-1})]^K$ of the input.

The objection to this method is that the accumulators must be very large if sufficient bits are to be provided to prevent their overflowing. This difficulty is overcome by employing modulo arithmetic [96]. Each accumulator and differentiator holds only sufficient bits to accommodate the length of an output word; no more than $K\log_2(N) + b$ bits are needed. The circuits are permitted to overflow naturally. It can be shown that this overflow does not affect the net output of the

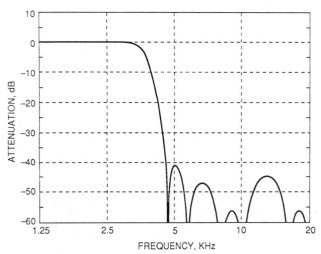

Fig. 42. The frequency response of a low-pass filter used for decimating 4-kHz telephone signals from 40 to 8 kilowords/s. Its transfer function is given by

$$(1 - 3/2z^{-1} + z^{-2})/(1 - 11/8z^{-1} + 5/8z^{-2})$$
$$\cdot(1 - 5/4z^{-1} + z^{-2})/(1 - 101/64z^{-1} + 7/8z^{-2})$$
$$\cdot(1 + z^{-1})(1 + z^{-2})(1 - z^{-5})/(1 - z^{-1})$$

decimator. The circuit can be further simplified by replacing one accumulator, one differentiator, and the resampling switch by an accumulate-and-dump circuit, as shown by the second circuit in Fig. 41.

### 4.4 The Low-Pass Filter

The low-pass filter in the final stage of the decimation in Fig. 34 is designed to meet the antialiasing requirements of the input signal. Its circuit can usually be very simple [93], and because the word rate $f_D$ is low, it can process bits serially rather than in parallel. Moreover, the word length of the coefficients are so short that dedicated multiplier circuits can be used rather than shared ones [90]. The accumulate-and-dump circuit that performs the final resampling at the Nyquist rate has a frequency response of the form given by Eq. (46). It contributes useful zeros to the decimating filter function.

Figure 42 shows the frequency response of a low-pass filter that is intended for decimating sampled 4-kHz telephone signals from an intermediate sampling frequency of 40 kHz to the Nyquist rate of 8 kHz. Zeros in the response near 4.5 and 6 kHz are provided by two recursive sections. Zeros at 10 and 20 kHz lie on the imaginary and negative axes of the z-plane. Zeros at 8 and 16 kHz are provided by an accumulate-and-dump circuit. The poles of the recursive sections are chosen to flatten the in-band response. These circuits are easy to build [90] because the word length of the coefficients is short.

### 5. Oversampling D/A Converters

#### 5.1 Demodulating Signals at Elevated Word Rates

The lower part of Fig. 2 showed an outline of an oversampled D/A converter. In this circuit, a digital filter was used

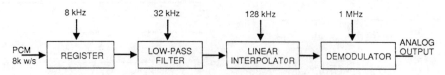

Fig. 43. An oversampling decoder for 4-kHz signals.

to interpolate sample values of the input signal in order to raise the word rate to a frequency well above the Nyquist rate [93]. A demodulator then truncates the words and converts the signal to analog form at the high sample rate. In most applications it is more efficient to raise the word rate of the signal in stages, in much the same way as it was decimated in stages, in the encoder. We illustrate the details of the over-sampling techniques for D/A conversion by an example that processes 4-kHz telephone signals encoded into 16-bit words at 8 kHz. Figure 43 shows an outline of this oversampled D/A converter. The input words are placed in a register from which they are fed to the low-pass filter at 32 kHz; each word is repeated four times. The output of this filter resembles the PCM encoding of the signal at 32 kHz. The next stage is a linear interpolator [90] that inserts three new words between each pair of 32-kHz words, raising the rate to 128 kHz. These words are placed in a register from where they are fed to a demodulator at 1 MHz; each word is repeated eight times. The demodulation rounds off the code to 1-bit words with a feedback quantizer, converts them to analog form, and smoothes them with an analog filter. The 1-MHz word rate is sufficiently high so that the quantization noise introduced into the signal is small, and the requirements of the analog smoothing filter are simple.

The filtering action [93] that is inherent in the processes that raise the word rate from 8 kHz to 1 MHz serves to smooth out the sampling images in the spectrum of the digital signal. Figure 44 illustrates the action of these interpolation processes: graph (a) represents the baseband signal, and (b) its spectrum when sampled at the Nyquist rate; (c) is the frequency response of the low-pass filter including that of the input holding register. Also (d) and (e) represent the output spectrum of the low-pass filter, drawn on different frequency scales, (f) is the sinc$^2$-shaped frequency response of linear interpolation, and (g) is the result of this interpolation. Finally, (h) is the frequency response of the last register which holds the signal between 128-kHz sample times, and (i) is the spectrum of its output, read at 1 MHz.

The requirements for attenuating the out-of-band images of the signal spectrum while decoding are usually less severe than were the requirements for attenuating signal components that alias into the signal while encoding. Consequently, the structure of the low-pass filter used in the encoder of Fig. 34 can conservatively be also used for the low-pass filter in the corresponding decoder of Fig. 43. The zeros in the filter response, which were provided by the accumulate-and-dump at the output of the A/D converter, are now provided by the holding register at the input of the D/A converter.

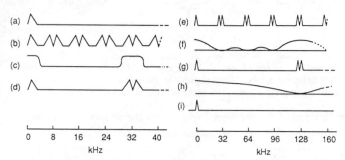

Fig. 44. Spectra of signals, and frequency responses of filters used for interpolating sample values. (a) the spectrum of the signal; (b) the spectrum of the sampled signal; (c) the low-pass filter response; (d) & (e) the spectrum of the filter output signal on different scales; (f) frequency response of linear interpolation; (g) the spectrum of the interpolated signal; (h) frequency response of the holding register; (i) the spectrum of the held signal.

### 5.2 Interpolating with Sinc$^K$-Shaped Filter Functions

The frequency responses associated with the linear interpolation circuit and the holding stages are sampled sinc$^K$ functions. The amount by which they attenuate images will next be calculated in order to determine good values for the intermediate sampling frequencies used in the circuit. For this purpose, let $f_I$ be the rate at which digital words are applied to an interpolator stage, and $Nf_I$ the rate at which words emerge at the output. Here, $N$ is defined as the interpolation ratio. The frequency response of this class of interpolation stages can be expressed as

$$I(f) = \frac{\text{sinc}^K\left(\pi f/f_I\right)}{\text{sinc}^K\left(\pi f/Nf_I\right)}, \tag{50}$$

where $K$ is the order of the interpolation. Images of the input signal will be situated adjacent to $f_I$ and its harmonics (i.e., in the frequency ranges $kf_I \pm f_0$, as illustrated in Fig. 44). The attenuation of the unwanted images is least at the frequency $f_I - f_0$. Its value can be read directly from the graph in Fig. 38 originally given for evaluating decimators. The relevant value of the abscissa is now the oversampling ratio of $f_I$ at the input to the stage of interpolation (i.e., $f_I/2f_0$).

As an example, some telephone applications require that the images be attenuated by at least 28 dB. For interpolation ratios $N \geq 4$ this is achieved by using a holding register, which corresponds to a sample-and-hold interpolation, $K = 1$, and an input oversampling ratio at least equal to 16 (i.e., sampling at or above 128 kHz for 4-kHz signals). Linear interpolation ($K = 2$) is satisfactory for oversampling ratios at least equal to 4 (i.e., sampling at or above 32 kHz for our example). These results are conservative because the low-pass

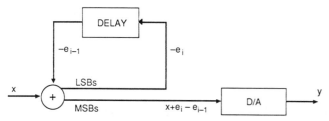

Fig. 45. A digital quantization with error feedback.

analog filter at the output of the decoder also contributes to the smoothing of the images, and the low-pass digital filter at the input can significantly attenuate the signal at the upper edge of its band from 3.5 to 4 kHz.

## 5.3 The Demodulator Stage

### 5.3.1 Quantizing the Digital Signal.
The first stage of the demodulator rounds off the long digital words to short ones, preferably all the way to a single-bit code which can be accurately converted to analog. Circuit structures of these quantizers resemble those of the modulators described in Section 2. The main difference between them is that the processing in the demodulator is digital instead of analog, and hence there is little trouble in achieving precision. The properties of the quantization noise derived in Section 2 can be used to determine the demodulation rates required to ensure adequately low quantization noise in the signal band. The need for this quantization distinguishes oversampling D/A converters from their conventional counterparts. Conventional D/A converters contain no such intentional source of error, but they are much more sensitive to analog circuit imprecision.

Just as there are several forms of oversampling modulators, so there are corresponding forms of demodulators. In general, there is no clear general advantage of one demodulator structure over another; the choice depends on the requirements of the application and on the technology available. Because much of the signal processing is analog in modulators and digital in demodulators, the trade-offs in their design differ. We next discuss the design of several demodulator circuits.

### 5.3.2 Quantization with Error Feedback.
The error feedback circuit shown in Fig. 23 is unsuitable for use as a modulator because of its sensitivity to errors in the analog subtractors, but this sensitivity is not a major concern in the design of the digital circuits of a demodulator. Figure 45 shows a digital implementation of an error feedback quantizer [111]. It uses digital codes that represent positive values. The sum generated by the adder is quantized by using only its most significant bits as the output. The remaining bits then represent the negative of the quantization error. These least significant bits are delayed and added to the next input sample for error correction. In the extreme case of single-bit quantization, only the carry signal from the adder constitutes the output, and all the sum bits are fed back.

The noise introduced into the signal by this quantization is the same as that resulting from first-order delta-sigma modu-

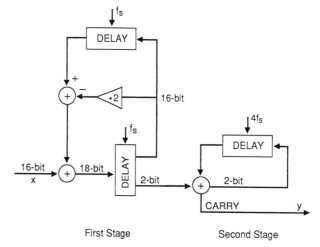

Fig. 46. A second-order error feedback digital quantizer with auxiliary first-order quantization to 1-bit code.

lation, and is given by Eq. (8). For busy signals, its spectrum is given by Eq. (9), and the noise power in the signal band by Eq. (11). For dc inputs the noise has the pattern structure illustrated in Fig. 8. For this and other reasons, first-order demodulation does not find wide application today, and higher-order circuits are usually preferred. The circuit was useful at times when it was important to have only very simple circuits in the demodulator [111].

Higher-order noise shaping can be achieved by replacing the delay in the feedback path by a predictor filter [97]. Such a system is shown in Fig. 46. The first stage of the circuit is a digital quantizer with second-order noise shaping. This circuit requires that its output words contain at least 2 bits if clipping of signal amplitudes in the feedback loop is to be avoided. Including a clipping circuit in order to get single-bit outputs causes significant increase in the noise, no matter where the clipping occurs. To avoid this penalty, the second stage of the circuit in Fig. 46 is included to perform a first-order quantization of the 2-bit words to 1-bit ones [108]. This stage is clocked four times faster than the first stage. The technique is useful only when this higher frequency is achievable, but then one also has the option of avoiding the need for the second stage altogether by clocking the first stage at a high enough rate to allow for the increased noise caused by clipping the feedback signal.

### 5.3.3 Cascaded Demodulators.
Demodulator circuits can be cascaded [105] to reduce their noise in the same way that modulators were cascaded, as discussed in Section 2.4. Figure 47 shows a cascade of two first-order quantizers in which the derivative of the second-stage output signal is added to the output of the first-stage. Even when the outputs of the individual stages are single-bit words, the net output contains 2-bit words. This is an advantage for cascaded modulators because it allows the output to oscillate between four levels and to avoid introducing excess noise into larger signals. But for demodulation this advantage is outweighed by the need for high precision in the four-level D/A circuits required.

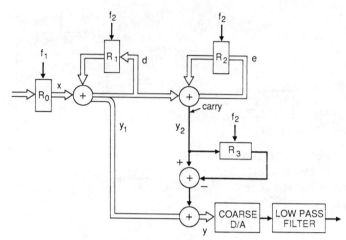

Fig. 47. A cascade of two first-order quantizers with digital combining of their outputs, and multilevel D/A.

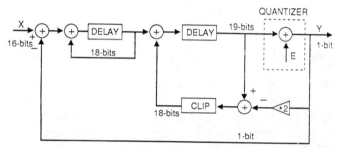

Fig. 48. Equivalent circuit of a second-order digital delta-sigma quantizer.

This difficulty is overcome to some extent by providing a separate D/A converter at the output of each quantizer. Then the signal passes only through the first-stage D/A, which should be a single-bit converter to avoid distorting the signal. Only the noise passes through the second converter, and misalignment of its levels result only in imperfect cancellation of noise and not in signal distortion. Reference [105] describes a demodulator in which the differentiation of the output of the second stage is performed in the analog circuit. Then both D/A converters can be single-bit ones. The precision required in matching the two D/A converters and the analog combining circuit is similar to the accuracy requirement of parameter $g$ in Eq. (24).

### 5.3.4 Circuit Design for Delta-Sigma Demodulation.
A digital implementation of an ordinary delta-sigma modulator can be used to quantize the signal in a demodulator. There are many ways of designing these circuits, and Fig. 48 gives one example of a second-order quantizer. This circuit introduces noise, which is described by Eqs. (19) and (20). Although it is not troubled by leakage in the accumulators, tones can be present in the noise. The randomness in the oscillation pattern of second-order modulators depends on avoiding signals that are rational multiples of $\Delta$ in the integrators. This is not usually possible in digital implementations, where the signal values are always rational. One method of ensuring randomness is to inject a relatively large dither signal. Another is to add a random bit-pattern to the

least significant bit of the input to mimic the effect of having irrational values stored in the first accumulator.

Structures used for higher-order feedback quantizers of the type described in Section 2.3.6 may also be used in demodulators. It may also be possible to use high-order digital delta-sigma quantizers with digital control circuits that prevent them from going into saturating limit cycles.

### 5.4 Design of Filters for Smoothing the Output

The analog filter at the output of an oversampling D/A converter serves mainly to remove the out-of-band quantization noise introduced by the demodulator. Digital interpolation filters have already removed the lower-frequency sampling images of the signal. Relatively simple filters can often be used. For example, with second-order delta-sigma demodulation the noise spectrum rises at about 12 dB per octave, and it is small at frequencies near the signal band. The requirements of 4-kHz telephone signals can be met by using a filter characteristic that falls 12 dB per octave at frequencies above 40 kHz. More complex filters are needed, however, when higher-order demodulators are used. In some cases, it is hard to achieve the required linearity and low harmonic distortion in the smoothing filter, and special design techniques must be used.

## 6. CONCLUDING REMARKS

Oversampling methods can provide very high resolution, even when relatively inaccurate analog components are used. For example, 20-bit performance has been reported for 20-kHz audio applications. The question that arises is, what are the factors that limit the performance of the oversampling converters? It appears that achieving adequate precision of circuit component values is not a serious problem, provided that the modulator and demodulator are properly designed.

The speed at which circuits can operate may be too limited currently for some applications. For example, even with a third-order loop filter, a 9-bit conversion of 3.5-MHz video signals requires modulation rates in excess of 100 MHz. Such speeds can be achieved only by using submicron silicon or GaAs technology, which is expensive and not widely available. We expect that such rates will be commonly available in the near future, and oversampling methods will find frequent use in these higher-frequency applications.

Increased sampling rates can, at least in theory, increase indefinitely the amplitude resolution of the lower-frequency applications, but the resolution will be limited by the noise of circuit elements. The natural increase in noise power with increased circuit bandwidth is not a major concern, because only the baseband noise passes through the digital filters. Indeed, the higher-frequency circuit noise can be a useful dither signal. However, transistors that are designed for high-frequency use tend to have higher noise densities at low frequencies than do lower-frequency transistors. Thus, the circuit noise will probably be the limiting factor on the performance of these converters. Even when very-high-frequency technologies are available, there will still be an incentive to find ways of lowering the oversampling ratio and so reduce circuit noise as well as extend the signal frequency range. This indicates that the search for practical configurations incorporating high-order loop filters and multibit converters will continue in the foreseeable future.

# Bibliography
## to Oversampling Methods
## for A/D and D/A Conversion

Reference numbers with asterisks refer to papers in this book.

## 1. Basic Concepts and Analysis

[1] D. Anastassiou, "Error diffusion coding for A/D conversion," *IEEE Trans. Circuits Sys.*, vol. CAS-36, pp. 1175–1186, Sept. 1989.

[2]* S. H. Ardalan and J. J. Paulos, "An analysis of nonlinear behavior in delta-sigma modulators," *IEEE Trans. Circuits Sys.*, vol. CAS-34, pp. 593–603, June 1987.

[3] W. R. Bennett, "Spectra of quantized signals," *Bell Sys. Tech. J.*, vol. 27, pp. 446–472, July 1948.

[4] L. M. Blumberg, "Invariants and the characteristics of sigma-differential pulse code modulation," *IEEE Proc. ISCAS'88*, pp. 1569–1572, June 1988.

[5]* J. C. Candy, "A use of limit cycle oscillations to obtain robust analog-to-digital converters," *IEEE Trans. Commun.*, vol. COM-22, pp. 298–305, Mar. 1974.

[6]* J. C. Candy and O. J. Benjamin, "The structure of quantization noise from sigma-delta modulation," *IEEE Trans. Commun.*, vol. COM-29, pp. 1316–1323, Sept. 1981.

[7]* W. Chou, P. W. Wong, and R. M. Gray, "Multistage sigma-delta modulation," *IEEE Trans. Inform. Theory*, vol. IT-35, pp. 784–796, July 1989.

[8] W. Chou and R. M. Gray, "Dithering and its effects on sigma-delta and multi-stage sigma-delta modulation," *IEEE Proc. ISCAS'90*, pp. 368–371, May 1990.

[9] C. C. Cutler, "Transmission systems employing quantization," 1960 U.S. Patent No. 2,927,962 (filed 1954).

[10] J. Das and P. K. Chatterjee, "Optimized ΔΔ modulation system," *Elect. Lett.*, vol. 3, pp. 286–287, June 1967.

[11] E. C. Dijkmans and P. J. A. Naus, "Sigma-delta versus binary weighted A/D conversion, what is the promise?" *Proc. 15th European Solid-State Circuit Conf.*, pp. 35–63, Sept. 1989.

[12] D. J. Goodman, "The application of delta modulation to analog-to-digital encoding," *Bell Sys. Tech. J.*, vol. 48, pp. 321–343, Feb. 1969.

[13] D. G. Goodman and L. J. Greenstein, "Quantizing noise of dm/pcm encoders," *Bell Sys. Tech. J.*, vol. 52, pp. 183–204, Feb. 1973.

[14]* R. M. Gray, "Oversampled sigma-delta modulation," *IEEE Trans. Commun.*, vol. COM-35, pp. 481–489, May 1987.

[15] R. M. Gray, "Spectral analysis of quantization noise in a single-loop sigma-delta modulator with dc input," *IEEE Trans. Commun.*, vol. COM-37, pp. 588–599, June 1989.

[16] R. M. Gray, W. Chou, and P.-W. Wong, "Quantization noise in single-loop sigma-delta modulation with sinusoidal input," *IEEE Trans. Commun.*, vol. COM-37, pp. 956–968, Sept. 1989.

[17]* R. M. Gray, "Quantization noise spectra," *IEEE Trans. Information Theory*, vol. IT-36, pp. 1220–1244, Nov. 1990.

[18] *Ning He, F. Kuhlmann, and A. Buzo, "Double-loop sigma-delta modulation with dc input," *IEEE Trans. Commun.*, vol. COM-38, pp. 487–495, April 1990.

[19] M. W. Hauser, "Principles of oversampling A/D conversion," *J. Audio Eng. Soc.*, vol. 39, pp. 3–26, Jan.–Feb. 1991.

[20]* H. Inose, Y. Yasuda, and J. Murakami, "A telemetering system code modulation—Δ – Σ modulation," *IRE Trans. Space Elect. Telemetry*, vol. SET-8, pp. 204–209, Sept. 1962.

[21] H. Inoise and Y. Yasuda, "A unity bit coding method by negative feedback," *Proc. IEEE*, vol. 51, pp. 1524–1533, Nov. 1963.

[22] J. E. Iwersen, "Calculated quantizing noise of single-integration delta-modulation coders," *Bell Syst. Tech. J.*, vol. 48, pp. 2359–2389, Sept. 1969.

[23] G. Lainey, R. Saintlaurens, and P. Senn, "Switched-capacitor second-order noise-shaping coder," *Elect. Lett.*, vol. 19, pp. 149–150, Feb. 1983.

[24] H. S. McDonald, "Pulse code modulation and differential pulse code modulation encoders," 1970 U.S. Patent, No. 3,526,855 (filed 1968).

[25] A. N. Netravali, "Optimum filters for interpolative A/D converters," *Bell Sys. Tech. J.*, vol. 56, pp. 1629–1641, Nov. 1977.

[26] G. R. Ritchie, "Higher order interpolation analog to digital converters," *Ph.D. Dissertation*, University of Pennsylvania, 1977.

[27]* T. Ritoniemi, T. Karema, and H. Tenhunen, "The design of stable high order 1-bit sigma-delta modulators," *IEEE Proc. ISCAS'90*, pp. 3267–3270, May 1990.

[28]* H. A. Spang III and P. M. Schultheiss, "Reduction of quantizing noise by use of feedback," *IRE Trans. Commun. Sys.*, pp. 373–380, Dec. 1962.

[29] R. Steele, *Delta Modulation Systems*, New York: Wiley, 1975.

[30]* S. K. Tewksbury and R. W. Hallock, "Oversampled, linear predictive and noise-shaping coders of order $N > 1$," *IEEE Trans. Circuits Sys.*, vol. CAS-25, pp. 436–447, July 1978.

## 2. Design, Simulation Techniques, and Architectures for Oversampling Converters

[31]* B. P. Agrawal and K. Shenoi, "Design methodology of $\Sigma\Delta M$," *IEEE Trans Commun.*, vol. COM-31, pp. 360–370, Mar. 1983.

[32]* T. R. Bishop, J. J. Paulos, M. B. Steer, and S. H. Ardalan, "Table-based simulator of sigma-delta modulators," *IEEE Trans. Circuits Sys.*, vol. CAS-37, pp. 447–451, March 1990.

[33]* B. E. Boser, K. P. Karmann, H. Martin, and B. A. Wooley, "Simulating and testing oversampled analog-to-digital converters," *IEEE Trans. Computer-Aided Design*, vol. 7, pp. 668–673, June 1988.

[34] B. Brandt, and B. Wooley, "A CMOS oversampling A/D converter with 12b resolution at conversion rates above 1 MHz," *ISSCC Dig. Tech. Pap.*, pp. 64–65, Feb. 1991.

[35] B. P. Brandt, D. E. Wingard, and B. A. Wooley, "Second-order sigma-delta signal acquisition," *IEEE J. Solid-State Circuits*, vol. SC-26, pp. 618–627, April 1991.

[36]* J. C. Candy, "A use of double integration in sigma-delta modulation," *IEEE Trans. Commun.*, vol. COM-33, pp. 249–258, Mar. 1985.

[37]* L. R. Carley, "An oversampling analog-to-digital converter topology for high resolution signal acquisition systems," *IEEE Trans. Circuits Sys.*, vol. CAS-34, pp. 83–90, Jan. 1987.

[38]* T. Cataltepe, G. C. Temes, and L. E. Larson, "Digitally corrected multi-bit $\Sigma - \Delta$ data converters," *IEEE Proc. ISCAS'89*, pp. 647–650, May 1989.

[39]* K. C. H. Chao, S. Nadeem, W. L. Lee, and C. G. Sodini, "A

higher order topology for interpolative modulators for oversampling A/D conversion," *IEEE Trans. Circuits Sys.*, vol. CAS-37, pp. 309–318, Mar. 1990.

[40] L. D. J. Eggermont, M. H. H. Hofelt, and R. H. W. Salters, "A delta-modulation to PCM converter," *Philips Tech. Rev.*, vol. 37, pp. 313–329, Nov.-Dec. 1977.

[41]* P. F. Ferguson Jr., A. Ganesan, and R. W. Adams, "One bit higher order sigma-delta A/D converters," *IEEE Proc. ISCAS'90*, pp. 890–893, May 1990.

[42]* A. Gosslau, and A. Gottwalk, "Optimization of a sigma-delta modulator by the use of a slow ADC," *IEEE Proc. ISCAS'88*, pp. 2317–2320, June 1988.

[43]* M. W. Hauser, and R. W. Brodersen, "Circuit and technology considerations for MOS delta-sigma A/D converters," *IEEE Proc. ISCAS'86*, pp. 1310–1315, May 1986.

[44]* M. W. Hauser, "Technology scaling and performance limitations in MOS delta-sigma analog-digital converters," *IEEE Proc. ISCAS'90*, pp. 356–359, May 1990.

[45]* P. J. Hurst and R. A. Levinson, "Delta-sigma A/Ds with reduced sensitivity to op amp noise and gain," *IEEE Proc. ISCAS'89*, pp. 254–257, May 1989.

[46]* L. E. Larson, T. Cataltepe, and G. C. Temes, "Multibit oversampled $\Sigma - \Delta$ A/D converter with digital error correction," *Elect. Lett.*, vol. 24, pp. 1051–1052, Aug. 1988.

[47] W. L. Lee and C. G. Sodini, "A topology for higher order interpolative coders," *IEEE Proc. ISCAS'87*, pp. 459–462, May 1987.

[48]* T. C. Leslie and B. Singh, "An improved sigma-delta modulator architecture," *IEEE Proc. ISCAS'90*, pp. 372–375, May 1990.

[49]* L. Longo and M. Copeland, "A 13 bit ISDN-band oversampled ADC using two-stage third order noise shaping," *IEEE Proc. Custom IC Conf.*, pp. 21.2.1–21.2.4, Jan. 1988.

[50]* Y. Matsuya, K. Uchimura, A Iwata, et al., "A 16-bit oversampling A-to-D conversion technology using triple-integration noise shaping," *IEEE J. Solid-State Circuits*, vol. SC-22, pp. 921–929, Dec. 1987.

[51]* J. J. Paulos, G. T. Brauns, M. B. Steer, and S. H. Ardalan, "Improved signal-to-noise ratio using tri-level delta-sigma modulation," *IEEE Proc. ISCAS*, pp. 463–466, May 1987.

[52] D. Ribner, R. Baertsch, et al., "A 16 b third-order sigma-delta modulator with reduced sensitivity to non-idealities," *ISSCC Dig. Tech. Pap.*, pp. 66–67, Feb. 1991.

[53] D. B. Ribner, "A comparison of modulator networks for high-order oversampling sigma-delta analog-to-digital converters," *IEEE Trans. Circuits Sys.*, vol. CAS-38, pp. 145–159, Feb. 1991.

[54]* J. Robert and P. Deval, "Second-order high-resolution incremental A/D converter with offset and charge injection compensation." *IEEE J. Solid-State Circuits*, vol. SC-23, pp. 736–741, June 1988.

[55]* Y. Shoji and T. Suzuki, "Improved double integration delta-sigma modulations for A/D and D/A conversions," *IEEE Proc. ISCAS'87*, pp. 451–454, May 1987.

[56] C. D. Thompson, "A VLSI sigma-delta A/D converter for audio and signal processing applications," *IEEE Proc. ICASSP'89*, Dec. 1988.

[57]* K. Uchimura, T. Hayashi, T. Kimura, and A. Iwata, "Oversampling A-to-D and D-to-A converters with multistage noise shaping modulators," *IEEE Trans. Acoust., Speech, Signal Processing*, vol. ASSP-36, pp. 1899–1905, Dec. 1988.

[58]* R. H. Walden, T. Cataltepe, and G. C. Temes, "Architectures for high-order multibit $\Sigma\Delta$ modulators," *IEEE Proc. ISCAS'90*, pp. 895–898, May 1990.

[59] L. A. Williams III, and B. A. Wooley, "Third-order cascade sigma-delta modulators," *IEEE Trans. Circuits Sys.*, vol. CAS-38, pp. 489–498, May 1991.

[60]* A. Yukawa, "Constraints analysis for oversampling A-to-D converter structures on VLSI implementation," *IEEE Proc. ISCAS'87*, pp. 467–472, May 1987.

## 3. Implementations and Applications of Oversampling A/D Converters

[61] R. W. Adams, "Companded predictive delta modulation: a low cost conversion technique for digital recording," *J. Audio Eng. Soc.*, vol. 32, pp. 659–672, Sept. 1984.

[62]* R. W. Adams, "Design and implementation of an audio 18-bit analog-to-digital converter using oversampling techniques," *J. Audio Eng. Soc.*, vol. 34, pp. 153–166, Mar. 1986.

[63] R. W. Adams, "An IC chip set for 20-bit A/D conversion," *J. Audio Eng. Soc.*, pp. 440–458, June 1990.

[64]* B. E. Boser and B. A. Wooley, "The design of sigma-delta modulation analog-to-digital converters," *IEEE J. Solid-State Circuits*, vol. SC-23, pp. 1298–1308, Dec. 1988.

[65] J. C. Candy, W. E. Ninke, and B. A. Wooley, "A per-channel A/D converter having 15-segment $\mu$-255 companding," *IEEE Trans. Commun.*, vol. COM-24, pp. 33–42, Jan. 1976.

[66]* L. R. Carley, "A noise-shaping coder topology for 15+ bit converters," *IEEE J. Solid-State Circuits*, vol. SC-24, pp. 267–273, April 1989.

[67] V. Comino, M. Steyaert, and G. C. Temes, "A first-order current-steering sigma-delta modulator," *IEEE Proc. CICC'1990*, pp. 6.3.1–6.3.4, May 1990.

[68] P. Defraeye, D. Rabaey, W. Roggeman, J. Yde, and L. Kiss, "A 3-$\mu$m CMOS digital codec with programmable echo cancellation and gain setting," *IEEE J. Solid-State Circuits*, vol. SC-20, pp. 679–687, June 1985.

[69] B. P. Del Signore, D. A. Kerth, N. S. Sooch, and E. J. Swanson, "A monolithic 20 b delta-sigma converter," *IEEE J. Solid-State Circuits*, vol. SC-25, pp. 1311–1317, Dec. 1990.

[70] J. D. Everard, "A single-channel PCM codec," *IEEE J. Solid-State Circuits*, vol. SC-14, pp. 25–37, Feb. 1979.

[71] P. Ferguson Jr., A. Ganesan, et al., "An 18 b 20 kHz dual sigma-delta A/D converter," *ISSCC Dig. Tech. Pap.*, pp. 68–69, Feb. 1991.

[72] H. L. Fielder and B. Hoefflinger, "A CMOS pulse density modulator for high-resolution A/D conversion," *IEEE J. Solid-State Circuits*, vol. SC-19, pp. 995–996, Dec. 1984.

[73]* V. Friedman, D. M. Brinthaupt, D.-P. Chen, T. Deppa, J. P. Elward, Jr., E. M. Fields, and H. Meleis, "Bit-slice architecture for sigma-delta analog-to-digital converters," *IEEE J. Solid-State Circuits*, vol. SC-24, pp. 274–280, April 1989.

[74]* M. W. Hauser, P. J. Hurst, and R. W. Brodersen, "MOS ADC-filter combination that does not require precision analog components," *ISSCC Dig. Tech. Pap.*, pp. 80–81, Feb. 1985.

[75]* T. Hayashi, Y. Inabe, K. Uchimura, and A. Iwata, "A multistage delta-sigma modulator without double integration loop," *ISSCC Dig. Tech. Pap.*, pp. 182–183, Feb. 1986.

[76]* T. Karema, T. Ritoniemi, and H. Tenhunen, "An oversampled $\Sigma\Delta$ A/D converter circuit using two-stage fourth order modulator," *IEEE Proc. ISCAS'90*, pp. 3279–3282, May 1990.

[77]* R. B. Koch, B. Heise, F. Eckbauer, E. Engelhardt, J. A. Fisher, and F. Parzefall, "A 12-bit sigma-delta analog-to-digital converter with 15-MHz clock rate," *IEEE J. Solid-State Circuits*, vol. SC-21, pp. 1003–1010, Dec. 1986.

[78]* B. H. Leung, R. Neff, P. R. Gray, and R. W. Brodersen, "Area-efficient multichannel oversampled PCM voice-band coder," *IEEE J. Solid-State Circuits*, vol. SC-23, pp. 1351–1357, Dec. 1988.

[79]* K. Matsumoto and R. W. Adams, "An 18b oversampling A/D converter for digital audio," *ISSCC Dig. Tech. Pap.*, pp. 202–203, Feb. 1988.

[80] T. Misawa, J. E. Iwersen, L. J. Loporcaro, and J. G. Ruch, "Single-chip per channel codec with filters utilizing $\Delta - \Sigma$ modulation," *IEEE J. Solid-State Circuits*, vol. SC-16, pp. 333–341, Aug. 1981.

[81]* S. R. Norsworthy, I. G. Post, and H. S. Fetterman, "A 14-bit 80-kHz sigma-delta A/D converter: modeling, design, and perfor-

mance evaluation," *IEEE J. Solid-State Circuits*, vol. SC-24, pp. 256–266, April 1989.

[82]  F. Op't Eynde, G. Yin, and W. Sansen, "A CMOS fourth-order 14 b 500 k sample/s sigma-delta A/D converter," *ISSCC Dig. Tech. Pap.*, pp. 62–63, Feb. 1991.

[83]* N. Rebeschini, van Bavel, et al., "A high resolution CMOS sigma-delta A/D converter with 320 kHz output rate," *IEEE Proc. ISCAS'89*, pp. 246–249, May 1989.

[84]* T. Ritoniemi, T. Karema, H. Tenhune, and M. Lindell, "Fully differential CMOS sigma-delta modulator for high performance analog-to-digital conversion with 5 V operating voltage," *IEEE Proc. ISCAS'88*, pp. 2321–2326, June 1988.

[85]  Y. Sakina, "Multi-bit $\Sigma\Delta$ A/D converters with nonlinearity correction using dynamic barrel shifting," *University of California, Berkeley Report*, Dept. of EECS, 1990.

[86]* J. W. Scott, W. Lee, C. Giancario, and C. G. Sodini, "A CMOS slope adaptive delta modulator," *ISSCC Dig. Tech. Pap.*, pp. 130–131, Feb. 1986.

[87]* D. R. Welland, B. P. Del Signore, E. J. Swanson, T. Tanaka, K. Hamashita, S. Hara, and K. Takasuka, "Stereo 16-bit delta-sigma A/D converter for digital audio," *J. Audio Eng. Soc.*, vol. 37, pp. 476–486, June 1989.

[88]  R. J. van de Plassche, "A sigma-delta modulator as an A/D converter," *IEEE Trans. Circuits Sys.*, vol. CAS-25, pp. 510–514, July 1978.

## 4. *Digital Filters for Oversampling A/D Converters*

[89]* J. C. Candy, Y. C. Ching, and D. S. Alexander, "Using triangularly weighted interpolation to get 13-bit PCM from a sigma-delta modulator," *IEEE Trans. Commun.*, vol. COM-24, pp. 1268–1275, Nov. 1976.

[90]* J. C. Candy, B. A. Wooley, and O. J. Benjamin, "A voiceband codec with digital filtering," *IEEE Trans. Commun.*, vol. COM-29, pp. 815–830, June 1981.

[91]* J. C. Candy, "Decimation for sigma delta modulation," *IEEE Trans. Commun.*, vol. COM-34, pp. 72–76, Jan. 1986.

[92]* S. Chu and C. S. Burrus, "Multirate filter designs using comb filters," *IEEE Trans. Circuit Sys.*, vol. CAS-31, pp. 913–924, Nov. 1984.

[93]* R. E. Crochiere and L. R. Rabiner, "Interpolation and decimation of digital signals—a tutorial review," *Proc. IEEE*, vol. 69, pp. 300–331, Mar. 1981.

[94]  E. Dijkstra, L. Cardoletti, O. Nys, C. Piguet, and M. Degrauwe, "Wave digital decimation filters in oversampled A/D converters," *IEEE Proc. ISCAS'88*, pp. 2327–2330, June 1988.

[95]  E. Dijkstra, M. Degrauwe et al., "A design methodology for decimation filters in signal delta A/D converters," *IEEE Proc. ISCAS'87*, pp. 479–482, May 1987.

[96]* E. Dijkstra, O. Nys, C. Piguet, and M. Degrauwe, "On the use of

modulo arithmetic comb filters in sigma delta modulators," *IEEE Proc. ICASSP'88*, pp. 2001–2004, April 1988.

[97]* D. J. Goodman and M. J. Carey, "Nine digital filters for decimation and interpolation," *IEEE Trans. Acoust., Speech, Signal Proc.*, vol. ASSP-25, pp. 121–126, April 1977.

[98]  M. W. Hauser, and R. W. Brodersen, "Monolithic decimation filter for custom delta-sigma A/D conversion," *IEEE Proc. ICASSP'88*, pp. 2005–2008, April 1988.

[99]  E. B. Hogenaur, "An economical class of digital filters for decimation and interpolation," *IEEE Trans. Acoust., Speech, Signal Proc.*, vol. ASSP-29, pp. 155–162, April 1981.

[100]  B. H. Leung, "Decimation filters for analog to digital converters based on quadratic programming," *IEEE Proc. ISCAS'90*, pp. 899–901, May 1990.

[101]* H. Meleis and P. LeFur, "Novel architecture design for VLSI implementation of a fir decimation filter," *IEEE Proc. ICASSP'85*, pp. 1380–1383, Mar. 1985.

[102]  T. Okamoto, Y. Maruyama, and K. Hinooka, "A 16 b oversampling codec with filtering DSP," *ISSCC Dig. Tech. Pap.*, pp. 74–75, Feb. 1991.

[103]  T. Saramaki, T. Karema, T. Rotoniemi, H. Tenhunen, "Multiplier-free decimator algorithms for superresolution oversampled converters," *Proc. ISCAS'90*, pp. 3275–3278, May 1990.

[104]* T. Saramaki and H. Tenhunen, "Efficient VLSI-realizable decimators for sigma-delta analog-to-digital converters," *IEEE Proc. IS-CAS'88*, pp. 1525–1528, June 1988.

## 5. *Theory and Implementations of Oversampling D/A Converters*

[105]* J. C. Candy and An-Ni Huynh, "Double interpolation for digital-to-analog conversions," *IEEE Trans. Commun.*, vol. COM-34, pp. 77–81, Jan. 1986.

[106]* R. Carley and J. Kenney, "A 16-bit 4th order noise-shaping D/A converter," *IEEE Proc. CICC.*, pp. 21.7.1–21.7.4, 1988.

[107]  B. Kup, E. Dijkmans, H. Naus, and J. Sneep, "A bitstream D/A converter with 18 b resolution," *ISSCC Dig. Tech. Pap.*, pp. 70–71, Feb. 1991.

[108]  H. G. Musmann and W. Korte, "Generalized interpolative method for digital/analog conversion of PCM signals," 1984 U.S. Patent, No. 4,467,316 (filed 1981).

[109]* P. J. Naus, E. C. Dijkmans et al., "A CMOS stereo 16-bit D/A converter for digital audio," *IEEE J. Solid-State Circuits*, vol. SC-22, pp. 390–395, June 1987.

[110]  M. Schouwenaars, W. Groeneveld, C. Bastiaansen, and H. Turmeer, "An oversampling multibit CMOS D/A converter for digital audio with 115 dB dynamic range," *ISSCC Dig. Tech. Pap.*, pp. 72–73, Feb. 1991.

[111]  G. R. Ritchie, J. C. Candy, and W. H. Ninke, "Interpolative digital to analog converters." *IEEE Trans. Commun.*, vol. COM-22, pp. 1797–1806, Nov. 1974.

# Part 1
# Basic Theory and Analysis

# An Analysis of Nonlinear Behavior in Delta-Sigma Modulators

SASAN H. ARDALAN, MEMBER, IEEE, AND JOHN J. PAULOS, MEMBER, IEEE

*Abstract* —This paper introduces a new method of analysis for delta-sigma modulators based on modeling the nonlinear quantizer with a linearized gain, obtained by minimizing a mean-square-error criterion [7], followed by an additive noise source representing distortion components. In the paper, input signal amplitude dependencies of delta-sigma modulator stability and signal-to-noise ratio are analyzed. It is shown that due to the nonlinearity of the quantizer, the signal-to-noise ratio of the modulator may decrease as the input amplitude increases prior to saturation. Also, a stable third-order delta-sigma modulator may become unstable by increasing the input amplitude beyond a certain threshold. Both of these phenomena are explained by the nonlinear analysis of this paper. The analysis is carried out for both dc and sinusoidal excitations.

## I. INTRODUCTION

RECENTLY, DELTA-SIGMA modulation has been receiving increased attention as an alternative to conventional A/D converters [1]–[3], [9]. Delta-sigma is one of a class of systems which use oversampling and 1-bit quantization to achieve high resolution A/D conversion at a lower rate. Oversampling is attractive for many systems in that the analog antialias filtering requirements are relaxed. In addition, delta-sigma modulators can generally be implemented with few precision circuits.

A scanning of the literature on delta-sigma modulation reveals a lack of comprehensive analyses of the modulator which fully account for the nonlinearity of the system. An exception is the fundamental work presented in [4], which clearly illustrates the influence of nonlinear phenomena on circuit behavior for first-order modulators and dc inputs. Thus, very few analytical design aids which predict nonlinear behavior can be found for delta-sigma modulation. Many researchers have observed experimental phenomena for which, to date, no analytical explanations exist. Some examples include the instability of high-order modulators as the input signal amplitude is increased, the actual decline of signal-to-noise ratio as the input amplitude is increased prior to saturation (see the experimental results of [5] and [12]), and the observation reported in [5], in which very high internal signal variances are encountered for large input amplitudes.

In this paper, higher order delta-sigma modulators are analyzed. These modulators are preferred because of their

Manuscript received August 14, 1986; revised February 2, 1987. This work was supported in part by Rockwell International, Dallas, TX, and by the Center for Communications and Signal Processing.
The authors are with the Center for Communications and Signal Processing, Department of Electrical and Computer Engineering, North Carolina State University, Raleigh, NC 27695-7914.
IEEE Log Number 8714311.

superior noise performance and relative freedom from harmonic quantization effects such as noise thresholding [4], [5]. These higher order systems can be unstable for certain values of the loop parameters and, in the case of third-order delta-sigma modulators, for certain ranges of the input signal.

This paper introduces a new method of analysis for delta-sigma modulators based on modeling the nonlinear quantizer with a linearized gain obtained by minimizing a mean-square-error criterion [7]. This method has been used to derive regions of stability for higher order modulators, including both parameter and signal dependencies. In addition, the method can be used to obtain more accurate solutions for quantization noise spectra and signal-to-noise ratios for several classes of input signals. In this paper, we first consider the case where the input signal is a dc amplitude. The approach is then extended to sinusoidal signals.

### Background on Methodology

The quasi-linear method for modeling the nonlinearity in nonlinear feedback systems was first introduced by Booton [13]. This technique, known as the describing function method, has been widely used and analyzed [7]. This method, however, is based on neglecting distortion components produced by the nonlinearity. The assumption here is that these components are filtered as they are fed back to the nonlinearity input. While this assumption yields adequate results in many nonlinear feedback systems, it will be shown that it is inappropriate for the analysis of delta-sigma modulators.

The method for modeling the nonlinearity presented in this paper is thus distinguished from the describing function technique in that the distortion components produced by the nonlinearity are not neglected. This paper will show that the distortion components affect the dynamic behavior of the modulator dramatically. This includes instability of higher order modulators and the increase in baseband noise as the input signal amplitude is increased.

In [8], Smith reviewed the different approaches for including the distortion components in the modeling of the nonlinearity. One method consisted of modeling the nonlinearity by a quasi-linear gain followed by an additive noise source representing the distortion components produced by the nonlinear element. For example, using this model West *et al.* [6] were able to predict baseband distortion in a nonlinear feedback system. Since the mod-

Reprinted from *IEEE Trans. Circuits and Sys.*, vol. CAS-34, pp. 593–603, June 1987.

ulator response is now seen to consist of both a signal related to the input to the modulator and random distortion components, the response of the nonlinearity to multiple inputs must be considered. To this end, the method of multiple linearized gains for multiple inputs has been used. This technique has been extensively studied in [7]. Using this method, a nonlinear feedback system can be represented as two interlocked linear systems.

In contrast to the work in [7], where the multiple inputs are applied externally, the random component in this paper is produced internally by the nonlinearity in the modulator. Smith has presented experimental results which show that under certain conditions the random distortion components appearing at the nonlinearity input may have a Gaussian probability density function (pdf). In the present work, this observation is exploited for the case of higher order modulators. Thus, the nonlinearity is modeled as two linearized gains, one for the input signal to the system and one for the random distortion components which are assumed to have a Gaussian pdf. The nonlinear system is then analyzed as two interlocked linear systems.

In this paper, with the nonlinear modeling approach outlined above as a basis, a novel approach for analyzing delta-sigma modulators is introduced. Using this approach, closed-form analytical expressions are obtained which relate, for example, the additive quantization noise to the input signal amplitude. Similarly, expressions are derived which relate the ratio of the amplitude of the signal component to the random noise component at the nonlinearity input, to the input signal amplitude. Such expressions have not been derived before. Using these expressions, it is then possible to compute the value of the linearized gains as a function of input amplitude by solving a set of simultaneous nonlinear and integral equations. From the gains, the signal-to-noise ratio of the modulator and other system parameters, including the variance and magnitude of various signals within the modulator, can be calculated.

## II. DELTA-SIGMA MODULATION

A simple representation of a first-order delta-sigma modulator is shown in Fig. 1. This circuit can be implemented with a differential integrator, a comparator, and a flip-flop or sample-and-hold amplifier. The output of this system is a bit stream whose pulse density is proportional to the applied input signal amplitude. (A more conventional digital representation of the input signal can be obtained by decimation and baseband filtering of the pulse stream.) In previous work [4], [5], [3], this system has been modeled by replacing the nonlinear element with a unity-gain linear element followed by an additive noise process. This simple model has proven to be inadequate for the accurate analysis of higher order modulator stability since it does not reflect the dependence of the nonlinear system on the input signal to the nonlinearity $e(t)$.

A second-order double-loop modulator is shown in Fig. 2. In order to analyze the modulator, fictitious sample-and-hold circuits are inserted into the circuit, as shown in

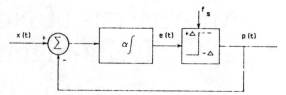

Fig. 1.   First-order delta-sigma modulator.

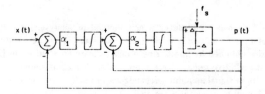

Fig. 2.   Second-order delta-sigma modulator.

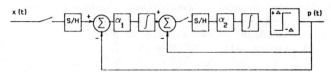

Fig. 3.   Second-order modulator with fictitious sample and holds.

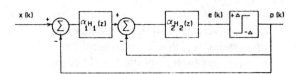

Fig. 4.   $z$-domain second-order modulator.

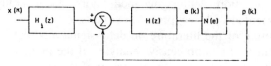

Fig. 5.   Block diagram of $z$-domain delta-sigma modulator.

Fig. 3. It is possible now to obtain a $z$-domain block diagram of the second-order modulator, as shown in Fig. 4. Since the sample-and-hold circuit and the integrator in the inner loop are cascaded with no sampler in between, they must be combined prior to obtaining the $z$-transform. Thus,

$$H_2(z) = Z\left[\frac{1 - e^{-T_s S}}{S}\frac{1}{S}\right] = (1 - z^{-1})Z\left[\frac{1}{S^2}\right] = \frac{T_s}{z-1}.$$

(1)

The $z$-transform of a simple integrator becomes

$$H_1(z) = \frac{z}{z-1}.$$

(2)

It can be shown that any continuous-time or multiloop system can be represented in the $z$-domain, as shown in Fig. 5, with the appropriate choice of the loop filters $H_i(z)$ and $H(z)$. The $z$-domain representation, however, is an approximation to the actual behavior of the continuous system. For a second-order loop, we have

$$H_i(x) = \frac{\alpha_1 H_1(z)}{1 + \alpha_1 H_1(z)}$$

(3)

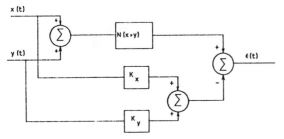

Fig. 6. Minimum mean square error linearized gain modeling of nonlinearity.

and

$$H(z) = [1 + \alpha_1 H_1(z)] \alpha_2 H_2(z). \quad (4)$$

## III. MODELING OF THE NONLINEAR QUANTIZER

### A. DC Input

For a dc input to the delta-sigma modulator, the signal at the input to the nonlinearity will consist of a dc term and feedback distortion components. Therefore, in order to model the nonlinearity, consider the random signal $z(t)$ with mean value $m_x$ and zero mean random component $y(t)$:

$$z(t) = m_x + y(t). \quad (5)$$

Consider the nonlinear mapping of $z(t)$ into $u(t)$:

$$u(t) = N(y(t) + m_x). \quad (6)$$

With reference to Fig. 6, we associate the linearized gain $K_y$ with the zero mean random component $y(t)$ and $K_x$ with the dc mean value or offset of the signal $z(t)$. In Fig. 6, $x(t)$ is a dc signal equal to $m_x$ in this case. The identification problem is to determine $K_x$ and $K_y$ such that the mean square error in modeling the nonlinear element using the linearized gains is minimized. Thus, we must minimize

$$\sigma_\epsilon^2 = E\{\epsilon^2(t)\} = E\left\{\left[u(t) - K_y y(t) - K_x m_x\right]^2\right\} \quad (7)$$

where $E\{\cdot\}$ denotes the expectation operator. Taking the partial derivatives and setting them to zero, we obtain

$$\frac{\partial \sigma_\epsilon^2}{\partial K_y} = -2E\{u(t)y(t)\} + 2K_y E\{y^2(t)\} = 0 \quad (8)$$

$$\frac{\partial \sigma_\epsilon^2}{\partial K_x} = 2K_x m_x^2 - 2m_x E\{u(t)\} = 0 \quad (9)$$

which yield

$$K_y = \frac{E\{u(t)y(t)\}}{E\{y^2(t)\}} \quad (10)$$

$$K_x = \frac{E\{u(t)\}}{m_x}. \quad (11)$$

In terms of the probability density function of $y(t)$, the

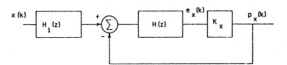

Fig. 7. Equivalent linearized system for dc and sinusoidal input.

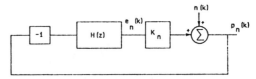

Fig. 8. Equivalent system for quantization noise.

results are

$$K_y = \frac{1}{\sigma_y^2} \int_{-\infty}^{\infty} y N(y + m_x) p(y) \, dy \quad (12)$$

$$K_x = \frac{1}{m_x} \int_{-\infty}^{\infty} N(y + m_x) p(y) \, dy. \quad (13)$$

An important consequence of using the above linear gains is that the error $\epsilon(t)$ becomes uncorrelated with $y(t)$:

$$E\{y(t)\epsilon(t)\} = E\{y(t)u(t)\} - K_y E\{y^2(t)\}$$
$$- K_x m_x E\{y(t)\} = 0. \quad (14)$$

In the application of the above modeling technique to the field of nonlinear control, the error $\epsilon(t)$ is usually neglected. This is based on the assumption that the error is filtered by the plant after feedback and forms a negligible part of the input signal to the nonlinearity [7]. For this reason, $y(t)$ is usually assumed to be dependent only on the input to the control system. In delta-sigma modulation, however, the nonlinearity introduces spectral components which cover a wide bandwidth, including the baseband. In this case, the error, $\epsilon(t)$, represents the noise due to quantization, which forms a major component of the modulator pulse stream. Furthermore, in many cases in nonlinear control, the output of the nonlinearity is the input to the plant. Hence, it is substantially filtered before feedback to the nonlinearity input. In contrast, the output of the nonlinear quantizer is the desired modulated pulse stream in delta-sigma modulation, which *is directly fed back* and subtracted from the input signal. For this reason, we must include the error term $\epsilon(t)$ in the modeling of the nonlinearity. We will show that the quantization noise has a major impact on stability for high-order modulators.

In this paper, we will assume that $\epsilon(t)$ has a white spectrum. The nonlinearity, in particular a 1-bit quantizer, produces harmonics of the input signal across the spectrum of interest, including baseband. Therefore, this assumption seems partly justified [8]. Thus, we replace the nonlinear quantizer in the modulator by the two linearized gains followed by an additive noise source representing the error $\epsilon(t)$. This is illustrated in Figs. 7 and 8, where we have separated the response due to the zero mean random component and the dc response. The additive noise is represented by $n(k)$. The input signal to the quantizer is $e(k)$. We are assuming that the input to the modulator is a dc level equal to $m_x$. From the figures, we have for the steady-state dc response,

$$m_e = \frac{H_i(1)H(1)}{1 + H(1)K_x} m_x \quad (15)$$

and for the response to the random noise component,

$$E_n(z) = N(z)\frac{H(z)}{1 + H(z)K_n}. \quad (16)$$

If we assume that $n(k)$ is white with variance $\sigma_n^2$, then

$$\sigma_e^2 = \frac{\sigma_n^2}{2\pi}$$
$$\int_{-\pi}^{\pi} \frac{|H(e^{j\omega})|^2}{1 + K_n[H(e^{j\omega}) + H^*(e^{j\omega})] + K_n^2|H(e^{j\omega})|^2} d\omega. \quad (17)$$

In most applications of delta-sigma modulation, the forward path includes an integrator. Therefore, $H(1) \to \infty$ and

$$m_e \to \frac{H_i(1)}{K_x} m_x. \quad (18)$$

Although we make no assumptions on the probability density distribution of the noise $n(k)$, the signal into the nonlinearity due to the noise, $e_n(k)$, is a double- or triple-integrated version of the noise $n(k)$ for second- and third-order loops. Since integration of a random variable tends to make its distribution approach a Gaussian distribution, we will assume a Gaussian distribution for $e_n(t)$. Substantial errors may result if this assumption is not true [8]. The linearized gains can be calculated based on (12) and (13) with a Gaussian distribution assumed for $p(y)$:

$$K_n = \frac{2\Delta}{\sigma_e\sqrt{2\pi}} e^{-m_e^2/2\sigma_e^2} \quad (19)$$

$$K_x = \frac{\Delta}{m_e} erf\left(\frac{m_e}{\sigma_e\sqrt{2}}\right). \quad (20)$$

In the above expressions, the nonlinearity is assumed to be a 1-bit quantizer with an output of $\pm\Delta$. Substituting for $K_x$ in (18), we obtain

$$m_e = m_x \frac{m_e}{\Delta}\left[erf\left(\frac{m_e}{\sigma_e\sqrt{2}}\right)\right]^{-1} \quad (21)$$

where we have assumed $H_i(1) \to 1$ if an integrator is used. Thus,

$$erf\left(\frac{m_e}{\sigma_e\sqrt{2}}\right) = \frac{m_x}{\Delta}. \quad (22)$$

Define

$$\rho = \frac{m_e}{\sigma_e\sqrt{2}}.$$

Then

$$\rho = erf^{-1}\left(\frac{m_x}{\Delta}\right). \quad (23)$$

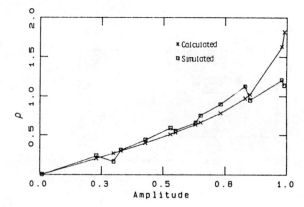

Fig. 9.  Calculated and simulated values of $\rho = m_e/\sqrt{2\sigma_e}$.

The above results show that $\rho$ is independent of the loop gains and is directly related to the dc input amplitude $m_x$. From Figs. 7 and 8, the delta-sigma pulse stream can be written as

$$p(k) = e(k)K_n + n(k) + m_eK_x. \quad (24)$$

Now, since $p(k)$ fluctuates between the two levels $-\Delta$ and $\Delta$, its power is constant and equal to $\Delta^2$. Hence,

$$E\{p^2(k)\} = E\{e^2(k)\}K_n^2 + \sigma_n^2 + m_e^2K_x^2 = \Delta^2 \quad (25)$$

where we have used the fact that the cross-correlation between $e(k)$ and $n(k)$ is zero since we are using the linearized minimum mean square error gains. From (25), we can derive the expression for the variance of the modeled additive noise as a function of input dc amplitude:

$$\sigma_n^2 = \Delta^2\left[1 - \frac{m_x^2}{\Delta^2} - \frac{2}{\pi}e^{-2[erf^{-1}(m_x/\Delta)]^2}\right]. \quad (26)$$

This expression shows that the noise variance depends only on the input dc level and is independent of the loop gains. When $m_x \to \Delta$, we have $\sigma_n^2 \to 0$.

We now examine the dependence of the gain $K_n$ on the input dc amplitude. From (23), we observe that as $m_x$ approaches $\Delta$, $\rho$ becomes very large. Based on (19), $K_n$ decreases exponentially with $\rho^2$. Hence, $K_n$ will decrease with increasing amplitude (although $\sigma_e \to 0$ as $m_x \to \Delta$, the exponential term in (19) decreases at a faster rate). As the gain $K_n$ decreases with amplitude, the noise shaping of the delta-sigma modulator changes. The implications of this nonlinear phenomenon on stability and signal to noise ratio will be examined later.

In order to test the theoretical results derived above, a second-order double-loop digital delta-sigma modulator was simulated with a dc input. The mean value $m_e$ and the variance $\sigma_{en}^2$ of the zero-mean random component $e_n(k)$ at the input to the nonlinearity were obtained after 1024 iterations. Also, the linear gains $K_x$ and $K_n$ were calculated from the simulation using time averages based on (10) and (11). From the computed gains, the additive noise $n(k)$ was obtained from the simulation by subtracting the quantizer output from $e_n(k)K_n + m_eK_x$. The loop gains were $\alpha_1 = 1.0$ and $\alpha_2 = 1.0$. The quantizer step size was

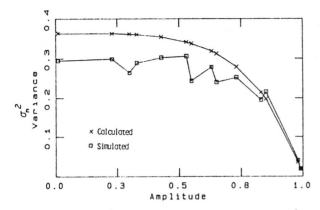

Fig. 10. Calculated and simulated additive noise variance $\sigma_n^2$.

$\Delta = 1.0$. In Fig. 9, the simulated and calculated value of $\rho$, based on evaluating (23) as a function of dc amplitude $m_x$, are plotted. In Fig. 10, the simulated and the calculated values of the additive noise variance $\sigma_n^2$, based on expression (30), are plotted. In both cases, the theoretical results agree reasonably well with the simulation results.

### B. Sinusoid Input and Nonlinearity Modeling

For a sinusoidally excited delta-sigma modulator, the signal at the quantizer input consists of a sinusoid at the same frequency as the input sinusoid, and feedback random distortion components. Therefore, in order to model the quantizer, consider the case where the input to the nonlinear quantizer consists of a sinusoid $x(t)$ and a Gaussian signal $y(t)$. Based on the previous discussion, we can associate the linear gains $K_x$ and $K_y$ with the sinusoidal and Gaussian inputs (see Fig. 6). Minimizing the mean square error between the linearized system and the actual nonlinearity, the following linearized gains are obtained:

$$K_x = \frac{1}{\sigma_x^2} \int_{-\infty}^{\infty} \int_{-\infty}^{\infty} x N(x+y) p(y) q(x) \, dx \, dy \quad (27)$$

$$K_y = \frac{1}{\sigma_y^2} \int_{-\infty}^{\infty} \int_{-\infty}^{\infty} y N(x+y) p(y) q(x) \, dx \, dy \quad (28)$$

where

$$q(x) = \frac{1}{\pi} \left( a_x^2 - x^2 \right)^{-1/2} \quad (29)$$

is the probability density function (pdf) of the sinusoid with amplitude $a_x$, and

$$p(y) = \frac{2}{\sigma_y \sqrt{\pi}} e^{-y^2 / 2\sigma_y^2} \quad (30)$$

is the pdf of the Gaussian input $y(t)$. The linear gains described by the equations above have been solved for the case of an ideal relay, which is equivalent to a 1-bit quantizer, by Atherton [7]. The results are

$$K_y = \left( \frac{2}{\pi} \right)^{1/2} \left( \frac{\Delta}{\sigma_y} \right) M(1/2, 1, -\rho^2) \quad (31)$$

$$K_x = \left( \frac{2}{\pi} \right)^{1/2} \left( \frac{\Delta}{\sigma_y} \right) M(1/2, 2, -\rho^2) \quad (32)$$

where

$$\rho = \frac{a_x}{\sqrt{2} \, \sigma_y} \quad (33)$$

is the square root of the ratio of the variance of the sinusoidal component to the Gaussian component at the nonlinearity input. The functions $M(\alpha, \gamma, x)$ are the confluent hypergeometric functions [10]:

$$\frac{\Gamma(\gamma - \alpha) \Gamma(\alpha)}{\Gamma(\gamma)} M(\alpha, \gamma, x) = \int_0^1 e^{xt} t^{\alpha - 1} (1 - t)^{\gamma - \alpha - 1} \, dt. \quad (34)$$

By replacing the nonlinear quantizer by the linearized gains followed by an additive noise source $n(t)$, the delta-sigma modulator can be separated into two interlocked linear systems, as illustrated in Figs. 7 and 8. In one system, the input forcing function is the sinusoid $x(k)$. In the other system, the forcing function is the additive noise source $n(k)$ produced by quantization. From Fig. 7, we have

$$\frac{E_x(z)}{X(z)} = \frac{H_i(z) H(z)}{1 + K_x H(z)}. \quad (35)$$

If integrators are used in the modulator, then in the baseband region and assuming that the frequency of $x(k)$ is small,

$$\frac{E_x(z)}{X(z)} \approx \frac{1}{K_x} \quad (36)$$

where $H_i(z) \approx 1$. It follows that

$$\sigma_{ex}^2 = \frac{1}{K_x^2} \sigma_x^2 \quad (37)$$

where, for a sinusoid, $\sigma_x^2 = a_x^2 / 2$.

As in the previous section, we assume that $n(k)$ is white. We also assume that the input to the nonlinearity due to the noise source $n(k), e_n(k)$, is Gaussian. From Fig. 8,

$$\sigma_{en}^2 = \frac{\sigma_n^2}{2\pi} \int_{-\pi}^{\pi} \frac{|H(e^{j\omega})|^2}{|1 + K_n H(e^{j\omega})|^2} \, d\omega. \quad (38)$$

The power at the output of the delta-sigma modulator is constant for a single-bit quantizer. Hence,

$$E\{ p^2(k) \} = \sigma_n^2 + K_x^2 \sigma_{ex}^2 + K_n^2 \sigma_{en}^2 = \Delta^2. \quad (39)$$

Substituting for the linearized gains and solving for the additive noise variance,

$$\sigma_n^2 = \Delta^2 \left[ 1 - \frac{2}{\pi} \rho^2 M^2\left(\tfrac{1}{2}, 2, -\rho^2\right) - \frac{2}{\pi} M^2\left(\tfrac{1}{2}, 1, -\rho^2\right) \right]. \quad (40)$$

Now, from (37) and (32),

$$K_x^2 = \frac{\sigma_x^2}{\sigma_{ex}^2} = \frac{2}{\pi} \frac{\Delta^2}{\sigma_{en}^2} M^2\left(\tfrac{1}{2}, 2, -\rho^2\right). \quad (41)$$

Hence, the ratio $\rho$ can be obtained as a function of the input sinusoid amplitude $a_x$. This is done by solving the

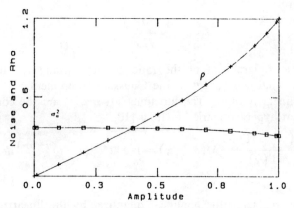

Fig. 11. Calculated additive noise variance $\sigma_n^2$ and $\rho$ sinusoidal input.

following nonlinear equation for $\rho$:

$$\rho^2 M^2\left(\tfrac{1}{2}, 2, -\rho^2\right) = \frac{\pi}{4}\frac{a_x^2}{\Delta^2}. \tag{42}$$

Using this expression, the additive noise variance can be found as a function of input sinusoid amplitude from (40):

$$\sigma_n^2 = \Delta^2\left[1 - \frac{a_x^2}{2\Delta^2} - \frac{2}{\pi}M^2\left(\tfrac{1}{2}, 1, -\rho^2\right)\right]. \tag{43}$$

An analysis of the above results shows that the gains $K_n$ and $K_x$ decrease as the input amplitude of the sinusoid increases.

Fig. 11 shows the calculated value of $\rho$ obtained by solving (42) using the Newton-Raphson technique plotted as a function of input sinusoid amplitude $a_x$ with $\Delta = 1$. In the same figure, the additive noise variance $\sigma_n^2$ is also plotted. Note that the additive noise variance remains almost constant, decreasing only slightly, and its value is close to that obtained if we assume that the noise is uniformly distributed (i.e., $\Delta^2/3$). However, the magnitude of $\rho$ is seen to increase as $a_x$ approaches $\Delta$. This implies that the variance of the input-signal-related component at the quantizer input becomes very large. This will be examined in the next section.

A few words are in order about the confluent hypergeometric functions. First of all, we note that

$$M\left(\tfrac{1}{2}, 1, -x^2\right) = e^{-x^2/2}I_0\left(\frac{x^2}{2}\right) \tag{44}$$

where $I_0(x)$ is the modified Bessel function of zero order. Furthermore,

$$M\left(\tfrac{1}{2}, 1\tfrac{1}{2}, -x^2\right) = \frac{\pi^{1/2}}{2x}\,erf(x). \tag{45}$$

This function is encountered for dc inputs. However, such nice closed-form solutions do not exist for the case $M(\tfrac{1}{2}, 2, -x^2)$. We note, however, that [10]

$$M(\alpha, \gamma, z) = \frac{\Gamma(\gamma)}{\Gamma(\gamma - \alpha)}(-z)^{-\alpha}\left[1 + O(|z|^{-1})\right] \tag{46}$$

when Real $(z) < 0$. $\Gamma(n)$ is the well-known gamma function. Using the above expression, we obtain the following approximations as $x$ becomes large:

$$M\left(\tfrac{1}{2}, 1, -x^2\right) \approx \frac{1}{\pi^{1/2}x} \tag{47}$$

$$M\left(\tfrac{1}{2}, 2, -x^2\right) \approx \frac{2}{\pi^{1/2}x} \tag{48}$$

$$M\left(\tfrac{1}{2}, 1\tfrac{1}{2}, -x^2\right) \approx \frac{\pi^{1/2}}{2x}. \tag{49}$$

Equation (48) is consistent with the exact relationship (45) as $x$ becomes large.

## IV. Noise Spectra and Signal-to-Noise Ratio

In this section, we will analyze the effects of input signal amplitude on the shaping of the noise spectra and the signal-to-noise ratio of the delta-sigma modulator. The noise transfer function can be written from the block diagram in Fig. 8 as

$$\text{NTF}(z) = \frac{P_n(z)}{N(z)} = \frac{1}{1 + K_n H(z)}. \tag{50}$$

For the second-order loop in Fig. 4,

$$P_n(z) = N(z)\frac{(1 - z^{-1})^2}{(1 - z^{-1})^2 + \alpha_2 K_n\left[(1 - z^{-1}) + \alpha_1\right]z^{-1}}. \tag{51}$$

The spectra of the noise can be obtained from (51) by noting that we have assumed $n(k)$ to be a white noise process. Thus,

$$N(e^{j\omega T_S})N^*(e^{j\omega T_S}) = \frac{\sigma_n^2}{f_S}. \tag{52}$$

Hence, the expression for the noise spectrum becomes

$$S_{nn}(f) = P_n(e^{j\omega T_S})P_n^*(e^{j\omega T_S})$$
$$= \frac{16\sigma_n^2 \sin^4\left(\pi\frac{f}{f_S}\right)}{16\sin^4\left(\pi\frac{f}{f_S}\right)[1 - \alpha_2 K_n] + 4\alpha_2 K_n \sin^2\left(\pi\frac{f}{f_S}\right)[\alpha_2 K_n - 2\alpha_2\alpha_1 K_n - 2\alpha_1] + (\alpha_1\alpha_2 K_n)^2}. \tag{53}$$

Clearly, the shaping of the noise spectra depends on the gain $K_n$. As the input signal amplitude increases, $K_n$ decreases and the noise moves in-band. This is shown in Figs. 12 and 13, where the noise spectra based on (53) are

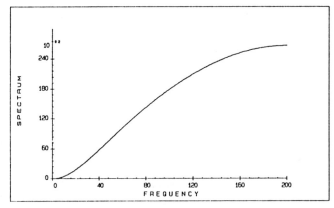

Fig. 12. Calculated noise spectrum, $K_n = 1.0$.

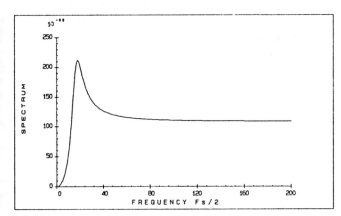

Fig. 13. Calculated noise spectrum, $K_n = 0.1$.

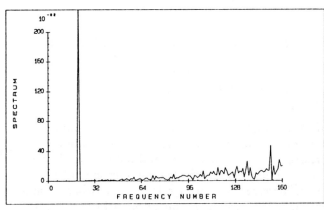

Fig. 14. Simulated amplitude spectrum, $a_x = 0.13$.

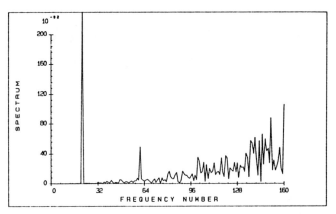

Fig. 15. Simulated amplitude spectrum $a_x = 0.93$.

The above result implies that the signal-to-noise ratio of the delta-sigma modulator is not a linear function of the variance of the input signal. This result cannot be obtained by the simple model used for the quantization process in [5] and [9].

To obtain an approximate expression for the signal-to-noise ratio, we note that within the baseband $f \ll f_s$. Thus,

$$S_{nn}(f) \approx \frac{16\sigma_n^2 \sin^4\left(\pi \dfrac{f}{f_S}\right)}{\alpha_1^2 \alpha_2^2 K_n^2}. \tag{54}$$

The in-band noise is calculated by integrating (54) over the baseband:

$$\sigma_{nb}^2 = \int_0^{f_b} S_{nn}(f)\, df \approx \frac{16\pi^4 \sigma_n^2}{[5\alpha_1 \alpha_2 K_n]^2}\left(\frac{f_b}{f_S}\right)^5. \tag{55}$$

Therefore, the signal-to-noise ratio becomes

$$\text{SNR} = \frac{\sigma_x^2}{\sigma_{nb}^2} \approx \frac{25\sigma_x^2 \alpha_1^2 [\alpha_2 K_n(\sigma_x)]^2}{16\pi^4 \sigma_n^2(\sigma_x)}\left(\frac{f_S}{f_b}\right)^5. \tag{56}$$

In the above expression, the dependence of the gain $K_n$ and the additive noise variance $\sigma_n^2$ on the input signal variance $\sigma_x^2$ has been indicated.

*Exact Numerical Calculation of Signal-to-Noise Ratio*

Based on the analytical results presented in Section III, we can compute the signal-to-noise ratio as a function of input signal amplitude for both dc and sinusoidal inputs. For dc inputs, (17), (19), (24), and (26) represent coupled, nonlinear, and integral equations. The numerical solution of these simultaneous equations, given a dc input amplitude $m_x$, yields the linearized gain $K_n$ and $\sigma_n^2$. Once $K_n$ and $\sigma_n^2$ are known, then the noise spectra $S_{nn}(f)$ can be numerically integrated over the baseband to yield the in-band noise component $\sigma_{nb}^2$. The SNR is then defined as

$$\text{SNR}_{dc} = \frac{m_x^2}{\sigma_{nb}^2}. \tag{57}$$

For sinusoidal inputs, the coupled, nonlinear, and integral equations (42), (43), (38), and (31) must be solved. In this

plotted for $K_n$ large and small. This nonlinear phenomenon, predicted theoretically above, is also observed in actual simulations. In Fig. 14, the amplitude spectrum of the delta-sigma modulator pulse stream, obtained by digital simulation of a second-order system, is shown; this highlights the baseband region. The sampling rate was 1.024 MHz. The spectrum was obtained using a 4096-point FFT and only the first 160 points are shown. The input signal consisted of a 5-kHz tone with an amplitude of 0.13. Again, $\Delta = 1.0$. Notice that the in-band noise is very small. However, in Fig. 15, where the amplitude is increased to 0.93, the in-band noise increases due to the shaping of the noise spectra as the gain $K_n$ decreases.

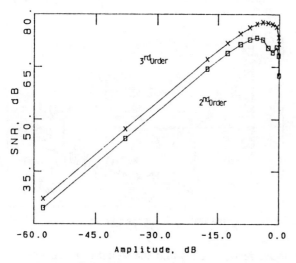

Fig. 16. Calculated SNR for second- and third-order modulator, $f_S/f_b = 256$.

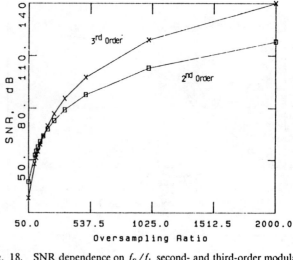

Fig. 18. SNR dependence on $f_S/f_b$, second- and third-order modulator.

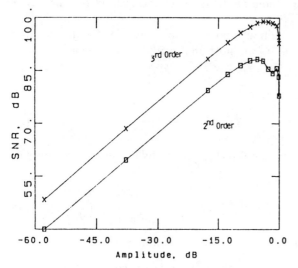

Fig. 17. Calculated SNR for second- and third-order modulator, $f_S/f_b = 512$.

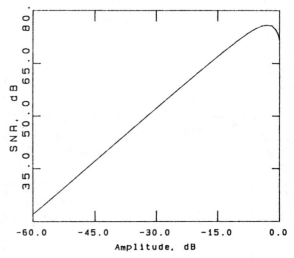

Fig. 19. Calculated SNR, sinusoidal input second-order modulator.

case, the SNR is defined as

$$SNR = \frac{a_x^2}{2\sigma_{nb}^2}. \tag{58}$$

Fig. 16 shows the calculated SNR of second- and third-order modulators plotted against the dc amplitude. The sampling rate was 1.024 MHz and the baseband bandwidth was 4 kHz. Notice that the SNR increases and then decreases as the input amplitude approaches to within 10 dB of the saturation point ($m_x = \Delta$). This theoretical result, which has been observed experimentally in [5] and [12], is caused by the decrease of the gain $K_n$ as the amplitude is increased. The decrease in $K_n$ in turn modifies the spectral noise shaping of the modulator, causing the noise to move in-band. The SNR decreases although based on (26) the additive noise variance decreases with increasing amplitude!

Another observation concerning Fig. 16 is that the gain in SNR by using a third-order loop is not substantial in this case. However, if the sampling rate is doubled, a

significant advantage can be gained by using a third-order loop over a second-order loop, as demonstrated in Fig. 17. The reason for this is that for a third-order loop, although the baseband noise is smaller than for a second-order loop, the noise increases at a larger rate as the baseband bandwidth is increased. In Fig. 18, the calculated signal-to-noise ratio of a second- and third-order modulator, with a fixed input amplitude, is shown as a function of the oversampling ratio $f_S/f_b$. From the figure, we observe that a third-order modulator loses its SNR advantage over a second-order modulator as the oversampling ratio is decreased.

The SNR of a second-order modulator with a sinusoidal input is shown in Fig. 19 as a function of input amplitude $a_x$. The sampling rate was $f_S = 1.024$ MHz and the baseband bandwidth was $f_b = 4$ kHz. Again, the SNR reaches a maximum and then declines as the sinusoid amplitude increases.

As was pointed out in Section III, the linear gain $K_n$ decreases as the input amplitude increases. This is shown in Fig. 20, where the calculated gain is plotted against

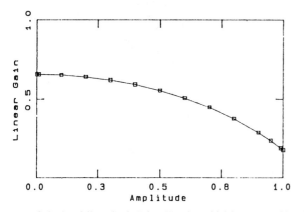

Fig. 20.   Calculated linearized Gain, $K_n$, sinusoidal input second-order modulator.

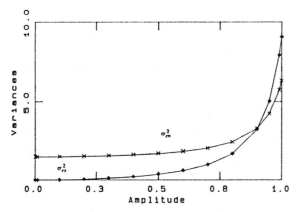

Fig. 21.   Calculated variances at quantizer input, second-order sinusoidal input.

input amplitude for a sinusoidal signal. Interestingly, the variance of the random noise at the quantizer input actually increases as the amplitude increases. The calculated value of this variance $\sigma_{en}^2$ is plotted in Fig. 21. Note that $\sigma_{en}^2$ increases well beyond its value at low amplitudes, where it constitutes the major component of the quantizer input. Also, the variance of the sinusoidal component at the quantizer input also increases, surpassing the noise variance $\sigma_{en}^2$. This is also shown in Fig. 21. This analytical observation, which has been observed in experimental circuits in [5], has important consequences on the design of actual circuits where the dynamic range of signals is limited. An important contribution of the present analytical results is that proper design considerations, accounting for the large variances, can be made when actual circuits are implemented.

## V.   STABILITY ANALYSIS

One of the major problems associated with higher order delta-sigma modulators is their stability. In this section, we present a stability analysis of the second- and third-order modulators. Expressions are derived which give bounds on the loop gains for stable operation. Furthermore, we show that a stable third-order modulator will become unstable as the input signal amplitude is increased beyond a certain threshold.

### A. Double-Loop System

The transfer function of the double-loop delta-sigma modulator between the input $x(k)$ and output $p_x(k)$ is

$$T(z) = \frac{K\alpha_2\alpha_1 H_1(z)H_2(z)}{1 + \alpha_2 K H_2(z)[1 + \alpha_1 H_1(z)]} \tag{59}$$

where $K$ is the linearized gain. The denominator of (59) is common for the input signal and noise transfer functions except that the appropriate gain must be substituted. As the amplitude of the input signal increases, the linearized gains decrease. In this section, we derive stability conditions as a function of loop parameters $\alpha_1$ and $\alpha_2$ and as a function of the gain $K$. To obtain the frequency response, we substitute $z = e^{j\omega T_S}$ in (59). The denominator can then be written as

$$D(j\omega T_s) = 1 + KGH(e^{j\omega T_S}) \tag{60}$$

where, if ideal integrators are used,

$$GH(e^{j\omega T_S}) = \frac{-\alpha_2}{4\sin^2\left(\dfrac{\omega T_S}{2}\right)}$$
$$\cdot [1 + \alpha_1 - \cos(\omega T_S) + j\sin(\omega T_S)].$$

In order to determine the relationship between $\alpha_1$, $\alpha_2$, and $K$ for a stable system, we set the imaginary part of $GH(e^{\omega T_S})$ to zero and its real part to $-1/K$. Thus,

$$\sin(\omega T_S) = 0. \tag{61}$$

Hence, the frequency of oscillation of the limit cycle just at the point of instability is

$$f = f_s/2. \tag{62}$$

Also, $\cos(\omega T_S) = -1$ for $f = f_s/2$. Hence,

$$\frac{\alpha_2 K}{4}(\alpha_1 + 2) = 1. \tag{63}$$

Therefore, when the above condition is met, the circuit will produce a sustained oscillation at a frequency half of the sampling rate. For stability,

$$\frac{\alpha_2 K}{4}(\alpha_1 + 2) < 1. \tag{64}$$

From this expression, it is clear that if $\alpha_1$ is chosen such that the system is stable, decreasing $K$ does not cause instability. Thus, the system will remain stable for increasing input signal amplitude. This is verified by the Nyquist plot for the second-order system in Fig. 22 ($\alpha_1 = 0.5$ $\alpha_2 = 1.0$). This is not the case for the third-order system, as will be shown below. Since increasing $\alpha_2$ directly increases the variance of the signal to the quantizer, $K$ will decrease accordingly. Therefore, the only degree of freedom is the choice of $\alpha_1$. To increase the signal-to-noise ratio, $\alpha_1$ must be increased. This leads to instability based on the expression above, where $\alpha_1 < 2$ must be satisfied assuming that $\alpha_2 K \approx 1$.

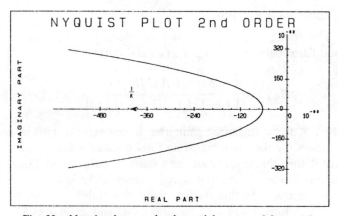

Fig. 22.  Nyquist plot second-order modulator ($\alpha_1 = 0.5, \alpha_2 = 1.0$).

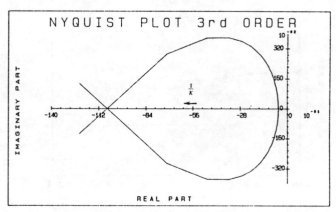

Fig. 23.  Nyquist plot third-order modulator ($\alpha_1 = 0.1, \alpha_2 = 0.1, \alpha_3 = 1.0$).

### B. Third Order System

The transfer function for the third-order system is

$$T(z) = \frac{K\alpha_3\alpha_2\alpha_1 H_1 H_2 H_3(z)}{1 + K\alpha_3 H_3(z)\left[1 + \alpha_2 H_2(z)\left[1 + \alpha_1 H_1(z)\right]\right]}. \tag{65}$$

Proceeding as in the second-order case, we obtain two solutions for the third-order system. The first solution is given by

$$\frac{K\alpha_3}{8}\left[4 + 2\alpha_2 + \alpha_1\alpha_2\right] \leqslant 1 \tag{66}$$

with a corresponding frequency of oscillation of $f = f_S/2$. From this expression, a stable third-order system can be designed by proper choice of $\alpha_1$ and $\alpha_2$. There are two degrees of freedom, since changing $\alpha_3$ causes $K$ to adjust accordingly. For stability, $\alpha_2$ must be small.

The frequency of oscillation for the second solution is determined by the following equation:

$$\sin^2\left(\frac{\omega T_s}{2}\right) = \frac{\alpha_1\alpha_2}{4}. \tag{67}$$

Setting the real part of $GH(e^{\omega T_S})$ equal to $-1/K$, we obtain

$$\frac{K\alpha_3}{8\sin^3\left(\frac{\omega T_S}{2}\right)}\left[4\sin^3\left(\frac{\omega T_S}{2}\right) + 2\alpha_2\sin\left(\frac{\omega T_S}{2}\right)\right.$$
$$\left. + \alpha_1\alpha_2\sin\left(\frac{\omega T_S}{2}\right)\right] \leqslant 1. \tag{68}$$

Substituting for $\sin(\omega T_S/2)$ from (67), we obtain the following condition for stability:

$$\alpha_3 K\left[1 + \frac{1}{\alpha_1}\right] \leqslant 1. \tag{69}$$

A Nyquist plot for the third-order system is shown in Fig. 23 ($\alpha_1 = 0.1, \alpha_2 = 0.1, \alpha_3 = 1.0$). We observe that as the linearized gain $K$ decreases, we approach the second solution given by (67) and (68). If $K$ decreases further, the third-order system will become unstable. In other words, increasing the amplitude of the input signal to a stable

third-order delta-sigma modulator will cause the modulator to become unstable. There is an amplitude region, however, for which the modulator is stable.

From (69), notice that $\alpha_1$ determines the range for which decreasing $K$ causes instability. If a small $\alpha_1$ is chosen, then the allowable amplitude range is increased at the expense of signal-to-noise ratio. Surprisingly, $\alpha_2$ only affects the frequency of oscillation. Hence, a large value can be used for $\alpha_2$ to increase SNR provided that (66) is satisfied for stability.

At this point, an interesting question arises. Which linearized gains, $K_x$ or $K_n$, cause instability in a third-order system? To answer this question, we evaluate the ratio $K_n/K_x$ as $m_x \to \Delta$. Thus, for a dc signal,

$$\lim_{m_x \to \Delta} \frac{K_n}{K_x} = \lim_{m_x \to \Delta} \frac{2}{\pi^{1/2}}\rho\frac{e^{-\rho^2}}{erf(\rho)} = 0 \tag{70}$$

since $\rho \to \infty$ as $m_x \to \Delta$.

Hence, $K_n$ is smaller than $K_x$. Thus, the distortion components will cause the modulator to go unstable. This supports the notion that the distortion components due to the quantizer must be included in the analysis of the modulator using linearized gains.

### VI. Conclusions

A new method of analysis for delta-sigma modulators based on modeling the nonlinear quantizer with minimum-mean-square error linearized gains followed by an additive noise source representing distortion components is described. Closed-form expressions have been derived which relate the quantizer additive noise variance to the input signal amplitude. The effects of increasing input signal amplitude on the shaping of the noise spectra and signal-to-noise ratio have been presented. The signal-to-noise ratio of the modulator is calculated directly as a function of input signal amplitude using analytical methods. The analysis is carried out for both dc and sinusoidal excitations.

Moreover, regions of stability for second- and third-order loops have been obtained, including bounds on the loop gains for stable operation. It is shown that the stability of third-order modulator is input-signal-amplitude dependent.

### REFERENCES

[1] M. W. Hauser, P. J. Hurst, and R. W. Brodersen, "MOS ADC-filter combination that does not require precision analog components," in *IEEE ISSCC Dig.*, Feb. 1985, pp. 80–82.

[2] B. P. Agrawal and K. Shenoi, "Design methodology for delta sigma modulation," *IEEE Trans. Commun.*, vol. COM-31, pp. 360–371, Mar. 1983.

[3] T. Misawa, J. E. Iwersen, L. J. Loporcaro, and J. G. Ruch, "Single-chip per channel CODEC with filters utilizing delta-sigma modulation," *IEEE J. Solid-State Circuits*, vol. SC-16, pp. 333–341, Aug. 1981.

[4] J. C. Candy and O. J. Benjamin, "The structure of quantization noise from sigma-delta modulation," *IEEE Trans. Commun.*, vol. COM-29, pp. 1316–1323, Sept. 1981.

[5] J. C. Candy, "A use of double integration in sigma delta modulation," *IEEE Trans. Commun.*, vol. COM-33, pp. 249–258, Mar. 1985.

[6] J. C. West, J. C. Douce, and B. G. Leary, "Frequency spectrum distortion of random signals in non-linear feedback systems," *IEE Monograph*, no. 419M, Nov. 1960, pp. 259–264.

[7] D. P. Atherton, *Nonlinear Control Engineering*. London: Van Nostrand, 1975.

[8] H. W. Smith, "Approximate analysis of randomly excited nonlinear controls," Research Monograph No. 34, MIT Press, Cambridge, Massachusetts, 1966.

[9] H. Meleis and P. Le Fur, "A novel architecture design for VLSI implementation of an FIR decimation filter," presented at Int. Conf. Acoustics, Speech, Signal Process., Orlando, FL, 1985.

[10] M. Abramowitz and I. A. Stegun, *Handbook of Mathematical Functions*. New York: Dover Publications, 1970.

[11] S. H. Ardalan and J. Paulos, "Stability analysis of high-order sigma-delta modulators," in *Proc. 1986 Int. Symp. Circuits Syst.*, (San Jose, CA), May 1986, pp. 715–719.

[12] M. W. Hauser and R. W. Brodersen, "Circuit and technology considerations for MOS delta-sigma A/D converters," in *Proc. 1986 Int. Symp. Circuits Syst.* (San Jose, CA), May 1986, pp. 1310–1315.

[13] R. C. Booton, Jr., "Nonlinear control systems with statistical inputs," M.I.T. Dynamic Analysis and Control Laboratory, Report No. 61, Mar. 1952.

# A Use of Limit Cycle Oscillations to Obtain Robust Analog-to-Digital Converters

JAMES C. CANDY, MEMBER, IEEE

*Abstract*—High quality analog-to-digital conversions are obtained using simple and inexpensive circuits that require no high-precision components. Samples of the analog signal are cycled rapidly through a coarse quantizer while the roundoff error is fed back and subtracted from the input. By means of this feedback, the coarse quantizations are caused to oscillate between levels, keeping their running average representative of the input. A binary coding of the quantized values, summed over Nyquist intervals, provides a high resolution PCM output. The precision is determined by a product of the cycle rate and the spacing of the coarse quantization levels. The system is surprisingly tolerant of inaccuracies in gains and threshold settings; indeed, it has many of the desirable properties of classical feedback servomechanisms. An 8-bit limit cycling converter intended for 1-MHz signal bandwidths has been fabricated of standard components that, in total, cost less than $150.

## I. INTRODUCTION

RECENT technical literature [1]–[11] is burdened with techniques for converting analog signals into digital form, yet there still is need to improve the methods of connecting analog sources to the powerful digital systems that are becoming available for processing, storing, and transmitting information. The difficulty of A/D conversions stems from the need for high precision in nonlinear circuits that must operate at fast rates. For example, resolutions of about 1 in 200 at 2 megasamples/s are desirable for videotelephone signals, and resolutions of about 1 in 10 000 at 8 kilosamples/s are needed for speech signals [8]. Providing reliable circuits having these orders of absolute precision is difficult and expensive. The technique to be described here eliminates much of the need for precision circuit components in A/D converters that are intended for use in communication systems. It takes advantage of tradeoffs that are peculiar to communications, in contrast to the requirements needed for instrumentation. Two percent tolerances in gain stability, linearity and dc level settings are usually adequate for video and speech transmission. On the other hand, there can be no appreciable discontinuities in the amplitude transmission characteristic; this fault is common of some well known A/D converter techniques.

The circuit described here is based on a technique proposed by Cutler [12] and elaborated by Brahm

Paper approved by the Associate Editor for Data Communications of the IEEE Communications Society for publication without oral presentation. Manuscript received September 24, 1973.
The author is with the Bell Laboratories, Holmdel, N. J. 07733.

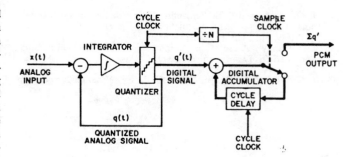

Fig. 1. Outline of the A/D converter in which digital connections are drawn with thicker lines than the analog connections.

[13] and by Miura *et al.* [14]. We will show that it is very tolerant of circuit imperfection. In essence, each input sample is coarsely quantized and digitally coded in a circuit that is not necessarily precise. An estimate of the quantization error, together with any spurious errors, is subtracted from the input value, and the difference becomes the input for the next cycle. After the process has repeated many times at high speed, an average of the digital outputs occurring in each sample time is a useful digital representation of the input signal. By this process, speed of operation replaces precision of circuit elements. The tradeoff is invaluable for designing integrated circuits. Simple inexpensive implementations of the converter have provided high resolution. For example, Picturephone® signals have been coded into 8-bit words at 2 megawords/s using standard circuit components that, in total, cost less than $150. A logarithmically companded coder suitable for speech is significantly less expensive. D/A converters based on related techniques have been implemented both for uniform and companded signals. These converters will be described in a later paper.

## II. A CIRCUIT FOR PULSE-CODE MODULATION (PCM)

The circuit, shown in Fig. 1, uses a negative feedback loop containing an integrator and a coarse quantizer that spans the signal range. A digital representation of the quantized signal increments the contents of an accumulator. After signals have cycled around the feedback loop

® Registered service mark of the American Telephone and Telegraph Company.

Reprinted from *IEEE Trans. Commun.*, vol. COM-22, pp. 298–305, March 1974.

a set number of times, an output is taken from the accumulator and its content set to zero before cycling for the next sample.

Table I illustrates the action of an algorithm using a quantized signal that assumes only integer values and an input amplitude equal to 8.2. It generates quantized values that repeat themselves with the pattern 88 988. In fact, the quantized signal oscillates between the two levels that bracket the input amplitude with duty ratio that attempts to keep the average quantized value equal to the input.

In order to describe the action of the circuit with linear equations, we will employ the technique described in [10]. The quantization will be approximated by addition of an error, $\epsilon$, that is assumed to be uncorrelated with the input, $x$. We then have the equation

$$q(N\tau) = \frac{1}{\tau} \int_0^{N\tau} (x(t) - q(t))\, dt + \epsilon(N\tau) \quad (1)$$

where $\tau$ is the cycle period, and $q(n\tau)$ the $n$th quantized signal. The integrator has unit output for a unit input persisting for $\tau$ seconds. Since $q(t)$ is a sampled function, its integral may be written as a summation; then reorganizing (1) and dividing by $N$, the number of cycles in a sample time, we get the result

$$\frac{1}{N} \sum_0^N q(n\tau) = \frac{1}{N\tau} \int_0^{N\tau} x(t)\, dt + \frac{\epsilon(N\tau)}{N}, \quad (2)$$

which shows that, for appropriate starting conditions, the average of the quantized signals represents the average input with a quantization error that is $N$ times less than the error associated with the coarse quantizer. High resolution is obtained by cycling the signals repeatedly around the feedback loop until $N$ is sufficiently large. For the application to Picturephone signals, the quantizer has sixteen levels and the cycle frequency is 32 MHz; an average of sixteen quantizations provides 8-bit resolution at 2-MHz sample rate.

The Appendix gives a more thorough analysis of the system. The following sections illustrate the more important properties with simple examples.

## III. OPERATING MODES

The initial condition for the integral in (2) is dependent on the application, and several different procedures are feasible. An obvious one uses a sampled and held input and sets both the integrator and the digital accumulator to zero content before processing each sample. Analysis given in the Appendix indicates that inaccuracy of loop gain or zero level can have significant influence on the digital output unless $N$ is especially large. A procedure that avoids the need for high precision connects the analog signal directly to the subtracting circuit in Fig. 1, and cycles data continuously around the feedback loop without ever resetting the integrator. The digital ouput is periodically read from the digital accumulator, which is then reset to zero content. Most of the following discussion

TABLE I
THE QUANTIZATION $q$ IS A REPRESENTATION OF THE SUM OF THE DIFFERENCES BETWEEN THE INPUT AND THE PREVIOUS QUANTIZATION

| Input $x_n$ | Error $\epsilon_n = (x_n - q_{n-1})$ | Integrand $I_n = I_{n-1} + \epsilon_n$ | Quantization $q_n = [I_n]$ |
|---|---|---|---|
| | | 8.4 | 8 |
| 8.2 | 0.2 | 8.6 | 9 |
| 8.2 | −0.8 | 7.8 | 8 |
| 8.2 | 0.2 | 8.0 | 8 |
| 8.2 | 0.2 | 8.2 | 8 |
| 8.2 | 0.2 | 8.4 | 8 |

is concerned with this continuous mode; its action resembles that of a servosystem because, for baseband frequencies, the effective gain of the integrator is large. Thus distortion and inaccuracies in the forward path of the feedback loop have only a second-order effect on the overall properties of the converter.

The continuous mode is suitable not only for single input signals, but also for time multiplexed inputs. Section VII will discuss this application.

## IV. RESOLUTION

This section is concerned with the noise introduced into the signal by the converter when its parameters are ideal. Effects of circuit inaccuracies will be discussed in the next section. As before, let $\tau$ be the cycle interval, and let $T = N\tau$ be the sample interval. Processing of the $K$th output sample commences after the $[(K-1)N]$th cycle and ends with the $(KN)$th cycle. Applying (1) at the start and finish of this sample, we have

$$\sum_0^{(K-1)N} q(n\tau) = \frac{1}{\tau} \int_0^{(K-1)T} x(t)\, dt + \epsilon(KT - T) \quad (3)$$

and

$$\sum_0^{KN} q(n\tau) = \frac{1}{\tau} \int_0^{KT} x(t)\, dt + \epsilon(KT). \quad (4)$$

Taking the difference of these equations and dividing by $N$ gives the important general result

$$\frac{1}{N} \sum_{(K-1)N}^{KN} q(n\tau) = \frac{1}{T} \int_{(K-1)T}^{KT} x(t)\, dt$$
$$+ \frac{\epsilon(KT) - \epsilon(KT - T)}{N}. \quad (5)$$

Equation (5) shows that the average of the quantized values over the sample period represents the average value of the input signal with error given by $1/N$ of the difference between the quantization error in the last cycle and the error in the last cycle of the previous sample.

It is well known [7] that quantization noise $\epsilon$ has uniform probability density spanning magnitudes up to half the quantization step size $\sigma$ and a flat spectral density,

$$E_{\text{RMS}} = \frac{\sigma}{(12)^{1/2}}. \quad (6)$$

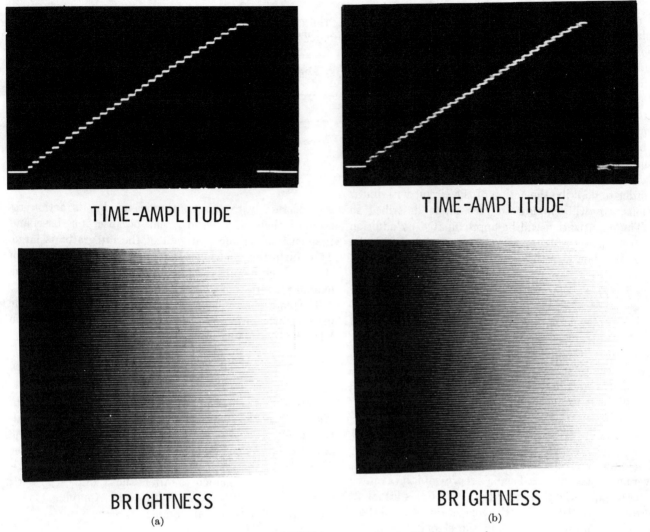

Fig. 2. (a) Response of PCM to a ramp input. (b) Response of the limit-cycle converter to a ramp input.

The noise introduced by the converter is the first difference of such quantization noise; it has a triangular probability density function spanning magnitudes up to a step size and baseband spectral density given by the first quarter cycle of a sinewave, i.e.,

$$\Delta E_{\mathrm{RMS}} = \frac{\sigma}{\sqrt{3}} \sin\left(\frac{\omega T}{2}\right). \qquad (7)$$

Thus, using a coarse quantizer with $L$ levels provides resolutions that are comparable to what is obtained from ordinary PCM converters having $NL$ levels. Although the rms noise from this converter is 3 dB greater than with ordinary PCM, it usually is less annoying because it is composed of higher frequency components. The pictures in Fig. 2 compare the visibility of the noise introduced by ordinary 5-bit PCM with the noise introduced by a 5-bit limit cycling converter. The input was a sawtooth voltage and in each case, the decoded output is displayed both as a trace on an oscilloscope and as a brightness modulation of a television display. The response of this converter resembles that of PCM with patterned dither [9].

Equation (5) shows that the output signal is representative of the average input signal over the entire sample interval, not just over a small aperture within the sample interval, as is the case with most ordinary PCM coders. Such averaging is equivalent to a low-pass filter with spectral response $[\sin(\omega T/2)/(T\omega/2)]$, which is plotted in Fig. 3. In many applications, this filtering action is useful, especially when signals have significant out-of-band energy. The effective gain of the filter falls 4 dB in the signal band, but this drop is easily compensated by preemphasis. Alternatively, it may be corrected by using a sampled and held input plus the technique for avoiding errors caused by transients that is described in Section VII.

## V. CIRCUIT PARAMETERS AND THEIR EFFECT ON THE SIGNAL

### A. Choice of Loop Gain and Circuit Stability

Stability is a primary consideration in the design of the circuit in Fig. 1. The example given in Table I shows that

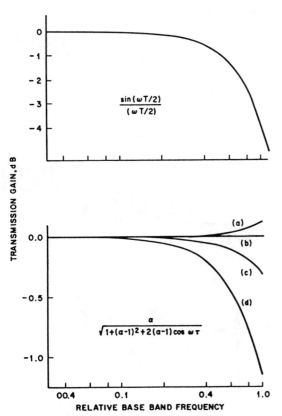

Fig. 3. Components of the transmission gain. (a) $\alpha = 1.9$. (b) $\alpha = 1$. (c) $\alpha = 0.75$. (d) $\alpha = 0.5$.

TABLE II
THE GAIN IN THE INTEGRATOR IS 0.2

| Input $x_n$ | Error $\epsilon_n = (x_n - q_{n-1})$ | Integrand $I_n = I_{n-1} + 0.2\epsilon_n$ | Quantization $q_n = [I_n]$ |
|---|---|---|---|
| | | 8.40 | 8 |
| 8.2 | 0.2 | 8.44 | 8 |
| 8.2 | 0.2 | 8.48 | 8 |
| 8.2 | 0.2 | 8.52 | 9 |
| 8.2 | −0.8 | 8.36 | 8 |
| 8.2 | 0.2 | 8.40 | 8 |

TABLE III
THE GAIN IN THE INTEGRATOR IS 1.3

| Input $x_n$ | Error $\epsilon_n = (x_n - q_{n-1})$ | Integrand $I_n = I_{n-1} + 1.3\epsilon_n$ | Quantization $q_n = [I_n]$ |
|---|---|---|---|
| | | 8.40 | 8 |
| 8.2 | 0.2 | 8.66 | 9 |
| 8.2 | −0.8 | 7.62 | 8 |
| 8.2 | 0.2 | 7.88 | 8 |
| 8.2 | 0.2 | 8.14 | 8 |
| 8.2 | 0.2 | 8.40 | 8 |

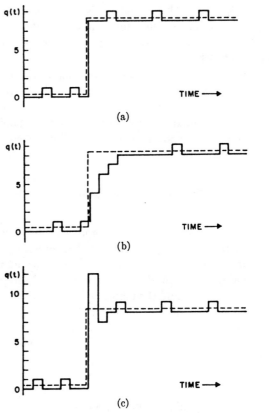

Fig. 4. Responses of the feedback loop to step changes of input, 0.33 to 8.20, with loop gains. (a) $\alpha = 1$. (b) $\alpha = 0.5$. (c) $\alpha = 1.5$.

the algorithm used in this converter is stable, in the sense that oscillations of the quantized value are bounded for a steady input, i.e., it has a limit cycle. This example assumes unity gain in all elements of the feedback loop. Tables II and III give the corresponding responses for gains 0.2 and 1.3 in the integrator. In all three examples, the initial state of the integrator is such that the first amplitude presented to the quantizer is 8.4. The algorithms illustrated in Tables I, II, and III are stable, and the average quantized signal equals the input amplitude. An investigation of a similar feedback loop, described in [10], demonstrates that the circuit is stable for values of gains less than 2. However, for gains in excess of 4/3, there is a tendency for the signal to oscillate between levels that are remote from the input amplitude. This may cause the quantizer to overload and thus increase the error $\epsilon$. It is recommended that loop gains be kept less than 1.5.

Using gains that differ from unity distorts the transient response of the circuit, as Fig. 4 demonstrates. These graphs show the quantized signal, $q$, responding to a step change of input for various gains, $\alpha$. The effective transmission through the system is given by (19) in the Appendix; it can be expressed as

$$| H(\omega) | = \frac{\alpha[\sin(\omega T/2)/(\omega T/2)]}{[1 + (\alpha - 1)^2 + 2(\alpha - 1)\cos(\omega\tau)]^{1/2}}. \quad (8)$$

Fig. 3 shows the two major components of this function

plotted against baseband frequency for $N = T/\tau = 16$, and gains $\alpha = 1.9, 1, 0.75,$ and $0.5$. Evidently, with $N \geq 16$, gain tolerances of $\pm 50$ percent should be acceptable for most communication applications. Thus designing an amplifier to perform the integrating and subtracting operations is simple, especially because its noise and dc offset can be related to equivalent signals added directly to the input. The amplifier specifications need be no more stringent than that of ordinary relay amplifiers.

## B. Threshold Settings

Providing accurately positioned voltage thresholds can be a very difficult task, especially when fast decisions are required. Ideally, the threshold levels should be placed centrally with respect to the quantization levels [7], but fortunately, use of the continuous mode of operation relaxes the requirement on threshold positioning.

Slow wholesale drift of the set of thresholds has insignificant effect on the signal, because the integrator can provide dc correction at negligible input cost.

Displacement of individual thresholds with respect to other ones can be studied by simulations similar to those described in Tables I, II, and III. A useful empirical method for predicting the consequences of displaced thresholds is to define a local gain of the quantizer as the ratio of the output level spacing to the corresponding threshold spacing. There usually is little practical difficulty in providing circuits for which the effective loop gains have a tolerance range of ±25 percent for all signal amplitudes.

In practice, circuit performances have been found to be very tolerant of threshold spacing. This is possibly because a decrease in the spacing of some thresholds is usually accompanied by an increase in the spacing of adjacent ones. Thus any tendency to oscillate at increased amplitude is localized.

## C. Alignment of the Quantization Levels

The quantized signal appears in Fig. 1 in two forms; the analog signal $q$ and the digital signal $q'$. We have seen, in (5), that the average of the analog values can be a good representation of the input, but it is an average of the digital values that produces the output. Discrepancies between the analog and the digital values affect the output directly, since mismatch is not protected by feedback. However, one of the attractive features of this A/D converter is that discrepancies in the coarse quantization levels result in a smooth nonlinearity of transmission, rather than introduction of noise or sharp discontinuity. Fig. 5 illustrates the effect diagramatically by tracing the quantized signal for ramp inputs. The course quantization levels are uniformly spaced in Fig. 5(a), and one step size is increased by 20 percent in Fig. 5(b). Fig. 5(c) shows the effective dc transmission characteristic for the two cases. This graph was obtained by assuming linear interpolation of values between the quantization levels.

It is advisable that the analog signal $q$ be constructed by adding together $L$ equal binary sources, one activated by each of the thresholds, rather than by permuting

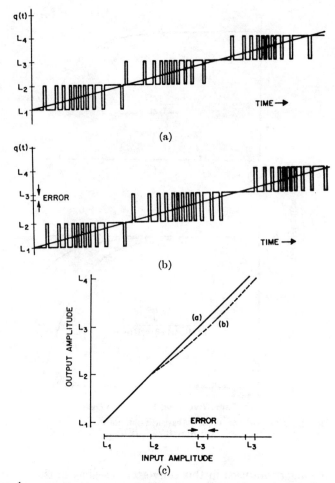

Fig. 5. (a) Graph of $q$ for a ramp input. (b) Effect of displacing level $L_3$ by 20 percent of a step size. (c) Effective dc transmission characteristics.

$\log_2 (L)$ geometrically weighted sources activated by the binary code. Use of weighted sources requires much greater precision in order to avoid large errors in step size at transitions of major bits of the code.

## VI. PRACTICAL CONSIDERATIONS OF TIMING CIRCUIT OPERATIONS

The circuit uses two clocks: a fast clock that times decisions of the coarse quantizer and the accumulation of digital numbers, and a slower clock that times the output gate. In order that these clocks be correctly phased, it has been advisable to generate them as part of the converter circuitry. Fig. 6 illustrates the method used. The fast clock $CF$ is divided in frequency by the factor $N$ to the desired sampling frequency $CS$. The phase of this slow clock is set by means of a timing input pulse $CH$, which resets the divider if there is significant phase discrepancy between $CS$ and $CH$. For television application, $CF$ has been a 32-MHz signal, $CS$ a 2-MHz signal, and $CH$ a horizontal drive signal.

## VII. TIME-SHARED OPERATION

The continuous mode of operation does not reset the integrator between samples. Therefore, data are carried over from sample to sample. This would be very un-

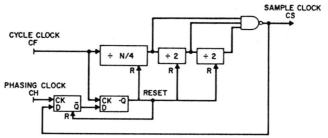

Fig. 6. Generating suitable clocks. The sample clock is automatically rephased if it deviates by more than ±1/16 of a sample interval from clock CH.

desirable for time-shared application where consecutive samples come from unrelated sources [8]. The circuit can be used for such applications, however, without having to actually reset the integrator. The cycle frequency would be increased in order to provide more cycles per sample. Then, the first few cycles are used to clear residual transients from the system. The quantized outputs for these cycles are ignored by holding the accumulator at zero content. When the accumulator is released, the action of the circuit is substantially the same as previously described. The noise, however, now has a flat frequency spectrum but still has triangular probability density and is 3 dB greater than the corresponding PCM noise. Ordinary PCM quality could be obtained by increasing N by a factor of two or more and "rounding off" the digital output words. Indeed, this is the method that was used to obtain the pictures shown in Fig. 2(a).

## CONCLUSION

The method of analog-to-digital conversion described has many of the desirable properties of the classical feedback method used in servomechanisms. Instead of making an accurate evaluation of each signal sample, or a succession of approximations requiring precise components and exacting adjustments, it uses the limit cycles in a coarsely quantizing feedback coder to give a precise determination of the average input value. The main limitation of the method described is the need for fast cycle times. There is little practical difficulty in using it for videotelephone, but broadcast television signals would challenge present circuit art. The technique is particularly attractive for speech signals, because the levels of the coarse quantizer can be companded. Then the interpolative action of the feedback loop provides uniform resolution between each pair of levels. Such quasi-linear companding is used for transmitting telephone signals [8].

The circuit in Fig. 7 is a practical design for an A/D converter; it is capable of 40-MHz cycle rates, providing 8-bit outputs at 2.5 megawords/s. Several of these circuits have been put to use in interframe coding experiments.

## APPENDIX

### A. Analysis of the System with Arbitrary Loop Gain

The diagram in Fig. 8 shows a linear representation of the basic system and notation for the signals. Small

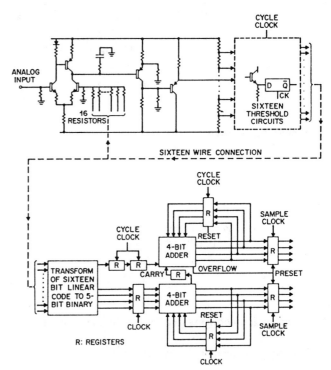

Fig. 7. Schematic circuit of an 8-bit A/D converter.

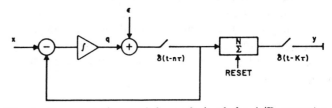

Fig. 8. Representation used for analysis of the A/D converter.

letters, $x(t)$, represent time functions and capitals, $X(\omega)$, represent the corresponding spectral function. $\tau$ will be the cycle time and $T = N\tau$, the sample time. We can describe the feedback by the following equation:

$$q(M\tau) = \frac{\alpha}{\tau} \int_0^{M\tau} [x(t) - q(t)]\, dt + A + \epsilon(M\tau) \qquad (9)$$

where $A$ accounts for the zero offset of the integrator with respect to the quantizer, and $\alpha$ is a gain constant. Now let

$$\frac{1}{\tau} \int_{t-\tau}^{t} x(t)\, dt = x'(t); \qquad (10)$$

then (9) can be written, in terms of sample functions, as

$$q(M\tau) = \alpha \sum_0^M \{x'(n\tau) - q(n\tau - \tau)\} + A + \epsilon(M\tau).$$

(11)

The equivalent spectral relationship [10], [14] is

$$Q(\omega) = \frac{\alpha[X'(\omega) - Z^{-1}Q(\omega)]}{(1 - Z^{-1})} + \frac{A}{j\omega} + E(\omega) \quad (12)$$

where $Z^{-1} = \exp(-j\omega\tau)$, a *cycle* delay. In order to complete the description of the entire system, we must introduce the action of the digital accumulator. It is convenient to regard its action as filtering followed by resampling. The filter is a simple running summation over $N$ previous cycles generating the function

$$\bar{q}(M\tau) = \frac{1}{N} \sum_{(M-N)\tau}^{M\tau} q(n\tau) \quad (13)$$

with spectral density

$$\bar{Q}(\omega) = \frac{(1 - Z^{-N})}{(1 - Z^{-1})} \frac{Q(\omega)}{N}. \quad (14)$$

Using (12), we have

$$\bar{Q}(\omega) = \frac{\alpha[(1 - Z^{-N})/(1 - Z^{-1})][X'(\omega)/N]}{1 + Z^{-1}(\alpha - 1)}$$

$$+ \frac{(1 - Z^{-N})(E(\omega) + A)}{[j\omega N(1 + Z^{-1}(\alpha - 1)]}. \quad (15)$$

This expression confirms that the system is stable [15] for $\alpha < 2$.

The initial transient described by the term

$$(1 - Z^{-N})A/j\omega(1 + Z^{-1}(\alpha - 1))$$

will decay rapidly to zero if $\alpha$ is near unity. For realistic values, it will influence the output only during the first few cycles of the first sample. It could be significant if the integrator were periodically reset, but here we are interested in the continuous mode of operation; $A$ will be henceforth ignored.

Now define

$$\bar{X}(\omega) = \frac{X'(\omega)}{N} \frac{(1 - Z^{-N})}{(1 - Z^{-1})}. \quad (16)$$

$\bar{X}(\omega)$ is the spectral density of the time function

$$\bar{x}(M\tau) = \frac{1}{N\tau} \int_{(M-N)\tau}^{M\tau} x(t)\,dt, \quad (17)$$

which represents a running average of the input over a Nyquist interval. Equation (15) becomes

$$\bar{Q}(\omega) = \frac{\alpha\bar{X}(\omega)}{1 + Z^{-1}(\alpha - 1)} + \frac{E(\omega)(1 - Z^{-N})}{N(1 + Z^{-1}(\alpha - 1))} \quad (18)$$

and it remains to sample this function at times $KN\tau = KT$.

Let us first consider the effects of resampling on the signal itself, considering the noise later. If the original input $x(t)$ is restricted to baseband, then the sampled functions $X'(M\tau)$ and $\bar{X}(M\tau)$ will contain energy only in the frequency bands

$$\left(\frac{2\pi i}{\tau} - \frac{\pi}{T}\right) < \omega < \left(\frac{2\pi i}{\tau} + \frac{\pi}{T}\right)$$

and

$$X(\omega) = X\left(\omega + \frac{i2\pi}{T}\right), \quad \text{for integer } i.$$

Resampling [14] these functions at intervals $N\tau$ will not affect the shape of their baseband spectral densities. Therefore, the baseband spectrum of the output signal is

$$Y(\omega) = \frac{\alpha[\sin(\omega T/2)/(\omega T/2)]X(\omega)}{1 + \cos\omega\tau(\alpha - 1) - j(\alpha - 1)\sin\omega\tau} \quad (19)$$

where

$$|\omega| < \frac{\pi}{T}.$$

The magnitude of this function is illustrated in Fig. 3. It should be emphasized that when the input is not band limited, the result (19) applies only for special cases. For example, using arguments similar to those given in the following, while discussing noise, it can be shown that using a sampled and held input makes the output more sensitive to gain $\alpha$.

Finally, we consider the effect of gain on the output noise. The noise $\epsilon(t)$ is not limited to baseband, therefore, the resampling operation is more easily studied in the time domain. From (18), the output noise can be written as

$$\frac{E(\omega)}{N}(1 - Z^{-N})(1 - Z^{-1}(\alpha - 1) + Z^{-2}(\alpha - 1)^2 - \cdots).$$

(20)

Therefore, the output noise samples are given by

$$\{\epsilon(KT) - \epsilon(KT - T)\} - (\alpha - 1)\{\epsilon(KT - \tau)$$
$$- \epsilon(KT - T - \tau)\} + (\alpha - 1)^2\{\epsilon(KT - 2\tau)$$
$$- \epsilon(KT - T - 2\tau)\} - \cdots. \quad (21)$$

The effect of using gains slightly different from unity, $(\alpha - 1)^N \ll 1$, is to increase the noise approximately by the factor

$$\frac{1}{[1 - (\alpha - 1)^2]^{1/2}} = \frac{1}{[\alpha(2 - \alpha)]^{1/2}}, \quad (22)$$

which, for practical values of $\alpha$, is less than 5 percent increase in the rms noise, and therefore, a negligible amount.

## ACKNOWLEDGMENT

The author would like to thank R. H. Bosworth and C. Vetrano for constructing circuits, and his associates in the systems Research Laboratory for helpful advice and encouragement.

## REFERENCES

[1] H. Schmid, *Electronic Analog/Digital Conversions.* New York: Van Nostrand–Reinhold, 1970.

[2] D. F. Hoeschele, Jr., *Analog-to-Digital/Digital-to-Analog Conversion Techniques.* New York: Wiley, 1968.

[3] D. H. Sheingold and R. A. Ferrero, "Understanding A/D and D/A converters," *IEEE Spectrum*, vol. 9, pp. 47–56, Sept. 1972.

[4] D. N. Kaye, "Focus on A/D and D/A converters," *Electron. Des.*, vol. 21, pp. 56–65, Jan. 4, 1973.

[5] D. Kesner, J. Barnes, and T. Henry, "Analog-to-digital converters," *Motorola Monitor*, vol. 11, pp. 23–27, Apr. 1973.

[6] R. C. Kime, "The charge-balancing A-D converter: An alternative to dual-slope integration," *Electronics*, vol. 46, pp. 97–100, May 24, 1973.

[7] K. W. Cattermole, *Principles of Pulse Code Modulation.* New York: American Elsevier, 1969.

[8] H. H. Henning and J. W. Pan, "D2 channel bank: System aspects," *Bell Syst. Tech. J.*, vol. 51, pp. 1641–1658, Oct. 1972.

[9] J. O. Limb, "Design of dither waveforms for quantized visual signals," *Bell Syst. Tech. J.*, vol. 48, pp. 2555–2582, Sept. 1969.

[10] R. C. Brainard and J. C. Candy, "Direct-feedback coders: Design and performance with television signals," *Proc. IEEE*, vol. 57, pp. 776–786, May 1969.

[11] H. Inose and Y. Yasuda, "A unity bit coding method by negative feedback," *Proc. IEEE*, vol. 51, pp. 1524–1535, Nov. 1963.

[12] C. C. Cutler, "Transmission system employing quantization," U. S. Patent 2 927 962, Mar. 8, 1960.

[13] C. B. Brahm, "Feedback integrating system," U. S. Patent 3 192 371, June 29, 1965.

[14] T. Miura, K. Shi, and J. Iwata, "Signal conversion system with storage and correction of quantization error," U. S. Patent 3 560 957, Feb. 2, 1971.

[15] P. L. Lindorff, *Theory of Sampled-Data Control Systems.* New York: Wiley, 1965.

# The Structure of Quantization Noise from Sigma–Delta Modulation

JAMES C. CANDY, FELLOW, IEEE, AND OCONNELL J. BENJAMIN

*Abstract*—When the sampling rate of a sigma–delta modulator far exceeds the frequencies of the input signal, its modulation noise is highly correlated with the amplitude of the input. We derive simple algebraic expressions for this noise and its spectrum in terms of the input amplitude. The results agree with measurements taken on a breadboard circuit.

This work can be useful for designing oversampled analog to digital converters that use sigma–delta modulation for the primary conversion.

## I. INTRODUCTION

QUANTIZATION noise introduced by delta modulation is easily analyzed when one assumes that overloading is avoided and that the noise is random [1]. The results of such analysis agree with practice provided the modulator has an active input, but the results are misleading for quiet inputs. It is the assumption of randomness that is faulty. When the input is steady the noise is highly colored, its spectrum appears as sets of distinct lines and the signal distortion is critically dependent on circuit conditions that determine whether or not strong lines lie in baseband.

The authoritative work on the structure of noise from delta modulation is that of Iwersen [2]. He gives a description of the noise spectrum and shows how it depends on imbalance between positive and negative step size and on the input signal.

Iwersen's work has been particularly useful for designing oversampled PCM encoders which make use of high speed delta modulation as the initial encoding [3]. The output of the modulator, when digitally filtered, is resampled at twice baseband frequency to provide the PCM. Deliberate imbalance of step size can greatly improve the idle channel resolution of these codecs [4].

The present work derives results for sigma–delta modulation that are closely related to those of Iwersen. The approach used in this analysis is different from Iwersen's and in some respects is more direct. We show that reasonable approximations provide very simple descriptions of the resolution of modulators that are sampled at a high frequency, and the results agree with practice. Before presenting this analysis, we describe measurements of noise from a real circuit. The main emphasis of this paper is on sigma–delta modulation because it is fast becoming the preferred modulation for oversampled codecs [4]-[8]. The Appendix shows how the analysis applies

Paper approved by the Editor for Data Communication Systems of the IEEE Communications Society for publication without oral presentation. Manuscript received January 21, 1981; revised April 29, 1981.

The authors are with Bell Laboratories, Holmdel, NJ 07733.

to ordinary delta modulation. The analysis and some of the results presented here also apply to the waiting-time problem described in [9].

## II. NOISE FROM SIGMA-DELTA MODULATION

Fig. 1 shows the circuit of a simple sigma–delta modulator, that accepts positive analog amplitudes and produces sequences of positive impulses. An impulse is generated, in time with the clock, whenever the integrated difference between the input and the output is positive. By this action the circuit regulates the rate at which impulses occur attempting to keep the average output equal to the average input. Zero input corresponds to no output impulses while maximum input corresponds to impulses generated at the clock rate. Then applied signals increase or decrease the rate its dynamic range where output impulses occur at one half the clock rate. Then applied signals increases or decreases the rate depending on whether they are positive or negative.

The output of the modulator is a digital representation of the signal, which can be demodulated by smoothing the impulses in a low-pass filter and removing the bias. In practical implementations the output pulses have finite duration but their shape can be allowed for in the design of the low-pass filter. Practical measurements described in this work came from a circuit that generates pulses lasting for a whole clock period, and they are smoothed in a low-pass filter that cuts off near 3.5 kHz.

When its input is steady, the modulator generates pulses in recurrent patterns that depend on the input level, and we expect the output to be noisy when repetition rates lie in baseband. Measured values of rms noise in baseband are plotted in Fig. 2 against input level, for three sampling rates. These graphs show that noise is largest when the modulator is biased near the ends of its range and next largest near the center. Indeed, most of the circuit's noise occurs in peaks adjacent to bias values that divide the dynamic range in small whole number ratios. Peaks of noise occur in pairs and are most prominent when the sampling rate is large compared to baseband. Arrows on the vertical axis of these graphs show average noise levels calculated from (35) of the Appendix.

For applications in communications it is advantageous to have low noise in the quiescent state; therefore, modulators should be biased to a quiet state away from the large peaks of noise. The advantage in doing so is demonstrated in Fig. 3 which shows the signal-to-noise ratio measured when the circuit was activated with a sinusoidal input for two conditions. One biased to the noisy state at the center of the dynamic range and the other to a relatively quiet state 1/20 of the range

Reprinted from *IEEE Trans. Commun.*, vol. COM-29, pp. 1316–1323, September 1981.

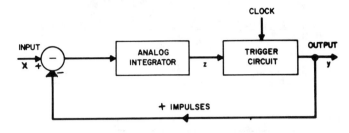

Fig. 1. An outline of a sigma–delta modulator, the trigger generates an impulse timed to the clock whenever $z$ is positive.

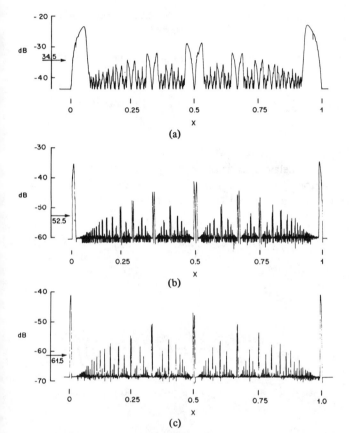

Fig. 2. Graphs of measured noise plotted against the level of dc input $X$. The active range is $0 < X < 1$ and for normal ac operation the modulator would be biased to $X = 0.5$. The arrow shows the expected average value of noise. Sampling rates are: (a) 64 kHz; (b) 256 kHz; (c) 512 kHz. Baseband is 3.5 kHz.

from center. The abscissa of these graphs is the amplitude of the input sine wave. Local depressions in the signal-to-noise ratio correspond to states where an extremity of the sine wave lies on a noise peak. For example, in Fig. 3(b), inputs that are a little larger than −20 dB touch the central noise peak.

The analysis we are about to present will enable these noise levels to be calculated, and because the structure of the noise is most noticeable at high sample rates, we will be concerned with sampling rates that far exceed baseband frequencies.

## III. A MODEL FOR SIGMA-DELTA MODULATION

We now develop a model that is a convenient basis for analyzing the noise generated by sigma–delta modulation. First, consider the asynchronous modulator that uses no tim-

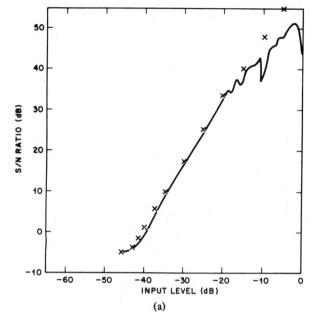

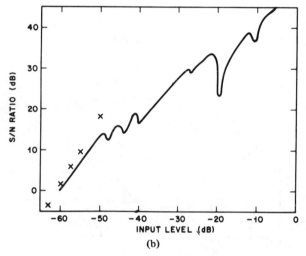

Fig. 3. Signal-to-noise ratio of a modulator with a 1.02 kHz sine wave input sampled at 256 kHz. $X$ marks the expected values of idle channel $S/N$: (a) The modulator biased at center of its range. (b) The modulator biased at 11/20 of its range.

ing clocks, as shown in Fig. 4. We let $x$ represent the input level and $z$ represent the integrated difference between the input and the output, $y$. An impulse of magnitude A is generated at the output whenever $z$ becomes positive. Fig. 5 shows representative waveforms of signals in this modulator. When $x$ is constant, the error $z$ is a regular sawtooth waveform and $y$ a regular stream of impulses, both of frequency $x/A$.

Now modify the modulator to synchronize its output to a clock of period $\tau$ as in Fig. 1, where output impulses can occur only when the clock is present and $z$ is positive. Waveforms in Fig. 6 show that the output impulses still occur at an average rate of $x/A$, (if $0 < x < A/\tau$), but each impulse is delayed from the corresponding asynchronous pulse in order to be aligned with the next clock. The synchronous sawtooth waveform $z$ overshoots positively but returns to its asynchronous value after the impulse occurs.

An impulse waveform identical to $y$ in Fig. 6 could be

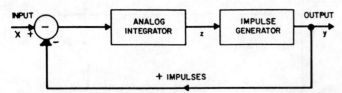

Fig. 4.   The asynchronous modulator, an impulse is generated whenever $z$ is positive.

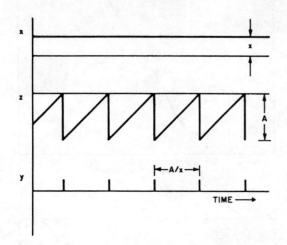

Fig. 5.   Waveforms in the asynchronous modulator with dc input $X$.

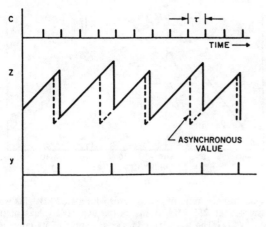

Fig. 6.   Waveforms in the clocked modulator, with dc input $X$.

generated directly by sampling a regular sequence of rectangular pulses of frequency $x/A$ and duration $\tau$ as shown in Fig. 7. This model would be inconvenient to implement, but it is very easy to describe mathematically; this we now do.

## IV. ANALYSIS OF THE SIMPLIFIED MODEL

The range of inputs that the synchronous modulator can accommodate without saturating is $0 < x < A/\tau$. In order to have unit dynamic range we scale amplitudes to make $A/\tau = 1$, then, the input would normally be biased to the state where $x = \frac{1}{2}$.

The train of impulses that is the clock signal is represented by the expression

$$C(t) = \tau \sum_i \delta(t - i\tau) = \sum_k \exp\left(\frac{2\pi j k t}{\tau}\right). \qquad (1)$$

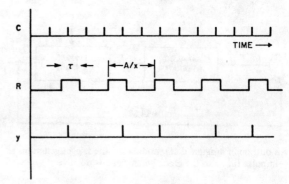

Fig. 7.   A sampling system that produces the same output as does the clocked modulator.

Similarly, the rectangular wave in Fig. 7 can be expressed as

$$R(t) = \sum_i \int dt \left\{ \delta\left(t - t_0 - \frac{i\tau}{x}\right) - \delta\left(t - t_0 - \tau - \frac{i\tau}{x}\right)\right\} \qquad (2)$$

where the delay $t_0$ is determined by initial conditions of the modulator. The Fourier series representation of the rectangular wave can now be written in the forms

$$R(t) = \frac{x}{\tau} \sum_l \int dt \left\{ \exp\left(2\pi j \frac{lx}{\tau}(t - t_0)\right) \right. \qquad (3)$$

$$\left. - \exp\left(2\pi j \frac{lx}{\tau}(t - t_0 - \tau)\right)\right\}$$

$$= \sum_l \frac{\sin(\pi lx)}{\pi l} \exp\left(2\pi j l \frac{x}{\tau}\left(t - t_0 - \frac{\tau}{2}\right)\right). \qquad (4)$$

The output, $R$ sampled by $C$, is given by the product of $C(t)$ and $R(t)$.

$$y(t) = \sum_l \sum_k \frac{\sin(\pi lx)}{\pi l} \exp\left(2\pi j \frac{(lx + k)}{\tau} t\right). \qquad (5)$$

In this last equation the constant delay $(t_0 + (\tau/2))$ has been ignored.

The result (5) represents the output signal as sets of spectral lines of frequency

$$f = \left(\frac{lx + k}{\tau}\right). \qquad (6)$$

Appendix II shows that this agrees with Iwersen's result for delta modulation [2]. Now recall that all the useful information in a sampled wave is contained in a band of frequency equal to half the sample rate

$$f \leqslant \frac{1}{2\tau}. \qquad (7)$$

In order to study this band we direct our attention to those values of $l$ and $k$ that satisfy

$$|lx + k| \leqslant \tfrac{1}{2} \tag{8}$$

and thereby eliminate the parameter $k$.

At this point it is useful to introduce the following notation: $I(v)$ will represent the nearest integer to the real number $v$ and $[v]$ will represent the fractional roundoff, that is,

$$[v] = (v - I(v)) \tag{9}$$

where

$$-0.5 < (v - I(v)) \leqslant 0.5.$$

For a component of (5) to lie in the band (7) requires that

$$k = -I(lx), \tag{10}$$

and its frequency can be expressed as

$$f_1 = \frac{[lx]}{\tau}. \tag{11}$$

Thus, the components of $y$ that lie in the half sample band of frequencies can be expressed as

$$y_1(x) = x + 2 \sum_{l=1}^{\infty} \frac{\sin(\pi lx)}{\pi l} \cos\left(2\pi[lx]\frac{t}{\tau}\right). \tag{12}$$

The first term in this expression represents the useful output; the second term is modulation noise.

## V. BASEBAND NOISE

For the noise component of $y$ at frequency $f$ to lie in baseband $f_0$, $0 \leqslant f < f_0 \leqslant \tfrac{1}{2}\tau$, requires that

$$f\tau = |[lx]| < f_0\tau. \tag{13}$$

The power associated with that component is given by the expression

$$P_1(lx) = 2 \frac{\sin^2(\pi[lx])}{(\pi l)^2} \tag{14}$$

which agrees with the result (3a) in [4]. This noise power will tend to be largest when $l$ is small, therefore, in order to examine the major properties of the noise, we want to locate those values of $x$ in the range $0 \leqslant x < 1$ that satisfy (13) with small values of $l$.

We shall assume that the net power is given by the sum of powers in individual components because situations where components have precisely the same frequency are very unusual.

Notice in Fig. 2 that the baseband noise occurs predominantly in narrow peaks adjacent to integer divisions of the signal

range, particularly when $\tau f_0$ is small. Let us now study the noise in the vicinity of the $m/n$ division by letting

$$x = \left(\frac{m}{n} + v\right) \tag{15}$$

where $m \leqslant n$ are incommensurate integers and $v$ a deviation. In order for $x$ to locate major peaks of noise, sensible ranges for these parameters are relatively coarse divisions compared to the oversampling ratio

$$\frac{1}{n} > f_0\tau \tag{16}$$

and $v$ small. Its true range will emerge as result (19).

Substituting (15) in (13) requires that

$$[lx] = \left[l\left(\frac{m}{n} + v\right)\right] < f_0\tau. \tag{17}$$

When $v$ is very small this condition can hold only when $m = 0$ or $l$ is a whole number multiple of $n$, otherwise

$$\left[\frac{lm}{n}\right] \geqslant \frac{1}{n} > f_0\tau.$$

We now put

$$l = in \tag{18}$$

where $i$ is a positive integer then

$$[lx] = [lv].$$

Then frequency components lie in baseband if

$$\tau f = [lx] = [lv] < f_0\tau. \tag{19}$$

We are particularly interested in the case where $l$ is small, that is,

$$l = in < \frac{f_0\tau}{|v|}; \tag{20}$$

then

$$[lv] = lv. \tag{21}$$

The range of $v$ that permits this condition to hold is

$$|v| < v_{max} = \frac{f_0\tau}{n}. \tag{22}$$

There are many values of $l$ that satisfy (19) without satisfying (20) but they are large. (For example, values near to $l = j/|v|$ where $j \geqslant 1$.) The noise power neglected by accepting condition (20) is of the order of one part in $(f_0\tau)^{-2}$. This is small because in most applications of sigma–delta modulation $f_0 \tau < 0.1$.

In order to obtain a simple expression for the noise power let us now consider modulators where $f_0 \tau$ is small enough that

$$\sin (\pi f_0 \tau) \approx \pi f_0 \tau. \qquad (23)$$

Then, using (21) and (14), the power in each baseband component that satisfies (20) may be written as

$$P_1(v) = 2v^2. \qquad (24)$$

To get an expression for the total noise power at a given value of $x$ we need to know the number of components that satisfy (21). This inequality may be written as

$$i < \frac{f_0 \tau}{n|v|} = \frac{v_{max}}{|v|} \qquad (25)$$

which sets the limit on the number of significant noise components in baseband. In particular, for $v$ in the range

$$\frac{v_{max}}{(i_1 + 1)} < |v| \leqslant \frac{v_{max}}{i_1}$$

there are $i_1$ components of noise and the net baseband power is given by

$$P(v) = (2i_1 v^2). $$

Fig. 8(a) shows a graph of calculated values of the net rms baseband noise $v\sqrt{2i}$, plotted against $v$. Fig. 8(b) shows the rms noise measured on a real circuit. The sharp corners in Fig. 8(a) are rounded in (b) because the practical filter used to define baseband does not have the abrupt cutoff that the theory assumes. The characteristic of the filter that was used is shown in Fig. 9. A useful approximation of the measured graph in Fig. 8(b) is obtained by assuming that $i$ is not only an integer but also a real number equal to $v_{max}/|v|$; then

$$P(v) = 2|v| v_{max} = \frac{2f_0 \tau}{n} |v|. \qquad (26)$$

A pair of maxima occur in the noise when $|v| = v_{max}$, their amplitude is given by

$$P_{max}(n) = 2v_{max}^2 = 2\left(\frac{f_0 \tau}{n}\right)^2. \qquad (27)$$

Fig. 10 shows a graph plotted against $1/n$ of the rms noise maxima, taken from Fig. 2. The value of $f_0$ calculated from the slopes of this graph is 3.0 kHz which is reasonably well in accord with the corner frequency of the filter characteristic in Fig. 9.

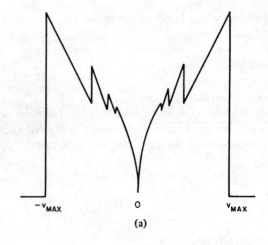

(a)

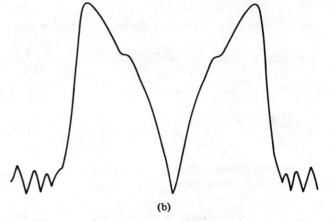

(b)

Fig. 8. An expanded view of a pair of peaks of noise. (a) Expected shape with ideal low-pass filter. (b) Measured shape with real baseband filter.

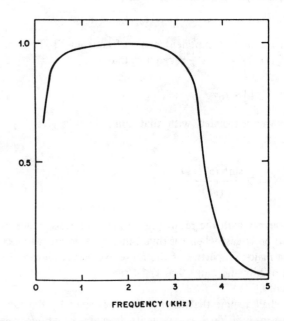

Fig. 9. Response of the low-pass filter that defines baseband in the experimental circuit. The analysis assumes an abruptly cutoff filter.

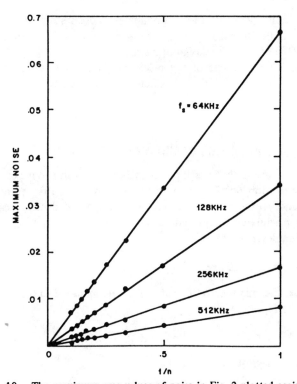

Fig. 10. The maximum rms values of noise in Fig. 2 plotted against $1/n$ for various sampling rates.

The total power in one peak of noise can be expressed as

$$P_T(n) = \sum_{i=1}^{\infty} \int_0^{\frac{v_{max}}{i}} 2v^2 \, dv \qquad (28)$$

$$= \frac{2}{3} v_{max}^2 \left( \sum_i^{\infty} \frac{1}{i^3} \right)$$

$$= 0.8 v_{max}^3 = 0.8 \left( \frac{f_0 \tau}{n} \right)^3.$$

Eighty-three percent of this noise is contributed by the primary component for which $i_1 = 1$ (i.e., $l = n$).

Results (25)–(28) are all independent of the parameter $m$ defined by (15); this fact agrees with the measurements presented as Fig. 2. Properties of the first peak at $x = 0$, for which $m = 0$, can be obtained by studying the final peak at $x = 1$, for which $n = 1$ because

$$\left\{ l \left( \frac{m}{n} + v \right) \right\} = [lv]$$

for all $l$ when $m = 0$ or $n = 1$.

## VI. APPLICATIONS OF THE RESULTS

The results derived in the previous sections apply to modulators that have steady dc input signals. They may be used, however, for dynamic inputs that change slowly compared with the sampling rate. This assertion was confirmed by re-plotting the graphs of Fig. 3 for signal frequencies in the range 100 Hz to 3 kHz and noting no appreciable change. We may therefore express the net noise power for any input that changes slowly by

$$N^2 = \int_0^1 P(x) p(x) \, dx. \qquad (29)$$

$P(x)$ is the noise for an input level $x$ as given by (26) and $p(x)$ is the fraction of time that the input has value $x$. For sinusoidal input having amplitude $A$ and dc offset $x_0$

$$p(x) = \frac{1}{\pi \sqrt{A^2 - (x - x_0)^2}}, \qquad |x - x_0| \leqslant A \qquad (30)$$

$$= 0, \qquad \text{otherwise.}$$

To calculate the noise contributed by the two peaks in the region of input level $x = m/n$ we use (26) to get

$$N_1^2(n) = \int_{-v_{max}}^{v_{max}} \frac{2}{\pi} \frac{v_{max} |v| \, dv}{\sqrt{A^2 - (x - x_0)^2}}. \qquad (31)$$

Now we consider three examples that relate to properties of the graphs in Fig. 3.

*Example 1:* We will calculate the noise contributed by the pair of large peaks at the center of the range where $x = 1/2$. The modulator is assumed to be biased to the center $x_0 =$ and the input is a sine wave of amplitude $A$. To describe the noise peaks we put $n = 2$, $v_{max} = (f_0 \tau)/2$ and $x = (1/2 + v)$, then (31) becomes

$$N_1^2(2) = 2 \int_0^{v_1} \frac{2}{\pi} \frac{v_{max} v \, dv}{\sqrt{A^2 - v^2}} \qquad (32)$$

where the range $v_1 = v_{max}$ if $A \geqslant v_{max}$ and $v_1 = A$, otherwise. It follows that

$$N_1^2(2) = \frac{4 v_{max}}{\pi} (A - \sqrt{A^2 - v_{max}^2}), \quad A \geqslant v_{max} \qquad (33)$$

$$= \frac{4 v_{max} A}{\pi}, \qquad A < v_{max}.$$

Selected values of the rms signal-to-noise ratio derived from (33) are shown on the graph in Fig. 3. We see good agreement for input amplitude below −20 dB. At larger amplitudes, noise from other peaks contribute strongly to the net impairment. We draw attention to the fact that some of the error predicted by (33) manifests itself as a loss of gain rather than as added noise. This loss is sometimes a significant part of the distortion and it has been included in the measurements reported in Fig. 3.

An interesting property of result (33) is the fact that when $A = v_{max}$ the signal-to-noise ratio is $\sqrt{\pi/8} = -4.06$ dB independent of the sampling rate.

*Example 2:* We now consider the case where the modulator is biased to center $x_0 = 1/2$ and the input sine wave is about −9 dB ($A \approx 1/6$) below saturation. In this condition the noise peaks at center have little effect on resolution (see Example 1). The dominant noise is that for which $n = 3$, it distorts both extremes of the sine wave where $x = \frac{1}{3}$ and $\frac{2}{3}$. The noise power from these peaks is largest when the maximum noise lies on the extreme of the sine wave. This noise can be expressed as

$$N_1{}^2(3) = 2 \int_{-v_{max}}^{v_{max}} \frac{2}{\pi} \frac{v_{max} v \, dv}{\sqrt{(\frac{1}{6} + v_{max})^2 - (\frac{1}{6} + v)^2}} . \quad (34)$$

For 256 kHz sampling and 3 kHz baseband

$$v_{max} = \frac{1}{3} \frac{3}{256} = \frac{1}{256}.$$

We neglect $v_{max}$ in comparison with $\frac{1}{3}$ to get

$$N_1{}^2(3) = \sqrt{3} \, \frac{4}{\pi} v_{max}$$

$$\cdot \left[ \int_0^{v_{max}} \frac{v \, dv}{\sqrt{v_{max} - v}} + \int_0^{v_{max}} \frac{v \, dv}{\sqrt{v_{max} + v}} \right]$$

$$= \frac{8}{\sqrt{3}\pi} (4 - \sqrt{2}) v_{max}{}^2 \sqrt{v_{max}}$$

and the signal-to-noise ratio = 35.8 dB. This compares with the measured value, 37 dB.

*Example 3:* With the modulator biased 1/20 of the range from center, we calculate the idle channel noise introduced by the two small peaks for which $n = 10$, $x = (11/20) + v)$, $x_0 = 11/20$, and $v_{max} = (f_0 \tau / 10)$. Following the same reasoning as that of Example 1 the noise is given by (33). The expected signal-to-noise ratio plotted in Fig. 3(b) agrees with the measured value for small inputs.

## VII. MULTILEVEL QUANTIZATION

Some sigma–delta modulators use multilevel quantization rather than the two level that we have considered. The analysis presented applies to such multilevel quantization provided that the net gain in the feedback loop [8] is unity and that signal amplitudes are scaled to make the step size unity.

Measurements on a real circuit indicate that changes of loop gain near unity have a weak effect on the noise. Indeed, modulators with gains in the range 0.7 − 1.3 all have similar noise structures.

With multilevel quantization the amplitude of the signal can far exceed the step size and the probability that the input has a particular value $p(x)$ is correspondingly reduced. This fact tends to whiten the net noise in multilevel modulators.

## CONCLUDING REMARKS

We have demonstrated that simple formulas give good predictions of the resolution that can be obtained from sigma-delta modulators. The method is particularly useful for determining the resolution at certain critical states where just a few noise peaks of Fig. 2 dominate. These states usually set the minimum resolution of the modulator and are thus of most interest in the design of a circuit.

Complete graphs of the resolution, such as those given in Fig. 3, would be tedious to calculate; they are better obtained by simulation.

## APPENDIX

### ANALYSIS BASED ON RANDOM NOISE

We assume that the input is sufficiently active to make the quantization noise so random that it may [8] be represented as additive white noise $E$ with power spectral density $\tau e_0{}^2$. The modulator is represented by the circuit in Fig. 11 and its action described by

$$Y_n = X_n + (E_n - E_{n-1}).$$

Thus, the rms spectral density of the noise added to the signal can be represented as

$$N(\omega) = \sqrt{\tau} e_0 (1 - \exp(-j\omega\tau))$$

and its magnitude expressed as

$$|N(\omega)| = 2\sqrt{\tau} e_0 \sin \left( \frac{\omega\tau}{2} \right).$$

The net rms noise in baseband $N_0$ is given by

$$N_0{}^2 = 4 e_0{}^2 \tau \int_0^{f_0} \sin^2 (\pi f \tau) \, df$$

$$= 2 f_0 \tau e_0{}^2 (1 - \mathrm{sinc} (2\pi f_0 \tau)).$$

For oversampled modulators $f_0 \tau \ll 1$, $N_0$ may be approximated by

$$N_0 = \frac{2\pi}{\sqrt{3}} e_0 (f_0 \tau)^{3/2}.$$

The quantization noise $E$ has uniform amplitude distribution in the range ±0.5. Therefore, $e_0 = \frac{1}{\sqrt{6}}$. Thus, the rms baseband noise from an active modulator is expected to be

$$N_0 = \frac{\sqrt{2}\pi (f_0 \tau)^{3/2}}{3}. \quad (35)$$

This value is marked on the graphs in Fig. 3.

Comparing $N_0{}^2$ with the noise power in each separate peak as given by (28) we find that the sum of the noise in the peaks is very nearly equal to $N_0{}^2$. About $\frac{3}{4}$ of this total is contained in the two peaks at the ends of the range. The two peaks at the center have about 0.1 of the total.

### NOISE FROM ORDINARY DELTA MODULATION

Fig. 12(a) is a diagram of a delta modulator and Fig. 12(b) is an equivalent circuit based on the sigma–delta modulator.

In order to obtain a constant input to the sigma–delta

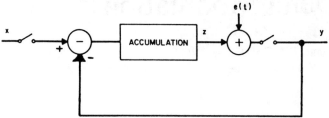

Fig. 11.   A linear sampled data model of a sigma–delta modulator.

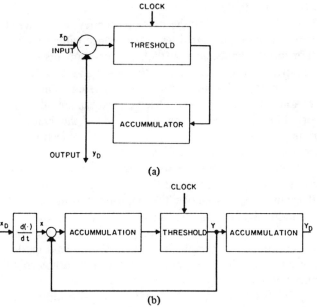

(a)

(b)

Fig. 12.   (a) Schematic of a delta modulator. (b) An analog of a delta modulator that is based on sigma–delta modulation.

modulator a constant rate of change is required at the input to the delta modulator. Then the output of the sigma–delta modulator, given by (5), can be accumulated to give the delta modulator output. Such accumulation may be represented by

$$y_D(t) = \frac{y(t)}{1 - \exp(-2\pi jf\tau)}$$

$$= \sum_l \sum_k \frac{\sin(\pi lx)\exp\left(2\pi j(lx + \kappa)\dfrac{t}{\tau}\right)}{\pi l(1 - \exp(-2\pi j(lx + \kappa)))}$$

$$y_D(t) = \sum_l \sum_\kappa \frac{\exp\left(2\pi j(lx + \kappa)\dfrac{t}{\tau}\right)}{2\pi jl(-1)^\kappa} .$$

Here, we have neglected the constant delays. This result agrees with Iwersen's when $x$ represents rate of change of input signal.

## ACKNOWLEDGMENT

The authors would like to thank W. H. Ninke and G. R. Ritchie for their helpful advice.

## REFERENCES

[1]   R. Steele. *Delta Modulation Systems.*   New York: Wiley. 1975. ch. 3.

[2]   J. E. Iwersen, "Calculated quantizing noise of single-integration delta modulation coders," *Bell Syst. Tech. J.*, vol. 48, pp. 2359–2389. Sept. 1969.

[3]   D. J. Goodman and L. J. Greenstein, "Quantizing noise of DM/PCM encoders," *Bell Syst. Tech. J.*, vol. 52, pp. 183–204, Feb. 1973.

[4]   T. Misawa, J. E. Iwersen, and J. G. Rush, "A single-chip CODEC with filters, architecture," in *Int. Conf. Commun. Rec.*, vol. 1, June 1980, pp. 30.5.1–30.5.6.

[5]   J. C. Candy, Y. C. Ching, and D. S. Alexander, "Using triangularly weighted interpolation to get 13-bit PCM from a sigma–delta modulator," *IEEE Trans. Commun.*, vol. COM-24, pp. 1268–1275. Nov. 1976.

[6]   J. D. Everhard, "A single-channel PCM codec." *IEEE J. Solid-State Circuits*, vol. SC-11, pp. 25–38, Feb. 1979.

[7]   J. C. Candy, B. A. Wooley, and O. J. Benjamin, "A voiceband codec with digital filtering," *IEEE Trans. Commun.*, vol. COM-29, pp. 815–830, June 1981.

[8]   R. C. Brainard and J. C. Candy, "Direct-feedback coders: Design and performance with television signals," *Proc. IEEE*, vol. 37, pp. 776–786, May 1969.

[9]   *Transmission Systems for Communications.* 5th ed.   Bell Telephone Laboratories, Inc., 1981.

# Multistage Sigma-Delta Modulation

WU CHOU, STUDENT MEMBER, IEEE, PING WAH WONG,
AND ROBERT M. GRAY, FELLOW, IEEE

*Abstract* —We provide a theoretical basis for multistage sigma-delta modulation (MSM), which is a cascade realization of several single-loop sigma-delta modulators with a linear combinatorial network. Equations are derived describing the output and the quantization noise of MSM for an arbitrary input signal and the noise shaping characteristic of MSM is investigated. The spectral characteristics of an $m$-stage sigma-delta modulator with both dc and sinusoidal inputs are developed. For both types of inputs the binary quantizer noise of the $m$th ($m \geq 3$) sigma-delta quantizer, which appears at the output as an $m$th order difference, is asymptotically white, uniformly distributed, and uncorrelated with the input level. It is also found that for an $m$-stage sigma-delta quantizer with either an ideal low-pass filter or a $\text{sinc}^{m+1}$ filter decoder, the average quantization noise of the system is inversely proportional to the $(2m+1)$th power of the oversampling ratio $R$. This implies that high-order systems are favorable in terms of the trade-off between the quantization noise and the oversampling ratio. Simulation results are presented to support the theoretical analysis.

## I. INTRODUCTION

ONE OF THE fundamental factors in the performance of digital signal processing systems is the precision of the digital and analog interface. Recently, oversampled sigma-delta modulators have received considerable attention as candidates for high-resolution analog-to-digital conversion because they are robust against circuit imperfections and well suited for VLSI implementation [1]–[9]. A sigma-delta modulator typically consists of a discrete time integrator and a one-bit quantizer inside a feedback loop. The incoming analog signal is sampled at many times the Nyquist rate, and the samples are digitized using a one-bit quantizer inside a feedback loop. The high rate bit stream is then processed by a digital filter and decimated to produce a high-resolution approximation to the analog signal at the Nyquist rate. Thus the requirement of a large number of high-precision thresholds in flash quantization is replaced by high accuracy in timing control of the high-speed circuit.

A recent exact analysis of the single-loop sigma-delta modulator [10]–[12] reveals that when the input is either dc or sinusoidal, the spectrum of the binary quantizer noise is discrete and highly colored. Moreover, the frequency components of the spectrum vary with the level of the input signal. It was reported that when an ideal low-pass filter or a triangular ($\text{sinc}^2$) filter is employed as the decoder, the

time average quantization noise power is inversely proportional to the third power of the oversampling ratio $R$, which is defined as the ratio of the actual sampling rate to the Nyquist rate.

Since the locations of the noise frequency components depend on the input signal level, it is conceivable that for certain input signals, strong frequency components of the noise spectrum will fall into the pass band and hence degrade the signal-to-noise ratio of the system. This phenomenon has already been observed experimentally and reported in [7]. Moreover, the spiky nature of the quantization noise spectrum might be subjectively objectionable when the oversampled quantizer is used in, for example, digital audio applications. Various schemes have been proposed in the past to preserve the advantages of oversampled quantizers, to remove the unpleasant properties of single-loop sigma-delta modulator, and to improve the trade-off between the oversampling ratio and the signal-to-noise ratio. For example, multi-loop sigma-delta modulation has been shown experimentally to have a less spiky quantization noise spectrum than its single-loop counterpart [4]. Also reported is that modulators with more than two loops can latch onto undesirable noisy modes due to overloading and hence require very careful design and fine tuning.

In [13] and [14], multistage sigma-delta modulation (MSM) was proposed. It has been shown experimentally for the cases of two and three stages that the quantization noise spectrum of such system is smooth and that it is free from overloading problems. In [15], a two-stage sigma-delta modulator was analyzed for both dc and sinusoidal input signals. It was shown that when the oversampling ratio is large (which is typical) or, equivalently, when the input signal frequency is much smaller than the actual sampling rate, the quantization noise is approximately white and uncorrelated with the input, where the approximation becomes exact when the input signal is dc.

We investigate the behavior of general MSM by solving the nonlinear difference equations describing the system. Since the analysis for two-stage sigma-delta quantizers has been reported previously [15], we focus our attention here on MSM with more than two stages. We point out, however, when a result in this paper agrees with or deviates from those of the two-stage sigma-delta quantizer. We study the mechanism of MSM and the noise shaping effect due to the linear combinatorial network. As in [15], we analyze the system for both dc and sinusoidal input signals.

Manuscript received December 12, 1987; revised August 1, 1988. This work was supported in part by the National Science Foundation under Grant ECS83-17981 and by a Stanford University Center for Integrated Systems Seed Grant.

The authors are with the Information Systems Laboratory, Department of Electrical Engineering, Stanford University, Stanford, CA 94305
IEEE Log Number 8929036.

Reprinted from *IEEE Trans. Inform. Theory*, vol. IT-35, pp. 784–796, July 1989.

In Section II we establish the input–output relation of the modulator and show that overloading problems do not exist in these systems, i.e., these systems are stable if the input is suitably bounded. In Section III we derive an explicit expression for the binary quantizer noise in terms of the input for both the dc and the sinusoidal cases. The time average moments and the spectral characteristics of the binary quantizer noise are derived in Section IV. The average quantization noise power and the signal-to-noise ratio of the system are considered in Section V. Specific examples for the performance of various decoding filters are given. Simulation results for the binary quantizer noise are presented in Section VI. Section VII summarizes our results.

## II. DIFFERENCE EQUATION AND THE LINEAR NETWORK

The discrete-time circuit model of a single-loop sigma-delta modulator is shown in Fig. 1. It consists of a digital integrator and a one-bit quantizer inside a feedback loop. The underlying idea of oversampled A/D conversion is that the modulator generates a high-rate digital output which consists of the original analog input signal plus high-frequency quantization noise components. The high-frequency quantization noise components are then removed by a low-pass filter at the output of the modulator, and the output signal is decimated to produce a high-resolution digital representation of the analog signal.

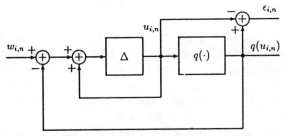

Fig. 1. Discrete time circuit model of single-loop sigma-delta modulator. $\Delta$ denotes unit delay and $q(\cdot)$ denotes nonlinear quantization function.

A block diagram of the MSM is shown in Fig. 2. It consists of a cascade of $m$ single-loop sigma-delta modulators. The analog input signal goes into the input of the first sigma-delta modulator. The binary quantizer error of each stage is fed into the input of the following stage. The outputs of all the stages are fed into a linear combinatorial network and processed to produce the output of the MSM.

The difference equations describing the $m$-stage sigma-delta modulator are as follows:

$$u_{i,n} = \begin{cases} u_i^*, & n=0 \\ u_{i,n-1} - q(u_{i,n-1}) + w_{i,n-1}, & n=1,2,\cdots \end{cases}$$
$$i = 1,2,\cdots,m \quad (1)$$

$$\epsilon_{i,n} = q(u_{i,n}) - u_{i,n}, \qquad n=0,1,\cdots; i=1,2,\cdots,m \quad (2)$$

$$w_{i,n} = \begin{cases} x_n, & i=1 \\ \epsilon_{i-1,n}, & i=2,3,\cdots,m \end{cases}, \quad n=0,1,\cdots \quad (3)$$

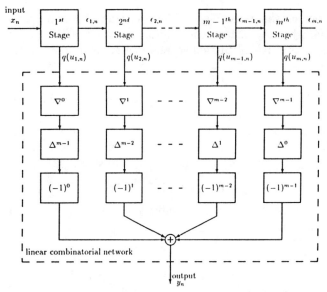

Fig. 2. Block diagram of $m$-stage sigma-delta quantizer. $\nabla$ denotes backward difference operator, and $\Delta$ denotes unit delay. Each stage consists of single-loop sigma-delta modulator.

where

$$q(u) = \begin{cases} b, & \text{if } u \geq 0 \\ -b, & \text{otherwise.} \end{cases}$$

$u_{i,n}$ is the $i$th integrator output, $u_i^*$ is the initial state of the $i$th integrator, $w_{i,n}$ is the input to the $i$th stage, $\epsilon_{i,n}$ is the binary quantizer error sequence of the $i$th stage, and $x_n \in [-b, b]$ is the input sequence to the system. From (1) and (2),

$$q(u_{i,n}) = w_{i,n-1} + \epsilon_{i,n} - \epsilon_{i,n-1}$$
$$= \begin{cases} x_{n-1} + \epsilon_{1,n} - \epsilon_{1,n-1}, & i=1 \\ \epsilon_{i-1,n-1} + \epsilon_{i,n} - \epsilon_{i,n-1}, & i=2,3,\cdots,m. \end{cases}$$
$$(4)$$

Define the backward difference operator as

$$\nabla^i \xi_n = \begin{cases} \xi_n, & i=0 \\ \nabla^{i-1}\xi_n - \nabla^{i-1}\xi_{n-1}, & i=1,2,\cdots. \end{cases}$$

For example,

$$\nabla^1 \xi_n = \xi_n - \xi_{n-1}$$

and

$$\nabla^2 \xi_n = \xi_n - 2\xi_{n-1} + \xi_{n-2}.$$

Note that $\nabla^i$ has the $z$ transfer function $(1-z^{-1})^i$. The linear combinatorial network is defined as

$$y_n = \sum_{i=1}^{m} (-1)^{i-1} \nabla^{i-1} q(u_{i,n-m+i}), \qquad m=2,3,\cdots \quad (5)$$

where $y_n$ is the output of the MSM. Equation (5) is a generalization of the circuit in the two-stage sigma-delta modulator [13], [14], where the quantization noise of the first stage is cancelled by the second stage. Here the quantization noise of each stage (except the last stage) is cancelled by its succeeding stage. Note that (5) suggests a natural implementation for the linear combinatorial network as shown in Fig. 2. The following Theorem shows that when the linear combinatorial network is defined as in (5), there is an alternative representation of the output sequence.

*Theorem 1:* For an $m$-stage sigma-delta modulator with the linear combinatorial network defined as in (5), the output sequence can be represented as a sum of the $m$-step delay of the input $x_n$ and an $m$th order difference of the binary quantizer error of the final ($m$th) stage, viz.,

$$y_n = x_{n-m} + (-1)^{m-1} \nabla^m \epsilon_{m,n}. \tag{6}$$

*Proof:* The proof is by induction on $m$. For $m = 2$, (5) reduces to

$$y_n = \nabla^0 q(u_{1,n-1}) - \nabla^1 q(u_{2,n})$$
$$= q(u_{1,n-1}) - q(u_{2,n}) + q(u_{2,n-1}). \tag{7}$$

Using (4), we have

$$y_n = x_{n-2} - \epsilon_{2,n} + 2\epsilon_{2,n-1} - \epsilon_{2,n-2}$$

which agrees with (6). Note that (7) agrees with previous studies of the two-stage sigma-delta modulator in [13]–[15]. Next, suppose (6) holds for $m$. Then using (5), we have for $m+1$ that

$$y_n = \sum_{i=1}^{m+1} (-1)^{i-1} \nabla^{i-1} q(u_{i,n-m-1+i})$$
$$= (-1)^m \nabla^m q(u_{m+1,n})$$
$$\quad + \sum_{i=1}^{m} (-1)^{i-1} \nabla^{i-1} q(u_{i,(n-1)-m+i})$$
$$= (-1)^m \nabla^m q(u_{m+1,n}) + x_{n-1-m} + (-1)^{m-1} \nabla^m \epsilon_{m,n-1}$$
$$= x_{n-(m+1)} + (-1)^m \nabla^m (q(u_{m+1,n}) - \epsilon_{m,n-1}). \tag{8}$$

Since the operator $\nabla$ is linear, we substitute (4) into (8) to get

$$y_n = x_{n-(m+1)} + (-1)^m \nabla^m (\epsilon_{m+1,n} - \epsilon_{m+1,n-1})$$
$$= x_{n-(m+1)} + (-1)^m \nabla^{m+1} \epsilon_{m+1,n},$$

which proves the theorem.

Observe that (5) and (6) are two different expressions describing the same output sequence. While (5) is more useful from the viewpoint of implementing the circuitry, (6) is more suitable for analysis because we only need the joint behavior of the input and the binary quantizer error of the final stage.

Let

$$N_n = y_n - x_{n-m} \tag{9}$$

be the overall quantization noise of the MSM. Note that the output at time $n$ is compared with the delayed input which is consistent with (6). Equation (6) implies that $N_n$ can be interpreted as the output of passing the binary quantizer noise of the final stage through a moving average filter with transfer function $(-1)^{m-1}(1-z^{-1})^m$. The frequency domain counterpart of (6) is the following result.

*Theorem 2:* For an $m$-stage sigma-delta modulator, if the binary quantizer noise of the final ($m$th) stage is a stationary or a quasi-stationary process (e.g., see [16, sec. 2.3]) with power spectrum $S_{\epsilon_m}(f)$, then the power spectrum of the overall quantization noise of the MSM is

$$S_N(f) = (2 \sin \pi f)^{2m} S_{\epsilon_m}(f). \tag{10}$$

*Proof:* From (6) and (9), the quantization noise of MSM is

$$N_n = (-1)^m \nabla^m \epsilon_{m,n}$$

which is the result of a linear operator acting on the binary quantizer noise $\epsilon_{m,n}$. By linear system theory,

$$S_N(f) = |H(f)|^2 S_{\epsilon_m}(f)$$

where

$$H(f) = (1 - z^{-1})^m \big|_{z = e^{j2\pi f}} = (j2 \sin \pi f)^m e^{-j\pi mf}.$$

Hence

$$|H(f)|^2 = (2 \sin \pi f)^{2m},$$

and (10) follows.

Equation (10) exhibits the noise-shaping effect of the MSM with $m$. As the number of stages $m$ increases, the attenuation of the transfer function $H(f)$ within the pass band also becomes greater. This implies a decrease of quantization noise power in the pass band. Although the amount of high-frequency noise increases due to a high transfer ratio in the high-frequency range, it can be removed easily by a good low-pass filter. This intuitive argument suggests that high-order MSM can provide better signal to noise ratios than low-order systems. We shall be more precise in a later section where we consider the average quantization noise for the cases of dc and sinusoidal inputs.

As was shown, all that is needed for an exact analysis of the MSM is the behavior of the quantization noise of the final ($m$th) stage. Since each stage of the MSM is a single-loop sigma-delta modulator, we can apply the results developed in [12]. In particular, if $w_{i,k} \in [-b, b]$ for all $i$ and $k$, then the normalized binary quantizer noise is given by

$$\frac{\epsilon_{i,n}}{2b} = \frac{1}{2} - \left\langle \sum_{k=0}^{n-1} \left( \frac{1}{2} + \frac{w_{i,k}}{2b} \right) \right\rangle,$$

$$n = 0, 1, \cdots; \ i = 1, 2, \cdots, m \tag{11}$$

where $\langle \alpha \rangle$ denotes the fractional part of $\alpha$, i.e., $\langle \alpha \rangle = \alpha \bmod 1$. Substituting (3) into (11) gives

$$\frac{\epsilon_{i,n}}{2b} = \begin{cases} \dfrac{1}{2} - \left\langle \displaystyle\sum_{k=0}^{n-1} \left( \dfrac{1}{2} + \dfrac{x_k}{2b} \right) \right\rangle, & \text{if } i = 1 \\[3ex] \dfrac{1}{2} - \left\langle \displaystyle\sum_{k=0}^{n-1} \left( \dfrac{1}{2} + \dfrac{\epsilon_{i-1,k}}{2b} \right) \right\rangle, & i = 2, 3, \cdots, m \end{cases}$$

$$n = 0, 1 \cdots. \tag{12}$$

Since $x_n$ is restricted in the range $[-b, b]$, (12) is valid for $i = 1$, and hence it implies $\epsilon_{1,n} \in [-b, b]$. Then (12) is also valid for $i = 2$ and hence $\epsilon_{2,n} \in [-b, b]$. Similar reasoning can be applied inductively to show that (12) is valid for $i = 1, 2, \cdots, m$. Observe that (12) also implies the multi-stage sigma-delta modulator is free from overloading since the dynamic ranges of all the signals are within the range $[-b, b]$. The following theorem gives the general form of the binary quantizer error of the $m$th quantizer in MSM.

*Theorem 3:* For an $m$-stage sigma-delta modulator with an arbitrary input $x_n \in [-b, b]$, the binary quantizer error of the final ($m$th) stage is

$$\frac{\epsilon_{m,n}}{2b} = (-1)^{m-1}\left[\frac{1}{2} - \left\langle \sum_{i_m=0}^{n-1} \cdots \sum_{i_2=0}^{i_3-1} \sum_{i_1=0}^{i_2-1}\left(\frac{1}{2} + \frac{x_{i_1}}{2b}\right)\right\rangle\right]. \tag{13}$$

*Proof:* We prove the result by induction on $m$. Let $m = 1$, then (13) coincides with (12). Assume (13) holds for $m$, then

$$\frac{\epsilon_{m+1,n}}{2b} = \frac{1}{2} - \left\langle \sum_{i_{m+1}=0}^{n-1}\left(\frac{1}{2} + \frac{\epsilon_{m,i_{m+1}}}{2b}\right)\right\rangle$$

$$= \frac{1}{2} - \left\langle \sum_{i_{m+1}=0}^{n-1}\left[\frac{1}{2} + (-1)^{m-1}\frac{1}{2} + (-1)^m\left\langle \sum_{i_m=0}^{i_{m+1}-1} \cdots \sum_{i_1=0}^{i_2-1}\left(\frac{1}{2} + \frac{x_{i_1}}{2b}\right)\right\rangle\right]\right\rangle \tag{14}$$

$$= \frac{1}{2} - \left\langle (-1)^m \sum_{i_{m+1}=0}^{n-1} \sum_{i_m=0}^{i_{m+1}-1} \cdots \sum_{i_1=0}^{i_2-1}\left(\frac{1}{2} + \frac{x_{i_1}}{2b}\right)\right\rangle$$

since the term $(1/2) + (-1)^{m-1}(1/2)$ does not contribute to the fractional part and

$$\langle\langle x\rangle \pm \langle y\rangle\rangle = \langle x \pm y\rangle$$

for all real numbers $x$ and $y$. For even $m$, we have $(-1)^m = 1$, and hence (14) becomes

$$\frac{\epsilon_{m+1,n}}{2b}$$

$$= (-1)^m\left[\frac{1}{2} - \left\langle \sum_{i_{m+1}=0}^{n-1} \sum_{i_m=0}^{i_{m+1}-1} \cdots \sum_{i_1=0}^{i_2-1}\left(\frac{1}{2} + \frac{x_{i_1}}{2b}\right)\right\rangle\right]$$

agreeing with (13). Similarly, $(-1)^m = -1$ for odd $m$ and (14) becomes

$$\frac{\epsilon_{m+1,n}}{2b} = \frac{1}{2} - \left\langle -\sum_{i_{m+1}=0}^{n-1} \sum_{i_m=0}^{i_{m+1}-1} \cdots \sum_{i_1=0}^{i_2-1}\left(\frac{1}{2} + \frac{x_{i_1}}{2b}\right)\right\rangle = \left\langle \sum_{i_{m+1}=0}^{n-1} \sum_{i_m=0}^{i_{m+1}-1} \cdots \sum_{i_1=0}^{i_2+1}\left(\frac{1}{2} + \frac{x_{i_1}}{2b}\right)\right\rangle - \frac{1}{2}$$

$$= (-1)^m\left[\frac{1}{2} - \left\langle \sum_{i_{m+1}=0}^{n-1} \sum_{i_m=0}^{i_{m+1}-1} \cdots \sum_{i_1=0}^{i_2-1}\left(\frac{1}{2} + \frac{x_{i_1}}{2b}\right)\right\rangle\right].$$

This completes the proof.

Observe that Theorems 1–3 apply to any arbitrary input signal. Two special cases of input signals that will be considered in depth are dc and a sinusoid. These are the two most common test signals considered in the analysis and comparison of oversampled analog-to-digital converters. The results can be extended to multiple sinusoidal inputs, and the details are given in Appendix I.

For a dc input, let $x$ be the input dc level, i.e., $x_n = x \in [-b, b]$ for $n = 0, 1 \cdots$. Define

$$\beta \triangleq \frac{1}{2} + \frac{x}{2b}$$

which is the normalized and shifted dc input value. The quantity $\beta$ plays a key role in the development of results in this paper (as it does in the single-loop case [11]). First, observe that $x \in [-b, b]$ implies $\beta \in [0, 1]$ and that $\beta$ and $x$ are related by a one-to-one mapping. Second, for a dc input, $\beta$ is a fixed constant. Third, we require that $\beta$ be an irrational number in $[0, 1]$, i.e., the input level $x$ is not a rational factor of $b$. This assumption is reasonable if we let $x$ be drawn from the set $[-b, b]$ according to some proba-

bility density function in which the condition of the assumption is satisfied with probability one. As discussed in [11], this can also be considered as an approximation to the case of a rational $x$ with a large denominator.

We also consider the sinusoidal input defined by

$$x_n = a\cos(2\pi nf/f_s) = a\cos n\theta, \qquad n = 0, 1, \cdots \tag{15}$$

where $a$ and $f$ are, respectively, the amplitude and the frequency of the sinusoidal input signal, and $f_s$ is the sampling frequency. Note that the amplitude $a$ is such that $0 \leq a \leq b$. We assume that the maximum bandwidth of the system is $W$, i.e., we only consider input signals of frequency no greater than $W$. The oversampling ratio $R$ is

defined as the ratio of the actual sampling frequency to the Nyquist frequency and hence $R = f_s/2W$. Since the oversampling ratio is assumed to be large, the quantity $\theta$ as defined in (15) is small and hence $x_n$ is a slowly varying sinusoidal sequence. Analogous to the dc input case, we require that the quantity $ka/b + lf/f_s$ be irrational for all integer pairs $(k, l) \neq (0,0)$. Again, this assumption is satisfied with probability one if we let $(a, f)$ be drawn from the set $[0, b] \times [0, W]$ according to some probability density function. The foregoing assumptions of irrationality allow us to apply results from the asymptotic distribution of fractional parts of sequences for evaluating the moments of the binary quantizer error. More detailed discussions are deferred to Section IV and the references therein.

Specializing Theorem 3 for dc and sinusoidal inputs, we obtain

$$\frac{\epsilon_{m,n}}{2b} = \begin{cases} (-1)^{m-1}\left[\dfrac{1}{2} - \left\langle \displaystyle\sum_{i_m=0}^{n-1} \cdots \sum_{i_2=0}^{i_3-1}\sum_{i_1=0}^{i_2-1} \beta \right\rangle \right], & \text{dc input} \\[2em] (-1)^{m-1}\left[\dfrac{1}{2} - \left\langle \displaystyle\sum_{i_m=0}^{n-1} \cdots \sum_{i_2=0}^{i_3-1}\sum_{i_1=0}^{i_2-1} \left(\dfrac{1}{2} + \dfrac{a\cos i_1 \theta}{2b}\right) \right\rangle \right], & \text{sinusoidal input.} \end{cases} \quad (16)$$

### III. Binary Quantizer Error of the $m$th Stage

We have established in Theorem 1 that the binary quantizer of the final stage completely determines the quantization noise at the output of the modulator. For this reason, we shall hereafter concentrate on this error sequence. Equation (16) provides an explicit expression of the binary quantizer error in terms of multiple sums. To proceed further, we need to evaluate the multiple sums in (16). Before doing so, it is convenient to introduce the following notation. Let $\mathscr{P}_k(\gamma)$ represent *generically* a polynomial in $\gamma$ of degree $k$. We emphasize the qualification 'generically' to indicate that $\mathscr{P}_k(\gamma)$ represents any polynomial provided its degree is $k$.[1] In particular, we can write

$$\mathscr{P}_k(\gamma) = \mathscr{P}_k(\gamma) + \mathscr{P}_l(\gamma), \qquad l < k$$

meaning that the addition of two polynomials of degrees $k$ and $l$ (with $l < k$), respectively, results in a polynomial of degree $k$. Similarly, we use $\mathscr{R}_k(\gamma)$ and $\mathscr{I}_k(\gamma)$ to represent generically polynomials of degree $k$ with rational and irrational coefficients, respectively. The reason for this notation is to avoid working out exactly all the coefficients of the polynomials involved in the development. As will be shown later, we only need to know the form of the polynomial to derive results, and for this reason we take the less tedious route.

*Lemma 1:* The $m$th order sum of a constant is a polynomial of degree $m$, i.e.,

$$\sum_{i_m=0}^{n-1} \cdots \sum_{i_2=0}^{i_3-1}\sum_{i_1=0}^{i_2-1} 1 = \mathscr{R}_m(n). \qquad (17)$$

[1] Note that the usage of $\mathscr{P}_k(\gamma)$ is analogous to the big-$O$ function in the asymptotic theory of mathematics.

*Proof:* The proof is by induction on $m$. For $m=1$, we have

$$\sum_{i_1=0}^{n-1} 1 = n$$

which is a polynomial of degree 1 with a rational coefficient. Assume (17) holds for $m$, i.e.,

$$\sum_{i_m=0}^{n-1} \cdots \sum_{i_2=0}^{i_3-1}\sum_{i_1=0}^{i_2-1} 1 = a_m n^m + a_{m-1}n^{m-1} + \cdots + a_0$$

where $a_0, a_1, \cdots, a_m$ are rational numbers. Then

$$\sum_{l=0}^{n-1}\sum_{i_m=0}^{l-1} \cdots \sum_{i_1=0}^{i_2-1} 1 = \sum_{l=0}^{n-1}\left(a_m l^m + a_{m-1}l^{m-1} + \cdots + a_0\right). \qquad (18)$$

Using an identity in [17, p. 1, 0.121], which is

$$\sum_{l=0}^{n-1} l^k = \frac{n^{k+1}}{k+1} + \mathscr{R}_k(n), \qquad \text{all } k \qquad (19)$$

and substituting (19) into (18) proves the lemma.

The following sums are useful for proving the next lemma [17, p. 30]:

$$\sum_{k=0}^{n-1}\cos k\theta = \frac{1}{2} + \frac{\sin\left(n - \dfrac{1}{2}\right)\theta}{2\sin\dfrac{\theta}{2}} \qquad (20)$$

$$\sum_{k=0}^{n-1}\sin k\theta = \frac{\cos\dfrac{\theta}{2} - \cos\left(n - \dfrac{1}{2}\right)\theta}{2\sin\dfrac{\theta}{2}}. \qquad (21)$$

*Lemma 2:* The following relation

$$\sum_{i_m=0}^{n-1} \cdots \sum_{i_2=0}^{i_3-1}\sum_{i_1=0}^{i_2-1} \cos i_1\theta$$

$$= \frac{n^{m-1}}{2(m-1)!} + P_{m-2}(n)$$

$$+ \frac{\cos\left[n\theta - \dfrac{m}{2}(\theta + \pi)\right]}{\left(2\sin\dfrac{\theta}{2}\right)^m}, \qquad m = 2, 3, \cdots \quad (22)$$

holds. The expression is also true for $m=1$ if we interpret $\mathscr{P}_{m-2}(n)$ as a zero polynomial when $m=1$.

*Proof:* The proof is by induction on $m$. Let $m=2$, then we can verify using (20) and (21) that

$$\sum_{i_2=0}^{n-1}\sum_{i_1=0}^{i_2-1}\cos i_1\theta = \frac{n}{2}+\frac{\cos\theta}{\left(2\sin\frac{\theta}{2}\right)^2}-\frac{\cos(n-1)\theta}{\left(2\sin\frac{\theta}{2}\right)^2},$$

agreeing with (22). Next, suppose that the lemma is true for $m$. Then

$$\sum_{l=0}^{n-1}\sum_{i_m=0}^{l-1}\cdots\sum_{i_1=0}^{i_2-1}\cos i_1\theta = \sum_{l=0}^{n-1}\frac{l^{m-1}}{2(m-1)!}+\sum_{l=0}^{n-1}\mathscr{P}_{m-2}(l)$$

$$+\sum_{l=0}^{n-1}\frac{\cos\left[l\theta-\frac{m}{2}(\theta+\pi)\right]}{\left(2\sin\frac{\theta}{2}\right)^m}.$$

Using (19), the first sum is

$$\sum_{l=0}^{n-1}\frac{l^{m-1}}{2(m-1)!}=\frac{n^m}{2(m!)}+\mathscr{P}_{m-1}(n).$$

Using (18) and (19), the second sum is seen to be a polynomial of degree $m-1$. For the third sum, we expand the cosine term in the numerator and then apply (20) and (21) to obtain

$$\sum_{l=0}^{n-1}\frac{\cos\left[l\theta-\frac{m}{2}(\theta+\pi)\right]}{\left(2\sin\frac{\theta}{2}\right)^m}$$

$$=\frac{\cos\left[n\theta-\frac{m+1}{2}(\theta+\pi)\right]}{\left(2\sin\frac{\theta}{2}\right)^{m+1}}+\mathscr{P}_0(n).$$

Collecting the results, we have

$$\sum_{i_{m+1}=0}^{n-1}\sum_{i_m=0}^{i_{m+1}-1}\cdots\sum_{i_1=0}^{i_2-1}\cos i_1\theta$$

$$=\frac{n^m}{2(m!)}+\mathscr{P}_{m-1}(n)+\frac{\cos\left[n\theta-\frac{m+1}{2}(\theta+\pi)\right]}{\left(2\sin\frac{\theta}{2}\right)^{m+1}},$$

proving the lemma.

*Theorem 4:* For an $m$-stage sigma-delta quantizer with dc input

$$x_n=x\in[-b,b],\qquad n=0,1,\cdots$$

where the ratio $x/b$ is an irrational number, the binary quantizer error of the final ($m$th) stage is

$$\frac{\epsilon_{m,n}}{2b}=(-1)^{m-1}\left[\frac{1}{2}-\langle\beta\mathscr{R}_m(n)\rangle\right]$$

$$=(-1)^{m-1}\left[\frac{1}{2}-\langle\mathscr{I}_m(n)\rangle\right],\qquad m=1,2,3,\cdots$$

where

$$\beta=\frac{1}{2}+\frac{x}{2b}.$$

*Proof:* Apply Lemma 1 to (16).

*Theorem 5:* For an $m$-stage sigma-delta quantizer with sinusoidal input

$$x_n=a\cos n\theta=a(\cos 2\pi nf/f_s),\qquad 0\le a\le b;\ n=0,1,\cdots$$

where the ratio $a/b$ is an irrational number, the binary quantizer error of the final ($m$th) stage is

$$\frac{\epsilon_{m,n}}{2b}=(-1)^{m-1}\left[\frac{1}{2}-\langle\lambda_n\rangle\right],\qquad m=2,3,\cdots$$

where

$$\lambda_n=\frac{an^{m-1}}{4b(m-1)!}+\mathscr{R}_m(n)+\mathscr{I}_{m-2}(n)$$

$$+\alpha_m\cos 2\pi\left(\frac{f}{f_s}n-\frac{mf}{2f_s}-\frac{m}{4}\right)$$

and

$$\alpha_m=\frac{a}{2b\left(2\sin\frac{\theta}{2}\right)^m}.$$

*Proof:* Apply Lemmas 1 and 2 to (16).

Theorems 4 and 5 give a closed-form representation of the binary quantizer error which permits the solution of time average moments in the next section. Before we close this section, we include a simple algebraic equality for ease of reference.

*Lemma 3:* Let $\psi(n)=a_mn^m+a_{m-1}n^{m-1}+\cdots+a_0$ be a polynomial of degree $m$ where $n$ is an integer. Then for all integers $i$ and $k$, the following equality is valid:

$$i\psi(n)+k\psi(n+l)$$

$$=\begin{cases}(i+k)a_mn^m+\mathscr{P}_{m-1}(n),& i+k\ne 0\\ kmla_mn^{m-1}+\mathscr{P}_{m-2}(n),& i+k=0.\end{cases}$$

If the $a_k$ are all rational or irrational numbers, then $\mathscr{P}$ can be replaced by $\mathscr{R}$ or $\mathscr{I}$, respectively. This lemma follows from simple algebraic manipulation.

## IV. Moments of The Binary Quantizer Error

We consider only deterministic input signals. Therefore, given an initial condition, the entire output sequence can be evaluated recursively from the difference equations in

Section II. Our objective here is to evaluate the first and second order time average moments. For a deterministic sequence $\xi_n$, define for positive integers $i$ and $k$ the time average moments as

$$\mathcal{M}\left\{\xi_n^i \xi_{n+l}^k\right\} = \lim_{N \to \infty} \frac{1}{N} \sum_{n=1}^{N} \xi_n^i \xi_{n+l}^k \qquad (23)$$

if the limit exists. Note that the right side does not depend on $n$ (as should be the case). To find the time average mean of the binary quantizer error $\epsilon_{m,n}$, we can apply (23) to the expressions in Theorems 4 and 5 and then evaluate the limits. In doing so, we need the following equalities.

1) Let $c(t) = a_0 + a_1 t + \cdots + a_k t^k$ be a polynomial with real coefficients. If among $a_1, a_2, \cdots, a_k$, at least one is irrational, then for any Riemann integrable function $f$,

$$\lim_{N \to \infty} \frac{1}{N} \sum_{n=1}^{N} f(\langle c(n) \rangle) = \int_0^1 f(r)\, dr. \qquad (24)$$

That is, the sequence $\{\langle c(n) \rangle\}$ is uniformly distributed modulo 1.

2) Let $\zeta(t) = \zeta_k t^k + \zeta_{k-1} t^{k-1} + \cdots + \zeta_0$ and $\eta(t) = \eta_l t^l + \eta_{l-1} t^{l-1} + \cdots + \eta_0$ be polynomials with real-valued coefficients. If among the coefficients (except the constant term) of the polynomial $i\zeta(t) + j\eta(t)$, there is at least one irrational number for each integer pair $(i, j) \neq (0, 0)$, then the following equality is valid for any Riemann integrable function $f$:

$$\lim_{N \to \infty} \frac{1}{N} \sum_{n=1}^{N} f(\langle \zeta(n) \rangle, \langle \eta(n) \rangle) = \int_0^1 \int_0^1 f(r, s)\, dr\, ds. \qquad (25)$$

The sequences $\{\langle \zeta(n) \rangle, \langle \eta(n) \rangle\}$ are jointly uniformly distributed modulo 1.

A discussion of the foregoing equalities can be found in [15]. The results follow from the mathematical literature on the distribution of the fractional parts of polynomials. (See, for example, [18]–[20], [24].)

To apply (24) and (25), we require the assumptions stated in the previous section, which we summarize here.

*1) DC input case:* The quantity $\beta$ is an irrational number in $[0, 1)$, i.e., the input level $x$ is not a rational factor of $b$.

*2) Sinusoidal input case:* The quantity $ka/b + lf/f_s$ is irrational for all integer pairs $(k, l) \neq (0, 0)$.

Note that these conditions can be dropped when a dithering signal is added to the input. We require these assumptions here, however, for deterministic input signals. Recall that the binary quantizer error of the final stage completely determines the quantization noise at the output of the modulator. We shall focus on its behavior in the remainder of this section.

*DC Input* $(m \geq 2)$

Using Theorem 4, the time average mean of the binary quantizer error $\epsilon_{m,n}$ is

$$\mathcal{M}\left\{\frac{\epsilon_{m,n}}{2b}\right\} = (-1)^{m-1}\left[\frac{1}{2} - \mathcal{M}\{\langle \psi_m(n) \rangle\}\right]$$

$$= (-1)^{m-1}\left[\frac{1}{2} - \lim_{N \to \infty} \frac{1}{N} \sum_{n=1}^{N} \langle \psi_m(n) \rangle\right].$$

Observe from Theorem 4 that for any fixed $m$, the generic polynomial $\mathcal{I}$ is actually a fixed polynomial of degree $m$ with irrational coefficients. Here we denote this polynomial by $\psi_m(n)$. Since the sequence $\{(1/2) - \langle \psi_m(n) \rangle\}$ is uniformly distributed in $[-1/2, 1/2]$, the mean and variance of $\epsilon_{m,n}/2b$ are 0 and $1/12$, respectively. To find the covariance, we use the characteristic function or transform method as in [12], [15]. Define the periodic function $g(\cdot)$ by

$$g(\gamma) = \langle \gamma \rangle$$

with its Fourier series

$$g(\gamma) = \sum_{k=-\infty}^{\infty} \hat{g}(k) e^{j2\pi k \gamma} \qquad (26)$$

where

$$\hat{g}(k) = \begin{cases} \dfrac{1}{2}, & \text{if } k = 0 \\[2mm] \dfrac{j}{2\pi k}, & \text{otherwise.} \end{cases}$$

Using (26), the autocorrelation of $\psi_m(n)$ becomes

$$\mathcal{M}\{\langle \psi_m(n) \rangle \langle \psi_m(n+l) \rangle\}$$

$$= \mathcal{M}\left\{\sum_{i=-\infty}^{\infty} \hat{g}(i) e^{j2\pi i \psi_m(n)} \sum_{k=-\infty}^{\infty} \hat{g}(k) e^{j2\pi k \psi_m(n+l)}\right\}.$$

If we can interchange the order of taking limits, then

$$\mathcal{M}\{\langle \psi_m(n) \rangle \langle \psi_m(n+l) \rangle\}$$

$$= \hat{g}(i)\hat{g}(k)\mathcal{M}\{e^{j2\pi(i\psi_m(n) + k\psi_m(n+l))}\}$$

$$\triangleq \sum_{i=-\infty}^{\infty} \sum_{k=-\infty}^{\infty} \hat{g}(i)\hat{g}(k)\Phi_l(i, k) \qquad (27)$$

where $\Phi_l(\cdot, \cdot)$ is the sample average joint characteristic function of the sequence $\psi_m(n)$. Let

$$\tilde{\Phi}(i) = \mathcal{M}\{e^{j2\pi i \psi_m(n)}\}$$

be the sample average characteristic function of $\psi_m(n)$. Observe that $\tilde{\Phi}(i) = \Phi_l(i, 0)$ for all $i$ and $l$. A sufficient condition which allows such an interchange is that both $\tilde{\Phi}(i)$ and $\Phi_l(i, k)$ exist and that $\lim_{i \to \infty} \tilde{\Phi}(i) = 0$ [12]. We shall verify that these conditions are satisfied after we have evaluated $\Phi_l(i, k)$.

Consider the expression $i\psi_m(n) + k\psi_m(n+l)$ for non-zero $l$. Application of Lemma 3 produces

$$i\psi_m(n) + k\psi_m(n+l) = \begin{cases} 0, & i = k = 0 \\ \mathscr{I}_{m-1}(n), & i + k = 0, i \neq 0 \\ \mathscr{I}_m(n), & i + k \neq 0 \end{cases}.$$

Further application of (24) results in

$$\Phi_l(i,k) = \begin{cases} 1, & i = k = 0 \\ 0, & \text{otherwise} \end{cases}, \quad m = 2, 3, \cdots. \quad (28)$$

Here we point out that when $m = 1$, i.e., for a single-loop sigma-delta modulator, the term $\mathscr{I}_{m-1}(n)$ becomes a constant independent of $n$ and hence $\Phi_l(i,k)$ would not be 2-valued as in (28). This leads to a discrete spectrum as reported in [11]. Equation (28) clearly implies both $\tilde{\Phi}(i)$ and $\Phi_l(i,k)$ exist and that $\Phi_l(i,0)$ goes to zero as $i$ goes to infinity. Hence the limit interchange in (27) is justified. Then the covariance of $\epsilon_{m,n}$ becomes

$$\mathscr{M}\left\{\frac{\epsilon_{m,n}}{2b}\frac{\epsilon_{m,n+l}}{2b}\right\} = \begin{cases} \frac{1}{12}, & l = 0 \\ 0, & \text{otherwise}. \end{cases}$$

It can also be deduced that the binary quantizer error and the input are uncorrelated since

$$\mathscr{M}\{x\epsilon_{m,n}\} = x\mathscr{M}\{\epsilon_{m,n}\} = 0.$$

Hence for a dc input, the binary quantizer error of the final stage is white and uncorrelated with the input. This agrees exactly with the corresponding result for the two-stage sigma-delta modulator [15].

*Sinusoidal Input ($m \geq 3$)*

Using Theorem 5, the time average mean of the binary quantizer error $\epsilon_{m,n}$ is

$$\mathscr{M}\left\{\frac{\epsilon_{m,n}}{2b}\right\} = (-1)^{m-1}\left[\frac{1}{2} - \mathscr{M}\{\langle\lambda_n\rangle\}\right]$$

where

$$\mathscr{M}\{\langle\lambda_n\rangle\} = \lim_{N\to\infty}\frac{1}{N}\sum_{n=1}^{N}\left\langle\left\langle\frac{an^{m-1}}{4b(m-1)!} + \mathscr{R}_m(n)\right.\right.$$

$$\left.\left. + \mathscr{I}_{m-2}(n)\right\rangle + \alpha_m\cos 2\pi\left\langle\frac{f}{f_s}n - \frac{mf}{2f_s} - \frac{m}{4}\right\rangle\right\rangle$$

$$\triangleq \lim_{N\to\infty}\frac{1}{N}\sum_{n=1}^{N}\langle\langle c(n)\rangle + \alpha_m\cos 2\pi\langle d(n)\rangle\rangle.$$

Note that $c(n)$ and $d(n)$ are fixed polynomials for any fixed $m$. Denote the indicator function by $1_{\{A\}}$ which takes on the value of 1 if condition $A$ is true and 0 otherwise. With assumption 2 mentioned before, we can apply (25) to obtain

$$\mathscr{M}\{\langle\lambda_n\rangle\} = \int_0^1\int_0^1\langle r + \alpha_m\cos 2\pi s\rangle\, dr\, ds$$

$$= \int_0^1\int_0^1\left(r + \langle\alpha_m\cos 2\pi s\rangle\right.$$

$$\left. - 1_{\{r + \langle\alpha_m\cos 2\pi s\rangle \geq 1\}}\right)dr\, ds$$

$$= \frac{1}{2}$$

and hence

$$\mathscr{M}\left\{\frac{\epsilon_{m,n}}{2b}\right\} = 0.$$

Similarly, the variance is

$$\mathscr{M}\left\{\left(\frac{\epsilon_{m,n}}{2b}\right)^2\right\} = \mathscr{M}\{\langle\lambda_n\rangle^2\} - \frac{1}{4}$$

$$= \int_0^1\int_0^1\left(r + \langle\alpha_m\cos 2\pi s\rangle\right.$$

$$\left. - 1_{\{r + \langle\alpha_m\cos 2\pi s\rangle \geq 1\}}\right)^2 dr\, ds - \frac{1}{4}$$

$$= \frac{1}{12}.$$

The covariance of the binary quantizer noise can be written as

$$\mathscr{M}\left\{\frac{\epsilon_{m,n}}{2b}\frac{\epsilon_{m,n+l}}{2b}\right\} = \mathscr{M}\{\langle\lambda_n\rangle\langle\lambda_{n+l}\rangle\} - \frac{1}{4}. \quad (29)$$

As before, we use the transform method to find the covariance of $\langle\lambda_n\rangle$. Assuming the validity of limit interchange for the time being, we have

$$\mathscr{M}\{\langle\lambda_n\rangle\langle\lambda_{n+l}\rangle\} = \sum_{i=-\infty}^{\infty}\sum_{k=-\infty}^{\infty}\hat{g}(i)\hat{g}(k)\Phi_l(i,k)$$

$$(30)$$

where

$$\Phi_l(i,k) = \mathscr{M}\{e^{j2\pi(i\lambda_n + k\lambda_{n+l})}\}.$$

Consider the expression $i\lambda_n + k\lambda_{n+l}$ for nonzero $l$. Using

Lemma 3 and Theorem 5, we have

$$
i\lambda_n + k\lambda_{n+l} = \begin{cases} 0, & i = k = 0 \\[2mm] \dfrac{kmlan^{m-2}}{4b(m-1)!} + \mathscr{R}_{m-1}(n) + \mathscr{I}_{m-3}(n) + i\alpha_m \cos 2\pi\left(\dfrac{f}{f_s}n - \dfrac{mf}{2f_s} - \dfrac{m}{4}\right) \\[4mm] \qquad - i\alpha_m \cos\left[2\pi\left(\dfrac{f}{f_s}n - \dfrac{mf}{2f_s} - \dfrac{m}{4}\right) + 2\pi l\dfrac{f}{f_s}\right], & i + k = 0,\ i \neq 0 \\[4mm] \dfrac{(i+k)an^{m-1}}{4b(m-1)!} + \mathscr{R}_m(n) + \mathscr{I}_{m-2}(n) + i\alpha_m \cos 2\pi\left(\dfrac{f}{f_s}n - \dfrac{mf}{2f_s} - \dfrac{m}{4}\right) \\[4mm] \qquad + k\alpha_m \cos\left[2\pi\left(\dfrac{f}{f_s}n - \dfrac{mf}{2f_s} - \dfrac{m}{4}\right) + 2\pi l\dfrac{f}{f_s}\right], & i + k \neq 0 \end{cases}
$$

$$m = 2, 3, \cdots. \quad (31)$$

Observe that for the case $i + k = 0$, $i \neq 0$, the first three terms on the right side of (31) reduce to a polynomial with rational coefficients when $m \leq 2$. This requires special handling, and the analyses for $m = 1$ and 2, i.e., for the single-loop and the two-stage sigma-delta modulator, have been reported in [12] and [15], respectively. Using (25), we have for $m \geq 3$ that

$$
\Phi_l(i,k) = \begin{cases} 1, & i = k = 0 \\[2mm] \displaystyle\int_0^1\int_0^1 e^{j2\pi[r + i\alpha_m \cos 2\pi s - i\alpha_m \cos 2\pi(s + (f/f_s)l)]}\,dr\,ds, \\ & i + k = 0,\ i \neq 0 \\[2mm] \displaystyle\int_0^1\int_0^1 e^{j2\pi[r + i\alpha_m \cos 2\pi s + k\alpha_m \cos 2\pi(s + (f/f_s)l)]}\,dr\,ds, \\ & i + k \neq 0 \end{cases}
$$

$$
= \begin{cases} 1, & i = k = 0 \\ 0, & \text{otherwise.} \end{cases} \quad (32)
$$

Equation (32) obviously implies that $\lim_{i \to \infty} \Phi_l(i, 0) = 0$, and hence the limit interchange in (30) is valid. Equations (29), (30), and (32) imply

$$
\mathscr{M}\left\{\frac{\epsilon_{m,n}}{2b}\frac{\epsilon_{m,n+l}}{2b}\right\} = 0, \qquad l \neq 0.
$$

Summarizing, we have

$$
\mathscr{M}\left\{\frac{\epsilon_{m,n}}{2b}\right\} = 0
$$

$$
\mathscr{M}\left\{\frac{\epsilon_{m,n}}{2b}\frac{\epsilon_{m,n+l}}{2b}\right\} = \begin{cases} \dfrac{1}{12}, & l = 0 \\[2mm] 0, & \text{otherwise.} \end{cases}
$$

This indicates that the first- and second-order moments of the binary quantizer error of the final stage ($m$th) are identical to those generated by an independent identically distributed (i.i.d.) random process uniformly distributed in $[-b, b]$. We next show that in addition to being white, the noise is asymptotically uncorrelated with the input.

First, we consider the mean of the input:

$$
\mathscr{M}\{\cos n\theta\} = \mathscr{M}\left\{\cos 2\pi\left(\frac{f}{f_s}n\right)\right\} = \int_0^1 \cos 2\pi r\,dr = 0.
$$

Then the cross moment between $x_n$ and $\epsilon_{m,n}$ is

$$
\frac{1}{2b}\mathscr{M}\{\epsilon_{m,n}\cos[(n+l)\theta]\}
$$

$$
= (-1)^m \mathscr{M}\{\langle\lambda_n\rangle\cos[(n+l)\theta]\}
$$

$$
= (-1)^m \mathscr{M}\{\langle\langle c(n)\rangle + \alpha_m\cos 2\pi\langle d(n)\rangle\rangle
$$

$$
\cdot \cos\left[2\pi\langle d(n)\rangle + \left(l - \frac{m}{2}\right)\theta + \frac{m}{2}\pi\right]\}.
$$

Application of (25) results in

$$
\frac{1}{2b}\mathscr{M}\{\epsilon_{m,n}\cos[(n+l)\theta]\}
$$

$$
= (-1)^m \int_0^1\int_0^1 \langle r + \alpha_m\cos 2\pi s\rangle
$$

$$
\cdot \cos\left[2\pi s + \left(l + \frac{m}{2}\right)\theta + \frac{m}{2}\pi\right]dr\,ds
$$

$$
= (-1)^m \int_0^1\int_0^1 \left(r + \langle\alpha_m\cos 2\pi s\rangle - 1_{\{r + \langle\alpha_m\cos 2\pi s\rangle \geq 1\}}\right)
$$

$$
\cdot \cos\left[2\pi s + \left(l + \frac{m}{2}\right)\theta + \frac{m}{2}\pi\right]dr\,ds
$$

$$
= 0.
$$

We have shown that given a sinusoidal input and $m \geq 3$, the first- and second-order moments of the binary quantizer error $\epsilon_{m,n}$ have sample averages completely consistent with the assumption of an additive, white, and uniformly distributed noise that is uncorrelated with the input signal. This result is not true for single-loop sigma-delta modulators [12] and only approximately true for two-stage sigma-delta modulators operating at large oversampling ratios [15].

## V. DECODING FILTERS AND SYSTEM PERFORMANCE

In this section, we study the overall system performance when a decoding filter is used to recover the original analog input from the digital output of the oversampled MSM. We only consider linear time-invariant decoding filters since they are easy to implement and they provide reasonably good performance. Provided the decoding filter has unity gain in the baseband and does not cause spectral distortion to the signal, the average noise power at the output of the decoding filter can be written as [15]

$$\Delta = \int_0^1 |H(f)|^2 (2\sin \pi f)^{2m} S_{\epsilon_m}(f) \, df$$

where $H(f)$ is the transfer function of the decoding filter. For both dc ($m \geq 2$) and sinusoidal inputs ($m \geq 3$),[2] the spectral density of the binary quantizer error $\epsilon_{m,n}$ equals $b^2/3$. Hence the average noise power becomes

$$\Delta = \frac{b^2}{3} \int_0^1 |H(f)|^2 (2\sin \pi f)^{2m} \, df. \qquad (33)$$

Then the signal-to-quantization-noise ratio (SQNR) of the MSM is given by

$$\text{SQNR} = \frac{\text{signal power}}{\Delta} = \begin{cases} \dfrac{x^2}{\Delta}, & \text{dc input} \\[2ex] \dfrac{a^2}{2\Delta}, & \text{sinusoidal input.} \end{cases} \qquad (34)$$

Observe that $\Delta$ is not a function of the input, and hence the SQNR of the system increases with the power of the input. As in [15], we point out that the foregoing results are derived assuming an idealized model. In particular, gain mismatch and input noise can push the input level outside the dynamic range of the system, and hence the SQNR might degrade. Hence it is advisable to leave a guard region at both ends of the dynamic range in order to avoid overloading effects. We now evaluate the average quantization noise power for several decoding filters.

### A. Ideal Low-Pass Filter Decoder

The ideal low-pass filter is given by

$$H(f) = \begin{cases} 1, & \text{if } |f| \leq 1/2R \\ 0, & \text{otherwise} \end{cases}$$

where $R$ is the oversampling ratio. Using the periodicity of the digital transfer function, the average quantization noise power is

$$\Delta = \frac{b^2}{3} \left( \int_0^{1/2R} + \int_{1-1/2R}^1 \right) (2\sin \pi f)^{2m} \, df$$

$$= \frac{2^{2m+1} b^2}{3} \int_0^{1/2R} (\sin \pi f)^{2m} \, df.$$

For large $R$ we only need to consider the integrand for small values of $f$. We use a Taylor series expansion to write $\sin x = x + O(x^3)$, where $O(\cdot)$ is the big-$O$ function. (See, for example, [21, sec. 1.2].) Then

$$\Delta = \frac{2^{2m+1} \pi^{2m} b^2}{3} \int_0^{1/2R} f^{2m} \, df + \int_0^{1/2R} O(f^{6m})$$

$$\approx \frac{\pi^{2m} b^2}{3(2m+1) R^{2m+1}}. \qquad (35)$$

Observe that the average noise power at the decoder output goes to zero approximately as $1/R^{2m+1}$. This approximation is very good because $O(1/R^{6m+1})$ decays much faster than $O(1/R^{2m+1})$. This quantifies the improvement in SQNR obtainable using higher order systems. Since the ideal low-pass filter can only be approximated in practice, it is of interest to find simpler filters which can provide the same trade-off.

### B. Sinc$^k$ Filter Decoder

A popular decoder for sigma-delta quantizers consists of a cascade of comb filters (sinc$^R$ filters) which down-samples the signal to four times the Nyquist frequency, followed by a baseband filter to further down-sample the signal to the Nyquist frequency [6], [15]. It was shown in [11] and [15] that for the single-loop and two-stage sigma-delta quantizers, the sinc$^2$ and sinc$^3$ filters, respectively, have identical performance to the ideal low-pass filter in terms of the trade-off between the quantization noise power and the oversampling ratio. Therefore, it is natural to investigate the general sinc$^k$ filter for the multistage sigma-delta quantizer. The transfer function of the sinc$^k$ filter is

$$H(f) = \frac{e^{-jk\pi fR/4}}{e^{-jk\pi f}} \left( \frac{4\sin(\pi fR/4)}{R\sin(\pi f)} \right)^k.$$

To find the average quantization noise behavior for a general sinc$^k$ filter, we make the further assumption that $m < R$, i.e., the number of stages is less than the oversampling ratio. This assumption is reasonable because the oversampling ratio is typically so large that the signal-to-noise ratio of a reasonably good system can be achieved by a small number of stages. It is shown in [15] that for this family of decoders, (33) should be replaced by an inequality, i.e., the integral on the right side of (33) upper bounds $\Delta$.

The details of the following integrations are given in Appendix II. First, we consider the case where $k \leq m$. Then

$$\Delta < \frac{2^{2m} b^2 4^{2k}}{3R^{2k}} \int_0^1 \sin^{2(m-k)}(\pi f) \sin^{2k}(\pi fR/4) \, df$$

$$= \frac{b^2 4^{2k}}{3R^{2k}} \binom{2(m-k)}{m-k} \binom{2k}{k} \qquad (36)$$

where $\begin{pmatrix} c \\ d \end{pmatrix}$ is the binomial coefficient. For $k = m+1$, we have

$$\Delta < \frac{2^{2m} b^2 4^{2(m+1)}}{3R^{2(m+1)}} \int_0^1 \frac{\sin^{2(m+1)}(\pi f R)}{\sin^2(\pi f)} \, df = \frac{b^2 4^{2m+1}}{3R^{2m+1}} \begin{pmatrix} 2m \\ m \end{pmatrix}. \tag{37}$$

This shows that with the $\text{sinc}^{m+1}$ decoding filter, we can achieve the same $\Delta$ versus $R$ trade-off as with the ideal low-pass filter. It can be shown that for $k$ larger than $m+1$ the dependence of $\Delta$ on $R$ remains the same as $1/R^{2m+1}$ although the proportionality constant changes. Hence the $\text{sinc}^{m+1}$ filter is a good choice because of its case of implementation and good performance. This result is consistent with those previously reported in [11] and [15] for the single-loop and the two-stage sigma-delta quantizer.

## VI. Simulation Results

In this section we present simulation results for the following inputs: dc, a single sinusoid, and the sum of two sinusoids. The initial conditions of all integrator outputs are set to zero. The dynamic range is normalized with $b$ equal to 1, i.e., both the input and output take on values in the range $[-1, 1]$. The values $x$, $a$, and $f$ described previously are chosen with 17 significant digits so that they resemble irrational numbers. Hence the simulation results will be approximately the same as those predicted by the analysis. One million samples of the output and the binary error sequences are generated from which the sample mean and covariance are calculated.

In evaluating the sample spectral density, we used a covariance sequence of length 2047, i.e., we have set $K_n = 0$ for $|n| \geq 1024$, where $K_n$ represents the covariance sequence of the binary quantizer noise. A window function is applied to the truncated covariance sequence before the transform is taken. We have chosen the "four-term window with continuous third derivative" reported in [22], which is

$$w(n) = 0.338946 + 0.481973 \cos\left(\frac{\pi n}{N}\right)$$
$$+ 0.161054 \cos\left(\frac{2\pi n}{N}\right) + 0.018027 \cos\left(\frac{3\pi n}{N}\right),$$
$$n = -N, \cdots, -1, 0, 1, \cdots, N.$$

The discrete-time Fourier transform of the windowed covariance sequence is taken with normalized frequency such that the sampling frequency equals one. The spectral density is evaluated at 1024 frequencies uniformly distributed in the range $[0, 1]$. The simulated spectral densities of the binary quantizer noise of the final stage for three- and four-stage sigma-delta modulators are shown in Figs. 3–5.

The simulated results support the facts predicted by the foregoing analysis. The spectral density of the binary quantizer noise of the final stage is "white," i.e., it has a zero mean and a flat spectrum, for both dc ($m \geq 2$) and sinusoidal inputs ($m \geq 3$).

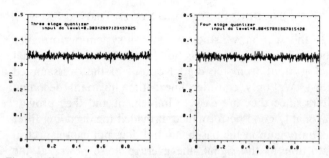

Fig. 3. Simulated power spectral density of binary quantizer error of final stage for three- and four-stage sigma-delta modulator driven by dc inputs.

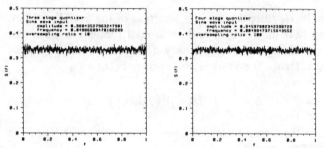

Fig. 4. Simulated power spectral density of binary quantizer error of final stage for three- and four-stage sigma-delta modulator driven by sinusoidal inputs.

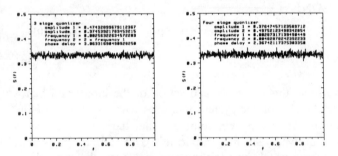

Fig. 5. Simulated power spectral density of binary quantizer error of final stage for three- and four-stage sigma-delta modulator driven by sum of two sinusoids.

## VII. Discussion

We have developed the sample moments of multistage sigma-delta modulation with more than two stages. The analysis was based on an exact solution of the nonlinear difference equations developed in [12]. These results, together with those in [11], [12], and [15], provide a complete understanding of idealized MSM with dc and sinusoidal inputs. The moments of the binary quantizer noise are found using results on the distribution of fractional parts of polynomials and the transform method, which leads to the spectral characteristic of the multistage sigma-delta quantizer (modulator plus decoder). We have shown that the binary quantizer noise of the final ($m$th) stage, which appears at the output as an $m$th order difference, is uncorrelated with the input signal and has first- and second-order moments identical to those of uniformly distributed white noise. This justifies the use of the additive white noise model for analyzing the spectral characteristic of the system for these two types of inputs. We have also shown in the appendix that the aforementioned results hold true for an input consisting of a finite sum of sinusoids. This gives rise to the interesting observation that although the MSM is a nonlinear system, its output is free of intermodulation products. This result is surprising given the highly nonlinear nature of the system.

Perhaps the most important aspect of an MSM is the noise-shaping characteristic provided by the linear combinatorial network. Although sigma-delta modulation is generally believed to have such noise-shaping characteristics, this behavior has only been proved using the linearized system with signal-independent white noise. We have shown that the behavior holds for multistage systems without such assumptions on the quantization noise, i.e., the behavior is exact for the ideal nonlinear system. In Section II, the noise-shaping function is shown to be $(2 \sin \pi f)^{2m}$, where the frequency is normalized so that the sampling frequency equals 1. At typical oversampling ratios, the bandwidth of the signal is a tiny fraction of 1, and hence the inband noise decreases exponentially with $m$. This is made precise in Section V where the average quantization noise for various decoding filters is calculated. It is shown that for an $m$-stage sigma-delta quantizer with either an ideal low-pass filter or a $\text{sinc}^{m+1}$ filter decoder, the average quantization noise decreases as $1/R^{2m+1}$, where $R$ is the oversampling ratio. This implies that high-order systems are favorable from the viewpoint of reducing the quantization noise. As discussed in [15], the bit rate of the system is not raised by going to more stages if the modulator and the decoder are fabricated on the same chip, which is usually done in practice [13], [14]. However, the complexity of the quantizer goes up with the number of stages.

Finally, we point out that in this analysis we have not considered effects such as gain mismatch of the stages, timing jitters, input noise, etc. For instance, gain mismatch of the stages can lead to only a partial cancelling of the spectral spikes in the single-loop system, and hence the performance of the MSM might degrade. These effects are probably more profound when the number of stages goes up. Investigation of these problems requires a more complicated model and perhaps more sophisticated mathematical tools.

## APPENDIX I

*Inputs with Multiple Sinusoids*

Consider an input which is a sum of two sinusoids.

$$x_n = a_1 \cos n\theta_1 + a_2 \cos(n\theta_2 + \gamma) \quad (38)$$

where

$$\theta_i = \frac{2\pi f_i}{f_s}, \quad i = 1, 2$$

represents the normalized frequencies of the two sinusoids and $\gamma$ is the phase delay. We restrict the dynamic range of the input so that $x_n \in [-b, b]$ for all $n$. Substituting (38) into (13) and following similar procedures as in Section III, we obtain

$$\frac{\epsilon_{m,n}}{2b} = (-1)^{m-1} \left[ \frac{1}{2} - \langle \lambda_n \rangle \right]$$

where

$$\lambda_n = \left( a_1 + \frac{a_2 \sin((\theta_2/2) - \gamma)}{\sin(\theta_2/2)} \right) \frac{n^{m-1}}{4b(m-1)!} + \mathcal{R}_m(n)$$

$$+ \mathcal{I}_{m-2}(n) + \alpha_{1,m} \cos 2\pi \left( \frac{f_1}{f_s} n - \frac{mf_1}{2f_s} - \frac{m}{4} \right)$$

$$+ \alpha_{2,m} \cos 2\pi \left( \frac{f_2}{f_s} n - \frac{mf_2}{2f_s} - \frac{m}{4} + \frac{\gamma}{2\pi} \right)$$

and

$$\alpha_{i,m} = \frac{a_i}{2b(2\sin(\theta_i/2))^m}, \quad i = 1, 2.$$

If there exists a rational number $r$ such that $f_1 = rf_2$, and we make the assumption 2 mentioned in Section IV, then the moments of the binary quantizer error can be evaluated by applying (25) and going through the same procedure as in Section IV. The moments turn out to be exactly the same as those for a single sinusoidal input signal. If $f_1$ and $f_2$ are not related by a rational factor, and the quantity

$$i \left( a_1 + \frac{a_2 \sin((\theta_2/2) - \gamma)}{\sin(\theta_2/2)} \right) + k \frac{f_1}{f_s} + l \frac{f_2}{f_s}$$

is irrational for all integer three-tuples $(i, k, l) \neq (0, 0, 0)$, then we can apply a three-dimensional analog of (25) to evaluate the moments of the binary quantizer noise. It turns out that the binary quantizer noise is again uncorrelated with the input signal and has first- and second-order moments identical to those of uniformly distributed white noise. The same result can be generalized to an input $x_n$ which consists of a finite sum of sinusoids.

## APPENDIX II

*Average Quantization Noise Power*

From [17, p. 25, 1.320.1],

$$\sin^{2n} x = \frac{1}{2^{2n}} \left[ \sum_{l=0}^{n-1} (-1)^{n-l} 2 \binom{2n}{l} \cos 2(n-l) x + \binom{2n}{n} \right]. \quad (39)$$

With this equality, the integrand in (36) becomes

$$\text{integrand} = \frac{1}{2^{2m}} \left[ \sum_{l=0}^{m-k-1} (-1)^{m-k-l} 2 \binom{2(m-k)}{l} \right.$$

$$\cdot \cos 2(m-k-l) \pi f + \left. \binom{2(m-k)}{m-k} \right]$$

$$\cdot \left[ \sum_{i=0}^{k-1} (-1)^{k-i} 2 \binom{2k}{i} \cos 2(k-i) \pi f R/4 + \binom{2k}{k} \right].$$

Using the fact that $k \leq m < R$ and $\int_0^1 \cos(2l\pi f) \, df = 0$ for $l \neq 0$, we can integrate (36) as

$$\Delta < \frac{2^{2m} b^2 4^{2k}}{3R^{2k}} \int_0^1 \frac{1}{2^{2m}} \binom{2(m-k)}{m-k} \binom{2k}{k} df$$

$$= \frac{b^2 4^{2k}}{3R^{2k}} \binom{2(m-k)}{m-k} \binom{2k}{k}.$$

Using a trigonometric identity in [23, p. 8, 1.9] we can write (37) as

$$\Delta < \frac{2^{2m}b^2 4^{2m+2}}{3R^{2m+2}} \int_0^1 \sin^{2m}(\pi f R/4) \left[ \frac{R}{4} + \sum_{l=1}^{(R/4)-1} 2\left(\frac{R}{4}-l\right)\cos(2l\pi f) \right] df$$

$$= \frac{2^{2m}b^2 4^{2m+1}}{3R^{2m+1}} \int_0^1 \sin^{2m}(\pi f R/4)\, df + \frac{2^{2m+1}b^2 4^{2m+2}}{3R^{2m+2}} \sum_{l=1}^{(R/4)-1} \left(\frac{R}{4}-l\right)$$

$$\cdot \int_0^1 \sin^{2m}(\pi f R/4)\cos(2l\pi f)\, df. \tag{40}$$

Applying (39) to (40), we see that every integral in the finite sum equals zero. Hence (40) becomes

$$\Delta < \frac{2^{2m}b^2 4^{2m+1}}{3R^{2m+1}} \int_0^1 \sin^{2m}(\pi f R/4)\, df = \frac{b^2 4^{2m+1}}{3R^{2m+1}} \binom{2m}{m}.$$

## REFERENCES

[1] H. Inose, Y. Yasuda, and J. Murakami, "A telemetering system by code modulation—$\Sigma - \Delta$ modulation," *IRE Trans. Space Electron. Telemetry*, vol. SET-8, pp. 204–209, Sept. 1962.

[2] H. Inose and Y. Yasuda, "A unity bit coding method by negative feedback," *Proc. IEEE*, vol. 51, pp. 1524–1535, Nov. 1963.

[3] J. C. Candy, "A use of limit cycle oscillations to obtain robust analog-to-digital converters," *IEEE Trans. Commun.*, vol. COM-22, pp. 298–305, Mar. 1974.

[4] ____, "A use of double integration is sigma delta modulation," *IEEE Trans. Commun.*, vol. COM-33, pp. 249–258, Mar. 1985.

[5] J. C. Candy, Y. C. Ching, and D. S. Alexander, "Using triangularly weighted interpolation to get 13-bit PCM from a sigma-delta modulator," *IEEE Trans. Commun.*, vol. COM-24, pp. 1268–1275, Nov. 1976.

[6] J. C. Candy, "Decimation for sigma delta modulation," *IEEE Trans. Commun.*, vol. COM-34, pp. 72–76, Jan. 1986.

[7] J. C. Candy and O. J. Benjamin, "The structure of quantization noise from sigma-delta modulation," *IEEE Trans. Commun.*, vol. COM-29, pp. 1316–1323, Sept. 1981.

[8] M. W. Hauser and R. W. Brodersen, "Circuit and technology considerations for MOS delta-sigma A/D converters," in *Proc. ISCAS*, San Jose, CA, 1986.

[9] W. L. Lee and C. G. Sodini, "A topology for higher order interpolative coders," *Proc. ISCAS*, Philadelphia, PA, May 1987.

[10] R. M. Gray, "Oversampled sigma-delta modulation," *IEEE Trans. Commun.*, vol. COM-35, pp. 481–489, May 1987.

[11] ____, "Spectral analysis of quantization noise in a single-loop sigma-delta modulator with dc input," *IEEE Trans. Commun.*, vol. COM-37, pp. 588–599, June 1989.

[12] R. M. Gray, W. Chou, and P. W. Wong, "Quantization noise in single-loop sigma-delta modulation with sinusoidal inputs," *IEEE Trans. Commun.*, to be published.

[13] T. Hayashi, Y. Inabe, K. Uchimura, and T. Kimura, "A multistage delta-sigma modulator without double integration loop," in *ISSCC Dig. Technical Papers*, Feb. 1986, pp. 182–183.

[14] K. Uchimura, T. Hayashi, T. Kimura, and A. Iwata, "Oversampling A-to-D and D-to-A converters with multistage noise shaping modulators," *IEEE Trans. Acoust., Speech, Signal Processing*, vol. ASSP-36, pp. 1899–1905, Dec. 1988.

[15] P. W. Wong and R. M. Gray, "Two stage sigma-delta modulation," *IEEE Trans. Acoust., Speech, Signal Processing*, to be published.

[16] L. Ljung, *System Identification: Theory for the User*. Englewood Cliffs, NJ: Prentice Hall.

[17] I. S. Gradshteyn and I. M. Ryzhik, *Table of Integrals, Series and Products*. New York: Academic, corrected and enlarged edition, 1980.

[18] H. Weyl, "Uber die Gleichverteilung von Zahlen mod Eins," *Math. Annal.*, vol. 77, pp. 313–352, 1916.

[19] F. J. Hahn, "On affine transformations of compact Abelian groups," *Amer. J. Math.*, vol. 85, pp. 428–446, 1963.

[20] A. G. Postnikov, "Ergodic problems in the theory of congruences and of diophantine approximations," in *Proc. Steklov Inst. Math.*, no. 82, pp. 1–128, Amer. Math. Soc. Translation, Providence, RI, 1966.

[21] N. G. de Bruijn, *Asymptotic Methods in Analysis*. New York: Dover, 1981.

[22] A. H. Nuttall, "Some windows with very good sidelobe behavior," *IEEE Trans. Acoust., Speech, Signal Processing*, vol. ASSP-29, pp. 84–91, Feb. 1981.

[23] F. Oberhettinger, *Fourier Expansions: A Collection of Formulas*. New York: Academic, 1973.

[24] F. J. Hahn, Errata for "On affine transformations of compact Abelian groups," *Amer. J. Math.*, vol. 86, pp. 463–464, 1964.

# Oversampled Sigma–Delta Modulation

ROBERT M. GRAY, FELLOW, IEEE

*Abstract*—Oversampled sigma–delta modulation has been proposed as a practical implementation for high rate analog-to-digital conversion because of its simplicity and its robustness against circuit imperfections. To date, mathematical developments of the basic properties of such systems have been based either on simplified continuous-time approximate models or on linearized discrete-time models where the quantizer is replaced by an additive white uniform noise source. In this paper, we rigorously derive several basic properties of a simple discrete-time single integrator loop sigma–delta modulator with an accumulate-and-dump demodulator. The derivation does not require any assumptions on the correlation or distribution of the quantizer error, and hence involves no linearization of the nonlinear system, but it does show that when the input is constant, the state sequence of the integrator in the encoder loop can be modeled exactly as a linear system in an appropriate space. Two basic properties are developed: 1) the behavior of the sigma–delta quantizer when driven by a constant input and its relation to uniform quantization, and 2) the rate-distortion tradeoffs between the oversampling ratio and the average mean-squared quantization error.

## I. INTRODUCTION

OVERSAMPLED sigma–delta ($\Sigma\Delta$) modulators were introduced by Inose and Yasuda [1] and have recently been much promoted as a promising candidate for implementing analog-to-digital converters [2]–[5]. A discrete-time model of a simple $\Sigma\Delta$ modulator or $\Sigma\Delta$ quantizer is shown in Fig. 1 and an equivalent model is shown in Fig. 2 for comparison. As in [2], the basic model has a single integrator in the encoder and an integrate-and-dump (or accumulate-and-dump) operation in the decoder. (By "decoder," we mean the operation of converting the binary sequence produced by the encoder into a sequence of digital approximations to the original sample. Only discrete time is considered here.) The system resembles a standard delta modulator as depicted in Fig. 3, except that in the encoder, the discrete-time integrator is before the quantizer and not in the feedback loop (or, equivalently, it is formed by adding an integrator to the signal before differencing) and there is no feedback loop in the decoder. While this is the simplest example of a $\Sigma\Delta$ modulator and more complicated systems having multiple encoder feedback loops or more sophisticated decoder filters are known to have superior performance, we confine attention to the basic system of Fig. 1 in order to develop the techniques and results in the simplest possible case. Possible extensions to more interesting systems will be described later.

The promise of this system and variations thereof is based on three basic properties. First, if the original information source is oversampled to, say, $N$ times the Nyquist rate and the decoder is then decimated by taking only every $N$th sample, then it is claimed that the overall result is a good approximation to the action of a uniform scalar quantizer at the

Paper approved by the Editor for Speech Processing of the IEEE Communications Society. Manuscript received April 4, 1986; revised August 20, 1986. This work was supported by the National Science Foundation under Grant ECS83-17981. This paper was presented at the 1986 IEEE International Symposium on Information Theory, Ann Arbor, MI, October 1986.

The author is with the Information Systems Laboratory, Department of Electrical Engineering, Stanford University, Stanford, CA 94305.

IEEE Log Number 8613930.

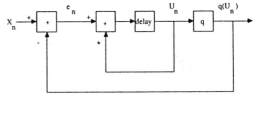

$$e_n = X_n - q(U_n)$$

$$U_n = e_{n-1} + U_{n-1}$$

**Encoder**

**Decoder**

Fig. 1. Sigma–delta quantizer.

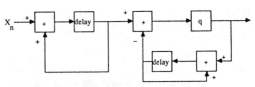

Fig. 2. Equivalent encoder.

original sampling rate. Hence, the system provides a good approximation to ideal PCM. Thus, for example, the tradeoff between the oversampling ratio $N$ and the average squared quantization error should be approximately the same as the tradeoff for PCM with an appropriate number of levels.

Second, the $\Sigma\Delta$ modulator is more robust against circuit imperfections than standard successive approximation or flash scalar quantizers because the single binary quantizer is less sensitive to such changes than are the multiple thresholds involved in the traditional implementations.

Third, there are intuitive reasons for preferring oversampled $\Sigma\Delta$ modulation to oversampled delta modulation for analog-to-digital conversion. For example, there is no error accumulation in the decoder because there is no feedback loop in the decoder. Thus, for example, the system is less sensitive to channel (binary) errors. Furthermore, arguments based on linearized models suggest that the spectral characteristics of the quantization error are better behaved for $\Sigma\Delta$ modulation than for delta modulation.

The above properties have been verified by simulations and experimental implementations, but the fundamental theoretical properties have not been rigorously demonstrated. We here focus on two fundamental questions and provide rigorous answers:

1) If a constant input is applied to the system and the integrator begins in an arbitrary state, how close is the output

Reprinted from *IEEE Trans. Commun.*, vol. COM-35, pp. 481–489, May 1987.

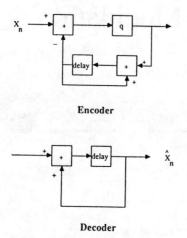

Fig. 3. Delta modulator.

to the input and how close is the output to that produced by a uniform scalar quantizer with a corresponding number of levels?

2) How does the mean-squared quantization noise vary with $N$, the oversampling ratio, for large $N$?

The assumption of a constant input over $N$ samples is consistent with the fact that every $N$th sample corresponds to the Nyquist rate, and hence the changes will be small. The first question is usually answered by simulation or experiment; several examples are provided in the literature demonstrating that specific fixed inputs yield a close output level. The basic intuition behind the operation is summed up by [5]:

The circuit tries to minimize this noise by having the quantized values oscillate between levels in such a way that the average of $q(t)$ approximates the average of $x(t)$.

It should also be noted that in the examples given, the quantizer input is always rational and oscillations or limit cycles about the input indeed take place. We shall see that such behavior is, in fact, a very special case. It is generally believed that whether or not the binary quantizer output is periodic, its time average should approximate a dc input. We shall see that this intuition is correct in a precise sense.

As to the second property, various results have been developed relating the oversampling ratio to the quantization error power based on the assumption that $N$ is large and these results have been compared to the performance of PCM [2], [3], [6]–[8]. The results vary in that different decoders are considered, but all are based on the assumption that the quantizer error sequence is uniformly distributed and independent of the input signal. Some of the developments also require that the quantizer error sequence is uncorrelated or independent. Candy and Benjamin [6] also use a simple approximate continuous-time model to obtain a detailed spectral analysis of the encoder quantization noise. Their approach is an extension of Iwersen's approach for the analysis of delta modulation systems [9]. We here derive an asymptotic (large $N$) expression for the basic discrete time $\Sigma\Delta$ modulator of Fig. 1 which resembles the traditional asymptotic quantization result of [10], [11] for uniform quantizers, but which does not require any assumptions on the behavior of the quantization error in the $\Sigma\Delta$ feedback loop. We also demonstrate an interesting connection between the simple $\Sigma\Delta$ modulator and uniform PCM that does not appear to have been previously noted. Given a constant input and an initial integrator state within half the input range of the input, then the $\Sigma\Delta$ output is within one least significant bit (LSB) of the output of an ideal uniform quantizer with the same number of possible output levels.

The time domain techniques for evaluating the average

squared quantization error do not immediately extend to more complicated (and more interesting) decoders such as the ideal low-pass filter and the cascade of two accumulate-and-dump filters that have been proposed and experimentally shown to provide better performance. The basic representation given in Theorem 1 of the single-loop encoder in terms of multiplication modulo 1 has, however, been combined with other techniques to analyze these and other decoders [12], [13], and these results will be mentioned in the final section.

In Section II, the basic equations describing a single-loop $\Sigma\Delta$ are given and used to compare the operation and performance of the $\Sigma\Delta$ modulator to a uniform quantizer. The assumptions and properties used in the derivation of the performance of a uniform quantizer are described and their applicability to analyzing the $\Sigma\Delta$ modulator is discussed. In Section III, a connection with ergodic theory is pointed out. This connection motivates the construction of an equivalent mathematical model which shows that the behavior of the $\Sigma\Delta$ modulator to a constant input is the same as the classical ergodic theory model of rotations on a circle or multiplication on the unit interval modulo 1. This has the perhaps surprising implication that the inherently nonlinear system can be modeled as a linear system in a different space. Difference equations are derived for the binary quantizer output, the integrator state, and the binary quantizer error. In the subsequent section, the basic properties are used to quantify the performance of the equivalent model, and hence of the $\Sigma\Delta$ modulator. The final section contains some general observations on the behavior of $\Sigma\Delta$ modulators with more complicated decoders.

## II. Sigma–Delta Modulation and Uniform Quantization

From Fig. 1, the $\Sigma\Delta$ modulation system can be described by the difference equations

$$U_n = e_{n-1} + U_{n-1}; \qquad n = 1, 2, \cdots$$
$$e_n = X_n - q(U_n) \qquad (2.1)$$

with an initial condition

$$U_0 = u_0$$

where the binary quantizer is defined by

$$q(u) = \begin{cases} +b & \text{if } u \geq 0 \\ -b & \text{if } u < 0. \end{cases}$$

We shall refer to $U_n$ as the state of the integrator at time $n$. The output is formed by decimating the sequence

$$Y_n = \frac{1}{N} \sum_{i=1}^{N} q(U_{n-i}),$$

that is, by forming the sequence $Y_{kN}$ for $k = 1, 2, \cdots$. The number of input samples per output sample $N$ is called the *oversampling ratio*. If the Nyquist frequency of the input data is $f_0$ and the sampling period is $\tau$, then $N = 1/(2f_0\tau)$ is assumed to be large (typically in the hundreds for audio). Note that $Y_{kN}$ can be thought of as the output of an integrate-and-dump (or comb) digital filter or as a special case of a low-pass filter.

Suppose now that the input is held fixed at a single sample value $X_n = x$ for times $n = 1, 2, \cdots, N$ and that the encoder integrator is in an initial state $U_0 = u_0$ (which depends on previous inputs). What is the behavior of the output of the decoder, that is, $Y_N$? Define the output for $x$ to be $Q(x)$. First observe that $Y_N = Q(x)$ has the form

$$Y_N = Q(x) = \frac{1}{N} \sum_{i=0}^{N-1} q(U_i)$$

and hence can take on any of the $N + 1$ different values

$$\frac{2b}{N} k - b; \qquad k = 0, 1, \cdots, N$$

where the $k$ above corresponds to the number of $+b$'s in the sequence of $N$ numbers. Call this collection the *reproduction alphabet* $\hat{A}(N)$. Since the output can only produce values in the interval $[-b, b]$, we assume that the input dynamic range is restricted to the same interval (at least approximately so). Thus, the binary quantizer in the feedback loop produces the maximum and minimum values of the inputs.

The reproduction alphabet $\hat{A}(N)$ is exactly the same as that produced by a uniform quantizer for the slightly larger interval $[-b(1 + 1/N), b(1 + 1/N)]$ having a bin width of $2b/N$ and $N + 1$ levels. Let $Q_{N+1}^*$ denote the uniform quantizer and recall that

$$|x - Q_{N+1}^*(x)| = \min_{a \in \hat{A}(N)} |x - a|,$$

that is, the encoder for a uniform quantizer is a minimum distortion or nearest neighbor mapping. Thus, *no* encoder (including the given $\Sigma\Delta$ quantizer being considered) can yield a smaller quantization error for the given reproduction alphabet than the uniform quantizer, and hence we have the simple bound

$$|x - Q(x)| \geq |x - Q_{N+1}^*(x)| \qquad (2.2)$$

for all inputs $x$. Observe that a $\Sigma\Delta$ modulator with oversamping ratio $N$ is best compared to a PCM system with $N + 1$ levels because of the common reproduction alphabet, rather than to a PCM system with $2N$ levels as in [2].

*Quantization Noise in Uniform Quantizers*

It is a classical result of asymptotic quantization theory that if the uniform quantizer is applied to a random variable $X$ with the given dynamic range and a smooth probability density function, then if $N$ is large, we have that

$$E[(X - Q_{N+1}^*(X))^2] \approx \frac{\Delta^2}{12} \qquad (2.3)$$

where

$$\Delta = \frac{2b}{N}$$

is the bin width of the quantizer, the distance between adjacent output levels. This result was first developed by Bennett in 1948 [11] and Widrow in 1956 [14]. (See Gersho [15] for a survey and history of such results.) This approximation can be made precise by approximating integrals by sums and vice versa, that is, no assumption that the quantization error is uniformly distributed and independent of the signal is *required* to prove this approximation. (See, e.g., [10].) A heuristic development of the above result [14], which can be made rigorous [11], shows that if the following properties are satisfied, then the quantization noise is approximately uncorrelated and uncorrelated with the input, and these facts can be used to provide a proof of (2.3):

1) successive input samples have small correlation,
2) the number of output points $N + 1$ is large,
3) the output points are very close to the midpoints of the set of inputs which yield these points.

From (2.2), we have an immediate lower bound for the mean-squared error when the $\Sigma\Delta$ quantizer is applied to a random variable $X$ (that is, when we assume that $X_i = X$ for $i = 0, 1, \cdots, N$):

$$E[(X - Q(X))^2] \geq E[(X - Q_{N+1}^*(X))^2] = \frac{\Delta^2}{12}.$$

It should be emphasized that this lower bound follows from the decoder filter used. As shall be discussed in the final section, different filters may produce a larger reproduction alphabet as a function of $N$ and that will yield a smaller lower bound.

*Quantization Noise in $\Sigma\Delta$ Modulation*

The mathematical developments of the quantization error power of discrete-time $\Sigma\Delta$ modulation all are patterned on the uniform signal-independent noise development for uniform quantization, that is, they assume that quantization error inside the loop is a sequence of uniform random variables that are independent of the input signal. Uncorrelation or independence of successive quantization errors is also usually assumed. This leads to simple developments of the quantizer error power for the given decoder and for others. To see this, we rewrite the difference equations to emphasize the quantizer error. Define the binary quantizer error

$$\epsilon_n = q(U_n) - U_n; \qquad n = 0, 1, 2, \cdots. \qquad (2.4)$$

The quantizer outputs can be expressed in terms of the binary quantizer error as

$$q(U_n) = U_n + \epsilon_n = U_{n-1} + x - q(U_{n-1}) + \epsilon_n$$
$$= x + \epsilon_n - \epsilon_{n-1}; \qquad n = 1, 2, \cdots, N, \qquad (2.5)$$

and hence we have the telescoping sum

$$Q(x) = \frac{1}{N} \sum_{i=1}^{N} (x + \epsilon_i - \epsilon_{i-1}) = x + \frac{\epsilon_N}{N} - \frac{\epsilon_0}{N}$$

and hence

$$x - Q(x) = \frac{\epsilon_0}{N} - \frac{\epsilon_N}{N}.$$

If $X$ is a random variable and one assumes that the $\epsilon_n$ are independent uniformly distributed random variables that are independent of the underlying signal, then the average mean-square quantization error would be $E[(X - Q(X))^2] = \Delta^2/6$, 3 dB or 1/2 bit worse than the corresponding PCM [7].

We now argue that this derivation cannot be made rigorous because the quantization noise sequence $\epsilon_n$ cannot be proved to satisfy the required three conditions. Most seriously, the quantizer has only one bit, which cannot be considered as a large number of levels. As a result, the asymptotic approximations usually used cannot be invoked. The quantizer is inside a feedback loop, and hence successive inputs to the quantizer cannot be assumed to be uncorrelated. In fact, they can be highly correlated because the original input $x$ is constant. To emphasize this point, observe from (2.5) that given a random input $X$,

$$\epsilon_n = q(U_n) - X + \epsilon_{n-1}.$$

While the relation does not prove that $\epsilon_n$ must be dependent of $X$ and $\epsilon_{n-1}$, the appearance of $X$ and $\epsilon_{n-1}$ in an expression for $\epsilon_n$ does not support the independence assumption.

Finally, it is not known whether the $\Sigma\Delta$ modulator has the property that the output levels form the centroids of all of the input values which map into the levels.

We shall see that the assumption of uniform independent quantizer noise is not required to analyze the given system, but we also admit that the independent and uniform assumption permits analyses of systems with more general decoders for which the techniques of this paper are not immediately applicable. Nonetheless, it should be emphasized that the assumption of such quantizer noise behavior cannot be proved using existing techniques because the basic conditions required for such a proof do not hold. In fact, it is shown in [12] that for the single-loop encoder and a dc input, the power spectrum of the binary quantizer noise can be evaluated exactly and (consistent with results in [6]) the spectrum is discrete and not even approximately flat.

## III. AN EQUIVALENT MODEL

The $\Sigma\Delta$ quantizer operation given an input $x$ and an initial condition $U_0 = u_0$ can be summarized by the difference equations

$$U_n = \begin{cases} U_{n-1} + x - b; & \text{if } U_{n-1} \geq 0 \\ U_{n-1} + x + b; & \text{if } U_{n-1} < 0 \end{cases} \quad n = 1, 2, \cdots.$$

(3.1)

Suppose for the moment that $x$ is nonnegative. (Similar conclusions follow for negative $x$ with obvious modifications.) Suppose also that $x \in [-b, b]$, the dynamic range of the quantizer. We wish to quantify how close $Q(x)$ is to $x$ and to the uniform quantized representation of $x$. We now present an intuitive argument to motivate the use of ergodic theory, but it is important to note that the eventual results will be stronger and not require the assumptions that we make shortly.

To begin, observe from (3.1) that once $U_n$ lies in the interval $[x - b, x + b]$, all future $U_n$ will lie within the interval. Furthermore, even if the initial condition $u$ is such that $U_0$ is not within the interval, $U_n$ will progress monotonically toward the interval and eventually lie within it. Thus, except possibly for a finite initial condition, all $U_n$ will lie within the interval. For simplicity, we can assume that $u$ is such that $U_0$ lies within the interval. This effectively assumes that the input does not change too rapidly. Observe that if $U_0 = 0$, then the condition is automatically satisfied if $x$ is within its assumed dynamic range.

Given that all $U_n$ lie within the interval, suppose that we replace the given initial condition $U_0 = u$ by a random initial condition $U_0$ that is uniformly distributed on this interval. One might think of this distribution as a steady-state distribution for the random variables $U_n$ that would result if the system were started at an arbitrary point and allowed to run for a long time. In particular, it is the quantizer input that is for the moment assumed uniform, not the quantizer error. If this is done, then the expected value of the binary quantizer output $q(U_n)$ can be computed to be

$$E[q(U_n)] = bPr(U_n \geq 0) - bPr(U_n < 0)$$

$$2bPr(U_n \geq 0) - b = 2b \frac{x+b}{2b} - b = x.$$

Recall that the output of the $\Sigma\Delta$ modulator is

$$Q(x) = \frac{1}{N} \sum_{i=0}^{N-1} q(U_i)$$

and the goal is to have this be close to $x$, which is the expectation of $q(X)$, that is, we wish to prove that

$$\frac{1}{N} \sum_{i=0}^{N-1} q(U_i) \approx E[q(U_0)]$$

in some sense for large $N$. This, however, is simply a law of large numbers or an ergodic theorem proving that a sample average approximates its expected value and hence will hold, for example, if the process $\{U_n\}$ is stationary and ergodic.

While it is true that the process as defined is both stationary and ergodic, we will not actually take this approach because it is undesirable to assume that $U_0$ is uniform (it is preferable to have an arbitrary initial state) and because the usual ergodic theorem is not strong enough to provide useful results here. In particular, it only states that large $N$ will yield a good approximation for a fixed $x$, but it does not say how large $N$ must be and the size of $N$, in general, depends on $x$. We shall see that the rate of convergence can be quantified and that the distribution of $U_n$ is asymptotically uniform regardless of the initial state.

In ergodic theory, a random process is modeled as an abstract dynamical system consisting of a transformation acting on a probability space. (See, for example, Peterson [16].) We will construct such a model for the $\Sigma\Delta$ quantizer and then obtain an equivalent model which will yield the desired results. Happily, both models are fairly simple. Define a sample space $\Omega = [x - b, x + b)$. As argued above, we can consider this to be the sample space for the random variables $U_n$. We will occasionally consider the probability space $(\Omega, B, m)$ formed with $B$ being the Borel sets and with a probability measure $m$ defined by a uniform probability density function on $\Omega$ ($m$ is the Lebesgue measure), but the principal results will not depend on this fact. Define a transformation $S: \Omega \to \Omega$ by

$$Su = \begin{cases} u + x - b; & \text{if } u \geq 0 \\ u + x + b; & \text{if } u < 0. \end{cases}$$

The transformation is invertible and measurable. Note that the transformation simply captures the operation of the difference equation of (3.1). The combination $(\Omega, m, S)$ is called an *abstract dynamical system* in ergodic theory. (A dynamical system is sometimes denoted by $(\Omega, B, m, S)$ to make the Borel field explicit.) The dynamical system can be related to the random process $U_n$ by defining

$$U_n = S^n U_0; \quad n = 1, 2, \cdots.$$

We can write the output of the $\Sigma\Delta$ quantizer now as

$$Q(x) = \frac{1}{N} \sum_{i=0}^{N-1} q(S^i U_0).$$

The dynamical system (or just the transformation) is said to be stationary if $m(S^{-1}F) = m(F)$ for all Borel sets $F$, and it is said to be ergodic if $S^{-1}F = F$ implies that $m(F)$ is 0 or 1. The dynamical system is stationary (ergodic) if and only if the discrete-time random process $\{U_n\}$ is stationary (ergodic) in the usual random process sense.

With the dynamical system model for the $\Sigma\Delta$ quantizer, we now map it into another model. Define the function $\alpha: [x - b, x + b) \to [0, 1)$ by

$$\alpha(u) = \frac{1}{2b} u + \frac{b-x}{2b} = \frac{1}{z} + \frac{u-x}{2b}. \quad (3.2)$$

Roughly speaking, $\alpha$ compresses and shifts the sample space, moving it from an interval of length $2b$ centered on $x$ to the unit interval centered on $1/2$. $\alpha$ is an affine, and hence continuous, measurable, and invertible function.

A transformation, and hence dynamical system can be defined on the unit interval to mimic the original transformation $S$:

$$Ty = \begin{cases} y + \dfrac{x-b}{2b}; & y \geq \dfrac{b-x}{2b} \\ y + \dfrac{x+b}{2b}; & y < \dfrac{b-x}{2b}. \end{cases} \quad (3.3)$$

It can easily be verified by direct calculation that $\alpha(Su) = T\alpha(u)$.

If we use a uniform density to define a probability measure on the unit interval (which we also call $m$), then we have another dynamical system $([0, 1), m, T)$. It follows from elementary ergodic theory that these two dynamical systems are isomorphic and that one will be stationary (ergodic) if and only if the other is. We shall draw on this fact in discussions, but it is not used in any proofs. Our focus is on the second system $([0, 1), m, T)$. We can relate the abstract dynamical system back to the concrete problem at hand by defining a binary quantizer $\bar{q}$ on $[0, 1)$ by

$$\bar{q}(y) = \begin{cases} +b; & y \geq \dfrac{b-x}{2b} \\ -b; & y < \dfrac{b-x}{2b} \end{cases}$$

and observe that

$$\bar{q}(\alpha(u)) = q(u)$$

and hence

$$Q(x) = \frac{1}{N} \sum_{i=0}^{N-1} \bar{q}(T^i \alpha(U_0)).$$

This shows that the output can be computed from the new model. The following provides another representation of the equivalent model that yields its useful properties.

*Claim:* Define

$$\beta = \frac{b+x}{2b} = \frac{1}{2} + \frac{x}{2b} \in [0, 1)$$

and let $\langle r \rangle$ denote $r$ mod 1, that is, the fractional part of $r$; then

$$Ty = \langle y + \beta \rangle.$$

Observe that the relation implies that the apparently nonlinear operation involving a binary quantizer can be expressed as a linear operation within the unit interval! This result forms the basis of the time domain techniques used here and in [13] and in the frequency domain techniques used in [12].

*Proof of Claim:* First note that if $y < (b - x)/(2b)$, then

$$y + \beta < \frac{b-x}{2b} + \frac{b+x}{2b} \leq 1$$

and since $\beta \geq 0$,

$$y + \beta \geq y \geq 0$$

and hence from (3.3),

$$Ty = y + \beta = \langle y + \beta \rangle.$$

If $y \geq (b - x)/(2b)$, then

$$y + \frac{b+x}{2b} \geq \frac{b-x}{2b} + \frac{b+x}{2b} = 1$$

and

$$y + \frac{b+x}{2b} < y + 1 < 2$$

and hence using (3.3),

$$\left\langle y + \frac{b+x}{2b} \right\rangle = y + \frac{b+x}{2b} - 1$$

$$= y + \frac{x-b}{2b} = Ty,$$

which proves the claim.

Repeated application of the claim implies that $T^n y = \langle y + n\beta \rangle$. The above development means that by studying the behavior of the transformation $T$, we can find the properties of the integrator state sequence $U_n$ and hence of $Q(x)$. For example, defining

$$Y_n = \alpha(U_n) = \frac{U_n}{2b} + \frac{b-x}{2b}$$

we have that

$$Y_n = T^n Y_0 = \langle \alpha(U_0) + n\beta \rangle$$

$$\langle \alpha(U_0) + \frac{n}{2} + \frac{nx}{2b} \rangle,$$

which is linear (mod 1) in the input $x$! The integrator state sequence can then be found by inverting (3.2) to be

$$U_n = 2b Y_n + x - b = 2b \langle \alpha(U_0) + n\beta \rangle + x - b \quad (3.4)$$

and hence the quantizer output is given by

$$Q(x) = \frac{1}{N} \sum_{i=0}^{N-1} \bar{q}(\langle \alpha(U_0) + i\beta \rangle). \quad (3.5)$$

Equations (3.4) and (3.5) have the form required for the basic results, but first a few observations are in order.

The transformation $T$ in this form has been much studied in ergodic theory. (See, e.g., [16]–[18] and their references.) It is equivalent to the transformation defined on the unit circle which rotates a point through an arc of length $\beta$. The system is stationary if the distribution on the interval (or circle) is uniform and it is ergodic if and only if $\beta$ is irrational. If $\beta$ is rational, then the transformation is periodic. In particular, if $\beta = k/K$ in lowest terms, then $T^K \beta = \beta$ and the sequence $T^i \beta$ is periodic with period $K$. As all of the examples in [2] are chosen with a rational input, the transformation is periodic, and indeed the quantizer output produces cycles as claimed. It should be noted, however, that with most probability densities, the probability of producing a rational number exactly with a truly continuous source is 0, and hence this behavior is not typical: with probability one, the output will not be periodic. (It will, however, be almost periodic in a precise sense.)

We now develop the key properties of the single loop $\Sigma\Delta$ encoder. The results are formally stated as a theorem for later reference.

*Theorem 1:* Given a $\Sigma\Delta$ encoder as in (2.1), an input $x \in [-b, b]$, and an initial state $U_0 \in [x - b, x + b)$, let $\beta = (b + x)/2b$ and $z = \alpha(U_0) = (U_0 + b - x)/2b$. Then

$$U_n = 2b\langle z + n\beta \rangle + x - b, \quad (3.6)$$

$$\frac{q(U_n)}{2b} = \langle z + n\beta \rangle - \langle z + (n+1)\beta \rangle + \frac{x}{2b}, \quad (3.7)$$

and

$$\frac{\epsilon_n}{2b} = \frac{1}{2} - \langle z + (n+1)\beta \rangle \quad (3.8)$$

where $z = \alpha(U_0) = U_0/2b + (b - x)/2b$.

*Proof:* Equation (3.6) is just (3.4). From (2.1), we have that

$$q(U_n) = X_n - e_n = X_n - (U_{n+1} - U_n) \qquad (3.9)$$

which, with (3.6), proves (3.7). Equation (3.8) follows by subtracting (3.7) from (3.6).

## IV. QUANTIZER PERFORMANCE

The first result is an easy corollary to Theorem 1.

*Corollary 1:* Given a $\Sigma\Delta$ quantizer with an oversampling ratio $N$, an input $x \in [-b, b)$, and an initial state $U_0 \in [x - b, x + b)$, then

$$x - Q(x) = \frac{2b}{N}(\langle z + N\beta \rangle - z) \qquad (4.1)$$

where $z = \alpha(U_0)$, and hence

$$|x - Q(x)| \leq \frac{2b}{N}.$$

*Comments:* The theorem states that the maximum quantization error for an input within the dynamic range of the quantizer and an initial condition within $b$ of the input is $2b/N$. Note that the bound does not depend on $u$ or $x$ provided $u$ is within $b$ of $x$. Note also that this bound is not quite as tight as that of a uniform quantizer which satisfies

$$|x - Q_{N+1}^*(x)| \leq \frac{b}{N},$$

a factor of two better resolution. Both, however, are inversely proportional to $N$. Unlike the motivating discussion, $x$ is not required to be irrational, no probability distribution is assumed for the initial state, and neither stationarity nor ergodicity are relevant.

*Proof:* Using (3.9), we have the telescoping sum

$$x - Q(x) = x - \frac{1}{N} \sum_{i=0}^{N-1} q(U_i) = x - \frac{1}{N} \sum_{i=0}^{N-1} (x - U_{i+1} + U_i)$$

$$= \frac{1}{N} \sum_{i=0}^{N-1} (U_{i+1} - U_i) = \frac{1}{N}(U_N - U_0) \qquad (4.2)$$

which, with (3.6), proves (4.1). Since the right-hand side has magnitude less than 1, the bound follows.

The bound has an easy corollary which is of interest. Suppose that $x$ is itself a member of the reproduction alphabet, and hence has the form $(2b/N)k - b$ for some $k = 0, 1, \cdots, N$. In this case, $\beta = k/N$, and hence $\langle z + N\beta \rangle = \langle z + k \rangle = z$, and hence from the theorem,

$$x - Q(x) = \frac{2b}{N}(\langle z + N\beta \rangle - z) = 0.$$

Thus, the $\Sigma\Delta$ quantizer has the property that if a reproduction symbol is used as an input, the output is the same symbol.

The following theorem provides a simple connection with the $\Sigma\Delta$ modulator and a uniform quantizer. Part of the result is a corollary to Corollary 1.

*Theorem 2:* Given the assumptions of Corollary 1, suppose that $Q$ is the $\Sigma\Delta$ quantizer and $Q_{N+1}^*$ is the uniform quantizer with the same number $N + 1$ of output levels, then

$$|Q(x) - Q_{N+1}^*(x)| \leq \frac{2b}{N}.$$

*Comments:* The theorem states that given an initial state

satisfying the conditions of Corollary 1, then the $\Sigma\Delta$ output is within one LSB of the uniform quantizer output.

*Proof:* Recall that the uniform quantizer selects the point of the form $(2b/N)k - b$; $k = 0, 1, \cdots, N$ that is closest to $x$. This can be restated by the rule

$$x \in \left[ \frac{2b}{N}k - b - \frac{b}{N}, \frac{2b}{N}k - b + \frac{b}{N} \right]$$

$$=> Q_{N+1}^*(x) = \frac{2b}{N}k - b; \qquad k = 0, 1, \cdots, N.$$

We claim that this is equivalent to the definition

$$Q_{N+1}^*(x) = 2b \frac{\left\lfloor N\beta + \frac{1}{2} \right\rfloor}{N} - b. \qquad (4.3)$$

To see this, observe that if $x \in [(2b/N)k - b - (b/N), (2b/N)k - b + (b/N))$, then

$$\beta = \frac{x + b}{2b} \in \left[ \frac{k}{N} - \frac{1}{2N}, \frac{k}{N} + \frac{1}{2N} \right)$$

and hence

$$N\beta + \frac{1}{2} \in [k, k+1)$$

so that $\lfloor N\beta + \frac{1}{2} \rfloor = k$. Observe from (4.3) that

$$Q_{N+1}^*(x) = \frac{2b}{N} \left( N\beta + \frac{1}{2} - \left\lfloor N\beta + \frac{1}{2} \right\rfloor \right) - b$$

$$= x + \frac{2b}{N} \left( \frac{1}{2} - \left\lfloor N\beta + \frac{1}{2} \right\rfloor \right) \qquad (4.4)$$

and therefore the error with uniform quantization is

$$Q_{N+1}^*(x) - x = \frac{2b}{N} \left( \frac{1}{2} - \langle N\beta + 1/2 \rangle \right)$$

which has magnitude less than $b/N$ as previously observed. Combining this with (4.1) yields

$$|Q(x) - Q_{N+1}(x)| = \frac{2b}{N} \left| z - \langle z + N\beta \rangle - \frac{1}{2} + \left\langle \frac{1}{2} + N\beta \right\rangle \right|$$

$$= \frac{2b}{N} \left| \lfloor z + N\beta \rfloor - \left\lfloor \frac{1}{2} + N\beta \right\rfloor \right|.$$

By assumption $z \in [0, 1)$ and hence the absolute value can take on only the values of 0 or 1, proving the theorem.

The exact representation of (4.1) can be used to derive the average mean-squared quantization error for large $N$ using calculus approximations similar to those used in asymptotic quantization theory for PCM (as in [10]). The following theorem summarizes the results.

*Theorem 3:* Given a $\Sigma\Delta$ modulator with oversampling ratio $N$, assume that the input for $N$ samples is a random variable $X$ described by a smooth probability density function. (The random variable is a sample of an input random process, but it is assumed that the sample does not change over $N$ consecutive samples.) We also assume that the modulator is tracking in the sense that if $U_0 = u$ is the initial state, then with probability one, $|X - u| \le b$. Let $f_X$ denote the density for $X$ (which can, in general, depend on $u$, but it is assumed to be smooth for any fixed $u$). Then

$$E[(X - Q(X))^2] \le \frac{(2\Delta)^2}{12}$$

where $\Delta = 2b/N$.

*Comment:* The average distortion is no greater than four times that for uniform quantization. Thus, if the same average distortion is required in both systems, $N$ for the oversampled $\Sigma\Delta$ modulation must be twice that of uniform quantization, an increase in rate of $\log_2 2 = 1$ bit.

*Proof:* Define the random variable $Y \in [0, 1)$ by $Y = (x + b)/2b$, that is, if $X = x$, then $Y = \beta$. Since the density for $X$ is smooth, so is that, say $f_Y$, for $Y$. If we define the random variable $W = \langle z + NY \rangle - z \in [-z, 1 - z)$, then we wish to estimate the second moment of the random variable $W$ since from Corollary 1

$$E[(X - Q(X))^2] = \left[\frac{2b}{N}\right]^2 E((W^2)).$$

Fix $-z \le a < b < 1 - z$ and observe that

$$\Pr(W \in [a, b)) = \Pr(\langle z + NY \rangle \in [a + z, b + z))$$

$$= \Pr\left(z + NY \in \bigcup_{k=0}^{N} [k + a + z, k + b + z)\right)$$

$$= \Pr\left(Y \in \bigcup_{k=0}^{N-1} \left[\frac{k+a}{N}, \frac{k+b}{N}\right)\right)$$

$$= \sum_{k=0}^{N-1} \Pr\left(Y \in \left[\frac{k+a}{N}, \frac{k+b}{N}\right)\right).$$

Since $f_Y$ is smooth and $N$ is large, we have as in asymptotic quantizer theory that $f_Y(y) \approx f_Y(k/N)$; $y \in [k/N, (k + 1)/N)$, and hence

$$\Pr\left(Y \in \left[\frac{k+a}{N}, \frac{k+b}{N}\right)\right) \approx f_Y\left(\frac{k}{N}\right) \frac{b-a}{N}.$$

With this approximation, the sum becomes

$$\Pr(W \in [a, b)) \approx (b - a) \sum_{k=0}^{N} \frac{1}{N} f_Y\left(\frac{k}{N}\right) \approx b - a$$

where we have used the fact that the final sum is approximately

$$\int_0^1 f_Y(y)\, dy = 1.$$

The above shows that the density for $W$ is approximately uniform on the interval $[-z, 1 - z)$. Thus, its second moment

is $1/12 + E(W)^2$, and hence

$$E[(X - Q(X))^2] \approx \left[\frac{2b}{N}\right]^2 \left(\frac{1}{12} + \left(\frac{1}{2} - z\right)^2\right)$$

$$\le \left[\frac{2b}{N}\right]^2 \left(\frac{1}{12} + \frac{1}{4}\right) = \left[\frac{2b}{N}\right]^2 \frac{4}{12}$$

which proves the result.

The key to the proof of the above approximation is the fact that we can prove that if $X$ has a smooth density, then the random variable

$$\left\langle z + N\left(\frac{X}{2b} + \frac{1}{2}\right)\right\rangle$$

has an approximately uniform distribution. This corresponds to the fact in ergodic theory that successive rotations of an irrational number are uniformly dense in the unit circle.

## V. COMMENTS

We have shown that the basic properties of a $\Sigma\Delta$ modulator can be derived without the assumption that the binary quantizer can be replaced by an independent additive noise sequence. The key idea is that the action of the quantizer on a constant input can be modeled as a multiplication modulo one. This equivalence permits an exact representation of the quantization error and a comparison to the operation of a uniform quantizer. The two quantizers are identical in output to a constant input if the $\Sigma\Delta$ encoder integrator is initialized to 0 at the beginning of every conversion cycle.

A natural question is whether or not the techniques used here can be applied to more interesting $\Sigma\Delta$ modulators having more complicated decoders or encoders. We here make a few remarks regarding more complicated decoders and encoders.

The basic accumulate-and-dump decoder takes $N$ binary numbers to produce a reproduction letter. Because these numbers are just added, there are only $N + 1$ possible letters, and hence the system can perform no better than the corresponding minimum distortion encoder, a uniform quantizer. If, however, the binary numbers are weighted, then a much larger reproduction alphabet can be obtained, and hence possibly better performance. One proposed decoder is to use a cascade of two integrators followed by a dump, that is, a digital filter with a $\text{sinc}^2$ transfer function. Thus, one would first form, as before,

$$Y_n = \frac{1}{N} \sum_{i=1}^{N} q(U_{n-i})$$

and use this sequence as input to another summer to form an output sequence

$$V_n = \frac{1}{N} \sum_{j=1}^{N} Y_{n-j} = \frac{1}{N} \sum_{j=1}^{N} \frac{1}{N} \sum_{i=1}^{N} q(U_{n-j-i}).$$

When sampled at time $2N$, this produces an output

$$V_{2N} = \frac{1}{N^2} \sum_{j=1}^{N} \sum_{i=1}^{N} q(U_{2N-j-i}) = \frac{1}{N^2} \sum_{k=2}^{2N} h_k q(U_{2N-k})$$

where

$$h_k = \begin{cases} k-1; & k = 2, 3, \cdots, N+1 \\ 2N-k+1; & k = N+2, \cdots, 2N. \end{cases}$$

This is a triangular impulse response with the well-known $sinc^2$ transfer function. This filter is known to provide better performance as a function of $N$. Observe that we have now doubled the number of samples used to produce a single output, but the number of possible outputs has been roughly squared. For example, if $b = 1$, the new output levels have the form

$$-1 + 2\frac{k}{N^2}; \qquad k = 0, 1, \cdots, N^2.$$

Thus, the performance of this quantizer is bounded below by that of a uniform quantizer having $N^2 + 1$ levels, not just $2N$ levels, that is, the lower bound to the mean-squared quantizer error goes down roughly as $N^{-4}$. If this decoder is indeed superior to the simple accumulate and dump, then the quantizer squared error must decrease at a rate of something between $N^{-2}$ and $N^{-4}$, but the techniques of this paper do not provide an easy answer. Some algebra yields the representation

$$x - Q(x) = \frac{2b}{N^2} \sum_{i=0}^{N-1} (\langle z + (i+N)\beta \rangle - \langle z + i\beta \rangle),$$

but the methods of this paper do not yield the behavior of this sum. (If the terms in the sum were uncorrelated analogous to the uncorrelated error assumption, then the expectation of the square could be analyzed.) Recently, Chou has used techniques from number theory to demonstrate that the absolute squared error tends to zero slightly slower than $1/N^3$ and that the average squared error exhibits similar behavior [13]. This behavior agrees with the analysis using the white quantization error assumption [8].

More generally, if arbitrary weights are permitted, one could, in principle, have a reproduction alphabet of $2^N$ possible values, the number of possible binary $N$ tuples. If one could obtain an asymptotic expression for the resulting expected squared error, one could optimize over the choice of coefficients. This may be possible when constraints are placed on the coefficients which make such an asymptotic expression tractable.

One can also consider an ideal low-pass filter (and hence a sinc function pulse response) instead of the FIR filters above (as in [1], [8]). While this yields simple results if the independent uniform quantization error is assumed (e.g., [1], [8]), it is not amenable to comparison to PCM because then the decoder filter does not have a finite impulse response and the reproduction alphabet is unbounded in size. In addition, the time domain techniques of this paper have not proved useful in evaluating the behavior of this decoder.

A definite advantage of the white noise assumption is that it permits the use of spectral arguments to evaluate the overall quantization noise. Recent work shows that Fourier techniques can be applied to the difference equations of Theorem 1 to obtain exact expressions for the power spectral density of the binary quantization noise, the binary quantizer output, and the overall quantization noise in terms of the decoder transfer function, and that these expressions can be evaluated in some examples and bounded in others [12]. Although limited to the simple encoder considered here and to dc inputs, such results indicate that the representation of Theorem 1 permits a spectral analysis for such nonlinear systems without having to assume whiteness of the binary quantizer noise, and hence linearization of the system. These results indicate that the binary quantizer noise is indeed not flat and that it consists of spikes, but that the overall mean-squared quantizer error is exactly that predicted by the white noise assumption.

More complicated encoders with multiple feedback loops and single-loop interpolative encoders which use more complicated filters have been shown by simulation to provide better performance than the single-loop encoder considered here [3], [7], [8], [19]–[21]. The assumption that the binary quantization noise can be replaced by an additive white noise source again permits a straightforward analysis of such systems, and it has been argued that this assumption is better justified in this case since the binary quantizer noise indeed appears to be uncorrelated [8]. The techniques of this paper have not yet been extended to multiple feedback loop encoders and research on the subject is continuing.

A final open problem concerns the source model. Although the classical analysis of $\Sigma\Delta$ modulators of Candy and Benjamin assumes dc signals, common test procedures focus on the behavior of such systems with sinusoidal inputs. Research is currently underway to see if the approach developed here can be extended to analyze such systems.

## ACKNOWLEDGMENT

The author thanks B. Boser, B. Wooley, and M. Hauser who introduced him to $\Sigma\Delta$ modulation, S. Boyd for helpful and insightful technical discussions, and J. Candy, W. Chou, C. Sodini, and W. Lee for many helpful comments. Extra thanks are due to B. Boser for many comments on the several drafts of this paper. Thanks also are due to the reviewers and colleagues who found several errors and typos in the earlier drafts.

## REFERENCES

[1] H. Inose and Y. Yasuda, "A unity bit coding method by negative feedback," *Proc. IEEE*, vol. 51, pp. 1524–1535, Nov. 1963.

[2] J. C. Candy, "A use of limit cycle oscillations to obtain robust analog-to-digital converters," *IEEE Trans. Commun.*, vol. COM-22, pp. 298–305, Mar. 1974.

[3] ——, "A use of double integration in sigma delta modulation," *IEEE Trans. Commun.*, vol. COM-33, pp. 249–258, Mar. 1985.

[4] ——, "Decimation for sigma delta modulation," *IEEE Trans. Commun.*, vol. COM-34, pp. 72–76, Jan. 1986.

[5] J. C. Candy, Y. C. Ching, and D. S. Alexander, "Using triangularly weighted interpolation to get 13-bit PCM from a $\Sigma\Delta$ modulator," *IEEE Trans. Commun.*, vol. COM-24, pp. 1268–1275, Nov. 1976.

[6] J. C. Candy and O. J. Benjamin, "The structure of quantization noise from sigma-delta modulation," *IEEE Trans. Commun.*, vol. COM-29, pp. 1316–1323, Sept. 1981.

[7] G. R. Ritchie, "Higher order interpolative analog to digital converters," Ph.D. dissertation, Univ. Pennsylvania, Philadelphia, 1977.

[8] B. Boser, personal communication.

[9] J. E. Iwersen, "Calculated quantizing noise of single-integration delta-modulation coders," *Bell Syst. Tech. J.*, pp. 2359–2389, Sept. 1969.

[10] H. Gish and J. N. Pierce, "Asymptotically efficient quantizing," *IEEE Trans. Inform. Theory*, vol. IT-14, pp. 676–683, Sept. 1968.

[11] W. R. Bennett, "Spectra of quantized signals," *Bell Syst. Tech. J.*, vol. 27, pp. 446–472, 1948.

[12] R. M. Gray, "Spectral analysis of sigma-delta quantization noise," submitted to *IEEE Trans. Commun.*, 1986.

[13] W. Chou, personal communication.

[14] B. Widrow, "A study of rough amplitude quantization by means of Nyquist sampling theory," *IRE Trans. Circuit Theory*, vol. CT-3, pp. 266–276, 1956.

[15] A. Gersho, "Principles of quantization," *IEEE Trans. Circuits Syst.*, vol. CAS-25, pp. 427–436, 1978.

[16] K. Petersen, *Ergodic Theory*. Cambridge, England: Cambridge Univ. Press, 1983.

[17] ——, "On a series of cosecants related to a problem in ergodic theory," *Compositio Mathematica*, vol. 26, pp. 313–317, 1973.

[18] E. Hecke, "Analytische Funktionen und die Verteilung von Zahlen mod. Eins," *Abh. Math. Semin. Hamburg Univ.*, vol. 1, pp. 54–76, 1922.

[19] C. H. Giancarlo and C. G. Sodini, "A slope adaptive delta modulator for VLSI signal processing systems," *IEEE Trans. Circuits Syst.*, vol. CAS 33, pp. 51–58, Jan. 1986.

[20] J. W. Scott, W. L. Lee, C. H. Giancarlo, and C. G. Sodini, "CMOS implementation of an immediately adaptive delta modulator," *IEEE J. Solid-State Circuits*, to be published.

[21] W. L. Lee and C. G. Sodini, "A topology for higher order interpolative coders," submitted to IEEE Circuits and Systems Conf.

# Quantization Noise Spectra

ROBERT M. GRAY, FELLOW, IEEE

*Abstract* —Uniform quantizers play a fundamental role in digital communication systems and have been the subject of extensive study for many decades. The inherent nonlinearity of quantizers makes their analysis both difficult and interesting. It usually has been accomplished either by assuming the quantizer noise to be a signal-independent, uniform white random process or by replacing the quantizer by a deterministic linear device, or by combining the two assumptions. Such linearizing approximations simplify the analysis and permit the use of linear systems techniques, but few results exist quantifying how good such approximations are for specific systems. These complications are magnified when the quantizer is inside a feedback loop, as for Sigma–Delta modulators. Exact descriptions of the moments and spectra of quantizer noise have been developed recently for the special case of single-loop, multistage and multiloop Sigma–Delta modulators. These results demonstrate that the white noise and linearization assumptions can be quite poor approximations in some systems and quite good in others. It turns out that many of the techniques used in the analysis were first applied to the analysis of quantizers by Clavier, Panter, and Grieg (1947) in pioneering (but often overlooked) work that preceded Bennett's (1948) classic study of quantization noise spectra. We take advantage of the benefit of hindsight to develop several results describing the behavior of quantization noise in a unified and simplified manner. Exact formulas for quantizer noise spectra are developed and applied to a variety of systems and inputs, including scalar quantization (PCM), dithered PCM, Sigma–Delta modulation, dithered Sigma–Delta modulation, two-stage Sigma–Delta modulation, and second-order Sigma–Delta modulation.

## I. INTRODUCTION

UNIFORM QUANTIZATION has long been a practical workhorse for analog-to-digital conversion (ADC) as well as a topic of extensive theoretical study. Deceptively simple in its description and construction, the uniform quantizer has proved to be surprisingly difficult to analyze precisely because of its inherent nonlinearity. The difficulty posed by the nonlinearity usually has been avoided by simplifying (linearizing) approximations or by simulations. Approximating the quantizer noise in system analysis by uniformly distributed white noise is extremely popular, but it can lead to inaccurate predictions of system behavior. These problems have been known since the earliest theoretical work on quantization noise spectra, but the modern literature often overlooks early results giving conditions under which the approximations

are good and providing exact techniques of analysis for certain input signals where the approximations are invalid.

Many of the original results and insights into the behavior of quantization noise are due to Bennett [1] and much of the work since then has its origins in that classic paper. Bennett first developed conditions under which quantization noise could be reasonably modeled as additive white noise. Less well known are the 1947 papers by Clavier, Panter, and Grieg [2], [3] that provided an exact analysis of the quantizer noise resulting when a uniform quantizer is driven by one or two sinusoids. The techniques of Clavier *et al.* are not as popular as the Bennett approximations, but they provided an important example in which the quantization noise is not white. More generally, their work also provides an interesting demonstration of a nonlinear system for which the output can be described exactly for a sinusoidal input using Fourier techniques.

Oversampled ADC's attempt to achieve the performance of high rate or high resolution quantizers by using low rate quantizers inside feedback loops along with linear filtering. When the analog waveform is sampled faster than the Nyquist rate, several successive bits from the low rate quantizer can be used to construct a single high rate approximation to the original signal sample at the Nyquist rate. The idea is to trade off the number of bits in the quantizer (and hence the required circuit tolerances in implementation) for timing accuracy (which is often easier to control). Alternatively, one uses a low rate quantizer often instead of a high rate quantizer once in order to produce each digitized output sample.

A variety of specific architectures for accomplishing this goal have been proposed in the literature: The Sigma–Delta or Delta–Sigma modulators [4]–[10] use only ideal integrators as linear filters, but there can be several feedback loops. These ADC's can also be cascaded to form multistage Sigma–Delta modulators [11], [12] or imbedded within additional feedback loops to form higher-order or multiloop Sigma–Delta modulators [7]. When only a single feedback loop around the quantizer is used, a general system can have linear filters in both the feedback and the feedforward paths. When the filtering is entirely in the feedback path, the system is called a *predictive coder* [13] and includes the classic example of a delta modulator [14]. When the filtering is entirely in the feedforward path, the system is called a *noise shaping*

Manuscript received November 21, 1988. This work was presented in part at the IEEE International Symposium on Information Theory, San Diego, CA, January 14–19, 1990.

The author is with the Department of Electrical Engineering, Stanford University, Durand 133, Mail Code 4055, Stanford, CA 94305.

IEEE Log Number 9036938.

Reprinted from *Trans. Inform. Theory*, vol. IT-36, pp. 1220–1244, Nov. 1990.

*coder* or an *interpolative coder* [13], [15]–[17]. The single-loop Sigma–Delta modulator can be considered as a special case of the interpolative ADC with the linear filter specialized to an ideal integrator. The multiloop Sigma–Delta modulators are not subsumed by the linear predictive or noise shaping coders models.

A longstanding problem with the application of feedback quantization systems has been the difficulty of analyzing and predicting their behavior; the complications of the nonlinear operation are aggravated by its presence within a feedback loop. Feedback systems with nonlinearities have long been a topic of study in the feedback control literature. (See, e.g., [18]–[20] and the references therein.) Two examples of particular importance are the ideal relay control system, which can be viewed as a predictive coder with a binary quantizer, and digital control systems, which can be considered as simply a feedback system containing a quantizer in the feedback loop.

Because of the difficulty of analyzing the behavior of nonlinear systems, a variety of approximations have been developed in both the control and communications literature in order to render the analysis tractable. Rarely can the validity of these approximations be tested analytically because of the paucity of systems amenable to exact solution, but many such techniques are known from experience and simulation to yield good results in some cases and bad results in others. There is a general belief that the higher order the system (the more feedback loops or stages or the higher the order of the linear filters), the more accurate the approximations. By far the most common assumption in the communications literature is to replace the quantization noise by a signal-independent, uniformly distributed, uncorrelated (white) additive noise source. This approximation linearizes the system and allows the computation of overall performance and signal-to-quantization noise (SQNR) ratios using ordinary linear systems techniques. Part of the model is obviously wrong since the quantization noise is in fact a deterministic function of the input signal (and hence cannot be signal independent), yet in many applications the uniform white quantization noise assumption yields reasonable results. Among other things, Bennett proved that under certain conditions the uniform white quantization noise assumption provides a good approximation to reality. In particular, if

1) the quantizer does not overload,
2) the quantizer has a large number of levels,
3) the bin width or distance between the levels is small, and
4) the probability distribution of pairs of input samples is given by a smooth probability density function,

then the white noise approximation yields good results in a precise sense.

The uniform white quantization noise assumption gained a wide popularity, largely due to the work of Widrow [21]. Unfortunately, the conditions under which the white noise approximation is valid are violated in

oversampled ADCs:

1) It is often not known whether or not the quantizer will overload for an input within a prescribed range,
2) the quantizers typically have only a few levels (usually two),
3) the bin width is typically not small (in fact it is usually the distance between the extremes of the input range), and
4) the feedback of a digital signal prevents the quantizer input from having smooth joint distributions.

Not surprisingly, it was found in some oversampled ADC's (such as the simple single loop Sigma–Delta modulator) that the quantizer noise was not at all white and that the noise contained discrete spikes whose amplitude and frequency depended on the input [22], [10]. Perhaps surprisingly, however, simulations and actual circuits of higher order Sigma–Delta modulators and of interpolative coders often exhibited quantizer noise that appeared to be approximately white. Unfortunately, these systems also often exhibited unstable behavior not predicted by the white noise analysis.

Another approach to linearizing the nonlinear system dominates the control field, that of describing function analysis. (See, for example, [18], [19], [20], [23], [24].) Here a specific input is assumed to the quantizer, usually a dc or a single sinusoid. The quantizer is then replaced by a linear gain chosen to minimize a mean-squared error between the outputs of the true quantizer and the linear approximation when the same sinusoid is input to both. Thus the true quantizer output when driven by a sinusoid is approximated by another sinusoid with an equivalent gain, where the gain depends on the assumed input signal. This can be viewed as approximating an infinity of harmonics at the quantizer output by the fundamental alone. Alternatively, one is implicitly assuming that all harmonics other than the fundamental are removed by the linear filtering in the loop so that only the single sinusoid remains as the input to the quantizer, which acts as a single-input input-dependent gain. Combining the linearized quantizer with the remaining linear filters in the feedback loop, one can determine limit cycles of the system and can find conditions under which the system is unstable by using root-locus or other criterion. The approach is clearly only an approximation, but it has served as a useful guide to design. An immediate problem with this approximation in oversampled ADC systems, however, is that the implicit assumption that the linear filters in the loop removes all harmonics is usually not satisfied. This is particularly true in ADC systems since the fundamental frequency within the loop can be dependent on the input signal and hence the loop filters cannot be assumed to remove the input-dependent harmonics.

The shortcomings of both the white noise assumption and the describing function approach can be partially ameliorated by combining the two: The describing function is taken as the fundamental component produced by the quantizer and the remaining components are then assumed to collectively behave as an additive white noise process with a gain that is determined in a least squares fit sense as is the describing function. This approach was introduced by Booton [25], [26] and was applied to Sigma–Delta modulators (with additional linear filtering permitted in the loops) by Ardalan and Paulos [27].

One can improve the approximations by using higher order terms in various expansions of the quantizer nonlinearity, but this approach has not been noticeably successful in the ADC application, primarily because of its difficulty. In addition, traditional power series expansions are not well suited to the discontinuous nonlinearities of quantizers. (See Arnstein [28] and Slepian [29] for series expansion solutions for quantization noise in delta modulators and DPCM.) The approach taken here does, in fact, resemble the series expansion technique because the nonlinear operation will be expanded in a Fourier series, but all terms will be retained for an exact analysis. This general approach to nonlinearities in communication systems is a variation the classical characteristic function method of Rice [30] and the transform method of Davenport and Root [31], who represented memoryless nonlinearities using Fourier or Laplace transforms. Here a Fourier series will play a similar role and the analysis will be for discrete time. A similar use of Fourier analysis for nonlinear systems was first applied to quantization noise analysis by Clavier, Panter, and Grieg [2], [3]. Subsequently the characteristic function method was applied to the study of quantization noise by Widrow [32] and Widrow's formulation has been used in the subsequent development of conditions under quantization noise is white, in particular by Sripad and Snyder [33] and Claasen and Jongepier [34]. Barnes and his colleagues used Sripad and Snyder's approach to analyze the similar problem of roundoff error in fixed rate multiplication [35], [36].

Since exact solution of the quantization noise behavior for general feedback quantization systems is unlikely, it is of interest to find simple systems which are amenable to exact analysis and, hopefully, are complicated enough to be of practical use. In the ideal relay feedback control problem, Hamel, Tsypkin, and others [37]–[40] provided an approach resembling the describing function approach that gives exact results for a very particular class of systems: Instead of assuming the input to the quantizer is a single sinusoid, the output is assumed to be periodic and hence to have a Fourier series expansion. If one then makes the additional assumptions of a particular form for the output (usually a simple square wave), it is sometimes possible to solve for conditions under which the assumed output exists. This approach has the drawback that it assumes a very simple output and then finds conditions under which it holds, but it does not provide a general solution.

One goal of this paper is to present a collection of results describing the quantizer noise behavior for a class of feedback quantization systems where exact results are possible and to compare these results and their implications with the common approximations. The basic class

considered is that of Sigma–Delta modulators: single-loop, multistage, and multiloop. The common attribute is that the linear filters within the loop are constrained to be ideal integrators, the forms first proposed by their originators [4], [5], [7], [8], [11], [12]. This paper surveys and in some cases extends several of these results. Most of these results have been reported in the recent literature [41]–[48], but this paper draws on hindsight to describe them in a unified fashion while providing an historical tour of the problem and emphasizing the underlying ideas.

The basic results do not require sophisticated mathematics, although the algebra is occasionally cluttered with Bessel functions (as one might expect when doing nonlinear modulation of sinusoids). Much of the original work in developing these results was in guessing the solutions to the nonlinear difference equations. Once done, it is easy to verify the solutions using simple induction. The basic stability properties of these systems will also be seen to follow from induction. The principal mathematical techniques required consist of 1) traditional linear systems theory, 2) "guessing" the solutions to certain nonlinear recursions, and 3) a straightforward discrete-time system extension of the transform method of Rice [30] and Davenport and Root [31] for memoryless nonlinearities.

Many of the basic ideas needed for analyzing uniform quantization within a feedback loop can be introduced in the simpler setting of a simple uniform quantizer not inside a feedback loop. This development provides results similar to those of Clavier, Panter, and Grieg and provides an interesting comparison with the results of Bennett and with the traditional white noise assumption and the describing function approach.

Section II develops a useful representation for the quantizer error in a uniform quantizer in the absence of overload (the no-overload quantizer noise is often called *granular noise*). A Fourier series expansion provides a representation of the error sequence that can then be used to express the mean, second moment, and autocorrelation functions of the quantizer noise in terms of characteristic functions. The moments and characteristic functions can be time-averages, probabilistic averages, or a combination of the two. Section III applies the formulas to the case considered by Clavier *et al.*: a single deterministic sinusoidal input. Section IV extends the application to the case of a general quasi-stationary input process with an additive memoryless noise or dither process. These two examples behave quite differently, the first having a complicated nonwhite spectrum and the second having a white spectrum. The two examples demonstrate extremes in behavior that are typical for more general architectures.

In Section V a single loop Sigma–Delta modulator is analyzed by using the results for a uniform quantizer to express the nonlinear difference equations describing the system in a convenient form and then by using induction to demonstrate a solution. The analysis techniques used for the memoryless quantizer are then applied to the solution to find the moments of the error process. The various time average moments are computed for a dc input and the development for a sinusoidal input is outlined. One form of dither process is also considered.

Sections VI and VII provide extensions to two-stage and two-loop Sigma–Delta modulators, respectively. Section VIII describes extensions to higher order loops and multistage quantizers and mentions several open problems.

## II. Uniform Quantization

The basic common component to all ADC's is a quantizer. The quantizer will be assumed to be uniform because *a priori* there is no knowledge of the statistical behavior of the quantizer input (or the original system input) and hence there is no way to optimize the quantizer levels for the input in a Lloyd–Max sense [49], [50]. Furthermore, a uniform quantizer is the simplest ADC and is almost universally used in oversampled ADCs. It is assumed that the quantizer has an even number, say $M$, of levels and that the distance between the output levels (the *bin width*) is $\Delta$. The special case of $M = 8$ is shown in Fig. 1.

The quantizer output level is always the closest (nearest neighbor) level to the input. This can be summarized as

$$q(u) = \begin{cases} \left(\dfrac{M}{2} - \dfrac{1}{2}\right)\Delta; & \left(\dfrac{M}{2} - 1\right)\Delta \le u \\[2mm] \left(k - \dfrac{1}{2}\right)\Delta; & (k-1)\Delta \le u < k\Delta; \\[2mm] & k = \left(-\dfrac{M}{2} + 2\right), \cdots, \left(\dfrac{M}{2} - 1\right) \\[2mm] \left(-\dfrac{M}{2} + \dfrac{1}{2}\right)\Delta; & u < \left(-\dfrac{M}{2} + 1\right)\Delta. \end{cases}$$

(1)

The quantizer error $\epsilon$ is defined by

$$\epsilon = q(u) - u. \tag{2}$$

The choice of the sign permits us to write the quantizer output as the sum of its input and the quantizer noise. Thus, in particular, if the input is a sequence of samples $u_n$, then we will be interested in the noise process $\epsilon_n = q(u_n) - u_n$ and we will wish to see if it resembles random white noise in some way.

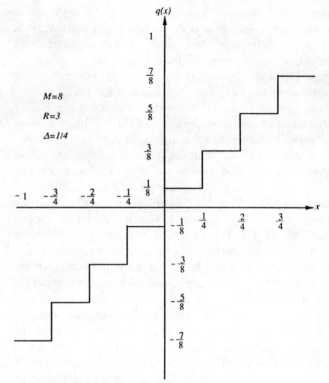

Fig. 1.   Uniform quantizer.

Note that except for inputs in the top or bottom regions of the input range, the magnitude of the error $\epsilon$ is bound above by $\Delta/2$. If the input is confined to the range $[-M\Delta/2, M\Delta/2]$, then the magnitude error is guaranteed to be no larger than $\Delta/2$. If the input is outside of this range, then the quantizer magnitude error will be larger than $\Delta/2$ and we will say that the quantizer is *overloaded*. The range $[-M\Delta/2, M\Delta/2]$ will be called the *no-overload range*. In an $M$ level quantizer with bin width $\Delta$, the no-overload range clearly has size $M\Delta$. In an ordinary PCM system we can prevent quantizer overload by preceding the quantizer by a saturator. In a feedback quantization system, however, the potential for overload makes even worse an already complicated analysis. It will be important, therefore, to determine conditions on a system under which the quantizer inside a feedback loop does not overload.

Normalize the quantizer output and input by the bin width $\Delta$ and use (1) to write

$$-\frac{\epsilon}{\Delta} = \frac{u}{\Delta} - \frac{q(u)}{\Delta}$$

$$= \begin{cases} \dfrac{u}{\Delta} - \left(\dfrac{M}{2} - 1\right) - \dfrac{1}{2}; & \left(\dfrac{M}{2} - 1\right) \le \dfrac{u}{\Delta} \\[2mm] \dfrac{u}{\Delta} - (k-1) - \dfrac{1}{2}; & (k-1) \le \dfrac{u}{\Delta} < k, \quad k = \left(-\dfrac{M}{2} + 2\right), \cdots, \left(\dfrac{M}{2} - 1\right) \\[2mm] \dfrac{u}{\Delta} - \left(-\dfrac{M}{2} + 1\right) - \dfrac{1}{2}; & \dfrac{u}{\Delta} < \left(-\dfrac{M}{2} + 1\right). \end{cases} \tag{3}$$

If the input is confined to the no overload region, then the previous formulas can be abbreviated to simply

$$-\frac{\epsilon}{\Delta} = \frac{u}{\Delta} - (k-1) - \frac{1}{2};$$

$$(k-1) \le \frac{u}{\Delta} < k, \quad k = \left(-\frac{M}{2}+1\right), \cdots, \frac{M}{2}. \quad (4)$$

This can be further abbreviated by introducing some notation. Every real number $r$ can be uniquely written in the form $r = \lfloor r \rfloor + \langle r \rangle$, where $\lfloor r \rfloor$ is the greatest integer less than or equal to $r$ and $0 \le \langle r \rangle < 1$ is the fractional part of $r$ (or $r \bmod 1$). From (4) we have that

$$-\frac{\epsilon}{\Delta} = \frac{u}{\Delta} - \left\lfloor \frac{u}{\Delta} \right\rfloor - \frac{1}{2}$$

and therefore

$$e = \frac{\epsilon}{\Delta} = \frac{1}{2} - \left\langle \frac{u}{\Delta} \right\rangle. \quad (5)$$

The normalized quantizer error $e$ is introduced here for convenience. The basic idea of this formula is well known, a similar function is drawn for the case of odd $M$ in Bennett [1], but its usual application is to quantizers with an infinite number of levels (an approximation to a very large number of levels). It is equally useful when the number of levels is small provided the quantizer is not overloaded. In addition, full advantage has not been taken of the representation as a fractional operator, an operator that has been much studied in ergodic theory.

Fig. 2 graphs $e$ as a function of the quantizer input $u$ for the special case of $M = 4$ and makes it clear that for $u$ constrained to the no-overload region, $e(u)$ is a periodic function of $u$ with period $\Delta$. Hence for $u$ in this region we can write a Fourier series for $e$ as

$$e = e(u) = \sum_{l \neq 0} \frac{1}{2\pi jl} e^{2\pi jlu/\Delta} = \sum_{l=1}^{\infty} \frac{1}{\pi l} \sin\left(2\pi l \frac{u}{\Delta}\right). \quad (6)$$

This series will hold for almost all values of $u$ and it is the key formula for all of the subsequent analysis. Note that if $u$ were not restricted to lie in the no-overload region,

then $e(u)$ would not be periodic and the Fourier series representation could not be used. (One could still use a Laplace transform as in Davenport and Root.) Note also for later reference that one can similarly write a Fourier series for $e^2$ as

$$e(u)^2 = \frac{1}{12} + \sum_{l \neq 0} \frac{1}{2(\pi l)^2} e^{2\pi jlu/\Delta}$$

$$= \frac{1}{12} + \sum_{l=1}^{\infty} \frac{1}{2(\pi l)^2} \cos\left(2\pi l \frac{u}{\Delta}\right). \quad (7)$$

Suppose now that a sequence $u_n$ is put into the quantizer, where we require that $|u_n| \le M\Delta/2$ so that there is no overload. We wish to study the behavior of the normalized error sequence $e_n = \epsilon_n/\Delta$. Here we will follow two different, but related, approaches. The first is that of Clavier, et al.: From (5)–(6) it is immediate that

$$e_n = \frac{1}{2} - \left\langle \frac{u_n}{\Delta} \right\rangle$$

$$= \sum_{l \neq 0} \frac{1}{2\pi jl} e^{2\pi jlu_n/\Delta} = \sum_{l=1}^{\infty} \frac{1}{\pi l} \sin\left(2\pi l \frac{u_n}{\Delta}\right). \quad (8)$$

For some specific examples of sequences $u_n$, (8) can be used to obtain a form of Fourier series representation directly for the sequence $e_n$. This representation then can be used to determine moments and spectral behavior. One problem with this approach is that it only works for certain simple inputs, e.g., sinusoids. Another difficulty is that ordinary Fourier series may not work (i.e., converge) since $e_n$ need not be a periodic function of $n$. For example, if $u_n = A\cos(2\pi f_0 n)$ and $f_0$ is not a rational number, then $u_n$ (and hence also $e_n$) is not periodic. (For example, $u_n = 0$ only if $n = 0$.) This problem can be handled mathematically by resorting to a generalization of the Fourier series due to Harald Bohr and the related theory of almost periodic functions, but we shall only make a few relevant observations when appropriate. (For a complete treatment of almost periodic functions, Bohr's classic [51] still makes good reading. See also [52], [53] and the discussion in [54] and the references therein.)

An alternative approach, which is also based on (6)–(7), is to instead focus on moments of the process rather than on a specific Fourier representation of the actual sequence. This will lead to the characteristic function method. Here the primary interest is the long term average behavior of the error sequence $e_n$. In particular we look at the time average mean, second moment, and autocorrelation function defined by

$$M\{e_n\} = \lim_{N \to \infty} \frac{1}{N} \sum_{n=1}^{N} e_n, \quad (9)$$

$$M\{e_n^2\} = \lim_{N \to \infty} \frac{1}{N} \sum_{n=1}^{N} e_n^2, \quad (10)$$

and

$$r_e(k) = M\{e_n e_{n+k}\} = \lim_{N \to \infty} \frac{1}{N} \sum_{n=1}^{N} e_n e_{n+k}, \quad (11)$$

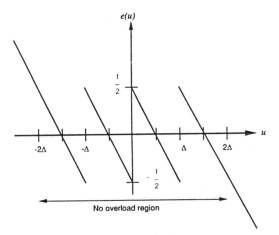

Fig. 2. Quantizer error.

respectively. Also of interest is the crosscorrelation,

$$r_{ue}(k) = M\{u_n e_{n+k}\} = \lim_{N \to \infty} \frac{1}{N} \sum_{n=1}^{N} u_n e_{n+k}. \quad (12)$$

If the input sequence $u_n$ is a stationary and ergodic random process, then the noise process $\epsilon_n$ will also be stationary and ergodic and these time averages will with probability one equal their expectations $E(e_n), E(e_n^2)$, $R_e(k) = E(e_n e_{n+k})$, and $R_{ue}(k) = E(u_n e_{n+k})$. While both forms of average are important, we follow the approach suggested for Sigma–Delta modulators in [48] and both generalize and unify the development for both deterministic and random inputs by using the idea of *quasi-stationary processes* considered by Ljung [55]. A discrete time process $e_n$ is said to be *quasi-stationary* if there is a finite constant $C$ such that

$$E(e_n) \leq C; \quad \text{all } n$$

$$|R_e(n,k)| \leq C; \quad \text{all } n,k$$

where $R_e(n,k) = E(e_n e_k)$, and if for each $k$ the limit

$$\lim_{N \to \infty} \frac{1}{N} \sum_{n=1}^{N} R_e(n, n+k) \quad (13)$$

exists, in which case the limit is defined as $R_e(k)$. For technical reasons we add the first moment condition that the limit

$$\overline{m}_e = \lim_{N \to \infty} \frac{1}{N} \sum_{n=1}^{N} E(e_n) \quad$$

exists to avoid the implicit assumption of a zero mean. Following Ljung we introduce the following notation: Given a process $x_n$, define

$$\overline{E}\{x_n\} = \lim_{N \to \infty} \frac{1}{N} \sum_{n=1}^{N} E(x_n), \quad (14)$$

if the limit exists. Thus for a quasi-stationary process $\{e_n\}$ the autocorrelation is given by

$$R_e(k) = \overline{E}\{e_n e_{n+k}\}, \quad (15)$$

the mean is defined by

$$m_e = \overline{E}\{e_n\} \quad (16)$$

and the average power is given by

$$R_e(0) = \overline{E}\{e_n^2\}. \quad (17)$$

A joint process $\{u_n, e_n\}$ is quasi-stationary if both coordinate processes $\{u_n\}$ and $\{e_n\}$ are quasi-stationary, if $|R_{ue}(n,k)| = |E(u_n e_k)|$ is uniformly bounded over $n$ and $k$, and if

$$R_{ue}(k) = \overline{E}\{u_n e_{n+k}\} \quad (18)$$

is well defined for all $k$. If $R_{ue}(k) = m_u m_e$ for all $k \neq 0$, the processes are said to be *asymptotically uncorrelated* with each other. The crosscorrelation between two processes becomes important, for example, if one wishes to do second moment analysis on linear combinations of the processes, for example, find the autocorrelation or spectrum of $u_n + e_n$.

These moments reduce to the corresponding time averages or probabilistic averages in the special cases of deterministic or random processes, respectively.

The fundamental common examples of quasi-stationary processes are the class of deterministic processes with convergent and bounded sample means and autocorrelations and the class of stationary random processes (with bounded mean and autocorrelation). The autocorrelation defined by (15) reduces to the sample average of (11) in the former case and the usual probabilistic autocorrelation in the latter. Useful but still simple examples are formed by combining simple deterministic processes with stationary random processes, e.g., adding a sinusoid to an independent identically distributed (i.i.d) process. More general examples of quasi-stationary random processes include processes that are second order stationary processes (second-order distributions do not depend on time origin) or asymptotically mean stationary processes (processes for which the ergodic theorem holds for all bounded measurements on the process [56]) provided that the appropriate moments are bounded.

The *power spectrum* of the process is defined in the general case as the discrete time Fourier transform of the autocorrelation:

$$S_e(f) = \sum_{n=-\infty}^{\infty} R_e(n) e^{-2\pi j f n}, \quad (19)$$

where the frequency $f$ is normalized to lie in $[0, 1]$ (corresponding to a sampling period being considered as a single time unit). We shall also use radian frequencies $\omega = 2\pi f$ when convenient. The usual linear system input/output relations hold for this general definition of spectrum (see Chapter 2 of Ljung [55]). In fact, the class of quasi-stationary processes can be viewed as the most general class for which the familiar formulas of ordinary linear system second-order correlation and spectral analysis remain valid.

We now proceed to apply the basic formulas (8) and (7) to find an expression for the basic moments of (15)–(18). Plugging (8) into (16) and (7) into (17) and assuming that the limits can be interchanged results in

$$\overline{E}\{e_n\} = \lim_{N \to \infty} \frac{1}{N} \sum_{n=1}^{N} \sum_{l \neq 0} \frac{1}{2\pi j l} e^{2\pi j l u_n / \Delta}$$

$$= \sum_{l \neq 0} \frac{1}{2\pi j l} \lim_{N \to \infty} \frac{1}{N} \sum_{n=1}^{N} e^{2\pi j l u_n / \Delta}$$

$$= \sum_{l \neq 0} \frac{1}{2\pi j l} \overline{E}\{e^{2\pi j l u_n / \Delta}\},$$

$$\overline{E}\{e_n^2\} = \frac{1}{12} + \sum_{l \neq 0} \frac{1}{2(\pi l)^2} \overline{E}\{e^{2\pi j l u_n / \Delta}\},$$

and for $k \neq 0$

$$R_e(k) = \sum_{i \neq 0} \sum_{l \neq 0} \frac{j}{2\pi i} \frac{j}{2\pi l} \overline{E}\{e^{2\pi j (i u_n / \Delta + l u_{n+k} / \Delta)}\}.$$

These expressions can be most easily given in terms of the one-dimensional characteristic function

$$\overline{\Phi}_u(l) = \overline{E}\{e^{2\pi j l u_n/\Delta}\} \qquad (20)$$

and a two-dimensional characteristic function

$$\overline{\Phi}_u^{(k)}(i,l) = \overline{E}\{e^{2\pi j (i u_n/\Delta + l u_{n+k}/\Delta)}\}; \qquad k \neq 0, \quad (21)$$

as

$$\overline{E}\{e_n\} = \sum_{l \neq 0} \frac{1}{2\pi j l} \overline{\Phi}_u(l), \qquad (22)$$

$$\overline{E}\{e_n^2\} = \frac{1}{12} + \sum_{l \neq 0} \frac{1}{2(\pi l)^2} \overline{\Phi}_u(l), \qquad (23)$$

and for $k \neq 0$

$$R_e(k) = - \sum_{i \neq 0} \sum_{l \neq 0} \frac{1}{2\pi i} \frac{1}{2\pi l} \overline{\Phi}_u^{(k)}(i,l). \qquad (24)$$

The interchange of the limits is an important technical point that must be justified in any particular application. For the purely deterministic case, it can be shown that if the one-dimensional time-average characteristic function of (20) exists (i.e., the limit converges to something finite), then so does the two-dimensional and the previous limit interchanges are valid [43], [48]. If the process is stationary, then the characteristic functions reduce to the usual probabilistic characteristic functions and a similar conclusion follows.

If the characteristic functions of (20)–(21) can be evaluated, then the moments and spectrum of the process can be computed from (22)–(24). This general approach to finding the moments of the output of a nonlinear system is a variation of the "characteristic function" approach described by Rice [30] and studied by Davenport and Root [31] under the name "transform method." Here discrete time replaces continuous time, Fourier series replace Laplace transforms, and stationary processes are replaced by quasi-stationary processes (and hence time-averages can replace probabilistic averages). The key idea of expanding the quantizer nonlinearity as a Fourier series was first developed by Clavier, Panter and Grieg (1947) and the extension of the approach to nonsinusoidal inputs was first accomplished by Widrow (1960) for stationary random processes.

The crosscorrelation does not have as nice a simple form in terms of characteristic functions as do the other moments, but a similar representation is given by

$$R_{ue}(k) = \overline{E}\{u_n e_{n+k}\} = \sum_{l \neq 0} \frac{1}{2\pi j l} \overline{E}\{u_n e^{j2\pi l (u_{n+k}/\Delta)}\}. \quad (25)$$

We next turn to specific applications of these formulas to ordinary memoryless quantization.

## III. PCM Quantization Noise: Deterministic Inputs

We first consider a purely deterministic input and hence the averages will all be time averages that is, $\overline{E}\{x_n\} = M\{x_n\}$. We will not consider the example of a dc input to

an ordinary uniform quantizer in any detail because the results are trivial: If the input is dc, say $u_n = u$, then the error is simply $e_n = \epsilon_n/\Delta = \epsilon(u)/\Delta = 1/2 - \langle u/\Delta \rangle$ and the mean is this constant value and the second moment is the constant squared (and hence the variance is 0). The autocorrelation is also equal to the second moment. This does point out, however, that the common white noise assumption is certainly erroneous for a dc input since $R_e(k)$ is not 0 for $k \neq 0$. This is well known and it is generally believed that the input to the quantizer must be "active" if the noise is to be approximately white. Hence we consider next a more interesting input, a sinusoid $u_n = A\sin(n\omega_0 + \theta)$ with a fixed initial phase $\theta$. The discrete time sinusoid can be related to the original continuous time signal by defining $\omega = \psi T$, where $T$ is the original sampling period and $\psi = 2\pi f$, where $f$ is the original continuous time frequency. We assume that $A \leq M/2$ so that the quantizer is not overloaded. Define also $f_0 = \omega_0/2\pi$. Here we can directly use (8) as did Clavier et al. to find an expression for the error sequence. For the given case,

$$e_n = \sum_{l \neq 0} \frac{1}{2\pi j l} e^{2\pi j l (A/\Delta)\sin(n\omega_0 + \theta)}. \qquad (26)$$

For brevity we denote $A/\Delta$ by $\gamma$. To evaluate this sum we use the Jacobi–Anger formula

$$e^{jz\sin\psi} = \sum_{m=-\infty}^{\infty} J_m(z) e^{jm\psi}, \qquad (27)$$

where $J_m$ is the ordinary Bessel function of order $m$ to obtain

$$e_n = \sum_{m=-\infty}^{\infty} e^{jn\omega_0 m} \left( e^{jm\theta} \sum_{l \neq 0} \frac{1}{2\pi j l} J_m(2\pi l\gamma) \right)$$

$$= \sum_{m=-\infty}^{\infty} e^{jn\omega_0(2m-1)} \left( e^{j(2m-1)\theta} \sum_{l=1}^{\infty} \frac{1}{\pi j l} J_{2m-1}(2\pi l\gamma) \right), \qquad (28)$$

where the even terms disappear because

$$J_m(z) = (-1)^m J_m(-z). \qquad (29)$$

Equation (28) provides a generalized Fourier series representation for the sequence $e_n$ in the form

$$e_n = \sum_{m=-\infty}^{\infty} e^{j\lambda_m n} b_m, \qquad (30)$$

where the qualifier "generalized" is used because the sequence $e_n$ need not be periodic. From this expression one can identify the power spectrum of $e_n$ as having components of magnitude $|b_m|^2$ at frequencies $\lambda_m = (2m - 1)\omega_0 \pmod{2\pi}$. Note that if $f_0$ is rational, then the input and hence also $e_n$ will be periodic and the frequencies $(2m - 1)\omega_0 \pmod{2\pi}$ will duplicate as the index $m$ takes on all integer values. In this case, the power spectrum amplitude at a given frequency is found by summing up all of the coefficients $|b_m|^2$ with $(2m - 1)\omega_0 \pmod{2\pi}$

giving the desired frequency. We will henceforth assume that $f_0$ is not a rational number (and hence that in the corresponding continuous time system, the sinusoidal frequency is incommensurate with the sampling frequency). This assumption simplifies the analysis and can be physically justified since a randomly selected frequency from a continuous probability distribution will be irrational with probability one. An irrational $\omega_0$ can also be viewed as an approximation to a rational frequency with a large denominator when expressed in lowest terms.

This direct approach indeed gives the spectrum for this special case. In particular, from the symmetry properties of the Bessel functions one has a power spectrum that assigns amplitude

$$S_m = \left( \frac{1}{\pi} \sum_{l=1}^{\infty} \frac{1}{l} J_{2m-1}(2\pi \gamma l) \right)^2 \qquad (31)$$

to every frequency of the form $(2m-1)\omega_0 \pmod{2\pi}$. Letting $\delta(f)$ denote a Dirac delta function, this implies a power spectral density of the form

$$S_e(f) = \sum_{m=-\infty}^{\infty} S_m \delta(f - \langle (2m-1)f_0 \rangle). \qquad (32)$$

We could use these results to evaluate the mean, power, and autocorrelation. Instead we turn to the more general transform method. The two methods will be seen to yield the same spectrum for this example.

For the given purely deterministic example, the one-dimensional characteristic function can be expressed as

$$\overline{\Phi}_u(l) = \lim_{N \to \infty} \frac{1}{N} \sum_{n=1}^{N} e^{j2\pi l \gamma \sin(n\omega_0 + \theta)}$$

$$= \lim_{N \to \infty} \frac{1}{N} \sum_{n=1}^{N} e^{j2\pi l \gamma \sin(2\pi \langle nf_0 + (\theta/2\pi) \rangle)}, \qquad (33)$$

where the fractional part can be inserted since $\sin(2\pi u)$ is a periodic function in $u$ with period 1. This limit can be evaluated using a classical result in ergodic theory of Hermann–Weyl (see, e.g., Petersen [54]): If $g$ is an integrable function, $a$ is an irrational number, and $b$ is any real number, then

$$\lim_{N \to \infty} \frac{1}{N} \sum_{n=1}^{N} g(\langle an + b \rangle) = \int_0^1 g(u) \, du. \qquad (34)$$

This remarkable result follows since the sequence of numbers $\langle an + b \rangle$ uniformly fills the unit interval and hence the sums approach an integral in the limit. Applying (34) to (33) yields

$$\overline{\Phi}_u(l) = \int_0^1 du \, e^{j2\pi l \gamma \sin(2\pi u)} = J_0(2\pi l \gamma). \qquad (35)$$

The mean and second moment of the quantizer noise can

then be found using the fact $J_0(r) = J_0(-r)$:

$$\overline{E}\{e_n\} = \sum_{l \neq 0} \frac{1}{2\pi jl} J_0(2\pi l \gamma) = 0, \qquad (36)$$

$$\overline{E}\{e_n^2\} = \frac{1}{12} + \frac{1}{\pi^2} \sum_{l=1}^{\infty} \frac{1}{l^2} J_0(2\pi l \gamma). \qquad (37)$$

Note that the result does not depend on the frequency of the input sinusoid and that the time average mean is 0, which agrees with that predicted by the assumption that $\epsilon_n$ is uniformly distributed on $[-\Delta/2, \Delta/2]$. The second moment, however, differs from the value of $1/12$ predicted by the uniform assumption by the right-hand sum of weighted Bessel functions. No simple closed form expression for this sum is known to the author, but other formulas useful for insight and computation can be developed. Before pursuing these, note that if $\gamma = A/\Delta$ becomes large (which, with $A$ held fixed and the no-overload assumption, means that the number of quantization levels is becoming large), then $J_0(2\pi l \gamma) \to 0$ and hence the second moment converges to $1/12$ in the limit. This is consistent with the asymptotic (high resolution) quantization theory developed by Bennett and others.

By expressing the Bessel function as an integral of a cosine and by applying the summation

$$\sum_{k=1}^{\infty} \frac{\cos kx}{k^2} = \sum_{k=1}^{\infty} \frac{\cos 2\pi k \left\langle \frac{x}{2\pi} \right\rangle}{k^2}$$

$$= \pi^2 \left( \frac{1}{6} - \left\langle \frac{x}{2\pi} \right\rangle + \left\langle \frac{x}{2\pi} \right\rangle^2 \right) \qquad (38)$$

(the second equality following, e.g., from Gradshteyn and Rhyzhik [57]) we obtain

$$\overline{E}\{e_n^2\} = \frac{1}{\pi} \int_0^{\pi} \left( \frac{1}{2} - \langle \gamma \sin \psi \rangle \right)^2 d\psi$$

$$= \frac{2}{\pi} \int_0^{\pi/2} \left( \frac{1}{2} - \langle \gamma \sin \psi \rangle \right)^2 d\psi. \qquad (39)$$

This result can also be obtained by applying Weyl's result directly to $\overline{E}\{e_n^2\}$ without using the transform approach. An alternative form can be found by observing that since $\langle \gamma \sin \psi \rangle = \gamma \sin \psi - (k-1)$ when $k > \gamma \sin \psi \geq k-1$, we have that if $K = \lfloor \gamma \rfloor$

$$\overline{E}\{e_n^2\} = \frac{2}{\pi} \sum_{k=1}^{K} \int_{b_{k-1}}^{b_k} \left( k - \frac{1}{2} - \gamma \sin \psi \right)^2 d\psi,$$

where $b_0 = 0$, $b_k = \sin^{-1}(k\Delta/A)$, $k = 1, 2, \cdots, K-1$, $b_K = \pi/2$. With some algebra this yields

$$\overline{E}\{e_n^2\} = \left( K + \frac{1}{2} \right)^2 + (\gamma)^2 + \frac{2}{\pi} \frac{A}{\Delta} - 2$$

$$- \frac{2}{\pi} \left( \sum_{k=1}^{K-1} \sin^{-1}\left( \frac{k\Delta}{A} \right) + \sum_{k=1}^{K-1} \cos\left( \sin^{-1}\left( \frac{k\Delta}{A} \right) \right) \right). \qquad (40)$$

The right-most term can be further simplified using the fact that $\cos(\sin^{-1}(a)) = \sqrt{1-a^2}$. Equation (40) has the advantage over (37) in that it is expressed as a finite sum and it more resembles the form of the results of Clavier *et al.* [2] for the harmonic distortion when a uniform quantizer with an odd number of levels is driven by a full range sinusoidal input.

Fig. 3 plots $\bar{E}\{e_n^2\}$ as given in (37) as a function of $\gamma = A/\Delta$ for $\gamma \le M/2$ (as required by the no overload condition).

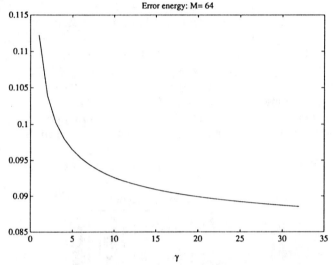

Fig. 3. Quantization error energy.

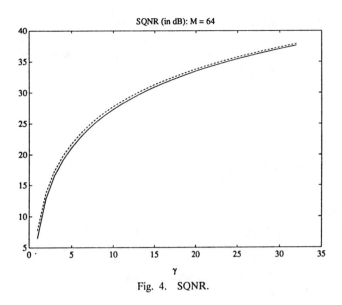

Fig. 4. SQNR.

A value of $M = 128$ is chosen, but the curve indicates the behavior for smaller $M$ since the points on the curve depend on $M$ only indirectly through the no-overload constraint. If $e_n$ where uniformly distributed, then one would expect a straight line at $1/12 = 0.0833$. The actual curve is above this value and is converging slowly for large $M$. Fig. 4 plots the resulting SQNR in dB, demonstrating that the inaccuracy of the average error has little effect on the resulting signal-to-noise ratio on a logarithmic scale. Thus the assumption of a uniform error distribution on the quantizer error gives a good approximation for the SQNR, but the approximation for the actual average distortion is not as accurate.

To compute the autocorrelation of the quantization noise, we use similar steps to find the joint characteristic function $\bar{\Phi}_u^{(k)}(i,l)$. We have that

Inserting (41) into (24) and using the fact that

$$J_{-n}(x) = (-1)^n J_n(x)$$

yields

$$R_e(k) = -\frac{1}{(2\pi)^2} \sum_{n=-\infty}^{\infty} (-1)^n \left( \sum_{l \ne 0} \frac{J_n(2\pi\gamma l)}{l} \right)^2 e^{jnk\omega_0}$$

$$= \sum_{n=-\infty}^{\infty} \left( \frac{1}{\pi} \sum_{l=1}^{\infty} \frac{J_{2n-1}(2\pi\gamma l)}{l} \right)^2 e^{jnk\omega_0}. \quad (43)$$

This has the form

$$R_e(k) = \sum_{n=-\infty}^{\infty} S_n e^{2\pi jk\lambda_n}, \quad (44)$$

where $\lambda_n = \langle (2n-1)\omega_0/2\pi \rangle$ are normalized frequencies in $[0,1)$ and

$$S_n = \left( \frac{1}{\pi} \sum_{l=1}^{\infty} \frac{J_{2n-1}(2\pi\gamma l)}{l} \right)^2 \quad (45)$$

are the spectral components at the frequency $\lambda_n$, and

$$\bar{\Phi}_u^{(k)}(i,l) = \lim_{N \to \infty} \frac{1}{N} \sum_{n=1}^{N} e^{2\pi j[l\gamma \sin(n\omega_0 + \theta) + i\gamma \sin((n+k)\omega_0 + \theta)]}$$

$$= \lim_{N \to \infty} \frac{1}{N} \sum_{n=1}^{N} e^{2\pi j[l\gamma \sin(2\pi\langle nf_0 + (\theta/2\pi)\rangle) + i\gamma \sin(2\pi\langle nf_0 + (\theta/2\pi)\rangle + k\omega_0))]}$$

$$= \int_0^1 du \, e^{2\pi j[l\gamma \sin(2\pi u + i\gamma \sin(2\pi u + k\omega_0))]}. \quad (41)$$

Using the Jacobi–Anger formula and the orthogonality of exponentials yields

$$\bar{\Phi}_u^{(k)}(i,l) = \sum_{n=-\infty}^{\infty} J_n(2\pi\gamma i) J_{-n}(2\pi\gamma l) e^{jnk\omega_0}. \quad (42)$$

hence

$$S_e(f) = \sum_n S_n \delta_{f-\lambda_n}, \quad (46)$$

exactly as found directly in (31)–(32).

The sequence $S_m$ is sometimes referred to as the Bohr–Fourier series of the autocorrelation because of Bohr's generalization of the ordinary Fourier series of periodic functions to the general exponential decomposition of (44), which implies that $R_e(k)$ is an almost periodic function of $k$. It suffices here to point out that the spectrum of the quantizer noise is purely discrete and consists of all odd harmonics of the fundamental frequency of the input sinusoid. The energy at each harmonic depends in a very complicated way on the amplitude of the input sinusoid. In particular, the quantizer noise is decidedly not white since it has a discrete spectrum and since the spectral energies are not flat. Thus here the white noise approximation of Bennett and of the describing function approach is invalid, even if $M$ is large.

An alternative form for (45) involving only a finite sum can be found in the same way (40) was found from (37). Since for odd $n$

$$J_n(z) = \frac{2}{\pi} \int_0^{\pi/2} \sin(n\theta) \sin(z \sin \theta) \, d\theta \qquad (47)$$

we have using the summation [57]

$$\sum_{l=1}^{\infty} \frac{\sin(xl)}{l} = \frac{\pi}{2} - \frac{1}{2}x; \qquad 0 \le x < 2\pi \qquad (48)$$

that for odd $n$

$$\sum_{l=1}^{\infty} \frac{J_n(2\pi\gamma l)}{l}$$

$$= \frac{2}{\pi} \int_0^{\pi/2} d\theta \sin(n\theta) \sum_{l=1}^{\infty} \frac{\sin(2\pi \langle \gamma \sin \theta \rangle l)}{l}$$

$$= \frac{2}{\pi} \int_0^{\pi/2} d\theta \sin(n\theta) \left( \frac{\pi}{2} - \pi \langle \gamma \sin(\theta) \rangle \right)$$

and hence as in (40)

$$\sum_{l=1}^{\infty} \frac{J_n(2\pi\gamma l)}{l}$$

$$= \frac{1}{n} - 2\gamma \frac{\pi}{4} \delta_{n-1}$$

$$+ 2 \sum_{k=1}^{K} \int_{b_{k-1}}^{b_k} d\theta \sin(n\theta) (\gamma \sin \theta - (k-1))$$

$$= \frac{1}{n} - 2\gamma \frac{\pi}{4} \delta_{n-1} + 2 \sum_{k=1}^{K} k \int_{b_{k-1}}^{b_k} \sin(n\theta) \, d\theta$$

$$= -\frac{2}{\pi n} \sum_{k=0}^{K-1} \cos\left( n \sin^{-1}\left( \frac{\Delta}{A} k \right) \right), \qquad (49)$$

where, as previously, $K \le M/2$ is the largest integer greater than or equal to $\gamma$. Thus (45) can also be written as

$$S_n = \left[ \frac{1}{n} - 2\gamma \frac{\pi}{4} \delta_{n-1} \right.$$

$$\left. + \frac{2}{\pi n} \sum_{k=0}^{K-1} \cos\left( (2n-1) \sin^{-1}\left( \frac{\Delta}{A} k \right) \right) \right]^2. \qquad (50)$$

Equation (50) strongly resembles the corresponding result of Clavier et al., for the case of odd $M$ and a full range sinusoidal input (that is, $\gamma = A/\Delta = (M-1)/2$).

Figs. 5 and 6 provide an example of a simulated waveform and spectrum together with the theoretical prediction. The waveform was generated from (8) with a randomly selected frequency and $N = 1024$ samples were computed. A $\gamma$ of 8 was chosen, which could correspond to a full range sinusoid with $M = 16$ levels ($M/2$ is the maximum value of $\gamma$ permitted by the no-overload condition). The spectral plot shows three quantities: The circles show the predicted spectrum using (50). It was found that the Bessel sums of (45) were painfully slow to converge

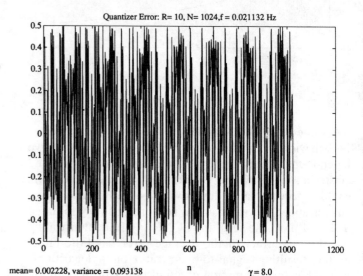

mean= 0.002228, variance = 0.093138     $\gamma = 8.0$

Fig. 5. PCM quantization error.

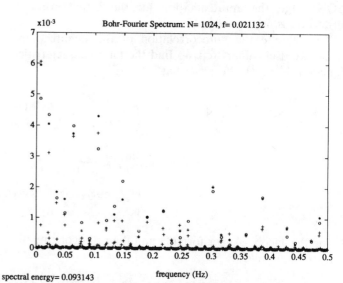

spectral energy= 0.093143

Fig. 6. PCM quantization error spectrum.

using Matlab, but the finite sums of (50) produced the desired plot in a few seconds. The asterisks show the corresponding spectral amplitudes computed at the predicted frequencies $\langle (2m-1)f_0 \rangle$, that is, the Bohr–Fourier coefficients at the odd harmonics of the input frequency. Finally, the addition signs show the spectrum as computed by an FFT, that is, the spectral components at 1024 uniformly spaced frequencies. The Bohr–Fourier coefficients computed from the actual waveform agree well with those predicted by the theory. (They should agree exactly in the limit as $N \to \infty$.) The FFT values are not always close. Intuitively this is because the spectrum is extremely "spikey" and the uniformly spaced Fourier frequencies are not the correct frequencies, that is, the odd harmonics of the input frequency. Hence the exponential sums are being evaluated for the wrong frequencies and the resulting values are not quite correct. Again in the limit as $N \to \infty$, the uniformly spaced frequencies will more closely approximate the true frequencies and the FFT spectrum should converge to the predicted spectrum.

The crosscorrelation is handled in a similar fashion. Beginning with (25) we have that

$$R_{ue}(k) = \sum_{l \neq 0} \overline{E}\{u_n e^{j2\pi l(u_{n+k}/\Delta)}\}$$

where

$$\overline{E}\{u_n e^{j2\pi l(u_{n+k}/\Delta)}\}$$
$$= \lim_{N \to \infty} \frac{1}{N} \sum_{n=1}^{\infty} A \sin(n\omega_0 + \theta) e^{j2\pi l\gamma \sin((n+k)\omega_0 + \theta)}$$
$$= A \int_0^1 \sin(2\pi u) e^{j2\pi l\gamma \sin(2\pi u + k\omega_0)} du.$$

Using the Jacobi–Anger formula again this becomes

$$\overline{E}\{u_n e^{j2\pi l(u_{n+k}/\Delta)}\}$$
$$= A \int_0^1 \sin(2\pi u) \sum_{m=-\infty}^{\infty} J_m(2\pi l\gamma) e^{jm(2\pi u + k\omega_0)} du$$
$$= A \sum_{m=-\infty}^{\infty} J_m(2\pi l\gamma) e^{jmk\omega_0} \int_0^1 \sin(2\pi u) e^{jm2\pi u} du$$
$$= jAJ_1(2\pi l\gamma) \cos k\omega_0.$$

Thus

$$R_{ue}(k) = A\cos(k\omega_0) \sum_{l=1}^{\infty} \frac{J_1(2\pi l\gamma)}{l}, \qquad (51)$$

which is not equal to the product of the means since the error mean is 0. Thus the error and the input are not asymptotically uncorrelated.

The basic procedure used above of computing characteristic functions that in turn yield the quantization error moments and spectra can be used with more complicated input signals to obtain exact formulas that can be evaluated numerically.

## IV. Dithering

We now consider a quantizer input process of the form

$$u_n = v_n + w_n, \qquad (52)$$

where $v_n$ is the possibly nonstationary original system input (such as the deterministic sinusoid previously considered) and $w_n$ is an i.i.d. random process which is called a *dither* process. A key attribute of the dither process is that it is independent of the $v_n$ process, that is, $v_n$ is independent of $w_k$ for all times $n$ and $k$. We still require that the quantizer input $u_n$ be in the no-overload region. This has the effect of reducing the allowed input dynamic range and hence limiting the overall SQNR. Dithering has long been used as a means of improving the subjective quality of quantized speech and images (see Jayant and Noll, Section 4.8, and the references therein [58]). The principal theoretical property of dithering was developed by Schuchman [59], who proved that if the quantizer does not overload and the characteristic function of the marginal probability density function of the dither signal was 0 at integral multiples of $2\pi/\Delta$, then the quantizer error $e_n = q(v_n + w_n) - (v_n + w_n)$ (at any specific time $n$) is independent of the original input signal $v_n$. It is generally thought that a good dither signal should force the quantizer error to be white even when the input signal has memory (the quantizer noise is trivially white if the original input is white). General results on this popular belief do not appear to have been explicitly published, although they can be derived from the work of Sripad and Snyder [33]. We explore this question using the techniques thus far developed.

Given an input as in (52) with $w_n$ a stationary random process, let

$$\Phi_w(j\alpha) = E(e^{j\alpha w_n})$$

denote the ordinary characteristic function of $w_n$ (which does not depend on $n$ since $\{w_n\}$ is stationary). Because of the independence of the processes, the one-dimensional characteristic function of (20) becomes

$$\overline{\Phi}_u(l) = \lim_{N \to \infty} \frac{1}{N} \sum_{n=1}^{N} E(e^{2\pi jl1/\Delta(v_n + w_n)})$$
$$= \Phi_w\left(j2\pi \frac{l}{\Delta}\right) \overline{\Phi}_v(l). \qquad (53)$$

The two-dimensional characteristic function of (21) is

$$\overline{\Phi}_u^{(k)}(i,l) = \lim_{N \to \infty} \frac{1}{N} \sum_{n=1}^{N} E(e^{2\pi j(1/\Delta)[i(v_n + w_n) + l(v_{n+k} + w_{n+k})]})$$
$$= \Phi_w\left(2\pi j\frac{1}{\Delta}i\right) \Phi_w\left(2\pi j\frac{1}{\Delta}l\right) \overline{\Phi}_v^{(k)}(i,l); \quad k \neq 0. \qquad (54)$$

Now suppose that the marginal distribution of $w_n$ is such that Schuchman's conditions are satisfied, that is, the quantizer is not overloaded and

$$\Phi_w\left(2\pi j\frac{1}{\Delta}l\right) = 0; \quad l = \pm 1, \pm 2, \pm 3, \cdots. \qquad (55)$$

(Recall that $\Phi_w(0) = 1$ for any distribution.) This is the condition shown by Schuchman to be necessary and sufficient for the quantization error to be independent of the original input $x_n$. The principal example is a dither signal with a uniform marginal on $[-\Delta/2, \Delta/2]$ (and an input amplitude constrained to avoid overload when added to the dither), in which case

$$\Phi_w(j\alpha) = \frac{2}{\alpha\Delta}\sin\frac{\alpha\Delta}{2}. \qquad (56)$$

For this case we have

$$\overline{\Phi}_u(l) = \begin{cases} 1; & l = 0 \\ 0; & \text{otherwise}, \end{cases} \qquad (57)$$

and for $k \neq 0$

$$\overline{\Phi}_u^{(k)}(i,l) = \begin{cases} \overline{\Phi}_v^{(k)}(0,0) = 1; & i = l = 0 \\ 0; & \text{otherwise}. \end{cases} \qquad (58)$$

Thus in this example we have from (22)–(24) that $e_n$ has zero mean, a second moment of $1/12$, and an autocorrelation function $R_e(k) = 0$ when $k \neq 0$, that is, the quantization noise is indeed white as is generally believed when Schuchman's condition is satisfied. Note that this is true for a general quasi-stationary input, including the sinusoid previously considered. If the input sequence is $A\sin(n\omega_0)$ as before and the dither sequence is an i.i.d. sequence of uniform random variables on $[-\Delta/2, \Delta/2]$, then the overload condition becomes $A + \Delta/2 \leq \Delta M/2$ or

$$\gamma = \frac{A}{\Delta} \leq \frac{M-1}{2}, \qquad (59)$$

which has effectively reduced the allowable $\gamma$. Thus, for example, in the $M = 16$ example of Fig. 7, the maximum allowable $\gamma$ is $15/2$. Fig. 7 shows an example of an error signal for this $\gamma$. The input sinusoid is the same as that of Fig. 5.

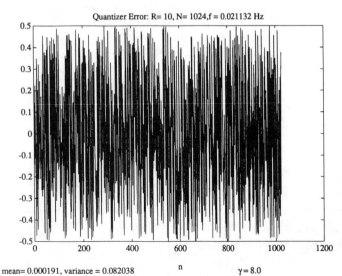

Quantizer Error: R= 10, N= 1024,f = 0.021132 Hz

mean= 0.000191, variance = 0.082038  $\qquad$ n $\qquad$ $\gamma = 8.0$

Fig. 7.  Dithered PCM.

We have already mentioned that Schuchman's conditions are sufficient to ensure that the original input and the quantization noise are independent, but for completeness we show that they are at least uncorrelated (or "linearly independent"). We have easily that

$$\overline{E}\{v_n e_{n+k}\} = \sum_{l \neq 0} \frac{1}{2\pi jl} \overline{E}\{v_n e^{j2\pi l((v_{n+k}/\Delta) + (w_{n+k}/\Delta))}\}$$
$$= \sum_{l \neq 0} \frac{1}{2\pi jl} \Phi_w\left(\frac{2\pi l}{\Delta}\right) \overline{E}\{v_n e^{j2\pi l((v_{n+k}/\Delta))}\},$$

which is 0 from Schuchman's conditions.

Note that when considering the crosscorrelation, it is often of most interest to consider the crosscorrelation of the quantization noise with the signal considered to be the original system input rather than the quantizer input, that is, with $v_n$ instead of $u_n$.

While the addition of a dither signal does turn the quantization noise into a simpler process, it should be remembered that this is not usually the goal of a quantizer. Adding a dither signal also adds distortion to the final reproduction since it corrupts the original signal in a noninvertible manner. If the decoded quantizer output is taken as the reproduction, then we can write

$$\hat{v}_n = q(v_n + w_n) = v_n + w_n + \epsilon_n,$$

showing that the overall error is

$$\hat{v}_n - v_n = w_n + \epsilon_n.$$

Intuitively, the overall quantization noise has increased because of the dither, but we cannot pin this down in general and evaluate the overall mean-squared error because $w_n$ and $\epsilon_n$ are correlated and hence

$$\overline{E}\{(\hat{v}_n - v_n)^2\} = E(w_n^2) + \overline{E}\{\epsilon_n^2\} + R_{w\epsilon}(0)$$

cannot be quantified without evaluating the final crosscorrelation term. If Schuchman's conditions hold, then there is a way to eliminate the performance loss, provided the receiver has a "side channel" and can be informed of the dithering signal exactly. If the receiver forms the reproduction process by subtracting out the dither signal from the decoded quantizer output, so called "subtractive dithering," to obtain

$$\hat{v}_n = q(v_n + w_n) - w_n = v_n + \epsilon_n,$$

then the reproduction is the input plus a signal-independent uniform white noise sequence. This result is nice mathematically, but has a basic drawback: The receiver does not usually know the dither signal exactly, unless the dithering signal was chosen in a deterministic way. In that case, however, the previous analysis is not valid—even if the dither is a pseudo-random signal.

A more serious problem with dithering than the corruption of the input signal is the fact that it reduces the SQNR achievable with a given quantizer since the input amplitude must be reduced enough so that the original signal plus the dither stays within the no-overload region. This loss may be acceptable (and small) when the number of quantization levels is large. It is significant if there are only a few quantization levels. For example, if $M = 2$, then a uniform dither on $[-\Delta/2, \Delta/2]$ can only avoid overload if the signal is itself confined to $[-\Delta/2, \Delta/2]$.

As a final remark, one could reverse the roles of the two processes and consider an original input to be a stationary process $w_n$ and $v_n$ to be a sinusoidal dither, as proposed by Jaffee [60]. Here the two-dimensional probabilistic characteristic function of the stationary process factors out of the time average of the sinusoidal dither. In general, however, the resulting $\Phi$'s will not be 0 for nonzero arguments and the noise will not be white.

## V. SINGLE-LOOP SIGMA–DELTA

The basic Sigma–Delta modulator can be motivated by an intuitive argument based on the dithering idea. Suppose that instead of adding an i.i.d. random process to the signal before quantization, the quantization noise itself is used as a dither signal, that is, i.i.d. signal independent noise is replaced by deterministic signal dependent noise which (hopefully) approximates a white signal independent process. Reversing the noise sign for convenience and inserting a delay in the forward path of the feedback loop (to reflect the physical delay inherent in a quantizer) yields the system of Fig. 8, where we preserve the $u_n$ notation for the quantizer input and label the original system input as $x_n$. The nonlinear difference equation governing this system is

$$u_n = x_{n-1} - \epsilon_{n-1} = u_{n-1} + x_{n-1} - q(u_{n-1}); \quad n = 1, 2, \cdots. \quad (60)$$

This difference equation can also be depicted as in Fig. 9, which is the traditional configuration for a single-loop Sigma–Delta modulator. The name follows from the fact that the system can be considered as a Delta modulator preceded by an integrator. The deterministically dithered form of Fig. 8 was introduced in 1960 by C. C. Cutler in Fig. 2 of [4], where he referred to the system as a quantizer "with a single step of error compensation." His system is identical except for the location of the system delay, which he placed in the feedback loop while most modern treatments place it in the feedforward loop with the quantizer. The form of Fig. 9 and the name of "Delta–Sigma" modulator were introduced by Inose and Yasuda in 1963 [5], who provided the first published of its basic properties. The modern popularity of these systems, much of the original analysis, and the name Sigma–Delta modulator is due to Candy and his colleagues [6]–[10].

One might hope that the deterministic dither might indeed yield a white quantization noise process, but unfortunately this circular argument does not hold for the simple system of Figs. 8 or 9, as will be seen.

Since $u_n = q(u_n) - \epsilon_n$, (60) yields the difference equation

$$q(u_n) = x_{n-1} + \epsilon_n - \epsilon_{n-1}, \quad (61)$$

which has the intuitive interpretation that the quantizer output can be written as the input signal (delayed) plus a difference (or discrete time derivative) of an error signal. The hope is that this difference will be a high frequency term that can be removed by low pass filtering to obtain the original signal. For convenience we assume that $u_0 = 0$ and we normalize the previous terms by $\Delta$ and use the definition of $\epsilon_n$ to write

$$e_n = \frac{\epsilon_n}{\Delta} = \frac{q(u_n) - u_n}{\Delta} = \frac{q(x_{n-1} - \epsilon_{n-1})}{\Delta} - \frac{x_{n-1}}{\Delta} - e_{n-1};$$
$$n = 1, 2, \cdots. \quad (62)$$

Since $u_0 = 0$, $\epsilon_0 = \Delta/2$.

We shall assume that the input range is $[-b, b)$, that is, $-b \le x_n < b$ for all $n$. Intuitively, we would like to make $\Delta$ small to keep the quantizer error small; but we dare not make it too small or the quantizer may overload ($M$ is considered fixed). In addition, we do not want to make $\Delta$ too large since then some of the quantizer levels may

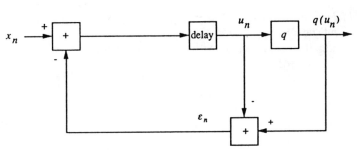

Fig. 8. Deterministic dithering.

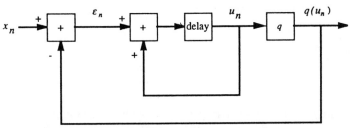

Fig. 9. Single-loop Sigma–Delta.

never be used and the bits will be wasted. The task is to pin down the relation between $\Delta$ and $b$, that is, given $b$, how do we choose $\Delta$? This is the key property of the system and is found using an induction argument. We wish to choose $\Delta$ so that the quantizer will never overload as long as the input stays within its range. A necessary condition is immediate: If $\epsilon_0 = \Delta/2$, then to ensure that $|u_1| \leq M\Delta/2$ we must have that

$$\left| x_0 - \frac{\Delta}{2} \right| \leq \frac{M\Delta}{2}$$

and therefore that

$$b + \frac{\Delta}{2} \leq \frac{M\Delta}{2}$$

or

$$b \leq (M-1)\frac{\Delta}{2}. \tag{63}$$

We now show that this condition is sufficient to ensure that the quantizer does not overload; that is, if $\Delta$ satisfies (63), then $|\epsilon_n| \leq \Delta/2$ for all $n$. We have already seen this is true for $n = 1$, so we proceed by induction. Assume that it is true for $k = 1, 2, \cdots, n-1$. If $u_n = x_{n-1} - \epsilon_{n-1}$ is in the no-overload region $[-M\Delta/2, M\Delta/2]$ then $|\epsilon_n| \leq \Delta/2$ by definition of no overload. Thus we need only prove that it is not possible for either $u_n > M\Delta/2$ or $u_n < -M\Delta/2$. We have, however, from the range of $x_{n-1}$ and the assumption on $\epsilon_{n-1}$ that

$$-b - \frac{\Delta}{2} \leq x_{n-1} - \epsilon_{n-1} \leq b + \frac{\Delta}{2}$$

and hence using (63) we have that

$$|u_n| = |x_{n-1} - \epsilon_{n-1}| \leq b + \frac{\Delta}{2} \leq \frac{M\Delta}{2}, \tag{64}$$

proving that the quantizer does not overload. Observe that the input bin for the largest quantizer level is $\{u: u \geq (M/2-1)\Delta\}$. Thus this level will be selected for $u_1 = x_0 - \Delta/2$ if $x_0 - \Delta/2 \geq (M/2-1)\Delta$ or $x_0 \geq (M/2 - 1/2)\Delta$. If $b < (M-1)\Delta/2$ this will never happen and the level will not be used. Combining this observation with (63) we are led to the following choice for $\Delta$:

$$\Delta = \frac{2b}{M-1}. \tag{65}$$

In the most common case of a binary quantizer, $\Delta = 2b$. We can now combine the fact that the quantizer does not overload with (5) for the quantizer error for a nonoverloading quantizer to obtain a recursion for the quantizer error sequence of a single-loop Sigma–Delta quantizer: $e_0 = 1/2$,

$$e_n = \frac{1}{2} - \left\langle \frac{u_n}{\Delta} \right\rangle = \frac{1}{2} - \left\langle \frac{x_{n-1}}{\Delta} - e_{n-1} \right\rangle; \qquad n = 1, 2, \cdots. \tag{66}$$

This formula can now be solved by induction so that the $e_n$ terms on the right are eliminated. It is simpler to express the recursion in terms of

$$y_n = \frac{1}{2} - e_n = \left\langle \frac{x_{n-1}}{\Delta} + \frac{1}{2} + y_{n-1} \right\rangle,$$

the solution to which is

$$y_0 = 0,$$

$$y_n = \left\langle \sum_{k=0}^{n-1} \left( \frac{1}{2} + \frac{x_k}{\Delta} \right) \right\rangle = \left\langle \frac{n}{2} + \sum_{k=0}^{n-1} \frac{x_k}{\Delta} \right\rangle; \qquad n = 1, 2, \cdots \tag{67}$$

as is easily verified by induction. Thus the quantizer error sequence is given by

$$e_n = \frac{1}{2} - \left\langle \frac{n}{2} + \sum_{k=0}^{n-1} \frac{x_k}{\Delta} \right\rangle, \tag{68}$$

which provides a comparison with the corresponding expression for the quantizer error sequence of an ordinary uniform quantizer operating on an input sequence $u_n$ of (5), that is,

$$e_n = \frac{1}{2} - \left\langle \frac{u_n}{\Delta} \right\rangle.$$

Thus when the quantizer is put into a feedback loop with an integrator, the overall effect is to integrate the input plus a constant bias before taking the fractional part. The overall nonlinear feedback loop therefore appears as an affine operation (linear plus a bias) on the input followed by a memoryless nonlinearity. Furthermore, the techniques used to find the time average moments for $e_n$ in the memoryless quantizer case can now be used by replacing $u_n$ by the sum

$$s_n = \sum_{k=0}^{n-1} \left( \frac{1}{2} + \frac{x_k}{\Delta} \right), \tag{69}$$

evaluating the characteristic functions of (20)–(21) and applying (22)–(24) for $\overline{\Phi}_s$ instead of $\overline{\Phi}_u$. Thus

$$\overline{\Phi}_s(l) = \overline{E}\{e^{\pi jln} e^{2\pi jl\sum_{i=0}^{n-1} x_i}\} \tag{70}$$

$$\overline{\Phi}_s^{(k)}(i,l) = e^{\pi jlk} \overline{E}\{e^{\pi j(i+l)n} e^{2\pi j(i+l)\sum_{m=0}^{n-1} x_m} e^{2\pi jl\sum_{m=n}^{n-1+k} x_m}\}. \tag{71}$$

Observe for later use that when $i = -l$, (71) simplifies to

$$\overline{\Phi}_s^{(k)}(-l,l) = e^{\pi jlk} \overline{E}\{e^{2\pi jl\sum_{m=n}^{n-1+k} x_m}\}. \tag{72}$$

We now evaluate these expressions for the special cases of deterministic dc and sinusoidal inputs.

### dc Inputs

Suppose that $x_k = x$ for all $k$, where $-b \leq x < b$ is an irrational number. A dc input can be considered as an approximation to a slowly varying input, that is, to the case where the Sigma–Delta modulator has a large oversampling ratio. In this case we replace $u_n$ by $s_n = n\beta$, where $\beta = (1/2 + x/\Delta)$, in (20)–(21) and evaluate

$$\overline{\Phi}_s(l) = \lim_{N \to \infty} \frac{1}{N} \sum_{n=1}^{N} e^{j2\pi l s_n} = \lim_{N \to \infty} \frac{1}{N} \sum_{n=1}^{N} e^{j2\pi l n\beta}$$

$$= \lim_{N \to \infty} \frac{1}{N} \sum_{n=1}^{N} e^{j2\pi \langle ln\beta \rangle}. \tag{73}$$

From (34) this is

$$\overline{\Phi}_s(l) = \int_0^1 du\, e^{j2\pi u} = \begin{cases} 0; & l \neq 0 \\ 1; & l = 0. \end{cases} \tag{74}$$

Similarly,

$$\overline{\Phi}_s^{(k)}(i,l) = \lim_{N \to \infty} \frac{1}{N} \sum_{n=1}^{N} e^{j2\pi(is_n + ls_{n+k})}$$

$$= \lim_{N \to \infty} \frac{1}{N} \sum_{n=1}^{N} e^{j2\pi(in\beta + l(n+k)\beta)}$$

$$= e^{2\pi lk\beta} \lim_{N \to \infty} \frac{1}{N} \sum_{n=1}^{N} e^{j2\pi \langle n(i\beta + l\beta) \rangle}$$

$$= \begin{cases} e^{2\pi lk\beta}; & i = -l \\ e^{2\pi lk\beta} \int_0^1 du\, e^{j2\pi u} = 0; & \text{otherwise}, \end{cases} \tag{75}$$

which is hence a special case of (72). Thus we have from (22)–(23) that

$$\overline{E}\{e_n\} = 0 \tag{76}$$

$$\overline{E}\{e_n^2\} = \frac{1}{12}, \tag{77}$$

which agrees with the uniform noise approximation, that is, these are exactly the time average moments one would expect with a sequence of uniform random variables. The second order properties, however, are quite different. From (24) and 1.443.3 of [57]

$$R_e(k) = \sum_{l \neq 0} \left( \frac{1}{2\pi l} \right)^2 e^{j2\pi lk\beta} = \frac{1}{2} \frac{1}{\pi^2} \sum_{l=1}^{\infty} \frac{\cos(2\pi lk\beta)}{l^2} \tag{78}$$

$$= \frac{1}{12} - \frac{\langle k\beta \rangle}{2}(1 - \langle k\beta \rangle). \tag{79}$$

This does not correspond to a white process. The exponential expansion implies that the spectrum is purely discrete having amplitude

$$S_n = \begin{cases} 0; & \text{if } n = 0 \\ \dfrac{1}{(2\pi n)^2}; & \text{if } n \neq 0 \end{cases} \tag{80}$$

at frequencies $\langle n\beta \rangle = \langle n(1/2 + x/\Delta) \rangle$. Thus the locations and hence the amplitude of the quantizer error spectrum depends strongly on the value of the input signal. Thus as in the simple PCM case with a sinusoidal input, the Bennett and describing function white noise approximations inaccurately predict the spectral nature of the quantizer noise process, which is neither continuous nor white. Figs. 10–17 provide examples of error signals

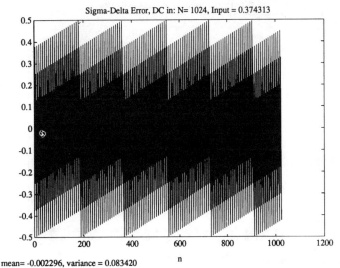

mean= -0.002296, variance = 0.083420

Fig. 10.   Quantizer error: dc input.

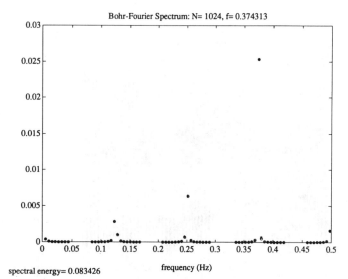

spectral energy= 0.083426

Fig. 11.   Quantizer spectra: dc input.

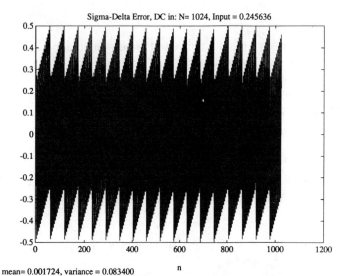

mean= 0.001724, variance = 0.083400

Fig. 12.   Quantizer error: dc input.

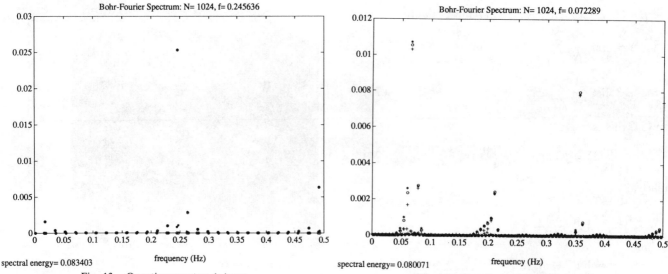

Fig. 13.   Quantizer spectra: dc input.

spectral energy= 0.083403

Fig. 15.   Quantizer spectra: Sin input.

spectral energy= 0.080071

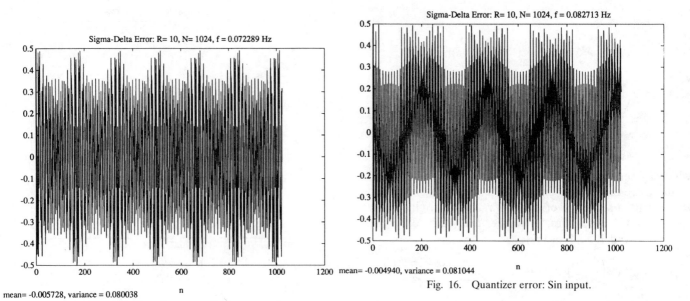

mean= -0.005728, variance = 0.080038

Fig. 14.   Quantizer error: Sin input.

mean= -0.004940, variance = 0.081044

Fig. 16.   Quantizer error: Sin input.

and their spectra for dc inputs and binary quantizers. As before, the circles denote the theoretically predicted spectral coefficients, the asterisks denote the computed exponential coefficients (Bohr–Fourier coefficients) at the same frequencies, and the pluses denote the FFT.

*Sinusoidal Inputs*

We now consider a more "active" input and set $x_n = A \cos n\omega_0$. We assume that $|A| \leq b$. From p. 30 of [57]

$$\sum_{k=0}^{n} \cos k\omega_0 = \frac{1}{2} + \frac{\sin\left(n\omega_0 + \frac{\omega_0}{2}\right)}{2\sin\frac{\omega_0}{2}}$$

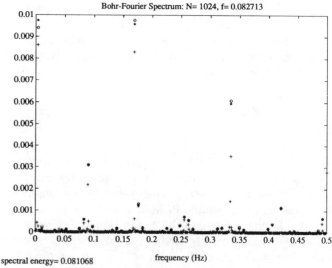

spectral energy= 0.081068

Fig. 17.   Quantizer spectra: Sin input.

and therefore

$$s_n = \sum_{i=0}^{n-1} \left( \frac{1}{2} + \frac{x_i}{\Delta} \right) = \frac{n}{2} + \frac{A}{2\Delta} + \alpha \sin\left( n\omega_0 - \frac{\omega_0}{2} \right),$$

where we have defined

$$\alpha = \frac{A}{2\Delta \sin \frac{\omega_0}{2}} = \frac{\gamma}{2 \sin \frac{\omega_0}{2}}.$$

Note the appearance of the sinusoid, which suggests that the final answers will be similar to those for the simple PCM case with a sinusoidal input. The answer will be complicated, however, by the additional $n/2$ term that will disappear inside the fractional operator when $n$ is even and add $1/2$ when $n$ is odd. The one-dimensional characteristic function is given by

$$\overline{\Phi}_u(l) = e^{j\pi l\gamma} \lim_{N \to \infty} \frac{1}{N} \sum_{n=1}^{N} e^{j\pi ln} e^{j2\pi l\alpha \sin(n\omega_0 - (\omega_0/2))}.$$

If $l$ is an even number, then $e^{j\pi ln} = 1$ and

$$\overline{\Phi}_u(l) = e^{j\pi l\gamma} \lim_{N \to \infty} \frac{1}{N} \sum_{n=1}^{N} e^{j2\pi l\alpha \sin(n\omega_0 - (\omega_0/2))}$$

$$= e^{j\pi l\gamma} \lim_{N \to \infty} \frac{1}{N} \sum_{n=1}^{N} e^{j2\pi l\alpha \sin 2\pi \langle n(\omega_0/2\pi) - (1/2)(\omega_0/2\pi) \rangle}.$$

Applying (34) yields

$$\overline{\Phi}_u(l) = e^{j\pi l\gamma} \int_0^1 du\, e^{j2\pi l\alpha \sin 2\pi u} = e^{j\pi l\gamma} J_0(2\pi l\alpha),$$

which bears a strong resemblance to the corresponding result (35) for the memoryless quantizer. The only difference is the lead term depending on the maximum input value and the replacement of $\gamma$ by $\alpha$. If $l$ is odd, application of (34) shows that $\overline{\Phi}_s(l) = 0$. In summary,

$$\overline{\Phi}_s(l) = \begin{cases} 0; & l \text{ odd} \\ e^{j\pi l\gamma} J_0(2\pi l\alpha); & l \text{ even.} \end{cases} \tag{81}$$

The two-dimensional characteristic function is somewhat messier. Some algebra and (34) yield

$$\overline{\Phi}_s^{(k)}(i,l) = \begin{cases} 0; & i+l \text{ odd} \\ e^{j\pi(i+l)\gamma} e^{j\pi lk} \int_0^1 du\, e^{j2\pi\alpha[i\sin 2\pi u + l\sin(2\pi u + k\omega_0)]}; & i+l \text{ even.} \end{cases}$$

Analogous to the development for the memoryless quantizer, the Jacobi–Anger formula can be inserted, the orthogonality of complex exponentials on the unit circle invoked, and the properties of Bessel functions used to find

$$\overline{\Phi}_s^{(k)}(i,l) = \begin{cases} 0; & i+l \text{ odd} \\ e^{j\pi(i+l)\gamma} e^{j\pi lk} \sum_m e^{jmk\omega_0} (-1)^m J_m(2\pi\alpha i) J_m(2\pi\alpha l); & i+l \text{ even.} \end{cases} \tag{82}$$

Thus from (14)–(22)

$$\overline{m}_e = \sum_{l \neq 0, l \text{ even}} \frac{1}{2\pi jl} e^{j\pi l\gamma} J_0(2\pi l\alpha) = \sum_{l=1}^{\infty} \frac{1}{2\pi l} \sin(2\pi l\gamma) J_0(4\pi l\alpha), \tag{83}$$

$$R_e(0) = \frac{1}{12} + \sum_{l \neq 0, l \text{ even}} \frac{1}{2(\pi l)^2} e^{j\pi l\gamma} J_0(2\pi l\alpha) = \frac{1}{12} + \sum_{l=1}^{\infty} \frac{1}{(\pi 2l)^2} \cos(2\pi l\gamma) J_0(4\pi l\alpha), \tag{84}$$

$$R_e(k) = -\frac{1}{(2\pi)^2} \sum_{m=-\infty}^{\infty} e^{jmk\omega_0} \left[ (-1)^m \sum_{i,l: i \neq 0, l \neq 0, i+l \text{ even}} \left( \frac{1}{i} e^{j\pi i\gamma} J_m(2\pi\alpha i) \right) \right.$$

$$\left. \cdot \left( \frac{1}{l} e^{j\pi l\gamma} J_m(2\pi\alpha l) e^{j\pi lk} \right) \right]. \tag{85}$$

The sum over $i$ and $l$ with even sum can be broken into separate sums with $i$ and $l$ both even or both odd. This eventually yields

$$R_y(k) = \sum_{m=-\infty}^{\infty} e^{jkm\omega_0} (-1)^m \left( c_e(m)^2 + (-1)^k c_o(m)^2 \right)$$

$$= \sum_{m=-\infty}^{\infty} e^{jkm\omega_0} (-1)^m c_e(m)^2 + \sum_{m=-\infty}^{\infty} e^{jk(m\omega_0 - \pi)} (-1)^m c_o(m)^2 \tag{86}$$

where

$$c_e(m) = \begin{cases} \dfrac{1}{2} - \dfrac{1}{\pi} \displaystyle\sum_{l=1}^{\infty} \dfrac{J_0(4\pi\alpha l)}{2l} \sin(2\pi l\gamma); & m = 0 \\[2ex] -\dfrac{1}{\pi} \displaystyle\sum_{l=1}^{\infty} \dfrac{J_m(4\pi\alpha l)}{2l} \sin(2\pi l\gamma); & m \text{ even} \\[2ex] \dfrac{j}{\pi} \displaystyle\sum_{l=1}^{\infty} \dfrac{J_m(4\pi\alpha l)}{2l} \cos(2\pi l\gamma); & m \text{ odd} \end{cases} \tag{87}$$

$$c_o(m) = \begin{cases} -\dfrac{1}{\pi} \displaystyle\sum_{l=1}^{\infty} \dfrac{J_m(2\pi\alpha(2l-1))}{2l-1} \sin\pi(2l-1)\gamma; & m \text{ even} \\[2ex] \dfrac{j}{\pi} \displaystyle\sum_{l=1}^{\infty} \dfrac{J_m(2\pi\alpha(2l-1))}{2l-1} \cos\pi(2l-1)\gamma; & m \text{ odd}. \end{cases} \tag{88}$$

Note that $c_e(0) = M\{y_n\}$.

Equations (86)–(88) give the sample autocorrelation function of the sequence $y_n$. This expression can be written in the form

$$R_y(k) = \sum_{l=-\infty}^{\infty} S_l e^{j2\pi k\lambda_l}. \tag{89}$$

The easiest means of indexing the frequencies and spectral amplitudes is to consider the indices $l$ in (87) to have the form $l = (m, i)$; $m = \cdots, -1, 0, 1, \cdots$, $i = 1, 2$, and

$$S_{(m,1)} = (-1)^m c_e(m)^2 \tag{90}$$

$$S_{(m,2)} = (-1)^m c_o(m)^2 \tag{91}$$

$$\lambda_{(m,1)} = \left\langle m\dfrac{\omega_0}{2\pi} \right\rangle \tag{92}$$

$$\lambda_{(m,2)} = \left\langle m\dfrac{\omega_0}{2\pi} - \dfrac{1}{2} \right\rangle. \tag{93}$$

Equation (89) defines the Bohr–Fourier series of the sequence $R_y(k)$. As in the PCM case, the spectrum of $y_n$ is purely discrete and has amplitude $s_l$ at the frequency $\lambda_l$. This spectrum is extremely nonwhite since it is not continuous and not flat. The output frequencies depend on the input frequency $\omega_0$ and comprise all harmonics of the input frequency $\omega_0$ as one might expect with a nonlinear device. It is interesting to observe that not only are all harmonics of the input frequency contained in the output signal, but also all shifts of these harmonics by $\pi$ (when computed in radians). These shifted harmonics are not present in the PCM case.

These results simplify in the special case where $A = b$, that is, a full scale sinusoid. In this case we have

$$\bar{m}_e = 0, \tag{94}$$

$$R_e(0) = \dfrac{1}{12} - \sum_{l=1}^{\infty} \dfrac{1}{(\pi 2l)^2} (-1)^l J_0(4\pi l\alpha), \tag{95}$$

$$c_e(m) = \begin{cases} \dfrac{1}{2}; & m = 0 \\[2ex] 0; & m \text{ even} \\[2ex] \dfrac{j}{\pi} \displaystyle\sum_{l=1}^{\infty} \dfrac{J_m(4\pi\alpha l)}{2l} (-1)^l; & m \text{ odd}, \end{cases} \tag{96}$$

$$c_o(m) = \begin{cases} \dfrac{1}{\pi} \displaystyle\sum_{l=1}^{\infty} \dfrac{J_m(2\pi\alpha(2l-1))}{2l-1} (-1)^l; & m \text{ even} \\[2ex] 0; & m \text{ odd}, \end{cases} \tag{97}$$

and hence

$$R_y(k) = \sum_{m=-\infty}^{\infty} S_m e^{j2\pi k\lambda_m}, \tag{98}$$

where

$$S_m = \begin{cases} \dfrac{1}{2}; & m = 0 \\[2ex] \left( \dfrac{1}{\pi} \displaystyle\sum_{l=1}^{\infty} \dfrac{J_m(2\pi\alpha(2l-1))}{2l-1} (-1)^l \right)^2; & m \text{ even} \\[2ex] \left( \dfrac{1}{\pi} \displaystyle\sum_{l=1}^{\infty} \dfrac{J_m(4\pi\alpha l)}{2l} (-1)^l \right)^2; & m \text{ odd} \end{cases} \tag{99}$$

98

and

$$\lambda_m = \begin{cases} \left\langle m\dfrac{\omega_0}{2\pi} - \dfrac{1}{2} \right\rangle; & m \text{ even} \\[2ex] \left\langle m\dfrac{\omega_0}{2\pi} \right\rangle; & m \text{ odd.} \end{cases} \quad (100)$$

It is likely that finite sums can be found for $S_m$ in the way that (40) was found from (37) and (50) was found from (45). The algebra is sufficiently messy and the results obtained by truncating (99) sufficiently good, however, that the author has not had the patience and stamina to derive the details. Figs. 14–17 provide examples of error signals and their spectra for sinusoidal inputs and binary quantizers.

Next consider the Sigma–Delta modulator with an input of the form

$$x_n = v_n + w_n, \quad (101)$$

where $w_n$ is an i.i.d. process that is independent of the quasi-stationary process $v_n$. This is again a form of dither, but now the i.i.d. signal is being added to the original signal rather than to the quantizer input. Denote by $\sigma_n$ the new sum process

$$\sigma_n = \sum_{k=0}^{n-1} \left( \frac{1}{2} + \frac{v_k}{\Delta} + \frac{w_k}{\Delta} \right), \quad (102)$$

and let $s_n$ denote the sum process in the absence of the dither,

$$s_n = \sum_{k=0}^{n-1} \left( \frac{1}{2} + \frac{v_k}{\Delta} \right). \quad (103)$$

We assume that the maximum amplitude of $x_n$ is, as before, no greater than $|b|$. As in the PCM dither analysis, we have since the dither process is independent of the signal that

$$\overline{\Phi}_\sigma(l) = \lim_{N \to \infty} \frac{1}{N} \sum_{n=1}^{N} E\left( e^{2\pi j l \sum_{k=0}^{n-1}(\frac{1}{2}+v_k)} \Phi_w\left( j2\pi\frac{l}{\Delta} \right)^n \right). \quad (104)$$

If $w_n$ does not have a lattice distribution (that is, if it does not have a discrete distribution on a lattice), then $\Phi_w(ju) < 1$ for $u \neq 0$ [48] and hence the terms in the previous sum go to 0 and

$$\Phi_\sigma(l) = 0; \quad l \neq 0, \quad (105)$$

which, as we have seen, implies a mean of 0 and $R_e(0) = 1/12$. The two-dimensional characteristic function is given by

$$\Phi_\sigma^{(k)}(i,l) = \lim_{N \to \infty} \frac{1}{N} \sum_{n=1}^{N} e^{2\pi j(i+l)\sum_{m=0}^{n-1}(\frac{1}{2}+v_m)}$$
$$\cdot e^{2\pi j l \sum_{m=n}^{n+k}(\frac{1}{2}+v_m)} \Phi_w\left( 2\pi\frac{i+l}{\Delta} \right)^n \Phi_w\left( 2\pi\frac{l}{\Delta} \right)^k. \quad (106)$$

If the distribution of $w_n$ is not a lattice distribution, then

unless $i = -l$ the terms in the above sum will tend to 0 as $n \to \infty$ and hence the limit will be 0. Thus for $k \neq 0$,

$$\Phi_\sigma^{(k)}(i,l) = \begin{cases} 0; & i \neq -l \\ \Phi_w\left(2\pi\frac{l}{\Delta}\right)^k \overline{\Phi}_s^{(k)}(-l,l); & i = -l. \end{cases} \quad (107)$$

Thus from (24), for $k \neq 0$

$$R_e(k) = \sum_{l \neq 0} \left( \frac{1}{2\pi l} \right)^2 \Phi_w^k\left(2\pi\frac{l}{\Delta}\right) \overline{\Phi}_s^{(k)}(-l,l). \quad (108)$$

This can be evaluated from the formulas for $\overline{\Phi}_s^{(k)}(-l,l)$ already found. In the case of a dc signal, we have from (75) that

$$R_e(k) = \sum_{l \neq 0} \left( \frac{1}{2\pi l} \right)^2 \Phi_w^k\left(2\pi\frac{l}{\Delta}\right) e^{2\pi l k \beta}, \quad (109)$$

which is consistent with the results in [48].

Observe that if $w_n$ is not a lattice distribution and hence if $|\Phi_w(v)| < 1$, then $R_e(k)$ is absolutely summable and hence its discrete-time Fourier transform converges to something finite, that is, the spectrum is purely continuous and does not have discrete components. Thus the addition of white noise to the input does result in a smooth spectrum. Analogous to the PCM dither case, the quantization noise would be white if the Schuchman condition could be satisfied and the $\Phi_w(2\pi l/\Delta)$ set to 0 for $l \neq 0$, but this cannot be accomplished in the Sigma–Delta example without increasing the input range past the allowed limits, that is, without overloading the quantizer.

Similar techniques can be used to consider the cross-correlation between the quantizer error and the input. As with the dithered PCM case, the "input" in question is usually taken to be the original system input, not the input to the quantizer itself. Before considering the cross-correlation, however, we consider its usefulness by considering the relationship between the quantizer error and the overall error. Recall (61):

$$q(u_n) = x_{n-1} + \epsilon_n - \epsilon_{n-1}.$$

As previously described, the quantizer output can be considered to be the input (delayed) plus the difference of two quantizer noise terms. Although these noise terms are in general dependent on each other and on the input, the intuition is that they behave individually like white noise and hence the difference behaves like high frequency noise. Thus the reproduction for $x_n$ is reasonably obtained by passing $q(u_n)$ through a low pass filter (or approximation thereto), the idea being that $x_n$ passes through relatively unscathed while the high frequency noise will be removed. If the reconstruction filter has pulse response $h_k$ and the oversampling ratio is $R$, then the filtered and downsampled reproduction can be expressed as

$$\hat{x}_{lN} = \sum_{i=0}^{\infty} q(u_{lN-i}) h_i. \quad (110)$$

One could measure the performance of the overall system

in terms of the average squared error $\hat{x}_{lN} - x_{lN}$, but this is somewhat misleading because the signal term in $q(u_n)$ has been passed through the filter $h_k$, which is assumed not to have much effect on the original signal $x_{lN}$. In particular, if $x_n$ is low-pass with the same bandwidth as the filter, then it should not be changed. Either by assuming that the filter has little effect on the signal or by simply admitting that it is the filtered signal which should be considered the ideal, a reasonable performance measure is the average squared error between the filtered quantizer output and the filtered signal, delayed by 1, that is, the average energy of the signal

$$\delta_{lN} = \hat{x}_{lN} - x_{lN} = \sum_{i=0}^{\infty} q(u_{lN-i})h_i - \sum_{i=0}^{\infty} x_{lN-i-1}h_i$$

$$= \sum_{i=0}^{\infty} \left( q(u_{lN-i}) - x_{lN-i-1} \right)h_i$$

$$= \sum_{i=0}^{\infty} \left( \epsilon_{lN-i} - \epsilon_{lN-i-1} \right)h_i.$$

The point is that the average of this signal will not depend explicitly on the crosscorrelation between the quantizer error and the original input (or the quantizer input), its form depends only on the first and second order moments of the quantizer error process. It is for this reason that crosscorrelations are only of secondary interest here, and it is the first- and second-order moments of the quantizer error process alone that are the primary consideration. Similar decompositions of the modulator output into a delayed signal and quantizer noise terms will hold for other Sigma–Delta architectures to be seen later.

Steps similar to those used for the characteristic function evaluations yield

$$\overline{E}\{x_n\epsilon_{n+k}\} = \sum_{l \neq 0} \frac{1}{2\pi jl} \lim_{N \to \infty} - \sum_{n=1}^{N} x_n e^{j\pi l(n+k)} e^{j2\pi l \Sigma_{i=0}^{n+k-1} x_i}.$$

In the case of a dc signal, the crosscorrelation is easily found to be 0 since all of the limiting terms are 0 (they are the same as $\overline{\phi}_s(l)$ for $l \neq 0$). Since $\overline{E}\epsilon_n = 0$, this implies that in this case the quantization noise and the input are in fact asymptotically uncorrelated. When the input is a sinusoid the formula can be evaluated with some effort and the input and quantizer noise seem to not be asymptotically uncorrelated.

## VI.  TWO-STAGE SIGMA–DELTA MODULATION

We now turn to the two-stage or cascaded Sigma–Delta modulator of Fig. 18. Here the error sequence $\zeta_n$ from the first loop is the input to the second loop, where the quantizer error is denoted $\epsilon_n$. From (60) the difference equations describing integrator states are

$$v_n = v_{n-1} + x_{n-1} - q(v_{n-1}); \qquad n = 1, 2, \cdots, \quad (111)$$

$$u_n = u_{n-1} + \zeta_{n-1} - q(u_{n-1}); \qquad n = 1, 2, \cdots. \quad (112)$$

The quantizers are exactly as in Section III, that is, uniform quantizers with $M$ levels and bin width $\Delta = \Delta/(M-1)$, where the input $x_n$ is between $-b$ and $b$. Note that this means that the first quantizer does not overload and hence $\zeta_n$ is in the range $[-b, b)$ and hence also the second quantizer does not overload. The output

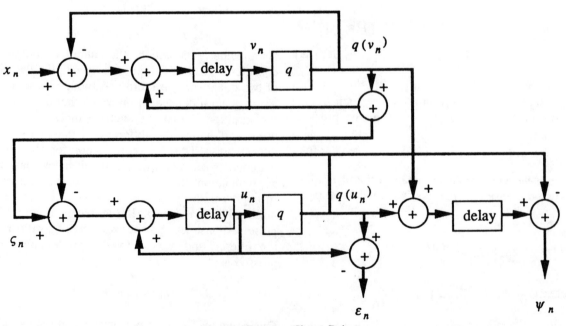

Fig. 18.  Two-stage Sigma–Delta.

of the two-stage sigma–delta modulator is defined by

$$\psi_n = q(v_{n-1}) - q(u_n) + q(u_{n-1}); \qquad n = 1, 2, \cdots,$$
(113)

a linear combination of the two quantizer outputs. As in (61) we have that

$$q(v_n) = x_{n-1} + \zeta_n - \zeta_{n-1},$$

$$q(u_n) = \zeta_{n-1} + \epsilon_n - \epsilon_{n-1}$$

and hence (113) becomes

$$\psi_n = x_{n-2} + \epsilon_n - 2\epsilon_{n-1} + \epsilon_{n-2}.$$
(114)

In contrast to (61), this has the interpretation of being the original signal plus a second-order difference (discrete time second derivative) instead of the single-order difference of the single-loop system. Note that although this signal depends on the outputs of both stages, the first stage quantization noise cancels out and only the second-stage noise remains. This fact is basic to the operation of the system. In particular, we need only know the behavior of the second stage quantization noise in order to find the behavior of the output. We shall see that the second stage noise is better behaved than the first-stage noise.

Equation (61) suggests that a natural means of producing the reproduction signal $\hat{x}$ is to pass $\psi_n$ through a linear filter with pulse response $h_k$ and downsample. As in the single-loop case, if we measure performance by comparing this reproduction with the original input passed through the same filter, than the overall average squared error will be the average energy in the signal

$$\delta_{lN} = \sum_{i=0}^{\infty} (\epsilon_{lN-i} - 2\epsilon_{lN-1-i} + \epsilon_{lN-2-i}) h_i.$$
(115)

This average is determined entirely by the first- and second-order moments of the quantizer noise $\epsilon$ of the second stage quantizer and it is not explicitly dependent on the crosscorrelation of this noise with either the second stage or original input signal.

Assuming that initially the integrator states (quantizer inputs) are $u_0 = v_0 = 0$, then applying (67) to both stages gives for $n = 1, 2, \cdots$

$$p_n = \frac{\psi_n}{\Delta} = \left\langle \frac{1}{2} - \frac{n}{2} + \sum_{k=0}^{n-1} \frac{x_k}{\Delta} \right\rangle,$$

$$e_n = \frac{\epsilon_n}{\Delta} = \left\langle \frac{1}{2} - \frac{n}{2} + \sum_{k=0}^{n-1} p_k \right\rangle$$

$$= -\frac{1}{2} + \left\langle \sum_{k=0}^{n-1} \sum_{l=0}^{k-1} \left( \frac{1}{2} + \frac{x_l}{\Delta} \right) \right\rangle$$

$$= -\frac{1}{2} + \left\langle \sum_{l=0}^{n-1} l \left( \frac{1}{2} + \frac{x_{n-l}}{\Delta} \right) \right\rangle.$$
(116)

As in the ordinary Sigma–Delta modulator, we can modify (20)–(24) by replacing the quantizer input $u_n$ by a sum

term

$$s_n = \sum_{i=0}^{n-1} \sum_{l=0}^{i-1} \left( \frac{1}{2} + \frac{x_l}{\Delta} \right) = \sum_{l=0}^{n-1} l \left( \frac{1}{2} + \frac{x_{n-l}}{\Delta} \right)$$
(117)

and then proceed exactly as before. We here illustrate the results only for the simple case of a dc input $x_n = x$. Define $\beta = 1/2 + x_n / \Delta$ as before and we have that

$$s_n = \frac{\beta}{2} n^2 - \frac{\beta}{2} n.$$
(118)

To evaluate the characteristic functions, we use the general version of Weyl's theorem [61]: If

$$c(n) = a_0 + a_1 n + a_2 n^2 + \cdots + a_k n^k$$

is a polynomial with real coefficients of degree $n \geq 1$ and if at least one of the coefficients $a_i$ is irrational for $i \geq 1$, then for any Riemann integrable function $g$

$$\lim_{N \to \infty} \frac{1}{N} \sum_{n=1}^{N} \langle g(c(n)) \rangle = \int_0^1 g(u) \, du.$$
(119)

We now use (119) exactly as (34) was used previously. We have with $c(n) = s_n$ in (119) that

$$\overline{\Phi}_s(l) = \lim_{N \to \infty} \frac{1}{N} \sum_{n=1}^{N} e^{2\pi j \langle c(n) \rangle} = \int_0^1 e^{2\pi j u} \, du = \begin{cases} 0; & l \neq 0 \\ 1; & l = 0, \end{cases}$$
(120)

and

$$\overline{\Phi}_s^{(k)}(i, l) = \lim_{N \to \infty} \frac{1}{N} \sum_{n=1}^{N} e^{2\pi j \langle d(n) \rangle}$$

where

$$d(n) = (i+1) \frac{\beta}{2} n^2 + (2il - i - l) \frac{\beta}{2} n + il(l-1) \frac{\beta}{2}$$

whence

$$\overline{\Phi}_s^{(k)}(i, l) = \begin{cases} 1; & i = -l = 0 \\ \int_0^1 e^{2\pi j u} \, du = 0; & \text{otherwise.} \end{cases}$$
(121)

These characteristic functions are identical to those of the dithered PCM case in (57) and (58) and hence the conclusions are the same: the quantization noise is indeed white and its marginal first and second moments agree with those of a uniform distribution! Thus the Bennett model is a good approximation for the second-stage quantization noise of a two-stage Sigma–Delta modulator driven by a dc (and only the second stage noise proves to be important in SQNR analysis). It is perhaps surprising that such a purely deterministic system with a fixed dc input can produce a sequence that appears to be uniformly distributed white noise when its first- and second-order moments are measured. Such a case of a deterministic process masquerading as a random process might lead the reader to suspect a connection with "chaos theory," but it turns out that the system is not chaotic by the standard definition (its Lyapunov exponent is 0). A slight variation on the foregoing analysis can be used to prove that the second stage quantizer noise and the original input are asymptotically uncorrelated.

The analysis can be extended to the case of a sinusoidal input, but the analysis is much more complicated and the noise is not white [47] and it is not asymptotically uncorrelated with the input. The analysis can also be extended to an i.i.d. input, in which case the noise is white [48].

## VII. SECOND-ORDER SIGMA–DELTA MODULATION

Another Sigma–Delta system is the so-called second order multiloop Sigma–Delta depicted in Fig. 19, introduced by Candy [7] and first rigorously analyzed by He, Buzo, and Kuhlmann [44], [46]. The shaded portion corresponds to a single-loop Sigma–Delta modulator, which emphasizes the fact that the second-order loop can be considered as a first-order loop with the original input replaced by the integrated error between the input and the quantizer output. From this viewpoint, the second-order Sigma–Delta is equivalent to Cutler's quantizer with "two steps of error compensation" of Fig. 3 of his 1960 patent application [4] except for the location of the delay.

An alternative block diagram for the same system is shown in Fig. 20, which shows that the system can, like the ordinary single loop system, be considered as a deterministically dithered quantizer. The difference is that now the quantizer error is passed through a linear filter before being added to the quantizer input. (This equivalent circuit was pointed out to the author by Randy Nash.) From either figure, the basic nonlinear difference equation for the quantizer input process $u_n$ is found to be

$$u_n = x_{n-1} - 2\epsilon_{n-1} + \epsilon_{n-2}, \qquad (122)$$

where, as before, $\epsilon_n = q(u_n) - u_n$ is the quantizer noise. Observe that the output of the second-order Sigma–Delta modulator is

$$q(u_n) = \epsilon_n + u_n = \epsilon_n - 2\epsilon_{n-1} + \epsilon_{n-2} + x_{n-1}, \quad (123)$$

a relation that bears a remarkable resemblance to (114) for the output of the two-stage Sigma–Delta modulator (and hence is capable of the same interpretation).

A well-known difficulty with the second- (and higher) order Sigma–Delta modulators is their potential stability. In particular, if one uses a binary quantizer with levels $\pm b$ in a second-order system, then it is easy to find an input within the range $[-b, b)$, which will overload the quantizer and hence will be capable of producing large errors. The potential overload also has the serious consequence for our purposes that it renders invalid a basic

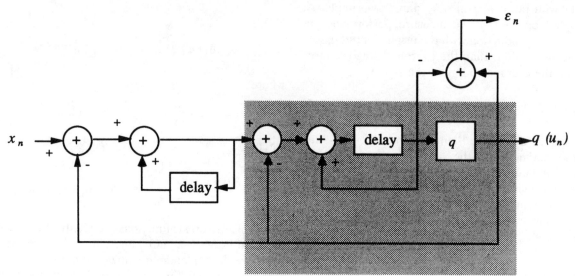

Fig. 19.  Second-order Sigma–Delta.

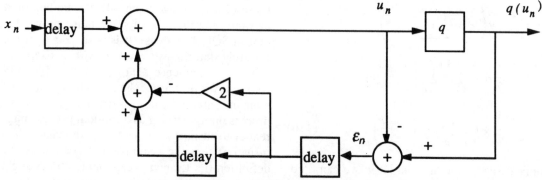

Fig. 20.  2nd-order loop: Deterministic dithering.

102

technique of the approach used here. No application of the techniques of this paper seems possible for the case of a binary quantizer, but the techniques do apply if we permit a two-bit (or higher) quantizer. To see this, we argue as in the choice of $\Delta$ in (32). Now suppose that the input maximum magnitude of $b$ is fixed and that we have $M$ quantization levels. As in the simple Sigma–Delta case, we want to choose $\Delta$ as small as possible (to minimize the quantization error) while still ensuring that no overload can occur, that is, the quantizer input $u_n$ must not exceed $M\Delta/2$ in magnitude and the quantizer error $\epsilon_n$ must not exceed $\Delta/2$. Abbreviating that argument, if the quantizer has not overloaded up to time $n-1$, then from (122) at worst the quantizer input at time $n$ will have magnitude no greater than $b+3\Delta/2$, which we require be less than $M\Delta/2$, that is, we need

$$b \leq (M-3)\frac{\Delta}{2}.$$

As in the ordinary Sigma–Delta case, choosing the smallest possible $\Delta$ with no overload yields

$$\Delta = 2\frac{b}{M-3}. \tag{124}$$

Clearly this result makes sense only if $M \geq 4$, that is, if the quantizer has at least two bits. We henceforth make this assumption and we now can apply the techniques of the previous section to this system. As before, consider the normalized process $e_n = \epsilon_n/\Delta$ and $y_n = 1/2 - e_n$. Some algebra reveals that the basic nonlinear difference equation is

$$y_n = \frac{x_{n-1}}{\Delta} + 2y_{n-1} - y_{n-2} - \frac{q(u_{n-1})}{\Delta}. \tag{125}$$

Analogous to (66)–(67) this is equivalent to the recursion

$$y_n = \left\langle \frac{x_{n-1}}{\Delta} + 2y_{n-1} - y_{n-2} + \frac{1}{2} \right\rangle \tag{126}$$

which has solution (provable by induction)

$$y_n = \left\langle \sum_{l=0}^{n-1} l\left( \frac{1}{2} + \frac{x_{n-l}}{\Delta} \right) \right\rangle. \tag{127}$$

This is *identical* to the solution (116) of the two-stage Sigma–Delta modulator with two one bit quantizers and hence the remaining analysis is also identical, proving that the quantizer noise for the second-order Sigma–Delta modulator with a two bit quantizer is also white.

## VIII. EXTENSIONS AND COMMENTS

The two stage Sigma–Delta results can be extended to multiple stages with binary quantizers, where it can be shown that dc inputs, sinusoidal inputs, and sums of sinusoidal inputs all yield white quantization noise [45]. Similarly, dithering multistage (two-stage or more) Sigma–Delta modulators with i.i.d. noise that does not cause overload also yields white quantization noise [48].

He, Buzo, and Kuhlman found the spectra for multi-bit higher-order Sigma–Delta modulators [44]. As with the second-order case considered here, the quantizer must have sufficient bits to avoid overload. They consider both dc and sinusoidal inputs. As with the two-stage and second-order systems considered here, the $M$-stage and $M$th order Sigma–Delta modulators yield the same quantization noise spectra.

The techniques provide a means of computing the spectra of quantizer noise whenever an expression can be found for the error sequence such that the one- and two-dimensional characteristic functions can be evaluated. Several examples where this is the case have been given. The results for PCM of single sinusoids can be generalized to multiple sinusoids, as was done by Clavier *et al.* The general result for dithered PCM is new, but the special case of stationary inputs follows from the work of Sripad and Snyder [33]. The results for simple Sigma–Delta modulation can easily be modified to hold for Delta modulation when the input slope is constrained (either by direct solution of the difference equations or by using the fact that Delta modulation is equivalent to Sigma–Delta modulation preceded by a difference operation). In this form the results were first found by Iwersen [62] and Misawa *et al.* [22] and a treatment along the lines taken here can be found in Section 6.7 of [56]. All of the systems developed and described have a key aspect in common that permitted solution: the linear filtering within the loop consisted only of ideal discrete time integrators, more complicated filters such as leaky integrators were not considered. A major open question is whether or not similar results can be developed for such more general systems. Kieffer [63]–[64] has extended the single-stage Sigma–Delta result to more general systems with dc inputs which include DPCM and leaky integrating Delta modulators, but his techniques are quite different and require deep ideas from the theory of Sturmian systems.

In the general case two issues become important: First, under what conditions is the system stable in the sense of not causing quantizer overload? Second, when can the nonlinear difference equations be solved exactly? An architecture of particular interest is the interpolative or higher order single-loop Sigma–Delta modulator of Fig. 21 where the ideal integrator is replaced by a general linear filter that includes at least one delay (to model the physical delay in the quantizer itself). One can use describing function theory and combine describing function theory with the white noise approximation to determine ordinary BIBO (bounded input/bounded output) stability and approximate the spectral behavior of the quantization noise, but it is known that this may give inaccurate results

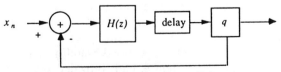

Fig. 21. Interpolative Sigma–Delta.

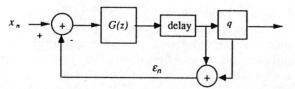

Fig. 22.   Interpolative Sigma–Delta: Deterministic dithering.

and yield systems that are unstable and have spikey noise spectra [65]. In fact, the no-overload stability condition found in the previous sections for specific examples is a much stronger condition than BIBO stability and it is not easily demonstrated in general. It is easy to find a simple sufficient condition for no-overload stability. If the feed-forward linear filter with causal transfer function $H(z)$ has pulse response $h_k$, then the system can also be depicted as in Fig. 16 as a form of deterministic dithered quantizer with a linear filter $G(z)$ in the feedforward loop, where

$$G(z) = \frac{H(z)}{1 + z^{-1}(H)z}. \qquad (128)$$

From Fig. 22 we have that

$$u_n = \sum_{k=0}^{n-1} g_k(x_{n-1-k} - \epsilon_{n-1-k}); \qquad n = 1, 2, \cdots. \quad (129)$$

A simple induction argument shows that the system is no-overload stable if

$$\sum_{k=0}^{\infty} |g_k| \le 1. \qquad (130)$$

A simple necessary condition for no-overload stability is that

$$|g_0| \le 1, \qquad (131)$$

since otherwise it is easy to construct an input for which the quantizer will overload. While (131) provides an easy test for eliminating nonstable filters, (130) seems too strong to be very useful. For example, simply replacing the integrator in a single-loop Sigma–Delta modulator with a leaky integrator yields a system that satisfies the necessary condition (131) with equality and hence violates the sufficient condition. Kieffer's results [63] (and earlier extensive simulations) indicate, however, that the system is no-overload stable. It is important for practical applications to predict stability of such a system (perhaps designed using the linearization techniques), but this still leaves open the problem of solving the difference equations and finding exact formulas for the noise spectra. This, too, is an open question. In particular, can $G(z)$ be chosen so that the quantization noise is white, or are more loops or stages necessary?

As a final observation, even though the second order Sigma–Delta modulator is known to not be no-overload stable if a binary quantizer is used, the system is still used. Lacking the property of no-overload stability does not necessarily mean the system is bad: provided the quantizer does not overload too often and the overload effects

are not catastrophic, the overall performance may still be good. The approach developed here depends in a crucial way on not overloading the quantizer; it is not clear whether the results can be modified to handle overload noise as well as granular noise.

## ACKNOWLEDGMENT

The author would like to thank Professor Yannis P. Tsividis for bringing the papers by Clavier, Panter, and Grieg to my attention, Professor Shlomo Schitz for bringing the transform method to my attention, Dr. Randy Nash for showing me the equivalent circuit for the second-order Sigma–Delta modulator, and my students Ping-Wah Wong and Wu Chou for their cooperation on much of the research reported herein and for their comments on this paper.

## REFERENCES

[1]  W. R. Bennett, "Spectra of quantized signals," Bell Syst. Tech. J., vol. 27, pp. 446–472, July 1948
[2]  A. G. Clavier, P. F. Panter, and D. D. Grieg, "Distortion in a pulse count modulation system," AIEE Trans., vol. 66, pp. 989–1005, 1947.
[3]  A. G. Clavier, P. F. Panter, and D. D. Grieg, "PCM distortion analysis," Elect. Eng., pp. 1110–1122, Nov. 1947.
[4]  C. C. Cutler, "Transmission systems employing quantization," 1960, U.S. Patent No. 2,927,962.
[5]  H. Inose and Y. Yasuda, "A unity bit coding method by negative feedback," Proc. IEEE, vol. 51, Nov. 1963, pp. 1524–1535.
[6]  J. C. Candy, "A use of limit cycle oscillations to obtain robust analog-to-digital converters," IEEE Trans. Commun., vol. COM-22, pp. 298–305, Mar. 1974.
[7]  ___, "A use of double integration in sigma–delta modulation," IEEE Trans. Commun., vol. COM-33, pp. 249–258, Mar. 1985.
[8]  ___, "Decimation for sigma–delta modulation," IEEE Trans. Commun., vol. COM-34, pp. 72–76, Jan. 1986.
[9]  J. C. Candy, Y. C. Ching, and D. S. Alexander, "Using triangularly weighted interpolation to get 13-bit PCM from a sigma–delta modulator," IEEE Trans. Commun., pp. 1268–1275, Nov. 1976.
[10] J. C. Candy and O. J. Benjamin, "The structure of quantization noise from sigma–delta modulation," IEEE Trans. Commun., vol. COM-29, pp. 1316–1323, Sept. 1981.
[11] K. Uchimura, T. Hayashi, T. Kimura, and A. Iwata, "VLSI-A to D and D to A converters with multi-stage noise shaping modulators," in Proc. 1986 ICASSP, Tokyo, 1986, pp. 1545–1548.
[12] Y. Matsuya, K. Uchimura, A. Iwata, T. Kobayashi, and M. Ishikawa, "A 16b oversampling conversion technology using triple integration noise shaping," in Proc. 1987 IEEE Int. Solid-State Circuits Conf., Feb. 1987, pp. 48–49.
[13] S. K. Tewksbury and R. W. Hallock, "Oversampled, linear predictive and noise-shaping coders of order n > 1," IEEE Trans. Circuits Syst., vol. CAS-25, pp. 436–447, July 1978.
[14] D. J. Goodman and L. J. Greenstein, "Quantizing noise of dm/pcm encoders," Bell Syst. Tech. J., vol. 52, pp. 183–204, Feb. 1973.
[15] W. L. Lee and C. G. Sodini, "A topology for higher order interpolative coders," Proc. ISCAS, pp. 459–462, May 1987.
[16] C. H. Giancarlo and C. G. Sodini, "A slope adaptive delta modulator for VLSI signal processing systems," IEEE Trans. Circuits Syst., vol. CAS-33, pp. 51–58, Jan. 1986.
[17] J. W. Scott, W. L. Lee, C. H. Giancarlo, and C. G. Sodini, "CMOS implementation of an immediately adaptive delta modulator," IEEE J. Solid-State Circuits, pp. 1088–1095, Dec. 1986.
[18] M. Vidyasagar, Nonlinear Systems Analysis. Englewood Cliffs, NJ: Prentice Hall, 1978.
[19] D. P. Atherton, Stability of Nonlinear Systems. Chichester: Research Studies Press, Wiley, 1981.
[20] D. P. Atherton, Nonlinear Control Engineering. New York: Van

Nostrand Theinhold, 1982.

[21] B. Widrow, "A study of rough amplitude quantization by means of Nyquist sampling theory," *IRE Trans. Circuit Theory*, vol. CT-3, pp. 266–276, 1956.

[22] T. Misawa, J. E. Iwersen, and J. G. Rush, "A single-chip CODEC with filters, architecture," *Int. Conf. Commun. Rec.*, vol. 1, pp. 30.5.1–30.5.6, June 1980.

[23] A. Gelbe and W. E. V. Velde, *Multiple-Input Describing Functions and Nonlinear Systems Design*. New York: McGraw-Hill, 1968.

[24] A. R. Bergens and R. L. Franks, "Justification of the describing function method," *SIAM J. Contr.*, vol. 9, pp. 568–589, 1971.

[25] J. R. C. Booton, "The analysis nonlinear control systems with random inputs," in *Proc. Symp. Nonlinear Circuit Anal.*, Polytechnic Institute of Brooklyn, Apr. 1953.

[26] ___, "Nonlinear control systems with statistical inputs," Dynamic Analysis and Control Laboratory Report No. 61, Mar. 1952.

[27] S. H. Ardalan and J. J. Paulos, "An analysis of nonlinear behavior in delta–sigma modulators," *IEEE Trans. Circuits Syst.*, vol. CAS-34, pp. 593–603, June 1987.

[28] D. S. Arnstein, "Quantization error in predictive coders," *IEEE Trans. Commun.*, vol. COM-23, pp. 423–429, Apr. 1975.

[29] D. Slepian, "On delta modulation," *Bell Syst. Tech. J.*, vol. 51, pp. 2101–2136, 1972.

[30] S. O. Rice, "Mathematical analysis of random noise," in *Selected Papers on Noise and Stochastic Processes*, N. Wax and N. Wax, Eds., pp. 133–294. New York: Dover, 1954. Reprinted from *Bell Syst. Tech. J.*, vol. 23, pp. 282–332, 1944 and vol. 24, pp. 46–156, 1945.

[31] W. B. Davenport and W. L. Root, *An Introduction to the Theory of Random Signals and Noise*. New York: McGraw-Hill, 1958.

[32] B. Widrow, "Statistical analysis of amplitude quantized sampled data systems," *Trans. Amer. Inst. Elect. Eng., Pt. II: Applications and Industry*, vol. 79, pp. 555–568, 1960.

[33] A. B. Sripad and D. L. Snyder, "A necessary and sufficient condition for quantization errors to be uniform and white," *IEEE Trans. Acoust. Speech Signal Processing*, vol. ASSP-25, pp. 442–448, Oct. 1977.

[34] T. A. C. M. Claasen and A. Jongepier, "Model for the power spectral density of quantization noise," *IEEE Trans. Acoust. Speech Signal Processing*, vol. ASSP-29, pp. 914–917, Aug. 1981.

[35] C. W. Barnes, B. N. Tran, and S. H. Leung, "On the statistics of fixed-point roundoff error," *IEEE Trans. Acoust. Speech Processing*, vol. ASSP-3, pp. 595–606, June 1985.

[36] I. Tokaji and C. W. Barnes, "Roundoff error statistics for a continuous range of multiplier coefficients," *IEEE Trans. Circuits Syst.*, vol. CAS-34, pp. 52–59, Jan. 1987.

[37] B. Hamel, "Contribution a l'etude mathematique des systems de reglage par tous-ou-rien," *C.E.M.V. Service Technique Aeronautique*, 1949.

[38] E. V. Bohn, "Stability margins and steady state oscillations in on-off feedback systems," *IRE Trans. Circuit Theory*, vol. CT-8, pp. 127–130, 1961.

[39] J. K.-C. Chung and D. P. Atherton, "The determination of periodic modes in relay systems using the state space approach," *Int. J. Contr.*, vol. 4, pp. 105–126, 1966.

[40] J. A. Tsypkin, *Theorie der relais systeme der automatschen regelung*. Munich: R. Oldenbourg-Verlag, 1958.

[41] R. M. Gray, "Oversampled sigma–delta modulation," *IEEE Trans. Commun.*, vol. COM-35, pp. 481–489, Apr. 1987.

[42] ___, "Spectral analysis of quantization noise in a single-loop sigma–delta modulator with dc input," *IEEE Trans. Commun.*,

vol. COM-37, pp. 588–599, 1989.

[43] R. M. Gray, W. Chou, and P.-W. Wong, "Quantization noise in single-loop sigma–delta modulation with sinusoidal inputs," *IEEE Trans. Inform. Theory*, vol. 35, pp. 956–968, 1989.

[44] N. He, A. Buzo, and F. Kuhlmann, "A frequency domain waveform speech compression system based on product vector quantizers," in *Proc. ICASSP*, Tokyo, Japan, Apr. 1986.

[45] W. Chou, P.-W. Wong, and R. M. Gray, "Multistage sigma–delta modulation," *IEEE Trans. Inform. Theory*, vol. 35, pp. 784–796, 1989.

[46] N. He, A. Buzo, and F. Kuhlmann, "Double-loop sigma–delta modulation with dc input," *IEEE Trans. Commun.*, vol. COM-38, pp. 487–495, 1990.

[47] P.-W. Wong and R. M. Gray, "Two-stage sigma–delta modulation," to appear in *IEEE Trans. Acoust. Speech Signal Processing*, 1989.

[48] W. Chou and R. M. Gray, "Dithering and its effects on sigma–delta and multistage sigma–delta modulation," to appear in *IEEE Trans. Inform. Theory*.

[49] S. P. Lloyd, "Least squares quantization in pcm," unpublished Bell Laboratories technical note. Presented in part at the Institute of Mathematical Statistics Meeting, Atlantic City, NJ, Sept. 1957. Published in the special issue on quantization, *IEEE Trans. Inform. Theory*, Mar. 1982.

[50] J. Max, "Quantizing for minimum distortion," *IEEE Trans. Inform. Theory*, pp. 7–12, Mar. 1960.

[51] H. Bohr, *Almost Periodic Functions*, Harvey Cohn, Translator. New York: Chelsea, 1947.

[52] S. Bochner, "Beitrage zur Theorie der fastperiodischen Funktionen I, II," *Math. Ann.*, vol. 96, pp. 119–147, 1927.

[53] S. Bochner and J. von Neumann, "Almost periodic functions in groups, II," *Trans. Amer. Math. Soc.*, vol. 37, pp. 21–50, 1935.

[54] K. Petersen, *Ergodic Theory*. Cambridge: Cambridge Univ. Press, 1983.

[55] L. Ljung, *System Identification*. Englewood Cliffs, NJ: Prentice-Hall, 1987.

[56] R. M. Gray, *Probability, Random Processes, and Ergodic Properties*. New York: Springer-Verlag, 1988.

[57] I. S. Gradshteyn and I. M. Ryzhik, *Table of Integrals, Series, and Products*. New York: Academic Press, 1965.

[58] N. S. Jayant and P. Noll, *Digital Coding of Waveforms*. Englewood Cliffs, NJ: Prentice-Hall, 1984.

[59] L. Schuchman, "Dither signals and their effects on quantization noise," *IEEE Trans. Commun. Technol.*, vol. COM-12, pp. 162–165, Dec. 1964.

[60] R. C. Jaffee, "Causal and statistical analysis of dithered systems containing three level quantizer," MS. thesis, M.I.T., 1959.

[61] F. J. Hahn, "On affine transformations of compact abelian groups," *Amer. J. Math.*, vol. 85, pp. 428–446, 1963. Errata in vol. 86, pp. 463–464, 1964.

[62] J. E. Iwersen, "Calculated quantizing noise of single-integration delta-modulation coders," *Bell Syst. Tech. J.*, pp. 2359–2389, Sept. 1969.

[63] J. C. Kieffer, "Sturmian minimal systems associated with the iterates of certain functions on an interval," in *Proceedings of the Special Year on Dynamical Syst., Lecture Notes in Mathematics*. New York: Springer-Verlag, 1988.

[64] ___, "Analysis of DC input response for a class of one-bit feedback encoders," *IEEE Trans. Commun.*, vol. COM-38, pp. 337–341, 1990.

[65] G. Temes, "Third-order single-loop delta–sigma modulator design," unpublished note.

# Double-Loop Sigma-Delta Modulation with dc Input

NING HE, STUDENT MEMBER, IEEE, FEDERICO KUHLMANN, MEMBER, IEEE, AND
ANDRES BUZO, MEMBER, IEEE

*Abstract*—Through an exact analysis of a discrete-time model having a two-bit (2-b) quantizer, we provide rigorous answers to two fundamental questions for a double-loop sigma–delta modulation system with dc input.

1) What is the long-term statistical behavior of the internal quantizer noise?

2) How does the asymptotic mean-squared sigma–delta quantization error vary as a function of the oversampling ratio?

## I. Introduction

OVERSAMPLED sigma–delta ($\Sigma\Delta$) modulation is currently receiving increased attention as an attractive alternative to conventional analog-to-digital (A/D) converters. It is amenable for VLSI implementation because circuit precision requirements can be significantly relaxed by oversampling input waveforms, as well as by coarse quantization and digital decoding [1]–[5], [8], [9], [13].

Single-loop $\Sigma\Delta$ modulation [3], [8] is the simplest example of oversampled quantization systems where only binary quantization is used. In order to implement high-resolution A/D conversion, single-loop $\Sigma\Delta$ modulators must be designed to operate at very fast rates (more precisely, very large oversampling ratios). However, small oversampling ratios are generally desirable because for a given maximum bandwidth of VLSI technology, smaller oversampling ratios allow one to handle signals of wider spectra, and hence increase the number of possible applications. This limitation of single-loop $\Sigma\Delta$ modulation can be compensated by using multiloop $\Sigma\Delta$ modulators and more sophisticated decoding filters. Due to the more favorable performance versus oversampling ratio tradeoffs, we can achieve a prespecified resolution by using much lower oversampling ratios.

The performance superiority of multiloop $\Sigma\Delta$ modulators has been observed by numerous simulation and experimentation results. Unfortunately, there are very few comprehensive theoretical results in the literature [4], [5]. The complex behavior of such highly nonlinear systems could more easily be understood through an exact mathematical description of an ideal model, and theoretical predictions of system performance could be useful aids for designing practical systems. Specifically, such an analysis should provide rigorous answers to the following fundamental questions.

1) What is the long-term statistical behavior of the internal quantizer noise, i.e., long-term time averages such as sample mean, power, autocorrelation, power spectrum, and sample distribution?

2) How does the mean-squared $\Sigma\Delta$ quantization noise vary with the oversampling ratio $N$ for large $N$, and with the number of loops in the $\Sigma\Delta$ modulator?

It should also be possible to provide worthwhile insight into other important issues, such as instability suffered in multiloop $\Sigma\Delta$ modulators.

Paper approved by the Editor for Quantization, Speech/Image Coding of the IEEE Communications Society. Manuscript received November 5, 1987; revised March 3, 1989. This work was supported by the Mexican National Utility (CFE) and the Electrical Research Institute (IIE). This paper was presented in part at the 1988 IEEE International Conference on Acoustics, Speech, and Signal Processing, New York, NY, April 1988.

The authors are with the Graduate School of Engineering, National University of Mexico, P.O. Box 70-256, 04510, Mexico D.F., Mexico.

IEEE Log Number 9034748.

Motivated by the above ideas, we present a rigorous derivation of various basic properties of multiloop $\Sigma\Delta$ modulators, and in particular, answer the aforementioned questions.

There are basically two different approaches for analyzing $\Sigma\Delta$ modulators: the additive white noise source approximation, and the rigorous or direct techniques. In the first approach, one tries to approximate the quantizer noise by choosing an input-independent additive noise source having a similar long-term average behavior as the actual quantizer noise, e.g., the same long-term sample distribution and power spectrum. The simplest noise model is the white noise with a uniform distribution. Under such an approximation, the nonlinear $\Sigma\Delta$ modulator is modeled as a linear system, and the performance can easily be derived by using well-known linear system techniques. Some of the properties derived by using this approach agree reasonably well with simulation results [4], [5].

The basis of the rigorous techniques is contained in the important papers on this subject by Gray [8], [9]. Instead of assuming the memoryless and uniformity characteristics, this approach tries to derive the true quantizer noise behavior by solving a system of nonlinear difference equations, and then actually determines the long-term spectrum and distribution.

Exact analysis was first applied successfully to discrete-time single-loop $\Sigma\Delta$ modulators with dc input [8], [9], and the major conclusion is that the quantizer noise, even though uniformly distributed, is definitely *not* white. In fact, the quantizer noise and output of single-loop $\Sigma\Delta$ modulators have discrete power spectra, which consist of spectral spikes whose frequency location and weight depend in a complex way on the system input [9]. This fact suggests that the single-loop system is unfavorable because the input-dependent character makes the design of decoding filters a difficult matter; besides, the discrete nature of the spectra may be subjectively undesirable for certain applications such as digital audio.

While the actual behavior of the quantizer noise in the single-loop system does not agree with the "white noise" hypothesis in the approximated analysis, we will show, following the rigorous approach, that the quantizer noise in *multiloop* $\Sigma\Delta$ modulators with dc input is (in the sense defined later) white, uniformly distributed, and independent of the input, a fact long believed to be true by the A/D researchers, but not rigorously proved yet.

To simplify our analysis of these highly nonlinear systems, we have made two basic assumptions. The first is that the modulator input is a constant sequence, i.e., a dc input. While sinusoidal waveforms are also commonly used to test the system performance, the dc input represents a useful and reasonable idealization of more general slowly varying waveforms.

The other basic assumption is that there is *no overload* in the internal quantizer. This can be accomplished by

1) limiting the allowable dynamic range of the modulator input, and

2) using sufficient quantization levels to accommodate the quantizer's input amplitude.

While 1) can be easily done, 2) seems restrictive for all but a few simple multiloop systems. The major drawback of using multilevel quantizers is, of course, that tighter circuit tolerance is required for implementation. For this reason, most practical modulators usually have only two quantization levels, and consequently are nearly always overloaded.

Reprinted from *IEEE Trans. Commun.*, vol. COM-38, pp. 487–495, April 1990.

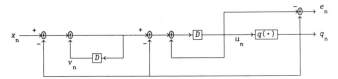

Fig. 1.  Discrete-time double-loop $\Sigma\Delta$ modulator. $D$: unit delay. $q(\cdot)$: quantizer.

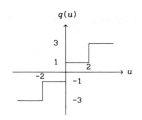

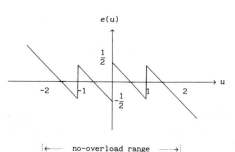

Fig. 2.  2-b uniform quantizer. (a) Input-output characteristic. (b) Quantizer error.

Unfortunately, the dc input and nonoverload assumptions we impose seem essential for this kind of analysis to work and for simple meaningful results to be derived. While for more general cases the problem still remains unsolved, the techniques used in this paper provide valuable quantitative predictions, as well as worthwhile insights into the performance tradeoffs involved in circuit design; these tradeoffs are obscured in more realistic and complicated situations.

In this paper, we will confine our attention to the special *double-loop* $\Sigma\Delta$ modulator with a *four-level* internal quantizer because it is the simplest multiloop modulator where these results and constraints can clearly be illustrated. The basic results developed for the double-loop system can be extended to general multiloop cases using essentially the same techniques, but the derivations are considerably more complex [12].

The paper is organized as follows. In Section II, we describe the discrete-time model and find an analytical expression for the quantizer noise sequence. Sections III and IV are devoted to answer the fundamental questions 1) and 2). We conclude the paper with some comments on possible future research directions.

## II. Discrete-Time Model

### A. Difference Equations

A discrete-time double-loop $\Sigma\Delta$ modulator is shown in Fig. 1 where $D$ represents a unit delay element and $q(\cdot)$ is a four-level (or 2-b) uniform quantizer (specified later).

From Fig. 1, one can write, by inspection, the following difference equations which describe the $\Sigma\Delta$ modulator:

$$v_{n+1} = x_n + v_n - q_n \qquad (2.1)$$

$$u_{n+1} = v_{n+1} + u_n - q_n \qquad (2.2)$$

where $x_n$ is the discrete-time input, $u_n$ and $v_n$ are the modulator states, and $q_n = q(u_n)$ is the quantizer output at time $n = 0, 1, 2, \cdots$.

It is desirable to handle only one basic process rather than several. A key process in our analysis is the quantizer error sequence defined by

$$e_n = e(u_n) = q(u_n) - u_n \qquad (2.3)$$

where the signs are chosen such that the quantizer output can be expressed as the sum of the quantizer input and a noise term, i.e.,

$$q_n = u_n + e_n. \qquad (2.4)$$

It is also convenient to define

$$d_n = q(u_n) - v_n. \qquad (2.5)$$

We find in Appendix A that the various processes can be expressed in terms of the input sequence and the error sequence only, as follows:

$$d_{n+1} = e_{n+1} - e_n \qquad (2.6)$$

$$q_{n+1} = x_n + e_{n+1} - 2e_n + e_{n-1} \qquad (2.7)$$

$$u_{n+1} = x_n - 2e_n + e_{n-1} \qquad (2.8)$$

and

$$v_{n+1} = x_n - e_n + e_{n-1}. \qquad (2.9)$$

In particular, the error sequence satisfies the following nonlinear equations:

$$e_{n+1} = e(x_n - 2e_n + e_{n-1}), \qquad n = 1, 2, \cdots \qquad (2.10)$$

$$e_0 = e(u_0), \quad e_1 = e(x_0 - e_0 - d_0), \text{ and } d_0 = q(u_0) - v_0 \qquad (2.11)$$

[recall that $e(\cdot)$ is defined by (2.3)].

### B. Stability Problems

Before proceeding to develop a simple analytical expression for the quantizer error sequence described by the difference equations (2.10)-(2.11), we need to determine bounds to the dynamic range of the various processes involved because they are useful to understand why the "no-overload" condition is essential for the techniques to work.

The internal signals of $\Sigma\Delta$ modulators are limited in practice by physical constraints such as the supply voltage. This fact suggests that a stability concept has to be introduced into the mathematical model; it is reasonable to say that the $\Sigma\Delta$ modulator is stable if for all $n$, the magnitudes of the states $u_n$ and $v_n$ are bounded by some predetermined constant $M$. Of course, a suitable value should be chosen for the parameter $M$ before any quantitative analysis can be carried out. While different $M$'s specify different degrees of system stability, for the purposes of this paper, the following concept will be particularly useful.

We will say that the $\Sigma\Delta$ is *no-overload stable* (or simply, stable) if the amplitudes of the internal states are bounded in the no-overload region of the quantizer.

To be more specific, consider the quantizer $q(\cdot)$ shown in Fig. 2. It has four output levels and unit bin width (our results can be extended in an obvious way for quantizers having arbitrary bin width). Its input-output relation is given by

$$q(u) = \begin{cases} m + 1/2, & \text{if } u \in [m, m+1), \text{ for } m = -2, -1, 0, 1 \\ \text{overload}, & \text{otherwise}. \end{cases} \qquad (2.12)$$

The function $e(u) = q(u) - u$ is also depicted in Fig. 2.

For this quantizer $q(\cdot)$, the stability is then defined by ($M = 2$)

$$-2 \leq u_n, v_n \leq 2 \text{ or } e_n \in [-1/2, 1/2], d_n \in [-1, 1]$$

for all but a finite number of $n$.

Finding stability conditions is thus an interesting problem; the following result provides a sufficient condition in that direction.

*1) No-Overload Stability Criterion:* If

$$x_n \in [-1/2, 1/2], \qquad n = 0, 1, 2, \cdots \qquad (2.13)$$

and

$$e_0 \in [-1/2, 1/2], \qquad d_0 \in [-1, 1], \qquad (2.14)$$

then

$$e_n \in [-1/2, 1/2], \qquad d_n \in [-1, 1] \qquad (2.15)$$

$$u_n \in [x_{n-1} - 3/2, x_{n-1} + 3/2] \subset [-2, 2] \qquad (2.16)$$

$$v_n \in [x_{n-1}, x_{n-1} + 1] \subset [-3/2, 3/2] \qquad (2.17)$$

for all $n = 1, 2, \cdots$.

*Proof (Outline):* Equations (2.13), (2.14), and (2.11) imply $e_1 \in [-1/2, 1/2]$; thus, (2.15) follows from (2.10) by induction. Equations (2.16) and (2.17) follow from (2.13), (2.8), and (2.9) with the aid of (2.15). □□□

*Comment:* This criterion states that if the input amplitude is confined to the interval $[-1/2, 1/2]$, then once the states reach the no-overload region (e.g., by resetting the system $u_0 = v_0 = 0$), all future values will lie in that range and the quantizer will be overload-free. Simulation results strongly suggest that resetting the states of the system is not necessary, i.e., internal saturation is just a transient phenomenon, and the modulator will eventually be stable, provided that the input amplitude is sufficiently small.

On the other hand, large input values can cause the modulator to be unstable. More precisely, suppose $x_n \geq 3/2 + \epsilon$ (or $x_n \leq -3/2 - \epsilon$) for some $\epsilon > 0$ and for all $n$; then

$$v_n \geq \mathcal{O}(n) \text{ and } u_n \geq \mathcal{O}(n^2) \text{ [or } v_n \leq -\mathcal{O}(n) \text{ and } u_n \leq -\mathcal{O}(n^2)].$$

*Proof:* Let $x_n \geq 3/2 + \epsilon$ (the other case can be handled similarly). Summing (2.1) and simplifying the resulting telescoping sum yields

$$v_n = v_0 + \sum_{k=0}^{n-1} (x_k - q_k).$$

Then

$$v_n \geq v_0 + \sum_{k=0}^{n-1} (3/2 + \epsilon - 3/2) = v_0 + \epsilon n.$$

Similarly, it follows from (2.2) that

$$u_n \geq u_0 + (v_0 - 3/2)n + \epsilon n(n+1)/2.$$

*C. Solution of the Difference Equation*

Now we proceed to solve the difference equations under the no-overload stability assumption. The following notation will be used throughout the development. Let $\lfloor r \rfloor$ and $\langle r \rangle$, respectively, denote the integral and fractional parts of any real number $r$, i.e., if $r \in [m, m+1)$ for some integer $m$, then $\lfloor r \rfloor = m$ and $\langle r \rangle = r - m = r - \lfloor r \rfloor$; thus, $r = \lfloor r \rfloor + \langle r \rangle$. Several useful properties of $\langle r \rangle$ follow from its definition and are summarized here for later use.

Let $m$ be an integer and $r$, $s$ be real numbers. Then

$$0 \leq \langle r \rangle < 1 \qquad (2.18)$$

and

$$\langle r + m \rangle = \langle r \rangle, \quad \langle r \pm s \rangle = \langle r \pm \langle s \rangle \rangle = \langle r \rangle \pm \langle s \rangle - K \qquad (2.19)$$

where

$$K = \begin{cases} 1, & \text{if } \langle r \rangle \pm \langle s \rangle \in [1, 2) \\ 0, & \text{if } \langle r \rangle \pm \langle s \rangle \in [0, 1) \\ -1, & \text{if } \langle r \rangle \pm \langle s \rangle \in (-1, 0). \end{cases}$$

The (granular) quantizer error can now be expressed in a very convenient form [cf. (2.12)]:

$$e(u) = \lfloor u \rfloor + 1/2 - u = 1/2 - \langle u \rangle, \qquad \text{if } u \in -[2, 2] \quad (2.20)$$

by limiting the quantizer input $u$ in the no-overload region. For an overload-free system, the difference equation (2.10) becomes

$$e_{n+1} = 1/2 - \langle x_n - 2e_n + e_{n-1} \rangle, \qquad n = 1, 2, \cdots. \quad (2.21)$$

The apparently difficult equation (2.10) expressed in this particular form can easily be solved by using simple induction (this is the real justification of the definition of no-overload stability). The solution is given by the following theorem (established in Appendix B).

*Theorem 1:* Under the assumption of *no-overload stability*, i.e., if $x_n \in [-1/2, 1/2]$, $e_0 \in [-1/2, 1/2]$, and $d_0 \in [-1, 1]$, then

$$e_n = 1/2 - \left\langle 1/2 - e_0 - d_0 n + \sum_{i=0}^{n-1} (n-i)(x_i + 1/2) \right\rangle. \quad (2.22)$$

Suppose, in addition, that the modulator is driven by a dc input

$$x_n = X \in [-1/2, 1/2], \qquad \text{for } n = 0, 1, 2, \cdots. \quad (2.23)$$

Then

$$e_n = 1/2 - \langle P_2(n) \rangle, \qquad n = 1, 2, \cdots \quad (2.24)$$

where

$$P_2(n) = 1/2 - e_0 - d_0 n + (X + 1/2)n(n+1)/2. \quad (2.25)$$

## III. LONG-TERM TIME AVERAGE PROPERTIES

*A. Preliminaries*

For a real-valued *deterministic* sequence $\{\xi_n, n = 1, 2, 3, \cdots\}$, we define its *sample mean* or time average mean as

$$M\{\xi_n\} = \lim_{n \to \infty} \frac{1}{N} \sum_{n=1}^{N} \xi_n \quad (3.1)$$

if the limit exists. One can similarly define other interesting long-term time averages, assuming beforehand the existence of the limits in question. It should be indicated that the time average is a property of the whole sequence, and thus any *finite* term effect will eventually disappear.

Let $\chi(\xi; I)$ denote the indicator function of any interval $I = [a, b)$, i.e.,

$$\chi(\xi; I) = \begin{cases} 1, & \text{if } a \leq \xi < b \\ 0, & \text{otherwise.} \end{cases} \quad (3.2)$$

We will call the following (set) function the *marginal distribution* of $\{\xi_n\}$:

$$\mu_\xi(I) = M\{\chi(\xi_n; I)\}. \quad (3.3)$$

$\{\xi_n\}$ is said to be *uniformly distributed* on $[c, d)$ if

$$\mu_\xi(I) = |I|/(d-c), \qquad \text{for each } I \subseteq [c, d). \quad (3.4)$$

Here, $|I|$ denotes the length of the interval $I$.

108

We can similarly define joint long-term averages for two sequences $\{\xi_n\}$ and $\{\zeta_n\}$. For example, we can define the *joint distribution* of $\{\xi_n\}$ and $\{\zeta_n\}$ by

$$\mu_{\xi\zeta}(I \times J) = M\{\chi[(\xi_n, \zeta_n); I \times J]\} \qquad (3.5)$$

for any Cartesian product of intervals $I$ and $J$.

$\{\xi_n\}$ and $\{\zeta_n\}$ are said to be *independent* whenever

$$\mu_{\xi\zeta}(I \times J) = \mu_\xi(I)\mu_\zeta(J) \qquad (3.6)$$

for every $I$ and $J$. They are *jointly uniform* on $[0, 1)^2$ if they are marginally uniform and independent, i.e.,

$$\mu_{\xi\zeta}(I \times J) = |I||J|. \qquad (3.7)$$

We define the *joint characteristic function* of $\{\xi_n\}$ and $\{\zeta_n\}$ by

$$\Phi_{\xi\zeta}(s, t) = M\{\exp[j2\pi(s\xi_n + t\zeta_n)]\} \qquad (j = \sqrt{-1}). \quad (3.8)$$

*Weyl's Theorem [14]:* Let $\{\xi_n\}$ and $\{\zeta_n\}$ be two $[0, 1)$-valued sequences. The following statements are equivalent.
1) $\{\xi_n\}$ and $\{\zeta_n\}$ are jointly uniform on $[0, 1)^2$.
2) The equality

$$M\{F(\xi_n, \zeta_n)\} = \int_0^1 \int_0^1 F(\xi, \zeta)\,d\xi\,d\zeta \qquad (3.9)$$

holds for *every Riemann-integrable* function $F: [0, 1)^2 \to \mathbb{R}$.
3) The joint characteristic function satifies

$$\Phi_{\xi\zeta}(s, t) = 0 \qquad (3.10)$$

for all *integer pairs* $(s, t) \neq (0, 0)$. Notice that

$$\Phi_{\xi\zeta}(0, 0) = 1. \qquad (3.11)$$

*Comment:* Because of (3.9), no limits need to be calculated. Equation (3.10) provides a convenient criterion to determine if two sequences are jointly uniform.

The time average *autocorrelation sequence* of $\{\xi_n\}$ will be defined by the correlation of $\{\xi_n\}$ and $\{\xi_n^{(k)}\}$:

$$R_\xi(k) = M\{\xi_n \xi_n^{(k)}\}, \qquad k = 0, \pm 1, \pm 2, \cdots \quad (3.12)$$

where

$$\xi_n^{(k)} = \xi_{n+k}, \qquad n = 1, 2, \cdots \qquad (3.13)$$

are the shifted versions of $\{\xi_n\}$.

The *power spectrum* of $\{\xi_n\}$ is then defined as the discrete-time Fourier transform of the autocorrelation sequence $\{R_\xi(k), k = 0, \pm 1, \pm 2, \cdots\}$, i.e.,

$$S_\xi(f) = \Sigma_k R_\xi(k)e^{-j2\pi fk}, \qquad f \in [0, 1). \qquad (3.14)$$

The average power $M\{\xi_n^2\}$ can be calculated by

$$R_\xi(0) = \int_0^1 S_\xi(f)\,df. \qquad (3.15)$$

A zero-mean sequence $\{\xi_n\}$ will be called a *white noise* if

$$R_\xi(k) = 0, \qquad \text{for all } k \neq 0 \qquad (3.16)$$

or, equivalently, the power spectrum has a constant value $M\{\xi_n^2\}$.

Let $\{\xi_n\}$ and $\{\zeta_n\}$ be the input and output, respectively, of a linear system having a finite impulse response (FIR) $\{h_n, n = 0, 1, 2, \cdots, N-1\}$ and transfer function $H(f) = \Sigma_n h_n \exp(-j2\pi fn)$. Then $\{\xi_n\}$ and $\{\zeta_n\}$ are related by

$$\zeta_n = \sum_{k=0}^{N-1} h_k \xi_{n-k}, \qquad n = N, N+1, \cdots. \qquad (3.17)$$

As in standard spectral analysis, the following formulas hold:

$$M\{\zeta_n\} = M\{\xi_n\}H(0) \qquad (3.18)$$

$$S_\zeta(f) = S_\xi(f)|H(f)|^2 \qquad (3.19)$$

and

$$M\{\zeta_n^2\} = \int_0^1 S_\xi(f)|H(f)|^2\,df. \qquad (3.20)$$

### B. Fundamental Result

We are now ready to determine the long-term statistical behavior of the quantizer error sequence in the double-loop $\Sigma\Delta$ modulator. Constant or dc input $x_n = X$ is assumed throughout. Recall that (Theorem 1)

$$e_n = 1/2 - \langle P_2(n) \rangle, \qquad n = 1, 2, \cdots$$

where $P_2(\cdot)$ is a polynomial of degree 2 with leading coefficient $(X + 1/2)/2$:

$$P_2(n) = 1/2 - e_0 - d_0 n + (X + 1/2)n(n+1)/2. \quad (3.21)$$

Recall also that $\langle r \rangle = r \bmod 1$, $\langle r \rangle \in [0, 1)$.

*Theorem 2:* Consider the sequence $\{\xi_n\}$ defined by

$$\xi_n = \langle P_2(n) \rangle, \qquad n = 1, 2, \cdots \qquad (3.22)$$

and its shifted versions

$$\xi_n^{(k)} = \langle P_2(n+k) \rangle, \qquad k = \pm 1, \pm 2, \cdots. \qquad (3.23)$$

If $X$ is an *irrational* number, then for all $k \neq 0$, the sequences $\{\xi_n\}$ and $\{\xi_n^{(k)}\}$ are jointly uniform on $[0, 1)^2$.

*Proof:* Fix $k \neq 0$ and let $\zeta_n = \xi_n^{(k)}$. Due to Weyl's theorem, the present theorem will be proved if we can show that $\Phi_{\xi\zeta}(s, t) = 0$ for all integers $s$ and $t$, except for $s = t = 0$. This can be easily done with aid of the following result from ergodic theory [6].

*Lemma:* Let $P_2(x)$ be a polynomial of degree two with an *irrational* leading coefficient $\alpha/2$. Form the polynomial $P_1(x)$ of degree one as follows:

$$P_1(x) = P_2(x+1) - P_2(x). \qquad (3.24)$$

Then the $[0, 1)$ sequences $\{\langle P_1(n) \rangle\}$ and $\{\langle P_2(n) \rangle\}$ are jointly uniform.

In order to simplify the calculation of

$$\Phi_{\xi\zeta}(s, t) = M\{\exp[j2\pi(s\langle P_2(n) \rangle + t\langle P_2(n+k) \rangle)]\},$$

observe from (3.21) and (3.24) that

$$P_2(n+k) = P_2(n) + P_1(n)k + P_0 k(k+1)/2$$

where

$$P_0 = P_1(x+1) - P_1(x) = \text{constant},$$

and hence it follows from (2.19) that

$$\langle P_2(n+k) \rangle = \langle P_2(n) \rangle + \langle P_1(n) \rangle k + \langle P_0 \rangle k(k+1)/2 \pmod 1$$

where, by the above lemma and the irrationality assumption, the fractional parts of the polynomials $P_1(n)$ and $P_2(n)$ are jointly uniform sequences. Thus, an application of Weyl's theorem to these processes yields the desired result:

$$\Phi_{\xi\zeta}(s, t) = \exp[j\pi t\langle P_0 \rangle k(k+1)]$$

$$\cdot \int_0^1 \int_0^1 \exp\{j2\pi[(s+t)\xi + tk\zeta]\}\,d\xi\,d\zeta$$

$$= \begin{cases} 1, & \text{if } s = t = 0 \\ 0, & \text{otherwise.} \end{cases}$$

□□□

The following result provides an answer to the first fundamental question.

*Corollary:* If the dc input $X$ to the $\Sigma\Delta$ modulator is a continuous random variable on $[-1/2, 1/2]$, then with probability one, the quantizer error process $\{e_n\}$ is

1) uniformly distributed on $(-1/2, 1/2]$, with $M\{e_n\} = 0$,
2) white noise, with power spectrum $S_e(f) = 1/12$,
3) independent of the input $\{x_n = X\}$ in the sense of (3.6),
4) furthermore, for each $k \neq 0$

$$M\{F(e_n, e_{n+k})\} = \int\!\!\!\int_{-1/2}^{1/2} F(\xi, \zeta)\,d\xi\,d\zeta \qquad (3.25)$$

holds for every Riemann-integrable function $F: (-1/2, 1/2]^2 \to \mathbb{R}$.

*Proof:* If $X$ is a continuous random variable, then it is *almost surely* an irrational number, and hence the conclusions of Theorem 2 will hold with probability one.

Statement 1) follows immediately since $e_n = 1/2 - \xi_n$, 2) follows because independence obviously implies whiteness, 3) can easily be checked by direct calculations, 4) is due to Weyl's theorem. Notice that

$$M\{F(e_n)\} = \int F(\xi)\,d\xi.$$

$\square\square\square$

The following example illustrates the fact that part 4) of the corollary cannot be arbitrarily generalized. Let $Q = \{\pm 1/2, \pm 3/2\}$ be the 2-b quantizer reproduction alphabet and consider the Riemann-integrable function

$$\chi(q; Q) = \begin{cases} 1, & \text{if } q \in Q \\ 0, & \text{otherwise.} \end{cases}$$

Recall that $q_{n+1} = X + e_{n+1} - 2e_n + e_{n-1} \in Q$. Clearly,

$$M\{\chi(X + e_{n+1} - 2e_n + e_{n-1}; Q)\} = M\{\chi(q_n; Q)\} = 1,$$

whereas

$$\int\!\!\int\!\!\int \chi(X + \alpha - 2\beta + \gamma; Q)\,d\alpha\,d\beta\,d\gamma = 0.$$

### C. Properties of the Various Processes

The long-term time average behavior of the various processes can be determined by using (3.25) through the quantizer error process as follows.

First consider the modulator output $\{q_n\}$. The mean is, as expected,

$$M\{q_{n+1}\} = M\{X + e_{n+1} - 2e_n + e_{n-1}\} = X. \qquad (3.26)$$

The autocorrelation and the spectrum are, respectively, given by

$$R_q(k) = M\{q_n q_{n+k}\} = \begin{cases} X^2 + 1/2, & k = 0 \\ X^2 - 1/3, & k = \pm 1 \\ X^2 + 1/12, & k = \pm 2 \\ X^2, & \text{else} \end{cases} \qquad (3.27)$$

$$S_q(f) = X^2\delta(f) + \frac{1}{12}|2\sin(\pi f)|^4, \qquad f \in [0, 1). \quad (3.28)$$

This output power spectrum is illustrated in Fig. 3.

Now recall that $v_{n+1} = X + e_{n+1} - e_n$. Consider

$$\Phi_v(t) = M\{\exp(j2\pi t v_n)\}$$

$$= M\{\exp[j2\pi t(X + e_n - e_{n-1})]\}$$

$$= e^{j2\pi tX} \int\!\!\int e^{j2\pi t(\alpha - \beta)}\,d\alpha\,d\beta = e^{j2\pi tX}\left(\frac{\sin(\pi t)}{\pi t}\right)^2$$

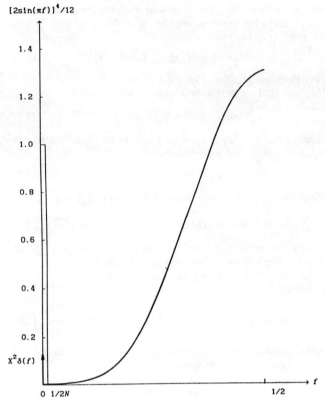

Fig. 3. Output power spectrum.

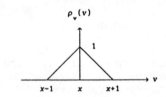

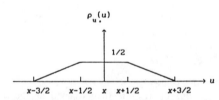

Fig. 4. Density functions. (a) $\rho_v(\cdot)$. (b) $\rho_u(\cdot)$.

which is the Fourier transform of the function [see Fig. 4(a)]

$$\rho_v(v) = \begin{cases} 1 - |v - X|, & \text{if } v \in [X - 1, X + 1] \\ 0, & \text{otherwise.} \end{cases} \qquad (3.29)$$

In a similar manner, one can compute $\Phi_u(\cdot)$ and $\rho_u(\cdot)$ for $\{u_n\}$, $u_{n+1} = X - 2e_n + e_{n-1}$ as follows:

$$\Phi_u(t) = e^{j2\pi tX}\frac{\sin(\pi t)}{\pi t}\frac{\sin(2\pi t)}{2\pi t}$$

$$\rho_u(u) = \begin{cases} 1/2, & \text{if } u \in [X - 1/2, X + 1/2] \\ 3/4 - |u - X|/2, \\ & \text{if } u \subset [X - 3/2, X - 1/2] \cup [X + 1/2, X + 3/2] \\ 0, & \text{otherwise.} \end{cases}$$

$$(3.30)$$

$\rho_v(\cdot)$ and $\rho_u(\cdot)$ can be interpreted as density functions in the sense that for $I \subseteq [X - 1, X + 1)$ and $J \subseteq [X - 3/2, X + 3/2)$

$$\mu_v(I) = \int_I \rho_v(v)\, dv \quad \text{and} \quad \mu_u(J) = \int_J \rho_u(u)\, du.$$

Now we are ready to see that for a $\Sigma\Delta$ modulator with a constant input $x_n = X$, the condition $X \in [-1/2, 1/2]$ is also necessary for no-overload stability; in other words, if $|X| > 1/2$, then the internal quantizer cannot be overload-free, independently of the initial states of the $\Sigma\Delta$ modulator. This can be proved by contradiction. Suppose that overload only occurs a finite number of times. Then the quantizer will eventually be overload-free, and hence the behavior of $u_n$ is given by $\rho_u(\cdot)$, as shown in Fig. 4(b). However, if $|X| > 1/2$, we have $\mu_u(\{|u| > 2\}) = \int_{|u|>2} \rho_u(u)\, du > 0$, and this implies that overload will still occur, which is a contradiction.

## IV. SYSTEM PERFORMANCE

### A. Sigma–Delta Quantization Noise

In this section, we will apply the results obtained so far to study the overall system performance when the double-loop $\Sigma\Delta$ modulator is used in combination with a suitable decoding filter.

The decoding filters are used to translate (decode) the modulated data $\{q_n\}$ into a more compact format suitable for transmission, storage, or further processing. The decoding filter should be designed to remove the noisy components $|2\sin(\pi f)|^4/12$ in the modulator output $\{q_n\}$ as much as possible. For simplicity, only linear time-invariant FIR filters will be considered here.

Let $\{h_n, n = 0, 1, 2, \cdots, N-1\}$ and $H(f) = \Sigma_n h_n \exp(-j2\pi f n)$ be the unit sample response and transfer function of a given filter. The number of taps $N$ is assumed to be equal to the oversampling ratio. We require that

$$H(0) = \Sigma_n h_n = 1 \qquad (4.1)$$

so that the filter has unit dc gain, and hence the desired signal component $X^2\delta(f)$ will pass without scaling.

Let $\hat{x}_n$ be the filter output $\hat{x}_n = \Sigma_k h_k q_{n-k}$; we define the $\Sigma\Delta$ quantization noise process by

$$\hat{x}_n - X = \sum_{k=0}^{N-1} h_k(q_{n-k} - X), \qquad n = N+1, N+2, \cdots.$$

$$(4.2)$$

It has zero mean and power given by [cf. (3.20), (3.28)]

$$\sigma^2 = M\{|\hat{x}_n - X|^2\} = \frac{1}{12}\int_0^1 |2\sin(\pi f)|^4 |H(f)|^2\, df. \quad (4.3)$$

Since the most popular performance measures are the mean-squared quantization error and the ubiquitous signal-to-quantization-noise ratio (SQNR), in the sequel we adopt the long-term time average $\Sigma\Delta$ quantization noise power in (4.3) as the (asymptotic) system performance measure.

We now consider some specific filters and evaluate their corresponding performances.

### B. Optimum FIR Filter

The following result, established in Appendix C, provides the answer to the second fundamental question:

*Theorem 3:* The solution to the following constrained optimization problem:

$$\text{minimize } \sigma^2 = \frac{1}{12}\int_0^1 |2\sin(\pi f)|^4 \left|\sum_{n=0}^{N-1} h_n e^{-j2\pi nf}\right|^2 df \quad (4.4)$$

over all $N$-tap linear filters $\{h_n, n = 0, 1, \cdots, N-1\}$ subject to the

unit dc gain constraint (4.1) is given by

$$h_n^* = 30\frac{(n+1)(n+2)(N-n)(N+1-n)}{N(N+1)(N+2)(N+3)(N+4)},$$

$$n = 0, 1, \cdots, N-1 \quad (4.5)$$

$$\sigma_*^2 = \frac{60}{N(N+1)(N+2)(N+3)(N+4)}. \quad (4.6)$$

*Comment:* The optimum FIR filter has also been found for single-loop and general multiloop systems [9], [11]. For large $N$, the following are useful approximations:

$$h_n^* \approx \frac{30}{N}\left(\frac{n}{N}\right)^2\left(1 - \frac{n}{N}\right)^2, \qquad n = 1, 2, \cdots, N-1$$

$$\sigma_*^2 \approx 60/N^5. \quad (4.7)$$

Observe also that $h_n^* = h_{N-n-1}^*$ (symmetry property).

### C. Sinc Filters

Although the optimum FIR filter gives the smallest mean-squared $\Sigma\Delta$ quantization error, simple decoding filters may be preferable for certain applications. Because of its simplicity for implementation and its excellent performance, the family of so-called sinc filters is widely used in practice [2], [5], [9], [11].

An $N$-tap sinc[1] filter (also called accumulate-and-dump filter or comb filter) is defined by

$$h_n = 1/N, \qquad n = 0, 1, \cdots, N-1.$$

and the magnitude of its transfer function is therefore

$$|H(f)| = \left|\frac{1}{N}\frac{\sin(\pi fN)}{\sin(\pi f)}\right|$$

(hence the name).

An $N$-tap sinc[2] filter can be viewed as the cascade connection of two $N/2$-tap sinc[1] filters. It is sometimes called a triangular filter because of the shape of its unit sample response. The magnitude of its transfer function is given by

$$|H(f)| = \left|\frac{2}{N}\frac{\sin(\pi fN/2)}{\sin(\pi f)}\right|^2. \quad (4.8)$$

An $N$-tap sinc[k] filter can be similarly defined by giving its transfer function magnitude as follows (suppose $N$ is a multiple of $k$):

$$|H(f)| = \left|\frac{k}{N}\frac{\sin(\pi fN/k)}{\sin(\pi f)}\right|^k. \quad (4.9)$$

In order to evaluate the power of the $\Sigma\Delta$ quantization noise, one can insert (4.9) into (4.3) and find the value of the integral

$$\sigma^2 = \frac{1}{12}\int_0^1 |2\sin(\pi f)|^4 \left|\frac{k}{N}\frac{\sin(\pi fN/k)}{\sin(\pi f)}\right|^{2k} df. \quad (4.10)$$

Simple examples correspond to $k = 2, 3$. The following results are proved in Appendix D:

$$\sigma^2 = 8/N^4 \quad \text{(with a sinc}^2 \text{ filter)} \quad (4.11)$$

$$\sigma^2 = 243/2N^5 \quad \text{(with a sinc}^3 \text{ filter)}. \quad (4.12)$$

Higher order sinc filters can clearly provide smaller $\Sigma\Delta$ quantization noise power. However, due to Theorem 3, the sinc[k] filters for $k > 3$ cannot be substantially better than the sinc[3] filter, which already achieves the best performance order:

$$\sigma^2 = \mathcal{O}(N^{-5}).$$

TABLE I
OVERSAMPLING RATIOS ($N$) VERSUS BIT RATES ($b$)

| PCM | | single-loop | | | double-loop | | | triple-loop | | |
|---|---|---|---|---|---|---|---|---|---|---|
| Bits | SQNR dB | opt. | $sinc^2$ | $sinc^1$ | opt. | $sinc^3$ | $sinc^2$ | opt. | $sinc^4$ | $sinc^3$ |
| 12 | 72.2 | 587 | 646 | 5793 | 104 | 120 | 202 | | | |
| 13 | 78.2 | 930 | 1024 | | 138 | 159 | 284 | | | |
| 14 | 84.3 | 1477 | 1626 | | 181 | 210 | 402 | 83 | 100 | 126 |
| 15 | 90.3 | 2345 | 2582 | | 238 | 276 | 568 | 102 | 120 | 159 |
| 16 | 96.3 | | | | 315 | 363 | 802 | 124 | 148 | 201 |
| 17 | 102.3 | | | | 416 | 480 | 1134 | 151 | 180 | 252 |
| 18 | 108.4 | | | | 549 | 633 | 1604 | 184 | 220 | 318 |
| 19 | 114.4 | | | | 723 | 834 | 2268 | 224 | 268 | 399 |
| 20 | 120.4 | | | | 955 | 1101 | | 273 | 324 | 504 |

|← —————————— oversampling ratios $N$ —————————— →|

Note: SQNR (dB) $\approx 6.02\ b$ (bits).

### D. Comparisons

Since oversampled $\Sigma\Delta$ modulation is intended to aid the implementation of high-resolution A/D converters, it is interesting to compare its performance to that of standard scalar uniform quantization. We thus ask: given a $\Sigma\Delta$ modulation system ($\Sigma\Delta$ modulator plus decoding filter), how large should the oversampling ratio be in order to achieve the same asymptotic mean-squared error of a uniform quantizer with a given (large) number of output levels?

To answer this question, recall the following classic result from asymptotic quantization theory [7]: if a random variable $X$ with an appropriate dynamic range, say $[-1/2, 1/2]$, and a "smooth" probability density is quantized by a uniform quantizer $q(\cdot)$ of bin width $\Delta$, then the mean-squared quantization error $E|X - q(X)|^2$ is approximately $\Delta^2/12$, provided that the number of output levels $M = 1/\Delta$ is large, i.e.,

$$E|X - q(X)|^2 \approx 2^{-2b}/12$$

where $b = \log_2 M$ is the bit rate of the uniform quantizer $q(\cdot)$.

On the other hand, the $\Sigma\Delta$ quantization noise power $\sigma^2$ has already been derived as a function of the oversampling ratio $N$ for several $\Sigma\Delta$ modulation systems; this can be summarized as follows.

*1) Single-Loop $\Sigma\Delta$ Modulation [9]:*

$$\text{Optimum FIR filter:} \quad \sigma^2 \approx 1/N^3$$

$$\text{Triangular filter:} \quad \sigma^2 = 4/3N^3$$

$$\text{Comb filter:} \quad \sigma^2 = 1/6N^2.$$

*2) Double-Loop $\Sigma\Delta$ Modulation [cf. (4.7), (4.11), (4.12)]:*

$$\text{Optimum FIR filter:} \quad \sigma^2 \approx 60/N^5$$

$$sinc^3 \text{ filter:} \quad \sigma^2 = 243/2N^5$$

$$\text{Triangular filter:} \quad \sigma^2 = 8/N^4.$$

*3) Triple-Loop $\Sigma\Delta$ Modulation [11], [12]:*

$$\text{Optimum FIR filter:} \quad \sigma^2 \approx 8400/N^7$$

$$sinc^4 \text{ filter:} \quad \sigma^2 = 81920/3N^7$$

$$sinc^3 \text{ filter:} \quad \sigma^2 = 1215/N^6.$$

By making the powers equal to $\sigma^2 = E|X - q(X)|^2$ and expressing $N$ in terms of $b$, one can get Table I.

It can be concluded that multiloop systems are far more efficient than single-loop systems. Observe also that the double-loop $\Sigma\Delta$ modulator with a $sinc^3$ filter provides a reasonable performance for most practical purposes.

### V. COMMENTS

In this paper, we have presented an exact analysis of the basic double-loop four-level $\Sigma\Delta$ modulator based on its discrete-time model. This analysis can be further extended to general $L$-loop systems [11], [12]. The conclusion is that the additive white noise source model does make correct assumptions on the long-term spectral characteristics of the quantizer error process, and hence yields correct predictions of the system performance which is of order $\sigma^2 = \mathcal{O}(N^{-(2L+1)})$, providing answers to the two fundamental questions stated in the Introduction of the paper.

Since the analysis is based on *asymptotic* assumptions, one can naturally ask for the size of the oversampling ratio for the various long-term properties to be accurate. Unfortunately, usual ergodic theorems are not strong enough to provide useful estimates of the oversampling ratios. Simulation results indicate that theoretical predictions are very good for most practical values ($\geq 100$) of oversampling ratios.

The cost of the highly efficient multiloop $\Sigma\Delta$ quantization is, as expected, an increased complexity and a reduced robustness of the system because multiple loops require multilevel quantization and tighter circuit accuracy to ensure system stability and good performance. For double- and triple-loop systems, however, this additional cost is generally moderate.

If binary quantization is desired or permitted, the only choice is the so-called *multistage* $\Sigma\Delta$ modulation [13], [15]. Multistage $\Sigma\Delta$ modulation could also be called multiquantizer $\Sigma\Delta$ modulation because it uses multiple binary quantizers, namely, one per stage. The two-stage $\Sigma\Delta$ modulator analyzed in [15] has the *same* performance as the double-loop $\Sigma\Delta$ modulator studied here, but instead of one 2-b quantizer, it uses two 1-b quantizers and some additional linear digital circuits.

Multiloop $\Sigma\Delta$ modulators are well known to suffer instability problems [4]; an important issue remaining unanswered is a quantitative description of the internal state saturation in multiloop $\Sigma\Delta$ modulators and to find conditions for system stability. To the authors' knowledge, there does not exist an exact performance prediction for overloaded systems.

Although interesting results have recently been derived for sinusoidal inputs [10], [15], the analysis for $\Sigma\Delta$ modulators with general input models is still an open problem.

### APPENDIX A

*Proof of (2.6)–(2.11):* Subtracting $q_{n+1}$ from both (2.1) and (2.2), and using the definition of $e_n$ and $d_n$, we find that

$$q_{n+1} = x_n + d_{n+1} - d_n$$

and (2.6). Equation (2.7) follows then by substituting (2.6) into the above equation.

Equation (2.8) is (2.7) with $e_{n+1}$ subtracted, and (2.9) is (2.1) with $q_n - v_n$ replaced by $e_n - e_{n-1}\ (= d_n)$.

Equation (2.10) follows by substituting (2.8) into (2.3), and (2.11) just expresses $e_0$ and $e_1$ in terms of the initial values $u_0$, $v_0$, and $x_0$.

### APPENDIX B

*Proof of Theorem 1:* (By induction.) Note that (2.22) is trivially true for $n = 0$, and for $n = 1$, (2.22) reduces to (2.11). Suppose that (2.22) is true for $n = 0, 1, \cdots, m$; then for $n = m + 1$, we have

$$e_{m+1} = 1/2 - \langle x_m + \langle e_{m-1} - e_m \rangle - e_m \rangle;$$

by substituting the induction hypothesis and using (2.19),

$$\langle e_{m-1} - e_m \rangle = \langle 1/2 - 1/2 + e_0 + d_0(m-1)$$

$$- \sum_{i=0}^{m-2} (m - 1 - i)(x_i + 1/2) - 1/2$$

112

$$\left\langle +1/2 - e_0 - d_0 m + \sum_{i=0}^{m-1}(m-i)(x_i + 1/2)\right\rangle$$

$$= \left\langle -d_0 + \sum_{i=0}^{m-1}(x_i + 1/2)\right\rangle;$$

thus, using (2.19) again,

$$\langle x_m + \langle e_{m-1} - e_m\rangle - e_m\rangle = \left\langle x_m - d_0 + \sum_{i=0}^{m-1}(x_i + 1/2)\right.$$

$$-1/2 + 1/2 - e_0 - d_0 m$$

$$\left. + \sum_{i=0}^{m-1}(m-i)(x_i + 1/2)\right\rangle$$

$$= \left\langle 1/2 - e_0 - d_0(m+1)\right.$$

$$\left. + \sum_{i=0}^{m}(m+1-i)(x_i + 1/2)\right\rangle,$$

and hence, (2.22) is also true for $n = m + 1$.

Equations (2.24)–(2.25) are immediate consequences of (2.22) and (2.23).

### APPENDIX C

*Proof of Theorem 3:* Expanding the sum in (4.4), we find that

$$\sigma^2 = \sum_{n=0}^{N-1}\sum_{k=0}^{N-1} h_n h_k R_{n-k}$$

$$= R_0 \sum_{n=0}^{N-1} h_n^2 + 2R_1 \sum_{n=0}^{N-2} h_n h_{n+1} + 2R_2 \sum_{n=0}^{N-3} h_n h_{n+2}$$

where $R_k$ is the autocorrelation of $\{q_n - X\}$ given by [cf.(3.27)]

$$R_k = \frac{1}{12}\int_0^1 |2\sin(\pi f)|^2 e^{j2\pi fk}\, df$$

$$= \begin{cases} (-1)^k \begin{pmatrix} 4 \\ 2+k \end{pmatrix}\dfrac{1}{12}, & \text{if } |k| \le 2 \\ 0, & \text{if } |k| > 2 \end{cases}$$

where $\begin{pmatrix} M \\ m \end{pmatrix}$ is the standard binomial coefficient. Introducing a Lagrangian multiplier $-2\lambda$, differentiating

$$\sigma^2 - 2\lambda\sum_{n=0}^{N-1} h_n$$

with respect to $h_n$, and setting the partial derivatives to zero yields the equations

$$\sum_{k=-n}^{2} R_k h_{n+k} = \lambda, \quad \text{if } 0 \le n \le 1$$

$$\sum_{k=-2}^{2} R_k h_{n+k} = \lambda, \quad \text{if } 2 \le n \le N-3$$

$$\sum_{k=-2}^{N-n-1} R_k h_{n+k} = \lambda, \quad \text{if } N-2 \le n \le N-1.$$

These $N$ equations plus the constraint (4.1) provide $N+1$ equations for the $N+1$ unknowns $h_n$ and $\lambda$, and the solution is given by (4.5)

and $\lambda^{-1} = 2\begin{pmatrix} N+4 \\ 5 \end{pmatrix}$. Finally,

$$\sigma_*^2 = R_0\Sigma_n h_n^2 + 2\Sigma_k R_k \Sigma_n h_n h_{n+k}$$

$$= \Sigma_n h_n(R_0 h_n + 2\Sigma_k R_k h_{n+k}) = \Sigma_n h_n \lambda = \lambda.$$

Thus, (4.6) follows.

### APPENDIX D

*Proof of (4.11) and (4.12):* Let $L = N/2$ be an integer. By definition of the $\text{sinc}^2$ filter,

$$x_N = (1/L)^2 \sum_{i=1}^{L}\sum_{k=1}^{L} q_{i+k}.$$

Since $q_n - X = e_n - 2e_{n-1} + e_{n-2}$, we have the following telescoping sum:

$$x_N - X = (1/L)^2 \sum_{i=1}^{L}\sum_{k=1}^{L}(e_{i+k} - 2e_{i+k-1}$$

$$+ e_{i+k-2}) = (e_N - 2e_L + e_0)/L^2,$$

and therefore,

$$\sigma^2 = M\{|x_{N+n} - X|^2\} = (1/L)^4 M\{|e_{N+n} - 2e_{L+n} + e_n|^2\}.$$

By recalling that $R_e(k) = 1/12$, $k = 0$; $R_e(k) = 0$, $k \ne 0$, we get (4.11).

Now let $L = N/3$ and $M = (L-1)/2$ be integers. Since

$$\frac{\sin(\pi fL)}{\sin(\pi f)} = \sum_{n=-M}^{M} e^{j2\pi fn},$$

$$(2\sin(\pi fL))^2 = j^2(e^{-j\pi fL} - e^{j\pi fL})^2$$

$$= (-1)e^{-j2\pi fL}\sum_{k=0}^{2}(-1)^k\begin{pmatrix} 2 \\ k \end{pmatrix}e^{j2\pi fLk}.$$

Thus,

$$\frac{\sin(\pi fL)}{\sin(\pi f)}(2\sin(\pi fL))^2 = (-1)e^{-j2\pi fL}$$

$$\cdot \sum_{n=-M}^{M}\sum_{k=0}^{2}(-1)^k\begin{pmatrix} 2 \\ k \end{pmatrix}e^{j2\pi f(Lk+n)}.$$

Applying the well-known Parseval's formula,

$$\int_0^1\left|\frac{\sin(\pi fL)}{\sin(\pi f)}\right|^2 |2\sin(\pi fL)|^4\, df = L\sum_{k=0}^{2}\begin{pmatrix} 2 \\ k \end{pmatrix}^2 = 6L = 2N.$$

Then (4.12) follows after some simple algebra.

### ACKNOWLEDGMENT

The authors are indebted to the reviewers for many helpful comments.

### REFERENCES

[1] J. C. Candy, "A use of limit cycle oscillation to obtain robust analog-to-digital converters," *IEEE Trans. Commun.*, vol. COM-22, pp. 298–305, Mar. 1974.
[2] J. C. Candy, Y. C. Ching, and D. S. Alexander, "Using triangular weighted interpolation to get 13-bit PCM from ΣΔ modulator," *IEEE Trans. Commun.*, vol. COM-24, pp. 1268–1275, Nov. 1976.
[3] J. C. Candy and O. J. Benjamin, "The structure of quantization noise from sigma-delta modulation," *IEEE Trans. Commun.*, vol. COM-29, pp. 1316–1323, Sept. 1981.
[4] J. C. Candy, "A use of double integration in sigma delta modulation," *IEEE Trans. Commun.*, vol. COM-33, pp. 249–258, Mar. 1985.
[5] ——, "Decimation for sigma delta modulation," *IEEE Trans. Commun.*, vol. COM-34, pp. 72–76, Jan. 1986.

[6] H. Furstenberg, *Recurrence in Ergodic Theory and Combinatorial Number Theory*. Princeton, NJ: Princeton Univ. Press, 1981.

[7] A. Gersho, "Principle of quantization," *IEEE Trans. Circuits Syst.*, vol. CAS-25, pp. 427–436, 1978.

[8] R. M. Gray, "Oversampled sigma–delta modulation," *IEEE Trans. Commun.*, vol. COM-35, pp. 481–489, May 1987.

[9] ——, "Spectral analysis of quantization noise in a single-loop sigma–delta modulator with dc input," *IEEE Trans. Commun.*, vol. 37, pp. 588–599, June 1989.

[10] R. M. Gray, W. Chou, and P.-W. Wong, "Quantization noise in single-loop sigma–delta modulation with sinusoidal input," *IEEE Trans. Commun.*, vol. 37, pp. 956–968, Sept. 1989.

[11] N. He, A. Buzo, and F. Kuhlmann, "Multi-loop sigma–delta quan-tization: Spectral analysis," in *Proc. IEEE ICASSP*, 1988, pp. 1870–1873.

[12] N. He, F. Kuhlmann, and A. Buzo, "Multiloop sigma–delta quantiza-tion," paper in preparation, 1989.

[13] K. Uchimura, T. Hayashi, T. Kimura, and A. Iwata, "Oversampling A-to-D and D-to-A converters with multistage noise shaping modula-tors," *IEEE Trans. Acoust., Speech, Signal Processing*, vol. 36, pp. 1899–1905, Dec. 1988.

[14] H. Weyl, "Über die Gleichverteilung von Zahlen mod. Eins," *Math-ematische Annalen*, vol. 77, pp. 313–352, 1916.

[15] P.-W. Wong and R. M. Gray, "Two-stage sigma–delta modulation," submitted to *IEEE Trans. Acoust., Speech, Signal Processing*, Oct. 1987.

# A Unity Bit Coding Method by Negative Feedback*

HIROSHI INOSE†, MEMBER, IEEE, AND YASUHIKO YASUDA‡

*Summary*—Signal-to-noise performances of a unity bit coding method and the characteristics of an experimental video encoder based upon the principle are described. The system contains a signal integration process in addition to the original delta modulator and features capability of transmitting dc component of input signal.

The characteristics of the quantizing noise to the signal frequency, the signal amplitude and the integrator time constant are obtained theoretically as well as experimentally. The characteristics of periodical noise which are inherent to the proposed system are also investigated.

The design and the characteristics of an experimental encoder for digital transmission of video signals are described as examples of the experimental equipment constructed to demonstrate the realizability of the principle. The experimental results show that considerably good reproduction of video pictures is obtained with sampling frequency as low as 30 Mc and suggest that the proposed system well fulfills the purpose.

## PRINCIPLE

DELTA MODULATION, characterized by simpler circuitry as compared with PCM, not always fulfills the purpose of digital transmission of signals. This is due to the fact that the output of the delta modulator carries the information which is the differentiation of the input signal. In other words, it is incapable of transmitting dc component, its dynamic range and SNR are inversely proportional to the signal frequency[1] and the inevitable integration at the receiving end causes an accumulative error when the system is subject to transmission disturbances. Consequently, delta modulation has been intended mainly for the transmission of such signals as speech which does not contain dc component and has less energy in higher frequencies. It has been felt that a different approach may be preferable for the transmission of telemetering signals, video signals and the like which generally have rather uniform frequency spectra with dc component, and for the transmission of signals through adverse transmitting conditions.

A modification of delta modulation has been proposed[2] to meet the requirements. To compensate for the inevitable differentiation of the input signal, the proposed system has a signal integration process added at the input of original delta modulator as is shown in

* Received July 24, 1963. Partly presented at the International Telemetering Conference, London, England; 1963.
† Faculty of Engineering, University of Tokyo, Japan.
‡ Institute of Industrial Science, University of Tokyo, Japan.
[1] F. de Jager, "Delta modulation—a method of PCM transmission using the one unit code," *Philips Res. Repts.*, vol. 7, pp. 442–466; 1952.
[2] H. Inose, Y. Yasuda, and J. Murakami, "A telemetering system by code modulation—Δ–Σ Modulation," IRE TRANS. ON SPACE ELECTRONICS AND TELEMETRY, vol. SET-8, pp. 204–209; September, 1962.

Fig. 1. The input to the pulse modulator $\epsilon(t)$ is the difference of the integrated input signal $\int s(t)dt$ and the integrated output pulses $\int p(t)dt$ so that $\epsilon(t) = \int s(t)dt - \int p(t)dt$. For the realization of this original configuration, an integrator with infinite dynamic range is needed. However, since the subtraction is a linear process, the above equation of $\epsilon(t)$ is rewritten as $\epsilon(t) = \int \{s(t) - p(t)\}dt$ so that the two integrators can be combined and replaced as shown in Fig. 2. In this modified configuration, the input to the integrator is the difference signal $\{s(t) - p(t)\}$ which stays within a certain limit if the system operates properly, and hence the integrator is practically realizable.

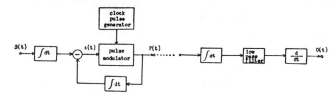

Fig. 1—Block diagram of a modified delta modulator.

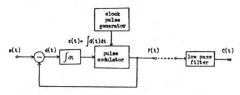

Fig. 2—Block diagram of the system.

The configuration of Fig. 2 has been named as $\Delta - \Sigma$ modulation and has been described with particular reference to the digital telemetering.[2] In the figure, the output pulses $p(t)$ are fed back to the input and subtracted from the input signal $s(t)$ which varies sufficiently slower than the sampling pulses. The difference signal $d(t) = s(t) - p(t)$ is integrated to produce $\epsilon(t) = \int d(t)dt$ and enters the pulse modulator. The pulse modulator compares the amplitude of the integrated difference signal $\epsilon(t)$ with a predetermined reference level and opens the gate to pass a pulse from the pulse generator when $\epsilon(t)$ is larger than the reference level and closes the gate to inhibit the pulse when $\epsilon(t)$ is smaller than the reference level. Through this negative feedback procedure, the integrated difference signal is always kept in the vicinity of the reference level of the pulse modulator, provided that the input signal is not too large. Hence if the amplitude of the input signal becomes large, the output pulses appear more frequently. In other words, the output pulses carry the information corresponding to the input signal amplitude. Demodula-

Reprinted from *Proc. IEEE*, vol. 51, pp. 1524–1535, November 1963.

115

tion in the receiving end is performed by reshaping the received pulses and passing them through a low-pass filter. Since no integration process is involved in the demodulation, no accumulative error due to transmission disturbances results in the demodulated signal.

The other feature of the system over delta modulation is the input-output characteristic. In the latter, the maximum allowable input signal amplitude that does not overload the system is inversely proportional to the input signal frequency, because the maximum slope of the local-decoder output waveform is constant under the predetermined step voltage and sampling frequency. On the other hand, in the former, the maximum allowable input signal voltage is equal to the product of the amplitude and the duty ratio of the output pulses and therefore is independent to the signal frequency.

The integrator in the forward path is not necessarily a single integrator but may be a double integrator, or more generally, be any signal processing networks favorable to the system performance.

The modulation system may be multiplexed on a time division basis as in other digital modulation systems. However, the pulse pattern of the information channels must be taken into consideration to the frame synchronization.

The frame synchronization pattern generally used in digital communications systems is successive ones. But this pattern in the modulation system is vulnerable because the same pattern appears stationary in the signal channel at the edge of the input dynamic range.

Similarly, such patterns as $\cdots 1010 \cdots$, $\cdots$ $001001 \cdots$ and $\cdots 110110 \cdots$ may appear stationary. Thus the synchronization pattern should be one of the patterns $\cdots 11001100 \cdots$, $\cdots$ $111000111000 \cdots$ etc., that do not appear stationary.

In the following sections, the considerations on the signal-to-noise performance of the proposed system as well as the design and the characteristics of an experimental video coder are described.

## Signal-to-Noise Characteristics

### Quantizing Noise

Fig. 3 shows a generalized block diagram of the system. The modulator output $p(t)$ is fed back to the input and subtracted from the input signal $s(t)$ to produce $d(t)$. The difference $d(t)$ enters a network having the transfer function $A(\omega)$ and comes out of it as $\epsilon(t)$. The network need not necessarily be an integrator. The comparator with characteristics $f\{\epsilon(t)\}$ forwards the sampling pulse $r(t)$ to the output according to $\epsilon(t)$. At the receiving end, the pulse $p(t)$ is fed through a demodulating network having the transfer function $B(\omega)$ and produces the demodulated output $o(t)$. The difference between $s(t)$ and $o(t)$, $n(t) = s(t) - o(t)$ is the error or the

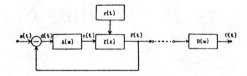

Fig. 3—Generalized block diagram of the system.

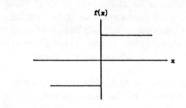

Fig. 4—Characteristic of the comparator.

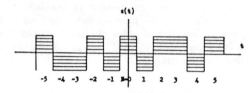

Fig. 5—Assumed waveform of $\epsilon(t)$.

noise of the system. Denoting the frequency domain characteristics of $s(t)$, $d(t)$, $\epsilon(t)$, $p(t)$ and $n(t)$ as $S(\omega)$, $D(\omega)$, $E(\omega)$, $P(\omega)$ and $N(\omega)$, respectively, the frequency domain equations of the system are expressed as

$$D(\omega) = S(\omega) - P(\omega) \tag{1}$$

$$E(\omega) = A(\omega) D(\omega) \tag{2}$$

$$O(\omega) = B(\omega) P(\omega) \tag{3}$$

$$N(\omega) = S(\omega) - O(\omega). \tag{4}$$

Rigorous solution of these equations is impossible since the comparator has such an abrupt nonlinearity as in Fig. 4. Hence, the following assumptions are in order to simplify the treatment.

1) The output pulse $p(t)$ has NRZ (Non-Return-To-Zero) waveform. In other words, the output pulse has the width equal to the sampling period.

2) The error voltage $\epsilon(t)$ is assumed to be rectangular waveforms whose amplitude ranges, with equal probability, from zero to $\Delta$, the peak of the response of the converting network to the output pulse $p(t)$, as shown in Fig. 5. If we assume that the amplitude of the rectangular waveforms is always $\Delta$,

$$\epsilon(t) = \sum_{\lambda=-\infty}^{\infty} \eta_\lambda f(t - \lambda T_r) \tag{5}$$

where $T_r$ is the sampling period and $\eta_\lambda$ is expressed by

$$\eta_\lambda = \begin{cases} +1, & \epsilon(t) > 0 \\ -1, & \epsilon(t) < 0. \end{cases} \tag{6}$$

However this obviously overestimates the value since $\epsilon(t)$ results from $d(t)$ which is the difference between

$p(t)$ and $s(t)$. With the assumption that the amplitude takes the value between zero and $\Delta$ with equal probability, the power of $\epsilon(t)$ is reduced by the following factor:

$$\frac{1}{\Delta^2}\int_0^\Delta y^2 \frac{dy}{\Delta} = \frac{1}{3}. \tag{7}$$

Therefore $\epsilon(t)$ is expressed as

$$\epsilon(t) = \frac{1}{\sqrt{3}} \sum_{\lambda=-\infty}^{\infty} \eta_\lambda f(t - \lambda T_r). \tag{8}$$

3) The demodulating network is assumed to have brickwall low-pass characteristics with cutoff angular frequency of $\omega_c$ as

$$B(\omega) = \begin{cases} 1 & |\omega| \leq \omega_c \\ 0 & |\omega| > \omega_c. \end{cases} \tag{9}$$

The Fourier transform of $f(t)$ is expressed as

$$F(\omega) = \Delta T_r S_a\left(\frac{\omega T_r}{2}\right) \tag{10}$$

where $S_a(x) = \sin x/x$. From (8) the Fourier transform of the portion of $\epsilon(t)$ between number $-m$ and $m$ is

$$E_m(\omega) = \frac{\Delta T_r}{\sqrt{3}} \sum_{\lambda=-m}^{m} S_a\left(\frac{\omega T_r}{2}\right) e^{-j\lambda\omega T_r}. \tag{11}$$

From (1), (2) and (9)

$$E(\omega)B(\omega) = A(\omega)\{S(\omega) - P(\omega)\}B(\omega)$$
$$= A(\omega)\{S(\omega) - O(\omega)\} = A(\omega)N(\omega). \tag{12}$$

Hence for the portion between $-m$ and $m$, the noise is expressed as

$$N_m(\omega) = \frac{B(\omega)}{A(\omega)} E_m(\omega). \tag{13}$$

The power spectrum of the noise is given as

$$W_n(\omega) = \lim_{m\to\infty} \frac{1}{(2m+1)T_r} N_m(\omega) N_m{}^*(\omega)$$
$$= \frac{B(\omega)B^*(\omega)}{A(\omega)A^*(\omega)} \frac{\Delta^2 T_r}{3} S_a{}^2\left(\frac{\omega T_r}{2}\right)$$
$$\cdot \lim_{m\to\infty} \frac{1}{2m+1} \sum_\lambda \sum_\mu \eta_\mu e^{j(\lambda-\mu)\omega T_r}. \tag{14}$$

Since the terms

$$\lim_{m\to\infty} \frac{1}{2m+1} \sum_\lambda \sum_\mu \eta_\lambda \eta_\mu e^{j(\lambda-\mu)\omega T_r}$$

is unity, (14) reduces to

$$W_n(\omega) = \frac{\Delta^2 T_r}{3} \frac{|B(\omega)|^2}{|A(\omega)|^2} S_a{}^2\left(\frac{\omega T_r}{2}\right). \tag{15}$$

Thus the noise power $N^2$ within the pass band of $B(\omega)$ is given as

$$N^2 = \frac{\Delta^2 T_r}{6\pi} \int_{-\omega_c}^{\omega_c} \frac{1}{|A(\omega)|^2} S_a{}^2\left(\frac{\omega T_r}{2}\right) d\omega. \tag{16}$$

On the other hand, the signal power $S^2$ when the input signal is sinusoidal is expressed as

$$S^2 = M^2 \frac{h^2}{2} \tag{17}$$

where $h$ is the one half of the dynamic range and $M$ is the ratio between the input signal amplitude and $h$.

The SNR of the system will be obtained if the characteristics of $A(\omega)$ is given.

*Single Integrator:* If $A(\omega)$ is characterized by a single integrator of Fig. 6 as

$$A(\omega) = \frac{G}{1 + j\omega\tau} \tag{18}$$

where $G$ is the gain of the amplifier and $\tau$ is the time constant CR, the peak of the response to a pulse with amplitude $h$ and duration $T_r$ is

$$\Delta = Gh(1 - e^{-T_r/\tau}). \tag{19}$$

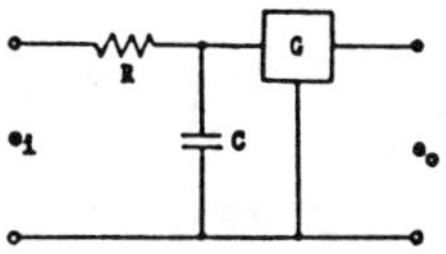

Fig. 6—Single integrator.

Hence

$$|A(\omega)|^2 = \frac{G^2}{1 - \omega^2\tau^2} = \frac{2}{h^2(1 - e^{-T_r/\tau})^2}. \tag{20}$$

Substituting this into (16),

$$N^2 = \frac{T_r h^2}{3\pi} (1 - e^{-T_r/\tau})^2 \qquad = 2 =$$
$$\cdot \int_0^{\omega_c} (1 + \omega^2\tau^2) S_a{}^2\left(\frac{\omega T_r}{2}\right) d\omega. \tag{21}$$

Since $(\omega T_r/2) \ll 1$, expanding $S_a{}^2(\omega T_r/2)$ as

$$S_a{}^2\left(\frac{\omega T_r}{2}\right) = 1 - \frac{1}{3}\left(\frac{\omega T_r}{2}\right)^2 + \cdots$$

and neglecting terms except the first, (21) becomes

$$N^2 = \frac{T_r h^2}{3\pi} (1 - e^{-T_r/\tau})^2 \omega_c \left(1 + \frac{\omega_c{}^2\tau^2}{3}\right). \tag{22}$$

Thus the SNR is expressed as

$$\frac{S^2}{N^2} = M^2 \frac{3\pi}{2T_r\omega_c(1 - \epsilon^{-T_r/\tau})^2 \left(1 + \frac{\omega_c{}^2\tau^2}{3}\right)} \quad (23)$$

Denoting $1/T_r = f_r$ and normalizing as $2\pi f_r/\omega_c = \alpha$ and $\omega_c\tau/2\pi = \beta$ and rewriting

$$\frac{S^2}{N^2} = M^2 \frac{3\alpha}{4(1 - \epsilon^{-1/\alpha\beta})^2 \left(1 + \frac{1}{3}\pi^2\beta^2\right)}, \quad (24)$$

$$\frac{S^2}{N^2} = M^2 \frac{3\alpha}{4\left(\dfrac{1 - \epsilon^{-1/\alpha\beta_1}}{1 - \gamma} + \dfrac{1 - \epsilon^{-1/\alpha\beta_1\gamma}}{\dfrac{1}{\gamma} - 1}\right)^2 \left\{1 + \dfrac{4}{3}\pi^2(1 + \gamma^2)\beta_1{}^2 + \dfrac{16}{5}\pi^4\beta_1{}^4\gamma^2\right\}} \quad (27)$$

and

$$\frac{S^2}{N^2} = \begin{cases} \dfrac{3}{4}\alpha M^2 & \text{for } \beta \to 0 \ \ (\text{or } \tau \to 0) \\[2ex] \dfrac{9}{16\pi^2}\alpha^3 M^2 & \text{for } \beta \to \infty \ \ (\text{or } \tau \to \infty). \end{cases} \quad (25)$$

This means that the SNR is proportional to $(f_r)^{3/2}$ for sufficiently large values of $\tau$. This relationship is similar to that of the delta modulation. However (24) indicates that the SNR is independent of the signal frequency. This is one of the major differences of the system from delta modulation in which both the SNR and the dynamic range are inversely proportional to the signal frequency. This characteristic enables the system's suitability to signals with rather uniform power spectrum such as in telemetering and video transmission.

Eq. (24) also indicates that the SNR is improved with the increase of $\alpha$ or the sampling frequency $f_r$, but is maximum with certain value of $\beta$ or the integrator time constant $\tau$. The results of calculations by Newton's iteration method indicate that, for $\alpha = 5$, 10 and 20, the maximum SNR are obtained when $\beta = 0.592$, 1.359 and 2.948, respectively. Fig. 7 shows the relation between the SNR and $\beta$ taking $\alpha$ as a parameter. As will be seen in the figure, the SNR decreases with the increase of $\beta$ only slightly, so that the integrator time constant would be better if it were larger. However, this decreases the integrator output, and hence reduces the

stability of operation. Therefore the time constant should be chosen to compromise the conditions.

*Double Integrator:* Fig. 8 shows a double integrator with an interstage amplifier of unity gain which is provided to isolate $R_1C_1$ and $R_2C_2$, so that the transfer function is expressed as

$$A(\omega) = \frac{G}{(1 + j\omega\tau_1)(1 + j\omega\tau_2)} \quad (26)$$

where $\tau_1 = R_1C_1$ and $\tau_2 = R_2C_2$. If the condition $\tau_1 + \tau_2 \gg R_1C_2$ exists, the transfer function of the circuit without the interstage amplifier is approximately the same as (26). Through the similar calculation as in the case of single integration, the SNR is expressed as follows.

where

$$\frac{2\pi f_r}{\omega_c} = \alpha, \quad \frac{\omega_c\tau_1}{2\pi} = \beta_1, \quad \frac{\omega_c\tau_2}{2\pi} = \beta_2 \quad \text{and} \quad \frac{\beta_2}{\beta_1} = \gamma.$$

If $\gamma \to \infty$,

$$\lim_{\gamma \to \infty} \frac{S^2}{N^2} = M^2 \frac{9\alpha^3}{16\pi^2\left(1 + \dfrac{12}{5}\pi^2\beta_1{}^2\right)(1 - \epsilon^{-1/\alpha\beta_1})^2} \quad (28)$$

and if $\gamma \to 1$ (or $\beta_2 \to \beta_1$)

$$\lim_{\gamma \to 1} \frac{S^2}{N^2} = M^2 \frac{3\alpha}{4\left\{1 - \left(1 + \dfrac{1}{\alpha\beta_1}\right)\epsilon^{-1/\alpha\beta_1}\right\}^2 \left(1 + \dfrac{8}{3}\pi^2\beta_1{}^2 + \dfrac{16}{5}\pi^4\beta_1{}^4\right)}. \quad (29)$$

For large values of $\alpha$, (27) is approximately reduced to the following.

$$\frac{S^2}{N^2} = M^2 \frac{3\alpha^5\beta_1{}^4\gamma^2}{1 + \dfrac{4}{3}\pi^2(1 + \gamma^2)\beta_1{}^2 + \dfrac{16}{5}\pi^4\beta_1{}^4\gamma^2}. \quad (30)$$

This indicates that the SNR is proportional to $(f_r)^{5/2}$.

Figs. 9 and 10 show the calculated SNR vs $\gamma$ taking $\alpha$ and $\beta_1$ as parameters. The figures indicate that the SNR is constant if either one of $\beta_1$ and $\beta_2$ is sufficiently large and that the SNR is maximum at an optimum value of $\gamma$ but does not decrease appreciably if $\gamma$ exceeds the optimum value.

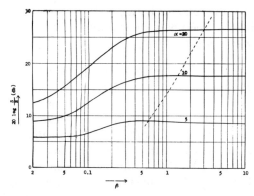

Fig. 7—Theoretical SNR vs normalized integrator time constant $\beta$ taking the normalized sampling frequency $\alpha$ as the parameter.

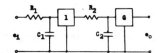

Fig. 8—Double integrator.

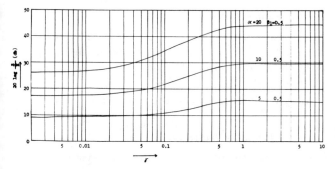

Fig. 9—Theoretical SNR vs the ratio of integrator time constants $\gamma$ taking normalized time constant $\beta_1$ and normalized sampling frequency $\alpha$ as parameters (double integrator).

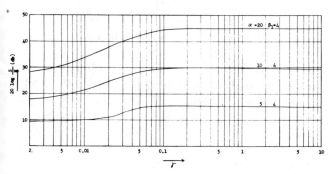

Fig. 10—Theoretical SNR vs the ratio of integrator time constants $\gamma$ taking normalized time constant $\beta_1$ and normalized sampling frequency $\alpha$ as parameters (double integrator).

### SNR and Signal Amplitude

The SNR of the system is maximum at an optimum signal amplitude. This is due to the fact that when the signal amplitude approaches from the center to the lower or the upper edge of the dynamic range, the equivalent repetition frequency of the output pulse reduces. Assuming that the repetition frequency reduces linearly with the departure of the signal amplitude from the center of the dynamic range, the equivalent normalized

repetition frequency $\alpha$ is

$$\alpha = \alpha_0\left(1 - \frac{|s|}{h}\right) \tag{31}$$

where $\alpha_0$ is the normalized repetition frequency when the signal amplitude is in the center of the dynamic range and $s$ is the instantaneous signal amplitude. Assuming also that the probability of the input signal taking the amplitude between $|s|$ and $|s+ds|$ as $p(s)ds$, the quantizing noise power increases by the following factor $H$.

$$H = 2\int_0^{s_0}\left(1 - \frac{s}{h}\right)^{-2k-1}p(s)ds \tag{32}$$

where $s_0$ is the maximum signal amplitude and $k$ takes the value of one for single integration and two for double integration, respectively.

Now, provided that the input signal has a uniform probability density of amplitude such as a triangular waveform, $p(s)=1/2s_0$ and the above equation is easily integrated to reduce to the following equation.

$$H = \frac{M(1-M)^{2k}}{1-(1-M)^{2k}}. \tag{33}$$

Taking this factor into consideration, the calculated SNR vs $M$ for $k=1$, $\beta=\infty$, $\alpha_0=20$ is shown in Fig. 11. When the input signal is sinusoidal, it is impossible to express $H$ as a simple function of $M$. But the results of numerical calculations reveal that the SNR vs $M$ curve for a sinusoidal input is not too different from that of a triangular input.

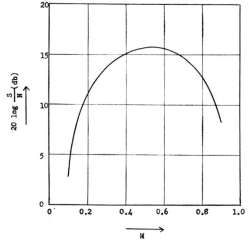

Fig. 11—Theoretical SNR vs normalized signal amplitude.

### Periodic Noise

This modulation system is subject to another sort of noise, that is, periodic noise which depends on the levels of dc input signal.

The aspects of generation of periodic noise are simply explained as follows. The modulation may be considered to be a sort of density modulation, if the output pulses are observed for a considerably long time. Therefore, the density $d$ of output pulses corresponds to the input signal amplitude.

Putting $d$ and $f_d$ as the density and the fundamental frequency of the output pulses, respectively, when the input signal amplitude is $s$,

$$d = \frac{1}{2}\left(1 - \frac{s}{h}\right) \qquad (34)$$

and

$$f_d = \frac{1}{2}\left(1 - \frac{s}{h}\right)f_r \qquad (35)$$

where the definition of $s$, $h$, $f_r$ is the same as in the previous sections.

If we assume that the distribution of pulses in a period $T_d = 1/f_d$ is as shown in Fig. 12, the series of pulses $P_d(t)$ is expressed as follows.

$$P_d(t) = hd + \frac{2h}{\pi}\sum_{n=1}^{\infty}\frac{1}{n}\sin n\pi d \cos 2\pi n f_d t. \qquad (36)$$

Hence the amplitude $A_1$ of the fundamental component of $P_d(t)$ normalized by $h$ is

$$A_1 = \frac{2}{\pi}\sin \pi d. \qquad (37)$$

The periodic component of $P_d(t)$ appears in the demodulated output when the fundamental frequency $f_d$ becomes equal to or smaller than the cutoff frequency $f_c$ of the low-pass filter.

Putting $f_d = f_c$, $d = f_c/f_r = 1/\alpha$. Then, (43) becomes

$$A_1 = \frac{2}{\pi}\sin \frac{\pi}{\alpha}. \qquad (38)$$

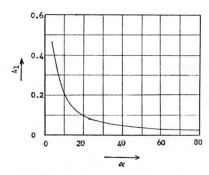

Fig. 12—An example of periodic output pulses.

Fig. 13—Fundamental component of periodic noise vs normalized pulse repetition frequency $\alpha$.

Fig. 13 show the relation between $A_1$ and $\alpha$.

Periodic noise becomes comparable with quantizing noise when $\alpha$ is small.

## EXPERIMENTAL RESULTS ON SYSTEM CHARACTERISTICS

Experimental equipment has been constructed to show the realizability of the principle. In order to be compared with the results of the theoretical calculations, the experimental characteristics will be described which have been obtained by the equipment having the signal bandwidth of dc to 4 kc and the variable sampling frequency from 10 kc to 500 kc. Fig. 14 shows the block diagram of the equipment.

### Input-Output Characteristics

Fig. 15 indicates the over-all input-output characteristics with dc input showing appreciable linearity. Characteristics to the sinusoidal inputs are essentially the same.

### Frequency Characteristics

Fig. 16 shows the over-all frequency characteristics of the modulation system compared with the low-pass filter in the demodulator. The similarity of both curves suggests that the frequency characteristics of the system itself may be independent of the signal frequency.

### Signal-to-Quantizing-Noise Ratio

Fig. 17 shows the measured SNR of the system with a single integrator against the pulse repetition frequency taking the signal frequency as the parameter. It can be seen that SNR is improved by the rate of 9 db/oct as the theoretical calculations indicate.

Fig. 18 shows the relation between the measured SNR of the system with a single integrator and the signal frequency. The dependence of the SNR on signal frequency may be considered to be resulted from the characteristics of the low-pass filter so that the SNR of the modulation system itself may be taken as independent of the signal frequency.

Fig. 19 shows the measured SNR of the system with single integration against integrator time constant, taking pulse repetition frequency as a parameter. As the theory of the previous section predicts, there exists the optimum value of time constant.

Fig. 20 is an example of the measured SNR of the system with single integration against the signal amplitude. The SNR has a certain maximum as has been shown by the calculation.

Fig. 21 shows the measured SNR of the system with double integration against the pulse repetition frequency. Since the time constants of the experimental integrators have been chosen smaller than the optimum value, the absolute value of the SNR is smaller, but the figure shows that the SNR improvement rate with respect to the pulse repetition frequency is 15 db/oct which conforms well to the theoretical calculation.

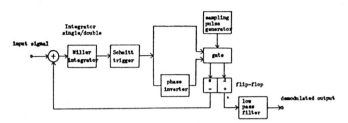

Fig. 14—Block diagram of the experimental modulation system.

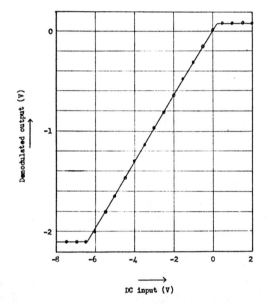

Fig. 15—Input-output characteristics (dc input).

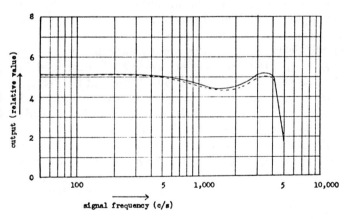

Fig. 16—Frequency characteristic of the experimental equipment.

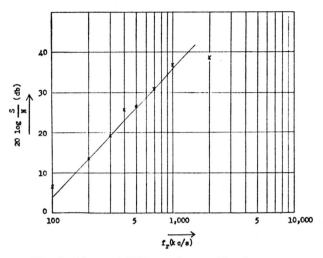

Fig. 17—Measured SNR vs pulse repetition frequency.

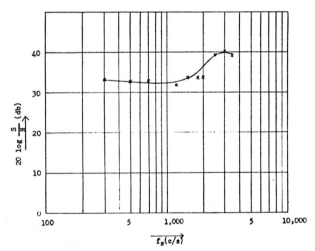

Fig. 18—Measured SNR vs signal frequency.

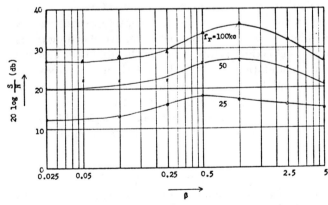

Fig. 19—Measured SNR vs noramlized integrator time constant $\beta$.

121

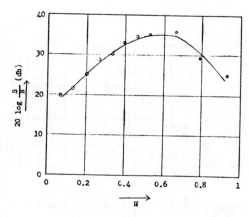

Fig. 20—Measured SNR vs normalized signal amplitude M.

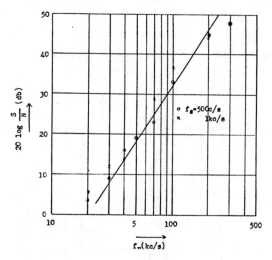

Fig. 21—Measured SNR vs pulse repetition frequency (double integration).

### Coding of Video Signals—An Example of the System

PCM[3-5] and delta modulation[6] techniques have been applied to the digital transmission of video signals. The PCM video encoder is considerably expensive and complicated because of the inevitable use of coding tubes or parallel coding arrangements. The application of the delta modulation to the video encoding simplifies circuitry and keeps the naturalness of the demodulated picture especially at low sampling rates.

The proposed system which possesses, in addition to the above mentioned features of delta modulation,

[3] W. M. Goodall, "Pulse code modulation for television," *Bell Lab. Record*, vol. 29, pp. 209–211; May, 1951.

[4] R. L. Carbrey, "Video transmission over telephone cable pairs by pulse code modulation," Proc. IRE, vol. 48, pp. 1546–1561; September, 1960.

[5] S. Oshima, H. Enomoto, and K. Amano, "High-speed logic circuit using Esaki diode," *J. Inst. Elec. Commun. Engrs.* of Japan, vol. 45, pp. 1541–1548; November, 1962.

[6] J. C. Balder and C. Kramer, "Video transmission by delta modulation using tunnel diodes," Proc. IRE, vol. 50, pp. 428–431; April, 1962.

such advantages that the dynamic range as well as SNR are independent of the signal frequency, may be considered to fit the purpose in view of the fact that the video signals have frequency spectra extending rather uniformly from very low frequency to a few megacycles. The design and characteristics of a video encoder are described here as examples of the experimental systems which have been constructed to demonstrate the feasibility of the principle.

Figs. 22 and 23 show schematically the basic composition and waveforms of the encoder. An Esaki diode is preferred as the pulse modulator. As shown in Fig. 24, the "set" and "reset" points of the Esaki diode are determined by the load line which is decided by the resistors $R_a$, $R_b$, $R_c$, $R_d$ and the supply voltage. Assuming that the Esaki diode is in the "reset" state, when the input signal $s(t)$ increases, the output of the integrator raises the load line upward. Thus a sampling pulse of positive polarity switches the Esaki diode to the "set" state. As the result, the current through the Esaki diode reduces and the potential drop across the resistor $R_a$ decreases. This is fed back to the input so that the increase of the input signal level is reduced. The Esaki diode is then reset by a succeeding sampling pulse of negative polarity. In other words, the resultant output $-p(t)$ of the Esaki diode is added to the input signal $s(t)$. The integrator integrates the signal $s(t) - p(t)$ and feeds the integrated difference signal $\epsilon(t) = \int \{s(t) - p(t)\} dt$ to the pulse modulator. If $\epsilon(t)$ is larger than the trigger level, the Esaki diode is switched again and reduces $\epsilon(t)$. If not, the Esaki diode remains unswitched. Through this negative feedback procedure, the integrated difference signal is always kept in the vicinity of the reference level at which the Esaki diode is triggered. Thus the output of the Esaki diode carries the information corresponding to the instantaneous amplitude of the input signal. The circuit is considerably simplified by using an Esaki diode. This is essential to reduce the circuit delay as well as the power consumption.

Although the principle of operation is described in the case that the comparison is performed at the "reset" state of the Esaki diode, the operation can be performed at the "set" state as well. In this case the Esaki diode provides output pulse $p(t)$ of positive polarity.

Figs. 25 and 26 show the modulator and demodulator circuits of the experimental system, respectively. The input of the modulator and the output of the demodulator are terminated by 75-$\Omega$ resistors. The dynamic range of the modulator is designed to be 75 mv, so that the video signal input of approximately 1 volt is divided by 20 and fed to the modulator. In the experimental system, the output of the modulator is directly connected to the demodulator through an emitter follower.

The modulator is composed of a summing network, a high-frequency amplifier, an integrator and a pulse modulator. The high-frequency amplifier is provided to

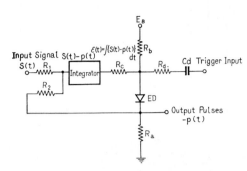

Fig. 22—Basic composition of the encoder.

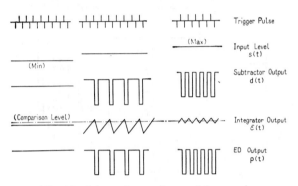

Fig. 23—Schematic waveforms of the encoder.

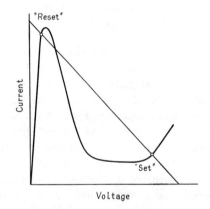

Fig. 24—Operating points of the Esaki diode comparator.

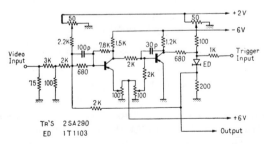

$T_R$'S  2SA290
ED  1T1103

Fig. 25—Experimental modulator circuitry.

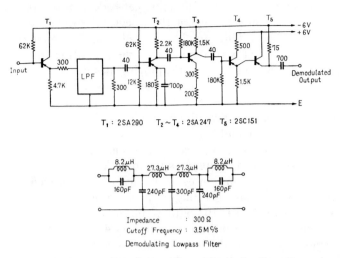

$T_1$ : 2SA290    $T_2 \sim T_4$ : 2SA247    $T_5$ : 2SC151

Impedance : 300 Ω
Cutoff Frequency : 3.5 MC/s
Demodulating Lowpass Filter

Fig. 26—Experimental demodulator circuitry.

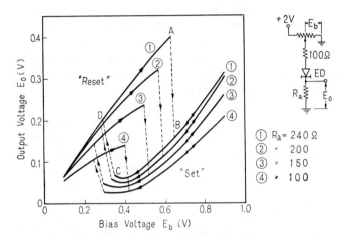

① $R_a$ = 240 Ω
② " 200
③ " 150
④ " 100

Fig. 27—dc characteristics of the comparator.

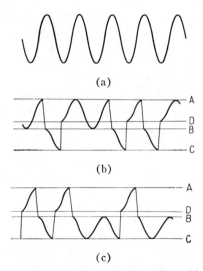

(a)

(b)

(c)

Fig. 28—Trigger and output waveforms. (a) Sinusoidal trigger. (b) Output waveform comparison at *A* (negative output). (c) Output waveform comparison at *C* (positive output).

amplify the output of the summing network in order to have sufficient output level of the integrator to trigger the Esaki diode. The amplifier should have enough bandwidth to deal with signals with the repetition frequency of sampling pulses. Employing negative feedback and peaking as shown, a common emitter amplifier with the transistor 2SA290 ($f_\alpha = 700$ Mc) has been constructed. The bandwidth and the gain of the amplifier are 0 to 50 Mc and 9, respectively. The integrator is a Miller integrator composed of a transistor 2SA290, a 2-k$\Omega$ resistor and a 30-pf capacitor. The frequency characteristics of the integrator have been proved satisfactory over the operating frequency range.

An Esaki diode 1T1103 with peak current of 2 ma, valley current of 0.3 ma, peak voltage of 40 mv, valley voltage of 300 mv and capacity of 6 pf, has been employed for the pulse modulator. Experiments have been carried out to choose optimum values of resistors so as to have larger output voltage with tolerable distortion of output waveforms. Fig. 27 shows the dc characteristics of the comparator. Figs. 28(b) and (c) show the output waveforms when triggered by sinusoidal wave of Fig. 28(a). The waveform of Fig. 28 (b) is obtained when the comparison level is chosen at point $A$ in Fig. 27, which corresponds to the "reset" point in Fig. 24. Assuming that the Esaki diode is in set state when the negative trigger exceeds point $C$, the Esaki diode is reset and jumps to point $D$. The circuit parameters are chosen so that the jump occurs at the negative peak. Following this the output waveform moves from point $D$ to point $A$ according to the increase of the trigger amplitude. At point $A$, if the integrator output voltage plus trigger amplitude is more than the voltage of point $A$, the Esaki diode is set and jumps to point $B$. The waveform of Fig. 27(b) results from this procedure and the negative output pulse is identified to be present when the waveform exceeds the level $B$, so that the voltage between $C$ and $B$ should preferably be made larger than the voltage between $B$ and $A$. The waveform of Fig. 28(c) is obtained when the comparison level is chosen at point $C$ in Fig. 27 which corresponds to the "set" point in Fig. 24. The polarity of output pulse is positive in this case. In the experimental system, the positive output resulted from the comparison at point $C$ is preferred because the output amplitude is larger.

The demodulator consists of a low-pass filter and a transitor amplifier. The low-pass filter has been designed to have the cutoff frequency of 4 Mc.

Fig. 29 shows the input-output characteristics of the experimental system for the signal frequency of 1 Mc and for the sampling frequency of 40 Mc. Fig. 30 shows the over-all frequency characteristics of the system. Figs. 31 (a), (b), (c), (d) and (e), on page 1534, show the output waveform of the modulator with dc input at the sampling rate of 10 Mc to 50 Mc. Fig. 32 (a) shows 500-kc sinusoidal input to the modulator and Figs. 32 (b),

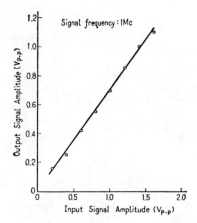

Fig. 29—Input-output characteristics of the experimental system.

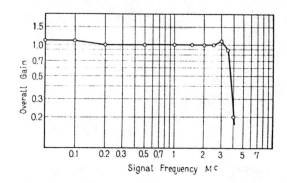

Fig. 30—Over-all frequency characteristics.

(c), (d), (e) and (f) show the demodulator output waveforms at the sampling rate of 10 Mc to 50 Mc.

Feeding standard test pattern signals to the experimental system, demodulated signals have been displayed on a television monitor. Fig. 33 (a) shows the original test pattern and Figs. 33 (b), (c), (d), (e) and (f) show the demodulated pattern at the sampling rate of 10 Mc to 50 Mc. The result of observations indicates that the system operating at the sampling rate as low as 30 Mc reproduces a considerably good picture and that even at the sampling rate of 10 Mc the reproduction is better than 1-bit PCM.

## Acknowledgment

Thanks are due to Professor Sakamoto and the members of the Laboratory of the University of Tokyo for their advice, and to J. Murakami, H. Fujita and H. Takano for their experimental contributions.

The authors also wish to acknowledge with thanks the grant of the High Speed PCM Research Project of the Ministry of Education under the direction of Professor T. Osatake, the supports of Dr. I. Sikiguchi of the Central Research Laboratories, Hitachi Limited and the Okabe Memorial Scholarship of the Institute of Electrical Communication Engineers of Japan.

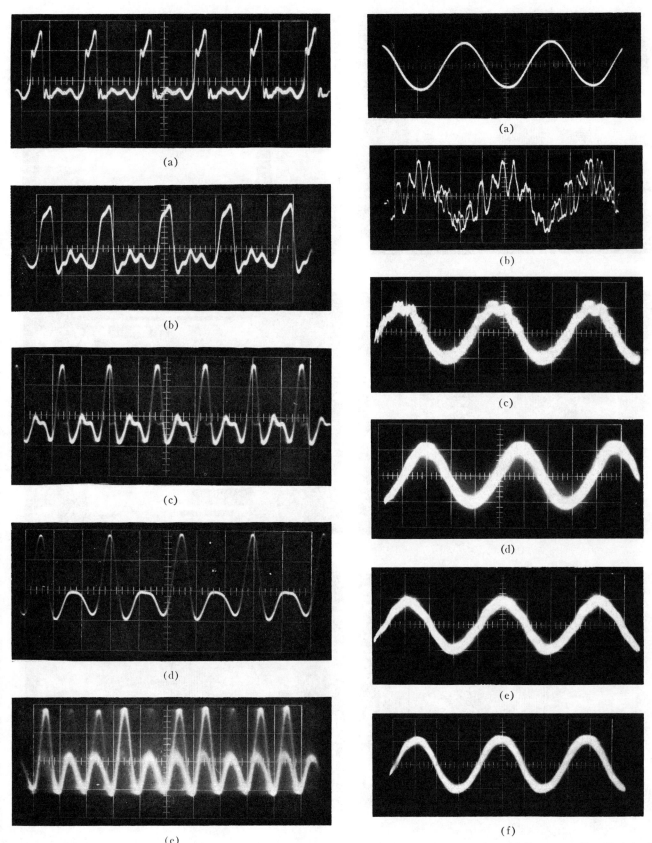

Fig. 31—(a) $f_r = 10$ Mc, $H$: 0.04 μsec/d.v., $V$: 0.1 v/d.v. (b) $f_r = 20$ Mc, $H$: 0.04 μsec/d.v., $V$: 0.1 v/d.v. (c) $f_r = 30$ Mc, $H$: 0.04 μsec/d.v., $V$: 0.1 v/d.v. (d) $f_r = 40$ Mc, $H$: 0.02 μsec/d.v., $V$: 0.1 v/d.v. (e) $f_r = 50$ Mc, $H$: 0.02 μsec/d.v., $V$: 0.1 v/d.v.

Fig. 32—(a) Input: 500 kc sinusoid, $H$: 0.5 μsec/d.v., $V$: 0.5 v/d.v. (b) $f_r$: 10 Mc, $H$: 0.5 μsec/d.v., $V$: 0.5 v/d.v. (c) $f_r$: 20 Mc, $H$: 0.5 μsec/d.v., $V$: 0.5 v/d.v. (d) $f_r$: 30 Mc, $H$: 0.5 μsec/d.v., $V$: 0.5 v d.v. (e) $f_f$: 40 Mc, $H$: 0.5 μsec/d.v., : $V$ 0.5 v d.v. (f) $f_r$: 50 Mc, $H$: 0.5 μsec/d.v., $V$: 0.5 v d.v.

Fig. 33— (a) Original. (b) $f_r = 10$ Mc. (c) $f_r = 20$ Mc. (d) $f_r = 30$ Mc. (e) $f_r = 40$ Mc. (f) $f_r = 50$ Mc.

# DESIGN OF STABLE HIGH ORDER 1-BIT SIGMA-DELTA MODULATORS

Tapani Ritoniemi, Teppo Karema and Hannu Tenhunen

Signal Processing Laboratory
Tampere University of Technology
P.O.Box 527, SF-33101 Tampere, Finland

*Abstract - In this paper, a method for designing stable 1-bit high order (≥ 3) sigma-delta modulators is presented. The stability analysis is based on the root locus and modeling the quantizer for each clock period at a time. The quantizer's gain in the modulator at the present clock period determines the modulator's stability for the next clock period. If the modulator is stable during each clock period, it is unconditionally stable and behaves as a linear A/D converter. Examples with 3rd, 4th, 5th and 6th order sigma-delta modulators are given to explore the use of the proposed method in practise. With the designed 6th order modulator it is possible to achieve 23 bit signal to quantization noise ratio at the oversampling ratio of 64.*

## I. INTRODUCTION

The need of high resolution analog-to-digital (A/D) converters has increased the use of the oversampling and sigma-delta modulation noise shaping technique. The resolution of sigma-delta A/D converter depends on the oversampling ratio (M) and the order of the noise shaping function, ie. the number and the location of the zeros in the noise transfer function (NTF). Using the second order modulator and oversampling ratio of 256 it is possible to reach 14.5-bit resolution [1]. Using third order modulator with M of 64 near 16-bit resolution has been demonstrated [2,3]. By using fourth order modulator over true 16-bit resolution has been reported [4]. The performance of the sigma-delta A/D converter is defined by the oversampling ratio and by the order of the modulator. For a given frequency range the uppermost performance is defined by the maximum sampling frequency on which the modulator is working properly. With current CMOS switched capacitor technologies the maximum sampling frequency is in the range of 5-20 MHz. Unfortunately, the maximum operating frequency doesn't give the best performance due the fact that the circuit noise increases. By raising the order of the noise shaping, improved resolution with the same CMOS technology can be obtained, or the Nyquist frequency can be increased respectively. For example, with the 6th order modulator over 23-bit theoretical resolution is possible with oversampling ratio of 64 and a 16-bit resolution with a modest oversampling ratio of 32. This makes the high order structures also ideal for wide frequency ranges. Unfortunately, such high order structures have never been demonstrated and are generally assumed to be unstable and impossible to design.

In this work, a method to design systematically high order modulator structures including 6th order topologies is presented. The design method is based on root locus analysis [7] and controlled internal node voltages for stable operation. 1-bit quantization is used solely because its linearity, which is difficult to realize with multi-bit quantization, especially for over 20 bit linearity range. The modulator's structure is straightforward direct connection of integra-

tors. With this approach, better performance can be obtained than using cascading techniques with first order modulators [2,3] or second order structures [5,6]. The proposed design methods can be used for designing switched capacitor analog modulators for A/D converters as well as digital modulator for D/A conversion. The method can be use also for multibit designs to increase their noise shaping order.

## II. MODULATOR ANALYSIS

The sigma-delta modulator can be divided into linear loop filter H(z) and a nonlinear quantizer shown in Fig. 1. The gain of the quantizer depends on the previous states of the modulator and the current input signal. The stability of the modulator is predicted by investigating of quantizer's gain $\lambda$, defined as voltage ratio of the quantizer's output and input voltages. The important property of the gain is that it is nonlinear because it is varies between the samples. Using this model we can define the signal transfer function (STF) as a function from modulator input to output using variable gain $\lambda$ in the feedback loop:

$$STF(z,\lambda) = \frac{H(z)}{1 + \lambda H(z)} \qquad (1)$$

For the additive noise produced by the quantization process we define:

$$NTF(z,\lambda) = \frac{1}{1 + \lambda H(z)} \qquad (2)$$

The output of the modulator is using (1) and (2):

$$D_{out} = STF \times V_{in} + NTF \times Q_n \qquad (3)$$

Where $V_{in}$ is the input voltage of the modulator and $Q_n$ is uniformly distributed white quantization noise. We can investigate the modulator's stability with chosen loop filter using root locus analysis [7] by varying the quantizer's gain $\lambda$. The stability is defined separately for each sample, and if the stability is found for all samples, the whole modulator is stable.

Looking the problem in the z-domain it is possible to find a linear stable range for the quantizer's input voltage using root locus analysis. This range can be considered as a safety region for the

Reprinted from *IEEE Proc. ISCAS'90*, pp. 3267–3270, May 1990.

quantizer's gain where modulator's operation can be predicted for the next clock period. When the voltage before the quantizer is smaller than the voltage in the linear stable region, the voltage is limited and the modulator is stable. The voltage before quantizer represents the sum of long term quantization error. When this voltage is low, which corresponds high quantizer gain, the output bit stream of the modulator corresponds the input signal and no information is lost. But when the quantizer's gain is smaller than in the linear stable region, the modulator has gone to unstable region and it can't recover because the negative feedback is changed to positive. If the voltage before quantizer is not limited, the modulator will start to oscillate when its input voltage exeeds the maximum input. The unstability can be avoided by limiting the voltages in the integrators in such way that the gain is higher than the minimum stable value and/or limiting the modulator's input range. This has been used as a design quide line for high order modulators presented in the next section.

## III. MODULATOR STRUCTURES

Sigma-delta modulators can be realized using feedforward structure as in Fig. 2 or multiple feedback structure as in Fig. 3 with the same root locus. This means that they behave almost identically. The small difference is found in the different voltage distribution in the integrators. The resulting NTF for even n-order modulator ($m=\frac{n}{2}$) is:

$$NTF(\lambda) = \frac{\prod_{i=1}^{m}( 1 + a_iH_{2i-1}H_{2i})}{\prod_{i=1}^{m}(1+a_iH_{2i-1}H_{2i})+\lambda\left[\sum_{i=1}^{m}[(b_{2i-1}H_{2i-1}+b_{2i}H_{2i-1}H_{2i})\prod_{j=i+1}^{m}(1+a_jH_{2j-1}H_{2i})\prod_{k=1}^{i-1}H_{2k-1}H_{2k}]\right]} \quad (4)$$

The transfer functions $H_1...H_n$ in SC implementation are delayed integrators, $H=\frac{z^{-1}}{1-z^{-1}}$. The resulting NTF is:

$$NTF(z,\lambda) = \frac{\prod_{i=1}^{m}( (z-1)^2 + a_i)}{\prod_{i=1}^{m}( (z-1)^2 + a_i)+\lambda\left[\sum_{i=1}^{m}[(z-1)^{i-1}(b_{2i-1}(z-1)+b_{2i})\prod_{j=i+1}^{m}((z-1)^2+a_j)]\right]} \quad (5)$$

It can be seen from the equations (4) and (5) that NTF zeroes are fixed and they don't vary with lambda as the poles do. Both the feedforward and the multiple feedback structure lead to stable structure with suitable coefficients $b_1...b_n$. Same kind formulas as Eq 4-5 can be derived for the odd order modulators.

The high order (≥3) modulator stability depends both on the input level and the feedback coefficients. Compared to the second and the first order structures which stability depends only on feedback coefficients, the useful input range of high order modulator is smaller. The maximum input voltage of the modulator can be found by simulation. The simulated results are in Fig. 4. The maximum input level for 3rd order structure is 0.7 when the feedback is 1 and maximum input for 6th order structure is 0.3 (with the reference voltage=±1).

## IV. OPTIMIZED NTF

With the architecture shown in Figs. 2-3 it is possible to adjust NTF zeroes and poles independently. The zeroes of the NTF can be optimized for maximum signal to quantization noise ratio (SQNR) on the base band. The poles of the NTF are optimized for maximum stability, because the poles of the NTF affects very little for SQNR. The design of the modulator is started by placing the zeroes of the NTF to such frequencies that equripple attenuation characteristics in baseband is achieved. The frequencies of the zeroes are controlled only by the feedback coefficients ($a_1, a_2 \cdots a_m$) of the biquads. The zeroes are located in $z_i = 1\pm\sqrt{a_i}$. This means that, when $H = \frac{z^{-1}}{1-z^{-1}}$, then accurate transmission zeroes can be only placed on 0-frequency. The low ferquency (≠0) transmission zeroes in realistic SC implementation can be made more accurate than the ideal integrators due to integrator's finite gain, which moves low-frequency zeroes towards to the unit circle.

The modulator's stability is controlled by the poles of the NTF. The transmission poles cannot be located exactly because the quantizer's gain changes dynamically through the modulator's operation. By adjusting the integrator weights ($b_1, b_2 \cdots b_n$), the modulator can be designed to be stable in the limited input range.

The stability of this architecture depends on the gain of the quantizer. This gain is controlled by the voltages of the integrators. The coefficient are defined in such way that the integrator voltages are unconditionally limited below the maximum of the stable operation range. The resulting the high order modulator structure will always be stable.

## V. EXAMPLES

The design examples are taken for the most promising high order modulator structures. The algorithm simulations are done using high level models. The performance is determined from FFT spectrum. The data is weighted by the Kaiser window with beta of 20. This window has 170 dB sideband attenuation which is enough for over 20 bit SQNR calculations.

### 1. 3rd order modulator

3rd order modulator gives about 90 dB SNR with oversampling ratio of 64 with $b_1 = 8$, $b_2 = 3$, $b_3 = 1$ and $a_1 = 0.001$. This structure is potential to replace second order structures. The modulator's maximum input is 0.7 compared to the reference voltage. This input voltage range is not practically low compared to the second order structure, because the voltage distribution of second integrator limits also the input voltage range in second order case. The implementation is less insensitive to process mismatches compared to cascaded structures. The simulated modulator's output spectrum is in Fig. 5.

### 2. 4th order SC modulator

The 4th order modulator can be realized using the feedforward architecture. One complex pair of the zeroes in the NTF is moved from the zero frequency using feedback coefficient $b_1$. Placing these zeroes optimally results only 2 dB improvement in SNR. This small improvement increases the complexity. By using the proposed design procedure the coefficients $b_1$, $b_2$, $b_3$, $b_4$ and $a_2$ can be determined. For this structure we get the NTF:

$$TF = \frac{[(1+a_2)(b_2-b_1)+b_4-b_3]z^4+[(3+a_2)b_1-2b_1+b_3]z^3+(b_2-3b_1)z^2+(b_1-4)z}{[\lambda((1+a_2)(b_2-b_1)+b_4-b_3)+1]z^4+[\lambda((3+a_2)b_1-b_2+b_3)-4]z^3+[\lambda(b_2-3b_1)+6]z^2+(\lambda b_1-4)z+1} \quad (6)$$

When the $b_1 = 45$, $b_2 = 37$, $b_3 = 4$, $b_4 = 1$ and $a_2 = 0.0015$ we get root locus shown in Fig. 6. Practical SC implementation is in Fig. 7. and corresponding spectrum in Fig. 8. For this modulator is possible to achieve over 18-bit signal to quantization noise ratio with oversampling ratio of 64. The 18-bit resolution is very near today's CMOS technology limit.

### 4. 5th order SC modulator

The odd order modulator can also be realized using the proposed procedure. The designed 5th order modulator gives about 20 bit resolution when using two NTF zeroes in the baseband. The simulated output spectrum of the modulator is presented in Fig. 9. and the root locus in Fig. 10.

### 5. 6th order SC modulator

For the sixth order modulator and we get same kind of results as for the fifth order. The structure offers better resolution with almost the same circuit complexity. However, it is much more difficult to find stability with $(b_1 \cdots b_6)$ coefficients than in fourth order case. Results of the successful design are in Figs. 11-12. The maximum modulator's input voltage is 0.3 relative to the reference voltage. This makes difficult to design such input stage which have the noise below quantization noise in any technology.

## VI. CONCLUSION

A method for designing stable high order 1-bit modulators was presented. The work shows that there are no limits in chosing the architecture for any order 1-bit modulators. The high order modulators push the quantization noise below the system noise level already with oversampling ratio of 64. The bandwidth of the high resolution A/D converters can be increased by using high order modulators and lower oversampling ratios. The presented design methodology can be applied also to D/A conversion.

## REFERENCES

[1] S.R. Norsworthy, I.G. Post and H.S. Fetterman, "A 14-bit 80-kHz Sigma-Delta A/D Converter: Modeling, Design, and Performance Evaluation", *IEEE Journal os Solid-State Circuits*, vol 24, pp. 256-266, Apr 1989.

[2] K. Uchimura, T. Hayashi, T. Kimura and A. Iwata, "Oversampling A-to-D and D-to-A Converters with Multistage Noise Shaping Modulators", *IEEE Trans. on Acoustics, Speech and Signal processing*, vol. 36, pp. 1899-1905, Dec 1988.

[3] M. Rebeschini, N. Bavel, P. Rakers, R. Greene, J. Caldwell and J. Haug, "A High-Resolution CMOS Sigma-Delta A/D Converter with 320 kHz Output Rate", in *Proc. IEEE Int. Symp. on Circuits and Systems*, May 1989, pp. 246-249.

[4] Crystal Semiconductor Corp., CS5528 preliminary product information.

[5] L. Longo and M. Copeland, " A13 bit ISDN-band oversampled ADC using two-stage third order noise shaping ," in *proc. IEEE Custom IC Conf.* Jan 1988, pp. 21.2.1-21.2.4.

[6] T. Karema, T. Ritoniemi H. Tenhunen, "Fourth Order Sigma-Delta Modulator Circuit for Digital Audio and ISDN Applications",in *Proc. IEE European Conference on Circuit Theory and Design,* Sep 1989. pp. 223-227.

[7] E. Stikvoort, "Some remarks on the stability and performance of the noise shaper or sigma-delta modulator," in *IEEE trans on Commun.* COM-36 pp. 1157-1162 Oct, 1988

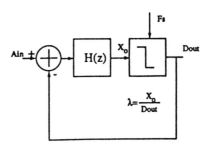

Fig. 1. General sigma-delta modulator structure

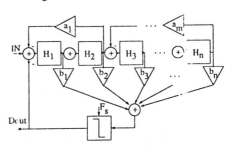

Fig. 2. Feedforward sigma-delta modulator structure

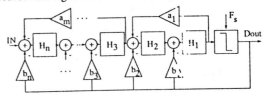

Fig. 3. Multiple feedback sigma-delta modulator structure

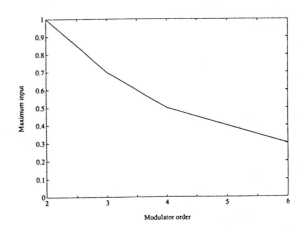

Fig. 4. Modulator's maximum relative input ( $b_n = 1$ and Vref = $\pm 1$)

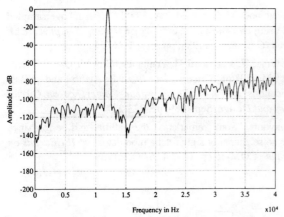

Fig. 5. Output spectrum of the 3rd order modulator, Fs=3.072 MHz, $2^{15}$ samples FFT weighted with Kaiser window (beta=20)

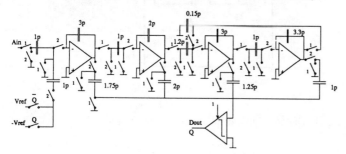

Fig. 6. Simplified SC implementation of the 4th order modulator structure

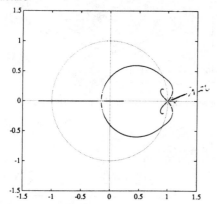

Fig. 7. Root locus for the 4th order modulator

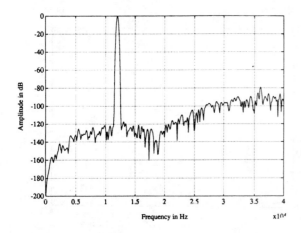

Fig. 8. Output spectrum of the 4th order modulator

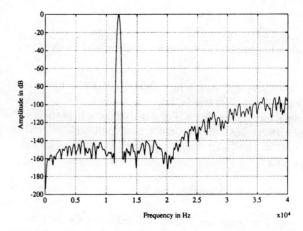

Fig. 9. Output spectrum of the 5th order modulator structure

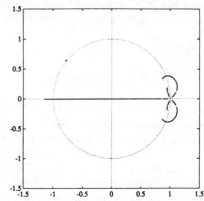

Fig. 10. Root locus for the 5th order modulator

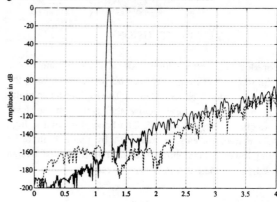

Fig. 11. Output spectrum of the 6th order modulator, solid line real valued NTF zeroes, dash line optimized complex valued NTF zeroes

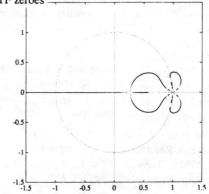

Fig. 12. Root locus of the 6th order modulator

# Reduction of Quantizing Noise by Use of Feedback*

H. A. SPANG, III†, MEMBER, IRE, AND P. M. SCHULTHEISS‡, MEMBER, IRE

*Summary*—Many information transmission systems use a discrete (digital) channel. Since most input signals are continuous, the conversion cannot be accomplished without an error which, for many cases, may be considered to have the characteristics of white noise. A method has been suggested to reduce this error by using linear feedback around the quantizer to shape the noise spectrum. Each output sample will then contain not only signal information but also information about the errors in the previous samples. Such a system is analyzed for random input signals of a rather general nature. Under assumptions allowing essentially no clipping in the quantizer and setting an upper bound on the coherence between samples of the input signal, the system can be represented by a simple model. A comparison is made of the mean-square error with and without feedback. It is shown that considerable reduction in noise power can be obtained by a slight increase in sampling rate. For example, an increase of 25 per cent in the sampling rate provides a 95 per cent decrease in error-noise power. This is equivalent to having about two additional bits per sample in the transmission channel.

## INTRODUCTION

MANY INFORMATION processing systems convert analog input signals into digital signals for transmission over a discrete channel. Such conversion (and subsequent reconstruction) cannot be accomplished without error because the continuum of possible input values must be represented by a discrete set of values in the channel. This error, commonly known as quantizing noise, is one of the major sources of inaccuracy in such systems. Its study has therefore assumed considerable importance,[1-4] and various schemes have been proposed to achieve noise reduction.[2-4] The present paper presents an analysis of a procedure invented by C. C. Cutler[4] of Bell Telephone Laboratories.

A block diagram of the system is shown in Fig. 1. The input signal is assumed to be sampled at regular intervals of τ seconds. A uniform quantizer with the characteristic shown in Fig. 2 is used for ease of computation. The difference between the output and input of the quantizer is fed back through a linear filter whose output samples are added to the input. If $x_4$ is sampled synchronously

* Received June 25, 1962. This paper is based in part on a dissertation by H. Austin Spang, III for the degree of Doctor of Engineering in the Yale School of Engineering. The dissertation was supported by Bell Telephone Laboratories, Murray Hill, N. J.
† GE Research Laboratory, Schenectady, N. Y. Formerly at Yale University, New Haven, Conn.
‡ Yale University, New Haven, Conn.

[1] W. R. Bennett, "Spectra of quantized signals," *Bell Sys. Tech. J.*, vol. 27, pp. 446–472; July, 1948.
[2] S. P. Lloyd, "Least Squares Quantization in P.C.M.," Bell Lab. Memo., Murray Hill, N. J. (unpublished).
[3] J. Max, "Quantizing for minimum distortion," IRE TRANS. ON INFORMATION THEORY, vol. IT-6, pp. 7–12; March, 1960.
[4] C. C. Cutler, "Transmission systems employing quantization," U.S. Patent No. 2,927,962; 1960.

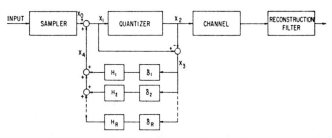

Fig. 1—The block diagram of the system.

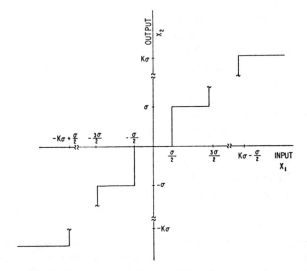

Fig. 2—Input-output characteristics of the quantizer.

with $x_0$, one can clearly represent the linear filter[5] by the combination of delay lines, $\delta_1, \delta_2, \cdots, \delta_R$ with time delays of $\tau, 2\tau, \cdots, R\tau$, and gain constants $H_1, H_2, \cdots, H_R$, respectively. Since $x_3 = x_1 - x_2$ and $x_1 = x_0 + x_4$ the input to the quantizer is a linear combination of the input signal and the previous $R$ quantizing errors.

An elementary example will suggest the manner in which the system reduces the quantizing noise. Suppose the input signal has the constant value $3\sigma/2$ [Fig. 3(a)]. Suppose further that at discontinuity points in Fig. 2 the quantizer output assumes the limiting value of $x_2$ as $x_1$ approaches the discontinuity from the left. In the absence of feedback the specified input signal results in a quantizer output constant at the value $\sigma$, or a constant quantizing error of $\sigma/2$ (Fig. 3b). With a feedback network using a single delay of $\tau$ and gain constant of unity ($H_1 = 1$, $H_r = 0$ for $r > 1$) the quantizer input at any

[5] For mathematical convenience, we have assumed that there is always some delay in the feedback path and that only the past $R$ samples of $x_3$ are of interest.

Reprinted from *IRE Trans. Commun. Systems*, pp. 373–380, December 1962.

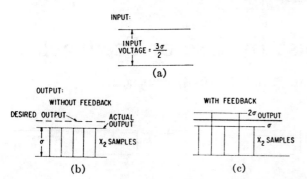

Fig. 3—The effect of feedback for a dc input to the quantizer.

sampling instant consists of the sum of the present input sample and the discrepancy between quantizer input and output at the last sample. $x_1$ therefore alternates between $3\sigma/2$ and $2\sigma$ while $x_2$ alternates between $\sigma$ and $2\sigma$. The quantizing error still has a magnitude of $\sigma/2$ at every sampling instant, but it now alternates in sign. The quantizing noise has been shifted in frequency from dc to a range where it can be eliminated to any desired extent by an averaging operation or low-pass filter.

It should perhaps be pointed out that the above example is artificial in one important respect. The bandwidth of the input signal is zero and therefore has no meaningful relation to the sampling frequency. One could obviously achieve large reductions in error by lowering the sampling frequency and employing proportionately finer quantization. In realistic situations one would use the sampling theorem as a guide in choosing an appropriate sampling frequency $(1/\tau)$. Roughly speaking one would choose $\tau$ such that most of the signal power falls into the range $0 \leq \omega \leq \pi/\tau$, *i.e.*, below half the sampling frequency. It will be shown that, under assumptions frequently met in practice,[6] any feedback network of the proposed form definitely increases the total error power in the frequency range $0 \leq \omega \leq \pi/\tau$. In order to reduce the error power by filtering, it is therefore necessary either to sacrifice some signal power or to increase the sampling frequency above the value which would have been used in the absence of feedback. The purpose of the feedback system is to redistribute the error power over the frequency band in a manner determined by the feedback network. With proper choice of this network a large reduction in quantizing noise can be achieved through a relatively slight increase in sampling frequency.

## The Error Spectrum

For the purposes of this paper the quantizing error $e^n$ at the $n$th sampling instant will be defined as

$$e^n = x_0^n - x_2^n. \qquad (1)$$

$x_0^n$ and $x_2^n$ are respectively the sampled input and the quantizer output at the $n$th sampling instant. (See Fig. 1.)[7]

Formal derivation of the error spectrum requires a great deal of rather cumbersome mathematical manipulation, most of which can be avoided by beginning with an intuitively reasonable assumption. In order to reach the results of primary interest as rapidly as possible the authors have chosen this intuitive approach. A more rigorous argument is outlined in the Appendix and a detailed derivation is available in a technical report.[8]

Widrow[9] has considered the noise spectra generated by uniform quantizers, such as the one shown in Fig. 2. For the case of moderately fine quantizing and sufficiently small input-signal power so that the quantizer rarely saturates,[10] he demonstrates that the quantizing noise has a total power of $\sigma^2/12$ and a spectrum essentially flat over the signal band. Under fairly weak restrictions on sample-to-sample correlation he further shows that self and joint moments of quantizer input and output can be obtained by representing the quantizer as a device which adds signal-independent white noise to the signal. Our interest centers on a closely related computation, that of the noise spectrum. It therefore appears reasonable that the same representation should be applicable, so that the feedback system can be analyzed in terms of the equivalent arrangement shown in Fig. 4. The formal analysis demonstrates that this is indeed permissible under the following two assumptions:

A) $$K\sigma - \frac{\sigma}{2}\sum_{r=0}^{R}|H_r| \geq M_1\sqrt{\psi_0} \qquad (2)$$

where $K\sigma$ is the saturation level of the quantizer, $\psi_0$ is the mean-square value of the signal, $M_1$ is a constant such that (for predetermined small $\epsilon > 0$)

$$\int_{M_1\sqrt{\psi_0}}^{\infty} dx_0 f_0(x_0) \leq \epsilon, \qquad (3)$$

and $f_0(x_0)$ is the probability density function of the input signal $x_0$.

B) $$\Omega\left(\frac{2\pi m_n}{\sigma}, \frac{2\pi m_{n-1}}{\sigma}, \cdots, \frac{2\pi m_0}{\sigma}\right) \leq \epsilon \qquad (4)$$

$m_p$ an integer $\neq 0$, and $p = 0, 1, \cdots, n$. $\epsilon$ is a preassigned small constant and $\Omega(\omega_n, \cdots, \omega_0)$ is the Fourier transform of the $n + 1$ dimensional probability density function of the quantizer input. Eq. (4) is, of course, closely related to Widrow's "multidimensional Nyquist condition".

---

[6] One of the assumptions sets an upper bound on the sample-to-sample correlation. The perfect correlation in our elementary example violates this assumption and accounts for the failure of the noise power to increase with feedback.

[7] Throughout this paper subscripts on $x$ identify the point of measurement on Fig. 1 while superscripts specify time in terms of sampling instants.

[8] H. A. Spang, III, "Quantizing Noise Reduction," GE Res. Lab., Schenectady, N. Y., Rept. No. 62-RL-(2999E); April, 1962.

[9] B. Widrow, "A Study of rough amplitude quantization by means of the Nyquist sampling theorem," IRE TRANS. ON CIRCUIT THEORY, vol. CT-3, pp. 266–276; December, 1956.

[10] Widrow actually works with an infinite quantizer, so that the saturation problem does not arise.

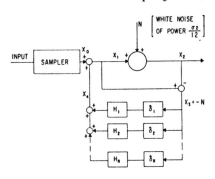

Fig. 4—Approximate model of the system.

In physical terms, assumption A) limits the probability that the quantizer input exceeds the saturation level to a value of order $\epsilon$. It therefore produces an effect similar to that of Widrow's infinite quantizer. Eq. (2) clearly becomes more difficult to satisfy as the number of elements in the feedback path (gain constants $H_r$) increases.

Physical interpretation of condition B) is more difficult in general. However, if the input process is Gaussian it can be shown that (4) is implied by the simpler relation

$$\frac{E[(x_0^n)^2 \mid x_0^{n-1}, \cdots, x_0^0]}{\sigma^2} \geq \frac{M_1^2}{4\pi^2}. \tag{5}$$

$E[(x_0^n)^2 \mid x_0^{n-1}, \cdots, x_0^0]$ represents the conditional variance of $x_0^n$ given $x_0^{n-1}, \cdots, x_0^0$ and $M_1$ is determined by (3). Hence (5) sets a minimum on the allowable spread of the conditional distribution, or a maximum on the allowable sample-to-sample correlation. In view of the coherence introduced by the feedback scheme it is reasonable that the criterion should compare $\sigma$ with the width of the conditional distribution rather than the *a priori* distribution of the input. If the input signal is not Gaussian, (5) becomes inapplicable. The authors have been unable to find a similarly useful replacement of (4) for any more general class of input distributions.

Using Fig. 4, it is a simple matter to calculate the autocorrelation function of the error. It is clear that

$$x_4^n = \sum_{r=1}^{R} H_r x_3^{n-r} = \sum_{r=1}^{R} H_r (x_1^{n-r} - x_2^{n-r}). \tag{6}$$

In the above equation we have assumed as a matter of convenience that $n \geq R$. From Fig. 4,

$$x_1^n = x_0^n + x_4^n = x_0^n + \sum_{r=1}^{R} H_r (x_1^{n-r} - x_2^{n-r}) \tag{7}$$

and

$$x_2^n = N^n + x_1^n \tag{8}$$

where $N^n$ is the $n$th sample of the noise. Since the noise is white, its autocorrelation function is

$$\psi_N(\beta \tau, \alpha \tau) = \frac{\sigma^2}{12} \delta[(\beta - \alpha)\tau] \tag{9}$$

where $\delta(\ )$ is the Kroneker delta function

$$\delta[(\beta - \alpha)\tau] = \begin{cases} 1 & \beta = \alpha \\ 0 & \beta \neq \alpha. \end{cases}$$

Substituting (8) into (7)

$$x_1^n = x_0^n - \sum_{r=1}^{R} H_r N^{n-r}. \tag{10}$$

Therefore, from (1)

$$e^n = x_0^n - x_2^n = x_0^n - x_1^n - N^n = -N^n + \sum_{r=1}^{R} H_r N^{n-r}. \tag{11}$$

Thus, the autocorrelation function of the error is

$$\psi_e[(\beta - \alpha)\tau] = \overline{e^\beta e^\alpha}$$
$$= \sum_{i=0}^{R} \sum_{s=0}^{R} H_i H_s \frac{\sigma^2}{12} \delta[(\beta - \alpha - i + s)\tau] \tag{12}$$

where by definition $H_0 = -1$ and $\beta \geq \alpha \geq R$. Written as a single sum,

$$\psi_e[(\beta - \alpha)\tau] = \begin{cases} \dfrac{\sigma^2}{12} \displaystyle\sum_{r=0}^{R-\beta+\alpha} H_r H_{\beta-\alpha+r} & 0 \leq \beta - \alpha \leq R \\ 0 & \beta - \alpha > R. \end{cases} \tag{13}$$

The coherence in the error extends only over $R$ adjacent samples because the noise is uncorrelated from sample to sample under the stated assumptions.

The mean-square value of the quantizing noise is obtained by setting $\beta - \alpha = 0$

$$\psi_e(0) = \frac{\sigma^2}{12} \left(1 + \sum_{r=1}^{R} H_r^2\right). \tag{14}$$

Thus, the quantizing noise without feedback is $\sigma^2/12$. The addition of feedback increases the over-all noise power. However, we shall now demonstrate that the frequency spectrum of the noise is no longer flat. It will therefore be possible to reduce the noise in a given frequency region by adding more noise to the signal outside this region.

J. Tou[11] has shown that the spectrum of a stationary sampled signal is

$$W_e(\omega) = \sum_{n=-\infty}^{\infty} \psi_e(n\tau) \exp[-jn\omega\tau]. \tag{15}$$

Since

$$\sum_{n=-\infty}^{\infty} \delta(n - i + s) \exp[-jn\omega\tau] = \exp[-j(i-s)\omega\tau], \tag{16}$$

it is clear from (12) that

$$W_e(\omega) = \frac{\sigma^2}{12} \mid H(\omega) \mid^2, \tag{17}$$

[11] J. Tou, "Digital and Sampled Data Control Systems," McGraw-Hill Book Co. Inc., New York, N. Y., p. 544; 1959.

where

$$H(\omega) = \sum_{r=0}^{R} H_r e^{-j\omega r \tau}. \tag{18}$$

Under the assumption of essentially no saturation the feedback around the quantizer, therefore, acts as a linear filter on the open-loop quantizing noise. It should be noted that the shape of this filter cannot be completely arbitrarily selected since by definition

$$H_0 = -1 = \frac{1}{\omega_s} \int_0^{\omega_s} H(\omega) \, d\omega, \tag{19}$$

where $\omega_s = 2\pi/\tau$, the sampling frequency.

Substituting (17) and (18),

$$E_n = \frac{\sigma^2}{12} \frac{\omega_1}{\omega_s} \sum_{r=0}^{R} \sum_{n=0}^{R} H_r H_n \frac{\sin (n - r)\omega_1 \tau}{(n - r)\omega_1 \tau}. \tag{21}$$

Observing that $H_0 = -1$, one can write (21) in matrix form.

$$E_n = \frac{\sigma^2 \omega_1}{12 \omega_s} [1 - 2H'S + H'N_R H] \tag{22}$$

where $H' = [H_1 \cdots H_R]$, the transpose of $H$;

$$N_R = \begin{bmatrix} 1 & \frac{\sin \omega_1 \tau}{\omega_1 \tau} & \cdots & \frac{\sin (R - 1)\omega_1 \tau}{(R - 1)\omega_1 \tau} \\ \frac{\sin \omega_1 \tau}{\omega_1 \tau} & 1 & & \vdots \\ \vdots & & 1 & \frac{\sin \omega_1 \tau}{\omega_1 \tau} \\ \frac{\sin (R - 1)\omega_1 \tau}{(R - 1)\omega_1 \tau} & \cdots & \frac{\sin \omega_1 \tau}{\omega_1 \tau} & 1 \end{bmatrix} \quad S = \begin{bmatrix} \frac{\sin \omega_1 \tau}{\omega_1 \tau} \\ \vdots \\ \frac{\sin R\omega_1 \tau}{R\omega_1 \tau} \end{bmatrix}. \tag{23}$$

## OPTIMIZATION OF THE SYSTEM

Eq. (17) indicates that the error spectrum can be controlled by adjusting the feedback network. The precise nature of the optimum network will clearly depend on the criterion used to measure system performance. For the purposes of this paper it will be assumed that the objective is to minimize the quantizing noise power in the frequency band $0 \leq \omega \leq \omega_1$, $\omega_1 \leq \pi/\tau$.[12]

It will be shown that the minimum realizable noise power in any given band is a monotone nonincreasing function of the number $R$ of feedback elements. However, as $R$ increases it becomes more and more difficult to satisfy (2) without increasing the saturation level $K\sigma$. In the limit, as $R \to \infty$, one must investigate the convergence of $\sum_{r=0}^{\infty} |H_r|$ in order to determine whether the assumption of low saturation probability can be satisfied for any finite saturation level. It is interesting to note that the condition $\sum_{r=1}^{\infty} |H_r| < \infty$ is a necessary and sufficient condition for the stability of the pulsed network in the feedback path. If and only if this network is stable can (2) be satisfied for some finite saturation level.

The entire stability problem can be eliminated by keeping $R$ finite. The subsequent numerical analysis therefore concerns itself with the minimization of noise power in a given band for a preassigned finite value of $R$.

The noise power in the band $0 \leq \omega \leq \omega_1$ is[13]

$$E_n = \frac{1}{\omega_s} \int_0^{\omega_1} W_e(\omega) \, d\omega. \tag{20}$$

A comparison of the quantizing noise power which occurs with feedback and the noise power which occurs without feedback is meaningful only if the two systems have the same number of quantizing intervals. In other words, the parameter $K$ in (2) must remain fixed while $\sigma$ can be adjusted within the limits permitted by (2). Otherwise, one could use a large number of feedback networks and in accordance with (2), adjust the number of intervals (hence $K$) such that

$$K\sigma \geq M_1 \sqrt{\psi_0} + \frac{\sigma}{2} \sum_{r=0}^{R} |H_r|. \tag{24}$$

For a constant value of $\sigma$, one would get an apparently large reduction in the noise power. However, without feedback, (24) becomes

$$K\sigma \geq M_1 \sqrt{\psi_0} + \frac{\sigma}{2} \tag{25}$$

which can be satisfied by a smaller $\sigma$ for the same value of $K$. Thus, one could quantize more finely without increasing the probability of saturation and achieve a reduction of noise power by these means. It is now evident that a portion of the noise reduction with feedback cannot be attributed to the effect of the feedback network but rather to the increased number of quantizing intervals demanded by (2) to avoid saturation.

For a fixed number of quantizing intervals (24) is satisfied if $\sigma$ varies such that

$$\sigma \geq \frac{2M_1 \sqrt{\psi_0}}{\left(2K - \sum_{r=0}^{R} |H_r|\right)}. \tag{26}$$

---

[12] An obvious modification would allow the assignment of arbitrary weight to the noise power density at any given frequency. The result would be a frequency weighted error criterion.
[13] J. Tou, *op. cit.*, p. 545.

Since the quantizing noise is proportional to $\sigma$, the smallest possible $\sigma$ should be used. Therefore, recalling that $H_0 = -1$,

$$\sigma = \frac{2M_1 \sqrt{\psi_0}}{\left[ (2K-1) - \sum_{r=1}^{R} |H_r| \right]}. \quad (27)$$

Without feedback,

$$\sigma = \frac{2M_1 \sqrt{\psi_0}}{(2K-1)}. \quad (28)$$

With this adjustment of $\sigma$, the quantizing levels are equally spaced over the effective amplitude range of the quantizer input, a range that increases with the number of feedback elements.

Substituting (27) into (22) one obtains the quantizing noise power.

$$E_n = \frac{M_1^2 \psi_0 \omega_1 [1 - 2H'S + H'N_R H]}{3\omega_s \left[ (2K-1) - \sum_{r=1}^{R} |H_r| \right]^2}. \quad (29)$$

Without feedback, $E_n$ reduces to

$$E_{n0} = \frac{M_1^2 \psi_0 \omega_1}{3\omega_s (2K-1)^2}. \quad (30)$$

We shall define an improvement ratio, $T(H)$, as the ratio of $E_n$ to $E_{n0}$.

$$T(H) \triangleq \frac{E_n}{E_{n0}} = \frac{[1 - 2H'S + H'N_R H]}{\left[ 1 - \frac{1}{2K-1} \sum_{r=1}^{R} |H_r| \right]^2}. \quad (31)$$

This ratio compares the noise power of the feedback system with that of an open-loop system having the same probability of saturation. It is therefore a measure of the change in quantizing noise power due to feedback.

We wish to minimize $T(H)$ with respect to $H$. The improvement ratio obviously has an upper bound of unity since $T(H) = 1$ can always be obtained by setting $H_r = 0$ for all $r$.

When $\omega_1 = \pi n/\tau$, $(n = 1, 2, \cdots)$, $N_R$ reduces to the identity matrix and the $S$ matrix vanishes. Therefore,

$$T(H) = \frac{1 + \sum_{r=1}^{R} H_r^2}{\left[ 1 - \frac{1}{2K-1} \sum_{r=1}^{R} |H_r| \right]^2}. \quad (32)$$

Eq. (32) is clearly minimized by

$$H_r = 0, \quad 1 \le r \le R. \quad (33)$$

Thus, if the passband of interest is equal to a multiple of half the sampling frequency, the least amount of quantizing noise is obtained with no feedback.

For general values of $\omega_1$ the minimization of $T(H)$ by analytical means does not appear to be feasible. The problem was therefore programmed for a digital computer.

Fig. 5 shows the minimum $T(H)$ for $K \to \infty$ as a function

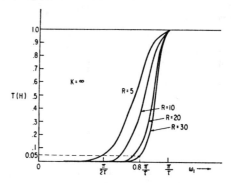

Fig. 5—The minimum $T(H)$ vs signal passband, $\omega_1$, for $K = \infty$.

of the passband ($\omega_1$) and the number of feedback elements ($R$). The case $K \to \infty$ (no saturation) was selected for detailed study because it yields the minimum value of $T(H)$ for any given $\omega_1$ and $R$. The curves indicate the possibility of substantial noise reduction through feedback as long as most of the signal power is concentrated below half the sampling frequency. Consider, for example, $R = 30$ and assume that most of the signal power is concentrated in the range $0 \le \omega \le 0.8 \pi/\tau$. With a sampling rate of $1/\tau$ samples/sec (25 per cent above the nominal Nyquist rate) it is possible, through use of feedback, to reduce the mean-square quantizing error by 95 per cent. This is equivalent to having more than two additional binary digits per sample available in the transmission channel.

The band ($0 \le \omega \le \omega_1$) over which substantial error reduction can be achieved increases monotonically with $R$. In the limit, as $R \to \infty$, its upper bound $\omega_1$ tends towards $\pi/\tau$, so that improvement is in principle obtainable with only slight increases of sampling frequency above the normal Nyquist rate. However, $\sum_{r=1}^{R} |H_r|$ also increases monotonically with $R$ and Fig. 6 indicates that the increase is quite rapid. Thus for large $R$ the assumption of small saturation probability can be satisfied only if $K$ is extremely large.

Successive terms of the sequence $H_1, H_2, \cdots, H_R$ of optimum gain constants alternate in sign in all cases for which computations were carried out. The magnitudes of the gain constants follow the pattern shown in Fig. 7. In qualitative terms, this behavior indicates that the feedback network is the sampled equivalent of a narrowband amplifier centered about half the sampling frequency. Reduction of noise power in the frequency range $0 \le \omega \le \omega_1$ is purchased at the expense of greatly increased noise power in the range $\omega_1 \le \omega \le \pi/\tau$. The noise spectrum shown in Fig. 8 ($R = 30$, $K = \infty$) confirms this general observation. It is interesting to note that the magnitude of the gain constants (Fig. 7) always tends to peak near $R/2$. No physical explanation for this phenomenon has been found.

In order to gain some insight into the dependence of $T(H)$ on $K$, computations were carried out for fixed $R$ and $\omega_1$. Fig. 9 shows the expected monotone decreasing behavior of $T(H)$ with increasing $K$ for values of $R$ and

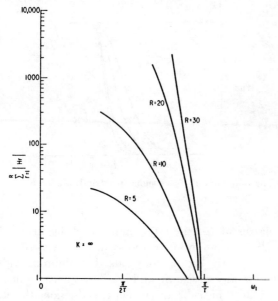

Fig. 6—$\sum^{R}_{r-1} |H_r|$ vs passband, $\omega_1$, for $K = \infty$.

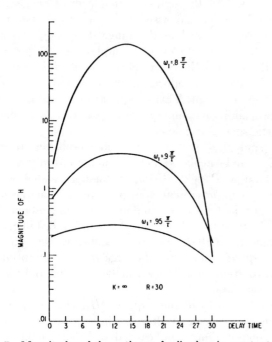

Fig. 7—Magnitudes of the optimum feedback gain constants.

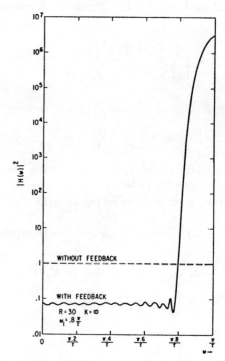

Fig. 8—Normalized spectrum of quantizing noise.

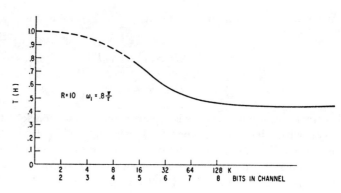

Fig. 9—The minimum improvement ratio as a function of $K$.

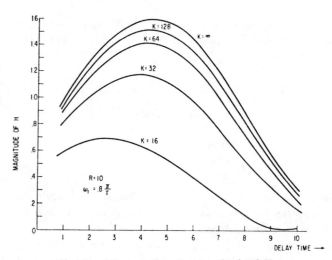

Fig. 10—Effect of $K$ on the magnitude of $H$.

$\omega_1$ within the range of practical interest. Due to computational difficulties, the dotted curve below $K = 16$ is only an approximation to the actual curve. As $K$ decreases one would expect the magnitude of the $H_r$ to decrease also. Fig. 10 lends computational support to this supposition. It also provides the interesting incidental information that the various curves do not differ greatly in shape. This suggests that the $K$ dependence of the optimum feedback network consists primarily of a gain constant multiplying the entire filter-transfer function.

The preceding discussion has ignored condition B) [see (4)]. Postulating a Gaussian signal and working with the more convenient (5) one obtains upon substituting the value of $\sigma$ from (27)

$$\frac{E[(x_0^n)^2 \mid x_0^{n-1}, \cdots, x_0^0]}{\psi_0}$$

$$\geq \frac{M_1^4}{\pi^2 \left[ (2K - 1) - \sum_{r=1}^{R} \mid H_r \mid \right]^2}. \qquad (34)$$

For values of $2K - 1$ large compared to $\sum_{r=1}^{R} \mid H_r \mid$ this requirement clearly presents no problem. More generally, suppose $M_1 = 3$ [from (3) this corresponds to $\epsilon = 0.0014$]. For the case described by Fig. 10, the right side of (34) has the value 0.0111 with $K = 16$ (5 bits per symbol) and the value $6.0 \times 10^{-4}$ with $K = 64$ (7 bits/symbol). Thus (34) does not appear to place any serious restriction on sample-to-sample correlation, at least for moderate values of $R$ and reasonably fine quantization.

## CONCLUSION

It has been shown that considerable reduction of noise power in the signal band can be obtained for a small increase in sampling frequency over the Nyquist rate. The quantizing noise can be further reduced by using a quantizer whose interval lengths depend on the probability density function at its input. However, the evaluation of such a system would be considerably more difficult.

In order to simplify the results a small probability of saturation was postulated. This imposes a fairly severe restriction on the feedback network. It is not known how much better the system could be if saturation were allowed. A second assumption set an upper bound on the amount of correlation between input samples. Intuitively, one would expect that highly correlated signals would lend themselves to more efficient processing. Thus one suspects that the present analysis gives an upper bound of $T(H)$ for signals so highly correlated that assumption B) is not satisfied.

## APPENDIX

## FORMAL DERIVATION OF THE ERROR SPECTRUM IN OUTLINE FORM

The initial problem is to state the statistical properties of the variables $x_1$, $x_2$, $x_3$, $x_4$ in the feedback loop of Fig. 1 in terms of the known statistical properties of the input variable $x_0$. This problem is clearly solved once the sta-

tistical properties of $x_1$ have been related to those of $x_0$ because $x_2$, $x_3$ and $x_4$ are completely determined by $x_1$. We therefore begin with

$$f_1(x_1^n, \cdots, x_1^0) = g_1(x_1^n \mid x_1^{n-1}, \cdots, x_1^0)$$
$$\cdot g_1(x_1^{n-1} \mid x_1^{n-2}, \cdots, x_1^0) \cdots g_1(x_1^1 \mid x_1^0) f(x_1^0) \qquad (35)$$

where $f_1(x_1^p, \cdots, x_1^0)$ is the joint probability density function of the variable $x_1$ at sampling instants 0 through $p$, while $g_1(x_1^p \mid x_1^{p-1}, \cdots, x_1^0)$ is the conditional probability density of $x_1$ at sampling instant $p$ given $x_1$ at all previous sampling instants.

At the initial sampling instant (index 0) the feedback signal $x_4$ is zero, hence $x_1^0 = x_0^0$. For $p > 0$

$$x_1^p = x_0^p + x_4^p. \qquad (36)$$

It is evident from Fig. 1 that $x_4^p$ is completely determined by $(x_1^{p-1}, \cdots, x_1^0)$. Hence, from (36) specification of $(x_1^{p-1}, \cdots, x_1^0)$ uniquely determines $(x_0^{p-1}, \cdots, x_0^0)$. Conversely, specification of $(x_0^{p-1}, \cdots, x_0^0)$ uniquely determines $(x_1^{p-1}, \cdots, x_1^0)$. It follows that

$$g_1(x_1^p \mid x_1^{p-1}, \cdots, x_1^0)$$
$$= g_0(x_1^p - x_4^p \mid x_1^{p-1} - x_4^{p-1}, \cdots, x_1^1 - x_4^1, x_1^0). \qquad (37)$$

Eq. (37) can now be substituted into (35) and since $x_4^p$ is a known function of $(x_1^{p-1}, \cdots, x_1^0)$ our first objective has been reached. There remains only the reduction of $f_1(x_1^n, \cdots, x_1^0)$ to a form more convenient for later manipulation. The result is

$$f_1(x_1^n, \cdots, x_1^0) = \sum_{k_0=-K}^{K} \cdots \sum_{k_{n-1}=-K}^{K} \prod_{r=0}^{n-1} SK[x_1^r, k_r \sigma]$$
$$\cdot f_0([I - H]\Phi + HX) \qquad (38)$$

where,

$$SK[x_1^r, k\sigma] = \begin{cases} U[x_1^r - (K - \frac{1}{2})\sigma] \; k = K \\ U[x_1^r - (k - \frac{1}{2})\sigma] - U[x_1^r - (k + \frac{1}{2})\sigma] \\ \qquad -K + 1 \leq k \leq K - 1 \\ 1 - U[x_1^r + (K - \frac{1}{2})\sigma] \; k = -K. \end{cases} \qquad (39)$$

$U(x)$ is the unit-step function, $f_0(\ )$ is the joint probability-density function of $(x_0^n, \cdots, x_0^0)$. $I$ is an $n + 1$ dimensional unit matrix and

$$X = \begin{bmatrix} x_1^0 \\ x_1^1 \\ \vdots \\ x_1^p \\ \vdots \\ x_1^n \end{bmatrix} \qquad \Phi = \begin{bmatrix} k_0\sigma \\ k_1\sigma \\ \vdots \\ k_p\sigma \\ \vdots \\ k_n\sigma \end{bmatrix} \qquad H = \begin{bmatrix} 1 & & & & & & 0 \\ -H_1 & 1 & & & & & \\ \vdots & & -H_1 & 1 & & & \\ -H_R & & & -H_1 & 1 & & \\ & -H_R & & & -H_1 & 1 & \\ 0 & & -H_R & & & -H_1 & 1 \end{bmatrix} \qquad (40)$$

The next step concerns itself with the statistical properties of the quantizing error. From (1) and the known parameters in Fig. 1, it is a straightforward matter to express $(e^n, \cdots, e^0)$ as a function of $(x_1^n, \cdots, x_1^0)$. A difficulty arises because the set of functional relations

$$e^p = g_p(x_1^p, \cdots, x_1^0) \qquad p = 0, \cdots, n \qquad (41)$$

does not represent a 1 : 1 transformation. This is physically obvious, for a change of any $x_1^p$ by an integral multiple of $\sigma$ clearly leaves the error unchanged. However, if one confines each $x_1^p$ to a particular quantizing interval $(k_p^* - \frac{1}{2})\sigma \leq x_1^p \leq (k_p^* + \frac{1}{2})\sigma$, then (41) has a unique inverse. This is clear since $e^0$ determines $x_1^0$, hence the additional specification of $e^1$ determines $x_1^1$ and so forth. Thus the $n$-dimensional rectangular volume $(e^n \leq k_n^*, \cdots, e^0 \leq k_0^*)$ maps in the $(x_1^n, \cdots, x_1^0)$ space into a compact volume whose boundaries are easily calculated. Integration over this volume with respect to the known probability density function $f_1(x_1^n, \cdots, x_1^0)$ yields the contribution to the error distribution function $F_e(k_n^*, \cdots, k_0^*)$ of the particular set of quantizing intervals selected. The complete error distribution is then found by summing over-all possible combinations of quantizing intervals. The result is

$$F_e(e^n, e^{n-1}, \cdots, e^0)$$

$$= \sum_{k_n=-K}^{K} \cdots \sum_{k_0=-K}^{K} \int_{-\infty}^{\infty} dx_1^n \cdots \int_{-\infty}^{\infty} dx_1^0$$

$$\cdot \prod_{r=0}^{n} SK[x_1^r, k_r\sigma] \prod_{s=0}^{n} U\left[e^s + \sum_{r=0}^{[R]_s} H_r(x_1^{s-r}\right.$$

$$\left. - k_{s-r}\sigma)\right] f_0([I - H]\Phi + HX) \qquad (42)$$

where

$$[R]_s = \begin{cases} s & 0 \leq s \leq R \\ R & s \geq R. \end{cases}$$

With the error distribution known, the autocorrelation function, $\psi_e(\beta\tau, \alpha\tau)$ of the error can be calculated from the following equation:

$$\psi_e(\beta\tau, \alpha\tau) = \int_{-\infty}^{\infty} \cdots \int_{-\infty}^{\infty} e^\beta e^\alpha \, dF_e(e^n, \cdots, e^0); \qquad (43)$$

$\beta$ and $\alpha$ are integers satisfying $n \geq \beta \geq \alpha \geq 0$.

We note that in (42) the variables $e^n, \cdots, e^0$ occur only as the argument of the step functions, $U(\ )$. Therefore, the Lebesgue-Stieltjes integral given in (43) will be zero except at the points

$$e^\beta = \sum_{r=0}^{[R]_\beta} H_r(x_1^{\beta-r} - k_{\beta-r}\sigma)$$

and

$$e^\alpha = \sum_{r=0}^{[R]_\alpha} H_r(x_1^{\alpha-r} - k_{\alpha-r}\sigma).$$

Thus,

$$\psi_e(\beta\tau, \alpha\tau) = \sum_{i=\beta-[R]_\beta}^{\beta} \sum_{s=\alpha-[R]_\alpha}^{\alpha} H_{\beta-i}H_{\alpha-i}\Lambda(i, s, H) \qquad (44)$$

where $\Lambda(i, s, H)$ is the rather complicated function

$$\Lambda(i, s, H) = \sum_{k_n=-K}^{K} \cdots \sum_{k_0=-K}^{K} \int_{-\infty}^{\infty} dx_1^n \cdots$$

$$\cdot \int_{-\infty}^{\infty} dx_1^0 (x_1^i - k_i\sigma)(x_1^s - k_s\sigma)$$

$$\cdot \prod_{r=0}^{n} SK[x_1^r, k_r\sigma] f_0([I - H]\Phi + HX). \qquad (45)$$

It should be observed that (44) is quite general. Nothing has been hypothesized concerning the input process except the existence of its $n + 1$ dimensional probability density function, $f_0(x_0^n, \cdots, x_0^0)$. Not even stationarity is required thus far. Similarly no significant use has been made of specific quantizer properties. A trivial generalization would remove the assumption that all quantizing intervals have the same width, $\sigma$.

In order to gain useful information from (44), it is necessary to reduce the expression for $\Lambda(i, s, H)$ to a simpler form. This can be done by making the two assumptions given in (2) and (4). It can then be shown[8] that

$$\Lambda(i, s, H) \cong \begin{cases} \dfrac{\sigma^2}{12} & i = s \\ 0 & i \neq s \end{cases} \qquad (46)$$

Substituting (46) into (44) one obtains the correlation function of the error

$$\psi_e(\beta\tau, \alpha\tau) = \sum_{i=\beta-[R]_\beta}^{\beta} \sum_{s=\alpha-[R]_\alpha}^{\alpha} H_{\beta-i}H_{\alpha-s} \frac{\sigma^2}{12} \delta(i - s) \qquad (47)$$

where

$$\delta(i - s) = \begin{cases} 1 & i = s \\ 0 & i \neq s \end{cases}.$$

If one does not make the assumptions implied by (2) and (4), the probability density functions of the input to the quantizer and the error are nonstationary even if the input signal is stationary.[8] This is reasonable, for with feedback all previous samples have some effect on the present input to the quantizer. On the other hand, it is clear that (47) depends only on the difference $\beta - \alpha$ as long as

$$\beta \geq \alpha \geq R. \qquad (48)$$

Thus, the assumptions are sufficient to make the error wide sense stationary after $R$ samples.

Eq. (47) can be simplified if one assumes that $\beta \geq \alpha \geq R$. Setting $i' = \beta - i$, $s' = \alpha - s$, and dropping the primes,

$$\psi_e((\beta - \alpha)\tau) = \sum_{i=0}^{R} \sum_{s=0}^{R} H_i H_s \frac{\sigma^2}{12} \delta[(\beta - \alpha - i + s)\tau]. \qquad (49)$$

This is the same expression as (12).

# Oversampled, Linear Predictive and Noise-Shaping Coders of Order $N>1$

STUART K. TEWKSBURY, MEMBER, IEEE, AND ROBERT W. HALLOCK

*Abstract*—First-order predictive coders (e.g., DPCM) and first-order noise shaping coders (e.g., interpolative coders) are familiar A/D conversion techniques. Using a feedback network containing an A/D and a first-order (single-pole) analog filter, they reduce the number of A/D output levels, for a given SNR requirement, at the expense of the additional analog filter complexity. Oversampling (i.e., sampling at higher than the Nyquist rate) provides excess bandwidth in the feedback loop, allowing further reductions in the number of A/D output levels at the expense of faster circuitry. This paper extends such first-order oversampled coders to include higher order analog filters under the constraint that the filters be independent of the statistical properties of the input analog signal. The resulting robust $N$th-order predictive and noise-shaping coders allow $N+1/2$ bits to be eliminated from the coder's A/D for each doubling of the sample rate. The design of such $N$th-order oversampled coders and an experimental third-order predictive coder are described.

## I. Introduction

OFTEN, a continuously varying analog signal must be converted at a uniform sample rate into a sequence of digital amplitude values from which, at least conceptually, a close replica of the input signal waveform can be reconstructed [1]. The fidelity of the reconstruction is measured by the average deviation of the reconstructed waveform from the input waveform (i.e., by a signal-to-noise ratio (SNR) requirement). Such an A/D converter may be called a "waveform coder" or simply a "coder." The number of discrete A/D output levels necessary in a waveform coder to achieve a given SNR performance may be reduced by operating the coder at a sample rate $f_s$ higher than the Nyquist rate $F_n = 2B$ (providing redundancy in the digital amplitude values) where $B$ is the input signal bandwidth and/or by imbedding the A/D converter in a linear analog feedback network (conditioning the input signal and the quantizing errors). Familiar examples of such techniques include i) delta modulation [2] (DM), ii) differential pulse-code modulation [3] (DPCM), iii) delta–sigma [4] ($\Delta$–$\Sigma$), and iv) direct feedback coders [5]. Such techniques are practical for waveform coding since the performance measure is the SNR, rather than the peak instantaneous error commonly used for data acquisition A/D converters. This paper extends conventional first-order oversampled (i.e., $f_s > F_N$) waveform coders to higher order waveform coders. The design of the higher order coders was constrained to obtain robust coders in the same sense as the coder examples

Manuscript received October 21, 1977; revised, February 6, 1978.
The authors are with Bell Laboratories, Holmdel, NJ 07733.

listed above. In particular, the linear analog feedback network in which the A/D is embedded is chosen to be independent of the statistical properties of the input signal.

The overall approach considered here for the A/D interface is shown in Fig. 1 [6]. An analog low-pass filter bandlimits the analog input signal to $0 < f < B$. The signal coder's A/D converter provides the digital output of the coder directly. Following the signal coder, a digital filter is used to correct known linear distortion of the signal due to the coder (and possibly also the analog low-pass filter) and to remove out-of-band coding noise. For "oversampled" A/D converters (i.e., $f_s \gg F_N = 2B$), the digital output of the digital filter is typically resampled at the Nyquist rate $F_N$. The components of the overall interface are designed to minimize the complexity of the A/D interface, reducing the complexity of the A/D converter circuitry at the expense of the circuitry for the linear analog feedback network.

In this paper, signal coders are classified as predictive coders and noise-shaping coders, with variations involving combinations of prediction and noise-shaping. The Appendices provide two optimization solutions for each type of coder. The conventional solutions, described in [7] for predictive coders and given in [8] for noise-shaping coders, generally give results which depend on the statistics of the input signal and of the quantizing errors, respectively. The solutions reported in this paper give feedback networks which are independent of the signal and error statistics (but, of course, a performance which depends on those statistics). Section III describes the results for predictive coders whereas Section IV describes the results for noise-shaping coders. Finally, Section V describes a third-order predictive coder which was built to demonstrate the performance that may be obtained.

## II. General Oversampled Linear Feedback Coder Structures

The linear feedback coder structures considered here are conveniently separated into two classes—predictive coders and noise-shaping coders according to their input/output $z$-transform response. The overall coder is modeled as a linear sampled data system by moving the sampling gate of its A/D converter ahead of the coder. The A/D converter is modeled as an additive noise

Reprinted from *IEEE Trans. Circuits and Sys.*, vol. CAS-25, pp. 436–447, July 1978.

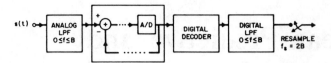

Fig. 1.   Overall A/D interface using oversampled coder.

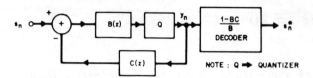

Fig. 2.   General linear feedback coder/decoder.

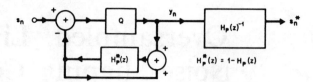

Fig. 3.   General predictive coder.

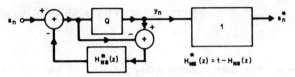

Fig. 4.   General noise-shaping coder.

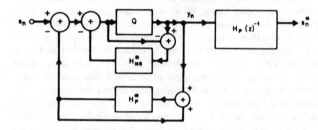

Fig. 5.   Nested predictive and noise-shaping coders.

source, adding the quantizing noise of the A/D converter to the converter's input sample. With this model, the general linear feedback coder has the form shown in Fig. 2 with a feed-forward filter $B(z)$ and a feedback filter $C(z)$. The z-transforms $Y(z)$, $S(z)$, and $Q(z)$ of the coder's output sample sequence $y_n$, input sample sequence $s_n$, and quantizing error sequence $q_n$, respectively (where $n$ is an integer denoting the sample time) are related by

$$Y(z) = [B(z)S(z) + Q(z)]/[1 + B(z)C(z)]. \quad (2.1)$$

To compensate for the linear distortion of the input signal $S(z)$ by the coder, the matched decoder with z-transform response

$$D(z) = [1 + B(z) \cdot C(z)]/B(z) \quad (2.2)$$

shown also in Fig. 2 is used. The decoder's output $S^*(z)$ is given by

$$S^*(z) = D(z) \cdot Y(z) = S(z) + Q(z)/B(z). \quad (2.3)$$

The general class of predictive coders are those, such as delta modulation and DPCM, which obey

$$Y(z) = H_p(z) \cdot \{S(z) + Q(z)\}. \quad (2.4)$$

The general structure of predictive coders is obtained by evaluating the constraints on $B(z)$ and $C(z)$ in (2.1) such that (2.4) is obtained, i.e.,

$$H_p(z) = B(z)/[1 + B(z) \cdot C(z)]$$
$$= 1/[1 + B(z) \cdot C(z)].$$

Therefore, $B(z) \equiv 1$, and from (2.1) through (2.4),

$$H_p(z) = 1/[1 + C(z)] \quad (2.5)$$

$$D(z) = [1 + C(z)] \quad (2.6)$$

and

$$C(z) = [1 - H_p(z)]/H_p(z). \quad (2.7)$$

Fig. 3 shows the general form of a predictive coder, such that only one filter is required.

The general class of noise-shaping coders are those, such as $\Delta$–$\Sigma$ and interpolative coders, which obey

$$Y(z) = S(z) + H_{NS}(z) \cdot Q(z). \quad (2.8)$$

The general structure of noise-shaping coders is obtained by using (2.8) and (2.1) to obtain the constraints on $B(z)$

and $C(z)$. In this case, the constraints are

$$B(z)/[1 + B(z) \cdot C(z)] = 1 \quad (2.9)$$

$$H_{NS}(z) = 1/[1 + B(z) \cdot C(z)]. \quad (2.10)$$

From (2.9) we obtain

$$B(z) = 1/[1 - C(z)] \quad (2.11)$$

and from (2.10),

$$H_{NS}(z) = 1 - C(z). \quad (2.12)$$

From (2.11) and (2.12),

$$C(z) = 1 - H_{NS}(z) \quad (2.13)$$

$$B(z) = 1/H_{NS}(z) \quad (2.14)$$

and

$$D(z) = 1. \quad (2.15)$$

Fig. 4 shows the general structure of a noise-shaping coder, again such that only one filter is required.

Predictive and noise-shaping coders can be nested [9], as shown in Fig. 5, to obtain the most general form of linear feedback coders. In this case

$$Y(z) = H_p(z) \cdot \{S(z) + H_{NS}(z) \cdot Q(z)\} \quad (2.16)$$

$$D(z) = 1/H_p(z). \quad (2.17)$$

From (2.16), the classification is evident. The frequency weighting of the input signal characterizes prediction and the frequency weighting of the quantizing error *relative* to the frequency weighted input signal characterizes noise shaping.

A final observation is relevant to the relationships between coder structures. Consider Fig. 6 which shows a predictive coder with filter response $H_{NS}(z)$ rather than $H_p(z)$. Then

$$Y(z) = [S'(z) + Q(z)] \cdot H_{NS}(z). \quad (2.18)$$

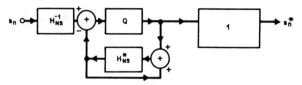

Fig. 6. Noise-shaping coder with predictive coder structure.

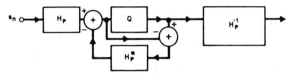

Fig. 7. Predictive coder with noise-shaping coder structure.

The preemphasis filter $P(z) = 1/H_{NS}(z)$ gives

$$S'(z) = S(z)/H_{NS}(z)$$

or

$$Y(z) = S(z) + H_{NS}(z)Q(z).$$

Therefore, a predictive coder whose filter is chosen to have the noise-shaping filter response can be converted to a noise-shaping coder by using a preemphasis filter. Similarly, a noise-shaping coder whose filter is chosen to be the optimum predictor filter giving

$$Y(z) = S'(z) + H_p(z) \cdot Q(z)$$

is converted to a predictive coder using a preemphasis filter $P(z) = H_p(z)$, as shown in Fig. 7. Therefore, predictive coder *structures* can be converted to noise-shaping coders and vice versa using preemphasis. An interesting special case occurs when the optimum prediction and noise-shaping filters have the same form (as is true for the higher order networks described below). In this case, preemphasis can convert a predictive *coder* into a noise-shaping coder and vice versa.

In the next section, "optimum" predictive and noise-shaping coders of order greater than 1 are presented. Furthermore, some particularly straightforward realizations are suggested. The discussions above, combined with the result that the "optimum" filters $H_p(z)$ and $H_{NS}(z)$ presented are equal, indicate the generality of the results and the structures.

## III. "Optimum" Oversampled Higher Order Linear Predictive Coder Structures

Fig. 3 showed the general form of a predictive coder. The coder output $Y(z)$ was given earlier as

$$Y(z) = H_p(z) \cdot \{S(z) + Q(z)\}$$

where the input signal is assumed bandlimited to $0 \leqslant f \leqslant B$ with total power normalized to unity, i.e.,

$$P_s = \int_0^B G_s(f)\, df \triangleq 1$$

where $G_s(f)$ is the power spectral density of $s(t)$. The prediction filter $H_p$ is assumed to be an FIR filter

$$H_p(z) = 1 - \sum_{m=1}^{N} a_m z^{-m}$$

where $N$ is the predictor order. The predictor coefficients $a_m$, $m = 1, 2, \cdots, N$ are chosen to reduce the prediction error and therefore to reduce the input full scale range of the A/D converter.

Appendix A gives two procedures for reducing the prediction error. The conventional approach minimizes the mean square prediction error $\sigma_\epsilon^2 = \langle \epsilon_n^2 \rangle$ by choosing the $a_m$, $m = 1, 2, \cdots, N$ according to the $N$ simultaneous equations $\partial \sigma_\epsilon^2 / \partial a_n = 0$, $n = 1, 2, \cdots, N$. The solutions are valid for any $f_s > 2B$, $f_s$ the signal sample rate. However, the resulting coder structure is not robust in the sense that the predictor coefficients depend on the statistics of the input signal and on the sample rate. The difficulty of using a single predictive coder for coding both voice-band speech and data at $f_s = 2B$ has been considered, for example, in [10].

The second optimization procedure in Appendix A requires that the signal be oversampled and uses a power series expansion of $\sigma_\epsilon^2$ in $(B/f_s)^2$. The $N$ predictor coefficients may be chosen such that the first $N$ terms of the expansion vanish, independent of the properties of the signal. The optimum prediction filter for an $n$th-order predictor obtained by this approach is

$$H_p(z) = (1 - z^{-1})^N. \tag{3.1}$$

The mean square prediction error which results is approximated by the lowest order nonvanishing term in the power series expansion of $\sigma_\epsilon^2$ giving (A.27)

$$\sigma_\epsilon^2 = (2\pi B/f_s)^{2N} \cdot \int_0^B G_s(f)(f/B)^{2N}\, df. \tag{3.2}$$

The predictive coder is robust in the sense that the predictor coefficients are independent of the input signal statistics and the sample rate. (However, it should be noted that for a given input signal and a given sample rate, the reduction in $\sigma_\epsilon^2$ obtained using (3.1) will be less than that which would be obtained using the conventional approach.) The reduction in the mean square prediction error, from (3.2), does of course depend on the signal statistics.

The output $S^*(z)$ of the decoder in Fig. 3 is given by

$$S^*(z) = S(z) + Q(z).$$

Therefore, the total output noise power at the output of the decoder is

$$P_Q(T) = \int_0^{f_s/2} G_Q(f)\, df \tag{3.3}$$

$G_Q(f)$ the quantizing noise power spectral density, and the total inband noise power at the output of the decoder is

$$P_Q(IB) = \int_0^B G_Q(f)\, df. \tag{3.4}$$

In the case of a uniform power spectral density $G_Q(f)$, (i.e., random quantizing errors)

$$P_Q(IB) = (2B/f_s) \cdot P_Q(T). \tag{3.5}$$

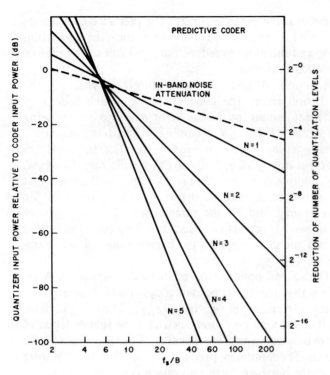

Fig. 8. Coder performance (series expansion design).

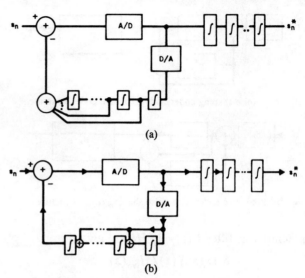

Fig. 9. (a) Form I of predictive coder $H_p^*(z) = [1-(1-z^{-1})^N]$.
(b) Form II of predictive coder $H_p^* = [1-(1-z^{-1})^N]$.

The extent to which the predictive coder simplifies the A/D converter is summarized in Fig. 8 for the case $G_s(f) = $ constant. For each 6-dB decrease in the prediction error power, one high order bit can be eliminated from the A/D. Fig. 8 (and (3.2)) shows that $\sigma_\epsilon^2$ decreases by $6 \cdot N$ dB per octave increase in the sample rate. Therefore, $N$ bits are eliminated for each doubling of the sample rate. Furthermore, the dotted line in Fig. 8 shows the decrease in the inband coding noise power assuming a constant $G_Q(f)$. Fig. 8 (and (3.5)) shows that the inband noise decreases by 3 dB per octave.

Therefore, the level separation can be increased by a factor of 1.4 for each doubling of the sample rate. Equivalently, one-half low-order bit is eliminated per octave increase in the sample rate if the out-of-band noise is removed using a low pass filter.

Therefore, as shown in Fig. 8, $N+1/2$ bits are eliminated by an $N$th-order predictive coder with out-of-band noise filtering for each doubling of the sampling rate.

This performance is achieved with a rather straightforward predictive coder structure. The overall feedback filter $C(z)$ of Fig. 2 is, from (2.7),

$$C(z) = [1 - H_p(z)] / H_p(z).$$

Using (3.1)

$$C(z) = [1 - (1-z^{-1})^N] / (1-z^{-1})^N.$$

Since

$$Z^{-1} \sum_{m=1}^{N} \frac{1}{(1-Z^{-1})^m} = (1 - [1 - z^{-1}]^N) / (1-z^{-1})^N$$

$$= C(z) \qquad (3.6)$$

the feedback network can be realized as a cascade of $N$ simple integrators, as in either Fig. 9(a) or (b). For the special case $N=1$, the conventional first order predictive coders of delta modulation and DPCM are obtained. The extension to higher order networks is, therefore, somewhat more complex than merely cascading integrators (i.e., double integration delta modulation is not a higher order predictor in the sense of this paper).

The decoder's $z$-transform response $D(z)$ is given by (2.6) as

$$D(z) = 1 + C(z) = (1-Z^{-1})^{-N}.$$

The decoder is therefore merely a cascade of integrators as shown in Fig. 9. The integrators of the decoder would normally be chosen somewhat leaky to give a finite gain at dc.

If an error $\delta(z)$ is associated with the D/A converter in a predictive coder (e.g., in the realizations of Fig. 9) and the coder's A/D conversion errors are $Q(Z)$, as before, then the output $S^*(z)$ of the decoder is readily shown to be

$$S^*(z) = S(z) + \left\{ Q(z) + [1-(1-Z^{-1})^{-N}] \cdot \delta(z) \right\}. \qquad (3.7)$$

Within the signal band, the D/A noise is amplified by an amplitude gain G (due to the $[1-Z^{-1}]^{-N}$ term in (3.7)) equal to the reciprocal of the signal attenuation by the coder (due to the $[1-Z^{-1}]^N$ term multiplying $S(z)$ in the coder output $Y(Z)$). In this sense, the predictive coder gives an A/D converter whose dynamic range is decreased without requiring a more accurate A/D but also gives a D/A converter whose dynamic range and allowed errors are decreased by equivalent amounts. The inherent precision of a conventional (nonfeedback) A/D converter operating at the Nyquist rate therefore appears in the coder's D/A converter. That precision requirement is relaxed only through the reduction of the inband D/A conversion noise by oversampling, i.e., by a factor $(f_s/2B)^{1/2}$.

Section V describes a realization of a third-order predictive coder and resolves the problem of overflow oscillations that characterize such higher order recursive structures [11].

## IV. "OPTIMUM" OVERSAMPLED HIGHER ORDER LINEAR NOISE-SHAPING CODER STRUCTURES

Fig. 4 showed the general form of a noise-shaping coder. The coder output $Y(z)$ was given in Section II as

$$Y(z) = S(z) + H_{NS}(z) \cdot Q(z).$$

The noise-shaping filter $H_{NS}(z)$ is assumed to be an FIR filter

$$H_{NS} = 1 - \sum_{m=1}^{N} a_m a^{-m}$$

where $N$ is the order of the noise-shaping coder. The filter coefficients $a_m$, $m = 1, 2, \cdots, N$, are chosen to reduce the inband coding noise power at the coder output, under the assumption that out-of-band noise is removed using a low-pass filter. As the inband coding noise power is reduced relative to the quantizing noise power of the A/D converter, it is possible to increase the A/D converter's decision level separation and thereby decrease the A/D complexity.

Appendix B gives two approaches for choosing the filter coefficients such that the inband coding noise power is reduced. The conventional approach minimizes the inband coding noise power $P_N$ by choosing the $a_m$, $m = 1, 2, \cdots, N$ such that the $N$ simultaneous equations $\partial P_N / \partial a_n = 0$, $n = 1, 2, \cdots, N$. In this case, the coefficients depend on the statistics of the quantizing errors of the A/D converter, rather than on the statistics of the input signal as was the case for the predictive coder. To the extent that the quantizing errors are random, a uniform noise spectral density can be used in (B.13) of Appendix B and the optimum coefficients $a_m$ then determined according to (B.16). The optimum coefficients for $N = 1$, 2, and 3 are shown in Fig. 10. The inband coding noise power $P_N$ can be obtained from (B.17) and normalized to the total quantizing noise power

$$P_Q(T) = \int_0^{f_s/2} G_Q(f)\, df$$

giving the results shown in Fig. 11. For each 6-dB decrease in $P_N / P_Q(T)$, the A/D decision level separation can be doubled. Since the A/D dynamic range remains essentially the full scale range of $s_n$, the result is the elimination of one lower order digit of the A/D converter for each 6-dB decrease of $P_N / P_Q(T)$. Such a coder design is valid so long as the quantizing errors remain essentially random but the specific filter coefficients do depend on the sample rate as shown in Fig. 10.

Appendix B also describes the selection of the optimum filter coefficients using the power series expansion method, giving (B.23)

$$H_{NS}(z) = (1 - z^{-1})^N \qquad (4.1)$$

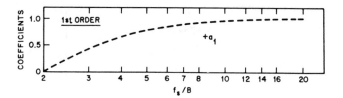
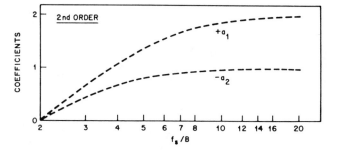
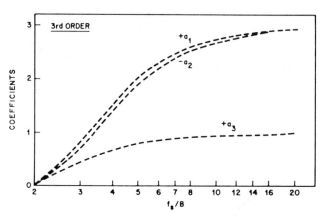

Fig. 10. Coefficients (mean square design).

for an $N$th-order noise-shaping coder. The inband coding noise power $P_n$ is approximated by the lowest order nonvanishing term in the series expansion of $P_N$ giving, by (B.25),

$$P_N = (2\pi B/f_s)^{2N} \cdot \int_0^B G_Q(f)(f/B)^{2N}\, df. \qquad (4.2)$$

Although the noise-shaping filter $H_{NS}(z)$ is independent of the quantizing error characteristics and of the sample rate, the reduction of the inband noise depends, of course, on the statistics of the quantizing errors. Assuming a uniform quantizing noise power spectral density, i.e.,

$$G_Q(f) = 2P_Q(T)/f_s$$

with total quantizing noise power $P_Q(T)$, equation (4.2) gives

$$P_N/P_Q(T) = (1/\pi)(2\pi B/f_s)^{2N+1}/(2N+1) \qquad (4.3)$$

which is plotted in Fig. 12. Since, as noted above, one low-order bit can be eliminated for each 6 dB decrease in $P_N$, $N + 1/2$ bits are eliminated for each doubling of the sample rate for an $N$th-order noise-shaping coder, as shown also in Fig. 12.

Using a heuristic extension of Candy's [12] interpolative coder, Ritchie [13] obtained a straightforward realization of an $N$th-order noise-shaping coder with the filter $H_{NS} =$

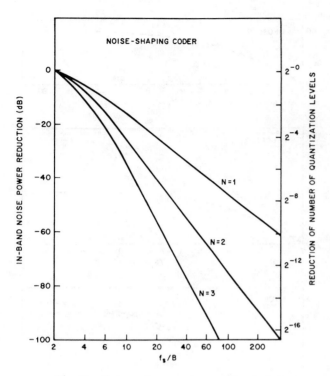

Fig. 11. Coder performance (mean square design).

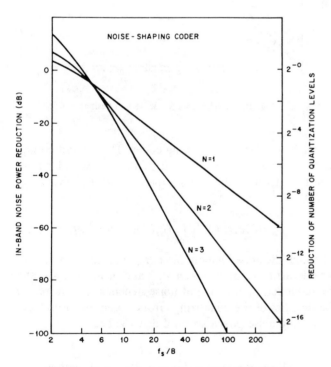

Fig. 12. Coder performance (series expansion design).

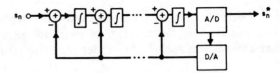

Fig. 13. Ritchie's form of noise-shaping coder.

before, then the coder's output signal $\delta^*(z)$ is readily shown to be

$$\delta^*(z) = S(z) + (1 - Z^{-1})^N$$

$$\cdot \left\{ Q(z) + \left[ 1 - (1 - z^{-1})^{-N} \right] \cdot \delta(z) \right\}. \quad (4.4)$$

Whereas the coder's inband A/D conversion noise is attenuated by the factor $(1 - Z^{-1})^N$, the inband D/A conversion noise is not attenuated. As a result, the inherent accuracy of a conventional (nonfeedback) converter operating at the Nyquist rate appears at the coder's D/A converter. That precision requirement, as was also the case for predictive coders, is relaxed only through the reduction of the inband D/A conversion noise by over sampling, i.e., by a factor $(f_s/2B)^{1/2}$.

## V. EXPERIMENTAL VERSION OF A THIRD-ORDER PREDICTIVE CODER

The higher order ($N > 1$) predictive and noise-shaping coder structures with $H(z) = (1 - z^{-1})^N$ were obtained for potential use in an A/D interface (producing a linear PCM coding of the 12-channel frequency division multiplexed group-band signal) of an all-digital, TDM/FDM translator [14], [15]. The group-band signal occupies a frequency band 60 kHz $< f <$ 108 kHz giving $B = 108$ kHz. An approximately 13-bit linear coding of the group-band signal is required.

The requirements on the group-band A/D interface specify not only a low background noise level (i.e., low quantizing noise) but also low harmonic distortion terms (i.e., a highly linear A/D characteristic), since harmonic distortion terms due to active channels may appear in idle channels, thereby acting as a background noise for that channel. A consequence of the linearity requirement is that the accuracy requirements on the decision levels of the A/D converter in a predictive coder (where the major A/D simplification is primarily achieved by reducing the A/D input full-scale range rather than improving resolution) are not as severe as in a noise-shaping coder (where the A/D simplification is fully achieved by improving the resolution). For similar reasons, the predictive coder is less sensitive to signal correlated timing jitter than the noise-shaping coder. From this perspective, the relative advantage of predictive coders over noise-shaping coders with respect to A/D complexity increases with increasing order $N$. Therefore, a predictive coder was chosen for experimental evaluation.

The sample rate required for an $N$th-order predictive or noise-shaping coder to act as an $N_T$-bit converter using only an $N_C$-bit converter is easily obtained, assuming a

$(1 - z^{-1})^N$ obtained above as the "optimum" filter. The general structure is shown in Fig. 13, using $N$ integrators in the feedforward loop, each integrator receiving an input from the direct feedback path. The special case $N = 1$ gives the direct feedback (i.e., interpolative) coder described by Candy [5], [12].

If an error $\delta(z)$ is associated with the D/A converter of a noise-shaping coder (e.g., in the realization of Fig. 13) and the coder's A/D conversion errors are $Q(z)$, as

TABLE I
SAMPLE RATE $f_s$ WITH $B = 108$ kHz

| ORDER OF CODER | SAMPLE RATE FOR 13-BIT ACCURACY WITH 3-BIT A/D | SAMPLE RATE FOR 12-BIT ACCURACY WITH 3-BIT A/D |
|---|---|---|
| N = 1 | 32.6 MHz | 20.6 MHz |
| N = 2 | 6.3 MHz | 4.7 MHz |
| N = 3 | 3.2 MHz | 2.6 MHz |
| N = 4 | 2.5 MHz | 1.9 MHz |
| N = 5 | 1.7 MHz | 1.5 MHz |

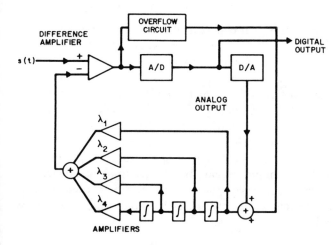

Fig. 14. Block diagram of third-order predictive coder realization.

uniform spectral density $G_s(f)$ for predictive coders and a uniform spectral density $G_Q(f)$ for both predictive and noise-shaping coders. In the case of predictive coders, the number of bits $N_B(1)$ eliminated due to prediction is merely

$$N_B(1) = \tfrac{1}{2} \log_2\left(P_s / \sigma_\epsilon^2\right)$$

and the number of bits $N_B(2)$ eliminated due to low-pass filtering the coder output is

$$N_B(2) = \tfrac{1}{2} \log_2\left(P_Q(T) / P_Q(\text{IB})\right)$$

with $N_T - N_C = N_B(1) + N_B(2)$. From (3.2) with $G_s(f) = P_s/B$,

$$\sigma_\epsilon^2 / P_s = \left(\frac{2\pi B}{f_s}\right)^{2N} \cdot \frac{1}{2N+1}.$$

Equation (3.5) gives $P_Q(\text{IB})/P_Q(T)$ directly. Therefore

$$-2(N_T - N_C) = \log_2\left[\left(\frac{2\pi B}{f_s}\right)^{2N+1} \cdot \frac{1}{\pi(2N+1)}\right]. \quad (5.1)$$

In the case of noise-shaping coders, the number of bits $N_B(3)$ eliminated is

$$N_T - N_C = N_B(3) = \tfrac{1}{2} \log_2\left(P_Q(T)/P_N\right)$$

or, using (4.3), the same result as given in (5.1).

The results from (5.1) for $N = 1, 2, 3$, and 4 with $N_T = 12$ or 13 and $N_C = 3$ are summarized in Table I. The relative reduction in $f_s$ as $N$ increases to $N+1$ is seen to decrease rapidly with increasing $N$. From these results, a third order predictive coder was chosen as the most interesting order and was realized using analog circuitry and a 3-bit A/D converter. The maximum sample rate of the experimental coder was about 1.4 M sample/s. The signal band $B$ was therefore adjusted, using (5.1), to correspond to the case $N_T - N_C = 10$ for $N = 3$.

Fig. 14 shows the block diagram of the third-order predictive coder which was realized. The saturation of the A/D converter was corrected by the overflow correction circuit to prevent overflow oscillations in the feedback loop. This correction circuit is probably the crucial factor which allowed the successful operation of the circuit. Fig. 14 also shows nonunity gains $\lambda_1$, $\lambda_2$, $\lambda_3$, and $\lambda_4$ in the connections of the integrator outputs to the summing node. The integrators are designed to give the desired step change over a time interval $T = f_s^{-1}$. However, due to nonzero delays through the A/D, D/A and the summing nodes, the input to the A/D converter will not have changed by the full desired amount when the next sample is taken. The gains $\lambda_1$, $\lambda_2$, $\lambda_3$, and $\lambda_4$ can be chosen to fully compensate known delays $\tau < T$.

145

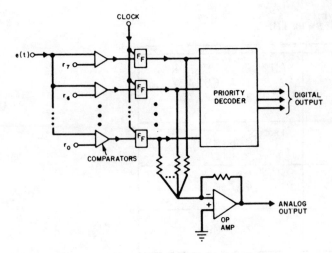

Fig. 15. 3-bit parallel A/D, D/A of realization.

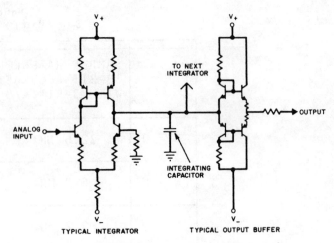

Fig. 16. Typical circuitry used in feedback circuit.

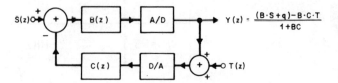

Fig. 17. Insertion and extraction of digital test signal.

The A/D and D/A converters were designed as a 3-bit parallel converter using comparators followed by clocked flip-flops with the flip-flop outputs summed using a resistive network to obtain the D/A function as shown in Fig. 15. This design avoids the need for a sample-and-hold circuit (so long as the A/D is relatively insensitive to the timing jitter of the comparator/flip-flop circuits) and provides a fast D/A output, minimizing the A/D/A delay time.

The predictive coder is particularly suitable for such an A/D since its timing jitter requirements are relaxed by a factor of $\sim 2^{60/7}$ due to the signal attenuation. The $\sim 150$ ps timing jitter requirements for a conventional 13-bit A/D converter, therefore, becomes $\sim 10$ to 20 ns for a 3-bit A/D of the predictive coder.

The integrator circuits were discrete realizations using an integrator design suggested by Candy[16]. Fig. 16 shows the integrator circuit and its buffer amplifier. The overall predictive coder worked reliably at sample rates up to $\sim 1.5$ MHz, short of the requirement for an actual group-band conversion application but adequate to evaluate the general technique.

The background noise was measured using a sinewave input, using the coder's D/A converter to drive a third-order analog integrator (simulating the decoder). This technique suppresses the D/A converter's errors (in the sense that they are not amplified relative to the A/D errors), allowing a direct evaluation of the performance of the A/D converter. The random background noise was found to agree closely with the theoretical predictions. The harmonic distortion products, for full-scale sinewaves, were 80 dB below the input levels as measured at the outputs. The harmonic distortion products decreased as the signal level decreased down to the limits of our measuring capabilities. This was not altogether surprising since the tolerance requirements on the A/D converter for such a performance are 6 to 8 percent. The high performance of this signal coder, achieved with straightforward analog feedback circuitry, suggests that it is well suited to such A/D conversion applications.

Finally, an inherent and potentially important advan-

tage of such signal coders as predictive and noise-shaping coders relative to conventional A/D converters involves the ability to perform in-service testing from a digital test facility. Such a procedure is shown in Fig. 17 where a digital tone is added into the coder's feedback loop and converted to an analog signal. The coder output will contain the response of the coder to that test tone, allowing in-service performance testing. In general, the test tone would have a frequency $f_0 > B$ to avoid contamination of the input signal. In the group-band conversion application, however, the test tone can be chosen to have a frequency $f_0 < 60$ kHz if its harmonics are sufficiently small.

## APPENDIX A
### PREDICTIVE CODER DESIGN

Fig. 3 showed the sampled data model of a linear feedback predictive coder. Regardless of the predictor used, the $\chi_n$ to the predictor in Fig. 2 is related to the input $s_n$ to the coder and the quantizing error $q_n$ associated with the A/D conversion of the prediction error $\varepsilon_n$ by

$$\chi_n = s_n + q_n. \qquad (A.1)$$

The linear predictor in Fig. 2 is assumed to have the general form

$$f_n = \sum_{j=1}^{N} a_j \chi_{n-j} = \left\{ \sum_{j=1}^{N} a_j s_{n-j} \right\} + \left\{ \sum_{j=1}^{N} a_j q_{n-j} \right\}. \qquad (A.2)$$

Letting $Y(z)$ be the z-transform of a general sequence of sampled values $y_n$, equation (A.2) may be expressed as

$$F(z) = H^*(z) \cdot X(z) = H^*(z) \cdot \left\{ S(z) + Q(z) \right\} \qquad (A.3)$$

where

$$H^*(z) = \sum_{j=1}^{N} a_j z^{-j} = 1 - H(z) \qquad (A.4)$$

is the $z$-transform of the $N$th-order predictor. The coder output $Y(z)$ is then

$$Y(z) = H(z) \cdot \{ S(z) + Q(z) \}. \qquad (A.5)$$

The multiplicative term $H(z)$ in (A.5) corresponds to a linear distortion of the input signal $s(t)$ which is removed by the matched decoder shown in Fig. 3 with $z$-transform

$$G(z) = H(z)^{-1} \qquad (A.6)$$

giving a decoder output $s_n^*$ with $z$-transform

$$S^*(z) = G(z) \cdot Y(z) = S(z) + Q(z). \qquad (A.7)$$

Finally, the prediction error $E(z) = > \epsilon_n$ is given by

$$E(z) = H(z) \cdot S(z) - \{ 1 - H(z) \} \cdot Q(z). \qquad (A.8)$$

Equation (A.7) shows that the error in the decoded output equals the quantizing error introduced by the A/D conversion of the prediction error. In this sense, prediction does not give an improved accuracy. However, by choosing $H^*(z)$ such that the contribution of $S(z)$ to the prediction error $E(z)$ is reduced, prediction does provide a reduced dynamic range over which the A/D converter must operate. The measure of the prediction error used below is the mean square prediction error $\sigma_\epsilon^2 = \langle \epsilon_n^2 \rangle$. Using the difference equation $\epsilon_n = s_n - f_n$ and (A.2), to relate $\epsilon_n$ to $s_n$ and $q_n$, gives

$$\sigma_\epsilon^2 = \langle \epsilon_n^2 \rangle = \psi_0 - 2 \sum_{m=1}^{N} a_m (\psi_m + \theta_m)$$

$$+ \sum_{m=1}^{N} \sum_{n=1}^{N} a_m a_n (\psi_{m-n} + \theta^*_{m-n} + \phi_{m-n}) \qquad (A.9)$$

where

$$\psi_m = \langle s_n \cdot s_{n-m} \rangle \triangleq \lim_{K \to \infty} \left\{ \frac{1}{2K+1} \right\}$$

$$\cdot \sum_{l=-K}^{K} (s_{n+l} \cdot s_{n+l-m}) \qquad (A.10a)$$

$$\theta_m = \langle s_n \cdot q_{n-m} \rangle \qquad (A.10b)$$

$$\theta^*_m = \theta_m + \theta_{-m} \qquad (A.10c)$$

$$\phi_m = \langle q_n \cdot q_{n-m} \rangle. \qquad (A.10d)$$

This can be conveniently expressed in matrix form. Letting $R$ be an $N \times N$ matrix with components

$$R_{mn} = \psi_{m-n} + \theta^*_{m-n} + \phi_{m-n} \qquad (A.11a)$$

$G$ be an $N$th-order vector with components

$$G_m = \psi_m + \theta_m \qquad (A.11b)$$

and $A$ be an $N$th-order vector with components

$$A_m = a_m$$

then (A.9) becomes

$$\sigma_\epsilon^2 = 1 - 2A^t \cdot G + A^t \cdot R \cdot A \qquad (A.12)$$

where we have assumed that the input signal is normalized to $\psi_0 = \langle s_n^2 \rangle = 1$.

The conventional minimum mean square optimization obtains the predictor coefficients $a_m$ by requiring [7]

$$\partial \sigma_\epsilon^2 / \partial a_m = 0, \qquad \text{for all } m, \text{ i.e., } m = 1, 2, \cdots, N. \qquad (A.13)$$

From (A.12) or (A.9), this gives

$$R \cdot \tilde{A} = G \qquad (A.14)$$

where $\tilde{A}$ is the $N$th-order vector of optimum coefficients. With $A = \tilde{A}$,

$$\sigma_\epsilon^2 = 1 - \tilde{A}^t \cdot G. \qquad (A.15)$$

Equations (A.14) and (A.15) define the standard predictive coder solutions, which are conventionally obtained assuming $\theta_m = \phi_m = 0$, for all integers $m$. From (A.14) it is apparent that the predictor depends on the assumed input signal statistics.

The optimization leading to the coder structures of this paper are based on a power series expansion method. Assuming again that $\theta_m = \phi_m = 0$ for all integer $m$, an expansion of $\sigma_\epsilon^2$ as a power series in $(B/f_s)$ is given below. This expansion, of course, presumes an oversampled coder (i.e., $f_s > 2B$) to obtain convergence.

$\psi_m$ of (A.10a) can be related to the power spectral density $G_s(f)$ of the input signal by

$$\psi_m = \int_0^B G_s(f) \cos (2\pi m f / f_s) \, df. \qquad (A.16)$$

The cosine term in (A.16) is expanded as a power series in $B/f_s$, giving

$$\psi_m = \sum_{k=0}^{\infty} \frac{(-1)^k (m)^{2k} (2\pi B/f_s)^{2k}}{(2k)!} \cdot \Gamma_k \qquad (A.17)$$

where

$$\Gamma_k \triangleq \int_0^B G_s(f) \cdot (f/B)^{2k} \, df. \qquad (A.18)$$

Using (A.17), the matrices $R$ and $G$ defined by (A.11a and b) can be expanded as

$$G = \sum_{k=0}^{\infty} g(k) \cdot \Gamma_k \cdot (B/f_s)^{2k} \qquad (A.19a)$$

and

$$R = \sum_{k=0}^{\infty} r(k) \cdot \Gamma_k \cdot (B/f_s)^{2k} \qquad (A.19b)$$

with $g(k)$ an $N$th-order vector with components

$$g_m(k) = \beta_{m,k}, \qquad m = 1, 2, \cdots, N \qquad (A.20a)$$

and $r(k)$ and $N \times N$ matrix with components

$$r_{m,n}(k) = \beta_{|m-n|,k}, \qquad m, n = 1, , \cdots, N. \qquad (A.20b)$$

147

The constants $\beta_{m,j}$ in (A.20a and b) are

$$\beta_{m,j} = \frac{(-1)^j (m)^{2j} (2\pi)^{2j}}{(2j)!}. \tag{A.21}$$

Using (A.19a and b) in (A.12) gives

$$\sigma_\epsilon^2 = \sum_{m=0}^{\infty} \Lambda_m \cdot \Gamma_m \cdot (B/f_s)^{2m} \tag{A.22}$$

where

$$\Lambda_m = \delta_{0,m} - 2A^t \cdot g(m) + A^t \cdot r(m) \cdot A \tag{A.23}$$

and

$$\delta_{0,m} = \begin{cases} 0, & \text{if } m \neq 0 \\ 1, & \text{if } m = 0 \end{cases}.$$

The power series expansion method for choosing optimum predictor coefficients in an $N$th-order predictor involves choosing the $a_k$, $k = 1, 2, \cdots, N$ such that $\Lambda_m = 0$, $m = 0, 1, \cdots, N-1$. The solution is

$$a_k = \frac{(-1)^{k+1} (N!)}{(N-k)! k!}, \qquad k = 1, 2, \cdots, N \tag{A.24}$$

or, equivalently,

$$H(z) = 1 - H^*(z) = (1 - z^{-1})^N. \tag{A.25}$$

The lowest order nonvanishing term for $\sigma_\epsilon^2$ is

$$\sigma_\epsilon^2 = \Lambda_N \cdot \Gamma_N \cdot (B/f_s)^{2N}. \tag{A.26}$$

or

$$\sigma_\epsilon^2 = (2\pi B/f_s)^{2N} \cdot \Gamma_N. \tag{A.27}$$

## APPENDIX B
### NOISE-SHAPING CODER DESIGN

Fig. 4 showed the sampled data model of a noise-shaping coder. Regardless of the feedback filter used in Fig. 4, the input to that filter is $q_n$, the current quantizing error of the A/D conversion of $\epsilon_n = s_n - f_n$. Letting $H(z)$ be the $z$-transform response of the feedback filter, the coder output signal $Y(z)$ is given by, with $H^*(z) \triangleq 1 - H(z)$,

$$Y(z) = S(z) + \{1 - H^*(z)\} \cdot Q(z) \tag{B.1}$$

where we shall assume below a transversal filter, i.e.,

$$H^*(z) = \sum_{j=1}^{N} a_j z^{-j} \tag{B.2}$$

giving

$$f_n = \sum_{j=1}^{N} a_j q_{n-j}. \tag{B.3}$$

The noise-shaping coder, according to (B.1), introduces a frequency weighting of the A/D converter's quantizing noise power spectral density $G_Q(f)$ at the coder output. Assuming $s(t)$ is bandlimited to $0 \leqslant f \leqslant B$ and that the signal is oversampled (i.e., $f_s \gg f_N = 2B$), the filter $H^*(z)$ is chosen such that the quantizing noise appearing in the

signal band is attenuated by $\{1 - H^*(z)\}$ in (B.1). In this sense, noise-shaping coders improve the resolution of the A/D converter but leave the A/D converter's dynamic range requirements basically unchanged.

As suggested by Fig. 4, the out-of-band coding noise power can be reduced, prior to the resampling operation in Fig. 1, using a digital low-pass filter with $z$-transform response $D(z)$. The low-pass filter acts as a decoder, giving an output

$$S^*(z) = D(z) \cdot Y(z) = D(z) \cdot S(z) + D(z) P(z) Q(z) \tag{B.4}$$

with

$$P(z) = 1 - H^*(z). \tag{B.5}$$

Letting $|p(f)|^2$ and $|d(f)|^2$ be the frequency response squared of $P(z)$ and $D(z)$, respectively, and letting $G_s(f)$, $G_s^*(f)$, and $G_Q(f)$ be the power spectral densities of $S(z)$, $S^*(z)$, and $Q(z)$, respectively, then

$$G_s^*(f) = |d(f)|^2 \{ G_s(f) + |p(f)|^2 G_Q(f) \} \tag{B.6}$$

where it has been assumed that $\theta_m = 0$ for all $m$ ($\theta_m$ is defined in (A.10b)). To prevent linear distortion of the signal $S(z)$, we require

$$|d(f)|^2 = 1, \qquad |f| < B$$

and (B.6) becomes

$$G_s^*(f) = G_s(f) + |d(f)|^2 |p(f)|^2 G_Q(f). \tag{B.7}$$

Identifying the second term on the right-hand side of (B.7) as the coding noise power spectral density $G_N(f)$, i.e.,

$$G_N(f) = |d(f)|^2 |p(f)|^2 G_Q(f) \tag{B.8}$$

the total coding noise power $P_N$ is

$$P_N = \int_0^{f_s/2} G_n(f) \, df$$

$$= \int_0^B |p(f)|^2 G_Q(f) \, df$$

$$+ \int_B^{f_s/2} |d(f)|^2 |p(f)|^2 G_Q(f) \, df. \tag{B.9}$$

The general noise-shaping coder design problem involves jointly choosing $D(z)$ and $H^*(z)$ such that $P_N$ is minimized. However, since the complexity tradeoff between digital filter complexity in $D(z)$ and analog circuit complexity in $H(z)$ is not well defined, the usual solution is to assume that $D(z)$ is an ideal low-pass filter, i.e.,

$$|d(f)|^2 = \begin{cases} 1, & 0 \leqslant f \leqslant B \\ 0, & B < f \leqslant f_s/2 \end{cases}.$$

Then

$$P_N = \int_0^B |p(f)|^2 G_Q(f) \, df. \tag{B.10}$$

The filter optimizations obtained below involved minimizing $P_N$ in (B.10) (i.e., minimizing the inband coding noise power).

With $H^*(z)$ of the form given by (B.2),

$$|p(f)|^2 = 1 - 2 \sum_{m=1}^{N} a_m \cos\left[2\pi mf/f_s\right]$$

$$+ \sum_{m=1}^{N} \sum_{n=1}^{N} a_m a_n \cos\left[2\pi(m-n)f/f_s\right]. \quad \text{(B.11)}$$

Using (B.11) in (B.10) gives

$$P_N = \psi_0 - 2 \sum_{m=1}^{N} a_m \psi_m + \sum_{m=1}^{N} \sum_{n=1}^{N} a_m a_n \psi_{m-n} \quad \text{(B.12)}$$

where

$$\psi_m = \int_0^B G_Q(f) \cos\left[2\pi mf/f_s\right] df. \quad \text{(B.13)}$$

Note the formal similarity between (B.12) (for $P_N$) and (A.9) (for $\sigma_e^2$) as well as the formal similarity between (B.13) (for $\psi_m$ of noise-shaping coder) and (A.16) (for $\psi_m$ used in predictive coder). Defining $R$ to be an $N \times N$ matrix with components

$$R_{m,n} = \psi_{m-n}/\psi_0 \quad \text{(B.14a)}$$

$G$ to be an $N$th-order vector with components

$$G_m = \psi_m/\psi_0 \quad \text{(B.14b)}$$

and $A$ to be the $N$th-order vector of filter coefficients

$$A_m = a_m \quad \text{(B.14c)}$$

equation (B.12) can be written

$$P_N = \psi_0\{1 - 2A^t \cdot G + A^t \cdot R \cdot A\}. \quad \text{(B.15)}$$

The conventional noise power minimization requires that the $a_m$, $m = 1, 2, \cdots, N$, be chosen according to

$$\partial P_N/\partial a_m = 0, \qquad m = 1, 2, \cdots, N$$

giving

$$R \cdot \tilde{A} = G \quad \text{(B.16)}$$

where $\tilde{A}$ is the $N$th-order vector of optimum coefficients. With $A = \tilde{A}$,

$$P_N = \psi_0\{1 - \tilde{A}^t \cdot G\}. \quad \text{(B.17)}$$

The power series expansion method involves expanding (B.13) as a power series in $B/f_s$. The procedure is identical to that described in Appendix A. In fact, even the results are identical with the substitutions $\sigma_e^2 \to P_N/\psi_0$, and $G_s(f) \to G_Q(f)$. In particular

$$P_N/\psi_0 = \sum_{m=0}^{\infty} \Lambda_m \cdot \Gamma_m \cdot (B/f_s)^{2m} \quad \text{(B.18)}$$

where

$$\Lambda_m = \delta_{0,m} - 2A^t \cdot g(m) + A^t \cdot r(m) \cdot A \quad \text{(B.19)}$$

$$\Gamma_m = \left[\int_0^B G_Q(f)(f/B)^{2m} df\right] \Big/ \psi_0 \quad \text{(B.20)}$$

$$\psi_0 = \int_0^B G_Q(f) df \quad \text{(B.21)}$$

and $g(m)$ and $r(m)$ are defined by (A.20a and b), respectively. The "optimum" noise-shaping filter coefficients are chosen such that $\Lambda_m = 0$, $m = 0, 1, \cdots, N-1$ giving

$$a_k = \frac{(-1)^{i+1}(N!)}{(N-k)!k!} \quad \text{(B.22)}$$

or

$$H^*(z) = 1 - (1 - z^{-1})^N. \quad \text{(B.23)}$$

The lowest order nonvanishhing term in (B.18) is

$$P_N = \psi_0\{\Lambda_N \cdot \Gamma_N \cdot (B/f_s)^{2N}\} \quad \text{(B.24)}$$

where $\Lambda_N = (2\pi)^{2N}$, giving

$$P_N = (2\pi B/f_s)^{2N}[\Gamma_N \cdot \psi_0]. \quad \text{(B.25)}$$

## ACKNOWLEDGMENT

The authors gratefully acknowledge the contribution of D. Cyganski and N. Krikelis who brought to our attention the importance of D/A errors on the overall coder performance.

## REFERENCES

[1] W. R. Bennett, "Spectra of quantized signals," *Bell Syst. Tech. J.*, vol. 27. pp. 446–472, July 1948.
[2] D. J. Goodman, "The application of delta modulation to analog-to-PCM encoding," *Bell Syst. Tech. J.*, vol. 48, pp. 321–343, Feb. 1969.
[3] C. C. Cutler, "Differential quantization of communications signals," U. S. Patent 2 605 361, July 29, 1952.
[4] H. Inose and Y. Yasuda, "A unity bit coding method by negative feedback," *Proc. IEEE*, vol. 51, pp. 1524–1535, Nov. 1963.
[5] J. C. Candy, "A use of limit cycle oscillations to obtain robust analog-to-digital converters," *IEEE Trans. Commun.*, vol. COM-22, pp. 298–305, Mar. 1974.
[6] S. K. Tewksbury, "A/D converters for digital filters," in *Proc. 1976 IEEE Int. Symp. Circuits and Systems*, pp. 602–605, Apr. 1976.
[7] R. W. Stroh, "Optimum and adaptive differential pulse code modulation," Ph.D dissertation, Dep. Elec. Eng., Polytechnic Institute of Brooklyn, Brooklyn, NY, 1970
[8] H. A. Spang and P. M. Schulteiss, "Reduction of quantizing noise by use of feedback," *IRE Trans Commun. Syst.*, pp. 373–380, Dec. 1962.
[9] C. C. Cutler, "Transmission systems employing quantization," U. S. Patent 2 927 962, 1960.
[10] J. B. O'Neal, Jr., and R. W. Stroh, "Differential PCM for speech and data signals," *IEEE Trans. Commun.*, vol. COM-20, pp. 900–912, Oct. 1972.
[11] D. Mitra, "Large amplitude, self-sustained oscillations in difference equations which describe digital filter sections using saturation arithmetic," *IEEE Trans. Acoust. Speech Signal Processing*, vol. ASSP-25, pp. 134–143, Apr. 1977.
[12] J. C. Candy, W. H. Ninke, and B. A. Wooley, "A per channel A/D converter having 15-segment μ-255 companding," *IEEE Trans. Commun.*, vol. COM-24, Jan. 1976.
[13] G. R. Ritchie, Private communication.
[14] S. L. Freeny, R. B. Kieburtz, K. V. Mina, and S. K. Tewksbury, "Systems analysis of a TDM/FDM translator/digital A-type channel bank," *IEEE Trans. Commun. Technol.*, vol. COM-19, pp. 1050–1059, Dec. 1971.
[15] C. Y. Kao and C. F. Kurth, "Method and apparatus for interfacing digital and analog copier systems," U. S. Patent 4 013 842, Mar. 22, 1977.
[16] J. C. Candy, private communication.

# Part 2
## Design, Simulation Techniques, and Architectures for Oversampling Converters

# Design Methodology for $\Sigma\Delta M$

BHAGWATI P. AGRAWAL, SENIOR MEMBER, IEEE, AND KISHAN SHENOI, MEMBER, IEEE

*Abstract*—The paper presents a design methodology based on correspondence between performance requirements, mathematical parameters, and circuit parameters of a sigma–delta modulator. This methodology will guide a design engineer in selecting the circuit parameters based on system requirements, in translating paper design directly into LSI design, in predicting the effect of component sensitivity, and in analyzing the operations of the sigma–delta modulator.

The sigma–delta modulator is viewed as a device which distributes the noise power, determined by peak SNR, over a much broader band, compared to signal bandwidth, shapes and amplifies it, and allows filtering of the out-of-band noise. The shaping and amplification are quantified by two parameters, $F$ and $P$, whose product is analogous to the square of step size of a uniform coder.

These two parameters are related, on one hand, to the time constants or location of zero and poles. On the other hand, inequalities are set up between performance parameters, like signal-to-noise ratio and dynamic range, and $F$ and $P$.

## I. INTRODUCTION

THE sigma-delta modulator belongs to the family of differential pulse code modulators which are designed on the principle of decorrelating a baseband signal or reducing the signal variance (power), prior to its quantization [1], [10]. For a "low-pass" signal the correlation between adjacent samples increases as the square of sampling frequency—a fact which suggests that the structure or order of the decorrelator will simplify with increasing sampling rate. It also suggests that the variance of the decorrelated signal will decrease as the square of sampling frequency and that the quantizer can be simplified or, equivalently, that the number of its levels can be reduced possibly to two. The two-level quantizer and decorrelator form, in principle, a delta modulator (DM). When preceded by a single integrator, the DM is called the first-order sigma–delta modulator ($\Sigma\Delta M$).

Although elegant, this intuitive viewpoint does not assist a design engineer in defining the circuit parameters, location of poles or zeros, or time constants for designing a second- or higher order $\Sigma\Delta M$. We, therefore, will present a design methodology for designing a $\Sigma\Delta M$. This methodology will establish correspondence between mathematical parameters and physical parameters. Such a correspondence will guide an engineer in translating paper designs into hardware designs, in predicting the effect of component sensitivity, and in analyzing stability.

Usually, the coder is designed to meet a certain signal-to-

Paper approved by the Editor for Data Communication Systems of the IEEE Communications Society for publication after presentation at the International Conference on Acoustics, Speech, and Signal Processing, Paris, France, May 1982. Manuscript received November 12, 1981; revised June 18, 1982.

B. P. Agrawal was with the ITT Advanced Technology Center, Shelton, CT 06484. He is now with GTE Business Communication Systems, Inc., McLean, VA 22102.

K. Shenoi is with the ITT Advanced Technology Center, Shelton, CT 06484.

noise ratio (SNR) for a specified dynamic range of signal. The specification of SNR determines the number of bits required for digital representation and the word rate is dictated by signal bandwidth. However, it is possible to trade word length or, equivalently, simplicity in quantizer design, for sampling rate. How is the tradeoff effected? One approach is to shape the quantization noise, which is white, so that most of the noise power lies outside the signal band.

The idea of noise shaping (shifting) immediately implies that the signal be sampled at a rate much higher than the Nyquist rate. It also implies that the noise will be filtered out prior to sample-rate reduction. However, for this scheme of noise shaping, measured in terms of a factor $F$, and subsequent filtering to be effective, it is necessary that the coder discriminate noise from signal. That is, the transfer functions seen by noise and signal ought to be distinct. For a "low-pass" signal, the signal transfer function (STF) should be, ideally, of low-pass-filter type and the noise transfer function be of high-pass-filter type. One way to achieve the two distinct functions by a single circuit is to embed the low-pass filter in a feedback loop around the quantizer.

Being high-pass, the noise transfer function (NTF) pushes the in-band noise out and reduces the proportion of in-band noise $F$. Simultaneously, the NTF may amplify the total noise power, measured as a mean power gain $P$. These two parameters, $F$ and $P$, are inversely related and can be identified directly with locations of poles and zero of the NTF and with the specifications of dynamic range and signal-to-noise ratio (SNR). That is, poles and zero can be systematically moved to meet system requirements on dynamic range and SNR.

Mathematically, the two parameters $F$ and $P$ are defined in Section II and their product $FP$ is shown to be inversely proportional to the peak SNR. The definitions themselves are based on a digital model of the $\Sigma\Delta M$. The design methodology is described in Section IV. It consists of two steps: system-level design and circuit-level design. At system-level design, the system specifications are transformed into mathematical parameters. At circuit-level design, the mathematical parameters are mapped into circuit parameters. Both these steps are illustrated with the design of a test chip in CMOS. The results obtained from the test chip are also included in this section. The idle channel behavior of the $\Sigma\Delta M$ is analyzed in Section V. The paper concludes with a summary of design philosophy and considerations.

The intent of the paper is, therefore, to present a methodology or tool for designing a $\Sigma\Delta M$ which has to meet specific system requirements. The methodology, it is expected, will eliminate the need for discrete implementation prior to large-scale integration. The most important feature of the methodology is the establishment of correspondence between

Reprinted from *IEEE Trans. Commun.*, vol. COM-31, pp. 360–370, March 1983.

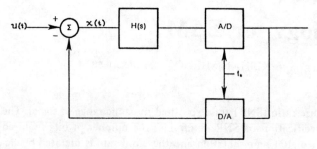

Fig. 1.   $\Sigma\Delta M$ configurations.

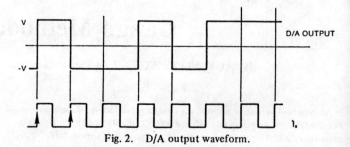

Fig. 2.   D/A output waveform.

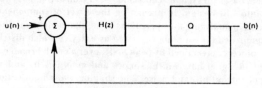

Fig. 3.   Discrete-time model of $\Sigma\Delta M$.

physical parameters, like time constants, mathematical parameters, like $F$ and $P$, and system requirements, like SNR and dynamic range.

## II. MODELING OF $\Sigma\Delta M$

The design methodology is based on a discrete-time model of the $\Sigma\Delta M$. This model is used both to define the mathematical parameters, $F$ and $P$, and to evaluate them via simulation. The product $FP$, it turns out, is a measure of quantization noise and permits us to draw an analogy between a $\Sigma\Delta M$ and a uniform quantizer. This analogy proves that it is not practical to simplify the quantizer by merely increasing the sampling rate.

### Discrete-Time Model

The basic configuration of a $\Sigma\Delta M$ is shown in Fig. 1. It consists of an analog filter $H(s)$, an A/D converter, and a D/A converter which is embedded in the feedback loop. In conventional, that is, 1-bit $\Sigma\Delta M$'s, the A/D converter is implemented as a comparator followed by a flip-flop and the D/A converter is a pulse shaper which outputs one of two precise waveshapes which correspond to the "high" and "low" states of "word" $b(n)$. For purposes of illustration, Fig. 2 shows a typical D/A output waveform with the levels $V$ and $-V$. For convenience, $V$ volts will be normalized to unity, hereon. The "order" of the $\Sigma\Delta M$ is determined by the order of $H(s)$. For a first-order $\Sigma\Delta M$, $H(s)$ is an integrator:

$$H(s) = \frac{g}{s}.$$

For a second-order $\Sigma\Delta M$, $H(s)$ takes the form

$$H(s) = \frac{g}{s}\frac{s+a}{s+b}.$$

We will restrict our attention to second-order $\Sigma\Delta M$'s although the analysis can be extended to higher order $\Sigma\Delta M$'s. Furthermore, the approach is valid for $\Sigma\Delta M$'s of configuration different than that of Fig. 1. For example, the configuration could be of a multiloop variety, as shown in Fig. 7. Further variations include cases where the "analog filters" are implemented by switched-capacitor techniques.

Denoting the sampling frequency by $f_s$ and the sampling interval by $T$ $(f_s = 1/T)$, and assuming that the input signal $u(t)$ is constant over the sampling interval $T$, a sampled-data equivalent of Fig. 1 can be generated, as shown in Fig. 3. For switched-capacitor implementations, Fig. 3 would be a natural description of the $\Sigma\Delta M$ and no assumption on the behavior of

the input signal between sampling epochs would be necessary. The comparator can be modeled as the addition of a "noise" signal $e(n)$, which yields the configuration shown in Fig. 4.

From Fig. 4, two transfer functions can be derived; one between the input $u(n)$ and the output $b(n)$, called the "signal transfer function" (STF), and the second between the noise $e(n)$ and output, called the "noise transfer function" (NTF). These are given by

$$\text{STF:}\quad W(z) = \frac{H(z)}{1 + H(z)}$$

$$\text{NTF:}\quad T(z) = \frac{1}{1 + H(z)}.$$

For the purposes of analysis, it will be assumed that "under normal operating conditions" the input $u(n)$ and the quantizing noise $e(n)$ are uncorrelated. Since the output of the $\Sigma\Delta M$ is $b(n) = \pm 1$ (normalized units), the $\Sigma\Delta M$ would be in "overload" state when the error exceeds unity. Under normal operating conditions, the error $e(n)$ is bounded by unity:

$$|e(n)| \leqslant 1.$$

The output of the $\Sigma\Delta M$, $b(n)$, has two components: first, the signal $u(n)$, modified by the STF, and second, a noise component $\eta(n)$ which is the additive noise $e(n)$ modified by the NTF.

Examination of the STF and NTF provides several insights into the behavior of $\Sigma\Delta M$. First, the input signal, or the desired components thereof, are confined to a small band around $dc$ $[-f_c, f_c]$. Consequently, the STF should be flat at low frequencies.

The noise transfer function, on the other hand, should be of a high-pass nature, effectively attenuating the quantization noise at low frequencies, albeit at the expense of amplifying the quantization noise at high frequencies.

Second, the $\Sigma\Delta M$ shapes the (flat) quantization noise spectrum so that, at the output, the noise spectrum is of a high-pass nature. Provided the STF does not distort the in-band signal,

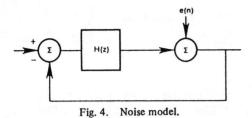

Fig. 4.   Noise model.

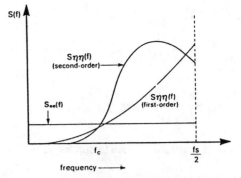

Fig. 5.   Noise spectra of first- and second-order ΣΔM.

the performance of the ΣΔM can be characterized, quantitatively, by the NTF. From the NTF, two parameters related to the noise-spectrum shaping are derived. First is the noise power gain $P$, which indicates the net noise power amplification. Second, is the noise shaping parameter $F$, which indicates what fraction of the total noise power is contained in the band of interest.

*Definitions of Parameters*

Under normal operating conditions, the noise power at the ΣΔM output is

$$\sigma_\eta^2 = E\{\eta(n)\}^2$$

$$= \frac{1}{2\pi j} \oint_{z=1} T(z)T(z^{-1})R_{ee}(z) \frac{dz}{z} \qquad (1)$$

where

$$R_{ee}(z) = \sum_k r_{ee}(k)z^{-k} \qquad (2)$$

is the $Z$-transform of the autocorrelation $r_{ee}(k)$ of the noise $e(n)$, i.e.,

$$r_{ee}(k) = E\{e(n)e(n+k)\}. \qquad (3)$$

In the above equations, $E$ is the expectation operator.

If $e(n)$ is a white noise sequence of variance $\sigma_e^2$, then

$$\sigma_e^2 = E\{e(n)\}^2 \leqslant 1 \qquad (4)$$

$$R_{ee}(z) = \sigma_e^2 \qquad (5)$$

and the output noise power $\sigma_\eta^2$ is

$$\sigma_\eta^2 = \sigma_e^2 P. \qquad (6)$$

The mean power gain $P$ is an important design parameter because it determines the total noise power at the ΣΔM output. The noise power is spread over the band $[-f_s/2, f_s/2]$ and only a small fraction of it lies in the band of interest $[-f_c, f_c]$. This fraction is computed from the spectrum of the output noise sequence $\eta(n)$. If $e(n)$ is white, then the spectrum of $\eta(n)$ is given by

$$S_{\eta\eta}(f) = R_{\eta\eta}(z)|_{z=e^{j2\pi f/f_s}} \qquad (7)$$

where

$$R_{\eta\eta}(z) = \sigma_e^2 T(z)T(z^{-1}).$$

Typical noise spectra of first- and second-order ΣΔM are shown in Fig. 5 (*not* to scale). Note that the effect of noise shaping is to move the quantization noise away from baseband, albeit at the expense of amplifying the noise power. The ratio of the area under the curve $S_{\eta\eta}(f)$ to the area under $S_{ee}(f)$ is the power gain $P$. The ratio of the area under the curve $S_{\eta\eta}(f)$ over the range $(0, f_c)$ to the area under $S_{\eta\eta}(f)$ over the range $(0, f_s/2)$ is the fraction $F$ of the noise power in the band of interest. An increase in the order of the ΣΔM reduces $F$ but increases $P$. Also, the required rolloff slope of the subsequent decimating filter increases with the increased order.

The fraction $F$ of noise power in the band of interest is given by

$$F = \frac{\displaystyle\int_0^{f_c} S_{\eta\eta}(f)\,df}{\displaystyle\int_0^{\frac{f_s}{2}} S_{\eta\eta}(f)\,df}. \qquad (8)$$

Thus, if the ΣΔM were followed by an ideal low-pass filter of cutoff frequency $f_c$, then the in-band noise power $\sigma_N^2$ would be

$$\sigma_N^2 = FP\sigma_e^2 \leqslant FP. \qquad (9)$$

Since the peak (sinusoidal) signal power is 1/2, the peak signal-to-noise ratio is

$$\text{SNR}_p \geqslant -10 \log (2FP) \text{ dB}. \qquad (10)$$

Although quite useful for design, the above relations (9) and (10) do not illustrate the effect of increasing or decreasing the sampling rate. To understand this effect, let us consider typical first- and second-order ΣΔM's.

A typical first-order ΣΔM would have a noise transfer function given by

$$T_1(z) = (1 - z^{-1})$$

yielding

$$S_{\eta\eta}(f) = 4\sigma_e^2 \sin^2 (\pi f/f_s). \qquad (11)$$

The total quantization noise in the band of interest $[-f_c,$

155

$f_c$] is given by

$$\sigma_N{}^2 = \frac{2}{f_s} \int_0^{f_c} 4\sigma_e{}^2 \sin^2(\pi f/f_s)$$

$$= \frac{2\sigma_e{}^2}{\pi} \left[ \frac{2\pi f_c}{f_s} - \sin(2\pi f_c/f_s) \right].$$

Using a two-term expansion for sin (assuming $f_s \gg f_c$),

$$\sigma_N{}^2 \doteq \tfrac{8}{3}\pi^2 \sigma_e{}^2 (f_c/f_s)^3. \tag{12}$$

Note that doubling of the sampling rate decreases the noise power by 9 dB.

A typical second-order $\Sigma\Delta$M would have a noise-transfer function given by

$$T_2(z) = \frac{(1 - z^{-1})^2}{1 + \alpha z^{-1} + \beta z^{-2}}.$$

Assuming, for the moment, that $\alpha = \beta = 0$,

$$S_{\eta\eta}(f) = 16\sigma_e{}^2 \sin^4(\pi f/f_s).$$

The in-band noise power is given by

$$\sigma_N{}^2 = \frac{2}{f_0} \int_0^{f_c} S_{\eta\eta}(f)\, df$$

$$\doteq 6 \cdot 4\pi^4 \sigma_e{}^2 (f_c/f_s)^5. \tag{13}$$

Note that doubling the sampling rate decreases the in-band noise power by 15 dB.

For the two cases considered above, the mean power gain $P$ is 2 for the first-order case and 6(!) for the second-order case. Furthermore, it was relatively simple to compute the in-band noise power $\sigma_N{}^2$ in a closed form. In practice, for example, when $\alpha$ and $\beta$ are nonzero, it is often more convenient to compute the fraction $F$ using numerical integration and then $\sigma_N{}^2$ as the product

$$\sigma_N{}^2 = FP\sigma_e{}^2. \tag{14}$$

*Analogy with Uniform Quantizer*

The product $FP$ is analogous to the square of step size of a uniform quantizer: to increase the peak signal-to-noise ratio, decrease the product $FP$. However, $F$ and $P$ are inversely related for a given order of NTF—variation of the design parameters to lower $F$ would increase $P$ by some factor. In concept, $F$ is reduced by increasing the sampling rate to distribute the noise power, fixed for a given number of bits over a larger band $(0, \tfrac{1}{2}f_s)$, and shaping the noise power density.

To emphasize the parallelism between the $\Sigma\Delta$M and the uniform quantizer, let us consider a uniform quantizer operating at the Nyquist rate $2f_1$, and with a step size $\Delta$. The instantaneous quantizing error $e(n)$ is the net quantizing error, since there is no error feedback. Equivalently, the power gain $P$ is

unity. The variance $\sigma_e{}^2$ is given by

$$\sigma_e{}^2 = k\left(\frac{\Delta}{2}\right)^2 \tag{15}$$

where $k$ is a constant determined by the probability distribution function (PDF) of the error. If a uniform PDF is assumed, $k = 1/3$. Also, note that for a 1-bit quantizer, $\Delta = 2$ (normalized units). The quantization noise power is spread (uniformly) over the frequency band $(-f_1, f_1)$ and only a fraction $F = (f_c/f_1)$ lies in the band of interest $(-f_c, f_c)$. The peak signal-to-noise ratio can be written as

$$\text{SNR}_P \text{ (uniform)} = -10\log_{10}\left(2k\left(\frac{\Delta}{2}\right)^2 FP\right)$$

$$= -10\log_{10}\left(2\left(\frac{f_c}{f_1}\right)k\left(\frac{\Delta}{2}\right)^2\right). \tag{16}$$

Clearly, the peak signal-to-noise ratio increases by 3 dB for each octave increase in sampling frequency. Unfortunately, $\text{SNR}_P$ decreases by 6 dB when the quantizer resolution is reduced by one bit. Consequently, increasing the sampling rate of a uniform quantizer is an inefficient way to reduce the requirements on resolution. In a $\Sigma\Delta$M, on the other hand, the use of feedback makes increasing the sampling rate an efficient way to ease the requirements on quantizer resolution. In particular, for first- and second-order $\Sigma\Delta$M's, the peak SNR increases at 9 dB/octave (12) and 15 dB/octave (13), respectively, with respect to the sampling frequency.

It then follows that, for example, a 13-bit, 8 kHz uniform quantizer can be replaced by a 1-bit (two-level) first-order $\Sigma\Delta$M operating at a sampling rate which is eight octaves higher than 8 kHz or a second-order $\Sigma\Delta$M operating at a sampling rate which is five octaves higher than 8 kHz. The actual choice of the order depends on the clock rate allowed by a particular technology (CMOS, NMOS, etc.).

## III. DESIGN METHODOLOGY

The design methodology consists of system-level and circuit-level designs. At system level the requirements, like signal-to-noise ratio and dynamic range, are translated into the mathematical parameters $F$, $P$, and $f_s$, which were defined in the previous section. These parameters are then mapped into the circuit parameters, such as location of poles and zero, of the NTF. The circuit-level design is also influenced by several other considerations which are described later on. The methodology of both the levels will be illustrated with a specific design of A/D and D/A converters.

*System-Level Design*

In telephony, the transmission and switching systems are typically required to meet the specifications shown in Fig. 8. That is, for signals between 0 and −30 dBm0, the SNR must exceed 35 dB. The "worst case" specification is the SNR of 25 dB at a signal level of −45 dBm0. When linearly extrapolated, the "worst case" corresponds to 73 dB at 3 dBm0.

The behavior of a sigma–delta modulator depends markedly

on the level of input signal. Therefore, we will divide the signal range into three distinct regions: high input or "overload," medium input or "normal," and low input or "idle-channel." The plot in Fig. 6, signal-to-noise ratio versus input signal power, delineates these three regions. The normal region is characterized by a linear relation between SNR and input power. The peak input signal is, arbitrarily, defined as 3 dBr. That is, a sine-wave of peak allowable amplitude $V_r$ has a power of 3 dBr. This peak amplitude is normalized to unity.

Note that in Fig. 6, the abscissa has two different "scales," dBr and dBm0. The dBm0 representation of signal level is a system specification and dBr is a component, that is, the ΣΔM, specification. They are related via 0 dBm0 = $-x$ dBr where $x$ ($\geqslant 0$) is also a design parameter. The designer of the ΣΔM has the freedom of choosing $x$, in effect shifting the specification template to lie entirely under the SNR curve of the ΣΔM, preferably with some margin to allow for nonideal behavior of circuit components.

The system requirements can be translated into two quantities: the peak SNR, $S$, 73 dB in the above example, and an overload point, 0, normally 0 dBm0 or 3 dBm0.

How does the overload point influence the design parameters $F$, $P$, and $f_s$? Since the output of the ΣΔM is bilevel, $\pm 1$, the output power is fixed, independent of input power. The output power must be shared between "signal" and "noise." If the share of the noise is large, the dynamic range, or overload point of the ΣΔM, will be small.

For elaboration, let us assume as a first approximation that the additive noise $e(n)$ has a uniform probability density function over $(-1, 1)$, corresponding to a variance $\sigma_e^2$, of 1/3. The output noise power would then be ($P/3$). Clearly, $P < 3$ is a necessary restriction on the noise gain. The above suggests that the ΣΔM is in an "overload" condition when the input signal power exceeds $(1 - P/3)$. Actually, the ΣΔM goes into overload when the comparator input exceeds 2 (normalized units), implying an instantaneous quantization error greater than 1. Simulation and experimental results have indicated that the input signal power is roughly 1/10 $(1 - P/3)$ when the ΣΔM just enters overload and, furthermore, that this overload condition is not catastrophic.

In the case of a switched-capacitor ΣΔM or a digital ΣΔM (discussed later), $P$ can be determined directly from the capacitor ratios or multiplier coefficients. When resistor–capacitor combinations are used to implement the analog filter $H(s)$, the sampling frequency does impact on $P$.

The determination of $F$, $P$, and $f_s$ is an iterative procedure and, since these parameters are interrelated, a consistency check is required. Furthermore, it is preferable to obtain a range for these parameters to account for any constraints imposed by circuit implementation.

A suitable first step is the choice of sampling frequency. For the example considered, which is derived from a telephony application, it is known that a 13-bit uniform coder at 8 kHz will meet the requirements. Assuming a second-order ΣΔM and using the rules of thumb of "15 dB/octave" and "6 dB/bit," the sampling frequency should be about 256 kHz (actually greater). In comparison, a first-order ΣΔM would need to operate at a sampling frequency in excess of 1 MHz.

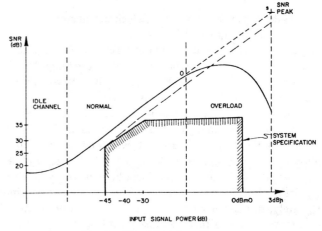

Fig. 6.   Signal-to-noise ratio curve.

The next step is to determine a ballpark figure for $P$. Assuming 6 dB margin, 0 dBm0 = $-6$ dBr ($x = 6$ in the example) and, as a first approximation,

$$\log_{10}\left(1 - \frac{P}{3}\right) = -6$$

or

$$P \doteq 2.25.$$

The next step is to calculate $F$ from

$$SNR_P = -10 \log_{10}(2FP).$$

For this example, the peak SNR must be $73 + x = 79$. The corresponding value of $F$ is

$$F \approx 0.2 \times 10^{-8}.$$

For a given implementation, the consistency of the triplet $(P, F, f_s)$ can be verified. The consistency is verified by checking sampling rate and dynamic range. The sampling rate should satisfy

$$F \approx k\left(\frac{f_c}{f_s}\right)^2.$$

The constants $k$ and $n$ were defined in the last section. If the consistency check fails, then one may increase the sampling rate, whose effect will be to slide the dotted SNR curve, in Fig. 6, upwards. Alternatively, one can reduce the headroom ($x$), whose effect is to move the system specification template (Fig. 6) to the right or, equivalently, reduce the overload point.

Once the triplet $(P, F, f_s)$ is selected, the circuit-level design can be undertaken. The circuit-level design involves evaluation of contour integrals. One may, therefore, determine the circuit-level parameters through simulation as explained below.

*Circuit-Level Design*

Apart from the peak SNR and dynamic range, there are a few other considerations which govern the choice of design parameters. These are idle channel noise and the requirements on the subsequent low-pass filter which attenuates the high frequency noise components.

To see how these requirements interact, consider the configuration of Fig. 1. The design consists of determining the circuit parameters $g$, $a$, and $b$. For convenience, it will be assumed that the sampling frequency is 1 MHz.

The zeros of the NTF should be kept as close to $z = +1$ as possible. The noise shaping parameter $F$ is minimized when all the zeros of the NTF are at $z = +1$, which corresponds to a (multiple) transmission zero at dc. If the NTF has zeros close to $z = +1$, the mean power gain $P$ can be reduced by moving the poles of the NTF closer to $z = 1$ (at the expense of $F$). Unfortunately, this causes an "early" peaking of the noise transfer function, requiring that the subsequent low-pass filter have a sharp rolloff. If the poles of the NTF are $re^{j\theta}$, then $s_{\eta\eta}(f)$ will peak close to $f = (\theta/2\pi) f_s$ and the peak will be sharp if $r$ is close to 1. If $r \geqslant 1$ the structure is unstable. Also, the idle pattern will have a fundamental frequency close to $f = (\theta/2\pi)f_s$, especially if $r$ is close to unity. From the viewpoint of low-pass filter requirements and idle channel behavior, it is advisable to keep $r$ as small as possible and $\theta$ as close to $\pi(180°)$ as possible. That is, the poles should lie well within the unit circle and away from the voiceband (at the expense of $P$ and, consequently, the dynamic range).

If $b$ is lowered, $F$ decreases but $P$ increases. That is, by lowering $b$, we shall reduce the overload point and degrade the stability. In contrast, if $a$ is decreased, $F$ increases and $P$ goes down, which corresponds to better stability. While choosing the circuit prameters, it is useful to remember the above correspondences.

A suitable strategy for choosing circuit parameters is necessarily dependent on the application. For example, consider the design of a $\Sigma\Delta M$ for use in telephony [11], [12]. The $\Sigma\Delta M$ is to use a 1 MHz clock rate and is followed by a simple finite impulse response (FIR) low-pass filter whose output is resampled at 32 kHz. After subsequent filtering, the signal is again resampled at 8 kHz. The chain is required to have the noise performance of a 13-bit uniform coder (at 8 kHz). One possible approach for selecting appropriate values for the design parameters $a$, $b$, and $g$ is as follows.

First, for any given $(g, a, b)$, the following are computed:
1) the poles $re^{j\theta}$
2) mean power gain $P$
3) the peak SNR from (10).

The given triplet $(g, a, b)$ is then accepted or rejected based on the following procedure:

1) reject if the angle of the pole is too small; a cutoff angle of 22.5° (64 kHz) was chosen
2) reject if $r > 0.75$
3) reject if $SNR_P \leqslant 78$ dB
4) reject if $P \geqslant 3$.

The guidelines chosen above are based on the following heuristic arguments. Since the ideal SNR performance of the $\Sigma\Delta M$ should exceed that of a 13-bit coder operating at 8 kHz, the peak SNR must exceed 78 dB. The acceptable pole locations are based on stability and the "knee" of the noise spectrum. Since $r < 1$ is necessary for stability, $r < 0.75$ provides some margin. The angle of the pole determines the knee of

the noise spectrum which, to ease requirements on the low-pass (decimating) filter, should be at as high a frequency as possible. Upon resampling the low-pass filter outputs at 32 kW/s, the noise energy centered around multiples of 32 kHz folds back, or aliases, into the baseband. If the angle of the pole is greater than 22.5° (64 kHz), the "early" peaking of the noise spectrum will be only a second-order effect.

Assuming $g = 1/T$, acceptable ranges for the pole and zero locations $b$ and $a$ are $2 \leqslant b \leqslant 4$ (kHz) and $75 \leqslant a \leqslant 100$ (kHz) as shown in Table I. Clearly, the $\Sigma\Delta M$ is relatively insensitive to the parameter values $a$ and $b$, permitting deviations of 15 percent of nominal.

Implementations of $\Sigma\Delta M$ can be classified into two broad categories identified with the type of feedback. These are the "single loop" and "multiloop" configurations. The configuration in Fig. 1 is of the single-loop type, wherein the digital output is fed back, in an analog form, at one point in the analog circuitry. A double loop version is depicted in Fig. 7. Note that in Fig. 7, the digital output is fed back at two "analog" nodes. Regardless of the configuration, it is possible to derive a signal transfer function and a noise transfer function. The performance of the $\Sigma\Delta M$ is then easily predicted via the power gain parameter $P$ and the noise shaping parameter $F$. The $\Sigma\Delta M$ can also be classified by the type of circuitry. Digital $\Sigma\Delta M$'s use (digital) arithmetic units, registers, etc., and are inherently discrete-time. Discrete-time "analog" $\Sigma\Delta M$'s can be implemented using switched-capacitor techniques. "True analog" $\Sigma\Delta M$'s would use analog components such as resistors and capacitors. For the discrete-time configurations, obtaining the signal and noise transfer functions is quite straightforward. Obtaining sample data equivalents for "true analog" $\Sigma\Delta M$'s requires some assumptions on the behavior of signals between sampling instants. In practice, however, the difference arising from these assumptions (for example, step invariance versus impulse invariance) will be masked by the nonideal behavior of the actual components.

### D/A or Digital $\Sigma\Delta M$ Design

Conventional D/A converters in telephony applications operate at an 8 kHz sampling frequency and require "sharp" postfilters and high-precision ladder networks. Additionally, they are quite expensive for the number of bits per sample encountered in telephony. An alternative conversion scheme consists of increasing the sampling rate by interpolation (digital filtering) and decreasing the number of bits per sample. The D/A converter would thus need fewer output levels and the requirements on the postfilter would be eased considerably.

Drawing a parallel with the "analog" $\Sigma\Delta M$ discussed earlier, if the sampling rate is high enough, the D/A converter need have only two output levels. The conversion of a word stream with a plurality of bits per word to a bit stream, or 1 bit per word, is accomplished by a digital $\Sigma\Delta M$. The design of a digital $\Sigma\Delta M$ closely parallels that of an "analog" $\Sigma\Delta M$. Referring to Fig. 3, the block $H(z)$ is implemented digitally and the comparator is simply a sign detector. In addition, considering the high sampling rate, it is necessary to ensure that the coefficients are "simple" and $H(z)$ can be implemented with a minimal number of add operations.

TABLE I
ΣΔM DESIGN GUIDELINES

| RATIO (a/b) | 10 | 15 | 20 | 25 | 30 | 35 | 40 | 45 | 50 |
|---|---|---|---|---|---|---|---|---|---|
| 2 KHz | REJECT BECAUSE ANGLE < 22.5° | | | | | $r\angle\theta = .45\angle 31°$ P = 2.5 SNR < 80.5 | $r\angle\theta = .49\angle 40°$ P = 2.6 SNR < 81.7 | $r\angle\theta = .52\angle 46°$ P = 2.7 SNR < 82.7 | $r\angle\theta = .54\angle 51°$ P = 2.85 SNR < 83.6 |
| 3 KHz | | | | $r\angle\theta = .47\angle 35°$ P = 2.6 SNR < 79.8 | $r\angle\theta = .51\angle 46°$ P = 2.7 SNR < 81.4 | $r\angle\theta = .56\angle 53°$ P = 2.9 SNR < 82.7 | M | | |
| 4 KHz | | | $r\angle\theta = .5\angle 39°$ P = 2.6 SNR < 78.9 | $r\angle\theta = .54\angle 51°$ P = 2.8 SNR < 80.9 | M | | | | |
| 5 KHz | | SNR < 78 | $r\angle\theta = .53\angle 50°$ P = 2.8 SNR < 79.6 | M | | | | | |
| 6 KHz | SNR < 78 | SNR < 78 | M | | REJECT BECAUSE P > 3 BOXES MARKED **M** ARE MARGINAL | | | | |
| 7 KHz | SNR < 78 | SNR < 78 | | | | | | | |
| 8 KHz | SNR < 78 | | | | | | | | |
| 9 KHz | SNR < 78 | | | | | | | **UNSTABLE** | |
| 10 KHz | SNR < 78 | | | | | | | | |

The simplicity of analysis of a discrete-time ΣΔM is illustrated here. Consider a double-loop digital ΣΔM depicted in Fig. 8. The behavior of the circuit is described by

$$y(n + 1) = y(n) + x(n) - b(n)$$

$$x(n + 1) = x(n) + u(n) - \tfrac{1}{2}b(n).$$

Treating the sign detector as an additive noise source,

$$b(n) = y(n) + e(n)$$

where $e(n)$ is the additive (quantization) noise. Some algebra yields the following signal and noise transfer functions:

$$\text{STF:} \quad W(z) = \frac{1}{z^2 - z + 0.5}$$

$$\text{NTF:} \quad T(z) = \frac{1}{z^2 - z + 0.5}.$$

Assuming $f_s/f_c = 2^8$, the fraction of noise power in the band $[-f_c, f_c]$ is

$$F = 0.91 \times 10^{-9}$$

and the noise power gain $P$ is

$$P = 2.4$$

which yields a peak signal-to-noise ratio, from (10), of 83.6 dB.

Note that the digital ΣΔM in Fig. 8 indeed has coefficients that are "simple." Furthermore, since the output is a one-bit word, the "adders" required to implement the feedback can be replaced by random logic of considerably less complexity than an adder.

*Test Chip Results*

This design methodology was used to design an A/D converter for central office application. The A/D was implemented as a switched capacitor ΣΔM and incorporated in a CMOS test vehicle. Several samples of this chip were tested, in the laboratory, for SNR, ICN, and dynamic range. The testing was carried out using an analog source, collecting the output data, and applying a "digital distortion analyzer," in nonreal time, on a high-speed array processor coupled to a PDP-11/60. The test results are shown in Fig. 9.

To check out the design methodology, a digital ΣΔM was built using the same capacitor ratios as that of the test chip. The digital ΣΔM was tested using a digitally generated sinusoid and an analog distortion analyzer.

The two implementations tracked each other. Indeed, as shown in Fig. 9, the peak SNR of both implementations is roughly 83 dB, which is in close agreement with the analytical prediction.

## IV. STABILITY AND NONLINEAR BEHAVIOR OF A ΣΔM

The behavior of a ΣΔM depends on the input signal level as shown in the typical SNR curve in Fig. 8. The three regions are discussed separately.

The overload region is characterized by the input power being greater than roughly $1/10 (1 - (P/3))$ where $P$ is the power gain. For such large input amplitudes (greater than $-15$ dBm0), the instantaneous quantization error $e(n)$ frequently exceeds unity (in magnitude). The total distortion components include harmonics of the input signal, indicating that the quantization error signal is no longer uncorrelated with the input.

The practical implication of the quantization error being greater than unity is that the amplifier voltage excursions, in analog implementation, could be greater than two (the ref-

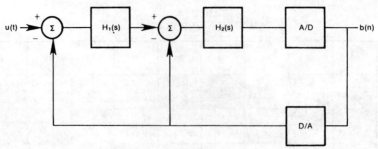

Fig. 7.   Double loop analog $\Sigma\Delta$M.

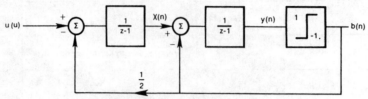

Fig. 8.   Double loop digital $\Sigma\Delta$M.

erence voltage $V_r$ being normalized to unity). The stability of the $\Sigma\Delta$M then depends on the manner in which the amplifiers come out of saturation and is dependent on the actual circuitry involved. In a digital $\Sigma\Delta$M, for example, the configuration in Fig. 8, $y(n)$ could be extremely large for a high level input. Consequently, the adders must be of the saturating variety. If the adders are allowed to overflow, and no corrective measures are taken, the $\Sigma\Delta$M in Fig. 8 could exhibit a wild oscillatory behavior.

In the "normal" region of operation, the SNR curve in Fig. 8 is substantially linear, indicating that the noise power is relatively constant. In this region, the assumption that the quantizing error $e(n)$ is not correlated with the input is valid. Consequently, the linearized representation of a $\Sigma\Delta$M, shown in Fig. 4, is applicable. The simple notion of stability, namely, that the poles should lie inside the unit circle, is easily applied. In particular, the poles of the circuit are the zeros of $(1 + H(z))$ and must lie within the unit circle.

In the "idle channel" region of operation, the SNR curve tends to flatten out, indicating that the distortion is proportional to the signal level. A heuristic explanation is as follows: in the idle channel condition, the quantization error signal $e(n)$ is no longer "white" but, rather, very coherent, with high frequency components in the range of an eighth to half of the sampling frequency, well outside the frequency band of interest. The effect of the input signal is to disrupt the "idle pattern" (discussed next) in a manner akin to frequency modulation. Consequently, at baseband are the fundamental (the input signal) and harmonics whose amplitudes are directly related to the amplitude of the fundamental. Since these harmonics are the primary components of the distortion, the SNR remains relatively constant.

Since the output of the $\Sigma\Delta$M is always ±1, even under zero input or idle conditions, the corresponding pattern is termed the "idle pattern" and is indicative of a limit cycle or oscillatory behavior. The fundamental frequency of oscillation is typically very high, one-half or a quarter of the sampling frequency, and usually introduces no in-band noise (idle channel noise) and is governed by the nonlinearity such as dead zone, hysteresis, and offset of the comparator. The presence (or ab-

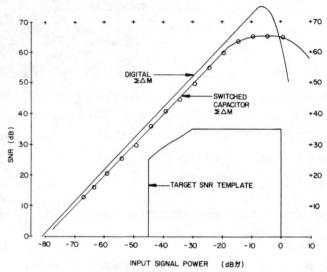

Fig. 9.   Test chip results.

sence) of a certain idling pattern can be determined in the following manner. Suppose an idling pattern with period $M$, $b(0)$, $b(1)$, $\cdots$, $b(M - 1)$ exists; then, with reference to Fig. 3, with the feedback loop broken and the above sequence as input to $H(z)$, the output $y(n)$ must be periodic, with period $M$, and satisfy $b(k) = \text{sgn } y(k)$. If $H(z)$ is written as

$$H(z) = \frac{z^{-1}(\alpha_0 + \alpha_1 z^{-1} + \alpha_2 z^{-2})}{1 + \beta_1 z^{-1} + \beta_2 z^{-2}} \qquad (17)$$

then the presence of the hypothesized idle pattern requires that

$$y(0) = -\beta_1 y(m - 1) - \beta_2 y(m - 2) + \alpha_0 b(m - 1)$$

$$+ \alpha_1 b(m - 2)\alpha_2 b(m - 3)$$

$$y(1) = -\beta_1 y(0) - \beta_2 y(m - 1) + \alpha_0 b(0)$$

$$+ \alpha_1 b(m - 1) + \alpha_2 b(m - 2)$$

$$\vdots$$

$$y(m-1) = -\beta_1 y(m-2) - \beta_2 y(m-3) + \alpha_0 b(m-2)$$

$$+ \alpha_1 b(m-3) + \alpha_2 b(m-4).$$

In matrix notation,

$$y = By + Ab \qquad (18)$$

where $B$ and $A$ are circulant matrices whose first columns are

$$[0 \ -\beta_1 \ -\beta_2 \ 0 \ 0 \ \cdots \ 0]^T$$

and

$$[0 \ \alpha_0 \ \alpha_1 \ \alpha_2 \ 0 \ \cdots \ 0]^T$$

respectively. The sequence $[y(0), y(1), \cdots, y(M-1)]$ is obtained as

$$y = (I-B)^{-1}Ab \qquad (19)$$

and the condition $b = \text{sgn } y$ can easily be verified.

The above analysis of an idle pattern provides additional information on the design of a $\Sigma\Delta$M. In analog $\Sigma\Delta$M's, the comparator usually has a "dead zone" of, say, $[-\epsilon, \epsilon]$. if the voltage at the comparator input has a magnitude of less than $\epsilon$, then the comparator cannot make a "hard" decision. Factors affecting $\epsilon$ are the comparator gain in its linear region, the clock rate, and the threshold behavior of the subsequent flip-flop. The condition $b = \text{sgn } (y)$ may be satisfied, but if min $|y(k)| < \epsilon$, then the idling pattern will not be stable. For example, if the hypothesized idle pattern is of alternating $+1$'s and $-1$'s, the observed idle pattern may contain, occasionally, two consecutive $+1$'s and two consecutive $-1$'s, although not necessarily immediately following. This jitter in the idle pattern could give rise to in-band idle channel noise.

In passing, it should be observed that the right-hand side of (19) can be computed efficiently even for very large values of $M$. In particular, the $M \times M$ matrices $(I-B)$ and $A$ are circulant matrices and can easily be diagonalized. Denoting the DFT (discrete Fourier transform) matrix by $D$, (19) can be written as

$$y = D^{-1}[\Lambda_\beta][\Lambda_A]Db$$

where $\Lambda_B$ and $\Lambda_A$ are diagonal matrices; $Db$ is the DFT of $b$. The elements of $\Lambda_B$ are the reciprocal of the eigenvalues of $(I-B)$ which are given by the DFT of the first column of $(I-B)$. Similarly, the elements of $\Lambda_A$ are obtained from the DFT of the first column of $A$. The DFT (and the inverse) are readily computed using fast Fourier transform (FFT) algorithms.

## V. CONCLUSIONS

Intuitively, the complexity of an A/D converter for a fixed peak SNR or number of bits can be reduced by increasing the ratio of sampling frequency to Nyquist frequency. This re-

duction in complexity of analog circuitry is achieved at the expense of digital circuitry in the form of a decimating filter which follows the A/D converter. However, this approach by itself is not practical because replacement, for example, of a 13-bit uniform quantizer by a two-level quantizer will necessitate an increase in sampling frequency of 24 octaves. It is, therefore, an imperative to imbed the quantizer in a mixed analog-digital feedback loop.

Early designs of such feedback encoders included delta modulators which exploited the high correlation between adjacent samples. Unfortunately, in delta modulators, both the signal transfer function and noise transfer function were of a high-pass variety and, therefore, could not be discriminated. These coders also required that the subsequent decimating digital filter had a suitable shape in the band of interest $(-f_c, f_c)$, as well as sufficient out-of-band rejection. A rearrangement of the functional blocks of a delta modulator, moving the integrator from the feedback path to the forward path just prior to the quantizer, created the sigma-delta modulator. Now the signal transfer function was relatively flat (low-pass type) in the frequency band of interest and the noise transfer function was high-pass. The decimating filter required similar out-of-band rejection, as for a $\Delta$M, but could be flat in the passband.

This "first-order" version of $\Sigma\Delta$M provided a 9 dB/octave increase in signal-to-noise ratio. The quantizer complexity could be reduced by more than one bit (at 6 dB/bit) by an octave increase in sampling frequency. Accordingly, a 13-bit uniform coder might be replaced by a 1-bit (two-level) first-order $\Sigma\Delta$M operating at a sampling rate of eight octaves higher, or a second-order $\Sigma\Delta$M operating at five octaves higher, than the sampling rate of the uniform order.

The logical trend was to increase the filter complexity in the feedback loop to obtain greater reduction in quantizer complexity for each octave increase in sampling frequency. However, such an approach raises several questions about the behavior of the $\Sigma\Delta$M. Is the circuit stable? What peak signal-to-noise ratio could be expected? Is there a systematic way to design these higher order $\Sigma\Delta$M's? What kind of idle pattern could be expected? How do the filter coefficients affect the dynamic range? How are the requirements on the subsequent decimating filter affected by the filter poles and zeros?

This paper attempts to answer these questions by extracting two parameters, $F$ and $P$, from the noise transfer function. The parameter $F$ is related to the shape of the noise spectrum and determines the fraction of quantization noise power in the frequency band of interest. In contrast, $P$, the power gain, measures the amplifications, due to feedback configuration, of noise introduced by the quantizer. The product $FP$ is analogous to the square of the step size of a (Nyquist rate) uniform quantizer. The power gain $P$ also determines the dynamic range of the $\Sigma\Delta$M. The noise transfer function also dictates the out-of-band rejection required in the decimating filter and, furthermore, assists in analyzing idle patterns for circuit verification.

The notion of trading complexity of analog circuitry for digital circuitry is easily applied to digital-to-analog conversion as well. A suitable digital interpolating filter is employed to increase the sampling rate and the digital $\Sigma\Delta M$ employed to reduce the number of bits/word required for signal representation. As is often the case, the digital $\Sigma\Delta M$ reduces the wordlength to 1 bit/word and the final D/A converter is required to generate one of just two precise waveforms per sampling period, as opposed to $2^N$ for an $N$-bit D/A converter. The design procedure presented here does not distinguish between "analog" and "digital," and provides a unified design methodology based on discrete-time formulation, for all forms of $\Sigma\Delta M$.

## REFERENCES

[1] R. J. van de Plassche, "A sigma–delta modulator as an A/D converter," *IEEE Trans. Circuits Syst.*, vol. CAS-25, no. 7, pp. 510–514, 1978.

[2] R. Steele, *Delta Modulation*. New York: Wiley, 1975.

[3] J. D. Everard, "A single channel PCM codec," *IEEE Trans. Commun.*, vol. COM-27, pp. 283–295, Feb. 1979.

[4] N. S. Jayant, "Digital coding of speech waveforms: PCM, DPCM, and DM quantizers," *Proc. IEEE*, vol. 62, pp. 611–632, 1974.

[5] ——, "A first-order Markov model for understanding delta modulation noise spectra," *IEEE Trans. Commun.*, vol. COM-26, pp. 1316–1318, Aug. 1978.

[6] J. Flood *et al.*, "Exact model for delta modulation process," *Proc. Inst. Elec. Eng.*, vol. 118, pp. 1155–1161, Sept. 1971.

[7] M. H. H. Hofelt, "On the stability of a 1-bit quantized feedback system," in *Proc. Int. Conf. Acoust., Speech, Signal Processing*, Washington, DC, 1979, pp. 844–848.

[8] M. H. H. Hofelt, D. J. G. Janssen, and V. D. Meeberg, "Integrated single-channel PCM encoder–filter combination," in *Proc. Int. Conf. Acoust., Speech, Signal Processing*, Hartford, CT, 1977.

[9] S. K. Tewksbury and R. W. Hallock, "Oversampled, linear, predictive and noise-shaping coders of order N 1," *IEEE Trans. Circuits Syst.*, vol. CAS-25, pp. 436–447, July 1978.

[10] J. L. Flanagan, "Speech coding," *IEEE Trans. Commun.*, vol. COM-27, pp. 710–736, Apr. 1979.

[11] B. P. Agrawal and K. Shenoi, "Telephone subscriber line unit with sigma delta D/A converter," U.S. Patent 4 270 027, May 1981.

[12] ——, "Digital gain control," to be published.

# Table-Based Simulation of Delta-Sigma Modulators

RICHARD J. BISHOP, JOHN J. PAULOS, MICHAEL B.
STEER, AND SASAN HOUSTON ARDALAN

*Abstract* —The program ZSIM (nonlinear $Z$-domain SIMulator) is used to explore the benefits and limitations of a table-based approach to the simulation of delta–sigma modulators with switched-capacitor integrators. Simulations demonstrating the effects of clock feedthrough and incomplete settling are presented, as well as simulations demonstrating the importance of the use of an accurate and charge conservative circuit simulator for the table point transient simulations. The methods used are appropriate for other discrete time systems where simulation of system-level performance is desired based on the results of transient circuit simulation.

## I. Introduction

Delta–sigma modulators [1] are one of a class of A/D converters that use oversampling to achieve a high level of precision at a lower sampling rate. Delta–sigma modulators can be implemented with few precision circuits and the component tolerances need not be precise [2], [3]. Also, delta–sigma modulators are easily implemented in digital MOS IC technologies through the use of switched capacitor integrators [4], [5].

A block diagram of a first-order delta–sigma modulator is shown in Fig. 1. The analog input signal $x(t)$ is encoded into a digital pulse stream $p(t)$. The feedback loop minimizes the error signal $e(t)$, where

$$e(t) = \int [x(t) - p(t)] \, dt. \qquad (1)$$

The error signal is quantized, sampled, and held for one clock cycle by the comparator to produce one output pulse. The system attempts to track the input signal $x(t)$ with the encoded output stream $p(t)$, so that the output $p(t)$ matches the input $x(t)$ in an integrated error sense. The pulsetrain $p(t)$ is a pulse density representation of $x(t)$, and the input can be reconstructed by passing the pulsetrain through a low-pass filter. In most applications, the pulsetrain is filtered and decimated to a lower sampling frequency to provide a more conventional PCM (pulse code modulation) representation of the input with high signal-to-noise ratio (SNR). The discrete time equivalent of Fig. 1 can be implemented using a switched-capacitor integrator. The error signal $e(t)$ is now represented by the following difference equation:

$$e[k] = e[k-1] + (x[k] - p[k]) \qquad (2)$$

where $p[k]$ is the sampled and held result of the previous comparator decision.

Delta–sigma modulators are difficult to simulate since tens of thousands of clock cycles are required in order to obtain mean-

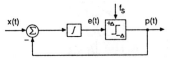

Fig. 1.  Block diagram of a first-order delta–sigma modulator.

ingful measures of SNR or signal-to-distortion ratios. Using accurate transistor-level circuit simulation, the transient analysis of a single clock cycle of the switched capacitor integrator may take several minutes of CPU time on a minicomputer or workstation. Detailed signal-to-noise curves would, therefore, take months or years of CPU time. Alternatively, difference equation models of the integrator can be used with dramatically reduced simulation time, but it is not easy to capture the actual nonidealities of the circuit in such a model. These nonidealities include the effects of finite bandwidth and slew rate of the operational amplifier, charge dump from the MOSFET switches, and nonlinearity of the amplifier transfer curve. Most of these effects can be modeled individually using difference equations. However, combinations of these effects are extremely difficult to model analytically, and the models obtained may only approximate the actual circuit behavior.

## II. ZSIM

ZSIM (nonlinear $Z$-domain SIMulator) [6], [7] combines the speed of difference equation simulation with the completeness of a transistor-level simulator. ZSIM uses tables (one table for each integrator) which are generated by a transistor-level simulator. Each table is a discrete representation of the nonlinear difference equation for a switched capacitor integrator which includes the effects of actual circuit nonlinearities. ZSIM uses these externally generated tables to quickly simulate the modulator over several thousand clock cycles.

ZSIM can be used to simulate delta–sigma modulators using either ideal difference equations or externally generated tables. ZSIM includes FFT-based signal analysis routines to compute SNR, signal-to-total harmonic distortion ratios (STHD), and signal-to-(noise + THD) ratios (SDR), as well as baseband spectra. ZSIM also generates a histogram of the table usage which allows the user to determine the most efficient table discretization.

The tables used by ZSIM allow for the discretization of the input and output data for the linear and nonlinear regions of operation of the integrator. The output of an ideal integrator can be represented by

$$y[k] = y[k-1] + \alpha(x[k] - p[k]) \qquad (3)$$

where $x[k]$ is the input signal during the current clock cycle, $y[k-1]$ is the integrator output from the previous clock cycle, $p[k]$ is the current output of the comparator which was sampled at the end of the previous clock cycle, and $\alpha$ is the integrator

Manuscript received February 23, 1989; revised May 24, 1989. This work was supported in part by a grant from Texas Instruments, Inc. and by the Center for Communications and Signal Processing. This paper was recommended by Associate Editor C.A.T. Salama.

The authors are with the Department of Electrical and Computer Engineering, North Carolina State University, Raleigh, NC 27695-7911.

IEEE Log Number 8933451.

Reprinted from *IEEE Trans. Circuits and Sys.*, vol. CAS-37, pp. 447–451, March 1990.

163

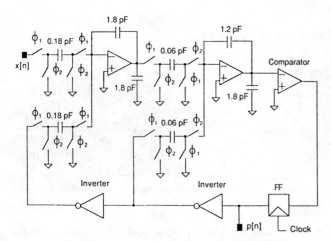

Fig. 2. Schematic of the second-order delta–sigma modulator.

gain. The output of the binary comparator, $p[k]$, is discrete, but the input signal $x[k]$ and the integrator output $y[k-1]$ are continuously valued and can, in general, range between the power supply rails. Therefore, the integrator tables are effectively divided into 2 two-dimensional tables (planes), with each plane corresponding to one value of $p[k]$. Within each plane, the rows represent the discretization of the input signal $x[k]$ and the columns represent the discretization of the previous integrator output $y[k-1]$. The table-based representation of the integrator can, therefore, be described by

$$y[k] = T(x[k], y[k-1], p[k]) \qquad (4)$$

where $T(x, y, p)$ denotes an interpolation of the table data. In its current form, ZSIM uses linear interpolation. Linear interpolation is considered to be adequate for delta–sigma modulators since the system components are designed to be approximately linear. The goals of this paper are: 1) to explore the limitations in table-based simulation as a result of finite precision in the transistor-level transient analyses used to construct the integrator tables and 2) to explore the effects of circuit limitations in delta–sigma modulators such as incomplete settling and charge dump from the MOSFET switches.

### III. CIRCUIT DESCRIPTION

The circuit used in this study is a second-order delta–sigma modulator using switched capacitor integrators. The schematic for this circuit is shown in Fig. 2. The switches are composed of an n-channel MOSFET in parallel with a p-channel MOSFET in order to minimize clock feedthrough (charge dump) [8]. These switched-capacitor integrators require the use of a two phase clock. The inputs to the modulator must be valid before the rising edge of clock phase $\phi_1$ and must remain valid for the duration of that clock phase. The output of the first integrator is valid at the end of clock phase $\phi_1$, and the output of the second integrator is valid at the end of clock phase $\phi_2$. Each clock has a rise/fall time of 4 ns. The clock phases are aligned such that there is a 2-ns delay between the falling edge of $\phi_1$ and the rising edge of $\phi_2$ and vice-versa.

The modulator is simulated for a 14-bit digital audio application, suitable for low-grade consumer electronics. For this application a clock rate of 5.6448 MHz is used with a decimation factor of 32, yielding a 176.4-kHz sampling frequency. In a typical application, this 4x oversampled data stream would be filtered and decimated again to obtain a 20-kHz signal bandwidth at a 44.1-kHz sampling rate.

The operational amplifier used to implement the switched capacitor integrators is a single-ended class AB amplifier operating from a single 5-V supply. The amplifier has an offset of 0.125 mV, a dc gain of 20 000, and a unity-gain bandwidth of 20 MHz. The first (inverting) integrator has an input gain of 0.1 and a feedback gain of 0.1. The second (noninverting) integrator has an input gain of 0.5 and a feedback gain of 0.05. With these gains, typical simulations show an output swing of about $-0.6$–$0.6$ V for the first integrator, and an output swing of about $-0.8$–$0.8$ V for the second integrator with respect to analog ground (2.5 V). The table discretization of the columns (input values) of the first integrator table was chosen to be 0, $-5$, $-10$, $-20$, and $-30$ dB relative to a full scale swing of 5 $V_{pp}$. These levels correspond to $\pm 2.5$, $\pm 1.4$, $\pm 0.8$, $\pm 0.25$, and $\pm 0.08$ V. The discretization of the rows (initial output values) of the first integrator table was chosen to accommodate the expected output swing of the integrator. The discretization points for $y[k-1]$ range from $-0.9$ to $+0.9$ V in increments of 0.2 V. The discretization of both the input (columns) and previous output (rows) for the second integrator table is identical to the discretization of the previous output of the first integrator. Discretization points at 0 V were avoided to prevent spurious distortion of small signals due to small slope discontinuities near zero.

### IV. GENERATING TABLES

In order for ZSIM to produce accurate results, the integrator tables used by ZSIM must be accurate. To accomplish this the transistor-level circuit simulator used must be capable of accurately simulating switched-capacitor integrators, the correct timing must be maintained during the table point transient analyses, and the table point transient analyses must start from the correct initial conditions. Tables were generated using both SPICE (version 2G.6) [9] and CAzM (Circuit Analyzer with Macromodeling) [10], [11]. In both cases the LEVEL 2 MOSFET model was used. CAzM is a table-based circuit simulator developed at the Microelectronics Center of North Carolina. CAzM has two advantages over SPICE. First, since CAzM is a table-based simulator, it is significantly faster than SPICE. Second, CAzM is a strictly charge-based simulator. Using SPICE, the switched-capacitor integrators operating at 5.6448 MHz exhibited errors of up to 20 mV relative to the ideal difference equations. Using CAzM, errors of less than 2 mV were observed. Fig. 3 shows the errors relative to the ideal difference equation for one row of a table using both SPICE and CAzM. Fig. 4 compares the SDR curves generated using both CAzM and SPICE tables to the SDR curve generated using ideal difference equations. These simulations were performed for a 2576.25-Hz input signal, and the noise calculations assume a 20-kHz bandwidth of interest. It is clear from Fig. 4 that the tables generated with CAzM predict more ideal performance than do the tables generated with SPICE. The greater accuracy seen in the CAzM tables is believed to be due to the fact that, unlike SPICE, CAzM is strictly charged-based and completely free from conservation of charge errors. In contrast, SPICE 2G.6 is capacitance-based, and it has been established that capacitance-based simulation is not accurate for nonlinear dynamic circuits since charge is lost as a result of numerical integration errors [12], [13]. It is especially important to keep track of charge in switched capacitor circuits, since small errors in charge can accumulate to produce an appreciable error over the transient analysis. Therefore, it is important to avoid the use of capacitance-based circuit simulators (such as SPICE) to generate the integrator tables.

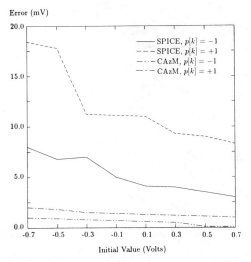

Fig. 3.  Errors of SPICE and CAzM tables relative to ideal difference equations.

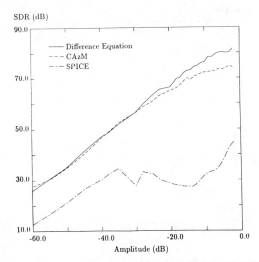

Fig. 4.  SDR versus amplitude using 5.6448-MHz CAzM tables, SPICE tables, and ideal difference equations.

Another important consideration when generating integrator tables is the timing used during the transient analysis. This timing must exactly mimic the operation of the integrators during normal operation. The table point simulations must begin just prior to the end of the previous cycle and then continue for one clock period; it is critical that the simulations begin when the assumed initial condition is valid. Referring to Fig. 2, the simulation of the first integrator must begin just prior to the end of clock phase $\phi_1$, and the simulation of the second integrator must begin just prior to the end of clock phase $\phi_2$.

Finally, it is critical when generating integrator tables to ensure that the transient analysis starts from the correct initial conditions. Since the points in the table are determined by simulating the circuit for a given input value and a given initial output for the integrator, it is important that the output of the integrator not drift before the first clock change. As an example, the table for integrator #1 was created using the circuit configuration shown in Fig. 5. The inductor (1 MH) acts as a short circuit at dc, allowing the circuit to converge to the proper initial condition at the virtual ground node for the given initial output voltage. The voltage source $V_s$ is varied from $-0.9$ to $0.9$ V in increments of 0.2 V to create the $y[k-1]$ initial conditions. As the transient

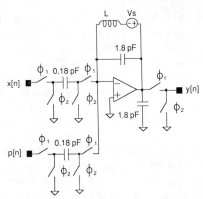

Fig. 5.  Circuit configuration for table generation.

analysis begins, the inductor effectively becomes an open circuit which disconnects the voltage source. The switches at the output of the integrator are necessary to account for the loading effects of the second integrator. These switches also can produce errors in the first integrator due to ac coupling of the switching transients back through the feedback capacitor. Therefore, it is important to include these switches when generating the table for integrator #1. The circuit used to generate the table for integrator #2 was identical to Fig. 5 except that there were no switches at the output of the integrator.

## V.  RESULTS OF ZSIM SIMULATIONS

This section discusses the results of ZSIM simulations for three variations of the delta–sigma modulator and some of the problems associated with table-based simulation. The first configuration is the nominal case, where the modulator is clocked at 5.6448 MHz. This circuit is nearly ideal and performs at the limit of the accuracy of CAzM's transient analysis. The second case is the same circuit clocked at 11.2896 MHz. The higher clock rate is interesting since it can provide true 16-bit digital audio performance with a decimation factor of 64. However, to facilitate comparison to the 5.6448-MHz case the ZSIM simulations for this case are identical to the 5.6448-MHz case, except that 11.2896-MHz integrator tables are used. This case is used to examine the effects of finite bandwidth and slew rate of the operational amplifier and provides an example of spurious errors which may occur during transient analysis. In the third case, the widths and lengths of the switches are increased by a factor of five to examine the effects of charge dump from the MOSFET switches.

### A. Nominal Case

In the nominal case the circuit is clocked at 5.6448 MHz, and the circuit errors are almost negligible. The SDR curve generated using CAzM tables tracks the SDR curve using ideal difference equations fairly well, as shown in Fig. 4, up to a signal level of $-30$ dB. In an actual design situation it would be critical to determine at this point whether the discrepancy above $-30$ dB is due to nonlinearity in the circuit or due to artifacts in the table generation process. Some indication can be obtained by examining the tables themselves.

The tables generated using CAzM (or SPICE) can be compared to tables generated using the ideal difference equations, and several error measures can be calculated. First, a table of point by point differences between each CAzM table and an ideal table was calculated. The average value of this difference table was then computed and removed, to eliminate the integrator output-referred offset. The offset error is due to the offset of the op-amp and charge dump from the switches and does not greatly affect the SNR or SDR performance. The slopes of the rows of the difference table was then calculated. These values were averaged, and the linear error was subtracted from the offset corrected difference table. This eliminates the linear gain error of the integrator, which again does not significantly affect the SNR or SDR performance. This linear gain error arises due to the finite dc gain and finite bandwidth of the operational amplifier. Finally, an overall rms error was calculated using the adjusted difference table with offset and linear gain errors removed. This error represents only the nonlinearities of the circuit and/or random errors arising from the finite precision of the circuit simulator. Spurious or random noise in the tables will cause the entire SDR curve to shift downward, while actual saturating nonlinearities in the circuit will cause the SDR curve to flatten out or fall off at higher amplitudes.

Using this methodology, the offset errors for the CAzM tables for the nominal circuit were found to be $-0.25$ mV for integrator #1 and $-0.44$ mV for integrator #2. The linear errors were $-0.035$ mV/V and $-0.141$ mV/V for integrators #1 and #2, respectively. The remaining rms nonlinear error was 0.14 mV for table #1 and 0.15 mV for table #2. (In comparison, the corresponding rms nonlinear errors for the SPICE tables were 12.8 and 37.0 mV.) The 14-bit digital audio application has a dynamic range approaching 85 dB. This application requires the table precision to be on the order of 0.05 mV. Notice that the rms error of 0.1–0.2 mV is not sufficient to obtain the 85-dB SDR required for this application. The errors in the adjusted difference table appear to be primarily random in nature, which would seem to indicate a limitation in the table point simulations rather than the actual circuit. The tolerances used in the CAzM simulations were $chgtol = 10^{-16}$, $abstol = 10^{-11}$, and $reltol = 10^{-3}$. Tightening these tolerances provides little improvement in the simulation results and leads to prohibitively long simulations.

### B. Increased Clock Rate

The simulations for the 11.2896-MHz case are identical to those for the 5.6448-MHz case except that the clock rate in the CAzM simulations was increased to 11.2896 MHz. By increasing the clock rate, it is possible to examine the effects of finite bandwidth and slew rate of the operational amplifier. The SDR curves for the 11.2896-MHz tables are shown in Fig. 6. While generating table #1 for this case, three spurious data points were obtained during transient analysis. These three errors occurred for an initial output value of $-0.1$ V and input values of $-0.25$, $-0.08$, and 0.08 V. A graphical representation of the spurious points is shown in Fig. 7. This figure shows the difference between the CAzM tables and ideal tables calculated from ideal difference equations. ZSIM simulations were performed both with the raw tables and with corrected tables where the three spurious points were replaced with linear interpolations of the surrounding points. The effect of these three spurious points can be seen in Fig. 6. The rms nonlinear error was 3.83 mV for table #1 and 11.88 mV for table #2. The rms nonlinear error for the corrected version of table #1 was 4.00 mV. Notice that although correcting the three spurious points in table #1 provides a dramatic improvement in SDR performance, it actually degrades the rms nonlinear error of the table. No conclusive explanation has been found for the spurious table points.

Another important issue raised by this case is the issue of incomplete settling. Errors can occur due to the fact that the

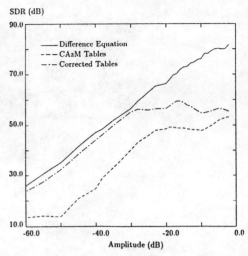

Fig. 6. SDR versus amplitude using 11.2896-MHz CAzM tables, corrected CAzM tables, and ideal difference equations.

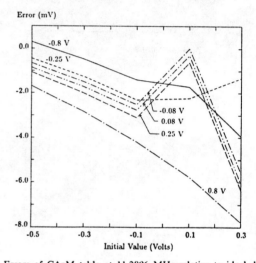

Fig. 7. Errors of CAzM table at 11.2896 MHz relative to ideal difference equation.

integrator response is underdamped and the residual ringing is sampled at the end of the clock cycle. These oscillations may be responsible for the increased noise observed in the 11.2896-MHz case as indicated in Fig. 6 by a downward shift in the SDR curve. There is also more distortion than in the 5.6448-MHz case as reflected by the flattening out of the SDR curve at a lower SDR level.

### C. Enlarged Switches

A final simulation was performed in which the widths and lengths of the switches were increased by a factor of five. By increasing the size of the switches it is possible to examine the exaggerated effects of charge dump from the MOSFET switches. Charge dump (or clock feedthrough) is a common problem in switched capacitor circuits. When a MOSFET is turned off, the charge stored in the channel flows out of the source and drain of the device. A common practice to reduce this problem is to put a p-channel device in parallel with an n-channel device, so that the stored charges will cancel. This configuration requires separate clocks for the p-channel and the n-channel devices, which must be 180 degrees out of phase. While this technique reduces the charge dump, it does not completely eliminate it. The charge feedthrough produces an output referred offset in the switched-

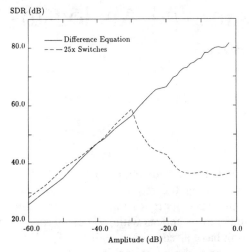

Fig. 8. SDR versus amplitude using 5.6448-MHz CAzM tables with 25x switches and ideal difference equations.

capacitor integrators which is weakly signal dependent. The SDR curves for this case are shown in Fig. 8. Although there is noticeable distortion due to charge dump from the switches, the noise level is similar to that of the nominal (5.6448 MHz) case.

## VI. CONCLUSIONS

A table-based method for simulating delta–sigma modulators has been described. The table-based method allows nonlinear circuit effects to be simulated within a reasonable amount of time. In order to generate accurate tables, capacitance-based circuit-level simulators (such as SPICE) should be avoided, and transient analyses of the switched-capacitor integrators must be properly initialized and the correct timing must be used. This method has been used to analyze the performance of a second-order delta–sigma modulator for a 14-bit digital audio application. The effects of the operational amplifier's finite bandwidth and slew rate on the performance of modulator were studied by increasing the clock rate of the modulator. The effects of charge dump were studied by increasing the size of the MOSFET switches. While these effects were successfully simulated, there are some circuit nonidealities which cannot be studied with this method, such as noise from the power supply and thermal noise. Also, while table-based methods are fast and accurate, the tables may not always provide good results. Problems can arise due to the finite precision of the circuit simulator or due to spurious errors during transient analysis.

## REFERENCES

[1] H. Inose, Y. Yasuda, and J. Murakami, "A telemetering system by code modulation—Delta-sigma modulation," *IRE Trans. Space Electron. Telem.*, vol. SET-8, pp. 204–209, Sept. 1962.

[2] M. W. Hauser, P. J. Hurst, and R. W. Brodersen, "MOS ADC-filter combination that does not require precision analog components," *IEEE Int. Solid-State Circuits Conf. Dig. Tech. Papers*, pp. 80–82, Feb. 1985.

[3] T. Misawa, J. E. Iwersen, L. J. Loporcaro, and J. G. Ruch, "Single-chip per channel CODEC with filters utilizing delta-sigma modulation," *IEEE J. Solid-State Circuits*, vol. SC-16, pp. 331–341, Aug. 1981.

[4] R. Koch, B. Heise, F. Eckbauer, E. Engelhardt, J. A. Fisher, and F. Parzefall, "A 12-bit sigma-delta analog-to-digital converter with 15 MHz clock rate," *IEEE J. Solid-State Circuits*, vol. SC-21, pp. 1003–1009, Dec. 1986.

[5] Y. Matsuya, K. Uchimura, A. Iwata, T. Kobyashi, and M. Ishikawa, "A 16b oversampling A/D conversion technology using triple integration noise shaping," *IEEE Int. Solid-State Circuits Conf. Dig. Tech. Papers*, pp. 48–49, Feb. 1987.

[6] G. T. Brauns, "ZSIM: A table-based Z-domain simulator for delta-sigma modulators," Master's thesis, North Carolina State Univ., 1987.

[7] G. T. Brauns, M. B. Steer, S. H. Ardalan, and J. J. Paulos, "ZSIM: A nonlinear Z-domain simulator for delta-sigma modulators," in *Proc. IEEE Int. Conf. on Computer-Aided Design*, pp. 538–541, Nov. 1987.

[8] W. B. Wilson, H. Z. Massoud, E. J. Swanson, R. T. George, Jr., and R. B. Fair, "Measurement and modeling of charge feedthrough in n-channel MOS analog switches," *IEEE J. Solid-State Circuits*, pp. 1206–1212, Dec. 1985.

[9] L. W. Nagel and D. O. Pederson, "SPICE—Simulation program with integrated circuit emphasis," presented at the 19th Midwest Symp. on Circuit Theory, Waterloo, Ont., Apr. 1973.

[10] W. M. Coughran, Jr., E. H. Grosse, and D. J. Roser. "CAzM: A circuit analyzer with macromodeling," *IEEE Trans. Electron Devices*, vol. ED-30, pp. 1207–1213, 1983.

[11] D. J. Erdman, "Duke release CAzM: Users guide," Tech. Rep. CS-1987-24, Duke Univ., June 1987.

[12] D. E. Ward and R. W. Dutton, "A charge-oriented model for MOS transistor capacitances," *IEEE J. Solid-State Circuits*, vol. SC-13, pp. 703–707, Oct. 1978.

[13] P. Yang, B. D. Epler, P. K. Chatterjee, P. Tuohyand, A. Gribben, A. J. Walton, and J. M. Robertson," Realistic worst-case parameters for circuit simulation," *IEEE Proc.*, vol. 134, pp. 137–140, Oct. 1987.

# Simulating and Testing Oversampled Analog-to-Digital Converters

BERNHARD E. BOSER, STUDENT MEMBER, IEEE, KLAUS-PETER KARMANN, HORST MARTIN, AND BRUCE A. WOOLEY, FELLOW, IEEE

*Abstract*—Quantities such as peak error and integral or differential nonlinearity are commonly used to characterize the performance of analog-to-digital converters. However, these measures are not readily applicable to converter architectures that employ feedback and oversampling. An alternative set of parameters for characterizing the linear, nonlinear, and statistical properties of A/D converters is suggested, and a new algorithm, referred to as the sinusoidal minimum error method, is proposed to estimate the values of these parameters. The suggested approach is equally suited to examining the performance of A/D converters by means of either computer simulations or experimental measurements on actual circuits.

## I. INTRODUCTION

SIGNAL PROCESSING systems can be divided into data acquisition and data processing components. While modern VLSI technology greatly simplifies implementation of the processing function by digital means, the same is not true for data acquisition, where analog signals must typically be conditioned and then converted to digital codes. A large number of transistors of small size and high speed are needed for digital processing, whereas conventional means of implementing the analog-to-digital (A/D) conversion function call for a variety of high-precision components. As a result, A/D converters are often implemented using special integrated circuit processes and are fabricated as separate chips. This approach is inefficient in that it both fails to take full advantage of VLSI technology and complicates system implementation by requiring multiple processes and chip sets. There is thus a pressing need for robust A/D conversion techniques that are insensitive to component variations and are compatible with VLSI technology. Oversampling is one approach to meeting this need.

In oversampled A/D converters coarse quantization at a high sampling rate is combined with negative feedback and digital filtering to achieve increased resolution at a lower sampling rate [1]–[3]. These converters thus exploit the speed and density advantages of VLSI while at the same time reducing the requirements on component accuracy. Digital speech processing systems and voice-band telecommunications codecs with A/D converters based on oversampling have already been realized [4]–[10], and the extension of the technique to performance levels required for digital audio has been demonstrated [11].

The performance of signal processing systems is typically specified by *mean squared error* (MSE) or *signal-to-noise ratio* (SNR), rather than by quantities such as peak error or integral and differential linearity that are commonly used to characterize classical Nyquist rate A/D converters. In this paper appropriate definitions of measures for both dynamic range and harmonic distortion based on the MSE are derived, and an algorithm, called the sinusoidal minimum error method, for estimating these parameters is suggested. In this algorithm the error at the output of a converter is determined by minimizing the power of the difference between the output signal and a template comprising a sinusoid at the frequency of the input signal, a dc offset, and explicitly evaluated harmonics. The applicability of the method to both the simulation and testing of oversampled converters is considered, and the algorithm is compared with alternative methods of evaluating A/D converter performance.

The sinusoidal minimum error method offers two advantages over spectral estimation techniques traditionally used for the simulation and testing of A/D converters [12], [13]. First, it is computationally more efficient because amplitude and phase of the converter output are computed only at the signal frequency and, optionally, any harmonics of interest. Second, no windowing of the data is necessary even in cases where the sampling frequency is not an integral multiple of the signal frequency. Thus, the difficult problem posed by spectral estimation of trading off accuracy against spectral resolution is avoided [14].

## II. PERFORMANCE OF OVERSAMPLED A/D CONVERTERS

In oversampled A/D converters, such as that shown in Fig. 1, the analog input $x(t)$ is sampled at a rate well above the Nyquist frequency and the amplitude is coarsely quantized [15]. The error introduced by the quantizer is spread out over the entire frequency band from zero to the sampling frequency $f_s$. Quantization noise above the signal band is then removed with a digital decimation filter wherein the signal is resampled at a rate $1/T$.

Manuscript received January 10, 1987; revised July 28, 1987, and January 4, 1988. The review of this paper was arranged by A. J. Strojwas, Editor.

B. E. Boser and B. A. Wooley are with the Center for Integrated Systems, Stanford University, Stanford, CA 94305.

K.-P. Karmann is with Siemens AG, Communication and Information Systems, Otto-Hahn-Ring 6, 8000 Munich 83, West Germany.

H. Martin is with Siemens AG, EWSD Basic Development System Peripherie, ETP 33, Boschetsriederstrasse 133, 8000 Munich 70, West Germany.

IEEE Log Number 8820200.

Reprinted from *IEEE Trans. Computer-Aided Design*, vol. 7, pp. 668–674, June 1988.

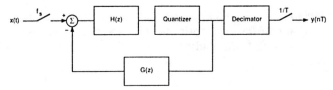

Fig. 1. Oversampled A/D converter.

Due to the feedback in the modulator, every digital output sample is a function of a sequence of analog input samples. Measures, such as peak error, that do not account for the history of the input signal are therefore meaningless for oversampled converters. Consequently, performance evaluation methods based on these parameters [16]–[18] generally cannot be used to characterize such converters. However, the mean squared error within the signal bandwidth of interest is an appropriate alternative basis for performance specification.

For every input signal $x(t)$, an A/D converter produces a sequence of digital output codes $y(nT)$. Errors in this conversion can be attributed to a variety of sources, including quantization in time (sampling) and amplitude, which are intrinsic to any A/D converter, linear errors such as gain and delay, nonlinear errors such as harmonic distortion, and additive errors such as thermal noise.

Errors due to the ideal sampling process itself are avoided simply by ensuring that the input signal is sampled at a rate that is at least twice the signal bandwidth. Quantization in amplitude maps the continuous amplitude range of the analog input signal onto a finite number of digital codes corresponding to a discrete set of amplitude levels. The linear errors describe the deviation of the gain and phase shift of the converter from ideal—usually unity gain and linear phase—and are often a function of the frequency of the input signal. Phase linearity and variation of the gain with frequency are important parameters in many applications, while the absolute value of the gain is often of less concern. Harmonic distortion is caused by nonlinearities in the A/D converter, and thermal noise is present in any analog circuit.

It is crucial that performance evaluation methods discriminate among the different types of error rather than simply lump them into a single parameter. For example, harmonic distortion has very different consequences for the operation of a system than a constant delay of the signal. Another important criterion for an evaluation procedure is its applicability to both computer simulations and measurements of actual devices, so as to simplify the comparison of expected with measured performance.

## III. SINUSOIDAL MINIMUM ERROR METHOD

In order to measure the performance of an A/D converter, test signals must be easy to define and generate. Sinusoids fulfill these requirements and are widely used for system characterization [19]. For a sinusoidal input

$$x(t) = A \cos 2\pi f_x t \qquad (1)$$

the output of the A/D converter is a sinusoidal signal at the input frequency, $f_x$, together with the error introduced by the quantization. In distortion measurement instrumentation [20], the total noise and distortion are commonly estimated by measuring the power of the difference between the output signal and a sinusoid at the frequency $f_x$ that has been fitted to the output so as to minimize the difference power. The *sinusoidal minimum error* algorithm is a generalization of this same principle. Specifically, the difference between the system output, $y(nT)$, and a template consisting of a sinusoid at the input frequency, a dc offset term, and harmonics that are to be explicitly evaluated is minimized. The template, $\hat{y}(nT)$, and a residual error, $e(nT)$, are defined such that

$$y(nT) = \hat{y}(nT) + e(nT) \qquad (2)$$

where

$$\hat{y}(nT) = \underbrace{a_0}_{\text{offset}} + \underbrace{a_1 \cos (2\pi f_x nT + \phi_1)}_{\text{signal}}$$

$$+ \underbrace{\sum_{k=2}^{K} a_k \cos (2\pi k f_x nT + \phi_k)}_{\text{harmonics}} \qquad (3)$$

and $e(nT)$ is an additive error comprising errors due to both amplitude quantization and thermal noise, as well as any harmonics not included in $\hat{y}(nT)$.

The offset, signal, and harmonics constitute spectral components of the system output and are combined in $\hat{y}(nT)$. In general, the number of harmonics present in the output is infinite. However, in practice only the power in the lower order harmonics is significant. Thus, an explicit parameter $K$ has been introduced in (3) to reflect that only a limited number of harmonics relevant to a particular application need be considered.

The amplitudes, $a_k$, and phases, $\phi_k$, of the output signal and its harmonics can be determined by fitting $\hat{y}(nT)$ to the system output $y(nT)$ so as to minimize the mean square (power) of the error $e(nT)$, $\sigma_{ee}^2$, where

$$\sigma_{ee}^2 = E\{e^2(nT)\} \triangleq \frac{1}{N} \sum_n e^2(nT). \qquad (4)$$

In this notation $E\{u(nT)\}$ is the expectation of the sequence $\{u(nT)\}$ and $N$ is the number of samples over which the expectation is evaluated.

The sequence $\hat{y}(nT)$ is a linear function of all $a_k$, but is nonlinear in $\phi_k$. However, if (3) is rewritten as

$$\hat{y}(nT) = \sum_{k=-K}^{+K} \hat{Y}(k) W^{kn} \qquad (5)$$

where

$$W = e^{j(2\pi f_x T)} \qquad (6)$$

and

$$\hat{Y}(k) = \frac{a_k}{2} e^{j\Phi_k} \qquad (7)$$

then the minimization of $\sigma_{ee}^2$ becomes linear in the complex variables $\hat{Y}(k)$. This can be demonstrated as follows.

169

A necessary condition for minimizing the power in $e(nT)$ is that

$$\frac{\partial}{\partial \hat{Y}(l)} \sigma_{ee}^2 = 0, \qquad -K \le l \le +K. \tag{8}$$

From substituting (2) and (4) for $\sigma_{ee}^2$ it follows that

$$\frac{\partial}{\partial \hat{Y}(l)} E\left\{ \left[ y(nT) - \hat{y}(nT) \right]^2 \right\} = 0. \tag{9}$$

Because the order of differentiation and taking the expectation is interchangeable, it is apparent from (5) and (9) that

$$E\left\{ -2\left( y(nT) - \sum_{k=-K}^{+K} \hat{Y}(k) W^{kn} \right) W^{ln} \right\} = 0 \tag{10}$$

and thus

$$E\left\{ y(nT) W^{ln} \right\} = E\left\{ W^{ln} \sum_{k=-K}^{+K} \hat{Y}(k) W^{kn} \right\},$$
$$-K \le l \le +K. \tag{11}$$

Equation (11) is a system of $2K + 1$ linear equations in the $2K + 1$ complex unknowns $\hat{Y}(k)$. The unknowns $a_k$ and $\phi_k$ are in turn related to the $\hat{Y}(k)$ by

$$a_k = \begin{cases} \hat{Y}(0) & \text{if } k = 0 \\ 2\left| \hat{Y}(k) \right|, & 1 \le k \le K \end{cases} \tag{12}$$

and

$$\phi_k = \arg \hat{Y}(k), \qquad 1 \le k \le K. \tag{13}$$

## IV. PERFORMANCE SPECIFICATION

The linear properties of an A/D converter can be characterized by its gain, $G = a_1/A$, and phase, $\phi_1$, both possibly functions of the signal frequency $f_x$. The signal power at the output of the A/D converter is

$$\sigma_{\text{out}}^2 = \frac{a_1^2}{2} \tag{14}$$

and the power in the harmonics is

$$\sigma_{hh}^2 = \frac{1}{2} \sum_{k=2}^{K} a_k^2. \tag{15}$$

Since $e(nT)$ is uncorrelated with $\hat{y}(nT)$, as proven in the Appendix, its power is given by

$$\sigma_{ee}^2 = \sigma_{yy}^2 - \sigma_{\text{out}}^2 - \sigma_{hh}^2 - a_0^2 \tag{16}$$

where

$$\sigma_{yy}^2 = E\left\{ y^2(nT) \right\}. \tag{17}$$

Additive errors are commonly expressed in terms of signal-to-noise ratio. Of the many possible definitions of this quantity, the following are particularly useful. The *signal-to-noise ratio*,

$$SNR = \frac{\sigma_{\text{out}}^2}{\sigma_{ee}^2} \tag{18}$$

accounts for the quantization error and thermal noise, and the *total signal-to-noise ratio* includes the harmonic distortion resulting from nonlinearities in the converter:

$$TSNR = \frac{\sigma_{\text{out}}^2}{\sigma_{ee}^2 + \sigma_{hh}^2}. \tag{19}$$

These formulas exclude the dc offset, $a_0$, which can be included as part of either the signal or error terms.

## V. PRACTICAL CONSIDERATIONS

Only a finite number, $N$, of output samples $y(nT)$ can be obtained from simulations or measurements on an A/D converter. Consequently, the expectation in (11) must be approximated by a sum over $N$ samples:

$$\sum_{n=0}^{N-1} y(nT) W^{ln} = \sum_{n=0}^{N-1} W^{ln} \sum_{k=-K}^{+K} \hat{Y}(k) W^{kn},$$
$$-K \le l \le +K. \tag{20}$$

The output $y(nT)$ and the signal frequency, $f_x$, relative to the sampling rate, $1/T$, can be determined either by a simulation or an actual measurement on the system. The amplitude $A$ of the input signal $x(t)$ of the converter need not be known other than to evaluate the gain $G$.

As an example, Fig. 2 presents simulation results obtained for an oversampled A/D converter of the form shown in Fig. 1, where $H(z)$ performs a double integration [1]. Both the *SNR* and the *TSNR* of the A/D converter are shown as a function of the normalized input signal power. For linear $H(z)$, the power in the harmonics is zero and the *SNR* and *TSNR* do not differ. Nonlinearities in $H(z)$ introduce harmonic distortion at high signal powers, and the *TSNR* thus decreases relative to the *SNR*.

The sinusoidal minimum error method presented here is equally suited to both simulations and measurements on actual circuits. A signal source with high stability and low harmonic distortion is the only precise analog component needed for experimental measurements. Consequently, very high signal-to-noise ratios can be determined accurately.

For simulation purposes, computational errors can be made arbitrarily small in digital algorithms and are therefore of no concern for the method presented here. However, both the finite record length, $N$, and the precision with which the signal frequency $f_x$ is known introduce systematic errors. Generally, these errors depend on the spectrum of the converter output $y(nT)$. The magnitude of the errors can easily be estimated by means of an example where, for the sake of simplicity, the output $y(nT)$ is assumed to be the sum of two sinusoids at frequencies $f_1$ and $f_2$:

$$y(nT) = A_1 \cos 2\pi f_1 nT + A_2 \cos 2\pi f_2 nT. \tag{21}$$

Such an output would be generated by a converter with a quadratic nonlinearity in response to a sinusoidal input with frequency $f_1$. When the sinusoidal minimum error

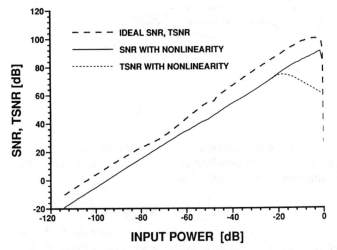

Fig. 2. *SNR* and *TSNR* of oversampled A/D converter.

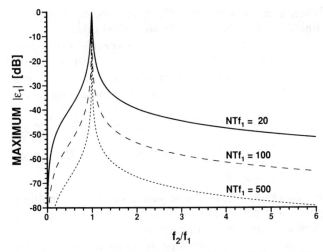

Fig. 3. Minimum sidelobe suppression for $NTf_1 = 20 \ 100$ and $500$.

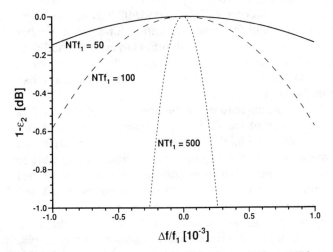

Fig. 4. Estimation error as a function of frequency error ($NTf_1 = 50 \ 100 \ 500$).

method is applied to determine the amplitude $A_1$ of the signal at $f_1$,

$$\hat{y}(nT) = a_1 \cos 2\pi f_1 nT. \qquad (22)$$

The estimated amplitude $a_1$ can be expressed in terms of $A_1$ and $A_2$ as

$$a_1 = A_1 + \epsilon_1 A_2. \qquad (23)$$

Ideally, the estimation of the amplitude, $a_1$, of the output signal at frequency $f_1$ is independent of the sine wave at frequency $f_2$; i.e., $a_1 = A_1$. For an infinite number of samples, $\epsilon_1 = 0$ for $f_1 \neq f_2$; that is, a signal at frequency $f_2$ is completely suppressed. For finite $N$, $\epsilon_1$ is bounded by

$$|\epsilon_1| \leq \left| \frac{1}{\pi NTf_1} \frac{f_2/f_1}{(f_2/f_1)^2 - 1} \right|, \qquad NTf_1 \gg 1. \quad (24)$$

Fig. 3 shows the upper bound on $|\epsilon_1|$ as a function of $f_2/f_1$ for various $NTf_1$. For example, with $N = 200$ and $f_1 = 1/10T (NTf_1 = 20)$, $\epsilon_1$ is less than $-38$ dB when $f_2 = 2f_1$.

Fig. 4 shows the estimation error that results when the signal frequency is not accurately known. For this calculation, the system output $y(nT)$ is assumed to be sinusoidal with frequency $f_2 = f_1 + \Delta f$; that is,

$$y(nT) = A_1 \cos 2\pi (f_1 + \Delta f) nT \qquad (25)$$

where $\Delta f$ is the inaccuracy of the signal frequency. If $A_1$ is to be estimated with the sinusoidal minimum error method, but due to measurement errors the frequency at which the estimation is performed is $f_1$ rather than $f_1 + \Delta f$, then

$$\hat{y}(nT) = a_1 \cos 2\pi f_1 nT. \qquad (26)$$

In this case, the estimated amplitude $a_1$ is given by

$$a_1 = (1 - \epsilon_2) A_1. \qquad (27)$$

The correct amplitude, $a_1 = A_1$, is obtained when $\epsilon_2 = 0$. The error in computing $a_1$ increases with $N$ but is negligible as long as $\Delta f NT \ll 1$. From Fig. 4 it can be seen that with $f_1 T = 1/10$, $N = 500$, and $\Delta f/f_1 = 0.05$ percent, $1 - \epsilon_2 = 0.996$; that is, the estimation error is less than 0.04 dB.

## VI. COMPARISON WITH OTHER TECHNIQUES

Several other techniques have been suggested to measure distortion in A/D converters. In this section the minimum sinusoidal error method is compared with two of these approaches.

Spectral techniques based on the Fourier transform [12], [21] are commonly used to determine the signal-to-noise ratio of A/D converters. One approach is to estimate the entire power density spectrum of the converter output and then separate the signal and the errors in the frequency domain [13]. The sinusoidal minimum error method is computationally more efficient than spectral methods based on estimating the entire spectrum since only the amplitude and phase of the signal and, optionally, the harmonics of interest need be estimated, rather than the entire spectrum. This savings in computational effort has proven crucial in simulation tools intended for the design and optimization of A/D converter structures.

While being computationally more efficient than estimating the entire spectrum, the sinusoidal minimum error method is otherwise very similar to spectral estimation techniques. In fact, the left-hand side of (20) is exactly the discrete Fourier transform (DFT) of $y(nT)$, whereas the right-hand side is the spectrum $\hat{Y}(k)$ multiplied by the matrix

$$\left[ \sum_{n=0}^{N-1} W^{n(l+k)} \right]_{kl}, \qquad -K \leq k, l \leq +K. \qquad (28)$$

If the duration, $NT$, of the sequence $y(nT)$ is a multiple of the signal period $1/f_x$, (28) is a unity matrix, and the sinusoidal minimum error method and the DFT give the same results. If this condition is violated—for example when the sampling frequency $1/T$ is not an integral multiple of the signal frequency—a window must be used for the DFT. Windows call for a tradeoff between sidelobe suppression and frequency resolution for a given number of samples, $N$. The sinusoidal minimum error method does not call for such tradeoffs. Consequently, in many cases fewer samples of $y(nT)$ are needed with this method than with the DFT.

An important reason for the popularity of the DFT is the existence of the FFT [12]. Because of the similarity of the sinusoidal minimum error method and the DFT, the same means of speeding up computations is applicable to both techniques. Equation (20) can be rewritten for real arithmetic and the matrix (28) can be precomputed and decomposed for a series of computations.

A computationally efficient technique for estimating only uncorrelated errors (which exclude, for example, harmonic distortion) has been suggested by Candy [22]. The same signal $x(t)$ is applied to two identical A/D converters as illustrated in Fig. 5. The signal and correlated errors at the outputs of the two systems will then be equal. If we further make sure that the uncorrelated errors are orthogonal by initializing the two oversampled converters to different states, the power in the difference of the two system outputs equals twice the power of the uncorrelated errors. While this procedure does not reveal any information about linear errors or harmonic distortion in an A/D converter, it is computationally more efficient than either the sinusoidal minimum error method or the discrete Fourier transform approach. Moreover, it is not restricted to periodic test signals, but allows for random input signals as well. The critical prerequisite for using this algorithm is the availability of two *identical* A/D converters. For example, only slight variations of the gains of the two systems could cause a difference of the signal at the outputs which far exceeds the uncorrelated errors. Consequently, the technique is practical only for computer simulations.

As pointed out previously, many of the techniques used to characterize classical A/D converters are not applicable to oversampled converters. Conversely, however, the sinusoidal minimum error method presented here can be used

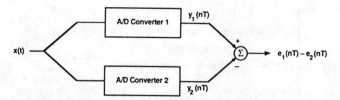

Fig. 5. Estimation of uncorrelated noise.

to evaluate both classical and oversampled A/D converters, thus making possible direct comparison between different conversion architectures.

## VII. CONCLUSIONS

A method for estimating the performance of A/D converters under real-time conditions has been proposed. The technique is applicable to both classical and oversampled converters and does not entail analog measurements that could limit its resolution. The tradeoffs between measurement time and accuracy of the algorithm have been explored, and estimates of the precision have been presented.

## APPENDIX

First it will be shown that $e(nT)$ and $\hat{y}(nT)$ are uncorrelated. This result will then be used to derive (16).

*Lemma:*

$$E\left\{ e(nT) W^{ln} \right\} = 0. \qquad (29)$$

*Proof:* It follows from (5) and (10) that

$$E\left\{ [y(nT) - \hat{y}(nT)] W^{ln} \right\} = 0, \qquad -K \leq l \leq +K \qquad (30)$$

and then from (2) that

$$E\left\{ e(nT) W^{ln} \right\} = 0, \qquad -K \leq l \leq +K \quad \text{QED.} \qquad (31)$$

*Proposition:* The cross-correlation of $e(nT)$ and $\hat{y}(nT)$ is zero:

$$E\left\{ e(nT) \, \hat{y}(nT) \right\} = 0. \qquad (32)$$

*Proof:* Substitution of (5) for $\hat{y}(nT)$ leads to

$$E\left\{ e(nT) \, \hat{y}(nT) \right\} = E\left\{ e(nT) \sum_{k=-K}^{+K} \hat{Y}(k) W^{kn} \right\}. \qquad (33)$$

Upon exchanging the summation and expectation operators,

$$E\left\{ e(nT) \, \hat{y}(nT) \right\} = \sum_{k=-K}^{+K} \hat{Y}(k) \, E\left\{ e(nT) W^{kn} \right\}. \qquad (34)$$

The value of the expectation in this expression is zero according to the above Lemma; hence

$$E\left\{ e(nT) \, \hat{y}(nT) \right\} = 0 \qquad \text{QED.} \qquad (35)$$

From (35) it follows that the power in the output sequence $y(nT)$ is

$$\sigma_{yy}^2 = E\{y^2(nT)\}$$
$$= E\{[\hat{y}(nT) + e(nT)]^2\}$$
$$= E\{\hat{y}^2(nT)\} + E\{\hat{y}(nT)e(nT)\}$$
$$\quad + E\{e(nT)\hat{y}(nT)\} + E\{e^2(nT)\}$$
$$= E\{\hat{y}^2(nT)\} + E\{e^2(nT)\}$$
$$= \sigma_{\hat{y}\hat{y}}^2 + \sigma_{ee}^2. \tag{36}$$

The power of $e(nT)$ is thus

$$\sigma_{ee}^2 = \sigma_{yy}^2 - \sigma_{\hat{y}\hat{y}}^2. \tag{37}$$

Since sine waves at different frequencies are orthogonal, the power of $\hat{y}(nT)$ is

$$\sigma_{\hat{y}\hat{y}}^2 = a_0^2 + \sigma_{\text{out}}^2 + \frac{1}{2}\sum_{k=2}^{K} a_k^2. \tag{38}$$

This last expression is exact only if either the sampling frequency $1/T$ is an integer multiple of the signal frequency, $f_x$, or the number of samples $N$ is infinite. The case when $N$ is finite is discussed in Section V.

The fact that $e(nT)$ is uncorrelated with $\hat{y}(nT)$ does not imply that $e(nT)$ and the system input $x(nT)$ are uncorrelated. Specifically, correlation between the input and the noise has been observed in some oversampled A/D converter architectures.

ACKNOWLEDGMENT

The authors would like to thank Dr. J. Tiemann and Prof. R. Gray for their comments and suggestions.

REFERENCES

[1] J. C. Candy, "A use of double integration in sigma delta modulation," *IEEE Trans. Commun.*, vol. COM-33, pp. 249–258, Mar. 1985.
[2] J. C. Candy, B. A. Wooley, and O. J. Benjamin, "A voiceband codec with digital filtering," *IEEE Trans. Commun.*, vol. COM-29, pp. 815–830, June 1981.
[3] J. C. Candy, "Decimation for sigma delta modulation," *IEEE Trans. Commun.*, vol. COM-34, pp. 72–76, Jan. 1986.
[4] G. L. Baldwin and S. K. Tewksbury, "Linear delta modulator integrated circuit with 17-MBit/s sampling rate," *IEEE Trans. Commun.*, vol. COM-22, pp. 977–985, July 1974.
[5] B. A. Wooley and J. L. Henry, "An integrated per-channel PCM encoder based on interpolation," *IEEE J. Solid-State Circuits*, vol. SC-14, pp. 14–20, Feb. 1979.
[6] J. D. Everhard, "A single-channel PCM codec," *IEEE J. Solid-State Circuits*, vol. SC-14, pp. 25–37, Feb. 1979.
[7] T. Misawa, J. E. Iwersen, L. J. Loporcaro, and J. G. Ruch, "Single-chip per channel codec with filters utilizing sigma-delta modulation," *IEEE J. Solid-State Circuits*, vol. SC-16, pp. 333–341, Aug. 1981.
[8] H. Kuwahara *et al.*, "An interpolative PCM codec with multiplexed digital filters," *IEEE J. Solid-State Circuits*, vol. SC-15, pp. 1014–1021, Dec. 1980.
[9] M. W. Hauser, P. J. Hurst, and R. W. Brodersen, "MOS ADC-filter combination that does not require precision analog components," in *IEEE Int. Solid-State Circuits Conf. Dig. Tech. Papers*, Feb. 1985, pp. 80–81.
[10] K. Yamakido, S. Nishita, M. Kokubo, H. Shirasu, K. Ohwada, and T. Nishihara, "A voiceband 15b interpolative converter chip set," in *IEEE Int. Solid-State Circuits Conf. Dig. Tech. Papers*, Feb. 1986, pp. 180–181.
[11] U. Roettcher, H. Fiedler, and G. Zimmer, "A compatible CMOS-JFET pulse density modulator for interpolative high-resolution A/D conversion," *IEEE J. Solid-State Circuits*, vol. SC-21, pp. 446–452, June 1986.
[12] A. V. Oppenheim and R. W. Schafer, *Digital Signal Processing*. Englewood Cliffs, NJ: Prentice Hall, 1975.
[13] F. Irons, "Dynamic characterization and compensation of analog to digital converters," in *Proc. IEEE Int. Symp. Circuits Syst.*, May 1986, pp. 1273–1277.
[14] F. J. Harris, "On the use of windows for harmonic analysis with the discrete Fourier transform," *Proc. IEEE*, vol. 66, pp. 51–83, Jan. 1978.
[15] D. J. Goodman, "Delta modulation granular quantization noise," *Bell Syst. Tech. J.*, vol. 48, pp. 1197–1218, May–June 1969.
[16] Hewlett Packard, "Dynamic performance testing of A-to-D converters," *Hewlett Packard Product Note 5180A-2*.
[17] J. Doernberg, H.-S. Lee, and D. A. Hodges, "Full-speed testing of A/D converters," *IEEE J. Solid-State Circuits*, vol. SC-19, pp. 820–827, Dec. 1984.
[18] M. Bossche, J. Schoukens, and J. Renneboog, "Dynamic testing and diagnostics of A/D converters," *IEEE Trans. Circuits Syst.*, vol. CAS-33, pp. 775–785, Aug. 1986.
[19] CCITT, "Digital networks—Transmission systems and multiplexing equipment," Tech. Report III.3, The International Telegraph and Telephone Consulative Committee, Nov. 1980.
[20] Hewlett-Packard, *Distortion Measurement Set 339A*, HP, 1984.
[21] L. R. Rabiner and B. Gold, *Theory and Application of Digital Signal Processing*. Englewood Cliffs, NJ: Prentice Hall, 1975.
[22] J. C. Candy, "Simulation of sigma-delta modulators based on uncorrelated noise," unpublished.

# A Use of Double Integration in Sigma Delta Modulation

JAMES C. CANDY, FELLOW, IEEE

*Abstract*—Sigma delta modulation is viewed as a technique that employs integration and feedback to move quantization noise out of baseband. This technique may be iterated by placing feedback loop around feedback loop, but when three or more loops are used the circuit can latch into undesirable overloading modes. In the desired mode, a simple linear theory gives a good description of the modulation even when the quantization has only two levels. A modulator that employs double integration and two-level quantization is easy to implement and is tolerant of parameter variation. At sampling rates of 1 MHz it provides resolution equivalent to 16 bit PCM for voiceband signals. Digital filters that are suitable for converting the modulation to PCM are also described.

## I. INTRODUCTION

THIS paper describes the design of a digital modulator that is intended for use in oversampled PCM encoders. These encoders modulate their analog inputs into a simple digital form at high speed; then digital processing transforms the modulation to PCM sampled at the Nyquist rate [1]–[9].

Properties of the preliminary modulation have strong influence on the design of the entire encoder. For example, the resolution of the PCM can be no better than that of the modulation, and the complexity and speed of the digital processor depends on the kind of modulation used and its resolution. The tolerance of the analog circuits employed in the modulator can determine the suitability of the design for integrated circuit implementation and the power consumed by these circuits can be a large part of the power used by the entire encoder. There is, therefore, much incentive to find an efficient modulator, one that provides high resolution (idle channel noise more than 80 dB below peak signal) at moderate sampling rates (less than 1 MHz for 4 kHz telephone signals) yet employs simple robust circuits (tolerances no tighter than ±3 percent).

Early work on oversampled encoders [1], [2] was mostly theoretical and based on delta modulation. Later, practical realization preferred sigma delta modulation but modified it to lower the sampling rate and simplify the digital processing. For a video application, multilevel quantization was used [3] to reduce the modulation rate. For telephone applications, some modulators [4], [5] achieve high resolution by biasing the modulator to an especially quiet state for idle channel operation. One [7] employed triple integration in the sigma delta modulator; another [8] employed digital accumulation and companded quantization levels in the feedback path.

Recent advances in digital integrated circuit technology have greatly reduced the need to have simple digital processing; indeed, now it is feasible [10] to have digital line equalization, echo canceling, digital hybrids, and conferencing on the chip with the codec. These applications, however, place stringent demands on resolution and dynamic range to be provided by the modulator.

The present work explores the advantages of having double integration in a sigma delta modulator. We demonstrate that a particular class of circuits can provide high resolution and be tolerant to imperfection. We explain the reasons for using

multiple integration, and calculate the signal-to-noise ratios. The results are confirmed by simulation and experimental measurements. We show that when more than two integrators are used, the circuit can latch into undesirable modes where its performance is ruined. Finally, we give a design for a digital processor for constructing PCM from this modulation.

## II. QUANTIZATION WITH FEEDBACK

Fig. 1(a) shows the circuit of a differential quantizer which is a form of the well-known delta modulator; we will use this circuit to explain our view of feedback modulation. Fig. 1(b) shows a sampled data model of the circuit; it assumes that the A/D and D/A conversion are ideal and that signals are random, so that the quantization may be represented by added noise $e$ and linear gain $G$ (level-spacing/threshold spacing). Accumulation $A$ represents the integration. Mathematical descriptions of related circuits have been presented in several places [1], [3], [4], [12], [14]–[16]. They show that the presence of feedback around the quantizer has three uses, which are summarized below.

*Prediction and Preemphasis:* The modulated signal $M$ comprises a noise component and a component that is proportional to the rate of change of input amplitude. Modulating a rate of change can be more efficient than modulating the amplitude directly, particularly for video and audio signals whose spectral densities fall with increasing frequency and whose sample values are highly correlated [15], [16]. This improved efficiency can result in decreased sampling rate or a reduction in the number of quantization levels needed for a given resolution.

*Control of Overloading:* Ordinary PCM quantization overloads by clipping signal amplitudes directly. When this happens to the signal applied to the A/D in Fig. 1, it is the derivative of the input signal that is clipped, resulting in slope overloading of the output signal. Distorting the slope of video [17] and audio signals can be less disturbing than clipping their amplitudes directly. This also can lead to improved efficiency of the modulation.

*Noise Shaping:* Placing the quantizer in a feedback loop with a filter shapes the spectrum of the modulation noise [3], [13], and at the same time it can decorrelate the noise from the signal. If we assume that the quantization noise $e$ in Fig. 1(b) is white, then the spectral density of the noise in the modulated signal rises with frequency; but after integration the noise is white again at the output. We will see that other circuit configurations [13]–[18] can shape the output noise spectrum to suit particular applications.

When higher order filters are used in place of simple integration [11]–[14], the properties of the modulation are modified. The modulated signal includes components that are proportional to high-order derivatives of the signal. Overloading limits not only the slope but also the rate of change of slope of the signal, and modulation noise rises more steeply with frequency. The restriction on the design of these high-order filters is the need to keep the feedback stable.

All these properties of feedback quantization can influence the design of modulators for oversampled codecs, and in some applications the requirements are in conflict. For example, optimum design of predictors usually calls for leaky integration, but optimum noise shaping calls for long-time constant inte-

Paper approved by the Editor for Signal Processing and Communication Electronics of the IEEE Communications Society for publication without oral presentation. Manuscript received March 28, 1984; revised October 16, 1984.

The author is with AT&T Bell Laboratories, Holmdel, NJ 07733.

Reprinted from *IEEE Trans. Commun.*, vol. COM-33, pp. 249–258, March 1985.

174

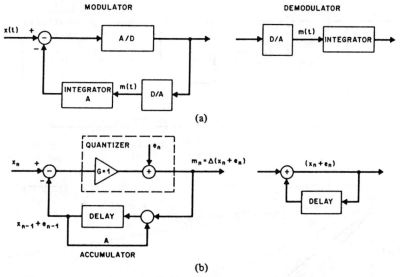

Fig. 1. (a) An example of a differential modulator and demodulator. (b) A sampled data representation of the differential modulation.

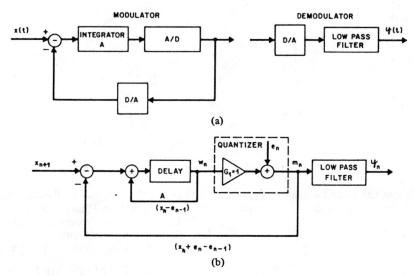

Fig. 2. (a) A sigma delta modulator and demodulator. (b) A sampled data representation of sigma delta modulation.

gration [15]. It is the different emphasis given to these separate properties that accounts for the different filters that have been proposed for feedback quantizers. Our design for the modulator will be based on the requirements of telephone toll-networks.

### III. REQUIREMENT OF GENERAL PURPOSE MODULATORS FOR TOLL NETWORK USE

Digital codecs used in the telephone toll-network must accept a wide range of signals, and their design may not rely on properties of restricted classes of signals nor properties of special receivers. We may not rely on there being high correlation between Nyquist samples, nor assume that slope overloading is any more acceptable than clipping amplitudes or that a colored noise is less objectionable than white noise.

We may take advantage of the fact that the signal is band limited, however, by moving quantization noise out of band where it can be removed by appropriate filters. The sigma delta modulator shown in Fig. 2 does this without differentiating the signal: it eliminates the need for integration at the receiver because low-frequency components of the modulated signal represent the input amplitude directly. This structure

has gained favor because it is very tolerant of imperfection and mismatch of the two D/A circuits [3]. The structure of Fig. 1 is not so tolerant because its D/A imperfections are multiplied by the large baseband gain of the integrating filter at the receiver.

The next section of this work will be directed at the task of generalizing the filter $A$ used in this modulator for the purpose of moving quantization noise out of the signal band. Applications that want to use signal prediction and special overload characteristics could do so by providing preemphasis and deemphasis filters external to the modulator [15].

### IV. SIGMA DELTA MODULATION

The modulator shown in Fig. 2 generates a quantized signal that oscillates between levels, keeping its average equal to the average input. It is easy to show [3] that for active inputs, the spectral density of the noise in the quantized signal is given by

$$N_M(f) = (1 - z^{-1})E(z) = \sigma \sqrt{\frac{2\tau}{3}} \sin{(\pi f \tau)} \qquad (1)$$

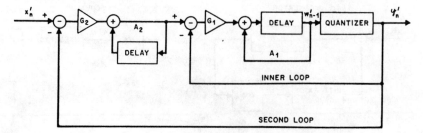

Fig. 3. A quantizer with two feedback loops around it. The gains of all elements are nominally unity.

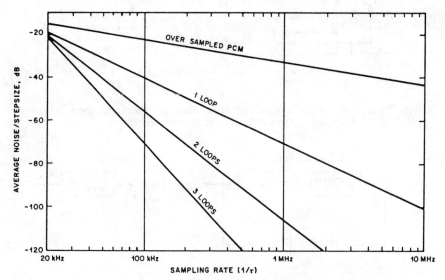

Fig. 4. Quantization noise plotted against sampling frequency for various numbers of feedback loops. Baseband is 3.5 kHz and the noise is referred to the step size: the noise of ordinary PCM is −10.8 dB on this scale.

where $\sigma$ is the quantization step size and $\tau$ the sampling period. The net noise in baseband $0 \leqslant f < f_0$ is then given approximately by

$$N_{M0} = \frac{\pi e_0}{\sqrt{3}} (2f_0\tau)^{3/2} \qquad (2)$$

provided that $f_0\tau \ll 1$ and $e_0 = \sigma/\sqrt{12}$ is the average noise generated by the quantizer alone. Thus, the resolution can be greatly increased by oversampling and feedback. For example, quantizing a 3.5 kHz signal at 16 MHz with this feedback quantizer reduces the noise $e_0$ by 96 dB, which is equivalent to a $2^{16}$ reduction of the step size.

This procedure for increasing the resolution with feedback can be reiterated as illustrated in Fig. 3. Here we have included gains $G_n$ in cascade with each integrator in order to describe circuit imperfections that cause the loop gains to be other than unity. Appendix A shows that the spectral density of the noise in the modulated signal generated by two feedback loops is given by

$$N_M(f) = (1 - z^{-1})^2 E(z) = 2e_0\sqrt{2\tau}(1 - \cos(2\pi f\tau)) \qquad (3)$$

when $G_n = 1$. The in-band noise is given approximately by

$$N_{M0} = \frac{\pi^2 e_0}{\sqrt{5}} (2f_0\tau)^{5/2}. \qquad (4)$$

This double feedback increases the resolution by 95 dB for a sampling rate of only 1 MHz and 3.5 kHz baseband. Fig. 4 compares the resolution of PCM to that obtainable with one, two, and three feedback loops. Measurements on circuits with active input signals agree with these calculated values.

The circuits that can be derived by reiterating simple feedback loops of this kind are a very useful class of feedback quantizers with high-order filters [14], [18]. Ritchie points out the penalties that must be paid for using feedback, and they are summarized in the next section.

## V. Penalties for Using Feedback

In Fig. 2 the signal applied to the quantizer can be expressed as the input less the noise from the previous cycle: $w_n = x_n - e_{n-1}$, and this noise uses up some of the dynamic range of the quantizer. Overloading may be avoided by adding one extra level to the quantizer because $e$ spans $\pm\sigma/2$. In a similar manner it may be shown that, when the number of feedback loops is $L > 0$, the range of signals applied to the quantizer is increased by $2^{L-1}\sigma$. If this signal exceeds the range of the quantizer, the modulation noise increases, as is illustrated by Fig. 5. For many applications, the increase in noise for large signal values can be tolerated, provided an adequate signal-to-noise ratio is maintained.

Besides requiring additional quantization range, feedback demands increased precision in the gains of the circuits. Gains in the range $\pm 10$ percent are usually acceptable for quantizers having a single feedback loop, but more precise gains are needed when additional feedback loops are used. In Fig. 6 we plot the calculated and measured signal-to-noise ratio against values of gains $G_n$ placed in series with each integrator for one, two, and three feedback loops. The measured change of noise with gain is larger and more variable than predicted by calculation. This is because the noise is correlated with signal amplitude in a way that depends on the gains $G_n$, and this correlation is ignored in the calculations.

The third penalty for using feedback concerns the depend-

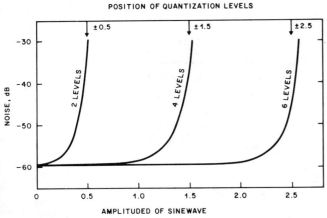

Fig. 5. The noise introduced into sinusoidal signals of various amplitudes by modulation with two feedback loops placed around a quantizer having the stated number of levels. Sampling is at 128 kHz.

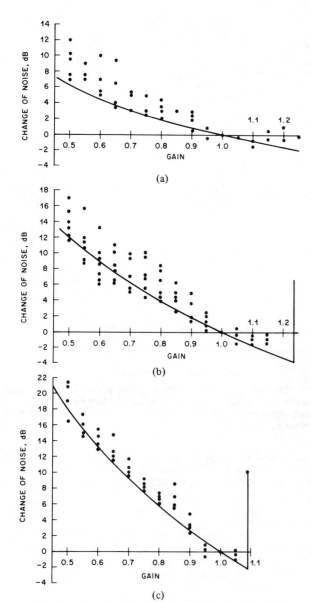

Fig. 6. The change in noise with gain placed in cascade with each integrator for multilevel quantization. The reference noise is that measured at unity gain value. The curve gives calculated values that apply for sampling rates that are at least eight times the Nyquist rate. (a), (b), and (c) are for one, two, and three feedback loops, respectively. These circuits become unstable at gains 2, 1.236, and 1.087.

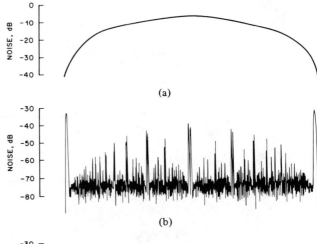

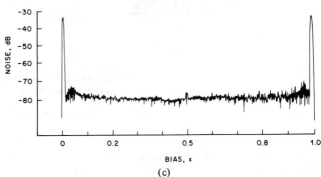

Fig. 7. Graphs of idle channel noise plotted against input bias. (a) Ordinary PCM. (b) Simple sigma delta modulation. (c) Two-level quantization with double feedback. Noise is referred to the step size. The sampling rate is 256 kHz.

ence of modulation noise on signal amplitude. A determination of this dependence for ordinary sigma delta modulation in [19] shows that it can have an important influence on the performance of oversampled codecs. Fig. 7 is a graph of noise plotted against the dc bias, $x$, applied to the modulator. Fig. 7(a) is the quantization error without feedback; the error is zero for $x = 0$ or 1 which corresponds to the position of two adjacent levels, and elsewhere the error is proportional to the distance to a level. Fig. 7(b) shows the idle channel noise of a quantizer with a single feedback loop. The noise is gathered into a series of narrow peaks, and the large peaks can be an embarrassment in the design of oversampled codecs.

## VI. TWO-LEVEL QUANTIZATION

We will demonstrate how the three penalties, described in the previous section, can be avoided by using just two feedback loops and degrading the quantizer to a single threshold circuit that generates a two-level output.

With a single threshold the inconvenience of establishing threshold spacing is removed, and the concept of gain of the quantizer becomes unreal unless other circuit properties provide a calibration for the amplitudes applied to the threshold. We find by experiment that signal levels adjust themselves so that the effective gain of the quantizer compensates for changes in the values of circuit gains $G_n$. The measured noise shown in Fig. 8 is almost independent of these gains and corresponds to the calculated value for loop gains of unity. Likewise, the change in the attenuation of the signal $Y(\omega)/X(\omega)$ is less than $\pm 0.05$ dB for values of gain $0.5 < G_n < 2$. We find that the penalty of having to establish quantizer gains in one and two feedback loops is substantially eliminated by the use of two-level quantization: signal levels in the circuit automatically adjust themselves to make the effective loop gains unity. With

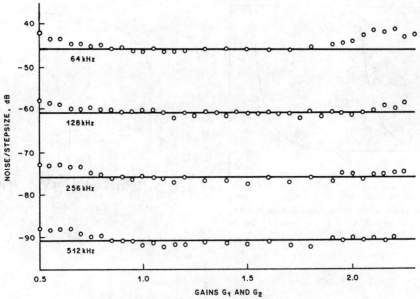

Fig. 8. The noise in double feedback, two-level modulation plotted against gains $G_n$ at various sampling rates. The input was excited with a random signal spanning $\pm 0.1\sigma$. The horizontal lines mark calculated values of noise for unity gain and multilevel quantization with the same step size.

more than two feedback loops, undesirable oscillations spoil the functioning of the circuit when the quantizer has only two levels. We discuss these oscillations in the next section.

The second penalty concerns the loss of quantization range because feedback increases the amplitudes that are applied to the quantizer. With a single threshold there is no limitation on the range of its input amplitude, only the output levels are defined, but we see in Fig. 5 that for two-level quantization the noise increases rapidly with signal amplitude. Fig. 9 presents similar data plotted on logarithmic scales.

The third penalty concerned the correlation of noise with input level. Fig. 7(c) shows a graph of modulation noise plotted against input bias for a modulator utilizing two feedback loops. Comparing it to the graphs in Fig. 7(b), we see that use of two loops substantially decorrelates the noise except at the ends of the range, where the modulation noise peaks in the same fashion that it does in modulators that utilize only one feedback loop. Reference [19] shows that the amplitude of these peaks of noise is given by

$$N_{max} = \sqrt{2}\,(f_0\tau)\sigma \tag{5}$$

which can be large compared to the calculated noise (4). But their width $v$ is narrow:

$$v = f_0\tau\sigma. \tag{6}$$

It appears that the use of two-level quantization and double integration could be the basis of a useful modulator. The next section explains why it is wise to use no more than two integrators.

## VII. LIMIT-CYCLES THAT OVERLOAD THE QUANTIZER

The feedback quantizers that we have described cause their outputs to oscillate between levels in a way that keeps their average value equal to the average input. In the desired mode of operation the signals held in the integrators are comparatively small, but when three or more feedback loops are present, other modes can be excited [14]. These modes are characterized by being very noisy and having large-amplitude, low-frequency oscillations in the integrators, which exceed the range

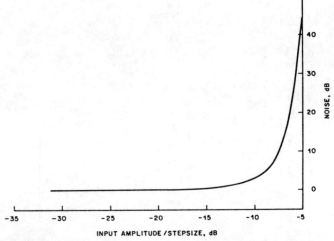

Fig. 9. The dependence of noise on the amplitude of a sinusoidal input signal for two-level modulation with double feedback. The noise is referred to its small signal value. This result applies to sampling rates that are eight or more times the Nyquist rate.

of the quantizer. In particular, when the quantizer has only two levels, the undesirable modes are easily excited and difficult to extinguish. The susceptibility of the circuit to enter undesirable modes has prevented measurement of the resolution of circuits that have triple integration and two-level quantization except for gains $G_2$, $G_3$ less than 0.55. Even for gains as low as 0.1, the unwanted modes were self-sustaining after being excited. When two feedback loops are used, however, the contents of the integrators always decay to a small value when excitations are removed, provided that the gains $G_n$ are less than 1.23 for multilevel quantization and less than 2 for two-level quantization.

When modulators function in an undesirable mode, the signals in the integrators are so large, compared to the largest quantization levels, that the inner feedbacks are ineffective. The behavior of the circuit is dominated by the outermost feedback, which, if it contains more than two integrations, is unstable. Stability is regained when the inner feedbacks are made effective by clipping the amplitudes of signals held in the integrators, or by nonlinear feedback [14]. But it is questionable whether the extra resolution obtained in practice can justify the use of these more complicated circuits and the tighter tolerances that they demand.

## VIII. IMPLEMENTATION OF THE MODULATORS USING TWO INTEGRATORS

The circuit shown in Fig. 3 is a sampled data model of the modulator; its signals are represented by impulse sample values. This circuit could well be implemented using switched capacitors for accumulation. Implementations in a bipolar technology, however, would prefer to use continuous signals such as those in the circuit of Fig. 10. The analysis of the switched and the continuous circuits in Appendix C shows that their operation is equivalent when the feedback signal $y(t)$ is held constant throughout the sample interval, the time constant $RC$ equals $1.5\tau$, and the two inputs are related by the expression

$$x_{n\tau}' = \int_{(n-1)\tau}^{n\tau} (2x(t) - x(t-\tau))\, dt. \tag{7}$$

The analog circuit in Fig. 10 is relatively easy to construct because there are only two main constraints on its design. There is the need to keep signals small in order to conserve power yet have signal levels large enough to swamp noise and imperfection in the threshold circuit, and there is the need to set the time constant $RC$ with sufficient precision. Fig. 11 shows graphs of the modulator's resolution plotted against input amplitude for three values of the time constant. Time constants changing in the range $1.2\tau$ to $1.8\tau$ give less than 1 dB variation in noise level; this should satisfy most applications.

Equation (7) can be used to define the frequency response of the filter that should be placed in cascade with the input of the sampled data circuit in Fig. 3 in order to make its response identical to that of the analog circuit in Fig. 8. That frequency response is given by

$$G(\omega) = \frac{X'(\omega)}{X(\omega)} = (2 - \epsilon^{-j\omega\tau})\operatorname{sinc}(f\tau). \tag{8}$$

This low-pass filter, inherent to the circuit of Fig. 10, is useful for reducing aliasing distortion. For example, when 3.5 kHz signals are being modulated at 512 kHz, any spurious signals in the range 508–516 kHz alias into band. But the distortion is small because $|G(508)| < -40$ dB. Table I lists the attenuation of the signals aliased into band for various sampling frequencies; also listed is gain introduced into baseband. This gain could be equalized in the digital processor.

Simulations show that signal amplitudes in the integrators of these two-level feedback modulators can be very large, and real implementations need to limit their size. Clipping their amplitudes speeds the recovery from overload but may increase the noise. Fig. 12 shows that essentially full resolution of two-level quantization is obtained by allowing the integrated signals to swing through at least $\pm 1.0$ step sizes. Variation of the modulator's net gain with input amplitude was less than $\pm 0.025$ dB.

Fig. 13 shows that leakage in the integration has negligible effect on noise, provided its time constant exceeds $1/2f_0$ seconds. The data in this figure agree with calculations that represent quantization by added noise, and assumes unity gain in the feedback loops.

## IX. DESIGN OF THE DIGITAL PROCESSOR

A digital processor will convert the output of the modulator into PCM by smoothing the signal with a digital low-pass filter

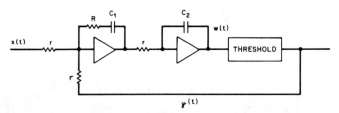

Fig. 10. An analog version of the double integrating modulator.

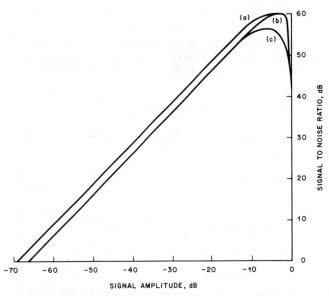

Fig. 11. The signal-to-noise ratio plotted against the amplitude of a sinusoidal input signal for the circuit in Fig. 10. The sampling rate is 256 kHz with 3.5 kHz baseband; the time constant $RC_1$ is (a) $1.5\tau$; (b) $2\tau$; (c) $\tau$.

TABLE I
GAINS OF THE FILTER $(2 - \exp(-j2\pi f\tau))\operatorname{sinc}(f\tau)$

| SAMPLING FREQUENCY $1/\tau$ kHz | GAIN AT 3.5 kHz dB | GAIN AT $(1/\tau - 3.5)$ kHz, dB |
|---|---|---|
| 64 | 0.87 | -23.9 |
| 128 | 0.24 | -30.8 |
| 256 | 0.06 | -37.1 |
| 512 | 0.015 | -43.2 |
| 1024 | 0.004 | -49.3 |
| 2048 | 0.001 | -55.3 |
| 4096 | 0.00 | -61.4 |
| 8192 | 0.00 | -67.4 |

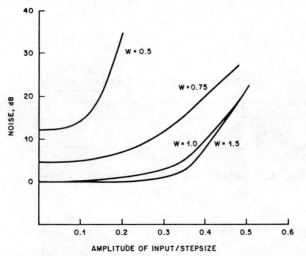

Fig. 12. The effect on noise of clipping the amplitude of signals in the two integrators for various amplitudes of a sinusoidal input. $W$ is the ratio of the clipping level to the step size. Quantization levels are at $\pm 0.5$.

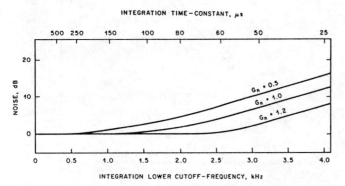

Fig. 13. The dependence of noise on leakage in both integrators for two-level quantization and double feedback. Baseband $f_0 = 3.5$ kHz.

and resampling it at the Nyquist rate. The filter attenuates spurious high-frequency components of the signal and the high-frequency components of the modulation noise, so that resampling does not alias significant noise into baseband.

Reference [9] describes the design of a digital processor that is suitable for use with ordinary sigma delta modulation. It reduces the sampling rate in stages, employing 32 kHz as an intermediate frequency for reaching the 8 kHz Nyquist rate of telephone signals. This technique leads to simple and efficient circuits. Use of double integration in the modulator does not influence the design of the filter with respect to out-of-band signal components, but it does influence the requirements for attenuating modulation noise, because the noise now rises more rapidly with increasing frequency.

The filter used in [9] for lowering the sampling rate to 32 kHz has a triangular-shaped impulse response, and frequency response given by

$$H(\omega) = \left(\frac{\sin fN\tau}{\sin f\tau}\right)^2. \tag{9}$$

Calculations of noise, in Appendix B, show that such filtering does not provide sufficient attenuation for modulations generated by means of double integration. It is shown that the modulation frequency would need to be raised from 1 to 2.5 MHz in order to make up for use of such inadequate filtering.

A filter that does provide adequate attenuation has fre-quency response

$$H(\omega) = \left(\frac{\sin fN\tau}{\sin f\tau}\right)^3 \tag{10}$$

and impulse response

$$h_n = \frac{n(n+1)}{2} \qquad \text{for } 1 \leqslant n < N$$

$$h_n = \frac{N(N+1)}{2} + (n-N)(2N-1-n)$$
$$\text{for } N \leqslant n < 2N$$

$$h_n = \frac{(3N-n-1)(3N-n)}{2} \qquad \text{for } 2N \leqslant n < 3N$$

where $N$ is the number of input sample values that occur in one period of the resampling. The duration of the filter impulse response is three resampling periods. Its frequency response has triple zeros at the 32 kHz resampling rate and all harmonics of it. When this filter is used, the noise, aliased into band, results in less than 0.5 dB loss of resolution. This filter can be implemented as an FIR structure that does not require full multipliers because the modulated signal is a 1 bit code. In the final stage of digital processing a sharp cutoff filter is needed. Analysis in Appendix C shows that at least 25 dB of attenuation is needed, which is provided by the filter described in [9]; it comprises two second-order low-pass sections followed by an accumulation and dump.

## X. CONCLUSION

We have demonstrated that a sigma delta modulator with double integration can be designed to provide resolution equivalent to that of 16 bit PCM when modulating 4 kHz signals at 1 MHz, with simple robust circuits. A somewhat higher modulation rate, 2.5 MHz, would permit use of simpler digital processing given by (9) for converting the modulation to PCM.

It is instructive to compare this modulator to the interpolating modulator described in [8]. They provide comparable resolution when the interpolating modulator generates 9 bit words at 256 kHz and the sigma delta modulator generates 1 bit words at 1 MHz. The interpolative modulator has more complex circuits, but the tolerances of the two modulators to imperfection are roughly equivalent. The overriding advantage of the modulator described here is the fact that its quantization is uniform. To obtain comparable resolution from the interpolating modulator without companding its quantization levels would require that it generate 4 bit words at 2 MHz.

There is need to have uniform quantization in order that the encoder can accept the sum of several independent signals without having interaction between them. For example, when digital hybrids and digital conferencing are to be provided, the sum of three or more speech signals may be present in the modulator at one time; then we require that the quantization noise be independent of the signal amplitude. The possible use of triple integration has been rejected because it can latch into noisy modes of operation.

## APPENDIX A
## MODULATION NOISE

Quantization in the modulator shown in Fig. 3 is represented by additive noise $e$. We assume it is white with spectral power density $2\tau e_0^2$, where $e_0$ is the noise power in the band of frequencies below the half sampling rate. When the gains $G_n$ are unity, we can describe the modulated signal by the

$z$-transform expression

$$Y'(z) = z^{-1} X'(z) + (1 - z^{-1})^2 E(z). \tag{11}$$

The noise in this signal has spectral density

$$N_M(f) = 2(1 - \cos(\omega\tau))e_0\sqrt{2\tau}. \tag{12}$$

If $N_{M0}$ is the component of noise in baseband, $0 \leqslant f \leqslant f_0$ and $\omega_0 = 2\pi f_0$, then

$$N_{M0}{}^2 = \int_0^{f_0} N_M{}^2(f) \, df = \frac{e_0{}^2}{\pi} [6\omega_0\tau$$

$$- 8 \sin(\omega_0\tau) + \sin(2\omega_0\tau)].$$

When $f_0\tau$ is small enough that $\sin(2\omega_0\tau)$ can be approximated by the first five terms in its Taylor expansion,

$$N_{M0} = \frac{e_0\pi^2}{\sqrt{5}} (2f_0\tau)^{5/2}. \tag{13}$$

This noise can be made very small by using high modulation rates. Subsequent digital processing, called decimation, lowers the rate without increasing the noise.

## APPENDIX B
### DESIGN OF THE DECIMATOR

The decimator uses low-pass filters to attenuate out-of-band components of the modulation that will be aliased into band by the resampling. To simplify the design of these filters, there is much advantage in lowering the sampling rate in stages. In the initial stages where the sampling rate is still large compared to the Nyquist rate, it is wise to place zeros of the filters at the new sampling rate and harmonics of it. Suitable sequences of evenly spaced zeros occur in trigonometric functions, and a particularly easy spectral response to implement is $(\sin \omega N\tau / \sin \omega\tau)$. In the time domain it is an averaging of $N$ samples with $\tau$ being the period of the input samples and $N\tau$ the period of the output samples.

In [9] a cascade of two such filters having response

$$\frac{1}{N^2} \left(\frac{\sin \omega N\tau}{\sin \omega\tau}\right)^2 = \frac{1}{N^2} \left(\frac{1 - z^{-N}}{1 - z^{-1}}\right)^2 \tag{14}$$

was used to lower sampling rates of sigma delta modulation to 32 kHz. This filter is inadequate for use with modulators employing double integration, as the following analysis will demonstrate. After filtering by (14), the modulation noise in (11) becomes

$$N_D(z) = \frac{e_0\sqrt{2\tau}}{N^2} (1 - z^{-N})^2. \tag{15}$$

Resampling with period $N\tau$ results in

$$N_D(Z) = \frac{e_0\sqrt{2N\tau}}{N^2} (1 - Z^{-1})^2 \tag{16}$$

where $Z^{-1} = z^{-N}$ represents a one-period delay at the new rate. Following arguments similar to (11)–(13), the in-band noise can be approximated by

$$N_{D0} = N^{1/2} \frac{e_0\pi^2}{\sqrt{5}} (2f_0\tau)^{5/2}. \tag{17}$$

Thus, the noise (13) is increased by the root of the decimation ratio $N$. This loss of resolution could be made up for by increasing the input modulation rate from $1/\tau$ to $N^{1/4}/\tau$.

A better decimating filter is

$$\frac{1}{N^3} \left(\frac{1 - Z^{-N}}{1 - Z^{-1}}\right)^3. \tag{18}$$

It modifies the modulation noise to be

$$N_D(z) = \frac{1}{N^3} (1 - z^{-N})^2 \left(\frac{1 - z^{-N}}{1 - z^{-1}} E(z)\right). \tag{19}$$

The spectral density of the noise following resampling can be obtained by reverting to the time domain. Let $E'(z)$ represent the accumulated noise

$$E'(z) = \left(\frac{1 - z^{-N}}{1 - z^{-1}}\right) E(z) \tag{20}$$

which is equivalent to

$$e'(i\tau) = \sum_{n=0}^{N} e((i - n)\tau). \tag{21}$$

When this is resampled with period $N\tau$, i.e., every $N$th sample is retained, there will be no correlation between the samples, if the original samples are uncorrelated. The resampled noise $e'(iN\tau)$ will be white with spectral density $Ne_0\sqrt{2\tau}$. It follows that the noise at the decimator output after resampling can be described as

$$\frac{1}{N^{5/2}} (1 - Z^{-1})^2 e_0\sqrt{2N\tau} \tag{22}$$

and its baseband component for $N\tau f_0 \gg 1$ is approximated by

$$N_{D0} = \frac{e_0\pi^2}{\sqrt{5}} (2f_0\tau)^{5/2}. \tag{23}$$

There is no change from (13). All of the penalty for decimating is represented by the in-band attenuation of the decimating filter which, when equalized, increases the noise.

This decimating filter can be used successfully to lower sampling rates to about four times the Nyquist rate, and its in-band attenuation will be less than 3 dB. Filters with much sharper cutoff characteristics are needed in the final stage of decimation. Their attenuation $R$ must be sufficient to make the modulation noise that is aliased into band small with respect to $N_{M0}$ in (13). That is,

$$\frac{2N\tau}{N^5} e_0{}^2 \int_{f_0}^{\frac{1}{2N\tau}} |R(\omega)|^2 (1 - Z^{-1})^4 \, df$$

$$\ll \frac{e_0{}^2\pi^4}{5} (2f_0\tau)^5 \tag{24}$$

which is satisfied if

$$|R(\omega)| \ll \frac{\pi^2}{\sqrt{30}} (2f_0N\tau)^{5/2}. \tag{25}$$

When $1/N\tau$ is 32 kHz and $f_0$ is 4 kHz, this requires $R \leqslant -25$ dB. This is easier than the requirement for antialiasing filters in $D$-channel banks, $R \leqslant -32$ dB. The low-pass filter described in [9] is adequate for use with double integrating modulators.

## APPENDIX C
### COMPARISON OF A MODULATOR THAT INTEGRATES ANALOG SIGNALS WITH ITS SAMPLED DATA EQUIVALENT

When the modulator in Fig. 3 has input samples $x_n{}'$ and output samples $y_n{}'$ and the signal applied to the quantizer is $w_{n-1}{}'$, a relationship between their values can be expressed as

$$w_N{}' = \sum_{n=0}^{N-1} \sum_{0}^{n} (x_n{}' - y_n{}') - \sum_{n=0}^{N-1} y_n{}' \qquad (26)$$

or

$$w_N{}' = \sum_{n=0}^{N-1} (N - n)x_n{}' - \sum_{n=0}^{N-1} (N - n + 1)y_n{}' \qquad (27)$$

provided all signals are initially zero.

An analog circuit that can have equivalent response is shown in Fig. 10. Here $x(t)$ is a continuous signal and the feedback signal $y(t)$ is held constant throughout each sample interval. To analyze this circuit we make use of the following relationships between integration and summation of sample values.

*Lemmas:* For $f(t) = 0$ when $t \leqslant 0$ and

$$f_{n\tau} = \frac{1}{\tau} \int_{n\tau}^{(n+1)\tau} f(t) \, dt, \qquad (28)$$

$$\frac{1}{\tau} \int_0^{n\tau} f(t) \, dt \equiv \sum_{n=0}^{N-1} f_{n\tau} = \sum_{n=0}^{N-1} (N - n)(f_{n\tau} - f_{(n-1)\tau}) \qquad (29)$$

and

$$\frac{1}{\tau} \int_0^{N\tau} dt \int_0^t f(t) \, dt = \sum_{n=0}^{N-1} (N\tau - t_n)f_{n\tau} \qquad (30)$$

where $t_n$ is the position, in time, of the center of area of the signal waveform during the $n$th sample interval, i.e.,

$$f_{n\tau}t_n = \frac{1}{\tau} \int_{n\tau}^{(n+1)\tau} tf(t) \, dt. \qquad (31)$$

When we assume that all signals in the circuit are zero for $t \leqslant 0$, we can express the amplitude that is applied to the quantizer in Fig. 3 at the $N$th sample time as

$$w(N\tau) = \frac{1}{r^2 C_1 C_2} \int_0^{N\tau} \left[ \int_0^t (x(t) - y(t)) \, dt \right. $$

$$\left. + RC_1(x(t) - y(t)) \right] dt. \qquad (32)$$

Applying results (13)–(15) and noting that because $y(t)$ is held constant, the center of area of its waveform lies mid-

way in the sample time $(n + \frac{1}{2})\tau$, we get

$$\frac{r^2 C_1 C_2}{\tau^2} w(N\tau) = \sum_{n=0}^{N-1} \left( \frac{\tau N - t_n}{\tau} + \frac{RC_1}{\tau} \right) x_{n\tau}$$

$$- \sum_{n=0}^{N-1} \left( N - \left( n + \frac{1}{2} \right) + \frac{RC_1}{\tau} \right) y_{n\tau}. \qquad (33)$$

We are primarily interested in modulators that are sampled at very high frequency compared to baseband so that the signal varies little during the sampling interval. To simplify the analysis, we shall now assume that the center of area of the input waveform for each sample lies midway in the interval $t_n = (n + \frac{1}{2})\tau$. Then

$$\frac{r^2 C_1 C_2}{\tau} w(N\tau)$$

$$= \sum_{n=0}^{N-1} (N - n) \left[ x_{n\tau} + \left( \frac{RC_1}{\tau} - \frac{1}{2} \right) (x_{n\tau} - x_{(n-1)\tau}) \right]$$

$$- \sum_{n=0}^{N-1} \left( N - n + \left( \frac{RC_1}{\tau} - \frac{1}{2} \right) \right) y_{n\tau}. \qquad (34)$$

Comparing this result to (27) we see that they can be equivalent, with the right initial conditions and

$$r^2 C_1 C_2 = \tau^2, \qquad RC_1 = 1.5\tau, \qquad y_n{}' = y_{n\tau}$$

and

$$x_n{}' = x_{n\tau} + \left( \frac{RC_1}{\tau} - \frac{1}{2} \right) (x_{n\tau} - x_{(n-1)\tau}).$$

This leads to the final result

$$x_n{}' = 2x_{n\tau} - x_{(n-1)\tau} \qquad (35)$$

which has spectral equivalent

$$X'(\omega) = (2 - e^{-j\omega\tau}) \operatorname{sinc}(f\tau)X(\omega). \qquad (36)$$

### REFERENCES

[1] D. J. Goodman, "The application of delta modulation to analog-to-digital PCM encoding," *Bell Syst. Tech. J.*, vol. 48, pp. 321–343, Feb. 1969.

[2] L. D. J. Eggermont, "A single-channel PCM coder with companded DM and bandwidth-restricting filtering," in *Conf. Rec., IEEE Int. Conf. Commun.*, June 1975, vol. III, pp. 40-2-40-6.

[3] J. C. Candy, "A use of limit cycle oscillations to obtain robust analog-to-digital converters," *IEEE Trans. Commun.*, vol. COM-22, pp. 298–305, Mar. 1974.

[4] J. D. Everhard, "A single-channel PCM codec," *IEEE J. Solid-State Circuits*, vol. SC-11, pp. 25–38, Feb. 1979.

[5] T. Misawa, J. E. Iwersen, and J. G. Rush, "A single-chip CODEC with filters, architecture," in *Conf. Rec., IEEE Conf. Commun.*, June 1980, vol. 1, pp. 30.5.1–30.5.6.

[6] J. C. Candy, Y. C. Ching, and D. S. Alexander, "Using triangularly weighted interpolation to get 13-bit PCM from a sigma-delta modulator," *IEEE Trans. Commun.*, vol. COM-24, pp. 1268–1275, Nov. 1976.

[7] L. vanDe Meeberg and D. J. G. Janssen, "PCM codec with on-chip digital filters," in *Conf. Rec., IEEE Int. Conf. Commun.*, June 1980, vol. 2, pp. 30.4.1–30.4.6.

[8] J. C. Candy, W. H. Ninke, and B. A. Wooley, "A per-channel A/D converter having 15-segment $\mu$-255 companding," *IEEE Trans. Commun.*, vol. COM-24, pp. 33–42, Jan. 1976.

[9]   J. C. Candy, B. A. Wooley, and O. J. Benjamin, "A voiceband codec with digital filtering," *IEEE Trans. Commun.*, vol. COM-29, pp. 815–830, June 1981.

[10]  R. Apfel, H. Ibrahim, and R. Ruebush, "Signal-processing chips enrich telephone line-card architecture," *Electronics,* vol. 55, pp. 113–118, May 5, 1982.

[11]  R. Steele, *Delta Modulation Systems.* New York: Wiley, 1975, ch. 3.

[12]  F. deJager, "Delta-modulation, a method of PCM transmission using the 1-unit code," *Philips Res. Rep.,* vol. 7, pp. 442–466, 1952.

[13]  H. A. Spang and P. M. Schultheiss, "Reduction of quantization noise by use of feedback," *IRE Trans. Commun. Syst.,* vol. CS-10, pp. 373–380, Dec. 1962.

[14]  S. K. Tewksbury and R. W. Halloch, "Oversampled, linear predictive and noise-shaping coders of order $N > 1$," *IEEE Trans. Circuits Syst.,* vol. CAS-25, pp. 436–442, July 1978.

[15]  R. C. Brainard and J. C. Candy, "Direct-feedback coders: Design and performance with television signals," *Proc. IEEE,* vol. 37, pp. 776–786, May 1969.

[16]  B. S. Atal, "Predictive coding of speech at low bit rates," *IEEE Trans. Commun.,* vol. COM-30, pp. 600–614, Apr. 1982.

[17]  J. C. Candy and R. H. Bosworth, "Methods for designing differential quantizers based on subjective evaluations of edge business," *Bell Syst. Tech. J.,* vol. 51, pp. 1495–1516, Sept. 1972.

[18]  G. R. Ritchie, "Higher order interpolative analog to digital converters," Ph.D. dissertation, Univ. Philadelphia, Pennsylvania, 1977.

[19]  J. C. Candy and O. J. Benjamin, "The structure of quantization noise from sigma delta modulation," *IEEE Trans. Commun.,* vol. COM-29, pp. 1316–1323, Sept. 1981.

# An Oversampling Analog-to-Digital Converter Topology for High-Resolution Signal Acquisition Systems

L. RICHARD CARLEY, MEMBER, IEEE

*Abstract* — High-precision analog-to-digital converters (ADC's) which periodically sample continuous time waveforms are required in many "signal acquisition" applications such as digital audio and instrumentation. This paper describes a specific topology for an oversampling differential pulse code modulation (DPCM)-type ADC that requires neither an analog anti-aliasing filter nor a sample-and-hold. Dithering techniques are presented which cause the quantization error to resemble an additive Gaussian white-noise source whose mean and variance are independent of the input. A smoothing technique is presented which improves the linearity of the ADC's transfer function. Results presented are based on both simulations and measurements of a test system constructed from discrete components. The potential for integration in a standard digital CMOS process is discussed.

## I. INTRODUCTION

ANALOG-TO-DIGITAL converters (ADC's) can be roughly divided into two categories: data acquisition ADC's and signal acquisition ADC's. Signal acquisition ADC's sample continuous-time signals converting them into discrete-amplitude discrete-time sequences suitable for digital signal processing, digital storage, and transmission over digital communication channels. Data acquisition ADC's are generally used to sample a single analog value of an input at a particular instant in time. Since a data acquisition system may be used to gather only a single sample of an input its accuracy is normally characterized by the maximum error of any conversion, whereas a signal acquisition system takes a large number of uniformly spaced[1] samples of an analog input waveform; therefore, its accuracy is better characterized by properties of many samples such as signal-to-noise ratio (SNR), the power spectral density (PSD) of the error, and the probability distribution function (PDF) of the error.

There are many advantages to sampling the input signal more frequently than the Nyquist rate (oversampling) when designing an ADC for signal acquisition. Oversampling can increase the resolution of the ADC [19]. The sharp[2]

analog anti-aliasing filter can be replaced by a simple analog filter and a digital anti-aliasing filter. This is an important advantage because a precision filter implemented in the digital domain does not require adjustments and an analog filter typically requires many circuit components which are difficult to implement in integrated circuit form (e.g., inductors and large capacitors). A third advantage of oversampling is that the dither signal (see Section III) commonly added to the input of an ADC to decorrelate successive quantization errors [3], [5], [1] can be high-pass filtered so that little dither signal power lies in the signal band. A final advantage of most oversampling ADC's, including the topology to be presented, is the elimination of the need for a sample-and-hold amplifier preceding the ADC. Designing sample-and-hold amplifiers with an accuracy greater than 13 or 14 bits is problematic due to dielectric absorption in capacitors; hence, their elimination is also important for high-resolution ADC's.

There are two additional advantages specific to the DPCM topology to be presented. First, by choosing the appropriate dither signal, the quantization errors at the output of the oversampling ADC will approximate a Gaussian white-noise whose mean and variance are independent of the analog input. Second, it is possible to decrease the ADC's nonlinearity by taking successive samples from different parts of its transfer function. This is done by adding an analog smoothing signal to the input and subtracting the equivalent digital signal from the output.

## II. OVERSAMPLING ADC TOPOLOGY

Two common categories of oversampling ADC's are noise shaping coders (e.g., delta–sigma modulators) and linear predictive coders (e.g., DPCM). Although the gain in resolution at a given oversampling ratio is higher for noise-shaping coders such as delta–sigma modulators [19], it has proven difficult to exceed about 13 bits of resolution using a delta–sigma modulator with a 1-bit DAC [13]. Multibit noise-shaping coders have been developed to achieve higher resolution [22], but their linearity is limited by the linearity of the DAC used in the feedback loop. There is one important advantage that the DPCM topology has over multibit noise-shaping coders: it is possible to

Manuscript received July 8, 1986. This work was supported in part by the National Science Foundation under Grant ENG-8451496.

The author is with the Department of Electrical and Computer Engineering, Carnegie Mellon University, Pittsburgh PA 15213.

IEEE Log Number 8611121.

[1] A discussion of nonuniform sampling is beyond the scope of this paper.

[2] In this context, a sharp filter is one in which the upper edge of the passband is close to the lower edge of the stopband and the attenuation in the stopband is large.

Reprinted from *IEEE Trans. Circuits and Sys.*, vol. CAS-34, pp. 83–90, January 1987.

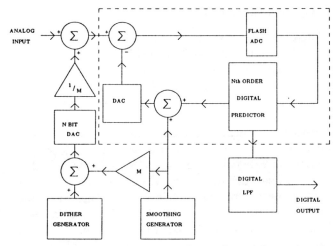

Fig. 1. DPCM ADC topology. Block diagram form of the topology for an oversampling signal acquisition system. The region contained within the dotted lines is the basic DPCM topology. In this implementation, the dither and smoothing signals are both digital signals.

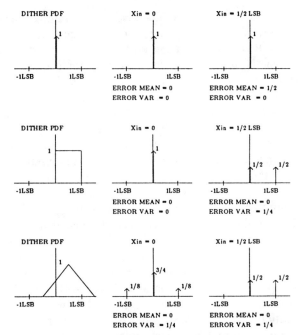

Fig. 2. Effect of dither PDF on ADC output's PDF. The PDF of the ADC's output is given for constant analog inputs of 0 and 1/2 LSB with a dither PDF which is an impulse at 0 (this corresponds to no dither input), a boxcar that is 1-LSB wide, and a 2-LSB-wide triangle.

accurately analyze the performance of the DPCM topology in the presence of dither noise. This is possible because, unlike noise-shaping coders, there is no analog filtering in the loop ahead of the nonlinearity (the quantizer). It is for this reason that a dither signal can be selected that results in a quantization error which is approximately Gaussian white noise and is independent of the input. In addition, the analysis also determines the effect of smoothing the DAC transfer function (see Section V) on the linearity of the ADC's transfer function.

The ADC topology consists of four main elements (see Fig. 1): 1) the DPCM loop which is enclosed by a dashed line in Fig. 1; 2) the digital anti-aliasing filter; 3) the digital dither noise generator and a digital-to-analog converter (DAC) to convert it to an analog signal; and 4) the smoothing signal generator and DAC (in Fig. 1, a single DAC is shown for both dithering and smoothing). For a more detailed discussion of the DPCM topology see [1], [19], and [20]. A discussion of digital anti-aliasing filters is beyond the scope of this paper. The basics of single- and multistage FIR decimating filters are presented in [10] and [11]. Recently, Chu and Burrus [12] developed decimating filters that combined IIR and FIR filter stages to further reduce the number of multiplications required per output point. The effects of dither will be discussed in Section III and the smoothing technique will be presented in Section V.

The primary limitation on the speed of the oversampling ADC topology is the DPCM loop. The sampling period must be longer than 1) the time required for the flash ADC to settle, plus 2) the predictor's worst-case propagation delay, plus 3) the worst-case time required for the DAC's output to settle to the required accuracy, plus 4) the settling time of the amplifier which takes the difference between the DAC's output and the analog input and amplifies that difference such that one DAC least significant bit (LSB) is equal to the LSB of the flash ADC. Since the input signal is oversampled, when the loop is tracking properly, the output of the DAC should only change by a

small fraction of full scale from one clock interval to the next. For some DAC topologies, this will result in a decrease in the settling time required.

The settling of the DAC and amplifier can occur concurrently with the computation of the next predictor output if the input to the DAC is latched. The cost of this increase in speed is a decrease in the accuracy of the predictor (it has to predict further into the future) which usually translates into an increase in the size of the flash ADC.

## III. DITHER

The resolution of the DPCM system alone is limited to the LSB size of the DAC. However, the resolution can be improved by adding a dither signal to the analog input [2]. Dither signals have been used for improving the resolution of images [3], speech signals [4], and audio signals [5].

### A. Dither Distribution Versus Quantizer Error

Fig. 2 illustrates how dither can increase the system resolution. With a constant quantizer input ($X_{in}$) and no dither added, which corresponds to a dither probability distribution function (PDF) that is an impulse of unit area at the origin, the error at every sample is constant. For example, with $X_{in} = 0$, there is no error, and, with $X_{in} = +1/2$ LSB, the error is 1/2 LSB at every sample.

With a constant input and a dither with a PDF which is a 1-LSB-wide boxcar (see Fig. 2), the error of successive samples will be different, and the mean value of the error will be zero. The 1-LSB-wide boxcar is the dither PDF with the smallest variance which results in zero mean quantization error for any constant analog input. When successive dither values are independent, successive quantization errors will also be independent. Therefore, if the digital output is averaged over a long enough period, the average will converge to a constant input with any desired resolution. Note, these results only hold for a quantizer with perfectly spaced transition points. In a real quantizer, there will be errors in the transition points; hence, there will be some dependence of the mean error on the input level even with a boxcar dither PDF added.

While a boxcar PDF dither results in zero mean error, the variance of the output still depends on the input. For example, the variance increases quadratically from 0 when $X_{in} = 0$ to $(1/4 \text{ LSB})^2$ when $X_{in} = 1/2$ LSB. With a constant input and a dither with a 2-LSB-wide triangular PDF (see Fig. 2), the variance of the error sequence is independent of the input. In fact, it has been shown that the dither PDF with the minimum variance which makes the first $N$ moments of the quantization error sequence independent of the input is the convolution of $N$ boxcar PDF's [6], [7]. For example, the 2-LSB-wide triangular PDF, which is the convolution of two 1-LSB-wide boxcar PDF's, is the minimum variance PDF which makes both the first and second moments (the mean and variance) of the error sequence independent of the input.

### B. Modeling Quantization Error

In an oversampling ADC, it is desirable to high-pass filter the dither signal before it is added to the flash ADC's input. Then after the DPCM output has passed through the digital anti-aliasing filter, only the fraction of the dither signal's energy in the signal passband will remain. Note, converting a sequence of values $(x_n)$ with a PDF which is a 1-LSB-wide boxcar and a frequency-independent power spectral density (PSD) into a sequence $(y_n)$ with a 2-LSB-wide triangular PDF and a high-pass filtered PSD can be done by letting $y_n = x_n - x_{n-1}$. The PSD of this high-pass filtered dither signal is proportional to $(1 - \cos(\omega))^2$. Assuming a 2-LSB-wide triangular dither PDF, the amount of dither power remaining at the output of a perfect low-pass filter with a cutoff frequency of $\pi/N$ is approximately $\text{LSB}^2/3.65\ N^3$. Thus, when oversampling by 4 or more, the dither power remaining in the output is negligible compared to the quantization error power remaining, $\text{LSB}^2/12\ N$, assuming that the quantization error is frequency independent.

To determine the exact PDF of the anti-aliasing filter's output is very difficult. The derivations in the preceding two subsections assumed that the input was a constant. Unfortunately, it is difficult to determine the spectrum of the quantization errors when the input is not constant and the dither is high-pass filtered. Therefore, extensive simulations were performed for a quantizer using a high-pass filtered dither with a 2-LSB-wide triangular PDF added to input signals of varying amplitudes that were sinewaves, sums of sinewaves, or Gaussian white noise. In all cases, the quantization error was very close to a 1-LSB-wide boxcar PDF, its PSD was nearly independent of frequency, and its mean and variance were independent of the input.

### C. Modeling the ADC System

There are four noise processes that appear at the input of the digital anti-aliasing filter: 1) dither noise, 2) quantization noise, 3) flash ADC transition point error noise, and 4) DAC level error noise.[3] As noted above, the high-pass filtered dither signal is removed from the output by the filter. Its only purpose is to make the mean and variance of the quantization error independent of the input and to make successive quantization errors independent. Analytical treatment of the flash ADC and DAC nonidealities is difficult due to their signal-dependent nature. Their effect is best explored via simulation. Therefore, both DAC and flash ADC will be idealized in the following discussion. Quantization noise is the only remaining noise source.

A digital decimation filter is typically composed of one or more stages of FIR filters [11]. Each output point from an $N$th-order FIR filter consists of a weighted sum of $N$ input points. Assuming that each of the input points is an independent sample of a 1-LSB-wide boxcar PDF,[4] the PDF for the output is the convolution of $N$ boxcar PDF's, each scaled on the voltage axis by the weight for the appropriate tap of the FIR filter. The central limit theorem predicts that, as $N$ becomes large, the PDF of the error at the filter's output approaches a Gaussian distribution. A typical FIR filter for oversampled A/D converters employs hundreds or even thousands of taps [13]. Since the PSD of the input sequence was independent of frequency and the FIR filter is approximately independent of frequency within its passband, the PSD of the error at the filter's output will also be independent of frequency up to the cutoff frequency of the FIR filter.

Therefore, the error at the filter's output is well modeled as a Gaussian white-noise process with constant mean, 0 (a nonzero mean error just corresponds to an input offset), and variance, approximately $\text{LSB}^2/12\ N$, where $N$ is the oversampling ratio. The discrete component version of this oversampling ADC topology (see Section VI) verified that, independent of the input, the distribution of the quantization errors at the filter's output was approximately Gaussian and that the PSD of the errors was independent of frequency.

### D. PSD of Quantization Error

Fig. 3 shows the PSD of the error of a quantizer with a sinusoidal input at a frequency of $3\pi/32$. Note that all of the noise power of the error is concentrated in odd harmonics of the input frequency and their aliases, whereas the error of the oversampling ADC is uniformly distributed in frequency (see Fig. 4). It is important to note that the perception of SNR may also depend on the PSD of the noise.

---

[3] The nonideal flash ADC can be modeled as an ideal ADC plus a signal-dependent additive noise source. Similarly, the nonideal DAC can be modeled as an ideal DAC plus a signal-dependent additive noise source.

[4] Since the decimating filter is a linear filter, superposition is valid and it is reasonable to ignore the dither noise at its input as it will be filtered out.

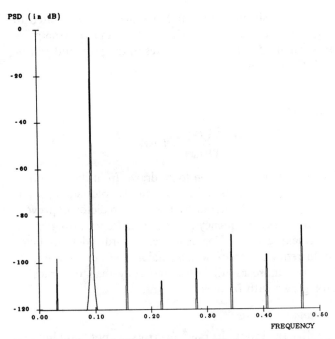

Fig. 3. PSD of normal PCM quantizer error. The input is a sinewave at a frequency of $3\pi/32$. Note, this is a simulation of the performance of an ideal quantizer with perfectly spaced transition points. The plot was made by taking the PSD of a 4096-point error sequence weighted by a Blackman window.

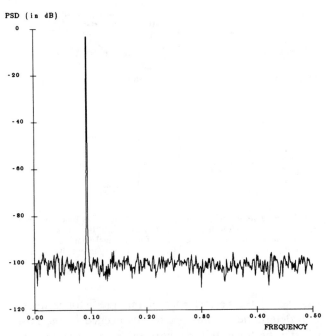

Fig. 4. PSD of PCM quantizer error with dither. The input is a sinewave at frequency of $3\pi/32$. Note, this is a simulation of the performance of an ideal quantizer with perfectly spaced transition points. The dither was generated with a random number generator and the plot was made as in Fig. 3.

In some applications, it is only the SNR of the ADC that is important. In other applications, tailoring the quantization errors to appear like additive Gaussian white noise may be important. For example, it is generally accepted that the SNR of an ADC for use in high-fidelity digital audio must be higher than the SNR of analog equipment [5]. Typically, 16-bit ADC's (96-dB SNR) are required for high-fidelity digital audio work. On the other hand, in preamps and other equipment, where the noise is predominantly Gaussian white noise, a SNR of 84 dB is acceptable. Using piano duets as test signals in a subjective study, Croll [14] concluded that, for equivalent perception, the SNR for normal quantization noise must be 12-dB higher than the SNR when the noise has a frequency independent PSD. One reason for this difference in perception may be that the Gaussian white-noise process, unlike quantization error, is independent of the input signal. Therefore, conclusions about the perception of a system's performance cannot be based only on its SNR. The nature of that noise and the task being performed must also be considered.

### E. Output Truncation and Dither

The digital filter may have product terms 32 bits in length or longer, while the noise level will typically be in the 13–16-bit range. Therefore, truncation of the output is desirable; however, truncating the number of bits in a digital signal is quantization. To maintain an ADC system in which the error is independent of the input, a second dither signal (a digital one) must be added before truncation. The choice of this digital dither is, as before, a dither PDF which consists of samples of the convolution of $N$

boxcars which make $N$ moments of the error independent of the input. Note, this digital dither cannot be filtered out because it must be added at the LPF's output; hence, it increases the noise power at the ADC's output. The number of bits kept is normally greater than the number of bits equivalent to the SNR at the filter's output to prevent the digital dither noise power from adding significantly to the total ADC noise. For example, if the SNR before truncation was 84 dB, which corresponds to 14 bits, then truncation to 16 bits might be selected.

### IV. THE DPCM LOOP (THE PREDICTOR)

The predictor is the most crucial element in determining the DPCM system's performance. If the predictor is more accurate, then the range of the errors which must be quantized by the flash ADC will be smaller. The more information one knows about the class of allowed input signals, the more accurate the prediction [1]. In the case when the only constraints on the input are that it is limited to a maximum amplitude and a maximum frequency, the most common predictor topology is based upon a Taylor-series expansion [18], [19]. Letting $\hat{x}_i$ be the prediction for $x_i$

$$\hat{x}_i = \begin{bmatrix} N \\ 1 \end{bmatrix} x_{i-1} - \begin{bmatrix} N \\ 2 \end{bmatrix} x_{i-2} + \cdots + (-1)^{N-1} \begin{bmatrix} N \\ N \end{bmatrix} x_{i-N}.$$

Alternatively, this $N$th-order predictor can be expressed in terms of $z$ transforms as

$$P(z) = \frac{\hat{X}(z)}{X(z)} = 1 - (1 - z^{-1})^N.$$

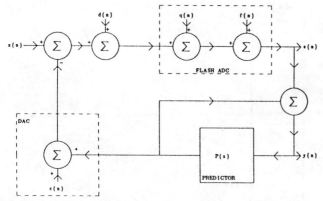

Fig. 5. Discrete-time model of oversampling ADC. This block diagram is arrived at by modeling the flash ADC as two additive noise sources, one for quantization error and a second for nonideal spacing of transition points, and modeling the DAC nonideal spacings by a third additive noise source. This is a discrete-time continuous-amplitude model.

Analysis of the performance of this predictor is necessary to determine the number of bits needed in the flash ADC. To perform this analysis, a linear discrete-time continuous-amplitude model of the ADC has been constructed (see Fig. 5). The flash ADC is modeled by an additive noise source representing quantization errors ($q_n$) and another additive noise source representing nonuniformities in the ADC transition point spacing ($f_n$). The nonideal DAC is replaced by an additive noise source ($c_n$) and an ideal DAC.

In previous work on DPCM coders, the decrease in the error power relative to the input power was presumed proportional to the decrease in the total noise power of the prediction error [18]–[20]. If the flash ADC size is based on the decrease of noise power only, there may be a nonzero probability that the range of the flash ADC will be exceeded. In that case, the quantization error is no longer bounded and the error of following predictions may increase. Therefore, to insure reliable operation of the DPCM loop, the size of the flash ADC should be based on the peak-to-peak prediction error and not on its power. Since the model is linear, determination of the peak prediction error can be divided into two parts: the peak error when tracking the analog input and peak error caused by all of the disturbances—dither, quantization, DAC nonlinearities, and ADC nonlinearities (see Fig. 5).

### A. Tracking Error

To determine the peak tracking error, it is first necessary to decide what input will cause the largest errors. For the Taylor-series predictor, the terms which have been left out are always a coefficient times a derivative of the input. It has been shown that, for amplitude and frequency limited waveforms, the signal with maximal derivatives is a sine-wave of maximum amplitude at the maximum frequency [8]. Therefore, the error must be determined as a function of frequency. Taking $z$ transforms and plugging in a predictor of order $N$

$$\frac{E(z)}{X(z)} = 1 - P(z) = \left(1 - z^{-1}\right)^N$$

which is consistent with [18]. Since the input is limited to frequencies below $\pi/M$, where $M$ is the oversampling ratio, when $M$ is large, frequencies in the passband will be small ($\omega \ll 1$) and

$$z^{-1} \equiv e^{-j\omega} \approx 1 - j\omega.$$

Therefore

$$\frac{E(j\omega)}{X(j\omega)} \approx (j\omega)^N.$$

One important conclusion to be drawn from the growth of the tracking error as $\omega^N$ is that the simple analog input filter must be of higher order than the predictor to prevent unwanted high-frequency input signals from causing excessive tracking error. For example, a third-order predictor would require a fourth-order analog prefilter to attenuate the input more rapidly with frequency than the tracking error grows with frequency.

### B. Random Errors

Unlike the input, the random errors are not band limited. Note that $y(n)$ is exactly the sum of $x(n)$, $c(n)$, $d(n)$, $q(n)$, and $f(n)$ (see Fig. 5). The error into the predictor is, therefore, just the sum of $c(n)$, $d(n)$, $q(n)$, and $f(n)$. Let the peak-to-peak (P-P) value of this sum be $\epsilon$. A conservative estimate for $\epsilon$ is the sum of the P-P range of the dither, plus the P-P range of the quantization noise, plus the P-P range of deviations from the ideal DAC transfer function, plus the P-P range of deviations from the ideal flash ADC transfer function. Given the alternation of sign in the predictor coefficients, the error sequence for which the largest predictor error occurs is the input sequence which alternates between the most positive error and the most negative error. Thus, the P-P error of an $N$th-order predictor is $(2^N - 1)\epsilon$.[5] Note that tracking error is proportional to full scale while the random errors are not. Fig. 6 shows total P-P errors versus oversampling ratio. At low oversampling ratios tracking error, which decreases at 6 dB times the predictor order per octave, predominates over the random errors which are independent of oversampling ratio. Note that the smallest error is not always obtained by the highest order predictor. In fact, the optimum predictor order decreases with increasing oversampling ratio. The optimum predictor order for a given oversampling ratio depends on the ratio of $\epsilon$ to FS.

### C. The Start-Up Problem

One potential problem with any feedback system which contains a predictor is that the predictor's output depends upon previous values. Therefore, first initiating proper tracking of the input, or reinitiating tracking after a transient overloads the flash ADC, can be very difficult, especially for higher order predictors. One method for initiating tracking is to detect an overload condition (this would require an extra comparator on either end of the flash

---

[5] This follows because the sum of the magnitude of the coefficients for an $N$th-order predictor is $2^N - 1$.

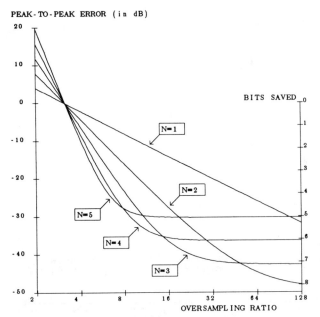

Fig. 6. Total prediction error versus oversampling ratio. These curves represent the P-P error of the predictor as a function of the oversampling ratio. The number $N$ refers to the order of the predictor. These curves are for the case when the full scale is 4096 LSB's (a 12-bit DAC) and the dither had a 2-LSB-wide triangular PDF. Flash ADC transition error and DAC level error are neglected.

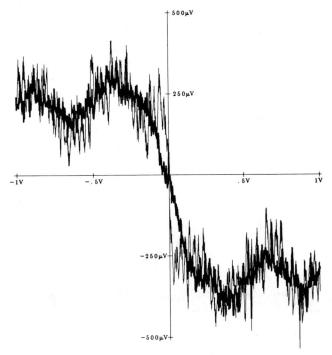

Fig. 7. ADC transfer function with and without smoothing. The two lines indicate the error from an ideal transfer function of the test system. Only the region from $-1$ V to $+1$ V is shown. The actual full scale range is $-10$ V to $+10$ V. The narrow line indicates the error without smoothing and the thick line is the error for a smoothing signal with a 0.3 V wide boxcar PDF.

ADC) and then strongly attenuate the input and set the stored values in the predictor to zero. Once the system is properly tracking the input, the attenuation is gradually removed. Another method is to add nonuniformly spaced comparators to the flash ADC making it impossible to overload the flash.

## V. SMOOTHING THE ADC CHARACTERISTIC

A general disadvantage of oversampled ADC's which employ multibit DAC's in feedback loops is that the linearity of the overall ADC is no better than that of the DAC, even though its resolution is increased by oversampling [19]. However, the linearity of the ADC can be improved by taking successive samples from different points on the DAC transfer function. This can be accomplished by adding a digital smoothing signal to the predictor's output and converting that smoothing signal to the analog domain with a separate DAC and adding it to the input (see Fig. 1). Assuming that the smoothing signal is not correlated with the input, the resulting DC transfer function is the original DC transfer function convolved with the smoothing signal PDF [9]. The smoothing signal decreases the ADC's differential nonlinearity over a range of voltages approximately equal to the width of the PDF of the smoothing signal. Practical considerations limit this width (as discussed below); therefore, while the local linearity of the ADC is increased by smoothing, the global linearity is not. Fig. 7 shows the error of the DC transfer function of the test system with and without smoothing. The smoothing signal has a boxcar PDF with a width of approximately 0.3 V and a constant PSD. As expected, local variations in the transfer function are attenuated while global variations remain unchanged. The improve-

ment in the local nonlinearity is significant. Without smoothing, a 100-mV P-P sinewave would suffer significant crossover distortion; however, with smoothing added, this sinewave only suffers a slight decrease in gain.

There are two limitations on the width of the smoothing signal's PDF. First, converting the smoothing signal to an analog value requires a DAC whose LSB is exactly matched to the LSB of the DPCM loop. Unfortunately, as the smoothing signal is made larger, the smoothing DAC requires precision approaching that of the DPCM DAC. Second, the smoothing signal decreases the dynamic range of the ADC; e.g., the maximum possible predictor output must be limited to the maximum DAC input minus the maximum possible smoothing signal. It is possible to modify the smoothing algorithm to prevent any loss of dynamic range. For example, one might simply limit the smoothing signal such that it is always less than the difference between the maximum input to the DAC and the predictor's output. However, this decreases the smoothing available near the upper and lower limits of full scale.

## VI. EXPERIMENTAL RESULTS

In order to verify that the oversampling topology functioned as expected, and to allow the study of different types of dither distributions, a test system was constructed using a 6-bit TRW flash ADC, a 12-bit Analog Devices DAC, and assorted MSI Schottky TTL. Only 5 bits of the flash ADC were used in the DPCM loop, and the MSB was used only to provide overload detection. The input was oversampled by a factor of 16. A general predictor capable of operating as either a first-, second-, or third-

189

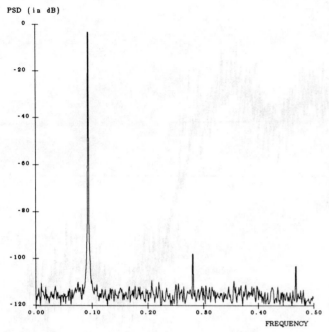

Fig. 8. PSD for maximal input with smoothing. The input is a 20 V P-P sinewave at a frequency equivalent to $3\pi/32$ in terms of the output sampling rate. The scale for the PSD is relative and 0 dB is not equal to FS. The smoothing signal has a 0.3 V wide boxcar PDF. The PSD was generated as in Fig. 3.

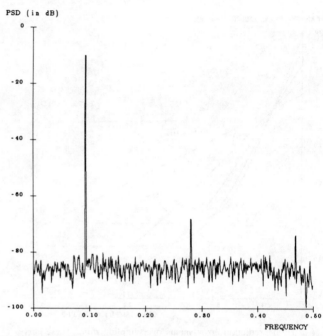

Fig. 9. PSD for low-level input without smoothing. The input is a 0.5 V P-P sinewave at a frequency equivalent to $3\pi/32$ in terms of the output sampling rate. The scale for the PSD is relative and 0 dB is not equal to FS. The PSD was generated as in Fig. 3.

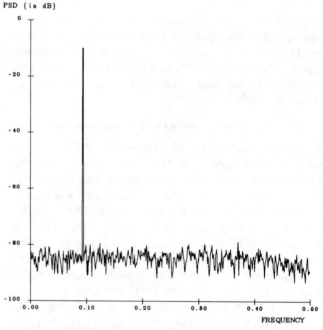

Fig. 10. PSD for low-level input with smoothing. The input is a 0.5 V P-P sinewave at a frequency equivalent to $3\pi/32$ in terms of the output sampling rate. The smoothing signal has a 0.3 V wide boxcar PDF. The scale for the PSD is relative and 0 dB is not equal to FS. The PSD was generated as in Fig. 3.

order predictor was constructed. All measurements presented were taken using the third-order predictor. The maximum DPCM clocking rate was 800 KHz and was limited predominantly by overload recovery time of the gain of 15 amplifier which boosted the analog difference between the input and the DAC such that 1 DAC LSB was equal to 1 flash LSB. The distribution of the dither signal was determined by a ROM for maximum flexibility in testing various dither distributions. A high-pass filtered dither signal with a 2-LSB-wide triangular PDF was used. The PDF for the smoothing signal was 128-LSB's wide and the PSD of the smoothing signal was nearly frequency independent. The smoothing signal and the addresses for the dither ROM were both generated by a pseudo-random sequence generator [3], [15], [16]. The feedback taps were chosen to make the pseudo-random sequence have a PSD approximately constant from 0 to $\pi$ [17]. In order to insure that the output was not affected by the dither generator's repetition frequency, the shift register was designed with a repetition period that was on the order of seconds. The output of the DPCM loop was fed into a general-purpose computer and a 377-tap single-stage FIR anti-aliasing filter was implemented in software.

The error from the ideal DC transfer curve, with and without smoothing, are shown in Fig. 7. The measured value of the quantization error was 82 dB below the maximum input signal which is very close to the 84 dB predicted, assuming that errors in the ADC transition points and DAC outputs were negligible. Fig. 8 shows the PSD of the output with a maximum amplitude input at $3\pi/32$. Note that the error is approximately independent of frequency as predicted. However, there is also notice-

able third- and fifth-harmonic distortion probably caused by the crossover distortion of the DAC.

To see the power of smoothing, consider the harmonic distortion of a low-level input. Fig. 9 shows the PSD for a low-level input without smoothing, and Fig. 10 shows the PSD for a low-level input with smoothing. As stated earlier, for a low-level input, smoothing converts a nonlinear crossover distortion into a linear gain error.

## VII. POTENTIAL FOR INTEGRATION

One important consideration is the potential of this DPCM topology for integration. By increasing the oversampling rate, the size of the flash ADC and the DAC can be decreased while maintaining a fixed SNR. For example, an SNR of 84 dB can be achieved at an oversampling ratio of 64 using an 11-bit DAC and a 4-bit flash ADC (see Fig. 6). An 11-bit DAC is difficult, but not impossible, to implement in a standard CMOS topology. For example, thin-film resistors could be deposited on top of the passivation and laser-trimmed to an accuracy on the order of 1 part in 10000. Another alternative is to use poly resistors or capacitors that are autocalibrated at power on [21]. Note, the design of the DAC should take advantage of the fact that the change in output between successive clock cycles will be small compared to full scale. This should allow a shorter settling time for the DAC, making oversampling by 64 possible even at digital audio speeds (DAC settling time about 300 ns).

## VIII. CONCLUSIONS

The effects of dither signals on the output distribution for an oversampling DPCM-type ADC were analyzed. A high-pass filtered dither signal with a 2-LSB-wide triangular PDF results in a quantization error which has a PDF that is approximately Gaussian with a mean and variance independent of the input without increasing the output noise power. The smoothing technique presented significantly decreased the harmonic distortion of the DPCM ADC. The relationship between oversampling rate, predictor order, and flash ADC size was determined. A discrete version of this topology was constructed and measurements agreed with analytic predictions. By increasing the oversampling ratio, it is possible that a high-resolution ADC system requiring neither an analog anti-aliasing filter nor a sample-and-hold can be fabricated on a single monolithic integrated circuit.

## REFERENCES

[1] N. S. Jayant and P. Noll, *Digital Coding of Waveforms*. Englewood Cliffs, NJ: Prentice-Hall, 1984.
[2] L. Schuchman, "Dither signals and their effect on quantization noise," *IEEE Trans. Commun. Theory*, vol. COM-12, pp. 162–165, Dec. 1964.
[3] L. G. Roberts, "Picture coding using pseudo-random noise," *IRE Trans. Inform. Theory*, vol. IT-8, pp. 145–154, Feb. 1962.
[4] N. Jayant and L. Rabiner, "The application of dither to the quantization of speech signals," *Bell Syst. Tech. J.*, vol. 51, pp. 1293–1304, July–Aug. 1972.
[5] B. Blesser, "Digitization of audio: A comprehensive examination of theory, implementation, and current practice," *J. Audio Eng. Soc.*, vol. 26, pp. 739–771, Oct. 1978.
[6] J. N. Wright, private communication, 1979.
[7] B. Widrow, "Study of rough amplitude quantization by means of Nyquist sampling theory," Ph.D. dissertation, Dept. Elec. Eng., MIT, 1956.
[8] A. Papoulis, *Signal Analysis*. New York: McGraw-Hill, 1977.
[9] G. Zames and N. A. Shneydor, "Dither in nonlinear systems," *IEEE Trans. Automat. Contr.*, vol. AC-21, pp. 660–667, Oct. 1976.
[10] J. H. McClellan, T. W. Parks, and L. R. Rabiner, "A computer program for designing optimum FIR linear phase digital filters," *IEEE Trans. Audio Electroacoust.*, vol. AU-21, pp. 506–526, Dec. 1973.
[11] R. E. Crochiere and L. L. Rabiner, *Multirate Digital Signal Processing*. Englewood Cliffs, NJ: Prentice-Hall, 1983.
[12] S. Chu and C. S. Burrus, "Multirate filter designs using comb filters," *IEEE Trans. Circuits Syst.*, vol. CAS-31, pp. 913–924, Nov. 1984.
[13] M. W. Hauser, "MOS ADC-filter combination that does not require precision analog components," in *Proc. ISSCC Conf.*, 1985, pp. 80–81.
[14] M. Croll, *Pulse Code Modulation for High Quality Sound Distribution: Quantizing Distortion at Very Low Signal Levels*. Great Britain: BBC Research Eng. Div. Monograph, 1970.
[15] J. H. Lindholm, "An analysis of the pseudo-randomness properties of subsequences of long *m*-sequences," *IEEE Trans. Inform. Theory*, vol. IT-14, pp. 569–576, July 1968.
[16] J. L. Manos, "Some techniques for testing pseudo-random number sequences," Lincoln Labs Tech. Note 1974-44, 1974.
[17] A. C. Davies, "Properties of waveforms obtained by nonrecursive digital filtering of pseudo-random binary sequences," *IEEE Trans. Comput.*, vol. C-20, pp. 270–281, Mar. 1971.
[18] D. Cyganski and N. J. Krikelis, "Optimum design, performance evaluation, and inherent limitations of DPCM encoders," *IEEE Trans. Circuits Syst.*, vol. CAS-25, pp. 448–460, July 1978.
[19] S. K. Tewksbury and R. W. Hallock, "Oversampled, linear predictive and noise-shaping coders of order $N > 1$," *IEEE Trans. Circuits Syst.*, vol. CAS-25, pp. 436–447, July 1978.
[20] J. B. O'Neal and R. W. Stroh, "Differential PCM for speech and data signals," *IEEE Trans. Commun.*, vol. COM-20, pp. 900–912, Oct. 1972.
[21] H. S. Lee, D. A. Hodges, and P. R. Gray, "A self-calibrating 15 bit CMOS A/D converter," *IEEE J. Solid-State Circuits*, vol. SC-19, pp. 813–819, Dec. 1984.
[22] J. W. Scott, W. Lee, C. Giancarlio, and C. G. Sodini, "A CMOS slope adaptive delta modulator," in *Proc. ISSCC Conf.*, 1986, pp. 130–131.

# DIGITALLY CORRECTED MULTI-BIT ΣΔ
# DATA CONVERTERS

T. Cataltepe*, A.R. Kramer**, L.E. Larson**,
G.C. Temes* and R.H. Walden**


\* Electrical Engineering Department, UCLA, Los Angeles, CA  90024
\*\* Hughes Research Laboratories, Malibu, CA 90265

## ABSTRACT

ΣΔ data converters using multi-bit internal A/D and D/A converters (noise-shaping converters) have lower quantization noise and are more stable than the usual single-bit systems. However, the required linearity of the internal multi-bit DAC is very difficult to achieve for an untrimmed integrated converter. This paper describes several novel multi-bit ΣΔ converters which use digital correction schemes to cancel the errors due to the nonlinearity of the internal DAC. Simulation and experimental results are given. They verify the high accuracy achievable with the proposed systems.

## INTRODUCTION

It is known [1] that both the accuracy and the stability of a ΣΔ data converter is improved if the internal A/D and D/A converters use N bits ($N > 1$), rather than a single bit as in most available ΣΔ systems. However, in such a system (called noise-shaping converter in [1]) the N-bit DAC must be linear to the full accuracy of the overall converter. For high-accuracy converters, this can only be achieved by using off-chip elements and/or trimming [1].

In this paper, we describe several novel ΣΔ data converters which utilize multi-bit internal subconverters and use digital correction schemes to cancel the effects of the internal DAC nonlinearity. This correction is performed by incorporating a memory (e.g., an EPROM) into the circuit. The basic concept is illustrated in Fig. 1, in which $H_1(z)$ is a linear dynamic two-port (typically, first or second order), and $N_1$, $N_2$ and $N_3$ are time-invariant memoryless two-ports with mildly nonlinear monotonic output/input characteristics $v_{out}(t) = N_i[v_{in}(t)]$ . The signals in the system may be all sampled-data analog, all digital, or mixed.

Assume that a frequency range exists where the loop gain of the feedback loop is much greater than 1, and that the system is stable. Then, for signal frequencies within this range, the feedback loop causes $v_f(t) \cong v_{in}(t)$ . Also, from Fig. 1, $v_f(t) = N_3[v_1(t)]$ and $v_{out}(t) = N_2[v_1(t)]$ . Now let $N_2[\cdot] \equiv N_3[\cdot]$ , i.e., let the nonlinear characteristics of two-ports $N_2$ and $N_3$ be matched. Then, $v_{out}(t) = N_2[v_1(t)] = N_3[v_1(t)] = v_f(t) \cong v_{in}(t)$ . If one or

more of the $N_i$ is an internal data converter, then this sytem can thus provide linear data conversion even with nonlinear internal subconverters. The matched compensating nonlinearity $N_2$ or $N_3$ can be realized by a programmable memory, e.g. by an EPROM. (An alternative system, in which the $N_2$ block precedes the feedback loop, is also possible.)

The described compensation technique may also be used for intentionally introduced nonlinearities, such as used in a companded converter . (suggested by Prof. K.W. Martin, UCLA).

In the next section, several applications of the basic principle of Fig. 1 to A/D and D/A conversion are described, along with some preliminary simulation and experimental results which verify the practical usefulness of the proposed systems.

## A DIGITALLY CORRECTED ΣΔ D/A CONVERTER

A ΣΔ D/A converter can be constructed using the system of Fig. 1, as shown in Fig. 2, where (as a typical example) $N = 4$ and an input signal with a word length of $M = 16$ bits were assumed. In the proposed system, the sampling rate $f_c$ of the digital input signal is raised by an oversampling factor $k$ . The interpolating filter at the input is used to suppress the signal spectrum everywhere except in narrow frequency bands centered around $f = 0$ , $\pm k f_c$ , $\pm 2k f_c$ ,...., where $f_c$ is the sampling rate at the input. The word length of the interpolated signal is then reduced to (say) $N = 4$ bits by the rounding stage containing a simple digital filter (e.g., an accumulator), a rounder and a memory. The rounding stage is followed by the N-bit DAC. The unavoidable nonlinearity of the DAC transfer function is cancelled by the EPROM which is programmed in a such a way that for any 4-bit input its output is the 16-bit equivalent of the analog DAC output. Also, as the analysis of the system of Fig. 1 indicated, the noise introduced by the 16-to-4 rounding operation is cancelled in the high-gain frequency region (baseband) of the H(z) block. Thus the system makes high-accuracy ( ~ 16 bit ) conversion possible, even if the linearity of the internal DAC is accurate only to, say, 5 bits.

Fig. 3 compares the simulated output spectra of two oversampled ΣΔ DACs at the inputs of their analog post-filters. The input signal was a sinewave with a peak amplitude equal to 0.577 $V_R$ , where $V_R$ is the reference voltage of the DAC.

Reprinted from *IEEE Proc. ISCAS'89*, pp. 647–650, May 1989.

The signal frequency was 2.875 kHz, and the fast clock frequency $k f_c = 1.024 \text{ MHz}$. Curve (a) is the output spectrum of a converter using an ideal 1-bit internal DAC, and curve (b) that of the proposed converter using a nonlinear 4-bit internal DAC and digital correction. The DAC nonlinearity was assumed to vary randomly from level to level, with a large RMS value equal to 25% of the 4-bit LSB. As the curves illustrate, the anticipated $(N - 1)6 = 18 \text{ dB}$ reduction in the noise and harmonic distortion due to the N-bit internal conversion was nearly fully achieved. The response of an *ideal* converter with a perfectly linear 4-bit internal DAC was found to be nearly indistinguishable from that of the nonlinear corrected one.

## A DIGITALLY CORRECTED A/D CONVERTER

The front end of a multi-bit $\Sigma\Delta$ ADC, also based on the system of Fig. 1, is shown in Fig. 4. Again, $N = 4$ and $M = 16$ were assumed, where M represents the bit accuracy of the corrected ADC. Since the output data of the EPROM appear at the high clock rate $k f_c$, the decimation filter normally following the analog front end would have to be a highly complex and fast circuit. To reduce the complexity of the post-filter, the EPROM output data are entered first into the rounding stage shown in Fig. 5. This consists of a feedback loop which is essentially a fully digital $\Sigma\Delta$ stage; it reduces the truncation noise in the high-gain frequency range of $H_2(z)$. The decimation filter which follows can now be much simpler.

As the results of both simulations and experiments indicate (cf. Figs. 8 and 10), the system achieves an improvement of $N - 1$ bits for the overall ADC accuracy, in agreement with the theoretical prediction.

Note that it is also possible to place the EPROM of the ADC system into the feedback loop of the rounding stage, as in the DAC of Fig. 2. Then, however, the nonlinearities of the DAC and the EPROM must be complementary, rather than matched. This makes the programming of the EPROM more difficult.

A digitally corrected ADC can also be realized by preceding the internal N-bit DAC of a conventional $\Sigma\Delta$ ADC front-end stage by the digitally corrected feedback loop shown in the DAC of Fig. 2. The resulting system [2] is somewhat more complicated than that of the ADC of Fig. 4; however, it does not require the truncation stage of Fig. 5 since the output word length is only N bits.

The proposed scheme is also applicable to the multi-stage $\Sigma\Delta$ ADCs proposed by Matsuya et al. [3]. Then, each stage requires digital correction. Simulations indicated (Fig. 6) that for a three-stage system, with 4-bit internal converters and an oversampling ratio $k = 128$, a 21-bit conversion accuracy may be feasible.

## SELF-CALIBRATION OF DIGITALLY CORRECTED CONVERTERS

The digitally corrected converters described above require a precise calibration procedure, which includes the measurement of the nonlinear DAC transfer characteristics and the storage of the resulting data in the EPROM. This process can be performed on-chip using the building blocks of the converter to be calibrated. The blocks are rearranged to form a single-bit $\Sigma\Delta$ ADC, as shown in Fig. 3 of ref. 2. The input signal of the ADC is obtained by generating a digital ramp with an N-bit counter, and feeding it to the N-bit DAC which must be corrected. Since the single-bit $\Sigma\Delta$ ADC is inherently linear, the M-bit output of its digital filter will provide the nonlinearity data on the N-bit DAC which must be stored in the EPROM. The calibration process requires approximately $2^{N+M}$ clock periods, and can be performed at the fast (oversampling) clock rate. Hence, it requires only a short time, and may be carried out automatically each time the converter is turned on.

## EXPERIMENTAL RESULTS

A breadboard model of a first-order digitally corrected ADC was built and tested to verify the concept. The internal word length was $N = 3$. The schematics of the system are shown in Fig. 7. The components employed were a 4-bit ADC (Analog Devices 9688), an 8-bit DAC (Analog Devices 9768), an OR gate, and a switched-capacitor integrator which realized H(z). In the linear mode, these components were wired in the usual way, but only the 3 MSBs of the ADC and DAC were used (Fig. 7a). In the nonlinear mode, a large nonlinearity ($\pm 1/2$ LSB of the 3-bit words) was introduced into the ADC by connecting its output to the OR gate. Also, a complementary nonlinearity was generated in the DAC by miswiring its input terminals (Fig. 7b). Since the inherent dc differential nonlinearity (DNL) of the ADC was less than $\pm 1/8$ LSB, and the DNL of the DAC only $\pm 1/256$ LSB, these artificially introduced nonlinearities dominated the operation. (The complementary nonlinearities were used to simulate the conditions anticipated for the planned implementation; similar nonlinearities can also be created using an AND, rather than an OR gate.)

Rather than using an EPROM directly to test the concept of digital correction, the simpler alternative of monitoring the DAC input word was employed. The 4-bit input word to the DAC is, of course, the digital representation of the DAC output, which is the same as the desired output of the EPROM. This approach was permissible since the artificially introduced DAC DNL of $\pm 1/2$ LSB was much greater than the intrinsic DAC DNL of $\pm 1/256$ LSB.

The breadboard $\Sigma\Delta$ ADC was operated with a 3.24 MHz clock and a 15.0 kHz sine wave input signal with a peak amplitude of $0.65 V_R$, where $V_R$ was the ADC reference voltage. A Tektronix DAS 9200 digital analyzer was used to capture sequences of samples, which in turn were analyzed using a Fast Fourier Transform (FFT) program. Three types of data were collected: (1) 3-bit linear mode operation with samples taken from the output of the ADC, (2) uncorrected 3-bit nonlinear mode data from the ADC/OR-gate combination output, and (3) corrected 3-bit nonlinear mode data with samples measured at the input to the DAC. In addition, for comparison purposes, both the linear and nonlinear $\Sigma\Delta$ ADCs were simulated using a special-purpose simulation program. The results are shown in Figs. 8-10.

Fig. 8 shows the measured spectrum corresponding to 3-bit linear-mode operation. The input generator produced signal harmonics that were only 55-60 dB below the fundamental and they are noticeable in the measured spectrum.

Fig. 9 displays the measured results for the uncorrected 3-bit nonlinear mode spectra. Here, it is apparent that the $\Sigma\Delta$ noise shaping has been lost and signal harmonics have been introduced that are larger than those attributable to the input generator. Fortunately, however, as is seen in Fig. 10, the noise-shaping characteristic has been recovered in the corrected 3-bit nonlinear data. The signal harmonics appear to have been attenuated relative to the curves of Fig. 9 but are still somewhat larger than those in Fig. 8. Finally, the signal-to-noise ratio (SNR) corresponding to an oversampling ratio (OSR) of 64 was computed from the experimental data in Figs. 8-10. The resulting values were 59 dB , 34 dB , and 55 dB , respectively. Thus, the corrected data shows a 21 dB improvement in SNR with respect to the uncorrected data, and comes within one bit of the linear mode result. This verifies the viability of the digital correction scheme.

The noise spectra obtained by simulation were nearly identical to the measured ones.

To obtain realistic results in the FFT computations, the frequency of the input sine wave was chosen to be in the form $lf_c/N$ , where N was the number of the FFT samples, $f_c$ the clock frequency, and $l$ an integer such that $1 < l < N-1$ and $l, N$ were relatively prime. (For radix-2 FFT, if only the $0 \le f \le f_c$ region is of interest, any odd $l$ value between 1 and $N-1$ is acceptable.) In addition, to obtain the true value of the quantization noise in the very low frequency region, the dc (average) value of the input sine wave had to be monitored, and (if necessary) subtracted from $v_{in}$ .

ACKNOWLEDGEMENTS

The research of T. Cataltepe and G. Temes was supported in part by the U.S. National Aeronautics and Space Administration under Grant NCC 2-374. The authors are grateful to B. Brandt of Stanford University, who also independently discovered the ADC scheme of Fig. 4, for useful discussions.

REFERENCES

[1]   R.W. Adams, "Design and implementation of an audio 18-bit ADC using oversampling techniques", *J. Audio Eng. Soc.* 34 (3), March 1986, pp. 153-166.

[2]   L.E. Larson, T. Cataltepe, and G.C.Temes, "Multibit oversampled $\Sigma\Delta$ A/D converter with digital error correction", *Electron. Lett.* 24 (16), August 4, 1988, pp. 1051-1052.

[3]   Y. Matsuya, K. Ichimura, A. Iwata, T. Kobayashi and M. Ishikawa, "A 16b oversampling A/D conversion technology using triple integration noise shaping", *ISSCC Digest of Techn. Papers*, New York, N.Y., Feb. 1987, pp. 48-49.

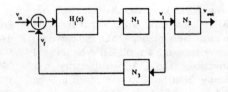

Fig. 1. The block diagram of the proposed digitally compensated nonlinear system.

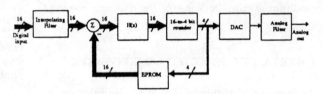

Fig. 2. Multi-bit $\Sigma\Delta$ DAC with digital correction.

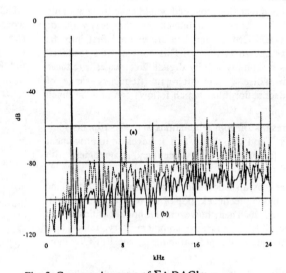

Fig. 3. Computed spectra of $\Sigma\Delta$ DAC's:
(a) using 1-bit internal DAC;
(b) using nonlinear 4-bit internal DAC and digital correction.

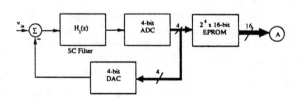

Fig. 4. Front end of the multi-bit $\Sigma\Delta$ ADC with digital correction.

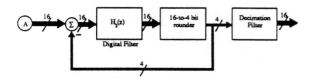

Fig. 5. Digital rounding and decimation stages.

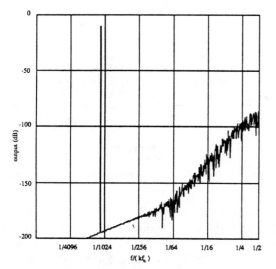

Fig. 6. The spectrum of the output signal of the three-stage multi-bit $\Sigma\Delta$ without digital filtering.

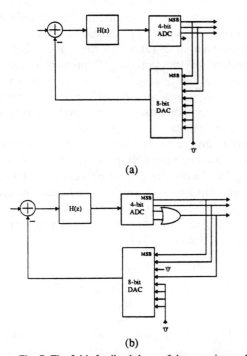

(a)

(b)

Fig. 7. The 3-bit feedback loop of the experimental oversampling A/D converter: (a) linear operation; (b) nonlinear operation.

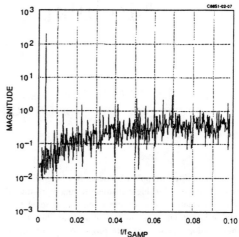

Fig. 8. Measured spectrum of the linear 3-bit $\Sigma\Delta$ ADC.

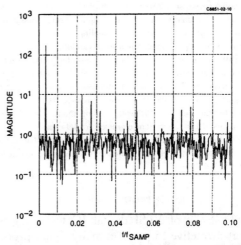

Fig. 9. Measured spectrum of the uncorrected nonlinear 3-bit $\Sigma\Delta$ ADC.

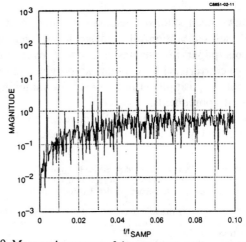

Fig. 10. Measured spectrum of the corrected nonlinear 3-bit $\Sigma\Delta$ ADC.

195

# A Higher Order Topology for Interpolative Modulators for Oversampling A/D Converters

KIRK C.-H. CHAO, MEMBER, IEEE, SHUJAAT NADEEM, WAI L. LEE, AND CHARLES G. SODINI, MEMBER, IEEE

*Abstract* —Oversampling interpolative coding has been demonstrated to be an effective technique for high resolution analog-to-digital (A/D) conversion that is tolerant of process imperfections. A novel topology for constructing stable interpolative modulators of arbitrary order is described. Analysis of this topology shows that with proper design of the modulator coefficients, stability is not a limitation to higher order modulators. Furthermore, complete control over placement of the poles and zeros of the quantization noise response allows treatment of the modulation process as a high-pass filter for quantization noise. Higher order modulators are shown to not only greatly reduce oversampling requirements for high resolution conversion applications, but also to randomize the quantization noise to avoid the need for dithering. An experimental fourth-order modulator breadboard demonstrates stability and feasibility, achieving a 90-dB dynamic range over the 20-kHz audio bandwidth with a sampling rate of 2.1 MHz. A generalized simulation software package has been developed to mimic time-domain behavior for oversampling modulators. Parameterized models are used for the system elements to determine the effects of circuit nonidealities on overall modulator performance. Circuit design specifications for integrated circuit implementation can be deduced from analysis of simulated data.

## I. INTRODUCTION

THE ADVENT of VLSI digital IC technologies has made it attractive to perform many signal processing functions in the digital domain placing important emphasis on analog-to-digital (A/D) conversion [1]. For high resolution, band-limited signal conversion applications, such as digital audio [2], a number of competing approaches have been studied, including floating-point conversion [3], self-calibration [4], and stochastic techniques [5], [6]. This paper investigates a class of oversampled interpolative converters [7]–[15] that can provide high resolution signal acquisition without precision matching of analog circuit components.

A generalized oversampling A/D converter (ADC) system is shown in Fig. 1. The input signal is passed through

Manuscript received August 30, 1988; revised August 3, 1989. This work was supported by Analog Devices, AT&T Bell Laboratories, DEC, GE, and IBM. This paper was recommended by Associate Editor T. T. Vu.

K. C.-H. Chao was with the Microsystems Laboratory, Massachusetts Institute of Technology, Cambridge, MA. He is now with Lotus Development Corporation, Cambridge, MA 02139.

S. Nadeem, W. L. Lee, and C. G. Sodini are with the Microsystems Technology Laboratory, Department of Electrical Engineering and Computer Science, Massachusetts Institute of Technology, Cambridge, MA 02139.

IEEE Log Number 8933455.

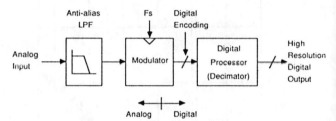

Fig. 1. Block diagram of an oversampling ADC system.

an anti-aliasing filter, sampled by the modulator at a rate much higher than the Nyquist rate, and converted into a high-speed low resolution digital signal. Further digital processing reduces the data rate down to the Nyquist rate, producing the desired high resolution digital output.

The major advantage of this system is that the modulator analog circuit complexity can be greatly reduced if the encoding is selected such that the modulator only needs to resolve a coarse quantization (frequently a single bit) [16]. Also, if oversampling rates are high, the baseband is a small portion of the sampling frequency. Consequently, constraints on the analog anti-aliasing filter can be relaxed, permitting gradual roll off and easy construction with passive components. The precision filtering requirement is now relegated to the digital domain, where a "brick-wall" anti-aliasing filter is needed to decimate the digital signal down to the Nyquist rate. Additional benefits can be gained with the digital processor which can also provide on-chip functions such as equalization, echo cancellation, etc. Thus this system can provide integrated analog and digital functions and is compatible with digital VLSI MOS technologies [17].

### 1.1. Delta–Sigma Modulators

Fig. 2 illustrates the simplest form of an oversampled interpolative modulator, which features an integrator, a 1-bit ADC and digital-to-analog converter (DAC), and a summer. This topology, known as the delta–sigma modulator [7], uses feedback to lock onto a band-limited input $X(t)$. Unless the input $X(t)$ exactly equals one of the discrete DAC output levels, a tracking error results. The integrator accumulates the tracking error over time and the in-loop ADC feeds back a value that will minimize the

Reprinted from *IEEE Trans. Circuits and Sys.*, vol. CAS-37, pp. 309–318, March 1990.

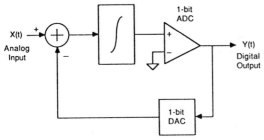

Fig. 2. Block diagram of the delta–sigma modulator.

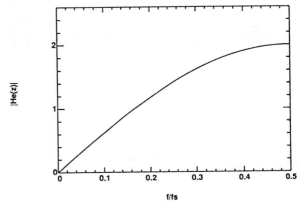

Fig. 4. Calculated quantization noise magnitude response $|H_E(z)|$ for delta–sigma loop.

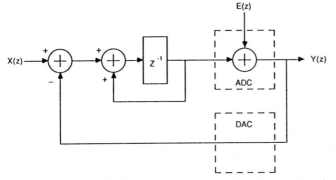

Fig. 3. Discrete-time equivalent of delta–sigma loop. The quantizer is modeled by an additive delay-free noise source. For a 1-bit quantizer, the DAC is replaced by a wire.

accumulated tracking error. Thus the DAC output toggles about the input $X(t)$ so that the average DAC output is approximately equal to the average of the input.

The operation of the delta–sigma modulator can be analyzed quantitatively by modeling the integrator with its discrete-time equivalent and the quantization process by an additive noise source $E(z)$ as illustrated in Fig. 3. $E(z)$ is assumed to be a white and statistically uncorrelated noise source [18]–[20], which has been shown to be reasonable under the usual conditions [21]. With this linearized model of the delta–sigma modulator, it can be shown that

$$Y(z) = X(z)z^{-1} + E(z)(1 - z^{-1}). \qquad (1)$$

Interpolative modulators are also called "noise-shaping" coders because of their effect on the quantization noise $E(z)$ as seen at the output $Y(z)$. A plot of the quantization noise response $|H_e(z)| = |1 - z^{-1}|$ is shown in Fig. 4. When the modulator is sampling much higher than the Nyquist rate, the baseband is in a region where the quantization noise will be greatly attenuated. Although only a coarse quantization is made by the modulator, the bulk of the quantization noise has been pushed to higher frequencies which can be removed by the subsequent digital processing stage. Thus the final output is a high resolution digital representation of the input.

The single-bit encoding scheme used by the delta–sigma modulator and similar interpolative modulators has a number of additional advantages: 1) the format is compatible for serial data transmission and storage systems, 2) subsequent digital processing can be simplified because multiplications and additions reduce to simple logic opera-

tions, and 3) highly linear converters are possible due to the inherent linearity of the in-loop 1-bit DAC.

The last point above deserves further explanation. The integral nonlinearity of the in-loop DAC often limits the harmonic distortion performance of many oversampling ADC's [11]. A multibit DAC has many discrete output levels that must be precisely defined to prevent linearity error. With only two discrete values, a 1-bit DAC always defines a linear transformation between the analog and digital domains. Gain and offset error can still exist in a 1-bit DAC, but linearity error is avoided without precision trimming of output levels.

However, many problems arise when implementing an ADC using a delta–sigma modulator. Most prominent is that the quantization noise actually is signal dependent [21] and not statistically uncorrelated as is usually assumed. This is related to the number of state variables in a system. For a delta–sigma modulator, the state of the system is determined by the integrator output value along with the input value. With only one state variable, the loop can lock itself into a mode where the output bit stream repeats in a pattern. Consequently, the spectrum of the output will contain substantial noise energy concentrated at multiples of the repetition frequency.

To counter this effect, dithering has been used with delta–sigma modulators to randomize the input so that repeating bit patterns will not form. However, this is not an attractive technique as an additional signal source is required and the input dynamic range is lowered.

### 1.2. Higher Order Loops

Candy [22] has done substantial work on a second-order interpolative modulator (the *double-loop*) based on the idea of embedding a delta–sigma loop within the main loop (Fig. 5). With two integrators in the loop, the quantization noise response rises as a quadratic function of frequency as compared to the linear relationship of the first-order topology. The higher order causes quantization noise to be further suppressed in the low frequency baseband and to rise more sharply in the higher frequencies. The net effect is that the total power of the quantization noise in baseband is further reduced, and therefore, a

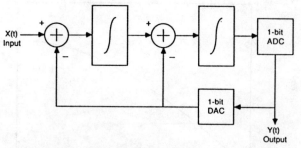

Fig. 5. The double-loop interpolative modulator. This is essentially a delta–sigma modulator within a delta–sigma modulator.

higher effective resolution can be achieved for the same oversampling ratio. The double-loop has been successfully demonstrated in a number of implementations [22], [23]. Also, due to the two integrators in the loop, repeating bit patterns are less likely to occur and, as a result, the quantization noise tends to be less signal dependent. Thus the double-loop increases the SNR possible for a given oversampling ratio while randomizing the output bit pattern so that noise spikes in the output spectra appear less frequently.

Extending the idea of multiple integration loops, higher order loops have proven quite difficult to stabilize [24], [25]. When more than one integrator is in the loop, it is possible for the system to be excited in a mode where a large amplitude, low frequency oscillation persists in the integrators. Approaches to stabilize the loops have fallen into two categories: 1) increasing the size (number of bits) of the in-loop ADC–DAC to allow a greater dynamic range of signals in the loop [26], and 2) cascading multiple stages of first-order delta–sigma modulators [25], [27]. In the first case nonlinearities in the inloop multibit ADC–DAC become the limiting factor for overall resolution while component mismatches in multistage modulators determine the resolution for the second case.

## II. ANALYSIS OF $N$TH-ORDER TOPOLOGY

The $N$th-order interpolative modulator to be described is shown in Fig. 6. It is important to note that linear analysis is used only to determine a starting point for modulator design. Further design refinements and verification is accomplished through software simulations and a breadboard circuit implementation.

While this paper concentrates on the results of a fourth-order implementation, the analysis and design techniques discussed here apply equally well to other order topologies. The salient features of this topology are the feedforward structure with loop coefficients $A_0, \cdots, A_N$ and a feedback structure with coefficients $B_0, \cdots, B_N$. The in-loop quantizer[1] is reduced to a single bit for previously cited reasons. As will be demonstrated, the large number of loop coefficients not only allows one to stabilize loops of any order,

[1] The term *quantizer* refers to a cascaded ADC–DAC combination. The input and output are both analog signals which differ by the error due to the quantization process.

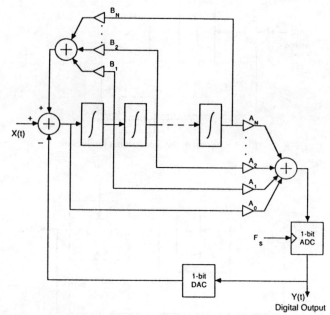

Fig. 6. Proposed $N$th-order loop topology with feedforward and feedback coefficients. The $A_i$ coefficients implement the poles of the quantization noise response, and the $B_i$ coefficients produce the zeros. A $z^{-1}$ delay is associated with the in-loop ADC.

but also allows optimizing the quantization noise shape to improve performance.

### 2.1. System Function

The system function for the topology of Fig. 6 is derived by replacing the integrators with their discrete-time equivalents and by modeling the in-loop ADC as an additive noise source $E(z)$ [22]. A linear model is used to facilitate design of the loop coefficients, which will be subsequently described. The DAC is assumed to be ideal and delayless, but a $z^{-1}$ delay is associated with the ADC. The resulting system function can be expressed as a combined response to the input $X(z)$ and the quantization noise $E(z)$;

$$Y(z) = H_X(z)X(z) + H_E(z)E(z) \qquad (2)$$

where

$$1H_X(z)$$

$$= \frac{\sum_{i=0}^{N} A_i (z-1)^{N-i}}{z\left[(z-1)^N - \sum_{i=1}^{N} B_i (z-1)^{N-i}\right] + \sum_{i=0}^{N} A_i (z-1)^{N-i}} \qquad (3)$$

$$H_E(z)$$

$$= \frac{(z-1)^N - \sum_{i=1}^{N} B_i (z-1)^{N-i}}{z\left[(z-1)^N - \sum_{i=1}^{N} B_i (z-1)^{N-i}\right] + \sum_{i=0}^{N} A_i (z-1)^{N-i}}. \qquad (4)$$

To gain an intuitive understanding of the system function, it is helpful to consider the case where the feedback coefficients $B_1, \cdots, B_N$ are set to zero. With this assumption, (2) can be simplified for the case of low frequency baseband signals by noting that $(z-1) \approx j\Omega$ and $|j\Omega| \ll 1$ where $\Omega = 2\pi(f/f_s)$; $f_s$ is the sampling frequency and $f$ is

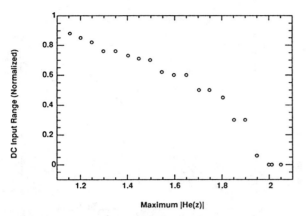

Fig. 7. Simulated dc input range of a fourth-order loop for various values of $|H_E(z)|$. With zero input, the loop becomes unstable when $|H_E(z)| > 2$. The vertical axis is normalized to quantizer step-size.

the frequency of interest. The system response reduces to

$$Y(z) \approx X(z) + \frac{E(z)(j\Omega)^N}{A_N} \qquad (5)$$

which shows that when $N$ is sufficiently greater than 1 and the oversampling ratio is large, i.e., $|j\Omega| \ll 1$, the quantization noise $E(z)$ is greatly attenuated. This demonstrates the accurate tracking aspect of the modulator for low frequency signals.

## 2.2. Stability

For a sampled-data system, stability requirements dictate that the poles be placed within the unit circle. Since the loop coefficients can be designed to have arbitrary values, the poles can be placed anywhere on the z-plane. The absolute control over the location of the poles will allow not only the design of stable loops but also the optimization of the loop response for maximum effective resolution. In most cases, requirements on the precision of the coefficients to obtain an optimized quantization noise response exceeds the requirements for a stable system.

Another mechanism for instability is due to the limited input range of the quantizer [15] which places further constraints on the design of the loop filter [28]. A signal at the input of the quantizer which exceeds the quantizer limits is reflected by an increase in the amount of quantization noise $|E(z)|$. This excess noise is circulated through the loop and can cause an even larger signal to appear at the quantizer input, eventually causing instability. For a fourth-order loop, it has been determined through simulations that $|H_E(z)| < 2$ for $|z| = 1$ at high frequencies is a necessary condition for stable operation with zero input. Applying an input to the system raises signal levels in the loop [28], [29], hence, $|H_E(z)|$ must actually be less than 2 for the modulator to remain stable (Fig. 7).

## 2.3. Design of Loop Coefficients

Treating the loop as a quantization noise filter allows linear filter design techniques to be used. The primary design criteria is to minimize the quantization noise energy

in baseband. However, suppressing quantization noise in the low frequency baseband produces the adverse effect of increasing the quantization noise level at higher frequencies. Also, the quantization noise response must not be allowed to approach 2 at high frequencies, otherwise the stability of the loop will be jeopardized.

Many methodologies can be used to determine the z-domain pole-zero values for $H_E(z)$. Once the pole-zero locations are known, it is a simple matter to determine the loop coefficients. We have chosen to use a Butterworth filter design because of its flat characteristics and its relative insensitivity to coefficient errors.

*2.3.1. $A_i$ Coefficients:* Again, it is useful to consider the case where the $B_i$ coefficients are set to zero and to first concentrate on the $A_i$ coefficients. The process involves computing the s-domain poles, then translating them into the z-domain with a bilinear transform. Once the desired values of the z-domain poles are known, values for $A_0, \cdots, A_N$ can be determined as follows. Let $H_D(z)$ be the desired $N$th-order response defined as

$$H_D(z) \equiv \frac{K(z-1)^N}{(z-p_1)(z-p_2)\cdots(z-p_N)} \qquad (6)$$

where $p_1, p_2, \cdots, p_N$ are the z-domain poles. Equation (6) can be related to $H_E(z)$ by the simple relationship

$$H_D(z) = zH_E(z). \qquad (7)$$

Expansion of the denominator terms gives

$$\frac{K(z-1)^N}{z^N + z^{N-1}C_{N-1} + \cdots + C_0} = \frac{z(z-1)^N}{z^{N+1} + z^N D_N + \cdots + D_0} \qquad (8)$$

where $C_i$ and $D_i$ are coefficients resulting from the expansion. Equating similar denominator terms produces a set of linear equations which determine the $A_i$ coefficients. For the case $N = 4$, $K = 1$, and s-domain poles at $-1.8074 \pm 0.7486j$, $-0.7486 \pm 1.8074j$ rad, the resulting $A_i$ coefficients are

$$A_0 = 0.8653$$
$$A_1 = 1.1920$$
$$A_2 = 0.3906$$
$$A_3 = 0.06926$$
$$A_4 = 0.005395.$$

*2.3.2. $B_i$ Coefficients:* In the section above, all the zeros of the quantization noise response $H_E(z)$ reside at dc ($z = 1$). As a consequence, the noise response rises monotonically out of baseband as an $N$-th order function. One finds that a small portion of the noise spectrum at the upper edge of baseband will dominate the total in-band noise energy. It is apparent that further shaping of the quantization noise will improve performance and this can be accomplished by using the $B_i$ coefficients to move the zeros away from dc.

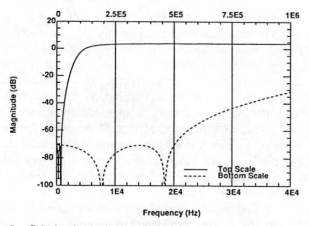

Fig. 8. Calculated quantization noise spectrum for a fourth-order loop. The loop coefficients implement a Butterworth high-pass response with Chebyshev rippling in the baseband region. Top axis is from 0 to 1 MHz; bottom axis shows details from 0 to 40 kHz.

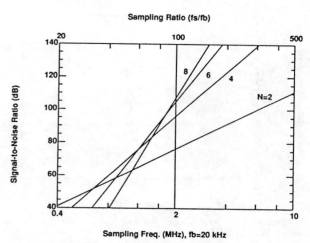

Fig. 9. Calculated effective resolution as a function of oversampling ratio and modulator order for various order loops.

The methodology developed here is to use the equal-ripple characteristics of the Chebyshev polynomials. These polynomials are defined by

$$T_0(x) = 1$$
$$T_1(x) = x$$
$$T_N(x) = 2xT_{N-1}(x) - T_{N-2}(x).$$

What makes them special is that $|T_N(x)| \leqslant 1$ for $|x| \leqslant 1$, and $|T_N(x)| > 1$ for $|x| > 1$. This property of the Chebyshev polynomials can be used to define a desired baseband response through proper scaling. For the case of large oversampling rates, i.e., $f_b \ll f_s$, where $f_b$ is the baseband frequency and $f_s$ is the sampling frequency, the desired $z$-domain zeros $z_i$ are located at

$$z_i = e^{j2\pi X_i f_b / f_s} \tag{9}$$

where $X_i$ are the roots of $T_N$. The $B_i$ coefficients can then be determined using the method outlined above for determining the $A_i$ coefficients. For $f_s = 2.1$ MHz and $f_b = 20$ kHz, the resulting $B_i$ coefficients are

$$B_1 = -3.540 \times 10^{-3}$$
$$B_2 = -3.542 \times 10^{-3}$$
$$B_3 = -3.134 \times 10^{-6}$$
$$B_4 = -1.567 \times 10^{-6}.$$

A plot of the quantization noise response for a $N = 4$ loop implementing the above coefficients is shown in Fig. 8.

From above and from (4), it is apparent that the zeros only depend on the $B_i$ coefficients. The zeros will usually be located near $z = 1$ because of high oversampling ratios, ensuring that the $B$ coefficient values will be small. The effect of $B$ coefficients in determining the $A$ coefficients is then negligible. Thus for large oversampling ratios, the $B$ coefficients determine the zeros and the $A$ coefficients determine the poles of the quantization noise response.

### 2.4. Quantization Noise

The effective resolution of the modulator can be calculated by evaluating $P^2$, the total quantization noise power

in baseband, which is given by

$$P^2 = \int_0^{f_b} |H_E(f)E(f)|^2 \, df \tag{10}$$

where

$$|H_E(f)| \approx \left| \left( \frac{\pi f_b}{f_s} \right)^N \left( \frac{2}{A_N} \right) T_N \left( \frac{f}{f_b} \right) \right| \quad \text{for } f \leqslant f_b \ll f_s. \tag{11}$$

It has been shown that for delta–sigma modulators $E(f)$ can be modeled as an additive white noise source of magnitude [18]–[20]:

$$|E(f)| = e_0 \sqrt{\frac{2}{f_s}} \tag{12}$$

where $e_0$ is the average quantization noise and is related to the quantizer step size $\sigma$ by

$$e_0 = \frac{\sigma}{\sqrt{12}}. \tag{13}$$

Equation (10) has been evaluated and is graphically presented in Fig. 9. As the modulator order increases, the lines become steeper, implying a greater payoff from oversampling. For 96-dB resolution across the 20-kHz audio range, Fig. 9 shows that a second-order loop requires a sampling rate of 5 MHz, whereas a fourth-order modulator requires only 2 MHz. In this case, the additional analog circuit complexity of two integrators can reduce the sampling rate by a factor of 2.5.

### III. SIMULATION RESULTS

Linear analysis is a useful tool for gaining intuitive understanding of modulator operation and for designing the loop coefficients. However, for nonlinear circuit effects, linear analysis is inappropriate. To tackle this problem, a generalized software simulator has been developed to study nonlinear circuit effects. The simulator mimics circuit behavior in the discrete-time domain [17] with

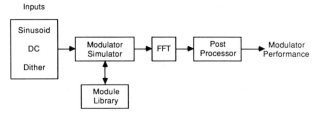

Fig. 10. Block diagram of the generalized oversampling modulator simulation system.

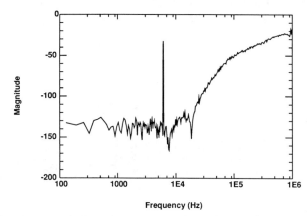

Fig. 11. Typical simulation spectrum of an ideal fourth-order loop. Sinusoidal input with amplitude = 0.1 (normalized to DAC output), dc offset = 0.02, and frequency = 6.1 kHz. The clock/sampling frequency is 2.1 MHz. Note the rippling in the baseband noise floor due to the optimized zeros.

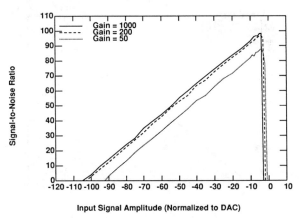

Fig. 12. Simulated SNR versus relative input amplitude for fourth-order loop showing effects of finite op-amp gain. Sinusoidal input dc offset = 0.02, frequency = 6.1 kHz, clock = 2.1 MHz, and oversampling ratio = 48.

modules that model block-level circuit functions. Each model contains associated nonidealities so that the effect of each nonideality can be examined. The modules can then be connected, via software, to simulate almost any topology for oversampling modulators. This simulator is especially useful for CMOS implementations of oversampling modulators, in which many analog functions are performed in discrete time.

Fig. 10 shows the different blocks of the simulation system. The user defines a modulator topology composed of modules from a library and sets parameters that control nonideal circuit behavior. With an input signal injected into the simulator, a long output stream from selected circuit nodes is generated. This output data can be processed with a fast Fourier transform (FFT) to reveal spectral content. Further post-processing yields performance measurements, such as SNR.

Discrete-time simulation of a sampled-data system can greatly simplify modeling of circuits, since only nodal values at sampling points are of interest. What happens at a node in-between clock edges is usually irrelevant, as long as the proper value at the sampling point is produced. This simulator contains models for various switched-capacitor integrators, quantizers, summers, gain blocks, etc. A representative sampling of nonidealities include finite op-amp gain, settling time, nonlinear capacitors for integrators, hysteresis of comparators, and nonlinear DAC's.

A typical spectrum of the modulator's output bit stream is shown in Fig. 11. For all simulations, the FFT's were performed on over 32 000 data points using a raised-cosine

window (von Hann). Various simulation results are presented in the following sections for the fourth-order modulator previously described.

### 3.1. Finite Op-amp Gain

Finite op-amp gain causes the inverting op-amp terminal to reflect the output voltage rather than behave as a virtual ground. Consequently, not all of the charge on the sampling capacitor will be transferred to the integrating capacitor, resulting in a "leaky" integration. This effect is still a linear process and can be incorporated into the linear S-C integrator model by the addition of leakage and integration gain coefficients. In the system function, integration gain errors are manifested as errors in the loop coefficients.

The effect of finite op-amp gain on fourth-order modulator performance can be seen in the SNR curves of Fig. 12. For op-amp gains less than 200, the performance has degraded by about 2 dB. However, with gains of 1000 or more, the performance is nearly ideal. This is especially important for MOS implementations, since amplifier gains of 1000 are easily achievable.

### 3.2. Integrator Settling Time

Typically, oversampled modulators are operated with clock frequencies over 1 MHz, thus settling becomes an issue in determining performance. This is especially true for switched-capacitor implementations, where each integrator can be expected to step to a new output level during every clock cycle. If the integrator is modeled as a single-pole system, then step responses will be exponential in nature. The fractional error introduced by this response is

$$\epsilon \equiv \frac{V_{\text{actual}} - V_{\text{step}}}{V_{\text{step}}} = -\exp\left(-T_s/\tau\right), \qquad \text{for } V_{\text{step}} \leqslant \tau S_R$$

(14)

where $T_s$ is sampling period, $\tau$ the time constant of the exponential response, and $V_{\text{step}}$ the output step size. Thus for a given $T_s$, $\epsilon$ remains constant. However, if the step size is sufficiently large, then slewing occurs and the integrator

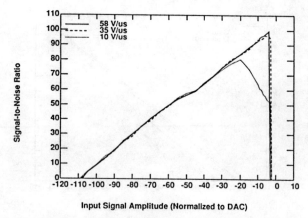

Fig. 13. Simulated integrator settling time results for various slew rates, $\tau = 13$ ns, $T_s = 100$ ns, and DAC step-size = 5. The worst case ($S_R = 10$ V/$\mu$s) performance is caused by slewing during the entire sampling period for large steps.

will settle with a combination of slew and exponential characteristics. The fractional error in this case is

$$\epsilon = -\frac{\tau S_R}{V_{\text{step}}} \exp\left[\frac{V_{\text{step}}}{\tau S_R} - 1 - \frac{T_S}{\tau}\right],$$

$$\text{for } \tau S_R < V_{\text{step}} \leqslant (\tau + T_S) S_R \quad (15)$$

where $S_R$ is the slew rate of the loaded op-amp. With this model, a set of SNR curves for various slew rates (Fig. 13) was produced. The parameters used in this experiment were calculated from SPICE simulation results of an integrator design operating under worst-case conditions. As can be seen, settling time for a first-order step-response only affects performance for lower slew rates. The slewing introduces nonlinear effects, yet the values of the percentage errors for the higher slew rates are so small as to be negligible.

### 3.3. Limited Op-amp Swing

The effect of limited output swing has already been documented [17], [30]. Because the integrators contain state information, clipping of output levels will cause loss of state information, thus performance will be degraded. For higher order loops, this is a serious problem because the last integrator has very large signal swings. A solution is to design the discrete-time integrators with time constant less than unity, thereby attenuating the signal levels at the integrator outputs. This attenuation is then compensated by multiplying the values of the $A_i$ and $B_i$ coefficients so that the overall loop transmission remains the same. Fig. 14(a) shows superimposed images of the four integrator outputs without the "gain compensation," and Fig. 14(b) shows the same outputs with the "gain compensation." Clearly, the last integrator now has signal levels well within the maximum swing bounds. Thus with proper system design, output swing is not a limitation for this topology.

### 3.4. $A_i$ and $B_i$ Coefficient Errors

As mentioned in the preceding section, the $A_i$ and $B_i$ coefficients determine the pole and zero locations of the

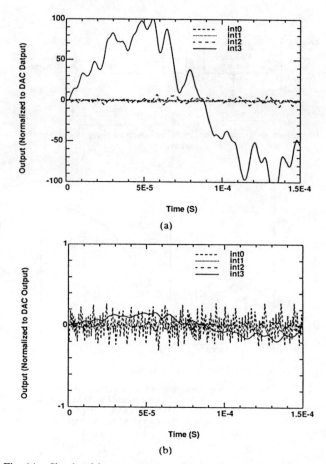

Fig. 14. Simulated integrator outputs for fourth-order loop. Sinusoidal input amplitude = 0.1, dc offset = 0.02, frequency = 6.1 kHz, and clock = 2.1 MHz. (a) Integrator gain = 1. (b) Gain = 0.2.

system. Errors from process mismatch change the coefficient values and cause performance degradation. Tolerances to each coefficient were determined and are listed in Table I. As can be seen, 5 to 30-percent errors in each coefficient can be tolerated for a 3-dB loss of dynamic range.

### 3.5. DC Idle-Channel Noise

One major problem with first- and second-order loops is that their quantization noise is highly dc-bias dependent [18], [31]. The first-order system contains only one state variable, thus it is very likely that repeating bit patterns will be produced. In a second-order system, somewhat better performance is possible because the states are more random, hence, quantization noise is less signal dependent. Presumably, for a fourth-order system, even better randomization should be achieved. This is indeed the case and is reflected in the flatness and lack of noise spikes with dc inputs as shown in Fig. 15. Thus no dither signal is necessary.

Thus from a system viewpoint, the fourth-order loop is quite immune to many nonidealities and does not suffer from the dc bias problems associated with lower order systems. These simulation results have been used to determine the design specifications for a CMOS implementa-

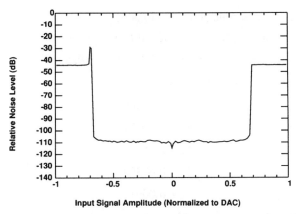

Fig. 15. Simulated fourth-order loop dc idle-channel noise. The lack of noise spikes within the valid input region indicates signal-independent noise.

TABLE I
MAXIMUM ERROR TOLERANCES OF $A_i$ AND $B_i$ COEFFICIENTS WHICH WILL CAUSE LESS THAN 3 DB LOSS OF DYNAMIC RANGE

| Coefficient | Butterworth | Elliptic | Coefficient | Butterworth | Elliptic |
|---|---|---|---|---|---|
| $A_0$ | 5% | 5% | | | |
| $A_1$ | 10% | 5% | $B_1$ | 30% | 30% |
| $A_2$ | 20% | 10% | $B_2$ | 10% | 10% |
| $A_3$ | 30% | 20% | $B_3$ | 30% | 20% |
| $A_4$ | 20% | 20% | $B_4$ | 30% | 30% |

tion of a fourth-order modulator for digital audio applications.

## IV. EXPERIMENTAL RESULTS

In order to verify the results of the theoretical analysis and demonstrate the robustness of the topology, an experimental fourth-order modulator has been constructed from discrete components. The breadboard consists of continuous-time integrators, summers, a comparator, and a 1-bit current DAC. This breadboard operates at a clock frequency of 2.1 MHz and implements the Butterworth–Chebyshev loop filter already described.

To make signal-to-noise measurements, a sinusoidal signal is fed into the modulator and the output is examined. However, in order to maintain the effectiveness of a spectrum analyzer, the square-wave output of the modulator must first be low-pass filtered to remove the high frequency components. In addition, limited dynamic range of the spectrum analyzer necessitates use of a notch filter to remove the signal component when making baseband noise measurements.

Fig. 16 is an output spectrum of the modulator. Notice the large amount of quantization noise, except in baseband where it is greatly suppressed by the response of the loop. A 10-kHz input signal component is also shown with no harmonic distortion visible, revealing the extreme linearity of the modulation process. As mentioned in the introduction, the excellent linearity is aided by the use of a 1-bit DAC which is inherently linear because it has only two discrete output levels. Measurement of the SNR as a function of input signal amplitude is shown in Fig. 17. It indicates a dynamic range of 90 dB with less than 0.1

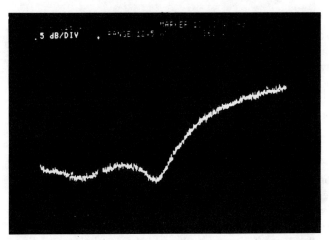

Fig. 16. Output spectrum of breadboard fourth-order loop showing quantization noise around the baseband region (0–40 kHz). Clock rate is 2.1 MHz. Input is a 10-kHz sinusoid with amplitude of −70-dB full scale.

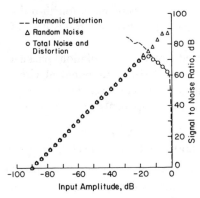

Fig. 17. Measured SNR of breadboard as a function of input amplitude. Random noise and harmonic distortion components are measured separately. Input frequency is 5 kHz.

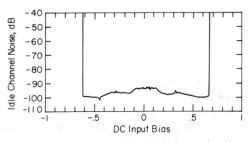

Fig. 18. DC idle-channel noise of breadboard as a function of input level (normalized to DAC output).

percent total harmonic distortion for inputs smaller than −3 dB and less than 0.01 percent total harmonic distortion for inputs smaller than −21 dB. Fig. 18 shows the modulator noise characteristic as a function of dc input level which does not exhibit the sharp increases in output noise at various dc input levels that are a problem for the first- and second-order loops. The reason the theoretical noise limit for the fourth-order modulator was not achieved is attributable to clock jitter. Since the DAC in the continuous-time implementation returns a current signal which is integrated by the loop, any jitter in the clock signal appears as noise at the input.

Using real audio sources as an input to this system has produced very good results. For this listener's ears, the sound quality was "excellent" with no discernible underlying tones, which would occur if signal-dependent nonharmonic noise were being generated. Stability was maintained, with the modulator faithfully tracking the input signal.

## V. Conclusions

The analysis of a novel $N$th-order interpolative modulator topology has been presented. It reveals that with proper design of feedforward loop coefficients, stability is not a limitation to higher order modulators. The loop stability is determined primarily by the feedforward coefficients, while feedback coefficients may be added to optimize quantization noise response in the baseband, thereby improving performance. A design methodology for a quantization noise-shaping filter has been described. Measured results of 90-dB SNR for an experimental fourth-order modulator demonstrates the feasibility and robustness of the topology.

A generalized software simulation package has been developed to mimic ideal and nonideal circuit behavior. With this package, system level design specifications for various modulator components may be determined.

## Appendix
### Elliptical Filter Implementation

The quantization noise response can also be designed using elliptical filter design techniques. An elliptical filter has the "optimal" characteristics of maximum attenuation and sharpness of transition region for a given filter order. The design is easily accomplished using a common elliptical filter design program [32], which gives the pole-zero locations of the filter.

To determine the loop coefficients, recall that for an $N$-th order filter with zeros $z_1, z_2, \cdots, z_N$ and poles $p_1, p_2, \cdots, p_N$, the desired system function $H_D(z)$ is

$$H_D(z) = \frac{(z - z_1)(z - z_2) \cdots (z - z_N)}{(z - p_1)(z - p_2) \cdots (z - p_N)}. \quad (16)$$

As before, $H_D(z)$ is related to $H_E(z)$ by the relationship

$$H_D(z) = z H_E(z) \quad (17)$$

$$\frac{(z - z_1)(z - z_2) \cdots (z - z_N)}{(z - p_1)(z - p_2) \cdots (z - p_N)}$$
$$= \frac{z[z^N + z^{N-1}F_{N-1} + \cdots + zF_1 + F_0]}{z^{N+1} + z^N D_N + \cdots + zD_1 + D_0} \quad (18)$$

$$\frac{(z - z_1)(z - z_2) \cdots (z - z_N)}{(z - p_1)(z - p_2) \cdots (z - p_N)}$$
$$= \frac{z^N + z^{N-1}F_{N-1} + \cdots + zF_1 + F_0}{z^N + z^{N-1}D_N + \cdots + D_1 + z^{-1}D_0} \quad (19)$$

where $F_i$ and $D_i$ are coefficients from expanding $H_E(z)$. For the case $N = 4$, an elliptical filter has been designed

with $z$-domain pole-zero locations at

$$p_i = 0.9138 \pm 0.2500j, \ 0.6970 \pm 0.1964j$$
$$z_i = 0.9985 \pm 0.0550j, \ 0.9997 \pm 0.0230j.$$

The resulting $A$ loop coefficients are

$$A_0 = 0.77486$$
$$A_1 = 1.07610$$
$$A_2 = 0.36610$$
$$A_3 = 0.07398$$
$$A_4 = 0.00912$$

and the $B$ coefficient values are

$$B_1 = -0.003558$$
$$B_2 = -0.003559$$
$$B_3 = -0.00000323$$
$$B_4 = -0.00000161.$$

From simulation results, the elliptical filter implementation was more susceptible to instability for a given perturbation of $A_i$ and $B_i$ coefficients as compared with the Butterworth–Chebyshev implementation. This fact has been summarized in the results of Table I, which show a tighter constraint on the allowable errors for the elliptical case.

## Acknowledgment

The authors gratefully thank Prof. H.-S. Lee for his insights throughout this project. Special thanks go to Prof. Bruce Musicus, Prof. John Wyatt, and Lloyd Clark for their helpful discussions.

## References

[1] D. A. Hodges, P. R. Gray, and R. W. Brodersen, "Potential of MOS technologies for analog integrated circuits," *IEEE J. Solid-State Circuits*, vol. SC-13, pp. 285–294, June 1978.
[2] B. Blesser and J. M. Kates, "Digital processing in audio signals," in *Applications of Digital Signal Processing*, A. V. Oppenheim, ed., Englewood Cliffs, NJ: Prentice-Hall, 1978, pp. 29–116.
[3] J. S. Kriz, "A 16-bit A-D-A conversion system for high-fidelity audio research," *IEEE Trans. Acoust. Speech, Signal Processing*, vol. ASSP-23, pp. 146–149, Feb. 1975.
[4] H. Lee, D. A. Hodges, and P. R. Gray, "A self-calibrating 15 bit CMOS A/D converter," *IEEE J. Solid-State Circuits*, vol. SC-19, pp. 813–819, Dec. 1984.
[5] L. R. Carley, "An oversampling analog-to-digital converter topology for high-resolution signal acquisition systems," *IEEE Trans. Circuits Syst.*, vol. CAS-34, pp. 83–90, Jan. 1987.
[6] ____, *Analog-Digital Conversion Handbook.* Englewood Cliffs, NJ: Prentice-Hall, 1986.
[7] H. Inose, Y. Yasuda, and J. Murakami, "A telemetering system by code modulation–Δ-Σ modulation," *IRE Trans. Space Electron. Telem.*, vol. SET-8, pp. 204–209, Sept. 1962.
[8] J. C. Candy, "A use of limit cycle oscillations to obtain robust analog-to-digital converters," *IEEE Trans. Commun.*, vol. COM-22, pp. 298–190, Mar. 1983.
[9] R. C. Brainard and J. C. Candy, "Direct-feedback coders: design and performance with television signals," *IEEE Proc.*, vol. 57, pp. 776–786, May 1969.
[10] M. W. Hauser, P. J. Hurst, and R. W. Brodersen, "MOS ADC-filter combination that does not require precision analog components," in *ISSCC Dig. Tech. Papers*, pp. 81–82, Feb. 1985.
[11] J. W. Scott, W. L. Lee, C. H. Giancarlo, and C. G. Sodini, "CMOS implementation of an immediately adaptive delta modulator," *IEEE J. Solid-State Circuits*, vol. SC-21, pp. 1088–1095, Dec. 1986.
[12] C. G. Giancarlo and C. G. Sodini, "A slope adaptive delta modulator for VLSI signal processing systems," *IEEE Trans. Circuits Syst.*, vol. CAS-33, pp. 51–58, Jan. 1986.

[13] S. K. Tewksbury and R. W. Hallock, "Oversampled, linear predictive and noise-shaping coders of order $N > 1$," *IEEE Trans. Circuits Syst.*, vol. CAS-25, pp. 436–447, July 1978.

[14] W. L. Lee and C. G. Sodini, "A topology for higher order interpolative coders," in *Proceedings 1987 Int. Symp. on Circuits and Systems*, pp. 459–462, May 1987.

[15] W. L. Lee, "A novel higher order interpolative modulator topology for high resolution oversampling A/D converters," Master's thesis, MIT, Cambridge, MA, 1987.

[16] J. R. Fox and G. J. Garrison, "Analog to digital conversion using sigma-delta modulation and digital signal processing," in *Proc. MIT VLSI Conf.*, pp. 101–112, Jan. 1982.

[17] M. W. Hauser and R. W. Brodersen, "Circuit and technology considerations for MOS delta-sigma a/d converters," in *Proc. 1986 Int. Symp. on Circuits and Systems*, pp. 1310–1315, May 1986.

[18] J. C. Candy, "The structure of quantization noise from sigma-delta modulation," *IEEE Trans. Commun.*, vol. COM-29, pp. 1316–1323, Sept. 1981.

[19] W. R. Bennett, "Spectra of quantized signals," *Bell Syst. Tech. J.*, vol. 27, pp. 446–472, July 1948.

[20] D. J. Goodman, "Delta modulation granular quantizing noise," *Bell Syst. Tech. J.*, pp. 1197–1218, May 1969.

[21] R. M. Gray, "Oversampled sigma-delta modulation," *IEEE Trans. Commun.*, vol. COM-35, pp. 481–489, May 1987.

[22] J. C. Candy, "A use of double integration in sigma delta modulation," *IEEE Trans. Commun.*, vol. COM-33, pp. 249–258, Mar. 1985.

[23] R. Koch and B. Heise, "A 120 kHz sigma-delta A/D converter," in *ISSCC Dig. Technic. Papers*, pp. 138–141, Feb. 1986.

[24] S. H. Ardalan and J. J. Paulos, "Stability analysis of high-order sigma-delta modulators," in *Proc. 1986 Int. Symp. on Circuits and Systems*, pp. 715–719, May 1986.

[25] Y. Matsuya, K. Uchimura, A. Iwata, T. Kobayashi, M. Ishikawa, and T. Yoshitome, "A 16-bit oversampling A-to-D conversion technology using triple integration noise shaping," *IEEE J. Solid-State Circuits*, vol. SC-22, pp. 921–929, Dec. 1987.

[26] M. J. Hawksford, "N-th order recursive sigma-ADC machinery at the analogue-digital gateway," presented at Audio Engineering Society Convention, May 1985.

[27] T. Hayashi, Y. Inabe, K. Uchimura, and T. Kimura, "A multistage delta-sigma modulator without double integration loop," in *ISSCC Dig. Tech. Papers*, pp. 182–183, Feb. 1986.

[28] S. H. Ardalan and J. J. Paulos, "An analysis of nonlinear behavior in delta-sigma modulators," *IEEE Trans. Circuits and Systems*, vol. CAS-34, pp. 593–603, June 1987.

[29] C. Wolff and L. R. Carley, "Modeling the quantizer in higher-order delta-sigma modulators," in *Proc. 1988 Int. Symp. on Circuits and Systems*, June 1988.

[30] B. E. Boser and B. A. Wooley, "Design of a CMOS second-order sigma-delta modulator," in *ISSCC Dig. Tech. Papers*, pp. 258–259, Feb. 1988.

[31] B. E. Boser and B. A. Wooley, "Quantization error spectrum of sigma-delta modulators," in *Proc. 1988 Int. Symp. on Circuits and Systems*, June 1988.

[32] A. H. Gray, Jr. and J. D. Markel, "A computer program for designing digital elliptic filters," *IEEE Trans. Acoust. Speech, Signal Processing*, vol. ASSP-24, pp. 529–538, Dec. 1976.

# ONE BIT HIGHER ORDER
# SIGMA-DELTA A/D CONVERTERS

P. F. Ferguson, Jr., A. Ganesan and R. W. Adams

Analog Devices
Wilmington, Massachusetts 01887
(508) 658-9400

## Abstract

A topology for higher order sigma-delta modulators is described and synthesis equations are given which allow for the arbitrary shaping of the signal and quantization noise transfer functions. The synthesis procedure is similar to that of switched capacitor filters and uses switched capacitor design techniques such as dynamic-range scaling, impedance scaling and circuit noise analysis [1]. Placement of pass band zeroes for the optimization of the quantization noise transfer function is discussed and some results measured on a third order test-chip are presented. Simulated results on non-linear stabilization of the loop will be discussed. The need for simulating the analog front-end along with the digital filters to find the true dynamic range of the system will be indicated.

## Introduction

Sigma Delta A/D converters have found widespread applications, particularly in signal processing application such as Digital Audio, ISDN, High Speed Modems, etc [2] - [4]. Most of the published literature either uses a single bit quantizer in a first or second order loop filter or uses a cascaded connection of single bit first/second order modulators resulting in a multi-bit digital output stream for further processing. The modulator in reference [5] is a notable exception to the above in that it deals with a single-bit multi-order loop and [6] is a commercial implementation of that architecture.

In this paper we will present a single-bit multi-order loop that does not require coefficient summers at the input and the output as in [5] and [6]. The complex transmission zeroes in the quantization noise shaping function are simply obtained by choosing appropriate resonator frequencies. The topology has excellent high frequency settling properties for two reasons. First, the interconnection of integrators and resonators is made without delay-free loops. Second, the amplifiers are connected so that the worst case settling occurs when two amplifiers settle in series. The errors resulting from this worst case settling only disturb the relative positioning of the noise shaping transmission zeroes for which the architecture is shown to be insensitive over a wide range.

The distributed nature of the quantized output feedback in this realization enhances the stability characteristics even under conditions where one or more of the integrators are in saturation. The architecture has the secondary advantage of requiring only modest open-loop gain from the amplifiers used for the integrators.

## Theory

Figure 1 shows the block diagram of the proposed architecture. Note that, in contrast to [5], the feedback path also has a shaping

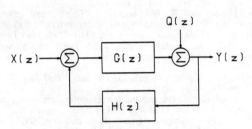

**Figure 1**
Block Diagram of the Proposed Architecture.

function in this architecture. However, the DC gain of the above architecture can still be determined by the ratio of two capacitors as in [5]-[6]. The ideal transfer function for the X input to the system is given by:

$$\frac{Y(z)}{X(z)} = \frac{G(z)}{1 + H(z)G(z)} \qquad (1)$$

and that of the input Q is given by:

$$\frac{Y(z)}{Q(z)} = \frac{1}{1 + H(z)G(z)} \qquad (2)$$

Note that the equations (1) and (2) can be arbitrarily synthesized by appropriate choice of G(z) and H(z). First, G(z) is chosen as a series connection of resonators and integrators to provide the appropriate shaping characteristic for the Q input. The feedback H(z) is then appropriately chosen to ensure the shaping characteristic of the X input and the stability requirements as derived in [5]. Feedforward terms can also be added to G(z) to reduce the amount of signal X seen at the output of intermediate opamps and to improve the transient response of the loop. Figure 2 shows the distributed nature of the G(z) (with feedforward added) and H(z) used to implement figure 1. Equations (3) and (4) are expanded forms of equations (1) and (2) to reflect the distributed architecture.

$$\frac{Y(z)}{X(z)} = \frac{\sum_{i=1}^{n} F_i(z) \prod_{j=i}^{n} G_j(z)}{1 + \sum_{i=1}^{n} H_i(z) \prod_{j=i}^{n} G_j(z)} \qquad (3)$$

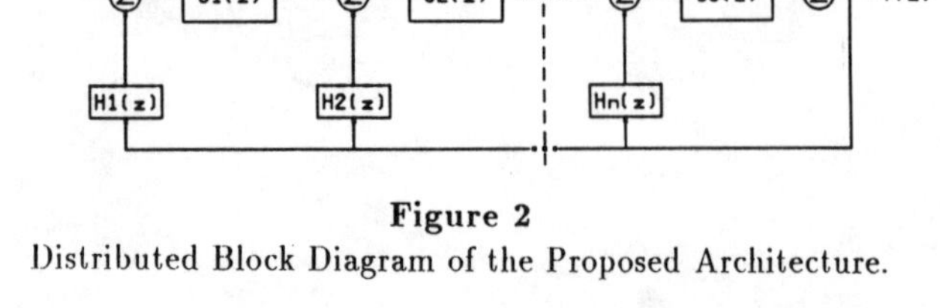

**Figure 2**
Distributed Block Diagram of the Proposed Architecture.

Reprinted from *IEEE Proc. ISCAS'90*, pp. 890–893, May 1990.

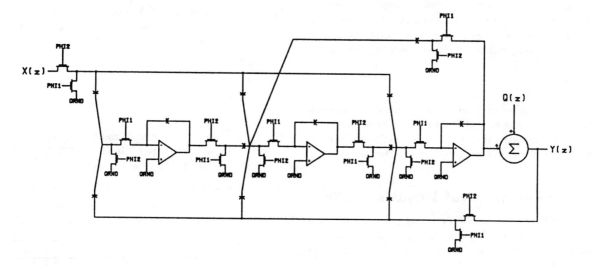

**Figure 4**
Third order sigma-delta modulator, linear model.

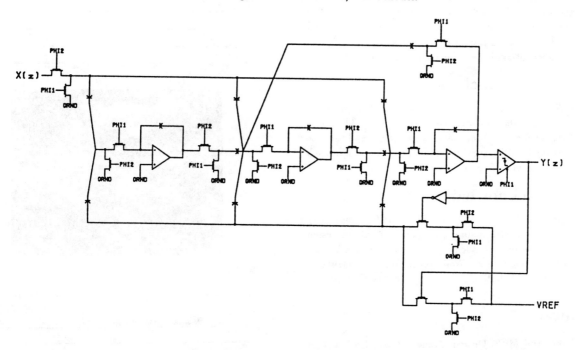

**Figure 7**
Third order sigma-delta modulator.

Figure 7 shows the sigma-delta modulator. Any negative signs in the feedback coefficients can be accommodated by swapping differential outputs in the filter implementation. In the conversion to a modulator, the complement of the comparator output can be used to perform these inversions. In figure 7 the comparator output is strobed at the end of $\phi_1$ and the whole of $\phi_2$ is used to ensure that the feedback elements are tied to the reference in an inverting or non-inverting manner. The feed-in elements during $\phi_2$ acquire the input and the integration results of the previous cycle. The inherent pipelined nature of the architecture due to the $z^{-1}$ delay in all the feedback elements helps to avoid extraordinary timing constraints one might encounter with other architectures.

Impedance scaling of all capacitor values was done by ratiometrically scaling all the capacitors incident at a summing node using traditional switched capacitor scaling techniques. Dynamic range scaling was accomplished by doing a direct transient simulation using the circuit of figure 7 with various "worst case" inputs and then scaling the maxima at the intermediate nodes by scaling the capacitors incident at the output of those integrators. The worst case input is not necessarily a full level sine wave at a certain frequency, since these converters tend to process signals with complex spectra and not simply discrete sinusoids. The transient simulations were run for an adequate length of time to ensure that the worst case maxima were indeed captured.

The above simulations were also used to find the maximum peak valued input signal that would cause instability in the circuit. A clamping circuit can be implemented at each opamp and simulations indicate that the modulator recovers gracefully after prolonged overload conditions. We suspect that the distributed nature of the quantized output being fed to all of the integrators is partly responsible for the resultant behavior.

The circuit noise optimization for $kT/C$ and opamp thermal noise was done using the linearized s-c filter model of figure 4 and using traditional switched capacitor noise analysis techniques [8]. Slew rate and settling requirements were optimized for the worst case output step encountered at the output of each integrator. While 1/f noise reduction was achieved in this realization by using large geometry devices, chopper stabilization techniques can also be used.

## Experimental Results

A fully differential third order modulator circuit was fabricated in a BiCMOS technology. The topology did not use feedforward terms, and did not feed back into the intermediate summing junction of the resonator; these improvements were discovered later. The modulator was first configured as a linear s-c circuit to verify the transfer functions. Figure 8 shows the simulated and measured X and Q transfer in the passband. The modulator was then configured as a sigma-delta modulator and figure 9 shows the passband response to a sine wave input. These results are in excellent agreement with transient simulation.

## Conclusion

A generalized procedure for synthesizing multi-loop single bit sigma-delta modulators has been presented. The close agreement between measured and simulated behavior and the ease of realization using traditional SC techniques makes this approach desirable for a multitude of applications.

## Acknowledgments

The authors would like to thank David Smart for timely help on modifications to our circuit simulator to ease the above design process. We deeply appreciate the help provided by Linda Dupuis and Susan Scott in layout, Tony Wellinger in circuit construction and the numerous contributions of the technical staff at Analog Devices.

## References

[1] K. R. Laker and M. S. Ghausi, Modern Filter Design. New Jersey: Prentice Hall, 1981.

[2] S. R. Norsworthy, et al., " A 14-bit 80 kHz Sigma-Delta A/D Converter: Modeling, Design and Performance Evaluation," IEEE Journal of Solid State Circuits, vol. SC-24, pp. 256 - 266, April 1989.

[3] Y. Matsuya, et al., " A 16 bit Oversampling A-to-D Conversion Technology Using Triple Integration Noise Shaping," IEEE Journal of Solid State Circuits, vol. SC-22, pp. 921-929, December 1987.

[4] R. Koch, et al., " A 12-bit Sigma Delta Analog-to-Digital Converter with a 15-MHz Clock Rate ", IEEE Journal of Solid State Circuits, Vol. SC-21, pp.1003-1010, December 1986.

[5] W. L. Lee and C. G. Sodini, " A Topology for Higher Order Interpolative Coders," ISCAS PROC.: 1987.

[6] D. Welland, et al., " A Stereo 16-bit Delta Sigma A/D Converter for Digital Audio," 85th Convention of the Audio Engineering Society, November 1988.

[7] B. E. Boser and B. A. Wooley, " The Design of Sigma-Delta Modulation Analog-to-Digital Coverters," IEEE Journal of Solid State Circuits, vol. SC-23, pp. 1298 - 1308, December 1988.

[8] J. H. Fischer, "Noise Sources and Calculation Techniques for Switched Capacitor filters," IEEE J. of Solid State Circuits, vol.SC-17, pp. 742-752, August 1982.

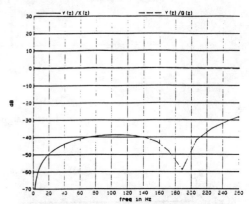

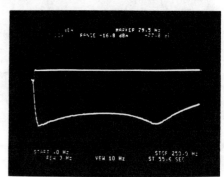

**Figure 8**
Simualted and measured x-path and q-path transfer functions.

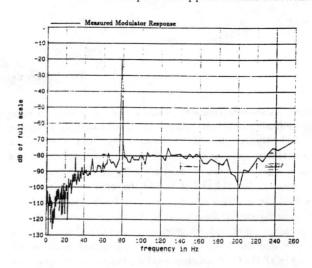

**Figure 9**
Sine wave response of the third order sigma-delta modulator.

# OPTIMIZATION OF A SIGMA-DELTA MODULATOR
# BY THE USE OF A SLOW ADC

A. Gosslau, A. Gottwald

Fakultaet fuer Elektrotechnik
Universitaet der Bundeswehr Muenchen
D-8014 Neubiberg, West Germany

## ABSTRACT

A 1 bit sigma-delta modulator ($\Sigma\Delta$M) with a loop delay $t_{dl}$ (conversion time of the ADC) is considered.
Its idle channel pattern - which is calculated by using the "modified z-transform" - is influenced by the loop delay time.
An optimization of this delay time yields a minimized signal power at the input of the quantizer and an improved signal-to-noise ratio at the output of the sigma-delta modulator.
Optimum performance is achieved for $t_{dl} = 0.25 \cdot T$ (T = sample period). Numerous simulations verify these results.
The use of a slow ADC can hence be advantageous.

## INTRODUCTION

Sigma-delta modulators ($\Sigma\Delta$M) (also referred to as analog feedback coders) are applied in digital signal processing systems (codecs) as analog-to-digital converters [1-6].
These coders can be treated as sampled-data control systems but - because of the strong nonlinearities - difficulties arise with the exact calculation of their behaviour [2,4,5].
Usually the loop delay - which mainly consists of the conversion time of the analog-to-digital converter (ADC) plus that of the digital-to-analog converter (DAC) - is not taken into account.
In [7] we developed a new model of a $\Sigma\Delta$M which allows a detailed investigation of the influence of the loop delay.
In [8] we demonstrated that the dynamic behaviour of a sigma-delta modulator can be improved by the use of an ADC with higher conversion time.
In this paper we show how the signal-to-noise ratio (SNR) at the output of an 1 bit sigma-delta modulator can be optimized by a proper choice of the loop delay time $t_{dl}$.

## MODEL OF THE SIGMA-DELTA MODULATOR

Fig. 1 shows the basic configuration of a sigma-delta modulator. It consists of an integrator with transfer function g/s, a clocked ADC, and a DAC which is embedded in the feedback loop.

Usually the quantizing inside the loop is coarse and the sampling frequency $f_s$ is very high compared to the bandwidth of the input signal $x_i(t)$ ("oversampling"). A digital lowpass filter ("decimation filter") which follows the sigma-delta modulator provides at its output a digital representation of $x_i(t)$ with high resolution and at a low sample rate.

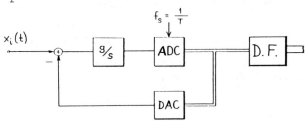

Fig. 1: Sigma-delta modulator with decimation filter

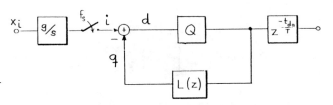

Fig. 2: Model of the sigma-delta modulator

A sigma-delta modulator can be described by the model of Fig. 2 [7] which consists of a cascade of a time continuous and a time discrete system. The amplitude quantization by the ADC plus DAC is described by the quantizer Q.
The block $z^{-t_{da}/T}$ stands for the delay by the conversion time of the ADC, while $t_{dl}$ is the sum of the conversion times of the ADC and DAC.
Because of the delay which is always present inside the feedback loop, L(z) must be calculated by the "modified z-transform" [9] (symbol $\mathbf{Z}_m$).

$$L(z) = \frac{z-1}{z} \cdot \mathbf{Z}_m \left\{ \frac{g}{s^2} \right\}$$

$$= gT \cdot z^{-1} \cdot \left( -t_{dl}/T + z/(z-1) \right) \qquad (1)$$

Reprinted from *IEEE Proc. ISCAS'88*, pp. 2317-2320, June 1988.

## MULTI-BIT AND 1 BIT SIGMA-DELTA MODULATORS

The behaviour of an analog feedback coder with a multi-bit quantizer can be analyzed by the use of a linearized model where the quantizer is modelled by an addition of quantizing noise [1].
We showed [7] that in this case the maximum achievable signal-to-noise ratio $SNR_{max}$ is independent on the loop delay $t_{dl}$ and increases with increased normalized gain $gT$. Stability considerations recommended an optimum loop delay $t_{dlopt} = 0.25 \cdot T$.
It should be noted, however, that the behaviour of a feedback coder with a multi-bit quantizer is amplitude dependent: For input levels which are smaller than the smallest quantizer output level the behaviour of a multi-bit sigma-delta modulator is that of an 1 bit $\Sigma\Delta M$!
In a $\Sigma\Delta M$ with a 1 bit ADC (i.e. a comparator) only the sign of the signal at the input of the quantizer is detected. Therefore the loop gain g has no influence on the behaviour of a 1 bit $\Sigma\Delta M$.
The influence of the loop delay $t_{dl}$ will be investigated in the following sections.

### THE IDLE CHANNEL PATTERN OF A 1 BIT SIGMA-DELTA MODULATOR WITH LOOP DELAY

In this section we investigate the idle channel pattern (ICP) at the input of the 1 bit quantizer Q. It is shown that only 3 different periods can exist which are dependent on the normalized delay $t_{dl}/T$ and on an initial value.
From Fig. 2 and with (1) we get the basic equations of a 1 bit $\Sigma\Delta M$ with delay time $t_{dl}$:

$$d_n = i_n - q_n \qquad (2)$$

$$q_{n+1} = q_n + sgn(d_n) \cdot g(T-t_{dl}) + sgn(d_{n-1}) \cdot gt_{dl} \qquad (3)$$

$$\text{where } sgn(d_n) = \begin{cases} 1 & \text{for } d_n \geq 0 \\ -1 & \text{for } d_n < 0 \end{cases}$$

With (2) and (3) we find for the idle channel pattern at the quantizer input ($i_n = 0$):

$$d_{n+1} - d_n = -g(T-t_{dl}) \cdot sgn(d_n) - gt_{dl} \cdot sgn(d_{n-1}) \qquad (4)$$

This leads to Table 1.

| $sgn(d_{n-1})$ | $sgn(d_n)$ | $d_{n+1} - d_n$ |
|:---:|:---:|:---:|
| 1 | -1 | $-gT$ |
| -1 | 1 | $-gT \cdot (1-2t_{dl}/T)$ |
| 1 | -1 | $+gT \cdot (1-2t_{dl}/T)$ |
| -1 | -1 | $+gT$ |

Table 1: Change of quantizer input as a function of previous inputs d

Only three different period lengths for the ICP are possible if an arbitrary initial value $d_o$ is taken. The length of the period depends on $d_o$ and on $t_{dl}/T$. This is demonstrated in the example of Fig. 3.

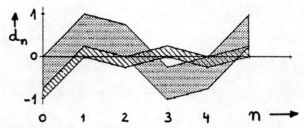

Fig. 3: Quantizer input (Idle channel pattern) $t_{dl}/T = 0.375$, $-1 < d_o < 0$, $gT = 1$, $d_{-1} < 0$

In this case only two different periods are possible:
period 2T for $2t_{dl}/T < |d_o| < 1$
period 4T for $0 < |d_o| < 2t_{dl}/T$
The probabilities P of these two patterns are different and depend on the loop delay.
We get (if all initial values $d_o$ are assumed to have equal probabilities):

$$P(2T) = 1 - 2t_{dl}/T \quad \text{and} \quad P(4T) = 2t_{dl}/T$$

The possible periods and their probabilities are given in Table 2 for different loop delays $t_{dl}/T$.

| | | loop delay | |
|---|---|:---:|:---:|
| | | $0 < t_{dl}/T \leq 0.5$ | $0.5 \leq t_{dl}/T < 1$ |
| period length | 2T | $1-2t_{dl}/T$ | 0 |
| | 4T | $2t_{dl}/T$ | $2-2t_{dl}/T$ |
| | 6T | 0 | $2t_{dl}/T-1$ |

Table 2: Probabilities of periods versus loop delay

### MINIMIZATION OF THE INPUT POWER OF THE QUANTIZER

In order to reduce the energy of the difference between the integrated input i and the quantized signal q the power of the difference signal d should be minimized.
The total power of the signal d at the quantizer input is the sum of the individual variances multiplied by the probabilities of the patterns. For gT=1 the total power is given in Fig. 4.

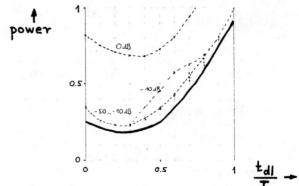

Fig. 4: Power at quantizer input versus loop delay
—— calculated for zero input (ICP)
---- simulated for different input levels

Simulations show that this characteristic is not only valid for zero input (ICP). We found a good agreement between this calculation and the simulations for higher input levels. The input signal was a sinewave with frequency $f = f_s \cdot 5/2048$.

210

We got the following results:
The power of the ICP at the quantizer input has its minimum for $t_{dlopt}$ = 0.25·T.
This optimum delay time yields a minimum power of the difference signal even for higher input levels (although at an increased power).
Therefore: In order to minimize the variance at the quantizer input of a 1 bit $\Sigma\Delta$M a loop delay of $t_{dlopt}$ = 0.25·T should be chosen.

## SHAPING OF THE NOISE POWER SPECTRUM

Only the values +1 and -1 are possible at the out- of an 1 bit $\Sigma\Delta$M, hence its output power is always 1. It is the sum of the signal power and the power of the quantizing noise. The SNR at the output of the digital filter (see Fig. 1) can only be increased if the filter bandwidth B is less than $f_s/2$.
A spectral shaping of the quantizing noise which re- duces the noise power in B is advantageous.
Iwersen [10] has shown that the spectrum of the quantizing noise consists of bessel spectra at the spectral lines of the idle channel noise (ICN).
The power spectrum of the ICN at the output of the $\Sigma\Delta$M is calculated from the ICPs. Their power spectra are multiplied by the individual probabilities and summed. This yields the power spectrum of the quan- tizer input.
The power spectrum at the quantizer output is found by multiplying the spectrum at the quantizer input by the inverse power transfer function $|1/L(z)|^2$ of the open loop.
The individual powers of the spectral lines are given in Table 3 (notice that the total power always equals 1).

| Frequency $f/f_s$ of spectral line | Loop delay | |
|---|---|---|
| | $0 < t_{dl}/T < 0.5$ | $0.5 < t_{dl}/T < 1$ |
| 0, 1 | 0 | 0 |
| 1/6, 5/6 | 0 | $(4/9)(2t_{dl}/T-1)$ |
| 1/4, 3/4 | $t_{dl}/T$ | $1-t_{dl}/T$ |
| 1/2 | $1-2t_{dl}/T$ | $(1/9)(2t_{dl}/T-1)$ |

Table 3: Power of spectral lines versus loop delay

Table 3 shows:
For small delays ($t_{dl} \approx 0$) the power spectral density (PSD) of the quantizing noise is concentrated at $f_s/2$. An increasing $t_{dl}/T$ shifts the maximum of the PSD towards $f_s/4$. For $t_{dl}/T \gg 0.5$ the PSD is concen- trated at $f_s/6$.
Numerous simulations have been made which show a very good agreement with the theory. See Fig. 5.
The input of the $\Sigma\Delta$M was a narrowband noise signal consisting of 9 equally spaced cosines (f=5...13) with equal amplitudes (20 dB below overload). Each of the spectra is the mean of 10 different spectra, calculated with random initial values. $f_s$=1024.

Fig. 5: Spectra of quantizer output between 0 and $f_s/2$. The loop delay time $t_{dl}/T$ is increased from 0.01 (top) to 0.99 (bottom) Vertical scale: 10 dB/div. Horizontal scale: $f_s/12$ per division

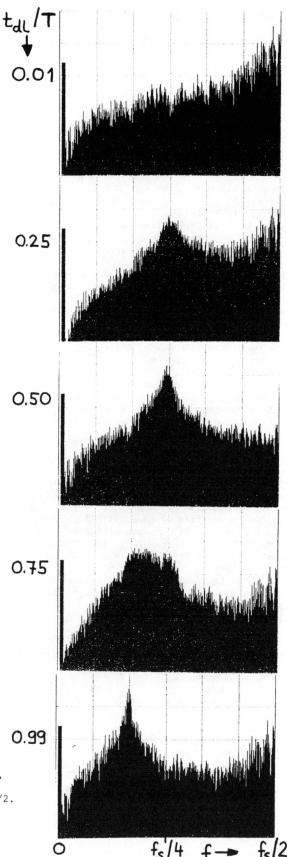

# OPTIMIZATION OF THE SIGNAL-TO-NOISE RATIO

Only that part of the noise power spectrum which lies inside of the filter bandwidth B deteriorates the SNR at the output of the digital filter.
A high SNR can be achieved either by a small bandwidth B, a high sampling frequency $f_s$, or a proper shaping of the noise power spectrum.
As it can be seen in Fig. 5, the PSD in vicinity of the input signal changes if $t_{dl}/T$ is varied.
A low PSD at low frequencies is achieved for a loop delay $t_{dl}$ = 0.25·T.
In order to investigate the influence of a loop delay on the signal-to-noise ratio we made numerous simulations similar to CCITT rec. O 132.
The input signal of the $\Sigma\Delta$M was a sinewave with frequency $f = f_s·9/8192$. For each input level 10 simulations with different (random) initial values inside the $\Sigma\Delta$M were made. The output spectra were calculated by a 8192-point-FFT. For the calculation of the SNR the mean of 10 different power spectra was taken. The displayed SNRs are those at the output of an ideal lowpass filter with a cutoff frequency $B = f_s·64/8192$ and $f_s·512/8192$, respectively.
In this paper only few representative curves can be shown (Fig. 6 and Fig. 7).
It is seen that the SNR can be improved by an optimization of the loop delay. For $t_{dlopt}$ = 0.25·T the improvement of the SNR at small input levels is almost 10 dB. At higher input levels (and also with higher bandwidths) the improvement is less.
This can be related to the fact that for higher inputs the bessel spectra (Fig. 5) are broader. Hence the noise power inside B increases and the signal-to-noise ratio decreases.

## CONCLUSIONS

We demonstrated how the performance of a sigma-delta modulator can be improved by proper choice of the loop delay. We found an optimum at $t_{dlopt}$ = 0.25·T. Our results are verified by numerous simulations.
As the loop delay mainly consists of the conversion time of the analog-to-digital-converter, the SNR at the output of a 1 bit $\Sigma\Delta$M (at the output of the decimation filter) can be improved by the use of a slow ADC. Thus the results of our calculations and simulations allow a simplification of sigma-delta modulators.

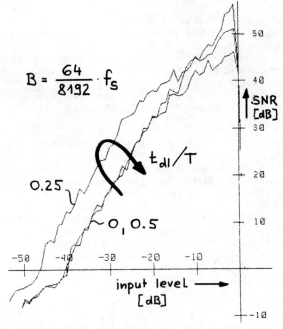

Fig. 6: Signal-to-noise ratio versus input level
Parameter: Loop delay    $B = f_s/128$

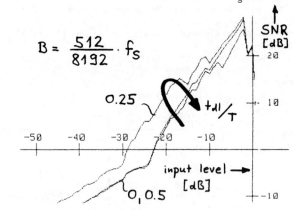

Fig. 7: Signal-to-noise ratio versus input level
Parameter: Loop delay    $B = f_s/16$

## REFERENCES

[1]  B.P. Agrawal and K. Shenoi: "Design methodology for $\Sigma\Delta$M", IEEE Trans. Commun., vol. COM-31, pp.360-370, March 1983.

[2]  C.H. Giancarlo and C.G. Sodini: "A slope adaptive delta modulator for VLSI signal processing systems" IEEE Trans. Circuits Syst., vol. CAS-33, pp.51-58, Jan. 1986.

[3]  D. Vogel et al.: "A digital signal-processing codec filter for PCM applications", IEEE J. Solid-State Circuits, vol. SC-21, pp.175-181, Feb. 1986.

[4]  R.M. Gray: "Oversampled sigma-delta modulation", IEEE Trans. Commun., vol. COM-35, pp.481-489, May 1987.

[5]  S.H. Ardalan and J.J. Paulos: "An analysis of non-linear behaviour in delta-sigma modulators", IEEE Trans. Circuits Syst., vol. CAS-34, pp.593-603, June 1987.

[6]  J.C. Candy: "A use of double integration in sigma delta modulation", IEEE Trans. Commun., vol. COM-33, pp.249-258, March 1985.

[7]  A. Gosslau and A. Gottwald: "Consideration of the loop delay for the dimensioning of analog feedback coders", FREQUENZ 41 (1987) 11/12, pp.329-333, Nov./Dec. 1987.

[8]  A. Gosslau and A. Gottwald: "The influence of a loop delay on the dynamic behaviour of a first-order sigma-delta modulator", Proceedings of the ECCTD 87, Paris, Sept. 1987.

[9]  E.I. Jury: "Sampled-data control systems", Robert E. Krieger Publishing Co., Huntington NY, 1977.

[10] J.E. Iwersen: "Calculated quantizing noise of single-integration delta modulation coders", Bell Syst. Tech. J., vol. 48, pp. 2359-2389, Sept. 1969.

# CIRCUIT AND TECHNOLOGY CONSIDERATIONS FOR MOS
# DELTA-SIGMA A/D CONVERTERS

*Max W. Hauser and Robert W. Brodersen*

Department of Electrical Engineering and Computer Science
University of California, Berkeley, CA 94720

## ABSTRACT

Delta-sigma, or one-bit oversample-and-decimate, analog-to-digital converters offer the promise of high-performance A/D interfaces in MOS IC technologies lacking high-quality analog components. Although the technique is potentially more fabrication-tolerant than other A/D methods, the MOS circuitry imposes performance limits not predicted in basic delta-sigma theory. Advantages of switched-capacitor delta-sigma conversion are discussed, as well as analysis and simulation of the effects of thermal noise, capacitor nonlinearity and operational-amplifier imperfections.

## 1. INTRODUCTION

In MOS IC technology, analog-to-digital converters based on delta-sigma modulation, or one-bit noise-shaping oversampling conversion [1-5], are becoming attractive for speech-processing, ISDN and other *signal-acquisition* applications. They promise high resolution without the component matching required of successive-approximation-type converters or the restricted speed of integrating converters. Moreover, delta-sigma A/D converters appear to fit future MOS fabrication trends: they consist largely of digital circuitry, and can therefore scale efficiently in area as digital MOS feature sizes shrink. Their few analog components can accommodate the large parasitics and imprecisely-defined element ratios associated with digital MOS technologies. Their resolution theoretically improves with oversampling ratio [1,2] and hence with circuit speed, implying that such converters should improve in performance as MOS transistors become smaller and consequently faster.

Yet recent experience with MOS delta-sigma A/D converters [3,4,5] has revealed performance constrained by analog-circuit imperfections, not oversampling ratio. Also, the analog circuitry, while consuming only a small fraction of total chip area, accounts for a large measure of design time. Although the system-level (z-transform) theory of delta-sigma converters has been elegantly developed by Candy [1], Agrawal and Shenoi [2], and others, to construct these systems effectively in MOS technology requires extending this theory to account for the particular circuit aberrations in the technology.

This paper examines some second-order effects and how they limit the performance of MOS delta-sigma A/D converters. Results of analysis and simulation bound the achievable converter SNR and the necessary op-amp and capacitor specifications. The preliminary sections discuss general advantages of oversampling converters in MOS technology, and one-bit (delta-sigma) converters in particular.

## 2. OVERSAMPLING CONVERTERS IN MOS

Figure 1 embraces a great many different forms of oversample-and-decimate* A/D converter, useful for signal-acquisition systems. Most of the oversampling converters demonstrated to date in IC form [3-6] are variations on Figure 1. Common to these different forms is a clocked feedback loop producing a coarse estimate that oscillates about the true value of the input, and a digital lowpass filter that averages this coarse estimate to obtain a finer approximation (accurate to more bits) at a lower (decimated) sampling rate.

Although trading off speed for resolution is a simple concept, the key to the utility of oversampling systems based on Figure 1 is that they enhance resolution faster than they reduce bandwidth. The feedback loop with integrating filter H(z) in Figure 1 forces the quantization error in the N-bit estimate to have a highpass spectrum, and therefore when the digital filter subsequently reduces the bandwidth by a certain factor, it excludes a more-than-linear proportion of the quantization noise. Typical theoretical signal-to-noise-ratio (SNR) enhancements, from the initial N-bit estimate, are [1]

$$\Delta SNR \ (dB) = 9L - 5.2 \qquad (1a)$$

$$\Delta SNR \ (dB) = 15L - 13 \qquad (1b)$$

where L is the number of *octaves* of oversampling, and the loop filter H(z) includes one (1a) or two (1b) integrations respectively.

Another feature of oversampling A/D circuits is that their digital lowpass filter can assume most of the input-antialias filtering burden, greatly relaxing the amount of analog prefiltering needed. They thus address a larger part of the analog-interface system than does a stand-alone successive-approximation A/D converter.

---

*The term "interpolative A/D" has also been used [8] since the averaging process interpolates between voltage levels of the N-bit estimate, but that term is now popularly identified with one particular form of multi-bit oversample-and-decimate converter developed by Candy and Wooley [6]

Reprinted from *IEEE Proc. ISCAS'86*, pp. 1310–1315, May 1986.

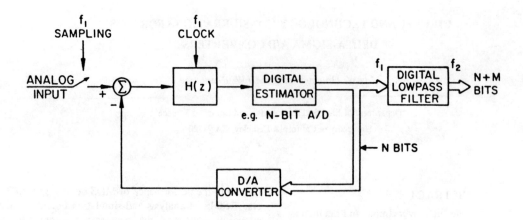

**Figure 1:** A basic oversample-and-decimate A/D converter. H(z) is an integrating filter.

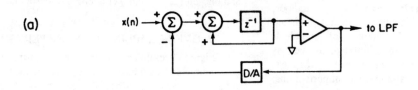

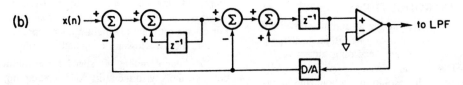

**Figure 2:** Typical delta-sigma modulators, using (a) one and (b) two integrators. The particular two-integrator configuration shown is that of Candy [1], with one delay in the forward path. The comparators are assumed memoryless; all delays are explicit in the $z^{-1}$ blocks.

Of special interest is the case N = 1 in Figure 1, where the coarse estimate and "D/A converter" have one-bit resolution. The loop of Figure 1 with N = 1 is a *delta-sigma modulator*, the term (originating in telemetry [7]) for systems that produce a clocked bit whose average value tracks an analog input. Delta-sigma converters, although requiring a larger oversampling factor than multi-bit versions to realize the same final resolution, admit a much wider variety of IC fabrication processes. Because they obtain resolution by interpolating linearly between two saturation levels, they do not require an array of precision binary elements for the D/A element in the feedback loop of Figure 1.

We remark that delta-sigma A/D converters have the further advantage of potentially unlimited SNR as oversampling factor increases. In contrast, multi-bit oversampling converters, such as earlier "codec" chips [6], employ an N-bit D/A converter in Figure 1, and rely on the input-output accuracy of the D/A converter for their ultimate linearity. Such multi-bit systems can display a *resolution* that is enhanced by oversampling, but their large-signal linearity (and hence, maximum SNR) is set by component matching in the D/A

converter, as with conventional A/D circuits. With a one-bit oversampling loop, the D/A element has only two output levels; uncertainty in those levels produces only a gain error or offset in the overall converter.

Figure 2 shows practical delta-sigma modulator configurations with both one and two integrators in the loop.

### 3. RATIO- AND PARASITIC-INSENSITIVITY

The MOS switched-capacitor integrator of Figure 3 can perform the discrete-time subtract-and-integrate function that makes up the loops of Figure 2. The configuration of Figure 3 is inherently insensitive to stray capacitances associated with the signal capacitors $C_1$ and $C_2$. We have constructed delta-sigma front ends in unmodified single-polysilicon digital MOS processes [3], using metal-to-polysilicon capacitors for $C_1$ and $C_2$. Such a capacitor structure exhibits a large parasitic capacitance from the bottom plate that may even considerably exceed the value of the signal-path capacitor. This does not appear to affect performance. Note that the *ratio accuracy* of $C_1/C_2$ is a second-order requirement, constrained by circuit stability to the order of 10% [1], and does not directly affect converter accuracy.

An incidental advantage of switched-capacitor circuits as the basis of an oversampling loop is that they, like the loop itself, are inherently discrete in time. After each clock transition, the node voltages settle to new values and stay there. While continuous-time integrators have also been used in delta-sigma loops [5], this causes the integrator gain in each clock cycle to be proportional to the clock period, effectively fixing the permissible clock rate. In a switched-capacitor topology the clock rate can be varied over a wide range to suit the application -- even in real time, such as for modem synchronization.

## 4. PERFORMANCE LIMITS DUE TO NOISE

Achievable dynamic range in monolithic A/D circuits is constrained by available signal swing at one extreme and noise sources at the other. Noise can arise from power-supply or substrate coupling, clock-signal feedthrough, and from thermal and 1/f noise generation in the MOS devices. Oversampling-converter front ends like that in Figure 1 are less sensitive to noise than they may first appear. Analog noise above the signal passband is blocked by the digital decimating filter, while in-band noise sources at points beyond the first stage of integration in the loop filter H(z) (for example, noise in the comparators in Figure 2) tend to be suppressed by the feedback action of the loop. The components of practical importance for circuit noise are the first op amp in H(z) and the MOS analog switches.

A lower bound on quantization noise in delta-sigma converters thus arises from low-frequency components of induced power-supply and substrate noise, primarily through the first op amp; 1/f noise in critical MOS transistors of the first op amp; and thermal noise in the op amp and MOS switches. While power-supply, substrate. and clock-feedthrough noise may be important, they depend heavily on layout and circuit topology, and can be mitigated with a fully-differential configuration. MOS 1/f noise is vulnerable to fabrication and possibly circuit measures, as with switched-capacitor frequency filters [9].

The thermal noise is a more fundamental problem. Because of thermal noise in the switch resistances, each voltage sampled onto a capacitance C exhibits an uncertainty of variance kT/C where k is Boltzmann's constant and T is absolute temperature. For an input circuit like Figure 3, the input capacitor sees two switch paths per clock cycle, for a noise variance of 2kT/C. The noise sources in the switches and op amp actually interact due to bandwidth effects, but

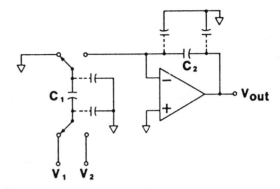

$$V_{out}(n)=V_{out}(n-1)+\frac{C_1}{C_2}[V_1(n-1)-V_2(n-1)]$$

**Figure 3:** A switched-capacitor integrator (accumulator). The expression holds even in the presence of large parasitic capacitors (dotted connections).

Castello [10] has established that a lower bound for the combined op-amp and switch thermal sources in this configuration is 2kT/C. If F is the factor by which the oversample-and-decimate process reduces signal bandwidth, the (uncorrelated) thermal noise variance falls by a similar amount. The maximum RMS value of a sinusoidal signal, within a power supply voltage $V_s$, is $V_s/(2\sqrt{2})$. Combining these factors gives a maximum signal-to-thermal-noise ratio for an oversampling front end based on Figure 3, expressed as an amplitude ratio, of

$$SNR_{MAX}=\frac{V_s}{4}\sqrt{\frac{FC}{kT}} \qquad (2)$$

($V_s$ is the total power-supply voltage, F is the oversampling factor, and C the magnitude of the input-sampling capacitor). Note that the result is independent of clock rate, for a given oversampling factor.

With a 5-volt power supply, 128:1 oversampling factor and 1-pF capacitors typical of contemporary designs, $SNR_{MAX}$ corresponds to about 17 bits, falling to 15 bits if the capacitors scale down to 0.1 pF.

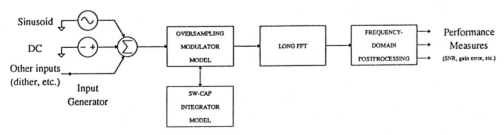

**Figure 4:** Software simulator for delta-sigma converters.

Extending the integrator of Figure 3 to a fully-differential form, based on a differential-output op amp, doubles the maximum signal swing but usually doubles the number of analog switches as well, affording a net SNR enhancement of only $\sqrt{2}$ (3 dB). The added complexity is justified, however, by greatly improved power-supply rejection.

## 5. SIMULATION

Because of the gross nonlinearity within delta-sigma loops, a comprehensive analytical description of them, useful for accurately predicting performance measures such as signal-to-noise ratio, has been elusive. The usual approach of analysis based on a linear model with additive quantization noise, although useful as a starting point, neglects just those peculiarities of delta-sigma loops that can determine ultimate performance.

An alternative is to solve the time-domain recursion equation for successive output samples, which is exact but must be repeated for each input waveform of interest. When done numerically, this amounts to a simulation of the system. We have used this technique to explore the effects of modifications to the delta-sigma loop.

Figure 4 shows a block diagram of the simulation software. Embedded in it is a switched-capacitor accumulator model that describes the transition of the node voltages from one clock cycle to the next, allowing for second-order effects such as nonlinear capacitors and incomplete settling. Copies of this accumulator are interconnected along with other elements to form an oversampling-modulator model, and this accepts input sources and parameters for each simulation. Some number of output samples from this modulator model, typically 100,000 or more, are expanded in a fast Fourier transform to complex frequency components. A long record of modulator output is necessary because the baseband signals chage very slowly with respect to the loop clock rate. Narrowband test signals such as one or more sinusoids can be collected in the frequency-domain output, and compared with the remainder of the baseband for an SNR measure.

A decimating filter can also be included after the modulator; however, analysis of the raw modulator output allows the modulator to be evaluated independently of any particular decimation filter, so that imperfections due to the modulator and filter remain distinct. We use some of the same software components evaluate actual modulators and A/D converters, in which case the circuit models are replaced by chips under test.

## 6. CAPACITOR LINEARITY

Figure 5 demonstrates how the SNR of a switched-capacitor delta-sigma converter can degrade when the capacitors display a voltage sensitivity. These curves were obtained from a discrete-time simulation of Candy's two-integrator delta-sigma modulator, Figure 2(b), using the integrator of Figure 3. According to these curves, the 2-10 ppm/V nonlinearity of good floating MOS capacitors permits a peak SNR equivalent to about 14 bits, but at 100 ppm/V this falls to the 12-bit level. This effect is due primarily to the input

capacitor of the first integrator. It has implications for the capacitor structures suitable for these A/D converters and may require further attention with future MOS fabrication technologies, whose thin oxide and oxide-nitride dielectrics experience increased electric fields and typically increased voltage coefficients.

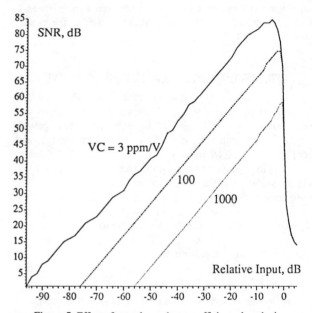

**Figure 5:** Effect of capacitor voltage coefficient when the integrator of Figure 3 is used in the loop of Figure 2(b). Input signal is a sinusoid with a frequency 0.1275 times the final sampling rate. 128:1 oversampling.

## 7. OP AMP REQUIREMENTS

In Figure 2, each comparator follows an integrating stage. A basic result of this is that the comparator's offset and noise are not as important as in other A/D configurations where comparators directly measure the analog input. The preceding integrator will adjust its output to the level required by the comparator's input threshold. Simple regenerative comparator circuits, which display excellent speed and resolution but poor DC offset, are suitable for delta-sigma A/D converters, and most of the analog design effort focuses on the op amp.

While op-amp settling time affects the maximum input clock rate, as in other switched-capacitor circuits [11], it can also limit the achievable SNR in oversampling converters. For a fixed output sampling-rate requirement, input clock rate sets the oversampling factor and hence the theoretical SNR of the A/D converter. However, simulations have shown that delta-sigma loops are extremely tolerant of *linear* settling error: allowing the op amp outputs to settle within only 10% of their final value in each clock cycle introduces negligible effect in the SNR curve at the 14-bit performance level. Practical speed limits are imposed by second-order effects rather than linear settling.

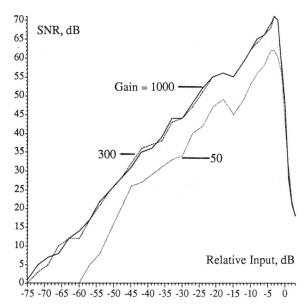

**Figure 6:** Effect of low op-amp open-loop gain in a dithered single-integrator delta-sigma converter. Input signal is a sinusoid with a frequency 0.1275 times the final sampling rate. Oversampling factor is 256.

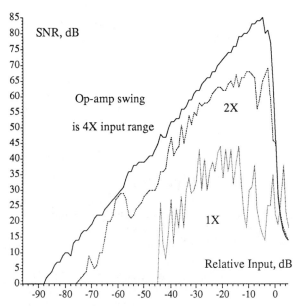

**Figure 7:** Effect of restricted op-amp swing. Input signal is a sinusoid with a frequency 0.1275 times the final sampling rate. Oversampling factor is 128.

Delta-sigma circuits can also give high resolution with low op-amp gain. The critical value for op-amp open-loop gain in a delta-sigma loop is approximately the oversampling factor, regardless of the number of integrators in the loop. Below this value of gain, in-band quantization noise rises rapidly. Even the simple single-integrator circuit of Figure 2(a) can maintain 12-bit resolution with an op amp gain as low as 300 (Figure 6). The A/D converter's *gain accuracy* and *DC offset* suffer, but resolution does not.

Limited op amp output swing can constrain converter input voltage range in two-integrator delta-sigma circuits. Stability in the loop dictates the product of gain constants in the integrators [1], and this means that the capacitor ratios that define these gains cannot be scaled at will to adjust signal swing at the op amp outputs. For Figure 2(b), the signals traversing the $z^{-1}$ delay elements correspond to op-amp output swings. The range of these voltage swings is approximately four times the value fed back by the D/A element, which is to say four times the range of the input signal x(n). Figure 7 shows the impaired performance that results when the op amps saturate at a lower value.

This phenomenon identifies a possible utility of the single-integrator delta-sigma converters of Figure 2(a): although yielding less resolution for a given oversampling factor than two-integrator versions, and requiring a dither signal [3] to avoid pathological quantization noise, they have the property that the integrator gain can be scaled at will, which can be exploited for rail-to-rail analog-input conversion capability.

## 8. CONCLUSIONS

The SNR of one-bit oversample-and-decimate A/D converters is not limited by component matching in the analog circuitry, unlike conventional A/D designs and oversampling converters based on a multi-bit initial conversion. This and the parasitic insensitivity of switched-capacitor integrators permits on-chip A/D conversion at high resolution (12 bits or better) even in inexpensive digital MOS fabrication processes that provide poor analog matching tolerances. Test results with an IC implementation of a one-bit oversampling converter confirm that op-amp-based switched-capacitor integrators allow a capacitor structure with a large parasitic component, such as metal-to-polysilicon. A switched-capacitor topology permits a flexible sampling rate as well, since the analog circuitry, like the converter itself, operates in discrete time.

The critical element to be designed in the oversampling converter is the MOS op amp, since it is both the slowest analog component and the one most vital to conversion accuracy. The design of the op amp, typically a single-gain-stage CMOS amplifier, requires a direct tradeoff between speed, gain and output swing. Simulation results suggest that open-loop gain may be sacrificed for output swing in order to maximize performance. A two-integrator delta-sigma loop, although nominally superior to a single-integrator loop in A/D performance, is more demanding of op-amp output swing since its input range is tied to the op-amp swing through a stability constraint.

Contrary to predictions from basic delta-sigma theory, the resolution of the A/D converter cannot always be improved by increasing the oversampling factor. Capacitor linearity imposes a technology-dependent practical limit on resolution. Thermal noise imposes a fundamental limit that gets worse as the capacitors scale down; this may dictate deliberately large capacitors and hence a restricted operating rate, since the converter input operates at a much higher speed than non-oversampling A/D circuits and can be limited by capacitor charging time.

## REFERENCES

[1]   James C. Candy, "A Use of Double Integration in Sigma Delta Modulation," *IEEE Transactions on Communications* vol. COM-33 no.3 pp. 249-258, March 1985.

[2]   Bhagwati P. Agrawal and Kishan Shenoi, "Design Methodology for $\Delta\Sigma M$." *IEEE Transactions on Communications* vol. COM-31 no. 3 pp. 360-370, March 1983.

[3]   Max W. Hauser, Paul J. Hurst and Robert W. Brodersen, "MOS ADC-filter Combination That Does Not Require Precision Analog Components," *1985 IEEE International Solid-State Circuits Conference Digest of Technical Papers* pp. 80-81, February 1985.

[4]   P. Defraeye *et al.*, "A 3-$\mu m$ CMOS Digital Codec with Programmable Echo Cancellation and Gain Setting," *IEEE Journal of Solid-State Circuits* vol. SC-20 no. 3 pp. 679-687, June 1985.

[5]   Rudolf Koch and Bernd Heise, "A 120kHz Sigma/Delta A/D Converter," *1986 IEEE International Solid-State Circuits Conference Digest of Technical Papers* pp. 138-139, February 1986.

[6]   James C. Candy, William H. Ninke and Bruce A. Wooley, "A Per-channel A/D Converter Having 15-segment $\mu$-255 Companding," *IEEE Transactions on Communications* vol. COM-24 no. 1 pp. 33-42, January 1976.

[7]   H. Inose, Y. Yasuda and J. Murakami, "A Telemetering System by Code Modulation -- $\Delta$-$\Sigma$ modulation," *IRE Transactions on Space Electronics and Telemetry* vol. SET-8 pp. 204-209, September 1962.

[8]   James C. Candy, "A Use of Limit Cycle Oscillations to Obtain Robust Analog-to-Digital Converters," *IEEE Transactions on Communications* vol. COM-22 no. 3 pp. 298-305, March 1974.

[9]   Rinaldo Castello and Paul R. Gray, "Performance Limitations in Switched-Capacitor Filters," *IEEE Transactions on Circuits and Systems* vol. CAS-32 no. 9 pp. 865-876, September 1985.

[10]  Rinaldo Castello, unpublished work.

[11]  David J. Allstot and William C. Black, "Technological Design Considerations for Monolithic MOS Switched-capacitor Filtering Systems," *Proceedings of the IEEE* vol. 71 no. 8 pp. 967-986, August 1983.

# TECHNOLOGY SCALING AND PERFORMANCE LIMITATIONS
# IN DELTA-SIGMA ANALOG-DIGITAL CONVERTERS

*Max W. Hauser*

School of Electrical Engineering
Cornell University
Ithaca, New York 14853

## ABSTRACT

Simple circuit models illuminate the underlying electrical tradeoffs between speed, resolution, power-supply voltage and fabrication technology when switched-capacitor delta-sigma A/D converters are realized using field-effect devices. These constraints are generic to oversampling A/D converters with switched-capacitor-integrator front ends, transcending the choice of modulator block diagram and even the type of FET technology (MOSFET, MESFET, *etc.*).

## 1. INTRODUCTION

A delta-sigma analog-to-digital converter consists of a digital decimating filter following a delta-sigma modulator (DSM). Because they are based on amplitude interpolation from a one-bit quantization, these converters exhibit analog resolution decoupled from element matching, and they also can be realized, using switched-capacitor circuitry, in mainstream digital MOS-VLSI technologies [1]. In recent years, electrical performance of these A/D converters has improved steadily due to technology scaling and to innovative circuit design, which presently achieves op amps of some 100 MHz unity-gain bandwidth [2] and DSMs exceeding 110 dB resolution [3] with relatively conservative MOS feature sizes.

However, designers of switched-capacitor delta-sigma A/D converters confront basic electrical tradeoffs when they seek to improve speed and resolution or to exploit scaled technologies. Faster sampling rates require greater capacitor-charging current from amplifier FETs, or else smaller capacitors; smaller capacitors limit dynamic range above the thermal ($kT$) noise floor; the noise-shaping/decimation process of delta-sigma A/D conversion interacts fundamentally with these factors in ways absent from purely-analog switched-capacitor circuits. This paper examines these tradeoffs. Section 2 identifies essential analog circuitry in switched-capacitor delta-sigma A/D converters and a representative circuit model; Section 3 develops simple bounds relating SNR, time constants, capacitor size and transistor feature size; Section 4 incorporates delta-sigma system theory with these results.

## 2. CORE CIRCUITRY

Many configurations for DSMs (that is, oversampling, noise-shaping, one-bit coders) are now in use, special proprietary topologies [4-6] as well as updated implementations [7-9] of the original first- and second-order "delta-sigma modulators" of Inose and Yasuda [10,11]. Pervading these many variations are a subtracting-integrator input structure and a voltage comparator (Figure 1).

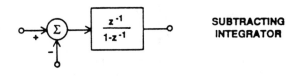

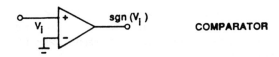

**Figure 1:** Essential subcircuits for a delta-sigma modulator.

Such an integrator operates on the analog input in a DSM, and therefore its analog aberrations will reflect directly to the A/D input. For linearity and parasitic insensitivity, the integrator is usually based on an op amp as in Figure 2, which captures essential characteristics of a wide class of circuits. The analog switches are field-effect devices of some kind, and the analysis may be easily modified for fully-differential or other variations. The circuit of Figure 2 leads to simple relations between dynamic range, time constants and technology parameters, independent of the specific topology of the rest of the modulator.

## 3. ANALOG CONSTRAINTS

**Dynamic range** in a switched-capacitor input circuit is limited most fundamentally by the thermal noise sampled onto the capacitors along with signal [12]. A white noise source, perfectly decimated by a factor D, falls in RMS value by a factor of $\sqrt{D}$. For the input circuit of Figure 2, with sinusoidal input-signal statistics, this leads to an upper bound on delta-sigma A/D RMS-signal to RMS-noise ratio of [1]

$$SNR \leq \frac{V_{SUP}}{4}\sqrt{\frac{DC_S}{kT}} \tag{1}$$

Reprinted from *IEEE Proc. ISCAS'90*, pp. 356–359, May 1990.

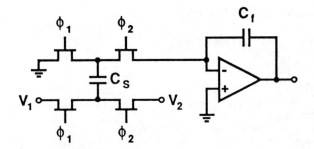

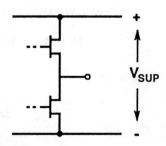

**Figure 2:** Switched-capacitor integrator prototype showing generic field-effect switches.

**Figure 3:** Minimal op-amp output circuit model with generic field-effect devices, not necessarily complementary.

where $V_{SUP}$ is the *total* power-supply voltage.

**Capacitor slewing time** originates in the finite current available from the amplifiers. If this current is $I_{PK}$ then the minimum time for a full-scale transition of the voltage on $C_S$ is

$$\tau_{SLEW} = \frac{C_S V_{SUP}}{I_{PK}} \qquad (2)$$

A lower limit on $\tau_{SLEW}$ may be derived for a specific amplifier circuit. For the minimal output stage of Figure 3, the output current is limited by the FET drain current at maximum gate-source voltage, regardless of the specific circuitry that drives the FET gates. Assuming that it is driven into velocity-saturated current flow, the maximum drain-current magnitude from a field-effect device can be expressed generically as [13]

$$I_D \leq \left[ \frac{v_l}{2L} \right] C_{IN} (V_{GS} - V_T) \qquad (3)$$

where $v_l$ is the limiting velocity of carriers in the channel, $L$ the length of the channel region, $C_{IN}$ the gate-to-channel capacitance, $V_{GS}$ the gate-to-source voltage, and $V_T$ the threshold voltage. With a maximum gate drive of $(V_{GS} - V_T) = V_{SUP}/2$, combining (3) and (2) gives

$$\tau_{SLEW} \geq 4 \left[ \frac{L}{v_l} \right] \frac{C_S}{C_{OA}} \qquad (4)$$

where now $C_{OA}$ denotes the input capacitance of the output-stage device. While in principle it might be possible to design with circuit capacitors $C_S$ much smaller than device capacitances, this would not in fact reduce the limiting value of $\tau_{SLEW}$ below $4L/v_l$ since the charging problem would just devolve to the circuit driving $C_{OA}$.

The **capacitor-switch time constant** arises from $C_S$ and the finite "on" conductance of the switch devices of Figure 2. This may be made arbitrarily small by scaling the device widths, but the parasitic gate-to-channel capacitance in the switches will scale accordingly. Therefore it is convenient again to express this time constant in a normalized form dependent on an extrinsic-intrinsic capacitance ratio.

$C_S$ charges nonlinearly after a clock transition even if the op amp is ideal, because switch channel conductance is

nonlinear. However as $C_S$ charges and $V_{DS}$ falls for each device in series with $C_S$, this conductance approaches a small-signal, ohmic value $g_{ds}$, and the time constant

$$\tau_{SW} = \frac{2C_S}{g_{ds}} \qquad (5)$$

describes the exponential charging.

A lower limit on $\tau_{SW}$ depends on the FET parameters. This situation entails no channel pinchoff or velocity saturation; classic prepinchoff FET characteristics around $V_{DS} = 0$ can be arranged to the form, analogous to (3),

$$g_{ds} = \left[ \frac{\mu}{L^2} \right] C_{IN} (V_{GS} - V_T) \qquad (6)$$

where $\mu$ is the carrier mobility in the device's channel and again $C_{IN}$ is the gate-to-channel capacitance. Single-channel FET switches exhibit a restricted analog signal range but assuming a favorable drive voltage $(V_{GS} - V_T) = V_{SUP}/2$, (5) and (6) imply

$$\tau_{SW} \geq \left[ \frac{4L^2}{\mu V_{SUP}} \right] \frac{C_S}{C_{SW}} \qquad (7)$$

where $C_{SW}$ is the renamed $C_{IN}$ value for the *switch* devices.

The remaining time constant characterizing Figure 2 embodies the **finite small-signal bandwidth** of the op amp. The simplest op amp, with single-pole response and a loaded unity-gain angular frequency of $\omega_U$, will exhibit a closed-loop settling time constant of $\tau_U = 1/\omega_U$. The magnitude of $\omega_U$ depends heavily on the details of op-amp design; an idealized value follows from the limits on a single transistor. From (3), the limiting value of small-signal transconductance $g_m$ can be expressed as

$$g_m \leq \frac{C_{IN} v_l}{2L} \qquad (8)$$

Driving a total load capacitance $C_L$, this device will yield a small-signal bandwidth

$$\omega_U = \frac{g_m}{C_L} \qquad (9)$$

Therefore,

$$\tau_U \geq 2 \left[ \frac{L}{v_l} \right] \frac{C_L}{C_{OA}} \qquad (10)$$

220

where again $C_{IN}$ has been renamed $C_{OA}$ for an op-amp device.

**Comparator** design, crucial in other (non-oversampling) types of A/D converters, is a minor issue in DSMs because these are extremely robust against most comparator imperfections, especially DC offset [1,7]. Therefore simple regenerative comparators suffice. Two cross-coupled field-effect devices obeying (8) will realize a right-half-plane real pole, causing exponential growth of an initial voltage-imbalance input, with growth time constant $\tau_1 = C_{IN}/g_m$. The total time required for a desired resolvable signal of $V_{RES}$ to grow to a logic level of $V_{LOGIC}$ is then

$$\tau_{COMP} = \tau_1 \ln \frac{V_{LOGIC}}{V_{RES}} \qquad (11)$$

and using (8), the bound on this time is

$$\tau_{COMP} \geq 2 \left[ \frac{L}{v_l} \right] \ln \frac{V_{LOGIC}}{V_{RES}} \qquad (12)$$

Boser [7] has reported successful DSMs with the ratio $V_{LOGIC}/V_{RES}$ on the order of 10.

The bracketed quantities in (4), (7), (10), and (12) are all measures of the channel transit time, which is the fundamental speed limit in a field-effect device and is the factor that constrains logic speed also [13]. This factor directly couples the analog speed bounds to the device fabrication technology. The capacitance ratios in (4), (7), and (10) relate explicit circuit capacitances to device capacitances, and will depend on other details of the circuit design that determine its immunity to the parasitic coupling, nonlinearity, and phase shift that the device capacitances introduce. These ratios will typically be in the range 10-1000, and cannot usefully be reduced below unity since other effects come into play and prevent the time constants from falling accordingly.

The $\tau_{SLEW}$ bound is a hard analog-speed constraint whereas, at least for first- and second-order DSMs, $\tau_{SW}$ and $\tau_U$ represent "soft" limits. $\tau_{SW}$ and $\tau_U$ are exponential-settling-time constants and the number of these that the circuit requires depends on other details; in particular it has been demonstrated that first- and second-order DSMs are insensitive to *linear* integrator errors such as finite op-amp DC gain and incomplete op-amp settling [1,7]. The comparator time constraint $\tau_{COMP}$ is normally small compared with $\tau_{SLEW}$.

It must be emphasized that these are technology-limited bounds under optimal conditions; second-order effects or more complex circuits can make the slewing and settling times worse. It is difficult to generalize about analog-versus-digital speed differences, but the multiplicands of transit time in (4), (7), and (10), and allowance for multiples of $\tau_{SW}$ and $\tau_U$, are sufficient to account for the frequent practical observation that well-designed MOS switched-capacitor circuitry clocks about an order of magnitude slower than digital logic in the same technology.

For n-channel silicon MOS devices, a nominal $v_l$ of $5 \times 10^{10}$ μm/sec implies that the bracketed factor in (4) is 20 psec with 1-μm effective channels (the bracketed factor in (7)

has similar magnitude when $V_{SUP} = 5V$). A safety factor $C_S/C_{OA}$ of 100 then yields an absolute minimum $\tau_{SLEW}$ of 8 nsec, and *this factor alone* would limit the sampling rate to 60 MHz in a two-phase system. This is a workable but not a comfortable limit in 1-micron technology.

## 4. MIXED ANALOG-DIGITAL CONSTRAINTS

The overall resolution and the decimating factor "D" in a delta-sigma A/D converter are also linked, independent of implementation, by the system-level theory of noise shaping, although the specific relationship depends on the modulator's topology. For example, when the modulator is a classic second-order DSM, the maximum signal-to-error ratio (loosely called SNR), in amplitude terms, obeys the empirical relation [14]

$$SNR_{MAX} \approx (0.11)D^{5/2} \qquad (13)$$

This is the peak A/D SNR with sinusoidal input if the modulator implementation and decimation are essentially ideal, and is about 86 dB (equivalent to a 14-bit linear A/D) when D = 128 (note that the widely-quoted linear-loop model for DSMs overestimates this SNR by some 14 dB [14]).

Combining (13) with the thermal-noise result of (1) yields a constraint on the **magnitude of the sampling capacitor** $C_S$ in order to preserve the full SNR capability of the modulator:

$$C_S \geq \frac{[16(0.11)^{2/5}kT]}{V_{SUP}^2}(SNR_{MAX})^{8/5} \qquad (14)$$

Similar expressions can be derived for other modulator topologies. At 70° C the bracketed quantity in (14) is $3.1 \times 10^{-20}$ J. With $V_{SUP} = 5$ V, this minimum capacitor value is only 3 fF for 80 dB maximum SNR ($D \approx 96$), but rises rapidly to 5 pF for an extreme SNR of 120 dB ($D \approx 600$).

Unfortunately the trend of (14) directly opposes the use of scaled technologies for high resolution. The lower bounds on analog cycle time (Section 3) scale roughly linearly with lateral dimensions and this suggests achieving a higher decimating ratio D at a given A/D output sampling rate. But the rapid rise of required capacitor value (as roughly the *square* of peak SNR, and this corresponds, through (13), to the *fourth power* of D) eventually intersects this speed improvement and may indeed compel *lower* A/D SNRs if the circuit capacitors are to scale along with the transistors. This does not preclude using scaled technologies in delta-sigma A/D conversion for enhanced output sampling rate.

## 5. AREA AND POWER

Castello and Gray [12] examined switched-capacitor linear filters and developed absolute-minimum bounds on power consumption and die area based on the capacitors. Some important differences exist, however, between DSMs and linear analog filters.

1. The DSM is associated with a large digital filter, which in contemporary and projected designs normally dominates the power consumption, and also dominates the die

area unless thick-oxide capacitors are used (in which case the capacitor area is deliberately nonminimum). The digital filter is also subject to minimum constraints, a separate topic [14].

2. The number of integrators in a DSM is modest, and unlike the integrators in linear filters, $C_F$ has similar magnitude to $C_S$ since DSM integrators are designed typically for unity-gain frequencies on the order of the clock rate. These factors account for the relatively small total capacitor area in DSMs; however it can still be linked to dynamic range through (14) and related expressions.

3. Signals entering the integrators in a DSM are not linearly related to the A/D input, owing to the large quantized feedback component, and this complicates calculation of the power consumed to charge capacitors.

## 6. CONCLUSIONS

Generic field-effect device characteristics applied to subcircuits that are sufficient to form a delta-sigma modulator suggest an absence of *fundamental* factors preventing analog clock rates in these circuits from continuing to increase with shorter-channel FET technologies. Four critical analog time constants, $\tau_{SLEW}$, $\tau_{SW}$, $\tau_U$, and $\tau_{COMP}$, are identified; these scale approximately linearly with lateral dimensions in the transistors. However, interaction of FET-channel thermal noise with the oversampling-decimation process imposes a minimum absolute circuit-capacitor size, independent of clock rate, for a particular modulator topology and SNR target. This capacitor requirement directly opposes the use of the higher speed of scaled technologies to achieve extreme resolution through the oversampling process. This capacitor bound also becomes more severe as the power-supply voltage decreases.

## REFERENCES

[1] Max W. Hauser and Robert W. Brodersen, "Circuit and Technology Considerations for MOS Delta-Sigma A/D Converters," *Proceedings of the 1986 IEEE International Symposium on Circuits and Systems* pp. 1310-1315, May 1986.

[2] K. Bult and G. Geelen, "A Fast-Settling CMOS Opamp with 90dB DC Gain and 116 MHz Unity-Gain Frequency," to appear in *1990 IEEE International Solid-State Circuits Conference Digest of Technical Papers*, February 1990.

[3] B. Del Signore *et al.* "A 20b Delta-Sigma A/D Converter," to appear in *1990 IEEE International Solid-State Circuits Conference Digest of Technical Papers*, February 1990.

[4] Kevin L. Kloker, Brett L. Lindsley and Charles D. Thompson, "VLSI Architectures for Digital Audio Signal Processing," Conference paper, Audio Engineering Society Seventh International Conference, May 1989.

[5] E. C. Dijkmans and P. J. A. Naus, "The Next Step Towards Ideal A/D and D/A Converters," Conference paper, Audio Engineering Society Seventh International Conference, May 1989.

[6] D. R. Welland *et al.*, "A Stereo 16-Bit Delta-Sigma A/D Converter for Digital Audio," *Journal of the Audio Engineering Society* vol. 37 no. 6 pp. 476-486, June 1989.

[7] Bernhard E. Boser and Bruce A. Wooley, "Design of a CMOS Second-Order Sigma-Delta Modulator," *1988 IEEE International Solid-State Circuits Conference Digest of Technical Papers* pp. 258-259, February 1988.

[8] Steven R. Norsworthy, Irving G. Post and H. Scott Fetterman, "A 14-bit 80-kHz Sigma-Delta A/D Converter: Modeling, Design, and Performance Evaluation," *IEEE Journal of Solid-State Circuits* vol. 24 no. 2 pp. 256-266, April 1989.

[9] Max W. Hauser, Paul J. Hurst and Robert W. Brodersen, "MOS ADC-filter Combination That Does Not Require Precision Analog Components," *1985 IEEE International Solid-State Circuits Conference Digest of Technical Papers* pp. 80-81, February 1985.

[10] H. Inose, Y. Yasuda and J. Murakami, "A Telemetering System by Code Modulation -- $\Delta - \Sigma$ modulation," *IRE Transactions on Space Electronics and Telemetry* vol. SET-8 pp. 204-209, September 1962. Reprinted in Jayant, editor, *Waveform Quantization and Coding*, IEEE Press, 1976.

[11] Hiroshi Inose and Yasuhiko Yasuda, "A Unity Bit Coding Method by Negative Feedback," *Proceedings of the IEEE* vol. 51 no. 11 pp. 1524-1535, November 1963.

[12] Rinaldo Castello and Paul R. Gray, "Performance Limitations in Switched-Capacitor Filters," *IEEE Transactions on Circuits and Systems* vol. CAS-32 no. 9 pp. 865-876, September 1985.

[13] James A. Cooper, Jr, "Limitations on the Performance of Field-Effect Devices for Logic Applications," *Proceedings of the IEEE* vol. 69 no. 2 pp. 226-231, February 1981.

[14] Max W. Hauser and Robert W. Brodersen, "Monolithic Decimation Filtering for Custom Delta-Sigma A/D Converters," *Proceedings of the 1988 IEEE International Conference on Acoustics, Speech and Signal Processing* pp. 2005-2008, April 1988.

# DELTA-SIGMA A/Ds WITH REDUCED
# SENSITIVITY TO OP AMP NOISE AND GAIN

## Paul J. Hurst and Roger A. Levinson

### DEPT. OF ELECTRICAL ENGINEERING AND COMPUTER SCIENCE
### UNIVERSITY OF CALIFORNIA
### DAVIS, CA. 95616

## ABSTRACT

Circuit noise and limited op amp gain are two major factors limiting the performance of integrated CMOS switched-capacitor delta-sigma modulators. New switching schemes which compensate for these limitations will be described.

## I. INTRODUCTION

In recent years, delta-sigma modulation has gained much interest as a technique for analog-to-digital (and digital-to-analog) conversion [1,2]. As an oversampling technique, delta-sigma modulation has several important advantages over common methods of conversion such as successive approximation, flash and serial. In theory, delta-sigma modulation can provide unlimited signal-to-noise ratio, or equivalently, an unlimited number of bits of resolution, by increasing the oversampling factor. Delta-sigma modulation requires no matched components and is insensitive to circuit offsets. Since the system is oversampled, minimal pre-filtering of the input is required. For these reasons, delta-sigma modulation suits the trend toward analog/digital integration on a single chip, and its performance will improve as process technology provides smaller, faster transistors.

In the pursuit of unlimited signal-to-noise ratio, three fundamental problems arise. First, increased signal-to-noise ratio requires an increase in the sampling rate, which is limited by op amp settling time. Second, circuit noise due to the op amps and the switches sets a noise floor in the signal band. And third, the op amp's finite gain limits the quantization noise shaping. In this paper, switched-capacitor techniques for increasing circuit speed, minimizing low-frequency circuit noise, and relaxing op amp gain requirements are presented.

## II. DSM BACKGROUND

The fundamental concepts involved in delta-sigma modulation are quantization noise shaping and oversampling. Noise shaping can be explained by referring to the second-order delta-sigma modulator (DSM) and its block diagram in Figures 1 and 2. In Figure 2, the comparator is linearized. It is modeled as an element that adds quantization noise to its input to produce its output.

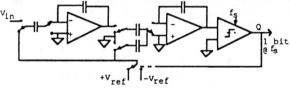

**Figure 1:** Second-order switched-capacitor DSM

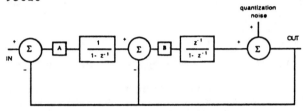

**Figure 2:** Linearized second-order DSM block diagram

For this DSM, the transfer function from the signal input to the output is

$$H_S(z) = \frac{OUT}{IN} = \frac{AB_z^{-1}}{1 + (AB + B-2)z^{-1} + (1-B)z^{-1}} . \qquad (1)$$

And the transfer function from the quantization noise to the output is

$$H_N(z) = \frac{OUT}{QUANT.NOISE} = \frac{(1-z^{-1})^2}{1 + (AB + B-2)z^{-1} + (1-B)z^{-2}} . \qquad (2)$$

The signal transfer function $H_S(z)$ is low-pass; the noise transfer function $H_N(z)$ is high-pass.

The noise power in the baseband for a second-order delta-sigma modulator is given by [2]

$$\sigma_N^2 = 0.2\ \pi^2\ \sigma_e^2 \left(\frac{2f_c}{f_s}\right)^5 \qquad (3)$$

where $f_s$ is the DSM sampling frequency, $f_c$ is the input signal bandwidth, $f_s/2f_c$ is the oversampling factor, and $\sigma_e^2$ is the quantization noise variance [1]. Using the above expression and scaling it with respect to empirical data yields the following equation for maximum signal-to-noise ratio (SNR) [2]:

$$SNR_{max} \cong 15N - 19\ dB \ \text{ where } N = \log_2 (f_s/2f_c). \qquad (4)$$

These results indicate that the SNR can be increased without limit by simply increasing the oversampling factor. However, a delta-sigma modulator implementation limits the achievable SNR by limiting the sampling frequency and introducing circuit noise. The circuit elements that cause the most problems are the operational amplifiers. In a CMOS process, high-gain, wide-bandwidth, low-noise operational amplifiers are not easily realized. Therefore, techniques must be found to relax the op amp specifications.

## III. PERFORMANCE LIMITATIONS DUE TO CIRCUIT NON-IDEALITIES

Since the thermal noise in the switches, kT/C noise, is inversely proportional to the capacitance associated with each switch, it can be made arbitrarily small by choosing a sufficiently large capacitor value [2].

This work was supported by the UC Micro Program, co-sponsored by Silicon Systems Inc.

Reprinted from *IEEE Proc. ISCAS'89*, pp. 254–257, May 1989.

In MOS op amps there are two major sources of noise, thermal noise and 1/f noise. However, at low frequencies (e.g. the audio band), the 1/f noise source dominates and becomes the limiting factor in achievable signal-to-noise ratio [2,3].

One major concern in the design of the analog delta-sigma loop is the effect of finite op amp gain on the signal-to-noise ratio. In a DSM, the integrators are in the feedback path for the transfer function from the quantization noise to the output. Therefore, the integrators' dc poles cause dc zeroes for the quantization noise transfer function (Equation 2). If the op amps have finite gain, the dc gain of the integrators will be limited to the op amp gain, and therefore the quantization noise shaping will be degraded. Previous work has shown that the op amp gain must be at least equal to the oversampling factor so as to introduce a negligible amount of noise in the baseband [2].

## IV. LOW-FREQUENCY NOISE COMPENSATION

One method for reducing the effect of low-frequency op amp noise is correlated double sampling, implemented in switched-capacitor circuitry with the auto-zeroing technique [4,5]. Figure 3 shows an implementation using auto-zeroed integrators in a delta-sigma modulator loop with the comparator replaced by it's linearized model. The transfer function from $v_{noise1}$ to the output of the DSM is

$$\frac{V_{out}(z)}{V_{noise1}(z)} = \frac{2(1-z^{-0.5})(1-z^{-1}-0.5z^{-0.5})}{3z^{-2}-6z^{-1}+4} \quad \text{for} \quad \frac{C_{in1}}{C_{f1}} = \frac{C_{in2}}{C_{f2}} = 0.5. \quad (5)$$

From the above transfer function, one can see that the input-referred noise of the first op amp is high-pass filtered. Equation 5 was verified by the SWITCAP [6] simulation shown in Figure 4.

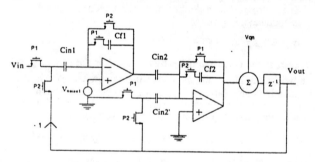

**Figure 3:** Second-order DSM with auto-zeroed integrators and linearized comparator

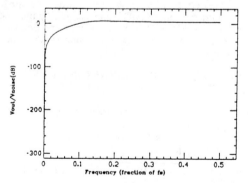

**Figure 4:** Transfer function of op amp input-referred noise in the auto-zeroed DSM with op amp gain=1000

## V. GAIN COMPENSATION

From an analysis of a standard switched-capacitor integrator, one concludes that the finite dc gain of the integrator (due to the finite op amp gain) results from a voltage change at the negative input of the op amp when the output changes. Therefore, the goal in designing a switched-capacitor integrator that is gain insensitive is to maintain a constant voltage at the negative op amp input node, independent of the output voltage.

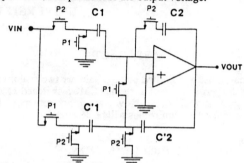

**Figure 5:** Gain-compensated integrator, C'1=2C1 and C'2=C2

An integrator that includes compensation for finite op amp gain was presented by Larson and Temes [7] and is shown in Figure 5. The transfer function for this integrator in the z-domain is

$$H(z) = \frac{-\frac{C1}{C2}[1 - \frac{\mu C1[z^{-1}-2z^{-0.5}]}{C2[1+\mu+3\mu(C1/C2)]}]}{1+\mu+\mu(\frac{C1}{C2})-z^{-1}[1+\mu+\mu(\frac{C1}{C2})(\frac{1+\mu(C1/C2)}{1+\mu+3\mu(C1/C2)})]} \quad , (6)$$

where $\mu$ is the inverse of the op amp gain. The dc gain of the compensated integrator is

$$H(z=1) = -\mu^{-2}\frac{1+\mu+4\mu(C1/C2)}{1+2(C1/C2)} . \quad (7)$$

The above equation shows that this switched-capacitor integrator will never achieve the ideal goal of infinite dc gain, but it can approach a dc gain equal to the negative of the op amp gain squared. It performs gain compensation by anticipating the output on Phase 1 and then giving the actual output on Phase 2, thereby causing the voltage at the negative input of the op amp to remain approximately constant from Phase1 to Phase2.

This gain-compensated integrator works well under the assumption that the input is moving slowly compared to the sampling rate. However, if the input is changing as fast as the sampling rate, $V_{out}$ on Phase 1 would not be a good approximation of $V_{out}$ on Phase 2. There are two integrator input signals in a second-order DSM that are changing significantly from one clock cycle to the next: the feedback signal and the output of the first integrator. Since DSMs are oversampled converters, the input signal is assumed to be a slow moving signal and therefore does not pose a problem. This gain-compensated integrator should be used for both integrators in a second-order DSM, and therefore a method for compensating for the fast moving integrator inputs is required.

In order to provide gain compensation when the integrator has a fast moving input signal, $V_x$, a half clock cycle delayed version of that signal, $V_x(n-0.5)$, is required as shown in Figure 6. Note that C'1 of Figure 5 has been split into two equal capacitors C"1 and C"'1. On Phase 1, the charge on C1 transfers onto C"'1 instead of C'2 and therefore does not contribute to the output voltage. This delayed input, $V_x(n-0.5)$,

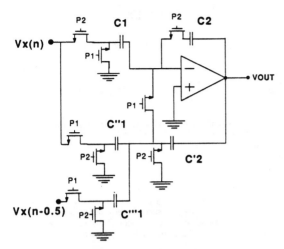

**Figure 6:** Gain-compensated integrator with half clock cycle input delay, C"1=C"'1=C1 and C'2=C2

is easily generated for the first integrator in the DSM since its fast moving input signal $V_x$ is the digital feedback signal. However, it is more complicated to implement for the second integrator, since one of its fast moving inputs is an analog signal - the output of the first integrator. The circuit in Figure 7 is an analog half-clock-cycle delay which is theoretically parasitic insensitive as long as the parasitics at node 1 are identical to those at node 2 [8]. This scheme was used to delay the output of the first integrator feeding the second integrator.

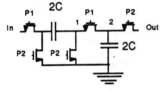

**Figure 7:** Half clock cycle delay; Out connects to an op amp virtual ground

The complete second-order DSM utilizing these gain-compensated integrators is shown (with a linearized comparator model) in Figure 8. Notice that the output of the first integrator on Phase 1 is used as the input to the second integrator during its anticipation cycle. This will introduce a small error in the second integrator since the output of the first integrator on Phase 1 is only an approximation of its actual output on Phase 2. Transient analyses using HSPICE were run on the final DSM implementation including the comparator (see Figure 11). Figure 9 shows a portion of the transient analysis output of the second integrator. Notice that the anticipation on Phase 1 is quite good.

Another benefit of this gain-compensated integrator is that there is a correlated double sampling effect because the input capacitors are always connected to the op amp inverting input. This reduces the effect of the op amp input referred noise as discussed in Section IV.

## VI. IC RESULTS

Test ICs containing the DSMs shown in Figures 1 , 10, and 11 were fabricated in a 3 μm digital CMOS process. Class AB op amps were used in the DSMs; a low-gain, higher-speed version was used in the gain-compensated circuit. Results from preliminary tests show that the delta-sigma modulators presented are functional, but design and modelling errors combined with widely varying process parameters severely limited the op amp

performance, causing the op amp settling time to be much longer than expected. The maximum SNR measured for each DSM with a sampling clock rate of 100 kHz and an oversampling factor of 100 is presented in Table 1. Figure 12 shows SNR versus input amplitude for an auto-zeroed DSM. SNR was measured by performing an FFT on the output signal and assuming an ideal decimation low-pass filter.

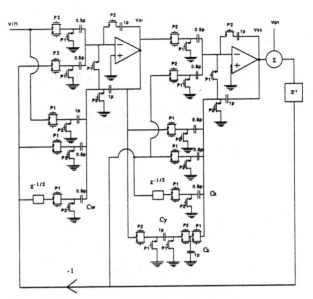

**Figure 8:** Gain-compensated DSM (shown with a linearized comparator), the 1/2 clock period delays are implemented digitally

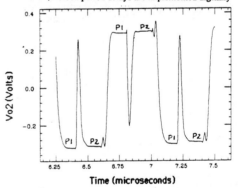

**Figure 9:** Transient analysis of gain-compensated DSM run on HSPICE showing second integrator output. Phase 1 is from 6.2 to 6.4 μs, and Phase 2 is from 6.4 to 6.6 μs, etc.

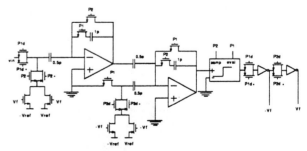

**Figure 10:** Complete auto-zeroed DSM schematic (P1* = P1 inverted, P1d = P1 delayed)

| DSM | Max. SNR | Op Amp Gain |
|---|---|---|
| Standard (Fig. 1 ) | 72dB | 180 |
| Auto-zero (Fig. 10) | 74.4dB | 180 |
| Gain Comp. (Fig. 11) | 70dB | 20-80 |

**Table 1**: Maximum SNR measured for each DSM. $V_{ref}=\pm1.23v$. Inputs were all approximately -5dBv. Supply voltage =5v.

A noise floor in our test set-up limited our measured SNR and prevented us from making a conclusion on the effectiveness of the auto-zeroed scheme. All of the DSMs were run at sampling rates higher than 100 kHz, however performance degraded significantly at these higher rates with the gain-compensated DSM functioning slightly faster than the others. Also, the gain-compensated DSM was operable with a single 3 Volt power supply and achieved a 60dB maximum SNR for an oversampling factor of 100. The results of the gain-compensated DSM are quite impressive considering the fact that the measured op amp gain in that DSM with a 3 V supply was between 10 and 20. Since none of the other DSMs functioned at all at 3 Volts, the gain-compensated DSM holds promise for future scaled technologies.

## VII. CONCLUSION

Switched-capacitor techniques have been presented which compensate for noise and finite gain limitations in second-order DSMs. Hand analysis and computer simulations show great promise for these new techniques; and preliminary tests conclusively showed that the gain compensation is extremely promising, especially for low power supply voltages. Because of these preliminary results, further investigation will be carried out.

## ACKNOWLEDGEMENTS

Technical discussions with Max Hauser and the assistance of Matthew Borg in the lab are gratefully acknowledged. In addition, we would like to thank Meta Software Inc. for the donation of HSPICE.

## REFERENCES

[1] J.C. Candy, "A Use of Double Integration in Sigma Delta Modulation," *IEEE Trans. on Communications*, Vol. COM-33 No. 3, March 1985.

[2] M.W. Hauser and R.W. Broderson, "Circuit and Technology Considerations for MOS Delta-Sigma Converters," *1986 IEEE International Symposium on Circuits and Systems*.

[3] B.E. Boser and B.A. Wooley, "The Design of Sigma-Delta Modulation Analog-to-Digital Converters," *IEEE J. Solid-State Circuits*, Vol. 23, December 1988.

[4] F. Krummacher, "Micropower Switched Capacitor Biquadratic Cell", *IEEE J. Solid-State Circuits*, Vol.17, June 1982.

[5] R.C. Yen and P.R. Gray, "A MOS Switched-Capacitor Instrumentation Amplifier," *IEEE J. Solid-State Circuits*, Vol. 17, December 1982.

[6] S.C. Fang, SWITCAP A.5R, Columbia University, Dept. of Electrical Engineering, 1986.

[7] L.E. Larson and G.C. Temes, "Switched Capacitor Building Blocks with Reduced Sensitivity to Finite Amplifier Gain, Bandwidth, and Offset Voltage," *1987 IEEE Int'l Symposium on Circuits and Systems*.

[8] P.E.Fleischer, A. Ganesan, K.A. Laker, "Parasitic Compensated Switched Capacitor Circuits," *Electronics Letters*, Vol.17, November 1981.

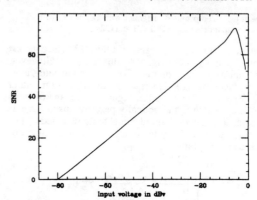

**Figure 12**: Auto-zeroed DSM SNR versus input amplitude

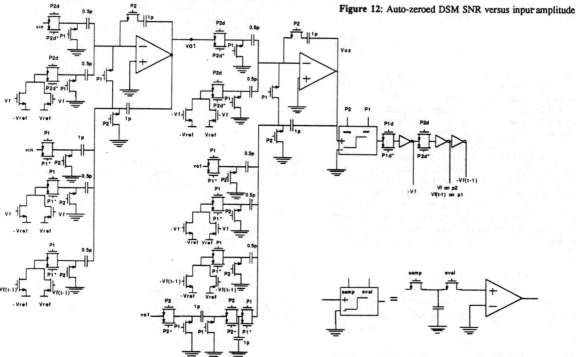

**Figure 11**: Complete DSM with gain-compensation schematic

# MULTIBIT OVERSAMPLED Σ-Δ A/D CONVERTOR WITH DIGITAL ERROR CORRECTION

L. E. LARSON

*Hughes Research Laboratories, Malibu, CA 90265, USA*

T. CATALTEPE
G. C. TEMES

*Department of Electrical Engineering, UCLA, Los Angeles, CA 90024, USA*

*Indexing terms: Analogue/digital conversion, Digital circuits, Error correction codes, Integrated circuits*

A new architecture is presented for a multibit oversampled Σ-Δ analogue/digital convertor (ADC). A novel digital correction scheme is employed to enhance the overall resolution without requiring increased precision of the analogue components.

*Introduction:* The Σ-Δ approach for analogue/digital conversion is of great current interest because of its applicability to VLSI signal processing and its insensitivity to process and component variations.[1-2] The typical Σ-Δ ADC utilises a single-bit ADC (a comparator) and a single-bit DAC (two switches) in a feedback configuration to produce a high-resolution digital output. However, multibit Σ-Δ ADCs exhibit a number of attractive features compared with the single-bit approach, including significantly lower quantisation noise for a given oversampling ratio, as well as improved stability characteristics.[1]

The principal drawback of the multibit Σ-Δ ADC is the stringent accuracy requirements placed on the feedback DAC. Though it has a relatively low resolution, its output voltage must be as accurate as the desired resolution of the overall ADC; typically 14 to 18 bits. Such linearity can only be accomplished through elaborate trimming or calibration schemes, and is hence not very practical for implementation on an integrated circuit.

This letter presents a new technique based on digital correction for the implementation of the multibit Σ-Δ ADC, which significantly reduces the accuracy restriction on the feedback DAC.

*Improved architecture:* The multibit Σ-Δ ADC is particularly sensitive to the static nonlinearity of the coarse DAC in the feedback loop, since the resulting error is directly added to the input. Hence, the DAC noise transfer function is identical to the input voltage transfer function, which is given by $H_1(z)/[1 + H_1(z)]$, where $H_1(z)$ is the transfer function of the loop filter. (This expression assumes that there is no delay through the ADC or DAC.)

The low-frequency error of the DAC can be improved by replacing the DAC itself with an oversampled DAC.[3] In principle, this requires an $M$ bit ADC in the feedback loop that is itself as accurate as the desired overall resolution of the convertor. However, this can be circumvented by precalibrating the DAC and employing a $2^N \times M$ bit memory to simulate the behaviour of the ADC. The resulting system is shown in Fig. 1. The output of the memory is identical to the output of the ideal ADC when presented with the output of the DAC.

Note that the circuit of Fig. 1 requires no analogue components to have a precision equal to that of the final conversion: the precision of the DAC, and hence the overall precision, is achieved by the feedback action. In fact the noise transfer function of the DAC is now approximately $[H_1(z)/(1 + H_1(z))] \cdot [1/(1 + H_2(z))]$, which corresponds to a greatly reduced noise gain.

The improved Σ-Δ ADC of Fig. 1 was simulated on a com-

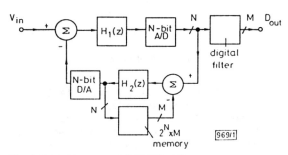

**Fig. 1** *Digitally corrected N-bit Σ-Δ A/D convertor*

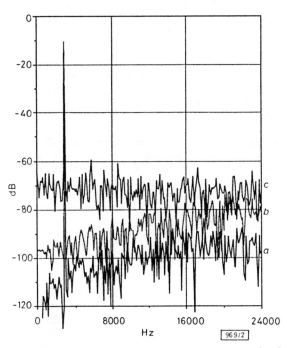

**Fig. 2** *Front-end output spectra for Σ-Δ convertor using four-bit internal ADC and DAC*

a Ideal case
b Nonlinear ADC and DAC
c Nonlinear ADC and DAC with digital correction

puter. For the purposes of simulation, $H_1(z)/2 = H_2(z) = z^{-1}/(1 - z^{-1})$ was chosen, although other filtering functions could be employed, consistent with stability requirements. Curve *a* in Fig. 2 is a plot of the FFT of the output of a traditional Σ-Δ ADC front-end using a four-bit ADC and DAC, where the DAC and ADC have 0 LSB error. In this case, a sinusoid of amplitude $V_{ref}/\sqrt{(3)}$ and frequency 2875 Hz was applied to the input; the clock rate was 1·024 MHz. Curve *b* shows the spectrum when the four-bit ADC and DAC both had a maximum randomly distributed error of 1/4 LSB. Finally, curve *c* illustrates the output spectrum when the proposed system of Fig. 1 was simulated under the same nonlinearity conditions as that of curve *b*. The dynamic range of the

Reprinted from *Electron. Lett.*, vol. 24, pp. 1051–1052, August 1988.

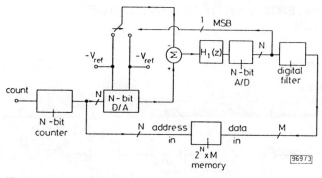

**Fig. 3** *Σ-Δ convertor rearranged for self-calibration*

output is now consistent with approximately 14 bits of resolution with an oversampling ratio of 128.

It should be noted that this technique only corrects static errors created by the $N$ bit DAC circuit. Any dynamic effects, such as settling errors or random noise are not reduced by this approach. However, these errors should not be larger than 1/2 LSB in most applications.[1] Finally, the output of the modulator is still of low resolution, and the digital filter can be the same as with the traditional approaches.

*Self-calibration approach:* The circuit of Fig. 1 requires the precise calibration of the $N$ bit DAC, and storage of the resulting digitised value in a $2^N \times M$ memory. This can be accomplished by the rearranged system illustrated in Fig. 3. This approach reconfigures the convertor as a single-bit $\Sigma$-$\Delta$ ADC, which is inherently insensitive to DAC errors at DC. Hence, the digital outputs of the circuit can be made arbitrarily precise, and they can be stored at memory locations determined by the input to the DAC. The major additional hardware required by this approach is a counter, which determines the DAC output level that is being calibrated. The cali-

bration can be carried out at the oversampling clock rate. The total number of clock periods required to perform the calibration is approximately $2^{N+M}$, plus whatever additional pulses are required for the system to settle.

An added advantage of this approach is that the minimum and maximum values of the overall ADC input and $N$ bit DAC output are intrinsically identical. This maximises the dynamic range of the circuit.

*Acknowledgments:* The authors would like to acknowledge valuable discussions with Dr. R. H. Walden, and the support of Dr. P. T. Greiling and Dr. R. E. Lundgren, all of the Hughes Research Laboratories.

**References**

1 ADAMS, R. W.: 'Design and implementation of an audio 18-bit analog-to-digital converter using oversampling techniques', *J. Audio Eng. Soc.*, 1986, **34**, pp. 153–166

2 MATSUYA, Y., UCHIMURA, K., IWATA, A., KOBAYASHI, T., and ISHI-KAWA, M.: 'A 16b oversampling A/D conversion technology using triple integration noise shaping'. ISSCC Digest of Technical Papers, 1987, pp. 48–49

3 CANDY, J. C., and HUYNH, A. N.: 'Double interpolation for digital-to-analog conversion', *IEEE Trans.*, 1986, **COM-34**, pp. 77–81

## Errata

LARSON, L. E., CATALTEPE, T., and TEMES, G. C.: 'Multibit over-sampled $\Sigma$-$\Delta$ A/D convertor with digital error correction', *Electron. Lett.*, 1988, **24**, (16), pp. 1051–1052

In Fig. 2 the labelling of curves *b* and *c* has been reversed

# AN IMPROVED SIGMA-DELTA MODULATOR ARCHITECTURE

T. C. LESLIE and B. SINGH

Plessey Research Caswell Ltd, Caswell, Towcester
Northants NN12 8EQ, England

## ABSTRACT

The performance of multi-bit sigma-delta coders is analysed and compared with that which can be obtained from the much more common single-bit implementations. It is demonstrated that the fundamental difficulty with multi-bit designs is the requirement for a high accuracy digital to analogue converter (DAC) in the feedback loop. A general expression for the reduction of in-band noise power as a function of coder order, quantiser resolution, and oversampling ratio is derived, together with a subsidiary condition imposed by the use of multi-bit DACs in the network. A new coder topology is proposed which uses multi-bit quantisation with single-bit feedback, overcoming the manufacturing or calibration difficulties inherent in the use of high resolution DACs.

## INTRODUCTION

Sigma-delta ($\Sigma\Delta$) converters continue to receive much attention in technical literature, because this component is vital for the successful operation of many wide dynamic range systems, ranging from ISDN U-interface transceivers, digital audio tape, digital radio, compact disc and sonar. In many of these applications the performance of the analogue-to-digital converter represents the fundamental limit for the operation of the system as a whole.

## BASIC THEORY OF OPERATION

If it is assumed that the quantisation error associated with any sample is independent of the sample value and that the reconversion of the digital signal back into the analogue domain is subject to an error which is also independent of the signal input to the DAC, then the k-th order sigma-delta modulator for which a generalised signal flow diagram is given in Fig.1 is characterised by the transfer relation [1]

$$s^*(z) = s(z) + e(z) + \left[\frac{z-1}{z}\right]^k * q(z) \qquad (1)$$

In this expression s(z) is the transform of the coder input, s*(z) that of the coder output q(z) that of the assumed white quantisation noise, and e(z) that of the DAC level accuracy noise. The premultiplying factor of this last term is often known as the noise transfer function (NTF).

The important feature to observe from eq.(1) is that whilst noise associated with the quantiser is ›shaped by the action of the feedback loop, that arising from the DAC is not. For this reason single-bit coders, which avoid the problem of DAC error, have been preferred.

If the sampling frequency (loop rate) is $f_s$ and the total (unshaped) quantisation noise power in the band $[-f_s/2, f_s/2]$ is given by $\sigma_q^2$, then the noise power spectral density, designated NPSD(f), at the coder output can be obtained from the noise transfer function of eq.(1) according to

$$\text{NPSD}(f) = 2^k * \left[1-\cos\frac{2\pi f}{f_s}\right]^k * \frac{\sigma_q^2}{f_s} \qquad (2)$$

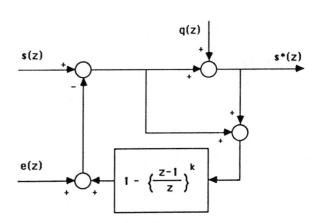

**Fig. 1: Generalised signal flow diagram of a k-th order noise-shaping ADC**

In order to calculate the total in-band noise power produced by the coder it is necessary to integrate eq.(4) over the baseband specified by $[-f_b, f_b]$. As $f_s >> f_b$ this gives

$$\text{TINP} = \sigma_q^2 * \frac{2\pi^{2k}}{2k+1} * \left\{\frac{2f_b}{f_s}\right\}^{2k+1} \qquad (3)$$

If the input signal to the quantiser is assumed to have an equiprobable amplitude distribution, which exactly spans the input range of this element, then the total unshaped noise power of the quantiser, $\sigma_q^2$, and its input signal power may be written in terms of its bit size and resolution. This can be use to derive a generally useful expression for the available dynamic range (DR) in dB as

$$\text{DR} = 6.02n + 3.01r * (2k+1) + 10\log\left\{\frac{2k+1}{\pi^{2k}}\right\} \qquad (4)$$

in which r is the oversampling ratio defined in octaves and n is the quantiser resolution. If a multi-bit DAC is required in the feedback loop, then its level accuracy noise is reduced only by the oversampling ratio (as is implied by eq.(1)), so that eq.(4) is subject to the subsidiary condition

$$\text{DR} \le E + 3r \qquad (5)$$

in which E is the signal to level accuracy noise ratio of the DAC in dB.

From eq.(4) three methods of decreasing the total in-band noise power can be determined, other factors remaining equal; either the coder order (k), the oversampling ratio (r) or the quantiser resolution (n) must be increased. Each of these approaches, however, has its own difficulties.

Increasing the order of the coder may lead to oscillation, since the stability margin of multiple loop $\Sigma\Delta$-coders is a relatively sensitive function of the number of loops [2]. Improving the available dynamic range by increasing the oversampling ratio is eventually limited by the process

Reprinted from *IEEE Proc. ISCAS'90*, pp. 372-375, May 1990.

technology; the dominant factor generally being the speed of the switched-capacitor integrators in the forward signal path. Decreasing the unshaped quantisation noise power term can only be achieved by utilising a multi-bit quantiser in the coder and, by implication, a multi-bit DAC at the end of the feedback loop. With this architecture, it is usually the DAC which limits the available dynamic range, through the condition given in eq.(5)

In attempting to extend the dynamic range of noise shaping coders, interest has concentrated recently on decreasing the quantiser noise power by increasing its word length, and finding methods of overcoming the level accuracy problem of the multi-bit DAC, either by using a dynamic element balancing technique [3] [4], or error-correction [5].

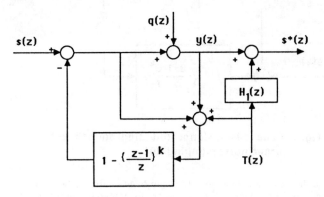

Fig.2: Generalised signal flow diagram of a k-th order multi-bit, noise-shaping ADC with single-bit feedback

## NEW CODER TOPOLOGIES

The essence of the new $\Sigma\Delta$-modulators, to be described in this section, is that the feedback signal comprises only the most significant bit (MSB) of the multi-bit quantiser. A signal flow diagram for one arrangement is shown in Fig.2. This will be referred to as offset-voltage feedback

since the action of the feedback loop is to subtract one of two fixed voltages, provided by the one-bit DAC from the input signal voltage. Intuitively this can be regarded as identical to the conventional single-bit approach, except that the quantiser also generates several lower-order bits, which are unused by the feedback network.

It should be stressed that the truncation signal T(z) in Fig.(2) is a known quantity, since, for any sample, it represents the quantiser output with the MSB removed. Simple loop analysis of Fig.2, assuming for the moment that the filter function $H_1(z)$ is identically zero, yields the transfer function for a k-th order coder, with a truncated feedback signal, as

$$y(z) = s(z) + q(z)\left[\frac{z-1}{z}\right]^k + T(z)\left\{1 - \left[\frac{z-1}{z}\right]^k\right\} \qquad (6)$$

The form of this expression immediately suggests that the effect of the feedback signal truncation on the coder output can be removed by an appropriate choice of $H_1(z)$,

$$H_1(z) = \left\{1 - \left[\frac{z-1}{z}\right]^k\right\} \qquad (7)$$

Substituting eq.(7) into eq.(6) gives an expression identical to eq.(1), so that the remainder of the analysis of section 2 remains valid. The significant improvement is that a multi-bit quantiser has been used, but only one bit is utilised for the feedback signal and thus only a single-bit DAC is required. This means that the feasible dynamic range is given by eq.(4) without the normally occurring subsidiary condition expressed in eq.(5). The problems of the multi-bit DAC have been avoided at the expense of a relatively trivial amount of extra digital signal processing. A signal flow diagram for an offset-voltage feedback coder of order two is shown in Fig.3

A similar approach can be applied to the cascade form of $\Sigma\Delta$-modulators. It has been established that, in this type of coder, only the quantisation noise of the final first-order section appears at the output of the coder as a whole. For this reason, only this particular quantiser in the architecture need be high resolution, in order to improve the dynamic range of the converter as a whole.

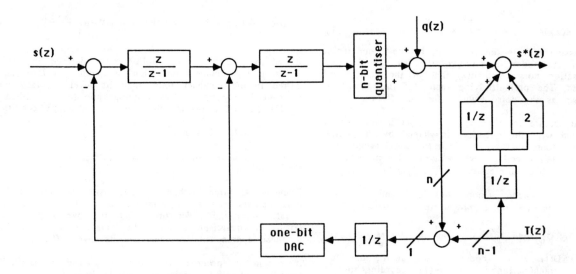

Fig. 3 Signal flow diagram for a second-order noise-shaping ADC with single-bit offset voltage feedback (multiloop realisation).

## SIMULATION PROGRAM

The coders described in the previous section were simulated using a proprietary software package known as GNS-CODER. This package permits the simulation of conventional multi-loop and cascade noise-shaping coders, as well as the topologies described in the previous section. Simulation of both analogue and digital predictive ADCs is also possible. Various component imperfections occurring within practical devices can be modelled: these include the voltage dependence of capacitors; broadband noise, flicker noise, and OPAMP parameters such as finite open-loop gain, slew rate, settling time and clipping.

GNSCODER performs time-domain simulations and provides results in either time- or frequency-domain. Spectral analysis is performed on the time-domain data using the FFT algorithm.

The program GNSCODER allows description of $\Sigma\Delta$-modulators in forms suitable for direct implementation. This ensures that each component is individually realisable, all voltage levels within the structure are tightly defined, and no component ever experiences overload conditions. The net result of these practical constraints is to reduce, irrespective of configuration, the available dynamic ranges of first, second, and third order coders by 6dB, 12dB and 36dB respectively, from the values which would be predicted by eq.(4).

## SIMULATION RESULTS

As a comparative test, the dynamic range available from three coder types was simulated as a function of input signal level. The selected coders were a single-bit second-order noise-shaper, a conventional 6-bit second order device and a device of the type described in the previous section with a 6-bit quantiser and single-bit feedback. The results are shown in Fig.4 which demonstrates the similarity in performance between a conventional 6-bit coder and the single-bit feedback approach and their common superiority over the conventional single-bit system.

Since only the MSB of the multi-bit quantiser is used in the feedback loop, an interesting question arises over whether inevitable manufacturing tolerances in the quantising element are reduced by the loop action in the same way as with the all multi-bit approach. To first order, differential non-linearities and even localised integral non-linearities can be treated in the same way as the quantisation error: both represent a deterministic distortion of the input signal but one which results in a relatively white error spectrum, which will be shaped by the loop action. Of more concern is the impact of large scale integral non-linearities.

In Fig.5, the output spectrum of a 6-bit coder with single bit feedback is shown, when the quantising element has been given a quadratic distortion equal to 10% of its nominal range. Such a transfer characteristic should result in

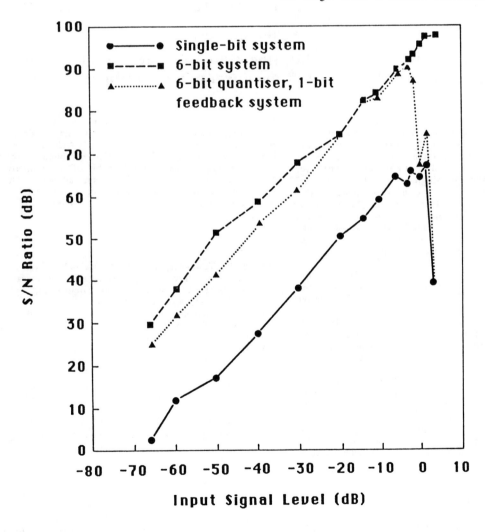

**Fig 4: Signal to noise ratio as a function of input signal level**

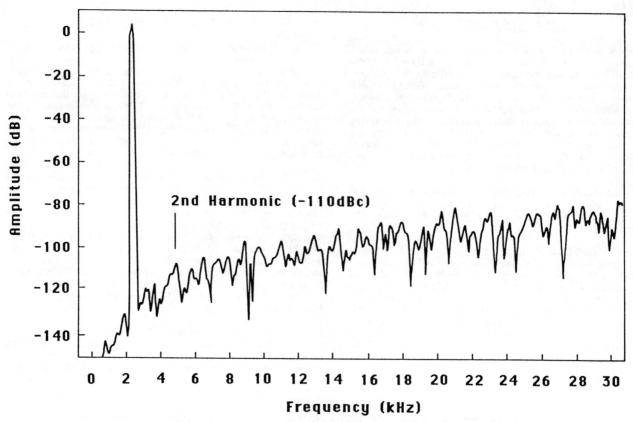

**Fig 5: Output spectrum for a 6-bit, single-bit feedback, second order noise-shaping ADC with 10% quantiser non-linearity**

the generation of substantial second harmonic distortion. For this simulation the notional baseband is 20kHz, the test tone is a 2.44kHz sine wave, 10dB down from the peak value. The loop rate is 2MHz, and the time record comprises 16384 points. The values of signal to (noise+distortion) over the baseband is 81dB for the single-bit feedback arrangement. The most significant feature of this result is that the level of second harmonic distortion for the coder is 110dB down from the carrier, and only just visible above the background quantisation noise level. This is an identical result to that which has been obtained from a fully 6-bit system with a similar quantiser non-linearity

Intuitively, it is possible to explain this result for the conventional multi-bit $\Sigma\Delta$-coder by analogy with a continuous-time system, and the normal operation of negative feedback, ignoring the quantisation error. Such an explanation cannot readily be applied to a system in which the feedback signal is truncated, since this last operation cannot easily be incorporated in a continuous amplitude, continuous time model. A direct calculation of the mechanism by which the harmonic distortion of the quantiser in the truncated feedback converter is reduced by the loop action would be very complex. Only recently [6]-[7], has the nonlinear operation of quantisation in noise-shaping coders been treated in a mathematically rigorous manner. The addition of a further two non-linear effects (the curvature in the quantiser characteristic and the truncation of the feedback signal) would seem to make an already complex calculation virtually intractable.

## CONCLUSIONS

Several new topologies for $\Sigma\Delta$-modulators have been proposed. It has been demonstrated by analysis and simulation that these combine the best aspects of multi-bit $\Sigma\Delta$-coders,

in terms of improved signal-to-noise ratio and harmonic distortion, without suffering their major drawback, the requirement for a high-accuracy DAC in the feedback loop. The omission of the latter component which would otherwise require calibration or be of a relatively complex architecture represents a significant advance. This benefit has been obtained at the expense of a very small amount of relatively trivial digital signal processing.

## REFERENCES

[1] S. K. Tewksbury and R. W. Hallock, "Oversampled, Linear Predictive and Noise-Shaping Coders of Order N>1", IEEE Transactions on Circuits and Systems, vol. CAS-25, pp. 436-447, July 1978.

[2] G. R. Ritchie, "Higher Order Interpolative Analog to Digital Converters", Ph. D Dissertation, University of Pennsylvania 1977.

[3] L. R. Carley, "A Noise-Shaping Coder Topology for 15+ Bit Converters", IEEE Journal of Solid State Circuits, vol. SC-24, pp.267-273, April 1989.

[4] R. J. Van De Plassche, "A Monolithic 14-Bit D/A Converter", IEEE Journal of Solid State Circuits, vol. SC-14, pp. 552-556, June 1979.

[5] L E Larsen, T Cataltepe, and G. C. Temes, "Multi-Bit Oversampled $\Sigma\Delta$ A/D Converter with Digital Error Correction", Electronics Letters, vol. 24, pp. 1051-1052, August 1988.

[6] R. M. Gray, "Spectral Analysis of Quantisation Noise in a Single-Loop Sigma-Delta Modulator with DC Input",IEEE Transactions on Communications, vol. 37, June 1989.

[7] R. M. Gray, W. Chou, and P. W. Wong, "Quantisation Noise in Single-Loop Sigma-Delta Modulation with Sinusoidal Inputs", IEEE Transactions on Communications, vol. 37, June 1989.

# A 13 bit ISDN-band Oversampled ADC using Two-Stage Third Order Noise Shaping

Lorenzo Longo
Bell-Northern Research
Ottawa, Ontario
Canada, K1Y-4H7
P.O. Box 3511,  Station C

Miles Copeland
Carleton University
Ottawa, Ontario
Canada

## Abstract

A 13 bit 80kHz baseband analog-to-digital converter suitable for use in applications such as the ISDN 'U'-interface will be described.  Two-stage third order noise shaping permits the use of a sampling frequency of only 2.56MHz.  The circuit has been implemented using conventional single-ended switched-capacitor techniques in a 1.5u CMOS process.

## Introduction

Recently, multi-stage noise shaping modulators have been designed for voiceband [1] and high quality audio [2] applications. These converters are composed of first order sigma delta modulators (SDM) in cascade and rely on the cancellation of quantization noise to achieve high resolution. Consequently, they require relatively high integrator gain and good capacitor matching compared to classical SDM's [3].

A method of reducing these requirements is to cascade a second  order SDM with a first order SDM. This architecture has been used to implement a 13 bit 80kHz baseband  ADC for ISDN.  The two-stage third order noise shaping modulator  has been implemented in a 2 level poly 1.5u CMOS process.

## Architecture

One of the major factors governing the theoretical signal-to-quantization noise ratio (SQNR) of an oversampled ADC is the order of noise shaping provided by the given architecture.  The benefit of higher order noise shaping is shown in Figure 1.  To achieve a given SQNR, the higher the order of noise shaping , the lower the oversampling ratio required.  In this case, the use of a third order system reduces the sampling frequency by a factor of approximately three compared to a second order system.

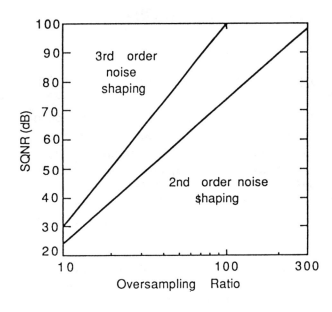

**Figure 1 :** SQNR  vs.  Oversampling ratio

Reprinted from *IEEE Proc. Custom IC Conf.*, pp. 21.2.1–21.2.4, January 1988.

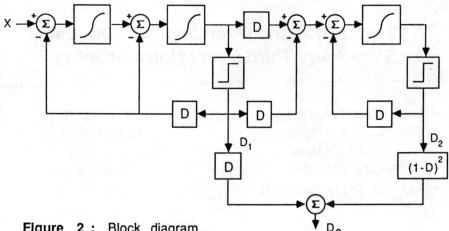

**Figure 2 :** Block diagram.

A block diagram of the ADC is shown in Figure 2. The architecture consists of a second order SDM followed by a first order SDM. The output of the first stage ($D_1$) is subtracted from second integrator output and used as the input to the second stage. This reduces the maximum input to the second stage. Assuming the quantizer can be modelled as an additive noise source [3], the outputs of the first and second stages can be represented as follows:

$$D_1(z) = X(z) + (1-z^{-1})^2 Q_1(z) \quad (1)$$

$$D_2(z) = -z^{-1} Q_1(z) + (1-z^{-1}) Q_2(z) \quad (2)$$

and the final output as :

$$D_o(z) = z^{-1} X(z) + (1-z^{-1})^3 Q_2(z) \quad (3)$$

where $Q_1(z)$ and $Q_2(z)$ are the first and second stage quantization noise sources. Consequently, this architecture provides third order noise shaping. The simulated performance curve, shown in Figure 3, predicts greater than 13 bits of resolution and linearity using a sampling rate of only 2.56MHz.

This architecture reduces integrator gain and capacitor matching requirements due to the fact that elimination of first stage quantization noise [$Q_1(z)$] requires

cancellation of two noise components which are relatively small [ nominally equal to $(1-z^{-1})^2 Q_1(z)$ ]. Oversampled ADC's which are composed of first order SDM's require cancellation of noise components which are significantly larger [ nominally equal to $(1-z^{-1}) Q_1(z)$ ].

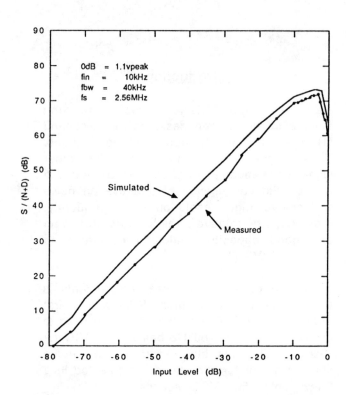

**Figure 3 :** Simulated and measured S / (N+D) curves.

234

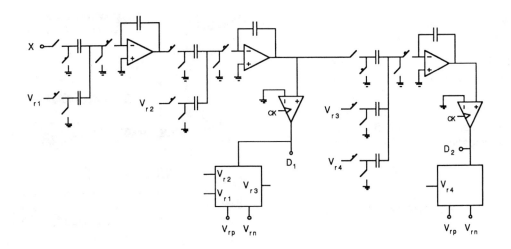

Figure 4 : Simplified circuit diagram.

## Implementation

Figure 4 shows a simplified circuit diagram of the ADC. Conventional single-ended switched-capacitor techniques are used. Capacitor ratios are chosen to maintain circuit stability, to ensure cancellation of first stage quantization noise and to minimize opamp saturation. Absolute capacitor values are selected to make switch thermal noise negligible compared to quantization noise. Capacitor matching requirements are approximately 5% and consequently are not a major concern. The maximum input signal level is 1.1v peak and the reference voltages ($V_{rp}$ & $V_{rn}$) are +/-1.5 volts. A folded-cascode operational amplifier with 65dB DC gain and 20MHz unity-gain bandwidth is used.

In addition to reducing integrator gain and capacitor matching requirements, this architecture does not exhibit high idle channel noise, as do converters based solely on sigma delta modulators. Consequently, a dither signal is not required.

## Performance

The ADC performance is evaluated by windowing the digital output with a four-term Blackman-Harris window [4] and performing a 32,768 point FFT. The noise and distortion is integrated over a bandwidth of approximately 40kHz. The measured performance curve [ S/(N+D) ], shown in Figure 3, is within 6dB of the simulated curve and indicates 13 bits of resolution and linearity. Figure 5 shows the output spectrum of the converter for a 10kHz sinusoidal input signal at -4dB. Harmonics are more than 80 dB below the fundamental. The performance is summarized in Table 1.

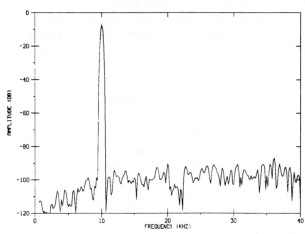

FIGURE 5 : MEASURED OUTPUT SPECTRUM.

| Resolution | 13 | bits |
|---|---|---|
| Sampling frequency | 2.56 | MHz |
| Baseband | 80 | kHz |
| Supply voltage | 5 | volts |
| Reference voltages | +-1.5 | volts |
| Power dissipation | 20 | mW |

**Table 1 :** Performance summary.

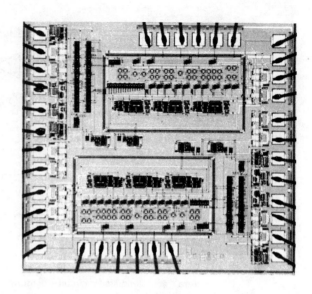

**Figure 6 :** Photograph of test chip.

A photograph of the test chip is shown in Figure 6. The ADC is implemented in a 2 level poly 1.5u CMOS process and occupies an area of 4.1mm$^2$, including bonding pads.

## Conclusions

In summary, two-stage third order noise shaping has been used to implement a 13 bit ADC suitable for use in applications such as the ISDN 'U'-interface. The architecture presented here reduces sampling frequency, opamp gain and capacitor matching requirements, and does not require a dither signal to achieve good idle channel noise performance.

## Acknowledgements

The authors would like to thank their management at Bell-Northern Research for providing a productive atmosphere for doing cooperative research. They would also like to thank Ed MacRobbie for opamp design and Patrick Goulet for his work on the data acquisition system.

## References

[1] Hayashi, T., Inabe, Y., Uchimura, K. and Kimura, T., "A Multi-Stage Delta-Sigma Modulator without Double Integration Loop", ISSCC Digest of Technical Papers, p. 182-183; Feb., 1986.

[2] Matsuya, Y.,Uchimura, K., Iwata A., Kobayashi T., Ishikawa M., "A 16b Oversampling A/D Conversion Technology using Triple Integration Noise Shaping", ISSCC Digest of Technical Papers, p. 48-49; Feb., 1987.

[3] Candy, J.C., "A Use of Double Integration in Sigma Delta Modulation", IEEE Transactions on Communications, Vol. COM-33, p.249-258; March, 1985.

[4] Harris, F. J., "On the Use of Windows for Harmonic Analysis with the Discrete Fourier Transform", Proceedings of the IEEE, Vol. 66, No. 1, p.51-83; Jan. 1978.

# A 16-bit Oversampling A-to-D Conversion Technology Using Triple-Integration Noise Shaping

YASUYUKI MATSUYA, KUNIHARU UCHIMURA, ATSUSHI IWATA,
TSUTOMU KOBAYASHI, MEMBER, IEEE, MASAYUKI ISHIKAWA, MEMBER, IEEE,
AND TAKESHI YOSHITOME

*Abstract* —A highly stable triple-integration noise-shaping technology is discussed which permits greater accuracy for monolithic audio A-to-D converters. Based on this new technology, using 2-μm CMOS technology, a 16-bit 24-kHz bandwidth A-to-D converter LSI with digital filters was successfully developed. An SNR (S/(N+THD)) of 91 dB and a total harmonic distortion (THD) of 0.002 percent at full-scale input were attained.

## I. INTRODUCTION

IMPROVING the accuracy of monolithic A-to-D converters is a never-ending research goal because the A-to-D converter is a key component of electrical systems, such as high-quality audio and high-accuracy measurement systems.

Conventionally, either the successive approximation or dual-ramp conversion technique is used for high-resolution A-to-D converters. In successive approximation, a means of trimming the weighting network is indispensable to achieving a conversion accuracy of over 15 bits [1]–[3]. This is because the conversion accuracy depends on the device matching tolerance of the weighting networks and is limited to 14-bit accuracy when using nontrimming weighting networks [4], [5]. Using the dual-ramp technique, high speed and accuracy are required in the integrator, current sources, comparators, and sample-and-hold (S/H) circuits. To realize these circuits, high $f_T$ bipolar process technologies must be used [6]. The development of an S/H circuit with over 16-bit accuracy is especially difficult, because the sampled charge in the sampling capacitor is leaked through the base impedance of the bipolar transistor.

Recently, oversampling has attracted considerable attention as a conversion technique for VLSI technology,

which does not require trimming or a precise S/H circuit [7], [8]. The noise reduction concept of the noise-shaping oversampling technique is to distribute quantization noise outside the signal band by setting the sampling rate much higher than the signal bandwidth and the noise shaping by an integration. This technique thus has the advantage that the device tolerance is relaxed, and it is possible to use a two-value quantizer which theoretically produces no distortion. By using a higher frequency as the sampling clock pulse and higher order integration noise shaping characteristics, higher accuracy is achieved.

The conventional noise-shaping oversampling technique, however, is limited to an accuracy of 80 dB in the audio signal bandwidth with CMOS process technologies, which are useful for VLSI's [9], [10]. This limitation stems from the fact that higher order integration noise-shaping characteristics over double integration cannot be realized due to the oscillation of the feedback loop.

To overcome this problem, we have developed a multistage noise-shaping technique which permits higher order, more than double integration, noise-shaping characteristics by using a new multistage configuration based on a stable first-order Δ-Σ quantizer [11]. Hereafter, this technique will be referred to as MASH. To confirm the feasibility of the MASH technique, we fabricated the double-integration MASH A-to-D conversion LSI [12].

In this paper, the realization of higher order integration noise-shaping characteristics, which are the essential features of the MASH technique, and the usefulness of MASH as a high-accuracy, over 16-bit, A-to-D conversion technique, will be described. In Section II, the accuracy limitations of the oversampling technique are discussed. Also in Section II, the necessity of triple integration noise shaping characteristics to realize 16-bit accuracy (which is required for high-quality audio encoding) using CMOS process technology is shown. In Section III, the operating principle of the MASH technique is described, and it will be shown theoretically that high-order, more than triple

Manuscript received June 18, 1987; revised July 26, 1987.
The authors are with the Linear Integrated Circuit Section, NTT Atsugi Electrical Communications Laboratories, 3-1, Morinosato Wakamiya, Atsugi-shi, Kanagawa Prefecture, 243-01 Japan.
IEEE Log Number 8717214.

Reprinted from *IEEE J. Solid-State Circuits*, vol. SC-22, pp. 921–929, December 1987.

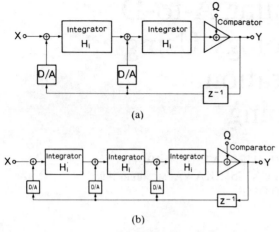

(a)

(b)

Fig. 1. (a) Conventional double-integration $\Delta$-$\Sigma$ quantizer signal flowchart. (b) Conventional triple-integration $\Delta$-$\Sigma$ quantizer signal flowchart.

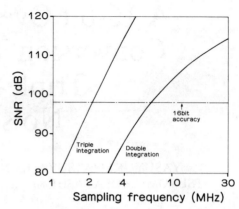

Fig. 2. Theoretical relationship between SNR and the sampling frequency.

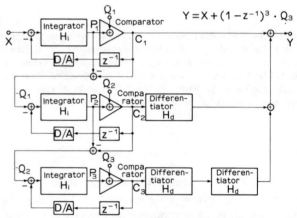

Fig. 3. Three-stage MASH signal flowchart.

integration, noise-shaping characteristics are easily realized. In Section IV, the practical accuracy limiting factors and the way these factors were optimized to realize the triple-integration MASH A-to-D converter are discussed. In Section V, the circuit configurations and operations of the developed 16-bit A-to-D conversion LSI with precise digital filters are described in detail. In Section VI, measurement results of the fabricated LSI with an SNR of 91 dB and a total harmonic distortion (THD) of 0.002 percent are presented. These results confirm the usefulness of MASH as a technique to enhance monolithic A-to-D converter accuracy.

## II. THE THEORETICAL ACCURACY LIMITING FACTORS OF THE $\Delta$-$\Sigma$ OVERSAMPLING TECHNIQUE

In the $\Delta$-$\Sigma$ oversampling technique which uses noise shaping characteristics, theoretical accuracy limiting factors are 1) oversampling frequency, and 2) the order of the integration of noise-shaping characteristics.

Signal flowcharts of the double- and triple-integration $\Delta$-$\Sigma$ quantizer using cascade integrators are shown in Fig. 1 (a) and (b), respectively, where $X$ is an analog input, $Y$ is a digital output, and $Q$ is a quantization noise [13]. The transfer function of an integrator $H_i$ is defined by $1/(1 - z^{-1})$. Therefore, the transfer functions of these signal flowchart are given by

$$Y = X + (1 - z^{-1})^2 Q \qquad (1)$$

$$Y = X + (1 - z^{-1})^3 Q. \qquad (2)$$

Fig. 2 shows the theoretical relationship between SNR and sampling frequency at the 24-kHz signal bandwidth. These are calculated using (1) and (2). To obtain better than 16-bit accuracy using double-integration noise-shaping characteristics, a sampling frequency of 6.6 MHz is required. If triple-integration noise-shaping characteristics are realized, the sampling frequency can be reduced to 2.0 MHz.

Using a CMOS switched-capacitor integrator, the maximum sampling frequency for high-accuracy integration is about 4 MHz [14], [15].

Even with low sampling frequencies under 4 MHz, 16-bit accuracy can be attained using triple-integration noise-shaping characteristics, but not with double-integration noise-shaping characteristics.

## III. PRINCIPLE OF THE THREE-STAGE MASH OPERATION

To realize an over 16-bit accuracy CMOS oversampling A-to-D converter, a triple-integration noise-shaping technique is required. However, the loop of the conventional triple integration $\Delta$-$\Sigma$ quantizer, as shown in Fig. 1(b), which includes a three-stage cascade integrator, oscillates because of a 270° phase shift.

Therefore, to realize stable triple-integration noise-shaping characteristics without the oscillation problem, a three-stage MASH configuration using stable first-order $\Delta$-$\Sigma$ quantizers (DSQ) was proposed. Fig. 3 shows a signal flowchart of this three-stage MASH. The analog output of the first DSQ is given by $P_1 - C_1$, which is equal to the quantization noise $-Q_1$, and is quantized by the second DSQ. The analog output $P_2 - C_2$ of the second DSQ, which is equal to the quantization noise $-Q_2$, is quantized by the third DSQ. When $H_i$ is $1/(1-z^{-1})$, outputs of the first, second, and third DSQ's are given by (3)–(5):

$$C_1 = X + (1-z^{-1})Q_1 \qquad (3)$$

$$C_2 = -Q_1 + (1-z^{-1})Q_2 \qquad (4)$$

$$C_3 = -Q_2 + (1-z^{-1})Q_3. \qquad (5)$$

Output $Y$ of the three-stage MASH is synthesized $C_1$, $C_2$, and $C_3$. $C_2$ is differentiated one time, $C_3$ is differentiated two times, and these are added to $C_1$. Thus, when $H_d = (1-z^{-1})$:

$$
\begin{aligned}
Y &= C_1 + (1-z^{-1})C_2 + (1-z^{-1})^2 C_3 \\
&= X + (1-z^{-1})Q_1 - (1-z^{-1})Q_1 + (1-z^{-1})^2 Q_2 \\
&\quad - (1-z^{-1})^2 Q_2 + (1-z^{-1})^3 Q_3 \\
&= X + (1-z^{-1})^3 Q_3 \qquad (6)
\end{aligned}
$$

Since quantization noise $Q_1$ can be suppressed by $C_2$ and $Q_2$ can be suppressed by $C_3$, (6) is equivalent to (2). The important point is that triple-integration noise-shaping characteristics can be obtained using three first-order $\Delta$-$\Sigma$ quantizers. This ensures stable operation.

The analog quantization noise $Q_1$ and $Q_2$ of $P-C$ for the next stage is generated easily without the analog subtractor. This operation is discussed in Section V.

## IV. PRACTICAL ACCURACY LIMITING FACTORS

Practical factors limiting conversion accuracy are the integrator gain, the settling speed, the capacitance tolerence, and the noise from the digital circuits.

### A. The Integrator Gain and Settling Speed

In Section II it was shown that 16-bit accuracy can be obtained by triple-integration noise-shaping characteristics and CMOS process technology at a sampling frequency of 2.0–4.0 MHz. Based on this range, we selected a sampling frequency $f_s$ of 3 MHz, which can easily generate the 48-kHz data output rate by $f_s/64$ for the 24-kHz signal bandwidth.

The integration gain accuracy $G_e$, which is defined as $V_o/V_{oi}$ ($V_o$: actual integrator output value; $V_{oi}$: ideal integrator output value), results from the finite amplifier gain of the integrator. This is because the virtual *GND* voltage, which is the negative input at the amplifier of the integrator, is varied depending on the integrator output voltage. When the amplifier gain is defined as $G_i$ ($G_i = V_o/V_g$; $V_o$: amplifier output voltage; $V_g$: virtual *GND* voltage), $G_e$ is given by

$$G_e = 1 - \frac{1}{G_i + 1}. \qquad (7)$$

To define the settling accuracy of the integrator as $G_s$, the

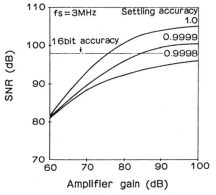

Fig. 4. SNR dependence on amplifier gain.

output of the integrator $P$ is given by

$$P = G_e \cdot G_s (X + Pz^{-1}). \qquad (8)$$

For the MASH configuration, the analog input and the feedback D/A output must be integrated individually. Therefore, the integrator is operated twice in each conversion operation. When the feedback D/A output is defined as $D_a$, the transfer function of the integrator considering integrating operations of two times is given by

$$
\begin{aligned}
P &= G_e \cdot G_s \big( D_a + G_e \cdot G_s (X + Pz^{-1}) \big) \\
P &= \frac{G_e \cdot G_s \cdot D_a}{1 - (G_e \cdot G_s)^2 z^{-1}} + \frac{(G_e \cdot G_s)^2 X}{1 - (G_e \cdot G_s)^2 z^{-1}} \qquad (9)
\end{aligned}
$$

Fig. 4 shows the SNR dependence on the amplifier gain at a settling accuracy of 1.0, 0.9999, and 0.9998 calculated by the three-stage MASH signal flowchart and (9). To obtain 16-bit accuracy at a 3-MHz sampling frequency, a gain of more than 85 dB and a settling accuracy of more than 0.9999 is required.

Focusing on the speed, the high-speed and low-distortion amplifier for the integrator is designed as shown in Fig. 5(a). It consists of two stages: a differential stage and a push–pull output stage. The measured characteristics of the amplifier are summarized in Table I. Where $V_{dd}$ is 5 V, the input signal swing is 2 $V_{pp}$, the input signal frequency is 1 kHz, the bandwidth for SNR measurement is 24 kHz, and the output swing for settling time measurement is 0.5 V. These characteristics are all satisfactory except for the gain, which was only 82 dB. To boost the gain, it was necessary to improve the integrator.

Usually, a switched-capacitor integrator is used for the $\Delta$-$\Sigma$ quantizer, as shown in Fig. 5(b), because its accuracy depends on only capacitance-matching tolerance. The integration gain accuracy $G_e$ of the switched-capacitor integrator is defined as $Q_i/(Q_i + Q_s)$ ($Q_i$: charge integrated in $C_i$; $Q_s$: charge remaining in $C_s$). $Q_i/(Q_i + Q_s)$ is given by

$$\frac{Q_i}{Q_i + Q_s} = \frac{C_i(V_o - V_g)}{C_i(V_o - V_g) + C_s \cdot V_g} \approx 1 - \frac{1}{\dfrac{C_i}{C_s} G_i + 1}. \qquad (10)$$

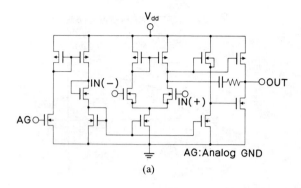

(a)

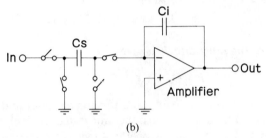

(b)

Fig. 5. (a) Configuration of the amplifier circuit. (b) Configuration of the integrator.

TABLE I
AMPLIFIER CHARACTERISTICS

| | |
|---|---|
| Gain | 82 dB |
| OdB bandwidth | 47 MHz |
| THD | 0.0003 % |
| S/N | 105 dB |
| Settling time (0.01%) | 100 ns |
| Power Dissipation | 8.5 mW |

Setting the capacitance ratio between the integrator capacitance $C_i$ and the sampling capacitance $C_s$ to 2:1, the amplifier gain considering $C_i$ and $C_s$ ($G_i \cdot C_i / C_s$) is improved to 88 dB. Using this design, a gain of 88 dB and an SNR of 99 dB were attained—values that are high enough to achieve 16-bit accuracy.

### B. Capacitance-Matching Tolerance

A gain mismatching of each DSQ causes the degradation of SNR. The gain of each DSQ is defined as digital output of the comparator/analog input. $C_1$, $C_2$, and $C_3$, shown in Fig. 3, are the ideal digital output at the DSQ gain of 1. $\alpha C_1$, $\beta C_2$, and $\gamma C_3$ are defined as the digital outputs at the DSQ when the DSQ gain is not equal to 1. Inserting $\alpha C_1$, $\beta C_2$, and $\gamma C_3$ into (6), we have the following equation:

$$Y = \alpha C_1 + \beta(1 - z^{-1})C_2 + \gamma(1 - z^{-1})^2 C_3$$
$$= \alpha X + (\alpha - \beta)(1 - z^{-1})Q_1 + (\beta - \gamma)(1 - z^{-1})^2 Q_2$$
$$+ \gamma(1 - z^{-1})^3 Q_3. \tag{11}$$

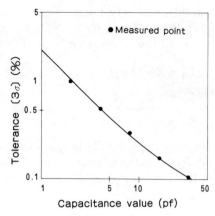

Fig. 6. Relationship between tolerance and capacitor value.

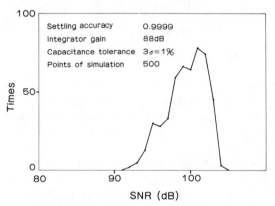

Fig. 7. Monte-Carlo simulated SNR results.

As can be seen from this equation, quantization noises $Q_1$ and $Q_2$ are not suppressed by the gain mismatching of each DSQ. This gain mismatching is caused by the capacitance ratio mismatching of $C_i$ to $C_s$.

The measured tolerance ($3\sigma$) dependence on the capacitance value [5] is shown in Fig. 6. Considering the parasitic capacitor, the circuit noise, and the integrator settling speed, a capacitance value of 2 pF is suitable. At this value, the capacitance tolerance is about 1 percent at $3\sigma$.

The SNR Monte-Carlo simulation results of the three-stage MASH A-to-D converter at a capacitance mismatching of 1 percent ($3\sigma$) among $C_s$, $C_i$, $C_{d1}$, and $C_{d2}$ of each DSQ, an amplifier gain of 88 dB, and a settling accuracy of 0.9999 are shown in Fig. 7. From these results, an SNR of 99 dB occupies 67 percent and an SNR of 92 dB occupies 99.7 percent.

### C. Reducing Noise from the Digital Circuits

The noise of the first-stage input is not noise shaped and must therefore be kept low. The main noise sources are the induced noise from the digital circuits and the $1/f$ noise of the integration amplifier. Using a wide gate area transistor of 2700 $\mu m^2$ as the amplifier input, an amplifier SNR of 105 dB is achieved. To reduce noise from the digital circuits, a differential configuration is adopted at the first DSQ, which can suppress the common-mode noise.

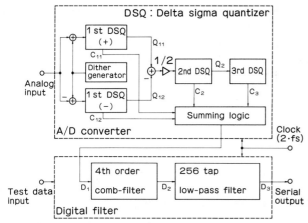

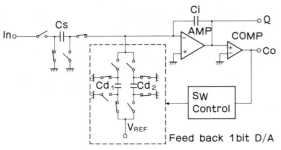

Fig. 8. Block diagram of the A-to-D conversion LSI.

Fig. 9. DSQ configuration.

## V. CIRCUIT CONFIGURATION

By incorporating the various optimizations discussed above in Section IV, we designed a three-stage MASH A-to-D conversion LSI with digital filters. In this section, we will describe the LSI configuration and circuit operations in some detail.

### A. Block Diagram

A block diagram of the A-to-D conversion LSI is shown in Fig. 8. It consists of an A-to-D converter using three-stage MASH, a fourth-order digital comb filter (FIR1), and a 256-tap low-pass digital filter with a dual-loop shift register (FIR2). In this figure, the DSQ's are first-order $\Delta$-$\Sigma$ quantizers. The digital-filter test data input function is added for chip selection and testing.

A first-order $\Delta$-$\Sigma$ quantizer generates discrete spectral lines and thus reduces the SNR for low input levels without dithering. Therefore, 64-kHz 0.5-$V_{pp}$ square waves are generated by the switched-capacitor circuits as a dither, and put in the differential stages as common mode. This dither vanishes by adding the first-stage differential outputs.

### B. Analog Circuit Configuration and Operation

Each DSQ can be constructed of the same simple switched-capacitor circuit, as shown in Fig. 9, consisting of an input switched-capacitor circuit, a feedback 1-bit D-to-A conversion circuit (DAC), an integrator, and a comparator.

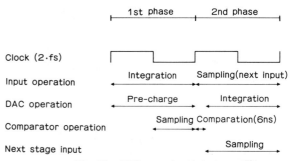

Fig. 10. DSQ operation timing.

In the $\Delta$-$\Sigma$ quantizer, the harmonic distortion of the DAC is one of the factors causing SNR degradation. The feedback 1-bit DAC consists of two switched-capacitor circuits. In these circuits, $C_{d1}$ outputs the positive full scale and $C_{d2}$ outputs the negative full scale. Theoretically, this DAC does not generate any harmonic distortions, because it outputs only two values. These modifications make it possible to fabricate a high-accuracy quantizer for the three-stage MASH configuration.

In the operation of this quantizer, the input voltage is first integrated by the input capacitor $C_s$. Simultaneously, feedback capacitors $C_{d1}$ and $C_{d2}$ are precharged to positive and negative full-scale charges, respectively. Ending the integration of the charge in $C_i$, the comparator compares the integrator output value and the *GND* level. Next, if the comparator output is positive, the negative full-scale charge, which is the output of the feedback D-to-A converter, is integrated. In the opposite case, the positive full-scale charge is integrated. The circuit operation timing of DSQ is shown in Fig. 10. The timing is divided into two phases: the input signal integration and DAC precharge phase, and the DAC output integration and next stage input switched-capacitor circuit sampling phase. The comparator acts at the end of the first phase for 6 ns. The SNR of the $\Delta$-$\Sigma$ quantizer is not degraded, even if the comparator has a conversion error and offset of more than $\pm 15$ mV, because the noise which is generated by the conversion error is also shaped. Therefore, for high-speed operation, the comparator consists of a simple positive feedback flip-flop without the preamplifier.

By using this operation timing, the integrator outputs the quantization noise $Q$ for the next stage input without any additional circuits. This can be shown as follows.

$P$, defined as the integrator output value at the end of the first phase, can be expressed as

$$P_1 = P_1 z^{-1} + X - C_1 z^{-1} \qquad (12)$$

where $X$ is the analog input value, and $C_1$ is the comparator output value of the first DSQ. The relationship between $X$ and $C_1$ was expressed by (3). The amplifier output value at the end of the second phase is given as $P_1 - C_1$ by transforming (12) to produce

$$P_1 - C_1 = \frac{X - C_1}{1 - z^{-1}}. \qquad (13)$$

241

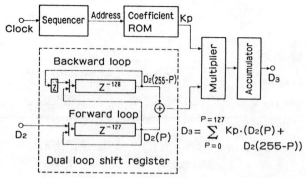

Fig. 11. Block diagram of the 256-tap low-pass transversal filter.

Therefore, $P_1 - C_1$ is given by (3) and (13):

$$P_1 - C_1 = -Q_1. \tag{14}$$

It is clear from (14) that the integrator output $P_1 - C_1$, which is the end of the second phase, equals the quantization noise $Q_1$. The configurations of the second and third stages are the same as that of the first stage. By this operation timing, the simple switched-capacitor quantizer without an analog subtractor for the three-stage MASH can be realized.

### C. Digital Filter Configuration and Operation

The digital filter consists of a 1/16 decimation fourth-order comb filter and a 1/4 decimation 256-tap low-pass transversal filter using a dual-loop shift-register technique, as shown in Fig. 8.

A block diagram of the 256-tap low-pass transversal filter is shown in Fig. 11. It consists of a dual-loop shift register with 127 stages and 129 stages, a multiplier, an accumulator, a coefficient ROM, and a sequencer.

With symmetrical coefficients, the number of multiplications can be cut by summing data before multiplication [16]. Also, by dividing the shift register into a forward and a backward loop, two symmetrical data can be output at the same time.

Fig. 12 shows how the dual-loop shift register operates using a 16-tap filter. The 16-tap filter consists of a seven-stage shift register as a forward loop and a nine-stage shift register as a backward loop. $Z_n$ is defined to be $D_2 z^{-n}$.

Fig. 12(a) is the initial condition of the data in the registers. The loop output data are $Z_9$ and $Z_8$. $Z_1$ is input into the forward loop, and $Z_8$ is input into the backward loop. $Z_{17}$ is thrown away. Next, the forward and backward loops are actually made into loops and shifted once, which is shown in Fig. 12(b). At this point, loop outputs are $Z_{10}$ and $Z_7$. Fig. 12(c) shows the results of shifting the forward and backward loops seven times. At that point, loop outputs are $Z_{16}$ and $Z_1$, and all pairs of symmetrical data from $Z_1$ to $Z_{16}$ are output by these shift operations. Then, when the forward and backward loops are shifted again, the results shown in Fig. 12(d) are produced. It is apparent that this is the same as the first condition plus one delay. This is the initial condition of the next operation.

The chief merits of this configuration are simplicity and controllability. Only, the shift registers are shifted, and the selectors are switched only once in each operation.

The use of a dynamic shift register reduces the area occupied by a one bit cell to that of a one bit RAM cell. The shift-register loops do not require a wiring area or any control logic. A precise on-chip digital filter is thus made possible.

## VI. Measurement Characteristics

The actual chip photomicrograph fabricated using 2-$\mu$m CMOS process technology is shown in Fig. 13. The capacitors are placed in the center of the A-to-D converter block in order to reduce mismatching. The two quantizers of the first differential stage are laid out symmetrically with the digital blocks, because cross-talk noise from within the digital blocks is canceled as common-mode noise by this layout.

The chip size of the converter is $2.7 \times 2.7$ mm, and the size of the whole chip is $9.0 \times 5.4$ mm. The digital blocks contain 16.6K gates and the number of analog elements is 1.6K.

A block diagram of the measurement system is shown in Fig. 14. In this system, the SNR at the bandpass filter output is more than 100 dB at 1 kHz. The analysis accuracy of the FFT analyzer with minimum window is more than 103 dB at nonsynchronization for the analog signal and there are 2K FFT points at 48 ksps. The clock pulse of the A-to-D conversion LSI is 3 MHz and the digital data output rate is 48 ksps.

The measured spectrum is shown in Fig. 15. The digital filter cutoff frequency is 24 kHz. The second- and third-order harmonic distortions are about −102 dB, and higher order harmonic distortions are below the noise floor. The THD from the second to tenth order is 0.002 percent.

The SNR including THD versus input level characteristics are shown in Fig. 16, where the input signal frequency is 1 kHz and the signal bandwidth is 24 kHz. The SNR is 91 dB at a full-scale input level and 93 dB at a small-signal input level. These characteristics are good enough for high-quality audio encoding.

The digital filter frequency response is shown in Fig. 17. This is measured from the input of the comb filter to the digital output, as shown in Fig. 8. The passband is 24 kHz and the sampling frequency is 3 MHz. We obtained a passband ripple of 0.001 dB and a stopband attenuation of 105 dB.

The performance of the LSI is summarized in Table II. The SNR + THD is 91 dB at full-scale input. The power dissipation of the whole LSI is 110 mW at 3 MHz and that of the analog block is 35 mW.

Fig. 18 compares the use configuration of our developed LSI with the conventional A-to-D converter. In conventional A-to-D converters, both an S/H circuit and a high-order antialiasing analog filter are required, both of which are difficult to make with current VLSI technology.

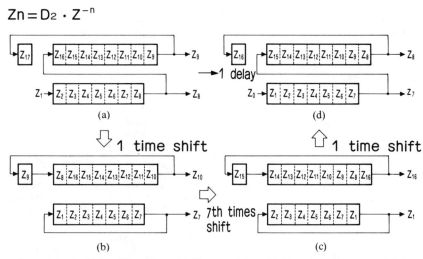

$$Zn = D_2 \cdot Z^{-n}$$

Fig. 12. Operations of the dual-loop shift register. (a) First condition. (b) Second condition. (c) Eighth condition. (d) First condition (next operation).

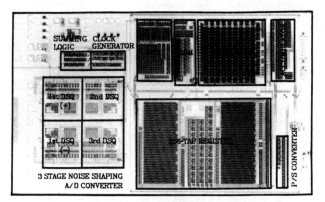

Fig. 13. Chip photomicrograph.

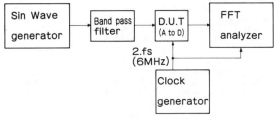

Fig. 14. Block diagram of the measurement system.

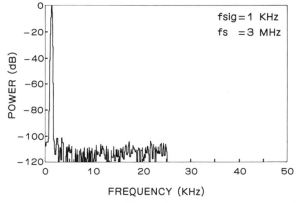

Fig. 15. Measured spectrum.

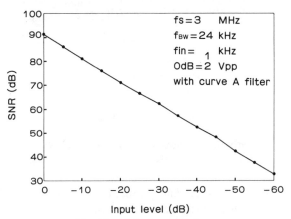

Fig. 16. SNR characteristics.

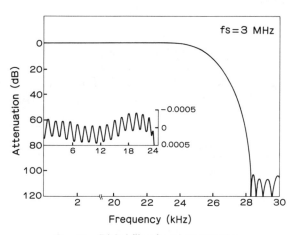

Fig. 17. Digital filter frequency response.

Our new LSI, however, does not require an S/H circuit, because the comparator acts only one time in each conversion operation. Also, a low-order *RC* filter can be used as an analog prefilter, because the sampling frequency is very high in comparison to the bandwidth (24 kHz).

IEEE JOURNAL OF SOLID-STATE CIRCUITS, VOL. SC-22, NO. 6, DECEMBER 1987

TABLE II
LSI PERFORMANCE

| Resolution | 16 bit |
|---|---|
| Sampling frequency(fs) | $\leqslant$ 3 MHz |
| Signal bandwidth | fs/128 |
| Signal/(Noise + T H D) | 91 dB |
| T H D | 0.002 % |
| Supply voltage | 5 V |
| Power Dissipation | 110 mW |

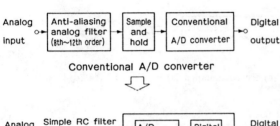

Conventional A/D converter

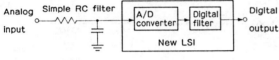

This LSI

Fig. 18. Use configuration of this LSI.

## VII. CONCLUSIONS

A triple-integration noise-shaping A-to-D conversion technology based on a multistage configuration of a delta-sigma quantizer has been developed. Applying this technology, a 16-bit 24-kHz bandwidth A-to-D converter with digital filters was integrated on a single chip utilizing a 2-$\mu$m CMOS process technology. An SNR of 91 dB and a THD of 0.002 percent at the full-scale input were successfully attained.

## ACKNOWLEDGMENT

The authors would like to thank T. Sudo, E. Arai, Y. Akazawa, T. Kimura, and T. Kaneko, for their helpful suggestions and encouragement.

## REFERENCES

[1] R. J. Van DePlassche and H. J. Schouwenaars, "A monolithic 14 bit A/D converter," *IEEE J. Solid-State Circuits*, vol. SC-17, pp. 1112–1117, Dec. 1982.
[2] J. R. Naylor, "A complete high-speed voltage output 16 bit monolithic DAC," *IEEE J. Solid-State Circuits*, vol. SC-18, pp. 729–735, Dec. 1983.
[3] Y. Matsuya, Y. Akazawa, and A. Iwata, "High linearity and high speed 1 chip A to D, D to A converter," *Trans. Inst. Electron. Commun. Eng. Japan*, vol. J69-C, pp. 531–539, May 1986.
[4] R. V. Plassche, "Dynamic element matching puts trimless converters on chip," in *Electronics*. Boulder, CO: Lake Publ. Corp., June 16, 1983, pp. 130–134.
[5] Y. Matsuya, Y. Akazawa, and A. Iwata, "Error analysis of weighting networks," paper of the Tech. Group on Semiconductors and Semiconductor Devices, *IECE* (Japan), vol. 81, SSD 81-58, pp. 25–32, Nov. 1981.
[6] T. Sugawara, M. Ishibe, H. Yamada, S. Majima, T. Tanji, and S. Komatsu, "A monolithic 14 bit/20 $\mu$s dual channel A/D converters," *IEEE J. Solid-State Circuits*, vol. SC-18, pp. 723–728, Dec. 1983.
[7] B. Agrawal and K. Shenoi, "Design methodology for $\Sigma\Delta$ M," *IEEE Trans. Commun.*, vol. CE-31, pp. 360–370, Mar. 1983.
[8] J. C. Candy, "A use of limit cycle oscillations to obtain robust analog to digital converters," *IEEE Trans. Commun.*, vol. COM-22, pp. 298–305, Mar. 1974.
[9] P. Koch, B. Heise, F. Eckbauer, E. Engelhardt, J. A. Fisher, and F. Parzefall, "A 12 bit sigma-delta analog to digital converter with 15 MHz clock rate," *IEEE J. Solid-State Circuits*, vol. SC-21, pp. 1003–1009, Dec. 1986.
[10] J. W. Scott, W. Lee, C. H. Giancarlo, and C. G. Sodini, "CMOS implementation of an immediately adaptive delta modulator," *IEEE J. Solid-State Circuits*, vol. SC-21, pp. 1088–1095, Dec. 1983.
[11] K. Uchimura, T. Hayashi, T. Kimura, and A. Iwata, "VLSI A to D and D to A converters with multi-stage noise shaping modulators," in *Proc. ICASSP*, Apr. 1986, pp. 1545–1548.
[12] T. Hayashi, Y. Inabe, K. Uchimura, and T. Kimura, "A multi stage delta-sigma modulator without double integration loop," in *ISSCC Dig. Tech. Papers*, Feb. 1986, pp. 182–183.
[13] S. K. Tewksbury and R. W. Hallock, "Oversampled, linear predictive and noise-shaping coders of order N > 1," *IEEE Trans. Circuits Syst.*, vol. CAS-25, pp. 436–447, July 1978.
[14] J. A. Guinea and D. Senderowicz, "A differential narrow-band switched capacitor filtering technique," *IEEE J. Solid-State Circuits*, vol. SC-17, pp. 1029–1038, Dec. 1982.
[15] T. Choi, R. T. Kaneshiro, R. W. Brodersen, P. R. Gray, W. B. Jett, and M. Wilcox, "High-frequency CMOS switched-capacitor filters for communications application," *IEEE J. Solid-State Circuits*, vol. SC-18, pp. 652–664, Dec. 1983.
[16] L. R. Rabiner and B. Gold, *Theory and Application of Digital Signal Processing*. Englewood Cliffs, NJ: Prentice-Hall, pp. 328–334.

# IMPROVED SIGNAL-TO-NOISE RATIO USING
# TRI-LEVEL DELTA-SIGMA MODULATION

John J. Paulos,   Gregory T. Brauns,   Michael B. Steer,   Sasan H. Ardalan

Center for Communications and Signal Processing
Department of Electrical and Computer Engineering
North Carolina State University
Box 7914, Raleigh, N.C. 27695-7914

## Abstract

A new approach to Delta-Sigma modulation is described which uses tri-level coding rather than the conventional binary coding. It is demonstrated through simulation that this approach dramatically reduces quantization noise, particularly for small input signals. The result is improved signal-to-noise performance and a relaxed digital filtering requirement. Circuit implementations for the tri-level modulator are discussed which demonstrate a minor increase in circuit complexity.

## Introduction

Delta-Sigma is one of a class of systems which use oversampling and 1-bit quantization to achieve high resolution A/D conversion at a lower rate [1]. Oversampling is attractive for many systems in that the analog anti-alias filtering requirements are relaxed  In addition, Delta-Sigma modulators can generally be implemented with few precision circuits [2,3]. A simplified implementation of a first-order Delta-Sigma modulator is shown in Fig. 1. This circuit can be implemented with a differential integrator, a comparator, and a flip-flop or sample-and-hold amplifier. The output of this system is a bit stream whose pulse density is proportional to the applied input signal. In analog-to-digital conversion applications, the basic Delta-Sigma modulator is followed by a decimating digital filter, which produces a more conventional digital representation of the signal at a much lower sampling rate.

Looking at the modulator in more detail, the operation of this circuit can be understood in terms of a feedback control system. At each sampling instant, the sign of the accumulated difference (or error) between the input and the output pulse stream is detected and held for one time period. If the comparator output is a logical "1", then a positive level is fed back to the differential integrator. If the comparator output is a logical "0", then a negative level is fed back to the integrator. Over time, the modulator will act to minimize the error between the input and the output pulse representation by constantly driving the integrated error to zero. For this discussion, it is convenient to consider the output of the flip- flop to be a sequence of pulses with amplitudes of precisely +1 V or -1 V. With this convention, the input can swing from -1 V to +1 V, and an input near zero is represented as an alternating +1, -1 sequence.

For small signals, most of the output power is actually noise, which will largely be removed by the decimating digital filter. However, some of this noise will fall into the signal band and will contribute to the quantization noise level of the overall system. Fortunately, the bit-by-bit quantization noise is shaped by the integrator so that a disproportionately small amount of the wideband noise falls into the signal band. If a 512 to 1 oversampling ratio is used, i.e., if the decimating digital filter has a passband of 1 part in 512 of the sampling rate, then the maximum signal-to-noise ratio of the filtered output can be in excess of 70 dB. Higher performance can be achieved either by using a larger oversampling ratio, or by using a higher-order loop filter in place of the integrator.

## Tri-level Coding

A rigorous analysis of quantization noise in Delta-Sigma modulators is presented in [4,5], and will not be reviewed in detail here. However, a key part of this analysis is a recognition of the fact that the power of the output sequence is unity, using the convention outlined above, independent of the output code sequence. Since the output is ideally a pulse density representation of the input, the quantization noise power at the output must be equal to unity less the input signal power. So, for small signals, the output quantization noise power is near unity, and the broadband signal-to-noise ratio is much less than one. This can be seen from Fig. 2, where the pulse stream response to a small positive input is shown. Although it is not proved here, it is reasonable to expect that if the broadband noise power can be reduced, the baseband signal-to-noise ratio (after filtering) will improve.

This can be achieved by using tri-level coding, as shown in Fig. 3. In this case, two comparator thresholds are used to establish a dead zone around zero. If the accu-

Reprinted from *IEEE Proc. ISCAS*, pp. 463-466, May 1987.

mulated error is large and positive, a code of +1 is produced. If the accumulated error is large and negative, a code of -1 is produced. However, if the accumulated error is small, a code of 0 is produced, and a zero-level signal is fed back to the integrator. In other words, if the accumulated error is small, defer feeding back a correcting signal until the accumulated error is significant. For low level inputs, rather than obtaining primarily an alternating +1, -1 sequence, the output will now be a sequence of 0 codes with occasional +1 or -1 entries to represent the small input signal. The pulse stream response of this circuit to a small positive input is shown in Fig. 4. The power of this output sequence is much less than unity, and therefore, the broadband quantization noise has been reduced.

Bi-level and tri-level Delta-Sigma modulators have been simulated using a new table-based simulation program described in the following section. An idealized integrator is used to demonstrate the theory of tri-level coding. For first-order systems, broadband quantization noise for tri-level coding was observed to be typically one third the level obtained with bi-level coding. A similar performance advantage was observed for in-band quantization noise, indicating an improved signal-to-noise ratio after digital filtering. Identical bi-level and tri-level Delta-Sigma modulators (with an oversampling ratio of 512) are compared in the signal-to-noise ratio versus input amplitude plot of Fig. 5. Over much of the range, the tri-level coding provides an improvement of greater than 15 dB.

The optimal comparator dead zone has been determined by running several signal-to-noise ratio analyses. Signal-to-noise ratio versus comparator threshold voltage with an input amplitude of -40 dB is displayed in Fig. 6 for the first-order Delta-Sigma modulator specified above. The bi-level signal-to-noise ratio is indicated as a reference point. Maximum signal-to-noise ratio is obtained for a threshold voltage of 0.2 V in the tri-level comparator, corresponding to a signal-to-noise ratio improvement of 18 dB over the bi-level case.

For second-order systems, broadband quantization noise for tri-level coding was observed to be typically one half the level obtained with bi-level coding. Identical bi-level and tri-level Delta-Sigma modulators are compared in Fig. 7. For small input signals, tri-level coding provides an improvement of only 3 - 4 dB. Signal-to-noise ratio versus comparator threshold voltage for an input amplitude of -40 dB is shown in Fig. 8 for the second-order Delta-Sigma modulator. The optimal comparator dead zone is observed to be 0.3 V, which corresponds to an improvement of 4 dB above the bi-level case. In this case, the choice of the comparator thresholds is problematic, and for several simulation samples the tri-level results are worse than those obtained with bi-level coding.

## Simulation

Simulation results have been made possible by the

development of a numerically efficient and accurate simulator for Delta-Sigma modulators. The simulator uses table look-up techniques to achieve high computation speeds, and high accuracy is maintained by using device-level simulation (e.g. using SPICE) to build the look-up tables. Delta-Sigma modulators are, as frequently implemented, sampled data systems, so it is only necessary for each table to model the performance of an individual subsystem at the sampling intervals.

While many analytic techniques [5,6,7] have been developed for predicting performance parameters such as signal-to-noise ratio, quantization noise spectra and stability thresholds, only numerical simulation can provide genuine system performance. Since the numerical simulation of Delta-Sigma modulators must be performed for a large number of clock cycles, often several tens of thousands, the use of general purpose circuit simulation programs such as SPICE would be prohibitively time-consuming. The development of ZSIM - a nonlinear Z-domain SIMulator - facilitates the incorporation of nonlinear component effects, slew-rate limiting, hysteresis, noise sources, and dead zoning of the comparator in a time efficient manner. In this approach, a Delta-Sigma modulator can be described as shown in Fig. 9, where each nonlinear function is represented by a table.

ZSIM consists of three modules - a Table Generator Module, a Simulator Module, and a Post-Processor Module with their relationship indicated in Fig. 10. The Table Generator Module develops the tables that describe the input-output characteristics of each subsystem. For this study, the table designed for the integrator of the modulator circuit is an ideal representation built from a difference equation. The Simulator Module implements the simulation of the entire system using the table description of each subsystem, and the Post-Processor Module calculates noise spectra and signal-to-noise ratio and displays results. The Post-Processor Module can work on either the modulator bit-stream output or the output of a decimating digital filter.

## Implementation of Tri-level Coding

The implementation of tri-level coding requires changes in both the analog circuitry and the decimating digital filter. In the analog section, a second comparator is required to divide the integrator output range into three segments for the tri-level coding. In addition, the integrator must be modified to allow tri-level feedback (+1, -1, and 0). Unfortunately, even-order harmonic distortion can occur if the +1 and -1 reference levels are not exactly matched (centered around zero). This can be seen by referring back to Fig. 3. In this system, the precision level generator is in the feedback path of the Delta-Sigma control loop. Therefore, the linearity of the input-output relationship depends directly upon the accuracy of the feedback reference levels. In contrast, for conventional bi-level Delta-Sigma systems, an imbalance between the

+1 and -1 levels produces only an offset in the input-output relationship.

Most recent Delta-Sigma implementations have used switched-capacitor techniques for the integrator [3,8]. A simple example of a differential switched-capacitor integrator for bi-level coding is shown in Fig. 11. This circuit could be extended to tri-level coding by adding a zero input that can be switched into the circuit in place of VREF or -VREF. However, this would lead to an error if VREF and -VREF are not perfectly matched. A better approach is to use a differential clocking scheme for the reference feedback as shown in Fig. 12. In this circuit, the clocking can be arranged so that the reference feedback is proportional to VREF-0, 0-VREF, and 0-0 for the codes +1, -1, and 0 respectively. This ensures that the integrator responses for the +1 and -1 codes are exactly symmetric around the integrator response for the 0 code.

The changes required in the decimating digital filter are primarily in the first FIR decimating stage, where integer arithmetic can be used. Although a complete tri-level decimator implementation has not yet been attempted, it does not seem that tri-level coding will require a dramatic increase in circuit complexity. In many early Delta-Sigma implementations [2], the decimation stage of the digital filter was implemented using an up/down counter. This technique can also be used for tri-level coding. For each clock period, the counter will either count up, count down, or do nothing. Although other more elaborate decimation schemes have been proposed to achieve higher performance filtering, it is believed that these techniques can be extended in a similar manner. In addition, since the tri-level coding significantly reduces the broadband quantization noise, it may be possible to achieve similar noise performance with simpler decimation techniques.

## Conclusion

The use of tri-level coding has been proposed as a means for improving signal-to-noise performance in Delta-Sigma converters. An improvement of more than 15 dB relative to conventional bi-level coding has been demonstrated through simulation for a simple first-order modulator. Second-order modulators show a relatively small improvement of only 4 dB relative to bi-level coding. Future work is planned to extend the analysis of quantization noise in Delta-Sigma modulators presented in [4,5] to the case of tri-level coding.

## References

[1] H. Inose, Y. Yasuda, and J. Murakami, "A telemetering system by code modulation - Delta-Sigma modulation," *IRE Trans. on Space Electronics and Telemetry*, vol. SET-8, pp. 204-209, Sept. 1962.

[2] R.J. van de Plassche, "A Sigma Delta modulator as an A/D converter," *IEEE Trans. on Circuits and Systems*, vol. CAS-25, pp. 510-514, July 1978.

[3] M.W. Hauser, P.J. Hurst, and R.W. Brodersen, "MOS ADC-filter combination that does not require precision analog components," *IEEE Inter. Solid-State Circuits Conf. Dig. of Tech. Papers*, pp. 80-82, Feb. 1985.

[4] S.H. Ardalan and J.J. Paulos, "An analysis of nonlinear behavior in Delta-Sigma modulators," To be published in *IEEE Trans. on Circuits and Systems*.

[5] S.H. Ardalan and J.J. Paulos, "Stability analysis of higher order Sigma-Delta modulators," *IEEE Int. Symp. Circuits and Systems Dig. of Tech. Papers*, pp. 715-719, May 1986.

[6] B.P Agrawal and K. Shenoi, "Design methodology for $\Sigma\Delta$M," *IEEE Trans. Comm.*, vol. COM-31, pp. 360-370, March 1983.

[7] J.C. Candy and O.J. Benjamin, "The structure of quantization noise from Sigma-Delta modulation," *IEEE Trans. Comm.*, vol. COM-29, pp. 1316-1323, Sept. 1981.

[8] T. Misawa, J.E. Iwersen, L.J. Loporcaro, and J.G. Ruch, "Single-chip per channel CODEC with filters utilizing Delta-Sigma modulation," *IEEE Journal of Solid-State Circuits*, vol. SC-16, pp. 333-341, Aug. 1981.

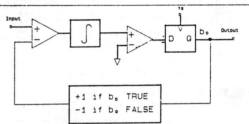

Fig. 1 : Simplified schematic of a first-order DSM

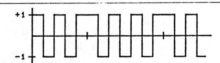

Fig. 2 : Typical output sequence for bi-level coding

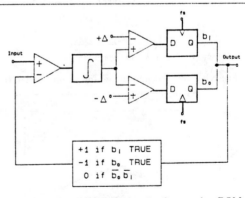

Fig. 3 : Simplified schematic of a first-order DSM using tri-level coding

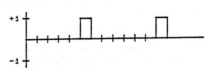

Fig. 4 : Typical output sequence for tri-level coding

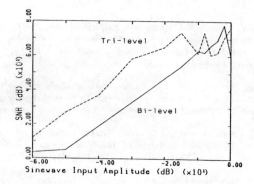

Fig. 5 : SNR vs. input amplitude for first-order DSM

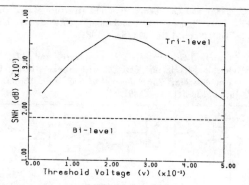

Fig. 6 : SNR vs. comparator threshold voltage of a
first-order DSM with an input amplitude of -40 dB

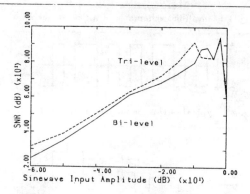

Fig. 7 : SNR vs. input amplitude for second-order DSM

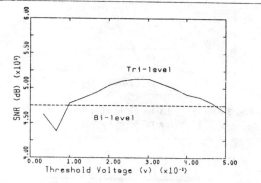

Fig. 8 : SNR vs. comparator threshold voltage of a
second-order DSM with an input amplitude of -40 dB

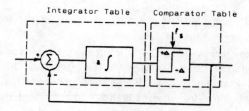

Fig. 9 : Subdivision of DSM for table representation

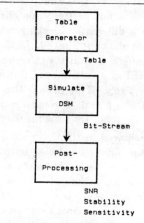

Fig. 10 : Simulation flowchart for ZSIM

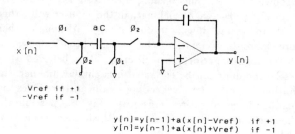

Fig. 11 : Switched-capacitor integrator for bi-level coding

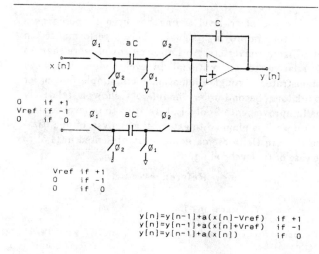

Fig. 12 : Switched-capacitor integrator for tri-level coding

248

# A Second-Order High-Resolution Incremental A/D Converter with Offset and Charge Injection Compensation

JACQUES ROBERT AND PHILIPPE DEVAL

*Abstract*—Sigma-delta modulation, associated with oversampling and noise shaping, is a well-known technique used in high-accuracy A/D converters. Such converters, required in telecommunications applications, are characterized by ac performance such as signal-to-noise ratio. Moreover, they are mainly dedicated to applications which can tolerate offset and gain errors. On the other hand, measurement and instrumentation applications require absolute accuracy, e.g., offset and gain errors cannot be tolerated. These applications are characterized by dc performance such as differential and integral nonlinearities, offset and gain errors, and they often require high resolution. The second-order incremental A/D converter, which makes use of sigma-delta modulation associated with a simple digital filter, is capable of achieving such requirements. Experimental results, obtained of circuits fabricated in a SACMOS 3-μm technology, indicate that 15-bit absolute accuracy is easily achievable, even with a low reference voltage.

## I. INTRODUCTION

RECENTLY, a first-order micropower incremental A/D converter was proposed [1], [2]. An absolute accuracy of 16 bits has been obtained, but the conversion time was large, of the order of 1 s for a consumption of 65 μW, restricting the applicability of such a circuit to very slowly varying signals. In this paper, the use of a second-order structure, allowing a considerable reduction of the conversion time, is first discussed. The optimal digital processing of the comparator outputs is derived with a simple temporal method. An offset and charge injection compensation is then presented. Finally, a comparison with sigma-delta converters will be carried out.

## II. FIRST- AND SECOND-ORDER SIGMA-DELTA MODULATOR

Fig. 1(a) represents a first-order sigma-delta modulator. The stability of this circuit is ensured because there is only one integrator in the loop. This structure has been used in a first-order incremental A/D converter [1], [2]. Fig. 1(b) represents a conventional second-order sigma-delta modulator. Since there are two integrators in the loop, their outputs can be very large and must be limited [4], which is

Manuscript received October 21, 1987; revised January 25, 1988.
The authors are with the Electronic Laboratory, Swiss Federal Institute of Technology, 1015 Lausanne, Switzerland.
IEEE Log Number 8820726.

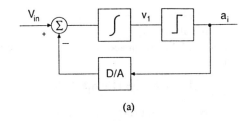

(a)

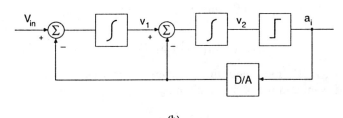

(b)

Fig. 1. (a) First-order sigma-delta modulator, and (b) conventional second-order sigma-delta modulator.

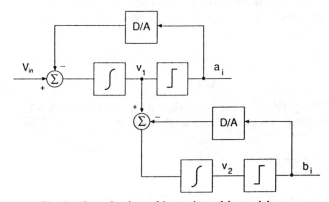

Fig. 2. Second-order multistage sigma-delta modulator.

not acceptable in the design of a second-order incremental A/D converter.

Fig. 2 shows a different structure of a second-order sigma-delta modulator, also called a multistage sigma-delta modulator [4], [5]. The stability of this modulator is ensured because there is only one integrator in each loop. This structure will be used in the second-order incremental A/D converter.

Reprinted from *IEEE J. Solid-State Circuits*, vol. 23, pp. 736–741, June 1988.

## III. First-Order Incremental A/D Converter

Fig. 3 represents the circuit diagram of the analog part of the first-order incremental A/D converter [1], [2]. It is composed of a stray-insensitive switched-capacitor integrator, a comparator, and switch control logic.

A four-phase nonoverlapping clock cycle constitutes one integration period (IP), as defined in Fig. 4. The integrator output voltage is designated by $v_1[i, j]$, where $i$ corresponds to the current integration period $IP_i$ and $j$ to the clock cycle ($j = 1, 2, 3,$ or $4$). Assuming ideal components, the circuit operation for one integration period is as follows.

During clock cycle $\phi_1$, $S_1$ and $S_4$ are closed, charging $\alpha_1 C_1$ to the input voltage $V_{in}$.

During $\phi_2$, $S_3$ and $S_5$ are closed, transferring the charge from $\alpha_1 C_1$ to $C_1$:

$$v_1[i, 2] = v_1[i, 1] + \alpha_1 V_{in}. \tag{1}$$

At the end of the charge transfer, the comparator output is denoted by $a_i$, where $a_i = 1 \ (-1)$ if $v_1[i, 2] > 0 \ (< 0)$.

During $\phi_3$, $S_3 \ (S_2)$ and $S_4$ are closed if $a_i = 1 \ (a_i = -1)$.

During $\phi_4$, $S_2 \ (S_3)$ and $S_5$ are closed if $a_i = 1 \ (a_i = -1)$, so that

$$v_1[i, 4] = v_1[i, 1] + \alpha_1(V_{in} - a_i V_R) \tag{2}$$

where $V_R$ represents the reference voltage. Note that the dynamic range of $V_{in}$ is given by

$$-V_R \leqslant V_{in} \leqslant V_R. \tag{3}$$

Now, consider a sequence composed of $p$ integration periods $IP_i$ ($1 \leqslant i \leqslant p$), preceded by the resetting of $v_1$ and a sample and hold of $V_{in}$. During $IP_1$, the comparator compares $v_1[1, 2] = \alpha_1 V_{in}$ with 0, resulting in a comparison level equal to

$$L_1 = 0 \Rightarrow \begin{cases} a_1 = 1, & \text{if } V_{in} > 0 \\ a_1 = -1, & \text{if } V_{in} < 0. \end{cases} \tag{4}$$

During $IP_2$, $v_1[2, 2] = \alpha_1(2V_{in} - a_1 V_R)$ is compared to 0, resulting in a comparison level equal to

$$L_2 = a_1 V_R/2 \Rightarrow \begin{cases} a_2 = 1, & \text{if } V_{in} > 0.5 a_1 V_R \\ a_2 = -1, & \text{if } V_{in} < 0.5 a_1 V_R. \end{cases} \tag{5}$$

Finally, during $IP_p$, $v_1[p, 2]$ is compared to 0:

$$L_p = \left( \sum_{i=1}^{p-1} a_i V_R \right) \Big/ p$$

$$\Rightarrow \begin{cases} a_p = 1, & \text{if } V_{in} > \left( \sum_{i=1}^{p-1} a_i V_R \right) \Big/ p \\ a_p = -1, & \text{if } V_{in} < \left( \sum_{i=1}^{p-1} a_i V_R \right) \Big/ p. \end{cases} \tag{6}$$

Fig. 5 represents each comparison level $L_i$ as a function of $i$ ($p = 8$). One can distinguish four "dead zones," two

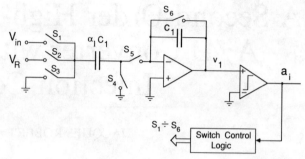

Fig. 3. Circuit diagram of the first-order incremental A/D converter.

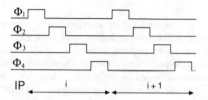

Fig. 4. Integration period (IP) associated with a four-phase nonoverlapping clock cycle.

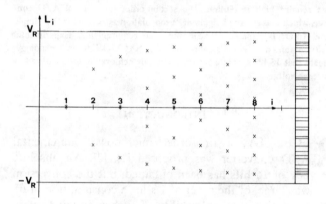

Fig. 5. Comparison levels $L_i$ as a function of $i$ ($p = 8$ in this case).

near 0, one near $+V_R$, and one near $-V_R$, which have widths equal to $V_R/p$. If a quantization error of less than $\pm 0.5$ LSB is desired, the maximum resolution is given by

$$n_1 = 1b[2V_R/(V_R/p)] = 1b(p) + 1 \text{ [bits]} \tag{7}$$

where $1b(p)$ denotes the binary logarithm of $p$. Hence, for a resolution of $n_1 = 16$ bits, a total number of $p = 32\,768$ integration periods is required for every conversion cycle, involving a very slow operation.

The output code $N_1$ is performed by using an up–down counter [2] which evaluates the quantity

$$N_1 = \left( \sum_{i=1}^{p} a_i \right) + \text{sign}(v_1[p, 4]) \tag{8}$$

where $\text{sign}(v_1[p, 4])$ is equal to 1 (0) if $v_1[p, 4]$ is positive (negative). As shown by (8), the digital processing of the comparator output corresponds to a very simple digital filter having a rectangular-shaped impulse response (each coefficient is equal to 1). Some different filters can be used [7]. Resolution may be increased, but *accuracy cannot*. This is due to the dead zones of Fig. 5. Fig. 6(a)–(c) represents

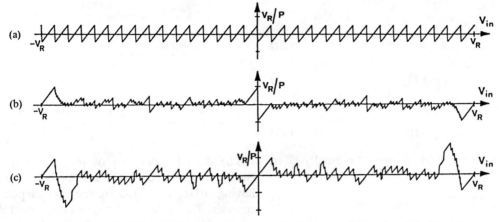

Fig. 6.  Simulation of the integral nonlinearity of a first-order sigma-delta modulator ($p = 16$): (a) rectangular window, (b) nonsymmetrical triangular window, and (c) symmetrical triangular window.

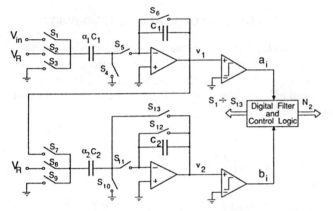

Fig. 7.  Circuit diagram of the second-order incremental A/D converter.

a simulation of the integral nonlinearity with $p = 16$ for three different filters: a) $C_i = 1$ (rectangular window as defined by (8)); b) $C_i = p + 1 - i$ (nonsymmetrical triangular window); and c) $C_i = i$ for $1 \leqslant i \leqslant p/2$, and $C_i = p + 1 - i$ for $p/2 + 1 \leqslant i \leqslant p$ (symmetrical triangular window [7]), respectively. One can see that resolution is increased in Fig. 6(b) and (c) (the total number of different codes is higher than in Fig. 6(a)), but the maximum integral nonlinearity errors which are associated with the "dead zones" of Fig. 5 are increased.

It has been demonstrated that the converter accuracy is independent of the capacitor ratio $\alpha_1$ [1], [2]. Moreover, the effect of the amplifier offset and the charge injection can be compensated by using a compensation described in Section VI.

## IV. SECOND-ORDER INCREMENTAL A/D CONVERTER

Fig. 7 represents the analog part of the second-order incremental A/D converter [5]. Assuming ideal components and $\alpha_1 = \alpha_2 = 1$, one conversion cycle is composed of the following.

1) A resetting of both integrator outputs and a sample and hold of $V_{in}$.

2) A first integration period $IP_1$ (note that during this integration period $v_2$ is not modified), where one obtains

$$v_1[1,4] = V_{in} - a_1 V_R \qquad (9)$$

$$v_2[1,4] = 0. \qquad (10)$$

3) $(p-1)$ integration periods; $V_{in}$ ($v_1[i-1,4]$) is integrated in the first (second) integrator during $IP_i$, so that

$$v_1[2,4] = 2V_{in} - (a_1 + a_2)V_R \qquad (11)$$

$$v_2[2,4] = V_{in} - a_1 V_R - b_2 V_R \qquad (12)$$

$$v_1[3,4] = 3V_{in} - (a_1 + a_2 + a_3)V_R \qquad (13)$$

$$v_2[3,4] = 3V_{in} - (2a_1 + a_2)V_R - (b_2 + b_3)V_R \qquad (14)$$

$$v_1[p,4] = pV_{in} - \sum_{i=1}^{p} a_i V_R \qquad (15)$$

$$v_2[p,4] = (p-1)pV_{in}/2 - \sum_{i=1}^{p-1} a_i(p-i)V_R - \sum_{i=2}^{p} b_i V_R. \qquad (16)$$

4) An integration period $IP_{p+1}$, involving

$$v_2[p+1,4] = p(p+1)V_{in}/2$$

$$- \sum_{i=1}^{p} a_i(p+1-i)V_R - \sum_{i=2}^{p+1} b_i V_R. \qquad (17)$$

Under the conditions $\alpha_1 = 1$, $\alpha_2 = 1$, it can be shown that the dynamic range of $v_2[p+1,4]$ is given by

$$-V_R \leqslant v_2[p+1,4] \leqslant V_R. \qquad (18)$$

With (17), (18) can be rewritten in the form of

$$-2V_R/[p(p+1)]$$

$$\leqslant V_{in} - \left\{ \sum_{i=1}^{p} a_i(p+1-i) + \sum_{i=2}^{p+1} b_i \right\} 2V_R/[p(p+1)]$$

$$\leqslant 2V_R/[p(p+1)]. \qquad (19)$$

Defining $D$ and $x$ as

$$D = \sum_{i=1}^{p} a_i(p+1-i) + \sum_{i=2}^{p+1} b_i \qquad (20)$$

$$x = 2V_R / [p(p+1)] \qquad (21)$$

(19) can be rewritten in the form of

$$-x \leqslant V_{in} - Dx \leqslant x. \qquad (22)$$

Ideal A/D converter quantization error is normally given by

$$-V_{LSB}/2 \leqslant V_{in} - NV_{LSB} \leqslant V_{LSB}/2 \qquad (23)$$

where $N$ is the digital representation of $V_{in}$ and $V_{LSB}$ is the analog voltage corresponding to the least significant bit. Equations (22) and (23) give

$$V_{LSB} = 4V_R / [p(p+1)]. \qquad (24)$$

The converter output code $N_2$ is thus given by

$$N_2 = \left( \sum_{i=1}^{p} a_i(p+1-i) + \sum_{i=2}^{p+1} b_i \right) \Big/ 2. \qquad (25)$$

Resolution is given by (note that $-V_R \leqslant V_{in} \leqslant V_R$)

$$n_2 = 1b(2V_R/V_{LSB}) = 1b[p(p+1)] - 1$$
$$= 21b(p) - 1 \text{ [bits]}, \quad \text{if } p \gg 1. \quad (26)$$

The above development does not take into account the error caused by the capacitor matching. Introducing $\alpha_1 = 1 + \Delta\alpha_1$ and $\alpha_2 = 1 + \Delta\alpha_2$, one can show that an additional quantization error $\epsilon_c$ results:

$$\epsilon_c = 0.25 p \Delta\alpha_1 \text{ [LSB]}. \qquad (27)$$

It is interesting to notice that the value of $\alpha_2$ does not affect the conversion accuracy. As an example, a 16-bit resolution, which requires from (26) that $p = 362$, associated with a standard technology ($\Delta\alpha_1 = 0.1 \sim 0.2$ percent) involves an error $\epsilon_c$ smaller than $0.1 \sim 0.2$ LSB.

An extra-bit accuracy can be obtained by detecting the sign of $v_2[p+1,4]$ at the end of the conversion cycle. This can be achieved without increasing significantly the conversion time. Equations (24)–(27) are replaced by

$$V_{LSB} = 2V_R / [p(p+1)] \qquad (28)$$

$$N_2 = D + \text{sign}(v_2[p+1,4]) \qquad (29)$$

$$n_2 = 21b(p) \text{ [bits]} \qquad (30)$$

$$\epsilon_c = 0.5 p \Delta\alpha_1 \text{ [LSB]}. \qquad (31)$$

Hence, for a $n_2 = 16$-bit resolution, a total number of $p + 1 = 257$ integration periods is required for each conversion cycle. Compared to a first-order structure, this yields an important reduction of the conversion time. It should be pointed out that (30) gives the *maximum resolution* corresponding to a quantization error of less than $\pm 0.5$ LSB.

## TABLE I

| Resolution | order | | | |
|---|---|---|---|---|
| | 1 | 2 | 3 | 4 |
| 12 | 2048 | 65 | 25 | 18 |
| 13 | 4096 | 91 | 31 | 21 |
| 14 | 8196 | 129 | 39 | 24 |
| 15 | 16368 | 182 | 48 | 28 |
| 16 | 32768 | 257 | 60 | 33 |

The digital processing of the comparator outputs, as shown by (20) and (29), corresponds to two digital filters having triangular- and rectangular-shaped impulse responses. The order of the sigma-delta modulator and the digital filter must therefore be equal if the quantization error is expected to satisfy (23).

## V. $M$TH-ORDER INCREMENTAL A/D CONVERTER

The developments of Section IV can be extended to a $m$th ($m > 2$) order incremental A/D converter. Due to the use of a multistage sigma-delta modulator, there are no overload effects even if $m > 2$ [5], [6]. It can be shown that resolution is given by

$$n_m = 1b\left\{ \left[ \prod_{i=1}^{m} (p+i-1) \right] \Big/ m! \right\} + 1$$

$$= m1b(p) - 1b(m!) + 1, \quad \text{if } p \gg 1. \quad (32)$$

Table I represents the total number of integration periods $N_{IP1} = p + m - 1$ required for each conversion cycle as a function of the resolution and the order. Each comparator output sequence is fed into a digital filter. Impulse response of the $i$th filter ($1 \leqslant i \leqslant m$) is of $(m - i + 1)$th order. Such filters can easily be implemented by using counters, adders, and registers, without requiring prohibitive die area. Converter output code is obtained by adding the filter outputs.

## VI. OFFSET AND CHARGE INJECTION COMPENSATION

Consider the general case of a $m$th-order incremental A/D converter. Each integrator $I_i$ ($1 \leqslant i \leqslant m$) can be characterized by an input-referred offset $V_{\epsilon i}$. This offset is caused by the charge injection of the switches which are situated on the right side of the capacitors $\alpha_i C_i$ ($1 \leqslant i \leqslant m$) and by the amplifier offset. Note that $V_{\epsilon i}$ can be made independent of $V_{in}$ [1]. In order to compensate the effect of these $V_{\epsilon i}$ terms, the conversion cycle is divided into three periods, preceded by a reset of each integrator output and a sample and hold of $V_{in}$.

1) During the first period, which requires $p + m - 1$ integration periods, the circuit operation corresponds to that described in Section IV. The integrator output voltage

TABLE II

| Resolution | order | | | |
|---|---|---|---|---|
| | 1 | 2 | 3 | 4 |
| 12 | 2049 | 94 | 41 | 30 |
| 13 | 4097 | 130 | 51 | 34 |
| 14 | 8197 | 184 | 63 | 40 |
| 15 | 16 369 | 258 | 77 | 48 |
| 16 | 32 769 | 366 | 97 | 56 |

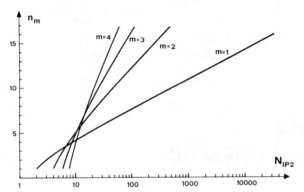

Fig. 8.  A graph of $n_m$ against $N_{IP2}$ for various values of the order $m$.

$v_m$ is given by

$$v_m[p+m-1,4]$$
$$= \left\{ \prod_{i=1}^{m}(p+i-1) \right\} V_{in}/m!$$
$$+ F_1(a,b,\cdots,V_R,p,m)$$
$$+ \sum_{i=1}^{m} \left\{ 2V_{\epsilon i} \prod_{j=i}^{m}(p+m-j) \right\} \Big/ (m-i+1)! \quad (33)$$

where $F_1(a,b,\cdots,V_R,p,m)$ represents the voltage component of $V_m$ related to the reference voltage.

2) During the second period, the output voltages of the first $m-1$ integrators are reset ($v_i[p+m,4]=0$, $1 \leqslant i \leqslant m-1$), while $v_m[p+m-1,4]$ is inverted [1]:

$$v_m[p+m,4] = -v_m[p+m-1,4]. \quad (34)$$

3) During the third period, which requires $p+m-1$ integration periods, the circuit operation is analogous to that of the first period, except that the voltage $v_1$ integrates $-V_{in}$ instead of $+V_{in}$. One obtains

$$v_m[2p+2m-1,4] = -\left\{ 2\left( \prod_{i=1}^{m}(p+i-1) \right) V_{in}/m! \right.$$
$$\left. + F_2(a,b,\cdots,V_R,p,m) \right\}. \quad (35)$$

One can see from this last equation that all error terms $V_{\epsilon i}$ have disappeared. It can be shown that resolution is now given by

$$n_m = 1b\left\{ \prod_{i=1}^{m}(p+i-1) \right\} \Big/ m! + 2$$
$$= m 1b(p) - 1b(m!) + 2 \text{ [bits]}, \quad \text{if } p \gg 1. \quad (36)$$

Table II represents the total number of integration periods $N_{IP2} = 2p + m$ required for each conversion cycle as a function of the resolution and the order. Fig. 8 presents a graph of $n_m$ against $N_{IP2}$ for various values of $m$. The ratio of conversion times with and without offset compensation is given by

$$N_{IP2}/N_{IP1} = 2^{(m-1)/m}, \quad m \geqslant 1, \ p \gg m. \quad (37)$$

It should be pointed out that the order of the digital filter is not modified by the offset and charge injection compensation.

## VII. Comparison with Sigma-Delta Converters

When the number of integration periods $p$ is doubled, (36) indicates a gain of $m$ bits. In terms of signal-to-noise ratio, this corresponds to an improvement of $6m$ dB per octave. Theory of sigma-delta modulation [3], [9] shows that the corresponding improvement of sigma-delta converters is equal to $6m + 3$ dB per octave of oversampling (9 dB (15 dB) for the first (second) order). Those results have been obtained under the condition that the input voltage is "sufficiently busy." It should be pointed out that these values correspond to the average noise level on the whole dynamic range [3]. It is well-known that certain quiet input levels produce a drastic increase of the in-band noise; such peaks of noise decrease only with a ratio of 6 dB (12 dB) per octave of oversampling for a first- (second-) order structure (see [8, fig. 2]). These peaks of noise are related to the "dead zones" of Fig. 5.

MOS technology involves large converter offset caused by the amplifier offset and charge injection. The incremental converter is very well suited to offset compensation, while sigma-delta converters are not. This difference excludes many sigma-delta converter applications.

Digital signal processing required by sigma-delta converters demands extensive digital filters, while the incremental converter needs only a simple one. As an example, a second-order filter designed for a 16-bit resolution involves a die area equal to only 1 mm² in a 3-μm SACMOS process. The design of incremental converters is therefore very versatile: filter design can easily be carried out for a wide range of applications.

The incremental converter requires a sample and hold, whereas sigma-delta converters sample continuously the input signal, which must be sufficiently busy. In many applications, the quantization of specific samples is desired. For example, some industrial process controls use multiplex operation. It also happens that the input signal is not always available. Therefore, as a sample and hold is necessary, incremental converters offer significant advantages over sigma-delta converters.

## VIII. Experimental Results

A second-order micropower incremental A/D converter has been integrated in a 3-μm low-voltage p-well SACMOS technology. The capacitances were equal to 10 pF ($\alpha_1 =$

Fig. 9. Photograph of the second-order incremental A/D converter.

TABLE III

| | |
|---|---|
| Supply voltage | $V_{DD} = - V_{SS} = 2.5$ V |
| Power consumption of the analog part | 325 $\mu$W |
| Resolution | 15 bits |
| Reference voltage | 0.82 V |
| $V_{LSB}$ | 50 $\mu$V |
| Conversion time | 10 ms |
| Active die area (including logic and pads) | 4.6 mm$^2$ |
| Offset | < 0.25 LSB |
| Gain error | < 0.3 LSB |
| Differential nonlinearity | < 0.3 LSB |
| Integral nonlinearity | < 0.3 LSB |
| PSSR | > 65 dB |

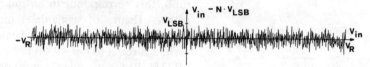

Fig. 10. Measured quantization error function. $V_R = 0.82$ V, $n = 15$ bits, and $T_c = 10$ ms.

$\alpha_2 = 1$), and the operational amplifiers were single-stage cascoded transconductance amplifiers. The photograph of the chip is shown in Fig. 9. Its die area is 4.6 mm$^2$.

Table III summarizes the experimental results while Fig. 10 presents a typical measured quantization error function.

## IX. Conclusions

A different use of sigma-delta modulation has been reported. Compared to a first-order incremental A/D converter, the use of a second-order structure allows a substantial reduction of the conversion time. No overload effects exist, even if multiple loop structures are used.

An optimal digital filtering has been derived. It was found that the order of both the filter and the sigma-delta modulator must be equal. Such filters are easy to implement and require modest die area. The presented A/D conversion principle is well suited to offset compensation.

Compared to some other high-resolution A/D converters, the incremental converter requires only a single reference voltage for bipolar operation. Moreover, it does not need laser trimming or self-calibration cycles. As with many other A/D converters, the incremental converter requires a sample and hold. This is its only drawback compared to sigma-delta converters.

## References

[1] J. Robert and V. Valencic, "Offset and charge injection compensation in an incremental analog-to-digital converter," in *Proc. European Solid-State Circuits Conf.* (Toulouse, France), Sept. 1985, pp. 45–48.

[2] J. Robert, G. C. Temes, V. Valencic, R. Dessoulavy, and P. Deval, "A 16-bit low-voltage CMOS A/D converter," *IEEE J. Solid-State Circuits*, vol. SC-22, no. 2, pp. 157–163, Apr. 1987.

[3] J. C. Candy, "A use of double integration in sigma-delta modulation," *IEEE Trans. Commun.*, vol. COM-33, no. 3, pp. 249–258, Mar. 1985.

[4] T. Hayashi, Y. Inabe, K. Uchimura, and T. Kimura, "A multistage delta-sigma modulator without double integration loop," in *ISSCC Dig. Tech. Papers*, Feb. 1986, pp. 182–183.

[5] J. Robert and P. Deval, "A second order high-resolution incremental A/D converter with offset and charge injection compensation," in *Proc. European Solid-State Circuits Conf.* (Bad Soden, Germany), Sept. 1987, pp. 109–112.

[6] Y. Matsuya, K. Uchimura, A. Iwata, T. Kobayashi, and M. Ishikawa, "A 16b oversampling A/D conversion technology using triple noise shaping," in *ISSCC Dig. Tech. Papers*, Feb. 1987, pp. 48–49.

[7] J. C. Candy, Y. C. Ching, and D. S. Alexander, "Using triangularly weighted interpolation to get 13-bit PCM from a sigma-delta modulator," *IEEE Trans. Commun.*, vol. COM-24, no. 11, pp. 1268–1275, Nov. 1976.

[8] J. C. Candy and O. J. Benjamin, "The structure of quantization noise from sigma-delta modulation," *IEEE Trans. Commun.*, vol. COM-29, no. 9, pp. 1316–1323, Sept. 1981.

[9] J. C. Candy, "Decimation for sigma-delta modulation," *IEEE Trans. Commun.*, vol. COM-34, no. 1, pp. 72–76, Jan. 1986.

# IMPROVED DOUBLE INTEGRATION DELTA-SIGMA MODULATIONS
## FOR A TO D AND D TO A CONVERSION

Yasuo SHOJI and Takao SUZUKI


Digital Communications Laboratories, Oki Electric Industry Co., Ltd.
4-10-3 Shibaura, Minato-ku, Tokyo 108, Japan

## ABSTRACT

This paper describes a new form of "double integration" delta sigma modulator as the analog-to-digital and digital-to-analog converter for the voiceband signals. It has the following three features compared with the earlier double integration delta sigma modulator.

1) The processing of the sampling rate can be executed at 1/2 clock speed.
2) The low-voltage power drive and expansion of the dynamic range are possible because the internal integral voltage is 1/2 or less.
3) The noise component out of the baseband is attenuated more effectively to improve the signal-to-noise ratios.

The properties of this new modulator and its performance are demonstrated with computer simulations.

## INTRODUCTION

The delta sigma modulation is a useful technique that employs integration and feedback with minimum complexity. The delta sigma modulator[1] and its interpolative type[2] have been proposed as the analog-to-digital and digital-to-analog converter for the voiceband used in the PCM.

The quantizer's output of the delta sigma modulator is basically two values, and the element value need not be adjusted during manufacture. In the analog-to-digital converter, the integrator can be composed of operational amplifiers and switched capacitors, and in addition, a digital-to-analog converter is not required in the feedback path. Therefore, delta sigma modulator features high performance, a small circuit scale, and the possibility of manufacturing an LSI of a small chip size.

In the analog-to-digital converter realized by the delta sigma modulator, the quantizer's output signal is passed through the digital filter equipped with the decimator function to eliminate the noise component out of the voiceband to reduce the sampling rate, and the digital signal having the specified dynamic range is realized at the

specified sampling rate.

In the digital-to-analog converter realized by the delta sigma modulator, on the contrary, the digital signal at the specified sampling rate is passed through the digital filter equipped with the interpolator function to increase the sampling rate up to the oversampling rate, the delta sigma modulator is performed to convert the signal into the digital signal, and then the signal becomes an analog signal.

The interpolative type of the delta sigma modulator requires an integrator after the quantizer. In the anlog-to-digital converter, therefore, there is a disadvantage that the circuit is complicated such as a digital-to-analog converter is required in the feedback path, and so on. Generally, the ordinary delta sigma modulator type is advantageous from the viewpoint of the circuit scale.

## DELTA SIGMA MODULATION

The delta sigma modulator which has one or two integrators is generally used in view of realizing the circuit. When more than two integrators are used, the behavior of the circuit becomes unstable. [3]

The dynamic range in the baseband of the single integration delta sigma modulator (SIΔΣM) and the double integration delta sigma modulator (DIΔΣM) are represented by equations (1) and (2) respectively.

$$DR|_{SI\Delta\Sigma M} = 10\,log\left\{ \frac{9}{16\pi^2}\cdot\left(\frac{f_S}{f_B}\right)^3\cdot A^2\right\} \qquad (1)$$

$$DR|_{DI\Delta\Sigma M} = 10\,log\left\{ \frac{15}{64\pi^2}\cdot\left(\frac{f_S}{f_B}\right)^5\cdot A^2\right\} \qquad (2)$$

where  $A$ : Input amplitude
$f_S$ : Oversampling frequency
$f_B$ : Baseband frequency

Figure 1 shows the graphs of equations (1) and

Reprinted from *IEEE Proc. ISCAS'87*, pp. 451–454, May 1987.

(2). It is evident from Figure 1 that, to ensure the dynamic range in the voiceband of 84 dB (equivalent to linear 14 bits) or more required for the PCM, so it is necessary to set the sampling rate about 2 or 1 MHz using the double integration delta sigma modulator.

To achieve the same characteristics in the voiceband using the single integration delta sigma modulator, it is necessary to set the sampling rate 4 MHz or more. Therefore, it seems difficult to apply the single integration delta sigma modulator to the analog-to-digial and digital-to-analog converter for the PCM in view of realization.

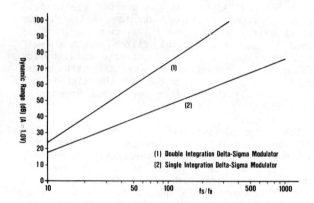

Figure 1.  Simulated dynamic range to $f_S/f_B$ ratios of delta-sigma modulators

The double integration delta sigma modulator has a high internal integral voltage, and in addition, it must be operated at a clock which is double the sampling rate. This was a big obstacle to realize the circuit.

IMPROVED DOUBLE INTEGRATION DELTA SIGMA MODULATION

Proposed in this paper is the improved double integration delta sigma modulator (IDIΔΣM) which has eliminated the aforementioned shortcomings of the double integration delta sigma modulator. The simulation results in consideration of its application to the voiceband of the PCM showed that it has better characteristics than the double integration delta sigma modulator.

Figure 2 shows the configuration of a coder of a general linear feedback type. The delta sigma modulator is realized by selecting the functions H(Z) and B(Z) appropriately. In the double integration delta sigma modulator, H(Z) and B(Z) should be determined as in equations (3) and (4).

$$H(Z) = \frac{1}{(1 - Z^{-1})^2} \qquad (3)$$

$$B(Z) = Z^{-1}(2 - Z^{-1}) \qquad (4)$$

Figure 3 shows the double integration delta sigma modulator configuration as a block diagram of the digital processing type obtained by applying the relationship shown in equations (3) and (4) to H(Z) and B(Z) in Figure 2. The relationship between the input and the output is represented by equation (5).

$$V_{OUT} = V_{IN} + (1 - Z^{-1})^2 \cdot Q(Z) \qquad (5)$$

where  $Q(Z)$ : Quantizer's quantizing noise
$V_{OUT}$ : Output voltage
$V_{IN}$ : Input voltage

In equation (5), the amplitude characteristic of the $(1 - Z^{-1})^2$ term shows the high-pass filter function that the attenuation is infinite at zero frequency and that it decreases monotonously as the frequency rises. Therefore, the output noise in voiceband which is sufficiently lower than the sampling rate is at a low level. So the delta sigma modulator is sometimes called the noise-shaping modulator.

Figure 4 shows the configuration of the improved double integration delta sigma modulator proposed in this paper. The improved double integration delta sigma modulator has two integrators and two quantizers. Each quantizer is connected to the output of each integrator. The final output is half the sum of the quantizer's outputs Y1, Y2. A delay of one sampling is inserted between integrator 1 and integrator 2. The outputs of quantizers 1, 2 are two values of +1, -1. The final output is three values of +1, -1, 0. When composing the analog-to-digital converter, a digital-to-analog converter is not required in the feedback path.

The relationships between the input VIN and the output Y1 of quantizer 1 and between the same input and the output Y2 of quantizer 2 are represented by equations (6) and (7) respectively.

$$Y_1 = \frac{V_{IN}(2 - Z^{-1}) + Q_1(Z) \cdot (2 - 3Z^{-1} + 2Z^{-2}) - Q_2(Z) \cdot (Z^{-1} - Z^{-2})}{2 - 2Z^{-1} + Z^{-2}} \qquad (6)$$

$$Y_2 = \frac{V_{IN} \cdot Z^{-1} - Q_1(Z) \cdot Z^{-1} + Q_2(Z) \cdot (2 - 3Z^{-1} + Z^{-2})}{2 - 2Z^{-1} + Z^{-2}} \qquad (7)$$

where  $Q_1(Z)$ : Quantizing noise of quantizer 1
$Q_2(Z)$ : Quantizing noise of quantizer 2

As the output is half the sum of Y1 and Y2, the relationship between the input VIN and the output VOUT is represented by equation (8).

$$V_{OUT} = F(Z) \cdot \{V_{IN} + (1 - Z^{-1})^2 \cdot (Q_1(Z) + Q_2(Z))\} \qquad (8)$$

where

$$F(Z) = \frac{0.5}{(1 - Z^{-1} + 0.5Z^{-2})} \qquad (9)$$

F(Z) in equation (8) shows the low-pass filter characteristic that the attenuation is zero at zero frequency. It shows a flat attenuation characteristic in the voiceband which is sufficiently lower than the sampling rate. Since the zero point of the denominator of F(Z) exists in the unit circle of the Z plane, therefore, F(Z) is a stable transfer function. The attenuation by F(Z) is about 14 dB at the frequency which is half the sampling rate. In the analog-to-digital converter, the decimator uses the low-pass filter to attenuate the noise component of the modulator that will be aliased into the baseband by resampling.

In the double integration delta sigma modulator shown in Figure 3, it is necessary to perform the operation of integrators 1, 2 in series at an interval of one sampling. In the improved double integration delta sigma modulator shown in Figure 4, meanwhile, it is enough to perform the operation of integrators 1, 2 in parallel like a pipeline at an interval of one sampling. Therefore, the improved double integration delta sigma modulator requires the clock speed which is half that of the double integration delta sigma modulator to execute the same sampling rate.

The improved double integration delta sigma modulator has a feature that its internal integral voltage is low. As this modulator has the three values of the final output, we find by simulations that the internal voltages adjust themselves to make the effective loop gains.

## SIMULATION RESULTS

Simulation was conducted by setting the sampling rate to 2,048 kHz in order to achieve a dynamic range of 84 dB (equivalent to linear 14 bits) in the voiceband required for the PCM.

Figure 5 shows the simulation results for the quantizing noise spectrum when the improved double integration delta sigma modulator's input level is 0 dB. The quantizing noise spectrum in the high frequency band is at a low level of -40 dB or less due to the aforementioned quantizing noise shaping filter's attenuation effect.

Figure 6 shows the simulation results for the relationship between the internal integral voltage and the input of the improved double integration delta sigma modulator (IDI$\Delta\Sigma$M) and of the double integration delta sigma modulator (DI$\Delta\Sigma$M). The internal integral voltage of IDI$\Delta\Sigma$M is less than about half that of DI$\Delta\Sigma$M, which implies the possibility of the low-voltage power drive and expansion of the dynamic range.

Figure 7 shows the simulation results for the signal-to-noise ratio in the voiceband when the sampling rate is decimated to 32 kHz as IDI$\Delta\Sigma$M and DI$\Delta\Sigma$M are used for the analog-to-digital converter. The transfer function of the digital filter for the decimator normally used the

function with triangular weights. The signal-to-noise ratio IDI$\Delta\Sigma$M is better than that of DI$\Delta\Sigma$M because of the noise shaping filter's attenuation effect on the noise component out of the voiceband.

From the above simulation results, it is evident that the improved double integration delta sigma modulator can achieve a dynamic range of 90 dB or more in the voiceband by operating at the sampling rate of 2,048 kHz. It can be applied as the analog-to-digital and digital-to-analog converter for the PCM. The attenuation of F(Z) in equation (7) is 0.0013 (dB) at the frequency point of 4 kHz. Therefore, fluctuations in attenuation in the voiceband are negligible.

When this modulator is operated as the analog-to-digital converter, it becomes evident from the simulation results that the influence on the signal-to-noise ratios by the leaky integral effect is negligible as long as the operational amplifier's gain is about 60 dB.

To check out the design of this modulator, the analog-to-digital converter was built using the same switched capacitor ratio. The experimental results have been verified by breadboard measurements in the laboratory.

## CONCLUSION

We proposed the improved double integration delta sigma modulator. We confirmed through simulations and measurements that this new modulator is very effective to improve the conventional double integration delta sigma modulator.

## ACKNOWLEDGEMENTS

The authors express their sincere thanks to Mr. Yoshio Masuda, Executive Director, Dr. Shigehisa Nakaya, Director of Research & Development Division and Dr. Atsushi Fukasawa, Director of Digital Communications Laboratories at Oki Electric Industry Co., Ltd.

## REFERENCES

[1] T. Misawa, et al., "Single-Chip per Channel Codec with Filters Utilizing $\Delta$-$\Sigma$ Modulation" IEEE J. Solid-State circuits, Vol. SC-16, No. 4, pp. 333-341, August 1981

[2] B.A. Wooley, J.L. Henry, "An Integrated Per-Channel PCM Encoder Based on Interpolation" IEEE J. Solid-State Circuits, Vol. SC-14, No. 1, pp. 14-20, Feb. 1979

[3] J.C. Candy, "A Use of Double Integration in Sigma Delta Modulation" IEEE Trans. Commun., Vol. COM-33, No. 3, pp. 249-258, March 1985

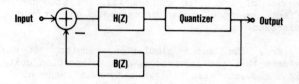

Figure 2.    General linear feedback coder

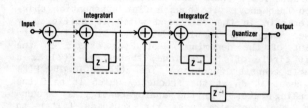

Figure 3.    Structure of double integration delta sigma modulator

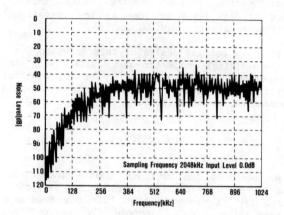

Figure 5.    Simulated quantization noise spectrum of IDIΔΣM

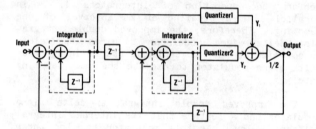

Figure 4.    Structure of improved double integration delta sigma modulator

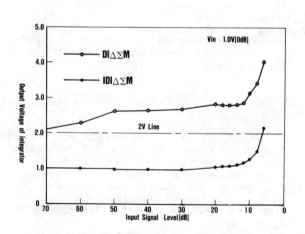

Figure 6.    Simulated output voltage of integrators of IDIΔΣM and DIΔΣM

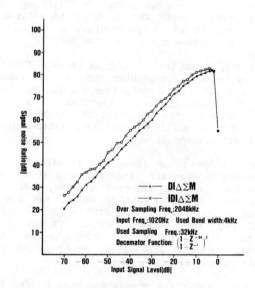

Figure 7.    Simulated signal-to-noise ratios of IDIΔΣM and DIΔΣM

# Oversampling A-to-D and D-to-A Converters with Multistage Noise Shaping Modulators

KUNIHARU UCHIMURA, TOSHIO HAYASHI, TADAKATSU KIMURA, AND ATSUSHI IWATA

*Abstract*—This paper proposes high resolution oversampling analog-to-digital and digital-to-analog converters. These converters utilize multistage noise shaping modulation techniques which can be implemented with VLSI MOS technology. The modulators consist of multiconnected single integration delta-sigma modulation loops. Quantization noise in the first delta-sigma loop is requantized by the next delta-sigma loop and cancelled by adding the requantized noise to the first stage signal. Quantization noise in the modulated signal is suppressed by double or multiple integration gain. Modulator resolution increases with the number of stages without feedback loop instability problems. A three-stage multistage noise shaping modulator with 1-bit quantization achieves the equivalent of 16-bit resolution for Hi-Fi audio band at 2 MHz sampling rate. Simulation results show that analog-to-digital and digital-to-analog converter circuits, constructed of switched capacitors, can be used without highly accurate components.

## I. INTRODUCTION

OVERSAMPLING analog-to-digital (A-to-D) and digital-to-analog (D-to-A) converters that employ modulation techniques, such as delta-sigma modulation [1]–[5], [15], adaptive delta modulation [6], [7], and interpolative modulation [8]–[11], reduce quantized levels by high sampling rate. They decrease both the number and variation sensitivity of analog components more effectively than conventional successive approximation converters. They can be used in combination with digital filters [12], therefore eliminating the need for precise analog pre- and postfilters. These features show that oversampling techniques are suitable for implementation using analog/digital compatible MOS VLSI and digital signal processing technologies.

However, delta-sigma modulation loop stability decreases when quantization noise is suppressed using second- or higher order integration [13], [14]. Furthermore, when analog components increase, device mismatch affects the signal-to-noise ratio (SNR) when multilevel quantization is used in such interpolative converters. Thus, these techniques cannot easily achieve the over 16-bit resolution required for Hi-Fi audio signal encoder/decoders.

This paper proposes high resolution oversampling A-to-D and D-to-A converters. These converters utilize multistage noise shaping (MASH) modulation techniques without employing higher order integrators. This paper

Manuscript received May 13, 1987; revised April 18, 1988.
The authors are with NTT Electrical Communications Laboratories, 3-1 Morinosato Wakamiya, Atsugi, 243-01, Japan.
IEEE Log Number 8824227.

also clarifies the suppression of baseband quantization noise by multiple high gain integration. Simulation results show that MASH modulators provide high resolution in two ways: 1) by eliminating the feedback loop instability problem caused by multiple order integrator phase shift; 2) by reducing D-to-A circuits required linearity. Analog circuit configurations suitable for VLSI implementation are also described.

## II. MULTISTAGE NOISE SHAPING MODULATOR QUANTIZATION NOISE

The noise shaping modulator formed by placing a quantizer and integrators in a feedback loop removes quantization noise from the baseband by integrator gain. Using 1-bit quantization achieves high resolution and high element mismatch tolerance by reducing analog components. A double integration delta-sigma modulator is shown in Fig. 1. The bypass in Integrator-1 improves feedback loop stability. The modulated output signal $V_{out}$ is given by the equation:

$$V_{out} = \frac{H_1 \cdot H_2 z^{-1} + H_2 z^{-1}}{(1 + H_1 \cdot H_2 z^{-1} + H_2 z^{-1})} V_{in}$$
$$+ \frac{Vq}{(1 + H_1 \cdot H_2 z^{-1} + H_2 z^{-1})}$$

for baseband frequencies ($H_1$ and $H_2 \gg 1$)

$$V_{out} \approx V_{in} + \frac{Vq}{(1 + H_1 \cdot H_2 z^{-1} + H_2 z^{-1})} \qquad (1)$$

where $V_{in}$ is the input signal, $Vq$ is quantization noise, and $H_1$ and $H_2$ are integrator gain equal to $1/(1 - z^{-1})$. Noise in the modulated signal is suppressed by $H_1$ and $H_2$ gain. Suppression is proportional to the number of integrators. However, it is difficult to maintain feedback loop stability when more than second-order integration is used.

This paper proposes using oversampling A-to-D and D-to-A converters incorporating MASH modulation techniques. These techniques provide high resolution and eliminate the feedback loop instability problem. The two-stage MASH modulator's basic configuration is shown in Fig. 2. Each stage consists of a first-order integrator, a quantizer, and a feedback circuit. The first stage quantizes the input signal in the same way as in a conventional delta-sigma modulator. The difference between input signal and quantizer output, that is, first stage quantization noise, is

Reprinted from *IEEE Trans. Acoust., Speech, Signal Processing*, vol. AASP-36, pp. 1899–1905, December 1988.

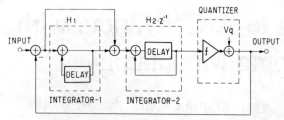

Fig. 1. Delta-sigma modulator with double integration.

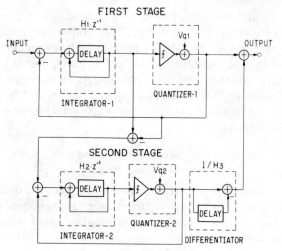

Fig. 2. Multistage noise shaping modulator with two stages.

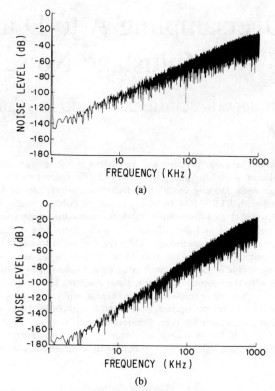

Fig. 3. Simulated quantization noise spectrum of multistage noise shaping modulator. (a) Two-stage. (b) Three-stage. Sampling rate is 2048 kHz.

accumulated by the integrator. The noise's low-frequency baseband components are amplified. Amplified noise is detected at the integrator output. The detected noise signal is requantized by the second stage. The differentiator shapes the quantizer output frequency spectrum to return it to the original signal spectrum. This signal cancels first stage quantization noise by adding it to the first stage output. If $H_1$ is assumed equal to $H_2$, the modulated output $V_{out}$ is given by the equation:

$$V_{out} = \frac{H_1 z^{-1}}{(1 + H_1 z^{-1})} V_{in} + \frac{(H_3 - H_1) z^{-1}}{(1 + H_1 z^{-1}) H_3} Vq_1$$

$$+ \frac{Vq_2}{(1 + H_1 z^{-1}) H_3}$$

for baseband frequencies ($H_1 \gg 1$)

$$V_{out} \approx V_{in} + \frac{(H_3 - H_1) z^{-1}}{(1 + H_1 z^{-1}) H_3} Vq_1 + \frac{Vq_2}{(1 + H_1 z^{-1}) H_3}$$

$$(2)$$

where $Vq_1$ and $Vq_2$ are the quantization noise of Quantizer-1 and Quantizer-2, respectively, and $1/H_3$ is the differentiator transfer function. If $H_1$ is designed to equal $H_3$, the second term in (2) is completely cancelled and the third term is suppressed by $H_1$ and $H_3$.

The performances of A-to-D and D-to-A converters with MASH modulators have been evaluated by computer simulation. To evaluate feedback loop stability and dynamic characteristics, analog waveforms and digital bit

streams were obtained by time domain simulation technique using extended Z-transform models. The required accuracy for analog components on VLSI chip were clarified by quantifying the performance degradation when nonideal models were used. The nonideal models include amplifier finite gain, dc offsets, comparator uncertainty, capacitance ratio errors, and noise sources. The quantization noise spectrum and the signal-to-noise ratio (SNR) were calculated from analog waveforms and digital bit streams by the FFT (Fast Fourier Transform) method.

The quantization noise spectrum of a two-stage MASH modulator simulated at a 2.048 MHz sampling rate is shown in Fig. 3(a). Noise is suppressed by the noise shaping characteristics of double integration. The noise spectrum slope has a 40 dB/decade gradient. The noise spectrum has the same shape as that of the double integration delta-sigma modulator. Further, second stage quantization noise is detected as the second stage integrator output, and three or more stage MASH modulators can be constructed without feedback loop instability problems. This three-stage MASH modulator makes highly effective triple integration feasible for the first time. The three-stage MASH modulator quantization noise spectrum is shown in Fig. 3(b). The slope has a 60 dB/decade gradient.

Simulated waveforms in the MASH modulator with 1-bit quantization are shown in Fig. 4. Amplitude is normalized by the maximum input level. Input waveforms of the second and third stages are exactly equal to the quantization noise of the first and second stages, respectively. Each stage output signal remains stable and does not exceed the maximum for each input signal range.

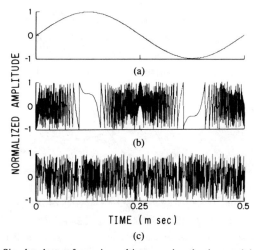

Fig. 4. Simulated waveforms in multistage noise shaping modulator. (a) First stage input. (b) Second stage input. (c) Third stage input. Quantization level is two. Sampling rate is 2048 kHz.

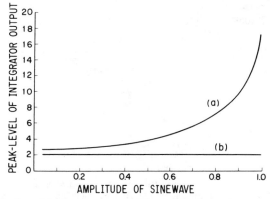

Fig. 5. Second-integrator output peak level versus input sine wave amplitude. (a) Delta-sigma modulator with double integration. (b) Multistage noise shaping modulator with two stages. Quantizer has two levels.

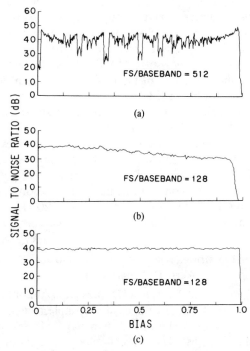

Fig. 6. Signal-to-noise ratios for a −40 dB input level and 1-bit quantization versus input bias. (a) Simple delta-sigma modulation. (b) Double integration delta-sigma modulation. (c) Two-stage MASH modulation.

## III. MULTISTAGE NOISE SHAPING MODULATOR PERFORMANCE

In a double integration delta-sigma modulator, the integrator output amplitude increases with the input level quantizer input range. The resultant feedback loops become unstable. The effect of input sine wave amplitude on second integrator output peak levels for 1-bit quantization is shown in Fig. 5. These are simulated results where the sampling frequency is 2048 times larger than the input frequency. The 1-bit quantizer feeds back two levels that are equal to positive and negative maximum levels of the input signal centering on ground level. The input signal differences from high and low quantizer outputs become unbalanced when the input level approaches its maximum. The output of the first integrator accumulating the input differences is shaped into a triangular waveform rising or falling with the input difference magnitude. When the input differences are in heavy unbalance and the rising or falling speed is very low, the triangular waveform has a long cycle time and includes low-frequency components. In the double integration delta-sigma modulator, the second integrator amplifies the first integrator output with gain inversely proportioning to the frequency. Therefore, the second integrator output level increases rapidly with input sine wave amplitude [Fig. 5(a)], and quantization noise also increases when the second integrator output level exceeds the quantizer input range. In the two-stage MASH modulator, integrator output peak level is unaffected by input sine wave amplitude and stays within quantizer input range [Fig. 5(b)].

In the simple delta-sigma modulator, input dc-bias affects quantization noise [5]. This dependence affects the performance of A-to-D and D-to-A converters and limits their use in many applications. SNR's have been graphed for a low level −40 dB input plotted versus input dc-bias in Fig. 6(a)-(c). The quantizer has a single threshold and generates two level outputs. Fig. 6(a) shows that the simple delta-sigma modulator is greatly affected and the SNR deteriorates at many bias levels. The double integration

delta-sigma modulator is affected only at the ends of input range [Fig. 6(b)]. Fig. 6(c) shows that the two-stage MASH modulator SNR are almost unaffected by input dc-bias. The MASH modulator can be used in the same way as a high resolution linear PCM converter over all input voltage ranges.

Feedback loop stability and sensitivity to input bias affect SNR. Simulated SNR characteristics plotted against input sine wave amplitude at 1-bit quantization and 16 kHz baseband are shown in Fig. 7. MASH modulator sampling rates are independent of the number of stages. This is because all MASH modulator feedback loop integrators operate in parallel, whereas delta-sigma modulator feedback loop integrators must operate in series. Therefore, the MASH modulator sampling rate is twice as high as the double integration delta-sigma modulator sampling rate.

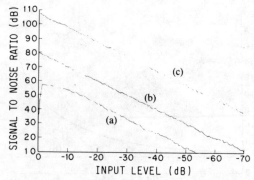

Fig. 7. Simulated signal-to-noise ratios at 1-bit quantization and 16 kHz baseband. (a) Double integration delta-sigma modulator at 1024 kHz sampling. (b) Two-stage MASH modulator at 2048 kHz sampling. (c) Three-stage MASH modulator at 2048 kHz sampling.

The SNR curve of the double integration delta-sigma modulator saturates above $-10$ dB [Fig. 7(a)]. The MASH modulator with SNR unaffected by input bias achieves ideal linear SNR curves [Fig. 7(b), (c)]. Fig. 7 also shows that the two-stage MASH modulator dynamic range is higher than that of the double integration delta-sigma modulator by 15 dB.

The relationship between two- and three-stage MASH modulator dynamic ranges and the ratio of sampling frequency to baseband frequency is calculated and shown in Fig. 8. Dynamic ranges increase with sampling frequency by 15 dB/octave and 21 dB/octave, respectively. The three-stage MASH modulator achieves high resolution of 16-bit at low 2 MHz sampling rate and wide 20 kHz baseband, using 1-bit quantization.

## IV. MASH MODULATOR CIRCUIT CONFIGURATION FOR A-TO-D CONVERTER

In a MASH modulator A-to-D converter, the integrators and feedback circuits can easily be implemented with switched capacitor technique. A two-stage MASH modulator suitable to MOS VLSI technology consists of two switched capacitor integrators, two 1-bit comparators, and a digital differentiator (Fig. 9). The input signal is sampled by capacitor CS1 and integrated to CI1. The quantized 1-bit signal is fed back to the input by charging a reference voltage into capacitor CD1. Analog switches and comparators are controlled by a sequence clock shown in Fig. 10. In a two-stage MASH modulator sample data model (Fig. 2), second stage input is generated from the difference of integrator and quantizer outputs. In the circuit of Fig. 9, the first stage integrator output has two phases a cycle. In the first half cycle, the integrator accumulates the input signal. In the second half cycle, it accumulates the quantized signal from the comparator. The value of the second half is equal to the difference of integrator and quantizer outputs in Fig. 2. The second stage capacitor CS2 should sample the first stage integrator output in the second half cycle.

For implementation in VLSI MOS technology, the MASH modulator must tolerate amplifier finite gain and capacitor ratio mismatch in the analog circuits. The first

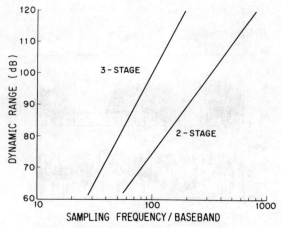

Fig. 8. Multistage noise shaping modulator dynamic range versus ratio of sampling frequency to baseband at 1-bit quantization.

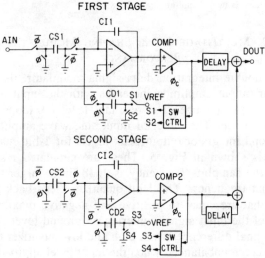

Fig. 9. Multistage noise shaping modulator circuit configuration for A-to-D converter.

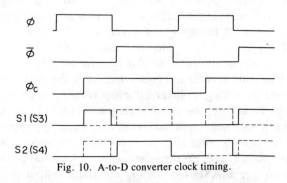

Fig. 10. A-to-D converter clock timing.

cause of SNR degradation is gain mismatch between the first and second stages. This gain mismatch is determined by the accuracy of the capacitor ratio of CS1/CD1 and CS2/CD2. Gain mismatch effects on SNR are shown in Fig. 11. To get an 80 dB dynamic range, equivalent to linear 13-bit resolution, about 5 percent gain mismatch is acceptable. Furthermore, capacitor ratio error of 1 percent achieves higher than 16-bit resolution [Fig. 11(a)]. This is because signal power transferred to the second stage is very small within the baseband, and the noise

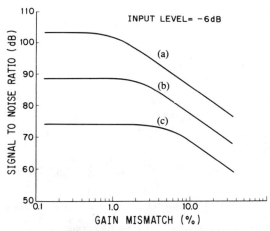

Fig. 11. Relation of gain mismatch between first and second stages to signal-to-noise ratios. Sampling frequency to baseband ratio is (a) 512; (b) 256; (c) 128.

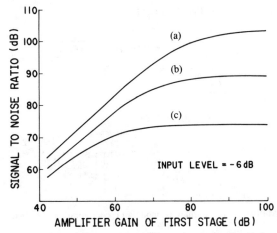

Fig. 13. Relation of first stage amplifier gain to signal-to-noise ratios. Sampling frequency to baseband ratio is (a) 512; (b) 256; (c) 128.

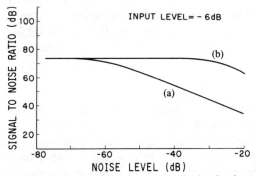

Fig. 12. Signal-to-noise ratios versus random noise level at integrator input. Noise voltages are from + VREF to − VREF at 0 dB. (a) First stage. (b) Second stage.

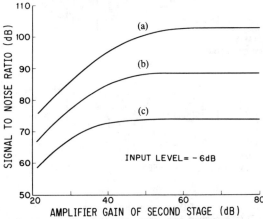

Fig. 14. Relation of second stage amplifier gain to signal-to-noise ratios. Sampling frequency to baseband ratio is (a) 512; (b) 256; (c) 128.

spectrum generated by the gain mismatch lies almost outside the baseband at the digital summing point.

The second cause of SNR degradation is circuit element noise such as MOS device $1/f$ noise, switched capacitor switching noise, and clock pulse crosstalk from digital circuits. Noise applied to integrator input is indistinguishable from the input signal or quantization noise. Fig. 12(a) shows that the integrator in the first stage must be comparatively low noise. In contrast to curve (a), curve (b) shows that the second stage is very tolerant of noise greater than 30 dB. Because the digital differentiator attenuates the second stage signal within the baseband before adding it to the first stage output, the second stage is more tolerant of circuit noise.

The third cause of SNR degradation is integrator leakage originating in amplifier finite gain. At lower frequencies, amplifier gain limits switched capacitor integrator gain. Low-frequency components of accumulated quantization noise are lost at a leakage time constant corresponding with the lower cutoff frequency. Simulated degradations caused by amplifier finite gain in the first and second stages are shown in Figs. 13 and 14. If the first stage amplifier gain is over 80 dB, SNR degradation is adequately small. Desired second stage amplifier gain is lower than first stage gain by 20–30 dB.

## V. MASH MODULATOR CIRCUIT CONFIGURATION FOR D-TO-A CONVERTER

The MASH modulator concept can be used to construct oversampling D-to-A converters. A two-stage digital MASH modulator for D-to-A converter is constructed from two digital quantizers, four digital integrators, and a differentiator as shown in Fig. 15. This configuration, where each stage has two integrators in a feedback loop, is designed for high element ratio mismatch tolerance. But feedback loop stability with two integrators is improved by reducing loop gain with a $1/4$ attenuator. When each stage has a digital integrator and a digital 1-bit quantizer in a feedback loop, the digital MASH modulator is equivalent to the reported interpolative configuration [15].

When a quantizer with three levels, 0, +1, and −1, is used in the first stage, the digital signal DOUT-1 has three levels: 0, +1, and −1. When a quantizer with two levels, +1 and −1, is used in the second stage, the second stage differentiator generates DOUT-2 having three levels: 0, +2, and −2. A 7-level local D-to-A circuit must be provided to convert these levels into the analog equivalent.

Two D-to-A conversion methods in Fig. 16 are investigated. The first method applies a binary digital signal generated by adding DOUT-1 and DOUT-2 to a conven-

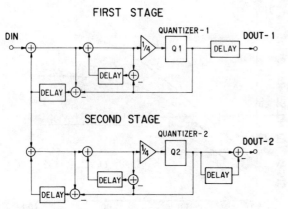

Fig. 15. Multistage noise shaping modulator configuration for D-to-A converter.

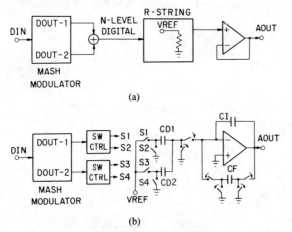

Fig. 16. (a) Conventional D-to-A conversion method using R-string circuit. (b) D-to-A conversion method using two individual switched-capacitor D-to-A circuits and an analog adder circuit.

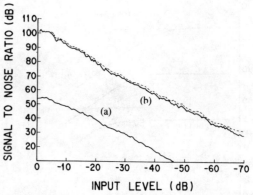

Fig. 17. Simulated D-to-A converter signal-to-noise ratios at 2048 kHz sampling and 16 kHz baseband, when element ratio error is 1 percent. (a) Using D-to-A circuit in Fig. 15(a). (b) Using D-to-A circuit in Fig. 15(b). Dashed line shows signal-to-noise ratios without element ratio error.

tional D-to-A circuit, such as the dividing voltage type with resistor string [Fig. 16(a)]. The second method applies the digital signal of each stage individually to switched capacitor D-to-A circuits [Fig. 16(b)]. In this circuit, the DOUT-1 and DOUT-2 signals select CD1 and CD2 switching modes from positive charge mode, negative charge mode, and discharge mode. An amplifier with a capacitor CI adds CD1 and CD2 charges. DOUT-1 DOUT-2 signals linearity are maintained and harmonic distortion does not occur because each analog output is generated using one capacitor and one reference voltage. But conventional D-to-A circuit nonlinearity generates harmonic distortion within baseband. CD1 and CD2 capacitor ratio error causes transfer gain mismatch between the first and second stages, just as in the MASH A-to-D converter.

The D-to-A converter's (Fig. 15) SNR characteristics simulated at a 1 percent element ratio error are shown in Fig. 17. Noises include harmonic distortion and quantization noise. The dashed line shows SNR characteristics without element ratio error. Quantization noise is suppressed by first integrator high gain in the first stage and by both integrators in the second stage. These 2-stage MASH modulator SNR characteristics are almost equiv-

alent to those of the 3-stage MASH modulator constructed by the single integration delta sigma loop. Compared to the dashed line, degradation for D-to-A circuit in Fig. 16(b) is negligible [Fig. 17(b)]. But degradation for conventional D-to-A circuits is significant [Fig. 17(a)].

## VI. CONCLUSION

This paper has proposed high resolution oversampling A-to-D and D-to-A converters utilizing multistage noise shaping modulation techniques. Multistage noise shaping modulators suppress quantization noise within baseband by double or multiple integration high gain without feedback loop instability problems. A three-stage MASH modulator with 1-bit quantization provides high resolution equivalent to 16-bit for Hi-Fi audio band at a 2 MHz sampling rate. Furthermore, the MASH modulator can be used like a high resolution linear PCM converter without being affected by input level and dc offset. Simulation results show that A-to-D and D-to-A converter circuits can be used without highly accurate components. These features are useful and suitable for high resolution digital signal processing applications in VLSI implementation.

### ACKNOWLEDGMENT

The authors wish to thank H. Mukai, T. Sudo, and E. Arai of the Electrical Communications Laboratories for their encouragement throughout this work.

### REFERENCES

[1] T. Misawa, J. E. Iwersen, L. J. Loporcaro, and J. G. Ruch, "Signal-chip per channel codec with filters utilizing delta-sigma modulation," *IEEE J. Solid-State Circuits*, vol. SC-16, pp. 333–341, Aug. 1981.
[2] S. Tewksbury and R. W. Hallock, "Oversampled, linear predictive and noise-shaping coders of order $N > 1$," *IEEE Trans. Circuits Syst.*, vol. CAS-25, pp. 436–447, July 1978.
[3] B. P. Agrawal and K. Shenoi, "Design methodology for sigma-delta-M," *IEEE Trans. Commun.*, vol. COM-31, pp. 360–370, Mar. 1983.
[4] H. L. Fiedler and B. Hoefflinger, "A CMOS pulse density modulator for high-resolution A/D converters," *IEEE J. Solid-State Circuits*, vol. SC-19, pp. 995–996, Dec. 1984.
[5] J. C. Candy and O. J. Benjamin, "The structure of quantization noise from sigma-delta modulation," *IEEE Trans. Commun.*, vol. COM-29, Sept. 1981.
[6] R. Gregorian and J. G. Gord, "A continuously variable slope adap-

tive delta modulation codec system,'' *IEEE J. Solid-State Circuits*, vol. SC-18, pp. 692–700, Dec. 1983.

[7] J. W. Scott, W. L. C. Giancario, and C. G. Sodini, ''A CMOS slope adaptive delta modulator,'' in *Proc. IEEE Int. Solid-State Circuits Conf.*, Feb. 1986, pp. 130–131.

[8] B. A. Wooley and J. L. Henry, ''An integrated per-channel PCM encoder based on interpolation,'' *IEEE J. Solid-State Circuits*, vol. SC-14, pp. 14–20, Feb. 1979.

[9] J. C. Candy, W. H. Ninke, and B. A. Wooley, ''A per-channel A/D converter having 15-segment $\mu$-255 companding,'' *IEEE Trans. Commun.*, vol. COM-24, pp. 33–42, Jan. 1976.

[10] A. Yukawa, R. Maruta, and K. Nakayama, ''An oversampling A-to-D converter structure for VLSI digital CODEC's,'' in *IEEE-IECEJ-ASJ ICASSP85 Proc.*, Mar. 1985, pp. 1400–1403.

[11] K. Yamakido *et al.*, ''A voiceband 15b interpolative converter chip set,'' in *Proc. IEEE Int. Solid-State Circuits Conf.*, Feb. 1986, pp. 180–181.

[12] L. van De Meeberg and D. J. G. Janssen, ''PCM codec with on-chip digital filters,'' in *Proc. IEEE Int. Conf. Commun.*, vol. 2, June 1980, pp. 30.4.1–30.4.6.

[13] J. C. Candy, ''A use of double integration in sigma delta modulation,'' *IEEE Trans. Commun.*, vol. COM-33, pp. 249–258, Mar. 1985.

[14] R. Koch and B. Heise, ''A 120 kHz sigma delta A/D converter,'' in *Proc. IEEE Int. Solid-State Circuits Conf.*, Feb. 1986, pp. 138–139.

[15] J. C. Candy and A. Huynh, ''Double interpolation for digital-to-analog conversion,'' *IEEE Trans. Commun.*, vol. COM-34, pp. 77–81, Jan. 1986.

# Architectures for High-Order Multibit ΣΔ Modulators*

R. H. Walden[†], T. Cataltepe[‡], G. C. Temes[‡]
[†]Hughes Research Laboratories, Malibu, CA 90265
[‡]Electrical Engineering Dept., UCLA, Los Angeles, CA 90024

**Abstract**
*ΣΔ modulators with higher-order noise shaping and multibit internal quantizers are described. A new realization method for single-stage high-order modulators is discussed. Also, a new multistage architecture with multibit quantizers and interstage scaling is described. A digital correction method for the effects of nonideal analog components of the multistage modulator is presented. Simulation results are included to verify the theoretical considerations.*

## I. Introduction

It was recently suggested to use multilevel internal quantizers to achieve better resolution from ΣΔ modulators [1, 2]. A digital correction method for such multibit ΣΔ modulators was reported earlier [3]. The resolution can further be improved by using higher-order noise-shaping modulators. One method of achieving high-order noise-shaping is to use a high-order filter in a single ΣΔ modulator, another method is to cascade simple ΣΔ modulators and combine their outputs. In this paper we describe a realization method for high-order single-stage ΣΔ modulators, and also a new architecture and digital correction method for multibit/ multistage ΣΔ modulators.

## II. Single-Stage High-Order ΣΔ Modulators:

Fig. 1 shows a the block diagram of a general ΣΔ modulator. In the figure, the quantizer is replaced by an additive noise source $E$ and a delay. $P_i$, $Q_i$, and $R_i$ are linear functions of $z^1$. The system is defined by the matrix equation

$$AX = z^1 RE + PX_0$$

where

$$X = [X_1, X_2, \ldots X_n]^T,$$

$$R = [R_1, R_2, \ldots R_n]^T,$$

$$P = [P_1, 0, \ldots 0]^T,$$

and

$$A = \begin{bmatrix} Q_1 & 0 & 0 \cdots & \cdots & -z^{-1}R_1 \\ -P_2 & Q_2 & 0 \cdots & \cdots & -z^{-1}R_2 \\ 0 & -P_3 & Q_3 & 0 \cdots & -z^{-1}R_3 \\ \vdots & 0 & \ddots & \cdots & \vdots \\ \vdots & \vdots & \vdots & \ddots & \vdots \\ 0 & 0 & 0 \cdots & -P_n & (Q_n - z^{-1}R_n) \end{bmatrix}$$

The output of the modulator $Y$ can be expressed as

---

*This work was supported by the Naval Ocean Systems Center, San Diego, CA, contract no. N66001-86-C-0193

$$Y(z) = \left[ \frac{1}{D(z)} \prod_{j=1}^{n} P_j \right] z^{-1} X_0 + \left[ \frac{1}{D(z)} \prod_{j=1}^{n} Q_j \right] z^{-1} E$$

where $D$, the determinant of the matrix $A$, is given by

$$D(z) = \det(A) = \left[ \prod_{j=1}^{n} Q_j \right] - z^{-1} \sum_{l=1}^{n} \left[ \prod_{j=1}^{l-1} Q_j \right] R_l \left[ \prod_{i=l+1}^{n} P_i \right]$$

The stability is determined by the roots of $D$. If we realize each stage (i.e. $P_i$, $Q_i$, $R_i$) by the switched-capacitor structure shown in Fig. 2, then it can be shown that if the desired $D$ is

$$D = 1 + \sum_{m=1}^{n} b_m z^{-m}$$

then the following $n$ equations can be solved for $a_i$

$$\sum_{k=n-m+1}^{n} \binom{k-1}{k-1-n+m} (-1)^{k-1} a_k = (-1)^n \left[ (-1)^m b_m - \binom{n}{m} \right],$$

$$m = 1, \ldots, n$$

where

$$a_l = \begin{cases} r_l \prod_{i=l+1}^{n} p_i, & l = 1, \ldots n-1 \\ \\ r_n, & l = n \end{cases}$$

As an example, a stable fifth-order system may have the roots of $D$ at $z = 0.5$, $0.5 \pm j0.25$, $0.5 \pm j0.5$. A simulation result of this system is shown in Fig. 3 for a signal amplitude of 0.577 $V_{REF}$. However, the simulations also indicate that the stability of the system depends on the input signal amplitude, and that the location of the roots of $D$ affects the dynamic range of the modulator.

## III. Multistage ΣΔ Modulator

Another method of achieving high-order noise shaping is to cascade simple first and second order ΣΔ modulators and then to combine the outputs of the individual stages where each of the stages is stable [4, 5]. The multistage architecture proposed here is shown in Fig. 4. The modulator shown consists of a cascade of two first-order stages, but it can easily be extended to more stages. The second stage input signal is the quantization error of the first stage. Since multibit quantizers are employed, this error signal is within 1 quantization level, thus it can be amplified to utilize the full dynamic range of the second stage. The output of the second stage must be divided by the same amplification factor before the (digital) addition of the outputs of both stages. One important advantage of the interstage scaling is that the the quantization error of the second stage is thus reduced by the scaling factor. The (digital) postprocessing blocks $H_A$ and $H_B$ are designed such that under

---

Reprinted from *IEEE Proc. ISCAS'90*, pp. 895–898, May 1990.

ideal conditions the quantization error of the first stage is completely cancelled. Therefore, due to the interstage scaling, the cascade of two first-order stages performs better than a single second order stage. A simulation result for the system shown in Fig. 4 with 4-bit internal quantizers and an interstage scaling factor of 10 is shown in Fig. 5.

## IV. Digital Correction For Multistage $\Sigma\Delta$ Modulators

One drawback of multistage aproach is that the cancellation of the first-stage quantization error depends on the precise matching between transfer functions of the analog components (the integrators and the interstage gain stage) and the digital processors $H_A$ and $H_B$. For example, if the digital processors are designed without taking into account the finite gain of the opamps in the analog stages, the S/N ratio of the final output signal deteriorates significantly as shown in Fig. 6 where the first and second stage opamp gains are assumed to be 100 and 150 respectively.

It can be shown that if we employ offset- and gain-compensated switched-capacitor integrators, as shown schematically in Fig. 7, then the finite gain of the opamps and the capacitor mismatches give rise mainly to a gain error, and the phase error can be ignored. Fig. 8 shows the simulation result for this system where the first and second stage opamp gains were again 100 and 150, respectively. The gain error can be corrected by modifying the digital processor function $H_B$ following the second stage to include two (variable) coefficients $c$ and $d$:

$$H_B = (z^{-1} - 1)(c + d\,z^{-1}).$$

By adjusting these coefficients, most of the error introduced by the opamps and capacitor mismatches can be eliminated. Fig. 9 shows the output spectrum of such a digitally corrected system. The coefficients $c$ and $d$ can be calculated to be 0.0136 and 1.0064 respectively.

## V. Conclusions

We described a new realization and analysis method for high-order single-stage $\Sigma\Delta$ modulators. We also described a new multistage approach to realize high-order noise-shaping. The new multistage modulator uses multibit internal converters and interstage scaling to utilize the full dynamic range of each stage. A digital correction method for the effects of nonidealities of the analog components in the multistage modulator is also introduced. All theoretical results are verified by simulations.

## Bibliography

[1] Adams, R. W., "Design and implementation of an audio 18-bit analog-to-digital converter using oversampling techniques," *J. Audio Eng. Soc.*, vol. 34, pp. 153-166, March, 1986.

[2] Paulos, J. J., G. T. Brauns, M. B. Steer, S. H. Ardalan, "Improved signal-to-noise ratio using tri-level delta-sigma modulation," *ISCAS' 87 Proceedings*, pp. 463-466, May, 1987.

[3] Cataltepe, T., G. C. Temes et al., "Digitally corrected multi-bit $\Sigma-\Delta$ data converters," *ISCAS'89 Proceedings*, pg. 647-650, May, 1989.

[4] Matsuya, Y., K. Uchimura, A. Iwata, et al., "A 16-bit oversampling A-to-D conversion technology using triple-integration noise shaping," *IEEE J. Solid-State Circuits*, vol. SC-22, pg. 921-929, Dec., 1987.

[5] Uchimura, K., T. Hayashi, et al., "VLSI- A to D and D to A converters with multi-stage noise shaping coders," *ICASSP' 86*, pg. 29.8.1-29.84, Apr., 1986.

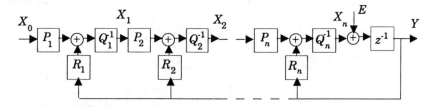

**Fig. 1.** Block diagram of a general single-stage $\Sigma\Delta$ modulator. The quantizer is shown as an additive noise source with delay.

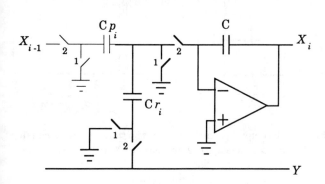

**Fig. 2.** A switched-capacitor realization of one of the stages of the modulator shown in Fig. 1.

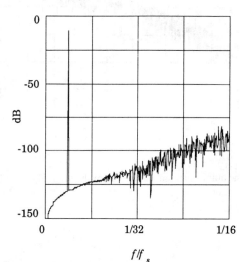

**Fig. 3.** The output spectrum of a 5th order single-stage $\Sigma\Delta$ modulator with 4-bit internal quantizer.

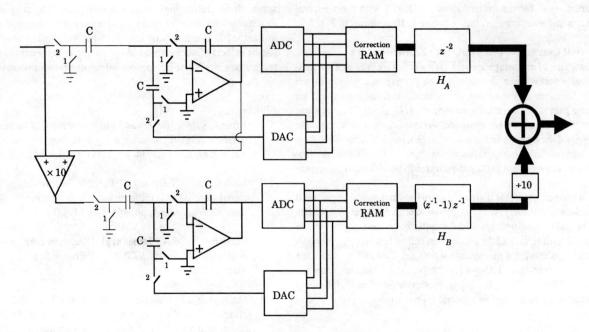

**Fig. 4.** The proposed multistage architecture with interstage scaling. In the figure, the scaling factor is chosen as 10.

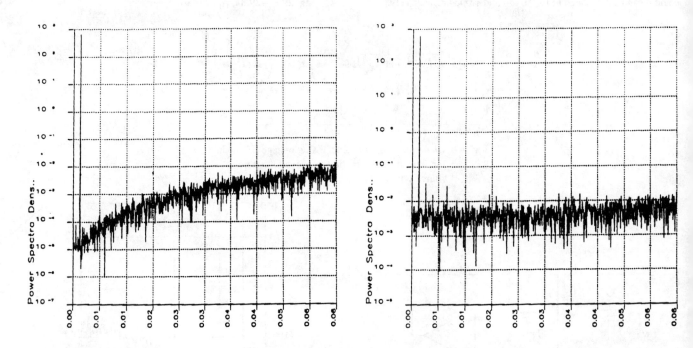

**Fig. 5.** The output spectum of the system shown in Fig. 4 with ideal opamps.

**Fig. 6.** The output spectrum of the system shown in Fig. 4 with first- and second-stage opamp gains of 100 and 150 respectively.

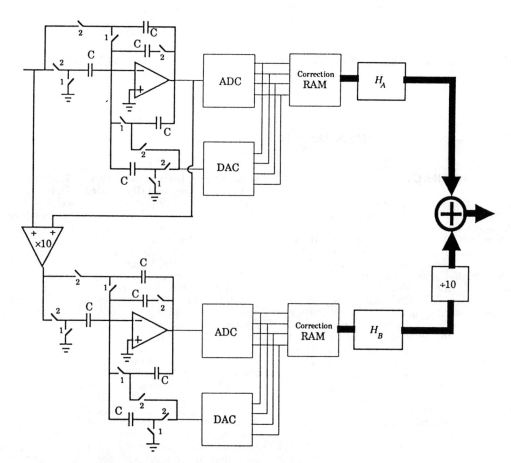

Fig. 7. The two-stage ΣΔ modulator with offset- and gain-compensated integrators.

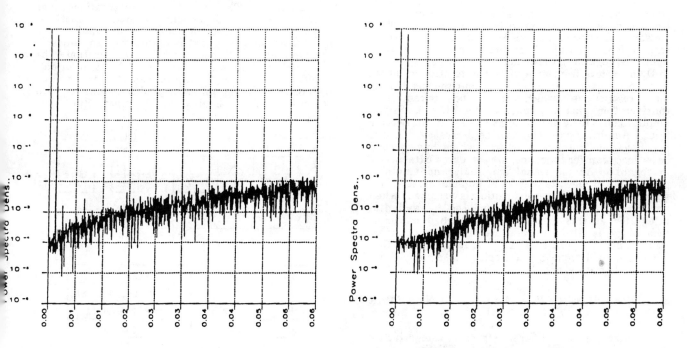

**Fig. 8.** The output spectrum of the system shown in Fig. 7 with first- and second-stage opamp gains of 100 and 150 respectively.

**Fig. 9.** The output spectrum of the system shown in Fig. 7 with first- and second-stage opamp gains of 100 and 150 respectively and with digitally corrected $H_B$.

# Constraints Analysis for Oversampling A-to-D Converter Structures on VLSI Implementation

*Akira Yukawa*

Microelectronics Research Labs.
NEC Corporation

## ABSTRACT

Requirements, applied to analog components and characteristics for oversampling A-to-D converter structures with second order loop transfer function, were analyzed through numerical and mathematical analyses. The results suggest that a first order predictive encoder, with first order noise shaping, is best suited to high resolution future VLSI implementation. It is most difficult to achieve a theoretical second order $\Delta$-$\Sigma$ modulation signal-to-noise ratio, because of the amplifier performance requirements. The operational amplifier gain should only be greater than $kR/\pi$, as long as the amplifier response is linear, where R is the oversampling ratio, $\omega$ is the angular stop band frequency and k is a parameter depending on structures and integrator circuit configuration, if a switched capacitor integrator is employed in a noise shaping technology. A second order delta modulation coder does not seem to be applicable to a precision converter, but to a high speed converter instead.

## 1. INTRODUCTION

Various oversampling A-to-D converter structures have been examined recently, for use as the front end of a precision A-to-D converter[1][2][3][4]. However, their performances are still below theoretical expectations. Although several reports have been published regarding the behaviours of oversampling A-to-D converters from analytical points of view[5][6][7], few reports have been published about the constraints and restrictions which must be taken into consideration when manufacturing a VLSI[8]. Signal amplitude has to be decreased as VLSI patterns become finer. Amplifier gain also will be decreased. In this paper, various encoder structures, which have second order loop transfer functions, are compared extensively in regard to analog signal amplitude and amplifier gain points of view, taking VLSI implementation into account. Here, digital substitution is assumed, when an analog integrator can be replaced by a digital integrator at this comparison. Switched capacitor technology is also assumed for analog processing. Requirements applied to an operational amplifier, when switched capacitor technology is employed for analog integrators, are also analyzed.

## 2. OVERSAMPLING A-to-D CONVERTER STRUCTURES

There are three different structure categories for oversampling encoders, namely; predictive, noise shaping and their combination. The first and second structures are usually called "Delta-Modulation"[9] and "$\Delta$-$\Sigma$ Modulation"[10], respectively. The name "Interpolative Encoder"[4] can be categorized as a special case of the last structure. A signal flow diagram can be drawn generally, as in Fig. 1, for these oversampling A-to-D converter structures with second order loop transfer function. Three input ports "A", "C" and "B", correspond to individual structures. Analog signal processing is carried out from the signal input port to the quantizer. Therefore, D-to-A converters are assumed to be placed just before the signal input nodes for the feedback signal. Signal and quantizing noise transfer functions are also summarized, as shown in the table in Fig. 1. A feedback signal bypass path, for loop stabilization, is the path from the 1st integrator to the 2nd integrator in Fig. 1.

There is an alternative configuration for the loop stabilization path position. It is a path bypassing the second integrator, as is shown in Fig. 2. There is no difference between the two configurations in regard to the transfer functions for "Delta Modulation." However, the first configuration needs less hardware for a digital accumulator than the second configuration requires, if a single bit quantizer is employed. Signal transfer functions are slightly different, for the other two structures. The second two configurations have input frequency dependent transfer functions. Although the frequency dependences are small in the signal band, this category should be avoided, unless a big advantage is obtained using the configuration. The first configuration dominates for "$\Delta$-$\Sigma$ Modulation". An efficient circuit configuration has been proposed[11] on CMOS LSI for the 1st order predictive encoder with 1st order noise shaping. Therefore, the first configurations are used for the analysis.

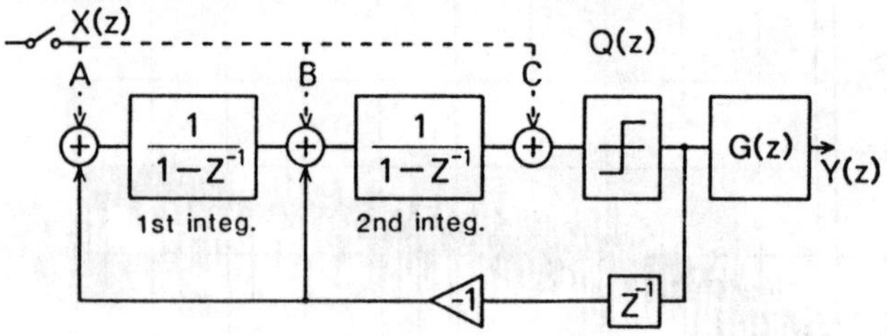

| Input | G(z) | Y(z) |
|-------|------|------|
| A | 1 | $X(z)+(1-Z^{-1})^2 Q(z)$ |
| B | $(1-Z^{-1})^{-1}$ | $X(z)+(1-Z^{-1})Q(z)$ |
| C | $(1-Z^{-1})^2$ | $X(z)+Q(z)$ |

[Fig. 1] Second order encoding structures.

Reprinted from *IEEE Proc. ISCAS'87*, pp. 467–472, May 1987.

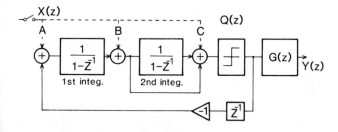

| Input | G(z) | Y(z) |
|-------|------|------|
| A | 1 | $(2-\bar{z}^1)X(z)+(1-\bar{z}^1)^2Q(z)$ |
| B | $(1-\bar{z}^1)$ | $(2-\bar{z}^1)X(z)+(1-\bar{z}^1)Q(z)$ |
| C | $(1-\bar{z}^1)$ | $X(z)+Q(z)$ |

[Fig. 2] Alternative structures to the second order coop.

## 2.1 Double Integration Delta Modulation

Assuming maximum input frequency $f_{max}$ and sampling frequency $f_s$ , feedback gain from the quantizer to the signal input port is expressed as $(f_s/2\pi f_{max})^2$. Therefore, the relation between maximum input amplitude A and step size $\Delta$ is

$$A = \Delta\left(\frac{f_s}{2\pi f_{max}}\right)^2 \qquad (1).$$

Since the resolution required for the D-to-A converter is the ratio between 2A and $\Delta$, it is approximately 12bits, assuming oversample ratio $f_s/2f_{max}$ as 128. The maximum voltage at the quantizer input is qualified through a simulation, based on the signal flow diagram shown in Fig. 1. This signal is the only analog signal processed using this configuration. The maximum voltage is about $2.6\Delta$ and is almost independent from the input signal amplitude, as shown in Fig. 3(a). Here, the amplitude is normalized by the step size. Although this structure can be realized without any amplifier, the quantizer always has to compare less than 1mV difference and D-to-A converter resolution has to be very fine. Therefore, this structure does not seem to be suitable to high resolution applications, but to less precision high speed uses, such as in a video A-to-D converter.

## 2.2 1st order predictive encoder with 1st order noise shaping

The relation between maximum input voltage and step size is the same as for the 1st order Delta Modulation coder,

$$A = \Delta\,\frac{f_s}{2\pi f_{max}} \qquad (2).$$

Therefore, the resolution required for the D-to-A converter is approximately 6bits, if the oversampling ratio is assumed to be 128. Maximum voltage and maximum voltage change, between two successive sampling periods at the quantizer inputs, are also qualified through a simulation based on the signal flow diagram shown in Fig. 1. The solid line indicates maximum amplitude and the dotted line indicates

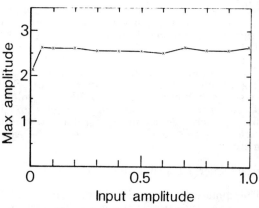

[Fig. 3a] Delta modulation.

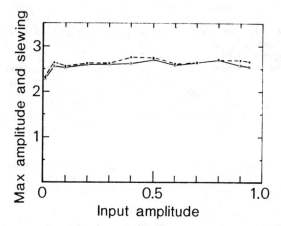

[Fig. 3b] 1st order predictive encoder with 1st order noise shaping.

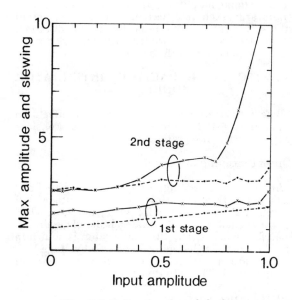

[Fig. 3c] Delta sigma modulation.

maximum difference between two successive sampling periods. This signal is the only analog signal processed using this configuration. Results are shown in Fig. 3(b). Both curves are also normalized by the step size. Maximum voltage and maximum voltage change are both about 2.6Δ and are almost independent from the input signal amplitude, as shown in Fig. 4(b). Since the step size is $2^6$ times smaller than full scale voltage, voltage swing and slew rate requirements to the amplifier, which forms an integrator, are mild. Therefore, an offset cancelling scheme can be easily implemented.

### 2.3 Second order Δ-Σ Modulation coder

The quantizing step size is the same for the maximum amplitude in this structure. The maximum voltages and maximum voltage differences at each integrator output are also simulated, based on the signal flow graph shown in Fig. 1. The maximum voltages at each integrator output increase as the input voltage is increased. The maximum voltage output capability, for the first integrator, has to be about 3 times the input voltage as shown in Fig. 3(c). The maximum voltage output capability, for the second integrator, has to be about 4 times the input voltage, even when input voltage is limited to less than 75% of the maximum input voltage. Slewing requirements are also very stringent. Therefore, a large penalty has to be paid, in order to employ offset cancelling techniques.

Theoretically obtained maximum signal to noise ratios, for the three structures, are plotted in Fig. 4. Maximum signal to noise ratios, for 1st order oversampling coders, are also shown in the same figure. Here, quantizing noise, generated by the quantizers, is assumed to be white and to be distributed equally between $+\Delta$ and $-\Delta$ values. The integrators are assumed to be ideal. Although a Δ-Σ Modulation coder is expected to show the best signal-to-noise ratio, the coder seems to have difficulty in handling maximum signal amplitude, if amplitude limitations in VLSIs are taken into account. Therefore, actual signal-to-noise ratios, obtainable by Δ-Σ Modulation, and 1st order predictive encoder with 1st order noise shaping, would be almost the same.

## 3. SWITCHED CAPACITOR INTEGRATOR MODELS

Before analyzing non-ideal integrator effects on oversampling A-to-D converters, let us consider switched capacitor integrator models.

### 3.1 Offset cancelling integrator

An offset cancelling integrator is beneficial through not only eliminating d.c. offset voltage, but also through eliminating low frequency noise, which is the dominant noise in MOS analog LSIs. The circuit, modeled for an offset cancelling integrator, is shown in Fig. 5. The amplifier has gain G and offset voltage $V_{off}$. The main parasitic effect is a parasitic capacitance at summing node S. The output voltage relation, between two successive sampling periods $V_{n-1}$ and $V_n$, is formulated using the charge conservation theory,

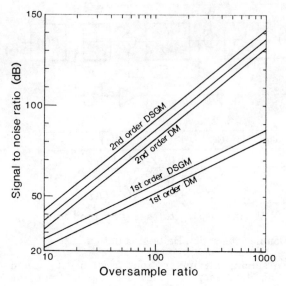

[Fig. 4] Theoretically obtained signal to noise ratio.

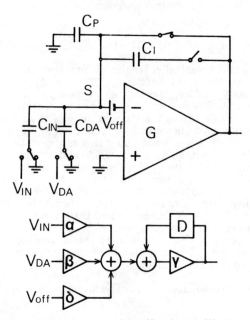

[Fig. 5] An integrator with off set cancelling model and signal flow diagram.

$$V_n = \frac{C_I V_{n-1} - \frac{G}{G+1}(C_{DA}V_{DA}+C_{IN}V_{IN})}{C_I + \frac{C_{DA}+C_{IN}+C_P}{A+1}} + \frac{\frac{G}{G+1}\frac{C_{DA}+C_{IN}+C_P}{A+1}}{C_I + \frac{C_{DA}+C_{IN}+C_P}{A+1}}V_{off}$$

(3).

Equation 3 can be expressed as the signal flow diagram, shown in Fig. 5. Here,

$$\alpha = \frac{G}{G+1}\frac{C_{IN}}{C_I}$$

(4),

$$\beta = \frac{G}{G \neq 1} \frac{C_{DA}}{C_I} \quad (5),$$

$$\gamma = \frac{1}{1 + \dfrac{C_{DA} + C_{IN} + C_P}{(G+1)C_I}} \quad (6),$$

$$\delta = \frac{G}{G+1} \frac{C_{DA} + C_{IN} + C_P}{(G+1)C_I} \quad (7).$$

If $G=100$ and $C_I = C_{DA} = C_{IN}$ are assumed and neglecting $C_P$, then $1-\gamma$, which is the integrater loss factor, is about 0.02.

### 3.2 Integrator without offset cancelling

The circuit modeled for an integrator without offset cancelling is shown in Fig. 6. The output voltage relation, between two successive sampling periods, can be formulated similarly to formulating the offset cancelling integrator as,

$$v_n = \frac{\{C_I + \dfrac{C_P}{A+1}\}v_{n-1} + \dfrac{G}{G+1}C_{IN}(V_{IN} - V_{DA}) + \dfrac{G}{G+1}V_{off}(C_{IN} + C_S)}{C_I + \dfrac{C_{IN} + C_P + C_s}{A+1}} \quad (8).$$

Equation 8 can be expressed as the same signal flow diagram, shown in Fig. 5, with slightly different parameters. Here,

$$a = \beta = \frac{G}{G+1} \frac{C_{IN}}{C_I + \dfrac{C_P}{A+1}} \quad (9).$$

$$r = \frac{C_{IN} + \dfrac{G}{G+1}}{C_I + \dfrac{C_{IN} + C_P + C_S}{G+1}} \quad (10).$$

$$\delta = \frac{G}{G+1} \frac{C_{IN} + C_S}{C_I + \dfrac{C_P}{A+1}} \quad (11).$$

If $G=100$ and $C_I = C_{IN}$ are assumed and neglecting $C_P$ and $C_S$, then $1-\gamma$ is about 0.01.

Terms $a$ and $\beta$ are attenuation terms for input signal and feedback signal and usually do not change noise spectra for different noise shapings. An integrator, having loss $1-\gamma$ inside the integration loop, has a pole at finite frequency $\omega T(1-\gamma)/\sqrt{\gamma}$, instead of at zero frequency for an ideal integrator, where T is the sampling interval.

### 4. OVERSAMPLING A-to-D CONVERTER ANALYSIS WITH LOSSY INTEGRATOR

If a digital circuit is prefered for the integrators, as much as possible, the 2nd order delta modulation coder requires no analog integrators. Therefore,

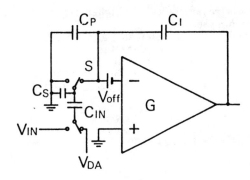

[Fig. 6] An integrator model without offset cancelling.

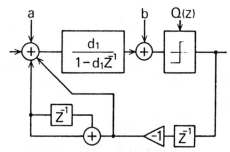

[Fig. 7] Signal flow diagram for a 1st order predictive encoder with 1st order noise shaping, including parasitic effects.

parasitic effects to be considered are distortion generated by the predictor D-to-A converter and noise generated by the comparator and the voltage subtraction network of predictor voltage from the input signal. The other two structures need analog integrators. Therefore, finite gain effect to the analog integrators has to be considered.

### 4.1 1st order predictive encoder with 1st order noise shaping

The 1st order predictive encoder, with 1st order noise shaping, can be modeled as shown in Fig. 7, if amplifier finite gain effect is included. Noise and offset voltages, generated at each node, are also included as "a" and "b". The system transfer equation is,

$$\{X(z) + a - (\frac{Z^{-1}}{1-Z^{-1}} + Z^{-1})Y(z)\}\frac{d_1}{1-d_1 Z^{-1}} + Q(z) + b = Y(z) \quad (12).$$

Solving Eq. 12, reconstructed signal $\tilde{X}(z)$ is expressed as,

$$\tilde{X}(z) = \frac{Y(z)}{1-Z^{-1}} = \frac{d_1\{X(z)+a\}+(1-d_1 Z^{-1})\{Q(z)+b\}}{1-(1-d_1)Z^{-1}} \quad (13).$$

Noise "a" appears at the reconstructed signal, similarly to the input signal. This means D-to-A converter nonlinearity, which is the main noise on this node, appears directly to the output spectrum. However, noise "b" is redistributed to higher frequencies as quantization noise.

Since $d_1$ is nearly equal to 1 and since the signal frequency is much smaller than the sampling frequency, the numerator in Eq. 13 can be approximated as 1. Therefore, the reconstructed input signal is attenuated by $d_1$. Quantizing noise $\bar{N}^2$ in the signal band $f_{max}$ is,

$$\bar{N}^2 = \frac{1}{2\pi} \int_{-\theta}^{\theta} |Q(\omega T)H(\omega T)|^2 d(\omega T) \qquad (14).$$

where,

$$H(\omega T) = 1 - d_1 e^{-j\omega T} \qquad (15).$$

$$\theta = 2\pi f_{max}/f_s \qquad (16).$$

Assuming $\theta$ to be much smaller than unity, and assuming $Q(\omega)$ to be white and distributed equally in $\pm \Delta$, the inband noise can be approximated as,

$$\bar{N}^2 = \frac{2}{9}\Delta^2(\frac{\theta}{2\pi})\{\theta^2 d_1 + 3(1-d_1)^2\} \qquad (17).$$

Since $d_1$ is nearly equal to 1, the first term is the noise for an ideal case. The second term is the excess noise, caused by the non-ideal integrator. Since $f_s/2f_{max}$ is the oversampling ratio R, $1-d_1$ is nearly $1/G$ or $2/G$, depending on the integrator configurations. The signal-to-noise ratio does not change, as long as $A >> R/\pi$. Assuming R as 128, the value $d_1$ becomes 0.91, if the second term is equal to the first term. Therefore, the gain required to the operational amplifier is less than 100, which is smaller than that expected from signal-to-noise ratio. In addition, since signal swing at the integrator output node is small, no distortion is generated in the integrator.

## 4.2 Second order $\Delta$-$\Sigma$ Modulation coder

The 2nd order $\Delta$-$\Sigma$ Modulation coder can be modeled, as shown in Fig. 8, if amplifier finite gain effect is included. Noise and offset voltages, generated at individual nodes, are also included as "a", "b" and "c". The transfer equation is,

$$\{\{X(z)+a-\tilde{X}(z)z^{-1}\}\frac{d_1}{1-d_1 z^{-1}}+b-\tilde{X}(z)z^{-1}\}\frac{d_2}{1-d_2 z^{-1}}+Q(z)+c=\tilde{X}(z)$$

$$(18).$$

Solving Eq. 18, reconstructed signal $\tilde{X}(z)$ is,

$$\tilde{X}(z)=\frac{d_1 d_2\{X(z)+a\}+(1-d_1 z^{-1})(1-d_2 z^{-1})\{Q(z)+c\}+d_2 b(1-d_1 z^{-1})}{1-d_1(1-d_2)z^{-1}}$$

$$(19).$$

Noise "a" appears at the reconstructed signal, similarly to the input signal. Noise "b" is redistributed in the first order. Noise "c" is redistributed to higher frequencies, in the second order, as quantized noise. Therefore, the noise generated in the 1st integrator plays the largest role in regard to the noise

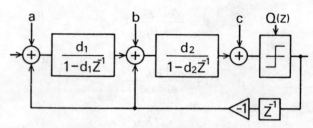

[Fig. 8]   2nd order noise shaping encoder signal flow diagram, including parasitic effects.

performance in this structure. The sources of "a" are amplifier noise and a distortion introduced by the first integrator nonlinearlity. Therefore, special care has to be taken with regard to the amplifier, which forms the integrator.

Since $d_1$ and $d_2$ are nearly equal to 1 and since the signal frequency is much smaller than the sampling frequency, the numerator for Eq. 13 can be approximated as 1. Then, the input signal, which appears at the reconstructed signal, is attenuated by $d_1 d_2$. Quantizing noise $\bar{N}^2$ in signal band $f_{max}$ is,

$$\bar{N}^2 = \frac{1}{2\pi} \int_{-\theta}^{\theta} |Q(\omega T)H(\omega T)|^2 d(\omega T) \qquad (20).$$

where,

$$H(\omega T) = (1-d_1 e^{-j\omega T})(1-d_2 e^{-j\omega T}) \qquad (21).$$

$$\theta = 2\omega f_{max}/f_s \qquad (22).$$

Under the same assumption as in Section 4.1, the inband noise can be approximated as,

$$\bar{N}^2 = \frac{\Delta^2}{3\pi}[(1-d_1)^2(1-d_2)^2\theta+\{d_1(1-d_2)^2+d_2(1-d_1)^2\}\frac{\theta^3}{3}+d_1 d_2\frac{\theta^5}{5}] \qquad (23).$$

The last term in Eq. 23 is the noise in an ideal encoder. The 1st and 2nd terms are excessive noises, caused by the non-ideal integrators. Assuming R as 128 and $d_1$ is equal to $d_2$, the value becomes 0.97, if the sum of the first two terms is equal to the last term. Therefore, the gain required in regard to the amplifiers, which form the integrators, is also not so large for noise redistribution character. However, harmonic distortion, introduced by the amplifiers, will be increased as the amplifier gain becomes smaller. Therefore, the amplifier gain has to be designed, taking this effect into account.

## 5.   INTEGRATOR SETTLING AND QUANTIZER UNCERTANITY EFFECTS

If the switches have their status changed, before the output voltage is settled to a final value in the circuit shown in Section 2, the integrator is regarded to have a loss. If this loss is linear, the loss can be dealt with like the finite gain effect for the integrating amplifier. However, if the output voltage does not settle, for some nonlinear reason, like amplifier slewing limitation, the quantizing noise $Q(z)$ distribution will be changed and linear approximation cannot be applied any more. Therefore, the amplifier has to have sufficient phase margin so that the response character is approximately linear.

If the quantizer generates random noise, the noise can be treated as mentioned in the previous section. If the quantizer has hysteresis $\delta$, the amplitude distribution for the quantizing noise becomes $\pm(\Delta+\delta)$, instead of $\pm\Delta$[12]. Therefore, the quantizing noise increases by this amount. This means that the requirement for the comparator is hardest to meet, in regard to the delta modulation coder, among the three structures.

## 6. CONCLUSION

Three structures, Delta modulation, 1st order predictive encoder with 1st order noise shaping and $\Delta$-$\Sigma$ modulation, were analyzed and compared extensively, in regard to the constraints applied to analog components.

The Delta modulation coder can be constructed without an analog integrator. However, since its step size is smallest among the three structures, the performance requirements applied to the quantizer are hardest to meet for precision A-to-D converters.

The $\Delta$-$\Sigma$ modulation requires very fast slewing and large dynamic range integrator. Therefore, it does not appear easy to attain theoretical signal-to-noise ratio, nor increasing sampling rate.

The 1st order predictive encoder, with 1st order noise shaping, stands between the above two. Since step size is moderate and integrator output signal swing is small, good performance at a high sampling rate is also expected.

## ACKNOWLEDGEMENT

The author would like to thank Dr. R. Maruta for informative discussions. The author appreciates the encouragement given by Drs. H. Shiraki, H. Sakuma and T. Enomoto.

## REFERENCE

1. Wooley, B.A. and Henry, J.L., "An integrated Per-Channel PCM Encoder Based on Interpolation", IEEE J. of Solid-State Cirucits, vol. SC-14, pp.20-25; 1979.

2. Misawa, T., et al, "Single-Chip per Channel Codec with Filters Utilizing $\Delta$-$\Sigma$ Modulation", IEEE J. of Solid-State Circuits, vol. SC-16, pp.333-341; 1983.

3. Yamakido, K., et al., "A Voiceband 15b Interpolative Converter Chip Set", ISSCC Digest of Tech. Papers, pp180-181; 1986.

4. Hayashi, T., et al., "A Multistage Delta-Sigma Modulator without Double Integration Loop", ISSCC Digest of Tech. Papers, pp.182-183; 1986.

5. Tewksbury, S.K. and Hallock, R.W., "Oversampled, Linear Predictive and Noise-Shaping Coders of N>1", IEEE Trans. on Circuit and Systems, vol. CAS-25, pp.436-447; 1978.

6. Agrawal, B.T. and Shenoi, K., "Design Methodology for $\Sigma\Delta$M", IEEE Trans. on Communications, vol. COM-31, pp360-369; 1983.

7. Candy, J.C., "A Use of Double Integration in Sigma Delta Modulation", IEEE Trans. on Communications, vol. COM-33, PP249-258; 1985.

8. Hauser, M.W. and Brodersen, R.W., "Circuit and Technology Consideration for MOS Delta-Sigma A/D Converters", ISCAS'86, pp.1310-1315; 1986.

9. Schouten, J.F., de Jager, F. and Greefkes, J.A., "Delta Modulation, A New Modulation System for Telecommunication", Philips Tech. Review, vol. 13, pp.237-268; 1952.

10. Inose, H., Yasuda, Y. and Murakami, J., "A Telemetering System by Code Modulation", IRE Trans. Space Electronics and Telemetry, vol. SET-8; 1962.

11. Yukawa, A., Maruta, R. and Nakayama, K., "An Oversampling A-to-D Converter Structure for VLSI Digital Codecs", ICASSP'85 Proceedings, pp.1400-1403; 1985.

12. Gotz, B., Ching, Y.C. and Baldwin, G.L., "Idle-Channel Noise Characteristic of a Delta Modulator with Step Imbalance and Hysteresis", IEEE Trans. on Communications, vol. COM-23, pp834-844; 1975.

# Part 3
# Implementations and Applications
# of Oversampling A/D Converters

# Design and Implementation of an Audio 18-Bit Analog-to-Digital Converter Using Oversampling Techniques*

**ROBERT W. ADAMS**

*dbx, Newton, MA 02195, USA*

An 18-bit analog-to-digital converter has been designed that uses the principle of oversampling rather than the successive-approximation technique of most converters. The design employs a two-stage digital filter/decimator to reduce the sampling rate. Unusual digital filter structures are used to reduce the computation rate to a practical level. Extension to 19 bits and beyond is discussed.

## 0 INTRODUCTION

There is an increasing need in the world of digital audio for high-resolution analog-to-digital (A/D) converters that are both inexpensive and easy to manufacture. Present-day 16-bit successive-approximation converters still represent a formidable design challenge and are considered to be state of the art. There are several practical as well as theoretical reasons why the successive-approximation technique of conversion is limited to 16 bits of resolution. For example, to design an 18-bit converter we would require:

1) An 18-bit monotonic digital-to-analog (D/A) converter with a settling time (to ½ LSB) of $< 1$ μs.

2) A comparator with a response time of 150 ns with an overdrive of only 40 μV and a peak-to-peak input noise voltage of $< 40$ μV.

3) A sample-and-hold circuit with an acquisition time of 1 μs and a droop specification of $< 40$ μV over the 20-μs sampling period.

Items 1) and 3) represent practical problems that could be overcome by advances in component technology or clever circuit design; item 2) is a theoretical limitation resulting from the conflicting requirements of high bandwidth and low noise.

It is apparent that if higher resolution is desired, a different approach must be taken. The technique used by the author and described in this paper employs the principle of oversampling followed by decimation.

This paper focuses on the two major elements of an oversampling converter: the fast "front-end" A/D converter and the digital decimator. We also discuss the D/A design and present some measurements made on the prototype A/D–D/A system. In later sections we discuss the feasibility of integration and methods of further resolution extension.

## 1 THEORY OF OVERSAMPLING CONVERTERS

Oversampling converters achieve increased resolution not by decreasing the error between analog input and digital output but by making the error occur more often. The frequency spectrum of the error function thus extends well beyond the band of interest, and the in-band noise power can be made very small even though the total noise power is quite large. The high bit rate of the oversampling converter can then be reduced by the process of decimation. The combined process is illustrated in Fig. 1.

The input signal $x(t)$ is assumed to be a sinusoid with a frequency of less than 20 kHz. It is sampled at the fast rate of $1/T$, where $1/T >> 48$ kHz, resulting in the spectrum of Fig. 1(a). This sequence is applied to a low-resolution converter running at the fast $1/T$ rate, which adds random noise to the signal, resulting in the spectrum in Fig. 1(b). The digital sequence is now applied to a digital low-pass filter with a cutoff frequency slightly higher than 20 kHz. This filter removes the out-of-band noise components and also acts as the system anti-aliasing filter. The signal can now be resampled at the lower rate of $1/KT$ (typically

* Presented at the 77th Convention of the Audio Engineering Society, Hamburg, 1985 March 5–8.

Reprinted with permission from *J. Audio Eng. Soc.*, vol. 34, pp. 153–166, March 1986.

48 kHz in our example) without excess noise being added due to aliasing.

Oversampling converters of this type have been known for many years and have found practical use in voice-grade converters for the telephone system. However, the technique has not been used for full-bandwith audio conversion because of the computational requirements of the digital low-pass filter. As an example, consider a system where the high-speed converter is running at 6 MHz (128 times faster than the final 48-kHz sample rate) and the cutoff frequency of the digital filter is 22 kHz. Due to the high ratio of sample frequency to cutoff frequency, the number of coefficients needed to achieve a reasonably steep cutoff characteristic is about 4000. To compute one output sample every 20 μs requires a multiply/accumulate time of 5 ns. This figure is beyond the state of the art by an order of magnitude. The solution to this problem was a major focus of the design effort and is described in detail later in this paper.

## 2 FRONT-END A/D DESIGN

Several different encoding schemes were examined for the front-end design. Some of these methods are:
1) Linear pulse-code modulation (PCM)
2) Feedback differential converters
   a) Delta modulation
   b) Differential pulse-code modulation (DPCM)
3) Feedback PCM converters
   a) Sigma-delta modulation
   b) Noise-shaped PCM converter.

The first of these, linear PCM, is unattractive because of the relationship between the sampling rate and the signal-to-band-limited-noise ratio (SNR),

$$\text{SNR (band-limited)} \propto \varepsilon \psi \sqrt{F_s/B_w} \qquad (1)$$

where

$F_s$ = sampling rate
$B_w$ = noise measurement bandwidth.

Eq. (1) tells us that a 16-bit converter, in order to achieve 18 bits of resolution in the audio band, would have to run at a sampling rate of 16 × 48 kHz. A 14-bit converter would have to run at 256 × 48 kHz. Clearly, this is not a practical approach to achieving increased resolution.

Eq. (1) is derived assuming that the noise power spectrum is relatively flat for frequencies below $F_s$. This assumption is justified if the input signal is sufficiently complex to decorrelate the error signal from the input signal. It can be shown, however, that the shape of the noise power spectrum of a quantized signal can be altered by employing frequency-selective negative feedback around the quantizer [1]. This technique can be used to lower the quantization noise in one portion of the frequency spectrum at the expense of increasing the noise power at other frequencies. In an oversampled system we can deliberately concentrate the quantization noise power above the band of interest with an appropriate choice of feedback filter. The result is a dramatic increase in low-frequency SNR.

Fig. 2 depicts some of the topologies used in feedback encoders. These topologies differ only in the location of the input summer and the resolution of the quantizer. The simplest feedback encoder is the delta modulator [Fig. 2(a)], which uses a comparator as a 1-bit quantizer. In this configuration the overload level of the modulator as a function of frequency follows the frequency response of the feedback filter $H(f)$, and the spectral characteristic of the noise floor at the decoded output is approximately white if the input is sufficiently random and the sampling rate is much higher than the band of interest [2]. While it may not seem that the noise floor

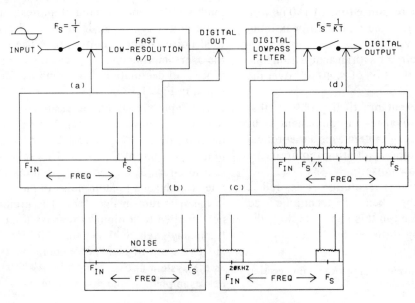

Fig. 1. Theory of oversampling converters. (a) Original sampled spectrum. (b) Spectrum with quantizing noise added. (c) Noise and signal above 20 kHz removed by digital low-pass filter. (d) Resampled at lower rate.

has been "shaped" in this example, we note that the overload level as a function of frequency is decreasing (assuming that $H(f)$ is a low-pass function), and hence the out-of-band SNR is reduced.

In Fig. 2(b) the input signal is now summed with the input of the feedback filter $H(f)$ rather than with the output of the filter, and the decoder is simply a low-pass filter rather than an integrator. The performance of this topology (known as a sigma-delta modulator) is equivalent to that of a standard delta modulator with a prefilter characteristic of $H(f)$ and a postfilter characteristic of $1/H(f)$. Thus we can conclude that the overload level as a function of frequency is flat and that the spectral characteristic of the noise floor rises with increasing frequency.

Fig. 2(c) depicts a DPCM encoder, which differs

from the delta modulation encoder of Fig. 2(a) only in that the error voltage $e(t)$ is quantized to more than 1 bit. The quantizer inside the feedback loop would ideally be a "flash" converter, which eliminates the need for a sample-and-hold circuit. The performance characteristics of this circuit are similar to those of the delta modulation encoder except that the dynamic range increases in direct proportion to the resolution of the quantizer.

Fig. 2(d) depicts a feedback encoder with the input signal summed with the input of the feedback filter $H(f)$ rather than with the output of the filter. This circuit is simply an extension of the sigma-delta modulator circuit in Fig. 2(b), but with a flash converter in place of the comparator. Therefore the performance will be identical to that of the sigma-delta modulator

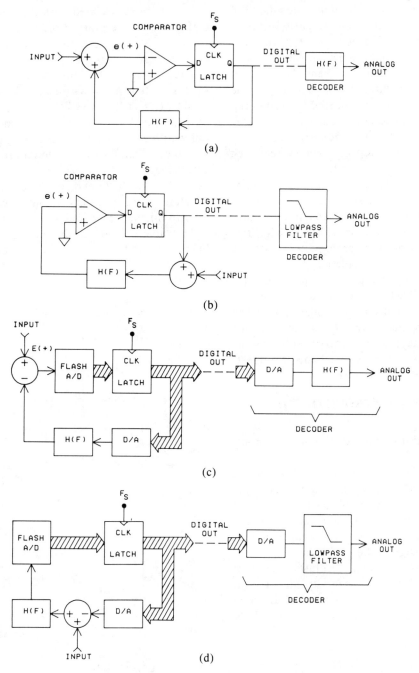

Fig. 2. (a) Delta modulator. (b) Sigma-delta modulator. (c) Differential pulse-code modulator. (d) Noise-shaped pulse-code modulator.

but with an extension of the dynamic range proportional to the resolution of the quantizer. The digital output of this converter is proportional to signal amplitude rather than signal slope, and therefore this circuit must be classified as a PCM converter. It differs from a standard PCM converter, however, in that the spectrum of the noise floor rises with increasing frequency. This converter is often referred to as a noise-shaping A/D converter.

Much work has been done in the area of finding an "optimum" feedback filter for the converters discussed above. An optimum feedback filter is one that minimizes the in-band noise at the expense of greatly increasing the out-of-band noise. Spang and Schultheiss [1] derived an improvement ratio comparing the in-band noise power of a feedback system with that of an open-loop system for stationary Gaussian input signals. The coefficients of the finite-duration impulse-response (FIR) feedback filter were then numerically optimized. Schindler [2] derived the optimum coefficients analytically by using certain assumptions, and then derived a method of testing the resulting idling patterns. Kimme and Kuo [3] derived an iterative procedure for finding the optimum coefficients of the feedback filter. This procedure is guaranteed to converge.

Unfortunately the derivation of optimum feedback filters is done without regard for the stability of the resulting system. While these filters may indeed minimize the noise power for Gaussian inputs, the filters that result in the most spectacular reduction of noise generally produce idling patterns with large high-frequency oscillations. In practice, analog filters of degree higher than 2 are apt to cause oscillations unless special nonlinear stabilization techniques are employed [4].

In summary, feedback converters are capable of achieving very high in-band dynamic-range figures when the sampling rate is much greater than the Nyquist frequency. This is accomplished by concentrating the noise power of the system above the band of interest, which minimizes the in-band noise.

## 3 PRACTICAL IMPLEMENTATION OF THE FRONT END

Several considerations went into the choice of front-end topology.

1) The bit rate of the converter must be low enough that the digital filter/decimator can run at reasonable speeds and use commonly available parts.

2) It is advantageous if the digital signal represents the input-signal amplitude rather than signal slope, as this simplifies the design of the decimator.

3) To achieve 18-bit-equivalent resolution, the front end should achieve a dynamic range (with dither added) of at least 105 dB. This is approximately 12 dB more than the 93 dB that can be achieved by a dithered 16-bit PCM system.

The first point imposes an upward limitation on the clock rate of approximately 10 MHz. The second consideration limits the choice to either a sigma-delta modulator or its close relative, the noise-shaped PCM

converter.

Theoretical expressions for the dynamic range of sigma-delta modulators can be found in the literature, and using these formulas we find that 18-bit resolution should be attainable using a sigma-delta modulator at reasonable (<10 MHz) sampling rates. There are two reasons, however, why we have decided to use a noise-shaped PCM system rather than the simpler sigma-delta modulator.

1) The sigma-delta modulator is extremely sensitive to waveform asymmetry at the latch outputs (discussed in Sec. 3.2). If 18-bit resolution is desired, the maximum tolerable mismatch between rise and fall times is only 30 ps.

2) Loop filters of order greater than two are highly desirable because they result in an error signal that is completely uncorrelated with the input signal. These higher-order filters are difficult to stabilize using a sigma-delta modulator, but are easily stabilized for the case of noise-shaped PCM coders.

The diagram of the converter finally adopted for use in this system is shown in Fig. 3. The clock rate is 6.144 MHz (128 × 48 kHz), and the error signal is quantized to 4 bits (16 levels). The measured dynamic range of the front-end converter (signal/20-kHz noise) is 106 dB, which exceeds the requirement for 18-bit-equivalent resolution. This excess dynamic range is needed in the front end, as there are several sources of additive noise in the decimator, which is analyzed later in this paper.

The quantizer is a flash converter composed of 15 high-speed (30-ns) comparators, each comparator re-

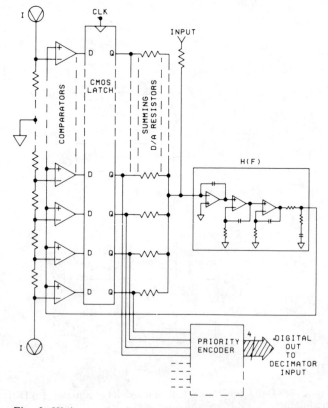

Fig. 3. High-speed noise-shaping front-end PCM converter.

ceiving the error signal on its positive input and a reference voltage from a resistive ladder network on its negative input. The loop D/A converter consists simply of 15 equal-value resistors connected to the latched output of each comparator and summed in a virtual-ground summer.

The loop filter is a fourth-order filter with four real poles (at the origin) and three coincident real zeros. Stabilization is accomplished by the use of nonlinear techniques similar to those described in [4].

The location of the zeros in the transfer function of the loop filter has a profound effect on the idling behavior of the system. As the frequency of the zeros is increased, the phase margin of the loop is reduced and the amplitude of the idling pattern is increased. We have adjusted the frequency of the zeros to the point where the idling pattern is just large enough to span three levels of the loop flash converter. This was done to eliminate any possibility of crossover noise, where the loop must jump discontinuously from one level to the next as the input signal changes. This technique is in many ways analogous to the use of crossover bias in a class-AB amplifier.

### 3.1 Linearity Considerations

The linearity of the converter is determined solely by the matching between the summing resistors. Each resistor controls the incremental gain over some specific part of the input voltage range, and hence the overall linearity is piecewise linear, with 16 different segments making up the total curve, as shown in Fig. 4. The transfer function is somewhat smoothed because the system does not jump from one level to the next but goes through a region where two adjacent comparators are active at once, which produces an averaging effect. This effect is analogous to the use of dither at the input to a quantizer, which has the effect of smoothing the staircase transfer curve [5, pp. 110–111].

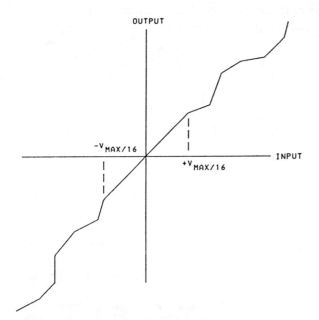

Fig. 4. Exaggerated linearity-error curve.

It should be noted that the linearity of the flash converter (determined by the resistive divider network feeding the comparators) has little effect on the linearity of the circuit. This is because the flash converter is within the feedback loop, and therefore its effect on the overall linearity is greatly reduced by the negative feedback.

The linearity within each piecewise-linear segment is essentially perfect due to the application of negative feedback, and therefore the linearity for small ($< V_{max}/16$ peak) input signals is extremely good. This is in sharp contrast to the linearity of a successive-approximation-type converter, where the distortion usually increases significantly at low signal levels.

The distortion performance at higher signal levels is determined almost entirely by the matching between the summing resistors in Fig. 3. We have derived the relationship between resistor mismatch and total harmonic distortion for the case where only the center resistor is in error. This produces a symmetrical transfer curve, as shown in Fig. 5. The Fourier transform of the output for a sine-wave input is given by

$$F(K) = \frac{4(M - 1)}{\pi}$$

$$\times \left( \frac{K \cos K\theta_1 \sin \theta_1 - \sin K\theta_1 \cos \theta_1}{1 - K^2} \right.$$

$$\left. + \frac{\sin \theta_1 \cos K\theta_1}{K} \right), \qquad K \text{ odd} \qquad (2)$$

where

$M$ = slope of center segment ($M$ = 1.01 for 1% slope error)
$K$ = frequency index
$\theta_1$ = radian angle at which the input signal crosses point $X_1$

This formula was used to compute the rms sum of the first 20 distortion components of the output waveform for various input levels and percentage center-slope errors. These results are shown in Table 1. Note that the distortion varies linearly with the slope error, and increases with decreasing input level. While this is a somewhat simplified case, it does allow us to estimate distortion based on resistor tolerances. In most cases 0.01% resistor-matching errors result in worst-case distortion figures of < 0.003%. This should be adequate for most applications.

### 3.2 Sensitivity to Waveform Symmetry

Several of our early prototypes of the oversampling converter suffered from large amounts of unexplained excess noise at the output of the decimator. This led to an investigation of the effects of waveform symmetry (at the latch outputs) on the noise floor of the system. We should bear in mind that the sum of these digital signals directly represents the input signal, and decoding may be accomplished by simply filtering the sum of the digital waveforms present at the latch outputs. It should therefore come as no suprise that the exact shape of the waveform at the latch outputs is critical, as we are attempting to achieve over 100 dB of dynamic range by summing 15 5-V digital signals switching at a 6-MHz rate.

A simple example of waveform asymmetry is illustrated in Fig. 6. In Fig. 6(a) a binary waveform is shown with nonsymmetrical rise and fall times. The average value of this waveform may be easily calculated. In Fig. 6(b) we have rearranged the waveform so that now the two 1s occur in a row. The average value of this waveform is clearly different from that of the previous waveform. This illustrates that in the case of nonsymmetrical binary waveforms, the effective value

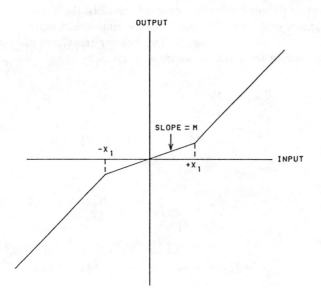

Fig. 5. Center-only nonlinearity.

Table 1. Total harmonic distortion versus input level and center-slope error.

| Error in center-segment slope (%) | Peak-to-peak input level expressed as number of segments | Total harmonic distortion (dB) |
|---|---|---|
| 0.01 | 3 | −97 |
| 0.01 | 9 | −108 |
| 0.01 | 15 | −113 |
| 0.032 | 3 | −87 |
| 0.032 | 9 | −98 |
| 0.032 | 15 | −103 |
| 0.1 | 3 | −77 |
| 0.1 | 9 | −88 |
| 0.1 | 15 | −93 |
| 0.316 | 3 | −67 |
| 0.316 | 9 | −78 |
| 0.316 | 15 | −83 |
| 1.0 | 3 | −57 |
| 1.0 | 9 | −68 |
| 1.0 | 15 | −73 |
| 3.16 | 3 | −47 |
| 3.16 | 9 | −58 |
| 3.16 | 15 | −63 |

of a particular bit depends on the value of adjacent bits. Because of the feedback nature of noise-shaping converters, an analog low-pass filter (a local decoder) connected to the nonsymmetrical latch output will show no excess noise in its output. However, when the decoder is a digital low-pass filter (as is the case in the decimator), the value of a 1 or 0 is assumed to be invariant. Thus nonsymmetrical latch output waveforms cause excess noise at the output of a "perfect" decoder.

The relationship between excess noise and waveform asymmetry may be derived mathematically. Since the signals to be analyzed may be described only in a statistical sense, we will use the theorem that equates the power spectrum of a signal to the Fourier transform of its autocorrelation function. To simplify the analysis, we will analyze the case of nonsymmetrical propagation delay, shown in Fig. 7. We assume that this signal is the latched output of a single-comparator feedback converter [Fig. 2(b)]. All 1-to-0 transitions are delayed by time $T_1$, where $T_1$ is the difference between positive and negative propagation delays. The resulting error signal, shown in Fig. 7(b), contributes excess noise to the signal.

To simplify the analysis, we must make some assumptions about the statistics of the error signal. We know that the noise produced in our front-end converter has a nonwhite spectral density function with most of its noise energy concentrated at higher frequencies. This implies that the autocorrelation function $R(U)$ of the binary signal shown in Fig. 7(a) is nonzero for $U > T$, which leads us to conclude that adjacent bits are not statistically independent. The error signal will therefore have some statistical relationship between samples, but this relationship is very weak since the errors occur only on the negative transitions of the binary signal. We will therefore treat the error signal as if it were a random binary pulse generator with a period of $T$, a pulse duration of $T_1$, and a minimum pulse spacing of $2T$. This last condition arises from the requirement that a binary signal cannot have two negative transitions in a row.

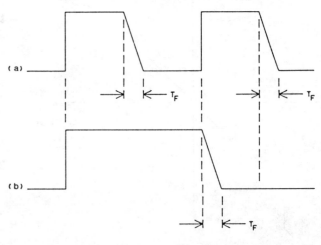

Fig. 6. Illustration of additive-noise mechanism for unequal rise and fall times. (a) 1–0–1 pattern. (b) 1–1–0 pattern with area different from (a).

284

The autocorrelation $R(U)$ of the error waveform can be found easily by inspection,

$$R(0) = 4 \frac{T_1}{T} P_{(1\to0)} \qquad (3a)$$

where

$T$ = 1/sampling rate
$T_1$ = error-pulse duration
$P_{(1\to0)}$ = probability of a negative-edge transition at the end of any time period $T$.

$$R(T) = 0 \qquad (3b)$$

due to the minimum pulse-spacing requirement of $2T$.

$$R(NT) = 4 \frac{T_1}{T} [P_{(1\to0)}]^2, \qquad N = \text{integer} > 2 . \qquad (3c)$$

This function is sketched in Fig. 7(c). We can find the power spectral density function by taking the Fourier transform of $R(U)$. Before doing this, however, we note that there is a strong periodic signal with repetition rate $T$, which will contribute spectral energy only at integer multiples of the sampling frequency. We can express $R(U)$ as the sum of the periodic waveform of Fig. 7(d) with the nonperiodic waveform since we are interested solely in the low-frequency noise contribution.

The transform of the nonperiodic part of $R(U)$ [Fig. 7(e)] can be further simplified by assuming that the noise bandwidth $B_w$ is much smaller than the sampling frequency (which is certainly true for $F_s = 6$ MHz). Applying the transform theorems of linearity and time delay, it is easy to show that the the negative triangle waves can be time shifted by $T$ so that they center on $U = 0$, with negligible effect on the low-frequency spectrum. When this is done, $R(U)$ becomes a single-triangle function, as shown in Fig. 7(f).

The transform of the autocorrelation function shown in Fig. 7(f) is well known and is given by

$$H(F) = AT_1 \left[ \frac{\sin(\pi FT_1)}{\pi FT_1} \right]^2 , \qquad (4)$$

$$A = \left[ \frac{4T_1}{T} P_{(1\to0)} \right] \left[ 1 - 3P_{(1\to0)} \right] .$$

For $F \ll F_s$ the power spectral density is nearly flat, with a value of

$$AT_1 = \left[ \frac{4T_1^2}{T} P_{(1\to0)} \right] \left[ 1 - 3P_{(1\to0)} \right] . \qquad (5)$$

To compute the SNR as a function of delay error $T_1$, we need to know the maximum rms sinusoidal signal that the binary signal of Fig. 7(a) can represent. Since the binary signal has levels of $\pm 1$, a 2-V peak-to-peak

sine wave (rms value of 0.707) can be reproduced without clipping. From Eq. (5) the SNR then becomes

$$\text{SNR} = \frac{1}{T_1 \sqrt{8 F_s P_{(1\to0)} [1 - 3P_{(1\to0)}] B_w}} . \qquad (6)$$

We can now substitute our minimum dynamic range requirements for SNR and calculate the maximum allowable timing error $T_1$. For the total system with 15 comparators, the dynamic range must exceed 104 dB. The dynamic range for a single-comparator system must therefore exceed $104 - 20 \log(16) = 80$ dB. If we assume that a "perfect" single-comparator system can achieve 82 dB of dynamic range, then the noise due to asymmetry alone must not exceed 84 dB below peak rms signal to meet the requirement of 80-dB minimum SNR. We can now evaluate Eq. (7) with SNR = 15,800, $F_s = 6$ MHz, and $B_w = 20$ kHz. For $P_{(1\to0)}$ we will use an estimated value of 0.25.

$$T_1 = 257 \text{ ps} . \qquad (8)$$

We conclude that the maximum allowable difference between positive and negative propagation delays is about one quarter of a nanosecond—a very small difference.

We can now readily extend this result to include the case of unequal rise and fall times, shown in Fig. 8(a). The error signal shown in Fig. 8(b) may be analyzed by adding the area under the positive error signal with the area under the negative error signal and equating the result with the area of the rectangular error function of Fig. 7(b). It can easily be shown that this procedure

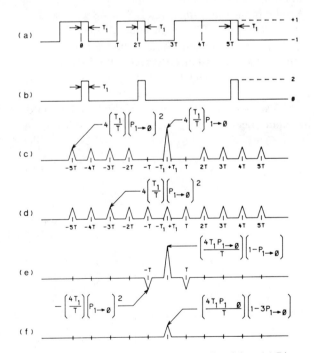

Fig. 7. Noise analysis of unequal propagation delay. (a) Binary waveform. (b) Error function. (c) Autocorrelation of error function. (d) Periodic part of autocorrelation. (e) Nonperiodic part of autocorrelation. (f) Equivalent low-frequency autocorrelation.

generates correct results for $F << F_s$. Using this method, we have

$$T_1 = \left| \frac{T_R - T_F}{2} \right| . \tag{9}$$

Substituting this result into Eq. (7), we arrive at

$$|T_R - T_F| < 514 \text{ ps} . \tag{10}$$

Thus the maximum allowable difference between rise and fall times is about one-half of a nanosecond.

These stringent symmetry requirements dictate the use of high-speed CMOS devices with tightly matched N and P device characteristics. We have found that the only commercial devices that meet these requirements are the 74HC series of CMOS digital integrated circuits, which are designed to serve as CMOS replacements for standard TTL parts. When these devices are used, less than 1 dB of excess noise due to asymmetry is present at the decimator output, when compared with a direct noise measurement of the front-end converter alone.

## 4 DESIGN OF THE DECIMATOR

The ideal decimator would be a steep linear-phase low-pass filter with a cutoff frequency $F_c$ less than $F_{so}/2$, where $F_{so}$ is the output sampling rate. As mentioned, this results in an excessive number of FIR filter coefficients due to the high ratio of input to output sampling rates (128:1). Our solution to this problem is to use two stages of decimation, with the first stage converting to an intermediate sampling frequency somewhere between the input sampling rate (128 × 48 kHz) and the final rate (48 kHz). This technique allows us to use a very simple high-speed filter structure for the first decimator and a standard FIR filter structure for the second decimator. The intermediate sampling frequency was chosen to be 2 × 48 kHz = 96 kHz.

Fig. 9 shows the requirements of the first decimator. The input spectrum is the noise floor of the front-end

converter and exhibits the rising spectral characteristic we expect from a noise-shaped PCM converter. If we wish to resample this signal at $2 \times F_{so}$ (96 kHz), the Nyquist criterion would dictate that all frequencies above $F_{so}$ need to be removed to prevent aliasing of the input spectrum. However, we note that only certain frequency regions will alias into the audio band, and to prevent alias components from appearing in the audio band, our filter need attenuate only these frequencies. These frequency bands are shown in Fig. 9(b) and are described mathematically by

$$F_{(alias)} = N \times 96 \text{ kHz} \quad \pm 20 \text{ kHz},$$

$$N \text{ any integer} > 0 . \tag{11}$$

We can now express the requirements of the digital filter in terms of how much out-of-band noise will be allowed to alias into the audio band. If we allow a maximum noise increase due to aliasing of 1 dB, we can write

$$\text{excess noise (dB)} = 10 \log$$

$$\times \left[ 1 + \frac{\sum_{N=1}^{32} \int_{NF_{si}-20 \text{ kHz}}^{NF_{si}+20 \text{ kHz}} N(F)^2 H(F)^2 \, dF}{\int_0^{20 \text{ kHz}} N(F)^2 \, dF} \right] < 1 \text{ dB} \tag{12}$$

where

$N(F)$ = input noise spectrum
$H(F)$ = response of digital filter (assumed to be unity for $F < 20$ kHz)
$F_{si}$ = intermediate sampling rate (96 kHz).

Various filter responses may be tested according to Eq. (12), using numerical integration techniques.

Clearly, our search for a suitable filter structure should concentrate on those filters that exhibit the highest attenuation in the frequency bands described by Eq. (11). One filter that satisfies this condition is the simple

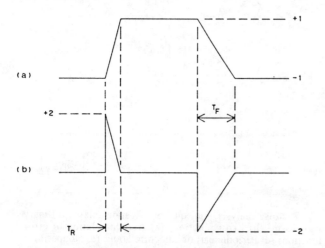

Fig. 8. Unequal rise and fall times. (a) Binary waveform at latch output. (b) Error function.

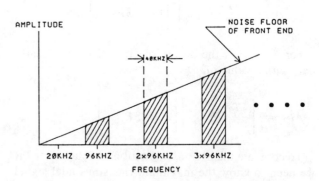

Fig. 9. Frequency regions which will result in audio-band aliasing components when resampled at 96 Hz.

moving-average filter shown in Fig. 10(a). This filter has tap weights of unity and the rectangular impulse response shown in Fig. 10(b). The frequency response (scaled for unity dc gain) is well known [6, pp. 90–91] and given by

$$\text{magnitude} \left[ \frac{H(F)}{H(0)} \right] = \left| \frac{\sin (\pi F P / F_s)}{P \sin (\pi F / F_s)} \right| \quad (13)$$

where

$P$ = number of samples averaged  
$F_s$ = sampling frequency.

This function is plotted in Fig. 11 for $P = 64$ and $F_s = 6.14$ MHz. Note that the maximum attenuation occurs at

$$F = N F_s / P , \qquad N \text{ any integer} > 0 . \quad (14)$$

Thus the frequencies where maximum attenuation occurs are precisely where we want them to be, according to Fig. 9. When the frequency response is evaluated according to the criteria given in Eq. (12), however, we find that we have insufficient alias rejection, resulting in excess audio-band noise.

The simplicity of this filter suggests that we try cascading several moving-average filters to obtain the required alias rejection. When this is done and the results are evaluated according to Eq. (12), we find that three series-connected averaging filters result in an excess noise figure of less than 1 dB, which is satisfactory. The frequency response of this combination is easily found to be

$$\text{magnitude} \left[ \frac{H(F)}{H(0)} \right] = \left| \frac{\sin (\pi F P / F_s)}{P \sin (\pi F / F_s)} \right|^3 . \quad (15)$$

This function is plotted in Fig. 11(b). Note that the response at 20 kHz is down by 1.89 dB. This error may be corrected using an external analog peaking low-pass filter or by designing the second stage of the decimator to compensate for this loss.

Unfortunately, the required computation rate of even this simple filter appears at first glance to be excessively high. Since we cannot throw away output samples (decimate) until the filtering operation is complete, the first two moving-average filters must compute one output sample for every input sample. Each output sample is the sum of the past 64 input samples, and therefore the computation rate for one moving-average filter becomes

64 additions per sample × 6.144 MHz

$$= 393 \text{ million additions per second.} \quad (16)$$

Fortunately, we can realize great savings in the computation rate if we use the recursive form of the moving-average filter. The derivation of the recursive form is

$$Y(N) = \sum_{M=N-P+1}^{M=N} X(M) = \left[ \sum_{M=N-P}^{WM=N-1} X(M) \right]$$

$$+ X(N) - X(N - P)$$

$$= Y(N - 1) + X(N) - X(N - P) \quad (17)$$

where

$Y(N)$ = output sequence  
$X(M)$ = input sequence  
$P$ = averaging length.

Eq. (17) indicates that each new output sample may be computed from the previous one by adding the new input value to the previous output value and then subtracting the input value that occurred $P$ samples ago. Therefore only one addition and one subtraction are required to compute each new output sample, and the required computation speed is only 80 ns per addition

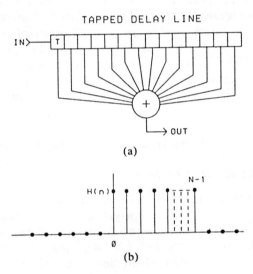

(a)

(b)

Fig. 10. (a) FIR implementation of moving-average filter. (b) Impulse response.

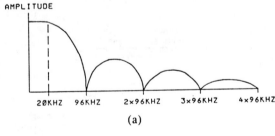

(a)

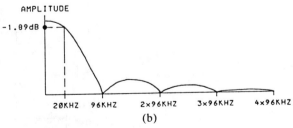

(b)

Fig. 11. (a) Frequency response of one moving-average filter. (b) Frequency response of three cascaded moving-average filters.

or subtraction. This is well within the capability of practical devices.

The hardware implementation of three cascaded moving-average filters is shown in Fig. 12. One input of the ALU (which is simply an add/subtract unit) is fed from a multiplexer that selects either the most recent input word or the output of a 64-stage shift register, which is used to delay the input data stream. The other input of the ALU is fed back from the latched output of the ALU. In operation the multiplexer is switched to the output of the shift register, and the delayed data are subtracted from the latched output of the ALU. The multiplexer is then switched to select the input data, and the input data word is added to the latched ALU output. This output is then used as the input to the next stage, which is operated out of phase with the first stage to eliminate the need for double buffering between stages.

Note that the width of the data path increases by 6 bits at each filter section. If we start with a 4-bit input word, this means that the final output of the three moving-average filters is 22 bits wide. This does not imply, however, that we have true 22-bit resolution; the dynamic range cannot exceed that of the front end.

This 22-bit word is truncated to 18 bits before it is fed to the next stage of the decimator. While this truncation does not appreciably increase the output noise power, it does have an effect on the spectral shape of the output noise floor. We shall discuss this effect later.

### 4.1 Second Stage of Decimator

The second stage of the decimator uses a very conventional FIR linear-phase low-pass filter. Its function is to remove all noise and signal components above 24 kHz and then to undersample (decimate) the output by a factor of 2. The filter should ideally be designed in such a way as to provide adequate attenuation of alias signals without introducing excessive amounts of "time smear," which in some situations may be audible [7].

If we choose the cutoff frequency to be exactly 24 kHz (one-quarter of the 96-kHz intermediate sampling rate), it becomes possible to reduce the computation rate of the final FIR filter by a factor of two. To see how this occurs, consider the impulse response of an ideal low-pass filter with a gain of unity for frequencies below 24 kHz and a gain of 0 above 24 kHz. The impulse response of such a filter is given by;

$$H(n) = \frac{\sin (2\pi n F_{co}/F_s)}{\pi n} \qquad (18)$$

where

$H(n)$ = impulse-response function
$F_{co}$ = cutoff frequency
$F_s$ = sample frequency.

Note that if we choose $F_{co}/F_s = 0.25$, then $H(n) = 0$ for even values of $n$ greater than 1. A window function can be applied to the infinite-length filter described by Eq. (18), resulting in a finite-length filter with half of its coefficients equal to zero. Filters of this type are known as half-band filters, and they can also be designed using standard linear programing techniques (such as the Remes exchange algorithm) if the passband and stopband ripple are made equal and the cutoff frequency is chosen correctly. Such techniques result in an optimum filter for a given number of coefficients and are preferred to the windowed sin (x)/x approach outlined above.

The final design uses a 61-tap FIR filter (half of which are zero) with a cutoff frequency of 24 kHz. Since there are only 31 nonzero coefficients, the required computation rate is only:

rate = 31 × 48 kHz

= 1.488 million multiplications per second.

$$(19)$$

The hardware implementation of this filter (Fig. 13)

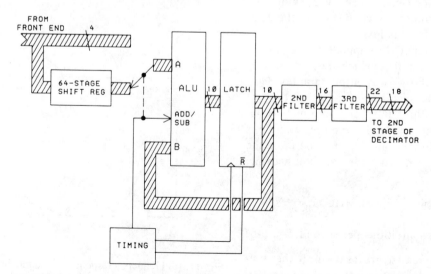

Fig. 12. Hardware to implement recursive form of moving-average filter. Second and third stages are identical to first stage except for data-path widths.

288

uses a shift-and-add algorithm to perform the multiplications. This decision was made in part because there are no inexpensive commercial multiplier integrated circuits with data paths wider than 16 bits, and we require an 18-bit part.

The frequency response of this filter is shown in Fig. 14. It attenuates audio-band alias components by at least 60 dB and has a passband ripple of ±0.0086 dB. Note that alias components do not appear below 20 kHz until the input frequency exceeds 28 kHz, and therefore 28 kHz is the frequency where the filter must achieve its desired stopband attenuation.

Fig. 14 shows that the filter response is mirrored about the intermediate sampling rate (96 kHz), as expected. This means that for input frequencies in the range of 96 ± 24 kHz we have only the attenuation of the first decimation filter. With the worst case of an input frequency of 72 kHz, Eq. (15) yields an attenuation figure of −31 dB for the first decimation filter. If large input signals are expected at these frequencies, external analog filters must be used to prevent aliasing. A simple 2nd-order peaked low-pass filter may be used at the input both to compensate for the high-frequency loss of the first decimator filter at 20 kHz and to provide additional attenuation at frequencies above 72 kHz.

FIR filters tend to generate output word lengths that are much longer than their input word lengths, so the output usually must be truncated. In our case the decimator output is truncated to 18 bits, and this becomes the final digital output of the system.

## 4.2 Effect of Truncation on the Output Spectrum

The effect of the two truncations on the output noise spectrum may be predicted by modeling each truncation as an additive white-noise source, as shown in Fig. 15(a). The analysis of the truncation error is identical to the analysis of the error introduced by a quantizer, and the ratio of signal power to noise power may be expressed by

$$SNR \ (dB) = 6.02 \times B + 1.76 \ dB \qquad (20)$$

where $B$ is the number of bits after truncation.

Each truncation contributes the same amount of total noise power, but the second truncation occurs at a 48-kHz sample rate and hence contributes more noise power in the audio band than the first truncation, which

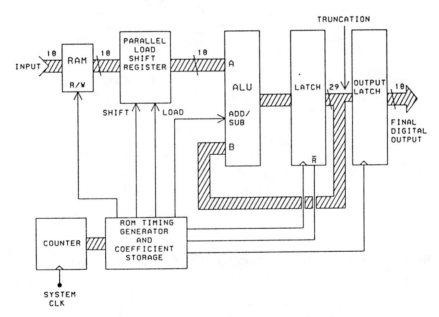

Fig. 13. FIR-filter hardware for final stage of decimator.

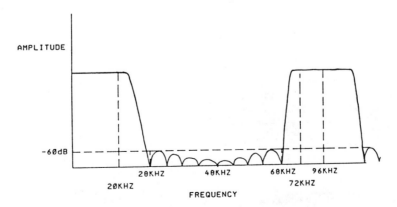

Fig. 14. Frequency response of FIR low-pass filter

occurs at a 96-kHz sample rate.

Fig. 15(b) shows how this additive noise changes the output noise spectrum. The input noise spectrum applied to the decimator rises with increasing frequency and contains very little low-frequency power. The noise added by truncation therefore has the effect of flattening the low-frequency portion of the spectrum.

In practice we find that there is more low-frequency output noise than can be attributed to truncation alone. This excess noise can be attributed to a variety of other sources, including front-end waveform asymmetry and poor linearity in the D/A converter, which must be used to measure the output noise spectrum.

## 5 D/A CONVERSION

D/A conversion is accomplished by using a conventional 16-bit D/A converter and time averaging between adjacent quantizing levels to achieve 18 bits of equivalent resolution. An example of this technique is shown in Fig. 16(a), where the desired output level is 1 LSB above a possible 16-bit quantizing level. We can easily see that the same average level can be achieved by switching to the next 16-bit level at the correct time, as shown by the dashed line in Fig. 16(a). It can be shown that this procedure results in the correct low-frequency spectrum. At higher frequencies the theory predicts that some excess noise will be generated, but the magnitude of this noise at 20 kHz is small enough so as not to increase the output noise power appreciably. However, if we attempt an increase in resolution of more than 2 bits, this excess noise may become unacceptable.

The theory behind time-averaged D/A converters is similar in many respects to the theory of noise-shaping A/D converters described in Sec. 1. A 16-bit quantizer has a fixed average error power for Gaussian inputs, and the only way to achieve increased resolution at low frequencies is to shape the noise power spectrum so that most of the noise is concentrated above the band of interest. It should therefore come as no surprise that the noise spectrum of a time-averaged D/A converter rises with increasing frequency. We can conclude that the time-averaging technique of D/A resolution extension is a simple feedforward form of noise shaping.

The hardware design is shown in Fig. 16(b). A 16-

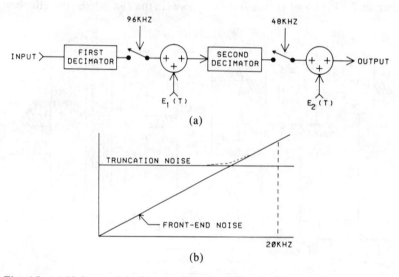

Fig. 15. (a) Noise model of truncation. (b) Effect on output noise spectrum.

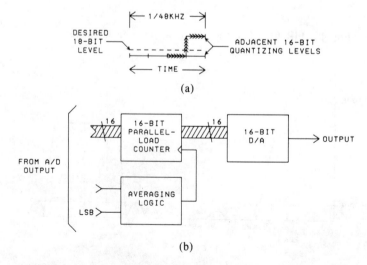

Fig. 16. (a) Time-averaging technique of resolution extension. (b) Hardware implementation.

bit parallel-load counter is used to latch the upper 16 bits of the digital output word. The lowest 2 bits are decoded to determine the correct time at which to increment the counter. The D/A output is then deglitched and applied to a $\sin(x)/x$ correction filter and output-reconstruction filter of conventional design.

The time-averaging technique of resolution extension generates additional quantizing levels that are evenly spaced between existing 16-bit levels. Therefore the linearity of the 18-bit D/A system cannot exceed the linearity of the 16-bit D/A converter itself. Poor linearity in the D/A converter will result in poor distortion performance at low signal levels. It may also result in modulation of the noise floor by the signal.

The time-averaging technique described above is not the only method of achieving 18-bit resolution in the D/A converter. A combination of time averaging and interpolation is currently used in Compact Disc players to achieve 16-bit dynamic range from a 14-bit D/A converter. This method of D/A conversion allows most of the output filtering to be done digitally, which eliminates the phase distortion inherent in analog reconstruction filters. When coupled with an oversampling A/D converter of the type described in this paper, this technique would allow us to implement an entire A/D–D/A system with minimum phase distortion.

## 6 PERFORMANCE OF PROTOTYPE

The following measurements were made on a system consisting of the oversampling A/D converter and the time-averaging 18-bit D/A converter. Distortion is measured at full signal level; distortion at lower levels is dominated by the nonlinearity of the D/A converter and is therefore difficult to measure directly.

Dynamic range (maximum rms signal to 20-kHz rms noise     105 dB

Total harmonic distortion (0.1% summing resistor matching)     < 0.03%

Total harmonic distortion (0.01% summing resistor matching)     < 0.003%

Frequency response     ± 0.1 dB, dc to 20 kHz

Relative 20-kHz phase shift (A/D only)     < 20°.

In addition to these measurements, informal listening tests were conducted at a local studio, using live source material and comparing the output of the A/D – D/A chain with the live signal. In single-blind tests no one could reliably distinguish between the processed and the unprocessed signals.

## 7 INTEGRATION CONSIDERATIONS

The prototype unit uses more than 100 MSI bipolar digital integrated circuits and consumes almost 40 W of power. Clearly, integration is necessary if the converter is to be used in commercial equipment. Integration

of the decimator portion of the oversampling converter has been studied and found to be quite feasible. The following requirements are found for the system:

3500 gates
4.5 kbits RAM, access time < 65 ns
1.25 kbits ROM, access time < 70 ns .

The access times indicate that 2-μm CMOS technology is required. It may be desirable that the ROM, which contains the FIR filter coefficients, be an external part. This would allow the user to customize the filter response to suit a particular application.

It may also be possible to integrate the front-end converter, with the exception of the analog feedback filter. The stringent waveform symmetry requirements discussed earlier would of course have to be met.

The decimator integrated circuit could be designed to accept large input wordlengths so that any desired resolution could be accommodated. For example, a 6-bit input word length would result in a 20-bit output word. When used with an 4-bit front end, the upper 2 bits of the decimator integrated circuit input would be disconnected, resulting in an output word length of 18 bits.

## 8 EXTENSION TO HIGHER RESOLUTION

There are two ways in which higher resolution (>18 bits) may be obtained. The first is simply to increase the resolution of the front-end quantizer. For example, if we double the number of comparators used in the circuit of Fig. 3, we gain 6 dB of dynamic range and obtain a 19-bit output. The decimator design changes very little; we need only widen the data paths by 1 bit.

The second way to increase the dynamic range is to design a more effective noise-shaping feedback filter. This causes more of the quantization noise to be pushed out of band, and hence decreases the in-band noise power. Alternatively, the same noise performance may be obtained with fewer front-end comparators. There is, however, a penalty to pay for this increased performance; the pulse-symmetry requirements discussed in Sec. 2 become even more severe.

While we have concentrated in this paper on achieving resolution beyond the 16-bit level, we should not forget that oversampling converters can be used to achieve 14-bit or 16-bit resolution as well. For example, a 16-bit converter would require a 2-bit front end with only three comparators. The resulting system would be extremely compact and cost-effective, with few external components.

## 9 CONCLUSION

Oversampling A/D converters of the type described in this paper are superior in all respects to the successive-approximation converters now in widespread use. Listed below are some of these advantages.

1) Digital anti-aliasing filters are precisely linear phase with no adjustments.

2) Low-level-distortion performance is superior to that of successive-approximation converters.

3) Resolution of up to 19 bits is practical, compared with the 16 bits attainable using successive approximation.

4) Cost when integrated is comparable with or slightly lower than that for present 16-bit converters.

Some of these advantages are lost when the total A/D–D/A system is examined, owing to poor D/A converter performance. It is logical to assume that similar improvements will be made in D/A converters, using the principles of oversampling, interpolation, and noise shaping.

We have demonstrated in this paper that resolution well beyond 16 bits is possible if oversampling/decimation techniques are employed. It is unclear whether the professional audio community wants or needs the additional dynamic range provided by this technique. Certainly the noise floor of digital recordings, while low, is not completely inaudible, especially when many tracks are combined in the mixing process. It is clear, however, that the dynamic range of PCM processors and recorders can be improved by at least 12 dB, with the added benefits of lower distortion and linear phase response.

## 10 REFERENCES

[1] H. A. Spang and P. M. Schultheiss, "Reduction of Quantizing Noise by Use of Feedback," *IRE Trans. Commun. Sys.*, pp. 373–380 (1962 Dec.).

[2] H. R. Schindler, "Linear, Nonlinear, and Adaptive Delta Modulation," *IEEE Trans. Commun.*, pp. 1807–1822 (1974 Nov.).

[3] E. G. Kimme and F. F. Kuo, "Synthesis of Optimal Filters for a Feedback-Quantization System," *IEEE Trans. Circuit Theory*, pp. 405–413 (1963 Sept.).

[4] R. W. Adams, "Companded Predictive Delta Modulation: A Low-Cost Conversion Technique for Digital Recording," *J. Audio Eng. Soc.*, vol. 32, pp. 659–672 (1984 Sept.).

[5] J. Vanderkooy and S. P. Lipshitz, "Resolution Below the Least Significant Bit in Digital Systems with Dither," *J. Audio Eng. Soc.*, vol. 32, pp. 106–113 (1984 Mar.).

[6] A. Oppenheim and R. Schafer, *Digital Signal Processing* (Prentice-Hall, Englewood Cliffs, NJ, 1975).

[7] R. Lagadec and T. G. Stockham, Jr., "Dispersive Models for A-to-D and D-to-A Conversion Systems," presented at the 75th Convention of the Audio Engineering Society, *J. Audio Eng. Soc. (Abstracts)*, vol. 32, p. 469 (1984 June), preprint 2097.

# The Design of Sigma-Delta Modulation Analog-to-Digital Converters

BERNHARD E. BOSER, STUDENT MEMBER, IEEE, AND BRUCE A. WOOLEY, FELLOW, IEEE

*Abstract* —Oversampled analog-to-digital (A/D) converter architectures offer a means of exchanging resolution in time for that in amplitude so as to avoid the difficulty of implementing complex precision analog circuits. These architectures thus represent an attractive approach to implementing precision A/D converters in scaled digital VLSI technologies. This paper examines the practical design criteria for implementing oversampled converters based on second-order sigma-delta ($\Sigma\Delta$) modulation. Behavioral models that include representation of various circuit impairments are established for each of the functional building blocks comprising a second-order $\Sigma\Delta$ modulator. Extensive simulations based on these models are then used to establish the major design criteria for each of the building blocks. As an example, these criteria are applied to the design of a modulator that has been integrated in a 3-$\mu$m CMOS technology. This experimental prototype operates from a single 5-V supply, dissipates 12 mW, occupies an area of 0.77 mm$^2$, and achieved a measured dynamic range of 89 dB.

## I. INTRODUCTION

THE emergence of powerful digital signal processors implemented in CMOS VLSI technology creates the need for high-resolution analog-to-digital (A/D) converters that can be integrated in fabrication technologies optimized for digital circuits and systems. However, the same scaling of VLSI technology that makes possible the continuing dramatic improvements in digital signal processor performance also severely constrains the dynamic range available for implementing the interfaces between the digital and analog representation of signals. A/D converters based on sigma-delta ($\Sigma\Delta$) modulation combine sampling at rates well above the Nyquist rate with negative feedback and digital filtering in order to exchange resolution in time for that in amplitude. Furthermore, these converters are especially insensitive to circuit imperfections and component mismatch since they employ only a simple two-level quantizer, and that quantizer is embedded within a feedback loop. $\Sigma\Delta$ modulators thus provide a means of exploiting the enhanced density and speed of scaled digital VLSI circuits so as to avoid the difficulty of implementing

complex analog circuit functions within a limited analog dynamic range.

A $\Sigma\Delta$ modulator consists of an analog filter and a coarse quantizer enclosed in a feedback loop [1]. Together with the filter, the feedback loop acts to attenuate the quantization noise at low frequencies while emphasizing the high-frequency noise. Since the signal is sampled at a frequency which is much greater than the Nyquist rate, high-frequency quantization noise can be removed without affecting the signal band by means of a digital low-pass filter operating on the output of the $\Sigma\Delta$ modulator.

The simplest $\Sigma\Delta$ modulator is a first-order loop wherein the filter consists of a single integrator [2], [3]. However, the quantization noise from first-order modulators is highly correlated [2]–[6], and the oversampling ratio needed to achieve resolution greater than 12 bits is prohibitively large. Higher order $\Sigma\Delta$ modulators, containing more than one integrator in the forward path, offer the potential of increased resolution. However, modulators with more than two integrators suffer from potential instability owing to the accumulation of large signals in the integrators [7], [8]. An architecture whereby several first-order modulators are cascaded in order to achieve performance that is comparable to that of higher order modulators has been suggested as a means of overcoming the stability problem [9]–[11]. These architectures, however, call for precise gain matching between the individual first-order sections, a requirement that conflicts with the goal of designing A/D converters that are especially insensitive to parameter tolerances and component mismatch. Second-order $\Sigma\Delta$ modulators are thus particularly attractive for high-resolution A/D conversion. The effectiveness of second-order $\Sigma\Delta$ modulator architectures has already been illustrated in a variety of applications. Digital speech processing systems and voice-band telecommunications codecs with A/D converters based on second-order $\Sigma\Delta$ modulation have been reported [12]–[15], and the extension of the performance achievable with such architectures to the levels required for digital audio [16] and higher [17] signal bandwidths has been demonstrated.

In this paper, the considerations faced in the design of second-order $\Sigma\Delta$ modulators are examined. First, analysis and simulation techniques for such modulators are introduced. Issues concerning the design and implementation of

Manuscript received May 20, 1988; revised August 8, 1988. This work was supported in part by the Semiconductor Research Corporation under Contract 87-DJ-112 and by a grant from Texas Instruments Incorporated.

The authors are with the Center for Integrated Systems, Stanford University, Stanford, CA 94305.

IEEE Log Number 8824187.

Reprinted from *IEEE J. Solid-State Circuits*, vol. SC-23, pp. 1298–1308, December 1988.

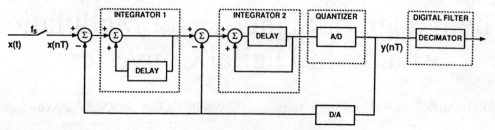

Fig. 1.  Block diagram of second-order $\Sigma\Delta$ modulator with decimator.

the building blocks comprising a $\Sigma\Delta$ modulator, as well as the interactions between these blocks, are then examined, leading to a set of functional design criteria for each of the blocks. In Section IV, the design criteria are applied to an example implementation. Measurement results for this experimental prototype are presented in Section V.

## II.  System Design Considerations

### A.  Second-Order $\Sigma\Delta$ Modulator

Fig. 1 shows the block diagram of a second-order $\Sigma\Delta$ modulator. The analog input signal $x(t)$ is sampled at the sampling frequency $f_s = 1/T$. A quantizer with only two levels at $\pm\Delta/2$ is employed so as to avoid the harmonic distortion generated by step-size mismatch in multibit quantizers. Out-of-band quantization noise in the modulator output is eliminated with a digital decimation filter that also resamples the signal at the Nyquist rate, $2B$. The power $S_N$ of the noise at the output of the filter is the sum of the in-band quantization noise $S_B$ together with in-band noise arising from other error sources, such as thermal noise or errors caused by jitter in the sampling time. An approximate expression for the quantization noise when the quantizer is modeled by an additive white-noise source [7] is

$$S_B = \frac{\pi^4}{5} \frac{1}{M^5} \frac{\Delta^2}{12}, \qquad M \gg 1. \tag{1}$$

The coefficient $M$ is the oversampling ratio, defined as the ratio of the sampling frequency $f_s$ to the Nyquist rate $2B$. For every octave of oversampling, the in-band quantization noise is reduced by 15 dB.

The performance of A/D converters for signal processing and communications applications is usually characterized in terms of the *signal-to-noise ratio*. Two definitions for this ratio will be used here. The TSNR is the ratio of the signal power to the total in-band noise, whereas the SNR accounts only for uncorrelated noise and not harmonic distortion. The useful signal range, or *dynamic range* (DR), of the A/D converter for sinusoidal inputs is defined as the ratio of the output power at the frequency of the input sinusoid for a full-scale input to the output signal power for a small input for which the TSNR is unity (0 dB). The dynamic range of an ideal Nyquist rate uniform PCM converter with $b$ bits is $DR = 3 \cdot 2^{2b-1}$. This definition of the dynamic range provides a simple means of comparing the resolution of oversampled and Nyquist

rate converters. For example, a 16-bit Nyquist rate A/D converter corresponds to an oversampled A/D with 98-dB dynamic range.

For the successful design and integration of a second-order $\Sigma\Delta$ modulator, it is important to establish the sensitivity of the system's performance to various circuit nonidealities. Functional simulation techniques must be used to examine the design trade-offs because the application of conventional circuit and system analysis methods to the study of higher order $\Sigma\Delta$ modulators has, to date, proven to be intractable. Circuit simulations alone are not an effective design approach, since they do not explicitly illustrate the fundamental trade-offs necessary in the design process. The approach taken here is based on the use of a custom simulation program that embodies quantitative models for the functional elements comprising a $\Sigma\Delta$ modulator that reflect nonidealities in the behavior of those elements. Descriptions of these elements are held in a generic form so that they can be mapped to a large variety of possible circuit implementations.

Because of the oversampling process and the long-term memory of second-order $\Sigma\Delta$ modulators, long data traces are necessary to accurately estimate the performance of such converters. MIDAS, a general-purpose simulator for mixed analog and digital sampled-data systems, has been used to generate these traces. MIDAS accepts a system description in the form of a net list and is thus flexible enough to accommodate a wide variety of architectural configurations. Estimates of dynamic range and signal-to-noise ratio, as well as distortion, are generated in MIDAS using the *sinusoidal minimal error method*, a computationally efficient algorithm suitable for both simulation and experimental measurement purposes [18].

Several types of nonidealities that are characteristic of analog circuit implementations of $\Sigma\Delta$ modulators have been studied. Signal range, electronic noise, and timing jitter are discussed below. The sensitivity of the modulator performance of the characteristics of the integrators and the comparator is then considered in Section III.

### B.  Signal Range

For the conventional second-order $\Sigma\Delta$ modulator architecture shown in Fig. 1, simulations reveal that the signal range required at the outputs of the two integrators is several times the full-scale analog input range, $\pm\Delta/2$. This requirement represents a severe problem in circuit technologies, such as CMOS VLSI, where the dynamic

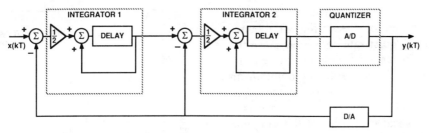

Fig. 2.   Modified architecture of second-order $\Sigma\Delta$ modulator.

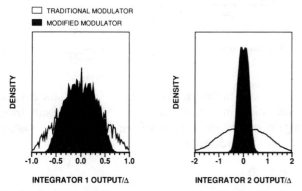

Fig. 3.   Comparison of integrator output probability densities for traditional and modified architectures with sinusoidal input 3 dB below overload.

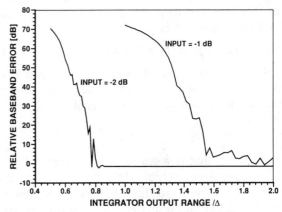

Fig. 4.   Simulated influence of integrator output range on baseband quantization noise for input sinusoids 1 and 2 dB below overload.

range is restricted. The modified modulator architecture shown in Fig. 2 calls for considerably smaller signal ranges within the integrators. This architecture differs from the traditional configuration in two respects: a forward path delay is included in both integrators, thus simplifying the implementation of the modulator with straightforward sampled-data analog circuits, and each integrator is preceded by an attenuation of 0.5. An extraction of the modified architecture from the conventional structure results in a configuration with an attenuation of 0.5 preceding the first integrator and a gain of 2 at the input of the second integrator. However, since the second integrator is followed immediately by a single-threshold quantizer, its gain can be adjusted arbitrarily without impairing the performance of the modulator [19].

Fig. 3 shows the probability densities of the outputs of the two integrators for both the traditional (Fig. 1) and the modified (Fig. 2) $\Sigma\Delta$ modulator architectures. Whereas the signals at the outputs of both integrators extend only slightly beyond the full-scale input for the modified modulator design, the signal ranges are considerably larger for the traditional architecture. The modified modulator architecture therefore requires a signal range in the integrators which is only slightly larger than the full-scale input range of the A/D converter. Fig. 4 shows the relative increase in baseband noise that results from clipping the integrator outputs for input signals 1 and 2 dB below overload. From these results it is apparent that for signals as large as 1 dB below overload, the performance penalty is negligible when the signal in both integrators is clipped to a range that is about 70 percent larger than the full-scale input range.

### C. Electronic Noise

In analog implementations of $\Sigma\Delta$ modulators, the signal is corrupted not only by quantization error, but also by electronic noise generated in the constituent circuits. Noise injected at the modulator input is the dominant contributor. Input-referred noise from the comparator undergoes the same second-order differentiation as the quantization noise, and noise injected at the input of the second integrator is subjected to a first-order difference function. Out-of-band noise is eliminated by the decimation filter, but high-frequency noise at multiples of the sampling frequency will be aliased into the baseband. The total input-referred noise power within the baseband does contribute to $S_N$ and thus ultimately limits the resolution of the A/D converter.

Offset is only a minor concern in many signal acquisition systems, as long as the quantization is uniform. The offset at the input to the first integrator is the only significant contributor because offsets in the second integrator and the comparator are suppressed by the large low-frequency gain of the integrators. In practice, excessive offsets should be avoided because of the consequent reduction in the effective signal range in the integrators.

### D. Sampling Jitter

The sampling theorem states that a sampled signal can be perfectly reconstructed provided that the sampling frequency is at least twice the signal bandwidth and that the sampling occurs at uniformly distributed instances in time. An anti-aliasing filter preceding the sampler ensures that the first of these requirements is fulfilled. Oversampled A/D converters put considerably less stringent requirements on this filter than Nyquist rate converters since the signal is sampled at a frequency which far exceeds its bandwidth. Sampling clock jitter results in nonuniform sampling and increases the total error power in the quantizer output. The magnitude of this error increase is a function of both the statistical properties of the sampling jitter and the input to the A/D converter. An estimate for this error is derived below.

The error resulting from sampling a sinusoidal signal with amplitude $A$ and frequency $f_x$ at an instant which is in error by an amount $\delta$ is

$$x(t + \delta) - x(t) \approx 2\pi f_x \delta A \cos 2\pi f_x t. \quad (2)$$

Under the assumption that the sampling uncertainty $\delta$ is an uncorrelated Gaussian random process with standard deviation $\Delta t$, the power of this error signal is

$$S_\delta = \frac{A^2}{2}(2\pi f_x \Delta t)^2 \quad (3)$$

with a spectrum that is a scaled and modulated (by the sinusoidal input signal) version of the timing jitter $\delta$. In an oversampled A/D converter, the decimation filter removes the content of this signal at frequencies above the baseband. Since the clock jitter is assumed to be white, the total power of the error is reduced by the oversampling ratio $M$ in the decimator process. The in-band error power $S_{\Delta t}$ is therefore

$$S\Delta_t \leqslant \frac{\Delta^2}{8}\frac{(2\pi B \Delta t)^2}{M}. \quad (4)$$

In this expression, the worst-case amplitude ($\pm \Delta/2$) and signal frequency ($B$) have been used in order to establish an upper bound on the error power.

The error caused by clock jitter is inversely proportional to the oversampling ratio $M$ and adds directly to the total error power $S_N$ at the output of the A/D converter. Since the in-band quantization noise $S_B$ is inversely proportional to the fifth power of $M$, the amount of clock jitter that can be tolerated decreases for an increase in oversampling ratio.

## III. INTEGRATOR AND COMPARATOR DESIGN

The two integrators in the forward path of a second-order $\Sigma\Delta$ modulator serve to accumulate the large quantization errors that result from the use of a two-level quantizer and force their average to zero. Ideally, for an integrator of the form used in the modulator architecture of Fig. 2 the output, $v(kT)$, is the sum of the previous output, $v(kT - T)$, and the previous input, $u(kT - T)$:

$$v(kT) = g_0 u(kT - T) + v(kT - T). \quad (5)$$

The constant $g_0$ represents the gain preceding the input to the integrator, which is 0.5 for each of the integrators in Fig. 2. The above equation corresponds to the following transfer function for an ideal integrator:

$$H(z) = \frac{g_0 z^{-1}}{1 - z^{-1}}. \quad (6)$$

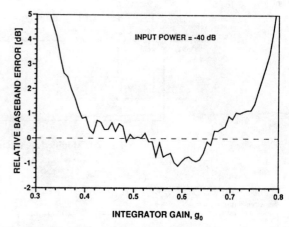

Fig. 5. Simulated influence of variations in integrator gain on baseband quantization noise.

Analog circuit implementations of the integrators deviate from this ideal in several ways. Errors which result from finite gain and bandwidth, as well as those due to nonlinearities, are considered below.

### A. Gain Variations

It was pointed out previously that a scalar preceding the second integrator in the $\Sigma\Delta$ modulator of Fig. 2 has no effect on the behavior of an ideal modulator because it is absorbed by the two-level quantizer. However, deviations in $g_0$ from its nominal value in the first integrator alter the noise shaping function of the $\Sigma\Delta$ modulator and consequently change the performance of the A/D converter.

Fig. 5 shows the change of the in-band quantization noise as a function of $g_0$. Gain variations of as much as 20 percent from the nominal value have only a minor impact on the performance of the A/D converter, confirming the general insensitivity of the $\Sigma\Delta$ modulator architecture to component variations. Larger $g_0$ means higher gain in the forward path of the modulator and consequently greater attenuation of the quantization noise. However, for gains larger than about 0.6, the signal amplitudes at the integrator outputs increase rapidly and the system becomes unstable.

### B. Leak

The dc gain of the ideal integrator described by (6) is infinite. In practice, the gain is limited by circuit constraints. The consequence of this "integrator leak" is that only a fraction of $P_0$ of the previous output of the integrator is added to each new input sample. The integrator transfer function in this case becomes

$$H(z) = \frac{g_0 z^{-1}}{1 - P_0 z^{-1}} \quad (7)$$

and the dc gain is $H_0 = g_0/(1 - P_0)$. The limited gain at low frequency reduces the attenuation of the quantization noise in the baseband and consequently, for the $\Sigma\Delta$ modulator of Fig. 2, results in an increase of the in-band

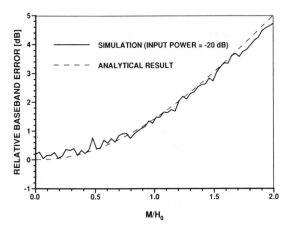

Fig. 6.  Influence of integrator leak on baseband quantization noise.

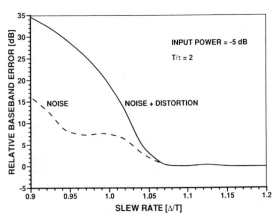

Fig. 7.  Simulated influence of integrator output slew rate on baseband quantization noise.

quantization noise $S_B$ that is given by

$$\frac{\Delta S_B}{S_B} = \frac{5}{\pi^4}\left(\frac{M}{H_0}\right)^4 + \frac{10}{3\pi^2}\left(\frac{M}{H_0}\right)^2. \qquad (8)$$

This relationship is plotted in Fig. 6, along with data obtained from simulations, for an input 20 dB below full scale. The performance penalty incurred is on the order of 1 dB when the integrator dc gain is comparable to the oversampling ratio.

### C. Bandwidth

In typical sampled-data analog filters, the unity-gain bandwidth of the operational amplifiers must often be at least an order of magnitude greater than the sampling rate. However, simulations indicate that integrator implementations using operational amplifiers with bandwidths considerably lower than this, and with correspondingly inaccurate settling, will not impair the $\Sigma\Delta$ modulator performance, provided that the settling process is linear.

For integrators with an exponential impulse response—as is observed for implementations which are based on an amplifier with a single dominant pole—the time constant of the response, $\tau$, can be nearly as large as the sampling period $T$. This constraint is considerably less stringent than requiring the integrator to settle to within the accuracy of the A/D converter. Simulation results indicate that for values of $\tau$ larger than the sampling period, the modulator becomes unstable. In Section V it will be argued that in practice $\tau$ must actually be kept somewhat smaller than $T$.

### D. Slew Rate

In the preceding subsection it was pointed out that a large time constant for the settling of the integrator output is acceptable, provided that the settling process is linear. In particular, the settling must not be slew-rate limited. The simulation results presented in Fig. 7 indicate a sharp increase in both quantization noise and harmonic distortion of the converter when the slew rate is less than $1.1\Delta/T$. These simulations are based on the assumption that if the integrator response is not slew-rate limited, the

impulse response is exponential with time constant $\tau$:

$$v(kT+t) = \frac{g_0}{1 - e^{-T/\tau}} u(kT)[1 - e^{t/\tau}] + v(kT). \quad (9)$$

The term $1 - e^{-T/\tau}$ has been included to separate the effects of finite slew rate from those due to variations in the equivalent gain. The peak rate of change in the impulse response occurs at $t = 0$ and is given by

$$\frac{dv(kT+t)}{dt} = \frac{g_0}{1 - e^{-T/\tau}} \frac{u(kT)}{\tau}. \qquad (10)$$

Slewing distortion occurs when this rate exceeds the maximum slew rate the integrator can support.

### E. Nonlinearity

The imperfections in analog circuit realizations of the integrators that have been considered above are either linear deviations from the ideal frequency response due to gain and bandwidth limitations, or large-scale nonlinearities, such as clipping and slewing. In this subsection the influence of differential nonlinearities on the modulator performance is examined. Such nonlinearities occur, for example, when the integrator implementation is based on capacitors that exhibit a voltage dependence, or on an amplifier with input-dependent gain. From simulations it has been observed that the consequence of these nonlinearities is harmonic distortion that limits the peak SNR achievable at large signal levels.

The following quantitative analysis of effects of integrator nonlinearity is based on representing the integrator by

$$\begin{aligned}
v(kT)&[1 + \beta_1 v(kT) + \beta_2 v^2(kT) + \cdots] \\
&= g_0 u(kT-T)[1 + \alpha_1 u(kT-T) \\
&\quad + \alpha_2 u^2(kT-T) + \cdots] \\
&\quad + v(kT-T)[1 + \beta_1 v(kT-T) \\
&\quad + \beta_2 v^2(kT-T) + \cdots]. \qquad (11)
\end{aligned}$$

This model has been found to be typical of a variety of possible integrator implementations. The parameters $\alpha_i$ and $\beta_i$ are the coefficients of Taylor series expansions of the integrator input and output and are associated with nonlinearities in the input and the storage elements, respectively. In switched-capacitor integrators, for example, these correspond to the voltage coefficients of the capacitors [20].

Fig. 8 shows simulation and analytical results obtained for evaluating the influence of integrator nonlinearities on the TSNR of the A/D converter, assuming a sinusoidal input. The performance degradation is proportional to the amplitude of the input for the first-order nonlinearity, and proportional to the square of the modulator input for second-order nonlinearity. The degradation is a consequence of harmonic distortion, rather than an increase in quantization noise; thus, it is possible to evaluate the harmonic distortion without taking the quantizer into consideration.

Only distortion introduced by the first integrator in a second-order $\Sigma\Delta$ architecture such as that of Fig. 2 need be considered, since errors introduced by the second integrator are attenuated by the feedback loop. The input to the first integrator is the difference between the modulator input, $x(kT)$, and the modulator output, which consists of the sum of the input delayed by two sampling periods and the quantization noise. The latter can be neglected in the distortion analysis, as has been pointed out above. For a sinusoidal modulator input with amplitude $A$ and frequency $f_x$, the input to the first integrator is approximately

$$x(kT) - x(kT - 2T) \approx 4\pi f_x TA \cos 2\pi f_x T, \qquad f_x T \ll 1.$$
(12)

The input-referred harmonics of the integrator for this signal can be determined either by distortion analysis with a circuit simulation program such as SPICE [21] or SWAP [22], or analytically, in a manner similar to the analysis presented in [20]. When the assumption is made that $\alpha_i = \beta_i$, the amplitudes of the first and second harmonics are

$$h_1 = \frac{\alpha_1}{2} A_1^2$$
(13)

and

$$h_2 = \frac{\alpha^2}{2} A_1^3$$
(14)

respectively. The power of the harmonic distortion in the output of the A/D converter due to integrator nonlinearity is then approximately $h_1^2/2 + h_2^2/2$, provided that the contribution of higher order harmonics is negligible. Harmonics at frequencies above the bandwidth of the converter are, of course, suppressed by the decimation filter. This result is in excellent agreement with simulations that do not include any simplifications.

### F. Comparator Hysteresis

The 1-bit quantizer in the forward path of a $\Sigma\Delta$ modulator can be realized with a comparator. The principle design parameters of this comparator are speed, which must be adequate to achieve the desired sampling rate,

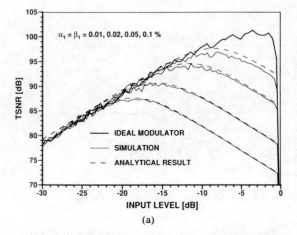

(a)

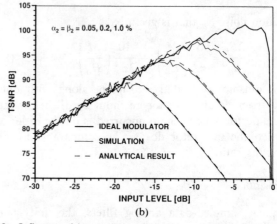

(b)

Fig. 8. Influence of integrator nonlinearity on A/D converter performance: (a) first-order nonlinearity, and (b) second-order nonlinearity.

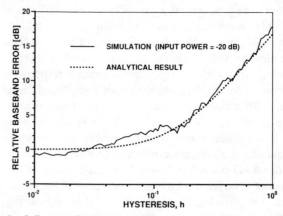

Fig. 9. Influence of comparator hysteresis on baseband quantization noise.

input offset, input-referred noise, and hysteresis. It has been pointed out already that offset and noise at the comparator input are suppressed by the feedback loop of modulator. Fig. 9 shows the performance of the A/D converter as a function of comparator hysteresis, defined as the minimum overdrive required to change the output. The power of the in-band noise, $S_N$, is virtually unchanged for hysteresis as large as 10 percent of the full-scale converter input, $\Delta$, and rises at 20 dB per decade above this point.

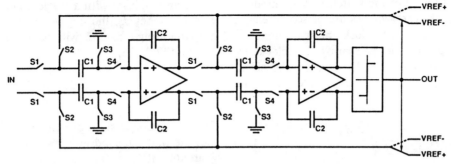

Fig. 10. Second-order ΣΔ modulator implementation.

The sensitivity of the A/D converter performance to comparator hysteresis is modeled quite accurately by an additive white noise with power $(h \cdot \Delta / 2)^2$, where $h$ is the magnitude of the comparator hysteresis relative to $\Delta$. The noise undergoes the same spectral shaping as the quantization noise. The sum of the quantization noise, $S_B$, and the hysteresis is therefore

$$S_N = \frac{\pi^{2L}}{2L+1} \frac{\Delta^2}{M^{2L+1}} \left[ \frac{1}{12} + 4h^2 \right], \qquad M \gg 1. \quad (15)$$

The factor 4 reflects the adjustment of the scalar preceding the second integrator from 2 to 0.5.

The sensitivity of ΣΔ modulators to comparator hysteresis is several orders of magnitude smaller than that of Nyquist rate converters. It is apparent from the model that this is attributable to the presence of negative feedback with high loop gain in a ΣΔ modulator.

## IV. IMPLEMENTATION

The considerations addressed above have been applied to the design of a second-order ΣΔ modulator that has been integrated in a 3-$\mu$m CMOS technology. The performance objective for this design was a dynamic range of 16 bits at as high a Nyquist rate as could be achieved within the constraints of the technology in which the circuit was integrated. The resolution of 16 bits corresponds to a dynamic range of 98 dB, which can be achieved with an oversampling ratio of $M = 153$ if the performance is limited only by the quantization noise. To allow for increased baseband noise due to circuit nonidealities, as well as to maintain an oversampling ratio that is a power of 2 so as to simplify the subsequent decimation to the Nyquist sampling rate, a sampling ratio of $M = 256$ was chosen. The modulator has been designed to operate from a single 5-V power supply.

### A. Circuit Topology

Since ΣΔ modulators are sampled-data systems, they are readily implemented in MOS technology with switched-capacitor (SC) circuits. Fig. 10 shows a possible topology. A fully differential configuration has been adopted in order to ensure high power supply rejection, reduced clock feedthrough and switch charge injection

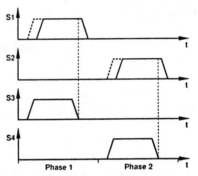

Fig. 11. Clock diagram for second-order ΣΔ modulator.

errors, improved linearity, and increased dynamic range. The two identical integrators in Fig. 10 each consist of an amplifier, two sampling capacitors $C_1$, and two integrating capacitors $C_2$. The ratio of $C_1$ to $C_2$ is chosen so as to realize the gain of 0.5 that precedes each integrator in the architecture of Fig. 2.

Operation of the modulator is controlled by a nonoverlapping two-phase clock. During phase 1 all of the switches labeled $S_1$ and $S_3$ are open, while those labeled $S_2$ and $S_4$ are closed, and the input to each integrator is sampled onto the capacitors $C_1$. In phase 2, switches $S_1$ and $S_3$ open, while $S_2$ and $S_4$ close, and charge stored on $C_1$ is transferred to $C_2$. During this phase, the closing of switches $S_2$ has the effect of subtracting the output of the two-level D/A network from the input to each integrator. The comparison of the outputs from the second integrator is performed during phase 1, and the comparator reset during phase 2. With this clocking arrangement, the time available for the integration and the time for the comparison are both one-half a clock cycle.

To first order, the charge injected by the MOS switches in the circuit of Fig. 1 is a common-mode signal that is canceled by the differential implementation of the modulator. Signal-dependent charge injection is further suppressed by opening switches $S_3$ and $S_4$ slightly before $S_1$ and $S_2$, respectively [23]. Since $S_3$ and $S_4$ are connected to either a ground or virtual ground node, they do not exhibit signal-dependent charge injection. Once $S_3$ or $S_4$ has opened, and before the other has closed, $C_1$ is floating; thus, the subsequent opening of $S_1$ or $S_3$ during the interval when both $S_3$ and $S_4$ are open will, to first order, not inject charge onto $C_1$.

A timing diagram for all of the switches in the modulator is given in Fig. 11. The switches are closed when the controlling clocks are high. The clocks must be nonoverlapping in order to prevent charge sharing. The clocks for switches $S_1$ and $S_2$ are generated by delaying the clocks for $S_3$ and $S_4$. An upper limit for the tolerable clock jitter follows from (4):

$$\Delta t \leqslant \frac{1}{2\pi B \cdot 2^b} \sqrt{\frac{2M}{3}}. \tag{16}$$

If the baseband error power induced by clock jitter is to be no larger than the quantization noise resulting from an ideal modulator, then it is necessary that $\Delta t \leqslant 630$ ps for $B = 20$ kHz.

The choice of the full-scale analog input range of the converter, which is equal to the quantizer step size $\Delta$, involves trade-offs among a number of design constraints. A large signal range is desirable due to the presence of electronic noise in the analog circuits. However, a large signal range results in increased harmonic distortion due to integrator nonlinearity. In addition, increasing the signal range calls for operational amplifiers with a higher slew rate. A differential full-scale input range of $\Delta = 4$ V has been chosen so as to limit the performance impairment due to electronic noise. The simulation results presented in Fig. 4 indicate that the signal range at the output of both integrators should be at least 50 percent larger than the full-scale analog input in order to avoid significant performance degradation. The output swing of the operational amplifiers should therefore be at least 6 V. This requirement is accommodated within a single 5-V supply through the use of the fully differential topology.

### B. Integrator Design

The design of the differential operational amplifier is key to the successful realization of the integrators. The specifications for this amplifier follow from the integrator performance requirements described in the previous section. A consideration of integrator leak mandates that the amplifier open-loop gain be at least equal to the oversampling ratio, $M = 256$. However, the gain must generally be somewhat larger than this in order to adequately suppress harmonic distortion.

An operational amplifier that is not slew-rate limited is essential in order to avoid slewing distortion. A class $AB$ configuration with a single gain stage similar to that described in [24] has been chosen to meet this constraint. The slew-rate requirement is more stringent for this implementation than was derived from Fig. 7 for two reasons. First, the integration is accomplished only during phase 2 and thus must be completed within one-half the clock cycle. Second, the signal swing at the integrator output in response to a step input is somewhat larger than anticipated in Fig. 7 owing to feedforward through the integrating capacitors $C_2$. Therefore, in the implementation of Fig. 10 the slew rate must be at least $3\Delta/T \approx 150$ V/$\mu$s.

For an amplifier with a single dominant pole and unity-gain frequency $f_u$, the impulse response of the integrator output during phase 2 will be exponential with a time constant [25], [26]

$$\tau = \frac{1 + C_1/C_2}{2\pi f_u}. \tag{17}$$

The simulation results presented in the previous section indicate that the condition $\tau \leqslant T$ must be met in order to guarantee stability of the modulator. This requirement corresponds to a lower limit for $f_u$ of

$$f_u \geqslant \frac{1 + C_1/C_2}{\pi T}. \tag{18}$$

The fact that only one-half of the clock period $T$ is available for the integration has been accounted for in this equation. From (18) it is apparent that the bandwidth $f_u$ must be greater than approximately one-half the sampling rate, provided that the step response is purely exponential. In practice, this latter requirement is not met precisely because of secondary effects such as nondominant poles and the dependence of the pole locations on the amplifier operating point, which in this design changes during transients.

Constraints on the dynamic range mandate that the sum of input-referred baseband noise of the first integrator and baseband noise in the output of the two-level D/A network be 104 dB below the power of a full-scale sinusoidal input to the converter if this noise is not to exceed the quantization noise. Sampling capacitors of 1 pF have been employed to reduce the level of thermal noise in the circuit, and large input transistors in the amplifier limit flicker noise.

Harmonic distortion limits the TSNR of the converter for large inputs. The main contributor to this distortion is nonlinearity in the first integrator, which is a consequence of the voltage dependence of the capacitors and the gain nonlinearity of the operational amplifier. In the previous section it was shown that it is possible to predict the impact of these nonlinearities through an analysis of the integrator alone. In these circumstances, the magnitude of the harmonics can be determined using a circuit simulator that includes distortion analysis.

An estimate of the tolerable voltage dependence of the capacitors can be extracted from the results presented in Fig. 8. For capacitors with a voltage dependence given by $C(v) = C_0(1 + \gamma_1 v + \gamma_2 v^2)$, it follows from charge conservation and comparison with (11) that $\alpha_1 = \beta_1 = \gamma_1 \Delta$ and $\alpha_2 = \beta_2 = \gamma_2 \Delta^2$. The first harmonic, $h_1$, will be smaller than predicted by (13) because of the differential configuration of the modulator. The peak TSNR will be reduced by about 6 dB for $\gamma_1 = 50$ ppm/V in a single-ended configuration; in the differential design the capacitor voltage coefficient can be several times larger for the same performance. Second-order harmonic distortion reduces the peak TSNR by 6 dB for $\gamma_2 = 30$ ppm/V$^2$ in a single-ended configuration.

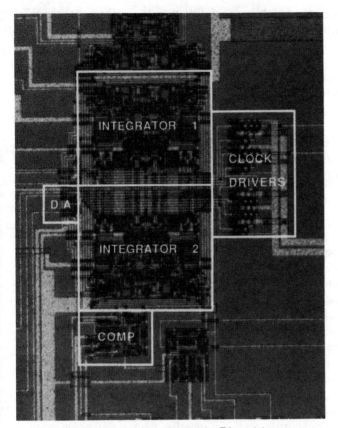

Fig. 12.   Die photo of second-order ΣΔ modulator.

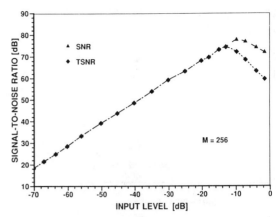

Fig. 13.   Measured SNR for a sampling frequency of 4 MHz and a signal frequency of 1.02 kHz.

## C. Comparator Design

Neither sensitivity nor offset considerations present stringent design constraints for the comparator in a second-order ΣΔ modulator. The two integrators provide preamplification of the signal, and due to the feedback the comparator offset is stored in the second integrator. The data in Fig. 9 imply that comparator hysteresis as large as 5 percent of the full-scale input range Δ has a negligible impact on the performance of the A/D converter. A simple regenerative latch without preamplification or offset cancellation, such as that presented in [27], fulfills the comparator requirements.

## V. EXPERIMENTAL RESULTS

The second-order ΣΔ modulator implementation of Fig. 10 has been integrated in a 3-$\mu$m CMOS technology. A photograph of the chip is shown in Fig. 12. Most of the 0.77-mm$^2$ die area is occupied by the two integrators, which have been laid out symmetrically in order to reduce component mismatch. The circuit dissipates 12 mW when operating from a single 5-V power supply.

For testing, the experimental ΣΔ modulator was connected to a high-quality sinusoidal signal source, and the initial stages of decimation were implemented with an off-chip digital filter. The output of this filter was then transmitted to a host computer for further filtering and an analysis of the performance. Amplitudes of both the out-

put signal and its harmonics, the quantization noise power and spectral density, and the signal frequency were estimated by the host using the same algorithms employed for performing the simulations [18]. The advantage of this approach over techniques that are based on the use of a high-precision D/A converter and analog test instruments is its insensitivity to the performance of analog test equipment. The only analog—and therefore potentially limiting—component in the measurement setup is the sinusoidal source.

Fig. 13 shows the SNR and TSNR measured for the experimental A/D converter. For this data the modulator was operated at its maximum clock rate of $f_s = 4$ MHz, and the oversampling ratio was $M = 256$. The corresponding Nyquist rate is 16 kHz. The frequency of the sinusoidal input signal was 1.02 kHz. From the data of Fig. 13, the measured dynamic range of the converter is found to be 89 dB, which corresponds to a resolution of 14.5 bits. Additional tests show that the modulator performance drops by less than 3 dB for signal frequencies up to half the Nyquist rate. The measured gain tracking is better than ±0.5 dB and is limited largely by the decimation filter design used.

For large input signals, the precision of the converter is limited by harmonic distortion rather than quantization noise, as is apparent from the divergence of the TSNR from the SNR in Fig. 13. In this design, the amplifier is the dominant source of distortion, a consequence of the large signal range in the integrators in comparison with the supply voltage. The capacitors in the experimental modulator were realized with double polysilicon layers, and their contribution to the total distortion is negligible.

The dynamic range of the ΣΔ modulator is plotted as a function of the sampling rate in Fig. 14. For sampling rates below 4 MHz, the dynamic range is independent of the clock rate; it then drops rapidly at higher operating frequencies. The pronounced decrease of the performance above 4 MHz has also been observed in simulation results and has been identified as resulting from instability. The unity-gain bandwidth of the operational amplifiers was measured to be 8 MHz with a phase margin of 80°. Thus, the modulator can be operated at speeds up to half the amplifier bandwidth without performance degradation. In

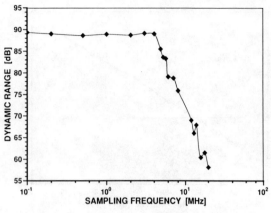

Fig. 14.   Maximum operating frequency.

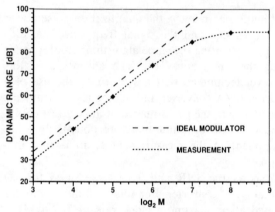

Fig. 15.   Dynamic range as a function of the oversampling ratio for a
sampling frequency of 4 MHz.

contrast, an order of magnitude higher bandwidth would be required if it were necessary for the amplifier outputs to settle to within the resolution of the overall converter.

Second-order effects are responsible for limiting the useful sampling frequency to a value which is smaller than that predicted by simulation of the modulator assuming integrators with a perfectly exponential impulse response. These effects include deviations in the impulse response from an exponential waveform because of the drastic change in bias currents that occur in the class $AB$ operational amplifier during large-signal transients. In addition, the time constants of the amplifier response during phase 1 and phase 2 differ as a result of changes in the equivalent load capacitance. The slew rate of the amplifier is greater than 200 V/$\mu$s, sufficient to prevent slewing distortion in all operating conditions of the integrators.

In Fig. 15 the dynamic range of the converter is plotted as a function of the oversampling ratio. The performance increases by 15 dB for every doubling of the oversampling ratio úp to $M \approx 64$, as is expected for an ideal $\Sigma\Delta$ modulator. In this regime, the dynamic range of the modulator is within 4 dB of that of an ideal modulator with perfect components and no error sources other than quantization noise. At higher oversampling ratios, the performance improvement obtained by increasing the oversampling ratio is reduced. Flicker noise in the input transistors of the first

integrator has been found to be the primary limitation when the signal is oversampled by a factor greater than 64. The dynamic range of 89 dB corresponds to fabrication of the modulator in a technology wherein the $1/f$ noise is characterized by a flicker-noise coefficient $K_1 = 5 \cdot 10^{-24}$ $V^2 \cdot F$ [28], a value consistent with typical CMOS technologies [29]. This flicker-noise limitation can be overcome by increasing the size of the input transistors, or through the use of a chopper-stabilized amplifier [30]. The size of the input devices was kept relatively small in this design because of concern for the amplifier frequency response.

## VI. CONCLUSION

Second-order $\Sigma\Delta$ modulators constitute an efficient architecture for implementing high-resolution A/D converters in scaled high-performance integrated circuit technologies. Both simulations and analytic results have been used to establish design criteria for the analog circuit blocks comprising such a modulator. Specifically, it has been found that integrator linearity has a crucial influence on the performance of these converters, whereas $\Sigma\Delta$ modulators impose only modest demands on integrator bandwidth and are relatively insensitive to offset and hysteresis in the comparator. The analysis presented here has also been used to identify mechanisms other than quantization noise that may limit the performance of $\Sigma\Delta$ modulators regardless of the oversampling ratio.

The limitations on both the speed and dynamic range of the experimental A/D converter reported herein can be readily overcome through the use of a higher performance technology. Scaled digital VLSI technologies are especially suitable for these types of converters because they provide the density and speed necessary to include the decimation filter, as well as other signal processing functions, on the same chip as the modulator. Conversely, oversampling architectures provide a means of exploiting the enhanced speed of scaled digital technologies so as to overcome constraints on the available dynamic range and the need for precision circuits and components. In a sampled-data CMOS implementation of a $\Sigma\Delta$ modulator, the only requirement imposed on the fabrication technology is the availability of capacitors. The accuracy of the capacitor ratios is not critical since the performance of $\Sigma\Delta$ modulators is not sensitive to the gain $g_0$ preceding the integrators. In comparison with Nyquist rate converters, relatively large capacitor voltage coefficients can be tolerated because the modulator feedback loop serves to reduce the resulting distortion.

## REFERENCES

[1] J. C. Candy, "A use of double integration in sigma delta modulation," *IEEE Trans. Commun.*, vol. COM-33, pp. 249–258, Mar. 1985.
[2] M. W. Hauser, P. J. Hurst, and R. W. Brodersen, "MOS ADC-filter combination that does not require precision analog components," in *ISSCC Dig. Tech. Papers*, Feb. 1985, pp. 80–82.

[3] B. H. Leung, R. Neff, and P. R. Gray, "A four-channel CMOS oversampled pcm voiceband coder," in *ISSCC Dig. Tech. Papers*, Feb. 1988, pp. 106–107.

[4] J. C. Candy and O. J. Benjamin, "The structure of quantization noise from sigma-delta modulation," *IEEE Trans. Commun.*, vol. COM-29, pp. 1316–1323, Sept. 1981.

[5] B. Boser and B. Wooley, "Quantization error spectrum of sigma-delta modulator," in *Proc. 1988 IEEE Int. Symp. Circuits Syst.*, June 1988, pp. 2331–2334.

[6] R. Gray, "Spectral analysis of sigma-delta quantization noise," to be published in *IEEE Trans. Commun.*

[7] J. C. Candy, "Decimation for sigma delta modulation," *IEEE Trans. Commun.*, vol. COM-34, pp. 72–76, Jan. 1986.

[8] S. Ardalan and J. Paulos, "An analysis of nonlinear behavior in delta-sigma modulators," *IEEE Trans. Circuits Syst.*, vol. CAS-3, pp. 593–603, June 1987.

[9] K. Uchimura, T. Hayashi, T. Kimura, and A. Iwata, "VLSI A-to-D and D-to-A converters with multi-stage noise shaping modulators," in *Proc. ICASSP*, Apr. 1986, pp. 1545–1548.

[10] T. Hayashi, Y. Inabe, K. Uchimura, and T. Kimura, "A multi-stage delta-sigma modulator without double integration loop," in *ISSCC Dig. Tech. Papers*, Feb. 1986, pp. 182–183.

[11] Y. Matsuya, K. Uchimura, A. Iwata, T. Kobayashi, and M. Ishikawa, "A 16b oversampling A/D conversion technology using triple integration noise shaping," in *ISSCC Dig. Tech. Papers*, Feb. 1987, pp. 48–49.

[12] J. C. Candy, Y. C. Ching, and D. S. Alexander, "Using triangularly weighted interpolation to get 13-bit PCM from a sigma-delta modulator," *IEEE Trans. Commun.*, vol. COM-24, pp. 1268–1275, Nov. 1976.

[13] T. Misawa, J. E. Iwersen, L. J. Loporcaro, and J. G. Ruch, "Single-chip per channel codec with filters utilizing $\Sigma$-$\Delta$ modulation," *IEEE J. Solid-State Circuits*, vol. SC-16, pp. 333–341, Aug. 1981.

[14] H. L. Fiedler and B. Hoefflinger, "A CMOS pulse density modulator for high-resolution A/D converters," *IEEE J. Solid-State Circuits*, vol. SC-19, pp. 995–996, Dec. 1984.

[15] P. Defraeye, D. Rabaey, W. Roggeman, J. Yde, and L. Kiss, "A 3-$\mu$m CMOS digital codec with programmable echo cancellation and gain setting," *IEEE J. Solid-State Circuits*, vol. SC-20, pp. 679–687, June 1985.

[16] U. Roettcher, H. Fiedler, and G. Zimmer, "A compatible CMOS-JFET pulse density modulator for interpolative high-resolution A/D conversion," *IEEE J. Solid-State Circuits*, vol. SC-21, pp. 446–452, June 1986.

[17] R. Koch et al., "A 12-bit sigma-delta analog-to-digital converter with a 15-MHz clock rate," *IEEE J. Solid-State Circuits*, vol. SC-21, pp. 1003–1010, Dec. 1986.

[18] B. Boser, K.-P. Karmann, H. Martin, and B. Wooley, "Simulating and testing oversampled analog-to-digital converters," *IEEE Trans. Computer-Aided Des.*, vol. CAD-7, pp. 668–674, June 1988.

[19] J. C. Candy, private communication, 1985.

[20] K-L. Lee and R. Meyer, "Low-distortion switched-capacitor filter design techniques," *IEEE J. Solid-State Circuits*, vol. SC-20, pp. 1103–1113, Dec. 1985.

[21] A. Vladimirescu, A. Newton, and D. Pederson, *SPICE Version 2G.1 User's Guide*, Univ. of Calif., Berkeley, Tech. Rep., Oct. 1980.

[22] *SWAP User Documentation*, Silvar-Lisco, Menlo Park, CA, 1986.

[23] T. Choi, R. Kaneshiro, P. Gray, W. Jett, and M. Wilcox, "High-frequency CMOS switched-capacitor filters for communications application," *IEEE J. Solid-State Circuits*, vol. SC-18, pp. 652–664, Dec. 1983.

[24] R. Castello and P. Gray, "A high-performance micropower switched-capacitor filter," *IEEE J. Solid-State Circuits*, vol. SC-20, pp. 1122–1132, Dec. 1985.

[25] G. C. Temes, "Finite amplifier gain and bandwidth effects in switched-capacitor filters," *IEEE J. Solid-State Circuits*, vol. SC-15, pp. 358–361, June 1980.

[26] K. Martin and A. S. Sedra, "Effects on the op amp finite gain and bandwidth on the performance of switched-capacitor filters," *IEEE Trans. Circuits Syst.*, vol. CAS-28, pp. 822–829, Aug. 1981.

[27] A. Yukawa, "A CMOS 8-bit high speed A/D converter IC," *IEEE J. Solid-State Circuits*, vol. SC-20, pp. 775–779, June 1985.

[28] P. R. Gray and R. G. Meyer, *Analog Integrated Circuits*. New York: Wiley, 1984.

[29] A. Abidi, C. Viswanathan, J. Wu, and A. Wikstrom, "Flicker noise in CMOS: A unified model for VLSI processes," in *1987 Symp. VLSI Technology, Dig. Tech. Papers*, May 1987, pp. 85–86.

[30] K.-C. Hsieh, "A low-noise chopper-stabilized differential switched-capacitor filtering technique," *IEEE J. Solid-State Circuits*, vol. SC-16, pp. 708–715, Dec. 1981.

# A Noise-Shaping Coder Topology for 15 + Bit Converters

L. RICHARD CARLEY, MEMBER, IEEE

*Abstract* —A novel topology for high-precision noise-shaping converters that can be integrated on a standard digital IC process is presented. This topology employs a multibit noise-shaping coder and a novel form of dynamic element matching to achieve high accuracy and long-term stability without requiring precision matching of components. A fourth-order noise-shaping D/A conversion system employing a 3-bit quantizer and a dynamic element matching internal D/A converter, fabricated in a standard double-metal 3-µm CMOS process, achieved 16-bit dynamic range and a harmonic distortion below −90 dB. This multibit noise-shaping D/A conversion system achieved performance comparable to that of a 1-bit noise-shaping D/A conversion system that operated at nearly four times its clock rate.

## I. Introduction

ONE-BIT noise-shaping converters, referred to as either delta–sigma modulators (ΔΣM's) or sigma–delta modulators, have recently achieved popularity for use in integrated circuit data converters, both A/D [1] and D/A [2]. Their attractiveness for IC systems that incorporate digital filtering and signal processing with A/D and/or D/A conversion is, in part, due to the fact that they employ a 1-bit internal D/A converter which does not require precision component matching. They can be implemented using a standard digital CMOS process without the economically costly addition of precision thin-film resistors or the use of laser trimming. However, experience has revealed that high integral linearity is very difficult to achieve, because it is not possible to avoid overloading the 1-bit loop quantizer. One obvious solution is to use a multibit quantizer in the loop employing sufficient levels to prevent quantizer overload. Eliminating quantizer overload, in addition to improving integral linearity, also facilitates the design of higher-order ( > 2) noise-shaping loops, because the low-frequency oscillations observed in higher-order ΔΣM loops [1] are a result of quantizer overload [7]. For these reasons, a multibit noise-shaping conversion system can achieve a signal-to-noise ratio (SNR) comparable to that of a second-order 1-bit noise-shaping (ΔΣM)

D/A conversion system operating at a much higher sampling rate. For example, the prototype noise-shaping D/A conversion system presented in this paper, which operated at 3.2 MHz and employed a 3-bit quantizer, achieved performance comparable to that of a ΔΣM-type conversion system that operates at 11.3 MHz [2]. The lower clock rate is particularly important since dynamic power consumption in the CMOS logic increases as the square of the clock frequency.

As stated above, many advantages result from using a multibit noise-shaping converter instead of a ΔΣM converter. However, there is one major disadvantage: the integral linearity of the noise-shaping conversion system is no better than the integral linearity of the multibit internal D/A converter [3], [4]. Achieving high integral linearity, hence low total harmonic distortion (THD), normally *requires* precision component matching. In this case, multibit noise-shaping converters cannot be fabricated using an inexpensive digital CMOS process, hence their relative lack of use as part of digital signal processing IC systems to date.

This paper presents a multibit noise-shaping coder topology that incorporates an internal D/A converter employing a novel form of "dynamic element matching" [5], [6] to achieve excellent integral and differential linearity, while requiring only modest component matching. For example, the prototype IC D/A conversion system has a peak element mismatch of nearly 0.2 percent, and a measured peak integral linearity error of only 0.0022 percent of full scale. This new topology allows multibit noise-shaping coders to employ internal D/A converters which achieve high linearity without requiring component trimming. Therefore, by employing a dynamic element matching internal D/A converter, multibit noise-shaping coders can be integrated with digital signal processing electronics using an inexpensive digital CMOS process as demonstrated by the prototype system.

## II. Noise-Shaping Coder Topology

Fig. 1 shows the topology for a multibit noise-shaping A/D conversion system and Fig. 2 shows the D/A conversion system. In the case of an A/D conversion system,

Manuscript received August 2, 1988; revised December 6, 1988. This work was supported in part by the National Science Foundation under Grant ENG-8451496 and by a grant from Analog Devices Corporation.
The author is with the Department of Electrical and Computer Engineering, Carnegie Mellon University, Pittsburgh, PA 15213.
IEEE Log Number 8826153.

Reprinted from *IEEE J. Solid-State Circuits*, vol. SC-24, pp. 267–273, April 1989.

304

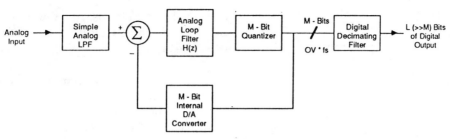

Fig. 1. Topology of noise-shaping A/D conversion systems. $OV$ is the oversampling ratio and $f_s$ is the sampling rate at the system's digital output. The number of bits at the output $L$ is a function of the decimating filter architecture and the SNR of the noise-shaping coder.

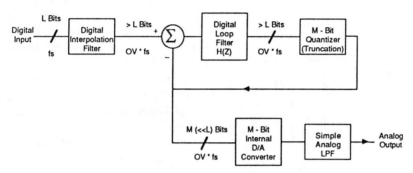

Fig. 2. Topology of noise-shaping D/A conversion systems. $OV$ is the oversampling ratio and $f_s$ is the sampling rate at the system's digital input. The number of bits at the interpolating filter output is greater than $L$ due to arithmetic operations in the filter, and is therefore a function of the interpolating filter architecture.

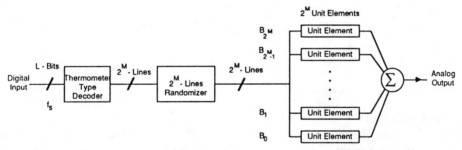

Fig. 3. Topology of dynamic element matching internal D/A converter. A thermometer-type decoder sets the number of output lines high that is equal to the digital input.

the quantizer is a true A/D converter—normally a bank of comparators, while in the case of a D/A conversion system, the quantizer merely corresponds to truncation. In both cases, a multibit internal D/A converter is necessary.

Assuming that quantizer overload does not occur, the design of multibit noise-shaping loops is quite simple compared to the design of $\Delta\Sigma M$ loops, because the gain of the quantizer is known (1 LSB/digital level), the quantization error is bounded between $+1/2$ and $-1/2$ LSB, and the quantization error in this case can be modeled as an additive noise that is independent of the input signal [12], [13]. Nonlinear numerical techniques have been applied to optimize the conversion system's performance for a given internal D/A converter through the choice of the loop-filter pole and zero locations [13].

Because the quantizer error is bounded, a worst-case bound can be computed for the range signal at the quantizer's input. The worst-case noise gain is determined by assuming that the quantization error will take on a se-quence of worst-case values, $+1/2$ or $-1/2$ LSB [12], [13]. The noise gain can be computed by summing the magnitude of the impulse response of the transfer function from the quantizer output back to its input. The quantizer must have enough levels to code the worst-case noise at its input plus the input signal range without overload. For example, if the $H(z)$ selected gives a worst-case noise gain of 4, then the noise due to quantization at the quantizer's input will be $\pm 2$ LSB's. If the input signal also occupied $\pm 2$ LSB's of range then the quantizer and the internal D/A converter would need eight levels.

## III. INTERNAL D/A CONVERTER TOPOLOGY

The dynamic element matching D/A converter topology consists of two parts: a parallel unit-element D/A converter structure and a digital "randomizer" which controls the connections between the input and the parallel D/A converter. First we will consider the parallel D/A con-

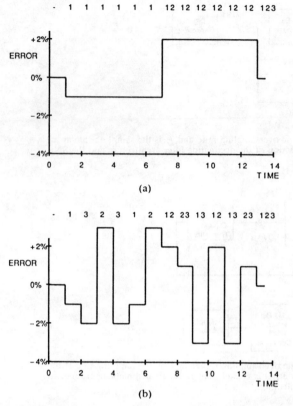

Fig. 4. How dynamic element matching works. (a) The error of a 2-bit parallel D/A converter implemented as the sum of three elements. (b) The error of the same D/A converter with the randomizer implementing dynamic element matching. Element 1 is 1 percent below desired value, element 2 is 3 percent above desired value, and element 3 is 2 percent below the desired value. The numbers above each trace indicate which elements are active. The input is 0, 1, 1, 1, 1, 1, 1, 2, 2, 2, 2, 2, 2, 3.

verter and the performance achievable by employing dynamic element matching in this case. Then we will consider possible implementations of the randomizer.

### A. Dynamic Element Matching Approach

The dynamic element matching D/A converter topology can be constructed using any D/A converter in which the $N$th output level is generated by activating $N$ approximately equal-valued elements, typically resistors or capacitors, and summing up their charge or current or voltage (see Fig. 3). Dynamic element matching is implemented by choosing different elements to represent the $N$th level as a function of time. The "randomizer" block varies which elements will be used to represent the $N$th level on each clock cycle (see Fig. 4). The goal of this approach is to convert the error due to element mismatch from a dc offset into an ac signal of equivalent power which, in an oversampling converter, can be partially removed by filtering. In Fig. 4, even when the input is constant, the error is a wide-band noise signal. With ideal randomization, a mismatch between the unit elements would be converted into a white-noise signal with zero mean error and a variance equal to the root-mean-square (rms) error between the

individual unit elements. In an oversampling data conversion system, which typically operates at oversampling ratios of 64:1 or more, nearly all of the error power can be filtered out.

First, let us consider the linearity of this D/A converter. For a dc input code of $N$, each element is active, on average, $N$ out of every $M$ clock cycles, where $M$ is the total number of elements. Therefore, each element of the D/A converter acts individually as a duty-cycle modulator and the integral linearity is limited only by the product of the fractional element mismatch $(\Delta E/E)$ and the fractional clock jitter $(\Delta T/T)$ [5], [6]. A second practical limit on the integral linearity results because there is normally a small change in the charge (or current) transferred by each element as a function of the number of elements active. With careful choice of D/A converter topology and the use of a precision clock, extremely high dc integral linearity can be achieved, even when the elements match very poorly. However, the element mismatch now appears as an ac noise signal added to the D/A converter output. The constraint on the element matching has been converted from a constraint based on the converter's linearity to a constraint based on the converter dynamic range.

If small scale factor errors are ignored, the maximum noise signal $n(t)$ varies in a parabolic fashion from zero at either zero or full scale to a maximum at half of full scale. At this maximum, $n(t)$, relative to the internal D/A converter's full scale $M$, is

$$\mathrm{rms}\left[\frac{n(t)}{M}\right] = \frac{\mathrm{rms}\left[\dfrac{\Delta E}{E}\right]}{2\sqrt{M}}.$$

Note, this expression is a factor of $\sqrt{2}$ smaller than the sum of $M/2$ random element errors because of the assumption that scale factor errors are negligible.

In an oversampling converter, only noise power in the signal passband is important. Assuming an ideal randomizer, the noise signal will be white and the rms fraction of the internal D/A converter's full scale for the noise signal in the passband $(n_p(t))$ will be

$$\mathrm{rms}\left[\frac{n_p(t)}{M}\right] = \frac{\mathrm{rms}\left[\dfrac{\Delta E}{E}\right]}{2\sqrt{M \times OV}}$$

where $OV$ is the oversampling ratio. Note, although the noise-shaping A/D conversion system shapes quantization noise in order to achieve a greater than 3-dB increase in SNR per octave, the internal D/A element mismatch noise is not in the feedback loop (see Fig. 2) and cannot be affected by feedback.

Since an $R$-bit D/A converter would require $2^R - 1$ elements, only a limited number of bits can be generated using this topology. The unusual nature of noise-shaping coders, that resolution is gained by oversampling the signal

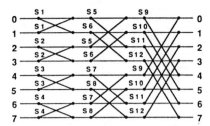

Fig. 5. Topology of the butterfly-type ramdomizer. Example of a three-stage eight-line butterfly randomizer. Each pair of switches marked with the same label is controlled to either exchange the two signal lines or pass them directly to the next stage.

## IV. DESIGN EXAMPLE

In order to validate the predictions for the linearity and SNR of the new noise-shaping coder topology, an experimental fourth-order noise-shaping D/A conversion system has been designed, optimized using CLANS [13], simulated, fabricated, and tested. Experimental versions using both resistor unit elements and capacitor unit elements in the internal D/A converter have been studied.

### A. System Topology

Fig. 2 shows the block diagram of the experimental noise-shaping D/A conversion system. Interpolation, the process of increasing the sampling rate, was performed in two stages. First, the input signal was interpolated by a factor of 2 and filtered using a standard 64-tap low-pass FIR filter topology. Then, the FIR filter output was interpolated by a factor of 32 and filtered using a comb filter [8]. The combined filter attenuates frequencies above 30 kHz by more than 80 dB with a 3.2-MHz clock. Extensive simulations indicate that the SNR at the output of the noise-shaping loop with this interpolator architecture is less than 3 dB below the SNR with an ideal interpolating filter. Although the filter complexity could be reduced still further, a dramatic decrease in the loop's SNR was observed for filter attenuations below 80 dB.

The noise-shaping loop filter ($H(z)$ in Fig. 2) employs a fourth-order filter with four poles at $z = 1$ and three zeros positioned to maintain system stability and to minimize the worst-case noise gain of the loop [9], [10]. CLANS [13] was used to optimize the locations of the three zeros in order to maximize the loop's SNR. Maximum tolerable noise gain is a constraint used in choosing the location of the zeros of the loop filter. A maximum noise gain of 4 was chosen as the constraint when placing the closed-loop poles for the noise-shaping loop of the prototype system. Therefore, the quantization error at the noise-shaping loop's output, which is in the range of $+1/2$ to $-1/2$ LSB, is amplified in the worst case by a factor of 4 at the quantizer input. Hence, the noise signal at the quantizer input will be in the range of $+2$ to $-2$ LSB. In order to prevent the noise-shaping quantizer from being overloaded, the 18-bit input signal is scaled so that its full-scale range is between $+2$ and $-2$ LSB, resulting in a total range of eight levels or 3 bits.

### B. 3-bit Internal D/A Converter

The capacitor version of the dynamic element matching 3-bit internal D/A converter is shown in Fig. 6. The D/A converter's $N$th output level is generated by charging all capacitors to a $+5$-V reference level and then switching $N$ of them into the summing junction of the $I^+$ operational amplifier and the rest into the summing junction of the $I^-$ amplifier during each clock cycle. Note that by always switching all of the capacitors between the same two voltages on every clock cycle many potential complications

processing but that linearity must be provided by the D/A converter, makes them a perfect application for this type of dynamic element matching which is limited to only a few bits but can provide extremely high accuracy. In addition, only a few levels, typically two to four, are necessary to prevent quantizer overload in noise-shaping coders; therefore, the number of parallel elements is not prohibitive for this application.

### B. The Randomizer

The randomizer connects the $M$ outputs from the decoder to the $M$ switching elements in a time-varying fashion. The number of possible connections is $M$ factorial. Therefore, when $M$ is small (on the order of 3 or 4) it is possible to randomly select between all possible connections. However, when $M$ is large (e.g., 8) the number of possible connections is so large that it may be necessary to select a subset of connections in order to conserve die area. For example, an ideal eight-level randomizer, one which connects each of the eight inputs to eight outputs, would have to include 40 320 possible connections.

One simple approach to randomizing over a subset of possible connections would be to have an $M$-port barrel shifter which rotates one increment after each clock. This represents only $M$ of the $M$ factorial possible permutations. This approach would completely decorrelate successive output errors only if the mismatch between elements were independent of the element's position on the die. Unfortunately, just the opposite is typically true. Adjacent elements are normally much more likely to match than distant elements due to gradients in process parameters across the wafer.

A compromise between these two extremes is the "butterfly" randomizer proposed by Kenney [14]. The butterfly randomizer circuit consists of a series of butterfly networks coupling the inputs to the outputs (see Fig. 5). In order that any input can be connected to any output, the number of butterfly stages should be at least equal to the number of bits in the internal D/A converter. More butterfly stages can be added if it is necessary to cover a larger fraction of possible connections. A pseudorandom sequence generator would normally be used to generate the random control sequences for the butterfly switches [15]–[17].

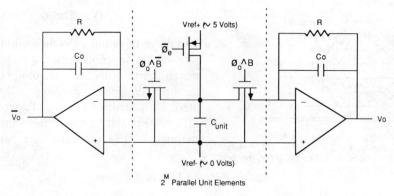

Fig. 6. Topology of the capacitive 3-bit internal D/A converter. $\phi_e$ and $\phi_o$ are the even and odd switch clocks, respectively, and $B$ is the digital input to that element. The unit $C$ was 2 pF.

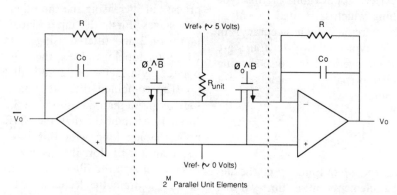

Fig. 7. Topology of the resistive 3-bit internal D/A converter. $\phi_e$ and $\phi_o$ are the even and odd switch clocks, respectively, and $B$ is the digital input to that element. The unit $R$ was 14K.

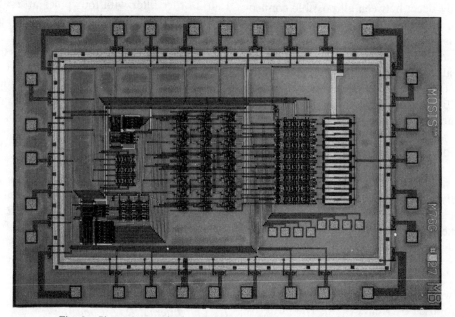

Fig. 8. Photomicrograph of randomizer and 3-bit capacitive D/A converter.

are avoided: varying current draw on the reference source, complications due to dielectric absorption in the capacitors, and even failure of the capacitor voltage to fully settle to either the reference voltage or ground. The primary nonlinearity error of this topology comes from changes in the transient at the op-amp summing junction with the number of elements selected.

A resistor version of the 3-bit D/A converter was also fabricated (see Fig. 7). As with the capacitor version, the $N$th output level was generated by switching $N$ resistors into the summing junction of the $I^+$ operation amplifier and the rest into the summing junction of the $I^-$ operational amplifier. The other end of each resistor was tied to the reference voltage. Note that even and odd switch

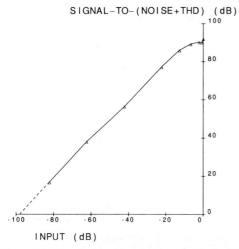

Fig. 9. Plot of SNR as a function of input amplitude. In this case distortion is *included* with the noise. Input signal is a 1-kHz sine wave.

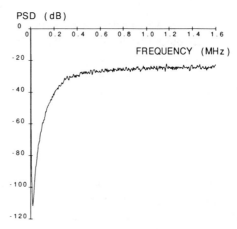

Fig. 10. PSD at D/A converter output. The input is a full-scale 1-kHz signal.

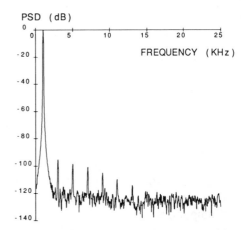

Fig. 11. PSD at analog filter output with a 0-dB 1-kHz input signal. The input is a full-scale 1-kHz signal. Note, the amplitude of the input signal has been artificially reduced by passing it through a 1-kHz notch filter in order to decrease the dynamic range to allow accurate measurement of the PSD.

clocks were used, even in the resistive case, to prevent varying propagation delays in the computation of $B$ from changing the ON period of the element.

Dynamic element matching was implemented by randomizing the elements activated by each code over time. Eight unit elements were used rather than the minimum of seven. At least one element was always OFF. The choice between using $2^M$ elements and $2^M - 1$ unit elements in the internal D/A converter is arbitrary, and is not important to its operation. In this case, eight levels were selected to make the design of the butterfly randomizer symmetric. The randomizer circuit consisted of a series of three "butterfly" networks coupling the inputs to the output (see Fig. 5). This randomizer implements 4096 of the available 40 320 combinations, about one-tenth of the possible connections. Simulations verified that this partial randomization resulted in good decorrelation of internal D/A converter errors in the face of both linear and quadratic variations in element value with position. Fig. 8 shows a photomicrograph of the capacitive 3-bit D/A converter, the decoder, and the digital randomizer circuitry.

The analog output from the internal 3-bit D/A converter (see Fig. 2) was filtered by a fifth-order Butterworth filter. The real axis pole was provided by the first op amp and the two complex pole pairs were implemented as a cascade of two Sallen–Key second-order sections. Because the SNR and the open-loop gain of the operational amplifiers limit the measurements of the D/A converter's performance, the analog filter was implemented off-chip using low-noise operational amplifiers.

Although all of the prototype system was not fabricated in IC form, other researchers have fabricated digital interpolating filters, digital noise-shaping loops, and analog output filters in CMOS processes equivalent in performance to those required in this system [2], [11]. For example, a higher-order (1:256) interpolating filter and a second-order noise-shaping loop were implemented by Naus *et al.* in approximately 18 mm² of die area [2]. Therefore, we conclude that by testing the novel portion of this

system, the dynamic element matching internal D/A converter, in IC form we have proven that integration of the entire system is viable.

### C. Results

In order to be compatible with a double-metal digital CMOS process, the capacitive version of the internal D/A converter used the field oxide between metal-1 and metal-2 as the unit capacitor. Unfortunately, this oxide was extremely irregular, the capacitor mismatch was over 1 percent. Although the integral linearity was excellent as expected, the noise resulting from element mismatch was unacceptably large. Therefore, the measured results reported in the rest of this section are for the resistive version of the internal D/A converter. The resistors in the resistive version of the internal D/A converter were $3\text{-}\mu\text{m} \times 2.7\text{-mm}$ polysilicon lines with a resistance of about 14K. On selected die rms element variations as low as 0.05 percent were observed.

Fig. 9 shows the SNR at the analog filter output as a function of the input amplitude for a 1-kHz sinusoidal input. The system achieves a dynamic range of about 98 dB. Fig. 10 shows the power spectral density (PSD) of the 3-bit D/A converter output without the analog filter for a 0-dB 1-kHz sinusoidal input. The flattening of the PSD of the noise at approximately 1 MHz is a result of the zeros incorporated into the noise-shaping loop's transfer function to control the noise gain. Note that this signal contains noise components from two sources: quantization noise from the noise-shaping loop and the approximately "white" noise resulting from the randomized element mismatch.

Fig. 11 shows the PSD at the output of the analog filter. Simulations with an ideal 3-bit D/A converter indicate that the PSD of the in-band noise is dominated, in the audio frequency range, by the component mismatch noise rather than the quantization noise from the noise-shaping loop. The dynamic range limited by the quantization alone was 108 dB for this noise-shaping coder. The harmonic distortion components are each more than 96 dB below the maximum input level.

## V. Conclusions

A new topology for multibit high-precision noise-shaping converters that incorporates a dynamic element matching internal D/A converter has been presented. This new topology makes possible the construction of multibit noise-shaping coders in a standard digital CMOS process. Experimental verification has been provided via a prototype 16-bit oversampling D/A conversion system that uses an all-digital noise-shaping loop followed by a dynamic element matching 3-bit D/A converter that was fabricated in 3-$\mu$m CMOS. The 16-bit D/A conversion system achieved a peak dynamic range of 98 dB and a THD of approximately $-94$ dB at a sampling rate of only 3.2 MHz. This represents almost a factor of 4 decrease in clock rate from a $\Delta\Sigma$M-type conversion system of comparable performance [2] and a potential decrease of nearly 16-fold in the dynamic power consumption of the CMOS logic implementing the interpolating filter and noise-shaping loop.

The multibit nature of the noise-shaping loop, which both increases the initial accuracy of the quantizer and facilitates the use of higher-order noise-shaping coders, makes it possible to attain performance equivalent to that of $\Delta\Sigma$M converters operating at much higher clock rates.

And it is the dynamic element matching technique which makes possible multibit noise-shaping coders that can be incorporated into a standard digital CMOS process with digital filters and signal processors.

## Acknowledgment

The author would like to thank J. Kenney and C. Wolff for their help in laying out and testing several versions of the experimental D/A converter reported herein.

## References

[1] J. C. Candy, "A use of double integration in sigma-delta modulation," *IEEE Trans. Commun.*, vol. COM-33, pp. 249–258, Mar. 1985.

[2] P. J. A. Naus *et al.* "A CMOS stereo 16-bit D/A converter for digital audio," *IEEE J. Solid-State Circuits*, vol. SC-22, no. 3, pp. 390–395, June 1987.

[3] J. W. Scott, W. Lee, C. Giancario, and C. G. Sodini, "A CMOS slope adaptive delta modulator," in *ISSCC Dig. Tech. Papers*, Feb. 1986, pp. 130–131.

[4] R. W. Adams, "Design and implementation of an audio 18-bit A/D converter using oversampling techniques," *J. Audio Eng. Soc.*, vol. 34, no. 3, pp. 153–166, Mar. 1986.

[5] R. J. Van De Plassche, "A monolithic 14-bit D/A converter," *IEEE J. Solid-State Circuits*, vol. SC-14, no. 3, pp. 552–556, June 1979.

[6] K. B. Klaassen, "Digitally controlled absolute voltage division," *IEEE Trans. Instrum. Meas.*, vol. IM-24, no. 3, pp. 106–112, June 1975.

[7] C. Wolff and L. R. Carley, "Modeling the quantizer in higher-order delta-sigma modulators," presented at the Int. Symp. Circuits Syst., Helsinki Finland, June 1988.

[8] S. Chu and C. S. Burrus, "Multirate filter designs using comb filters," *IEEE Trans. Circuits Syst.*, vol. CAS-31, pp. 913–924, Nov. 1984.

[9] J. Kenney, "Design methodology for *N*-th order noise-shaping coders," Masters Project Rep., Dept. of ECE, Carnegie-Mellon Univ., Pittsburgh PA, Jan. 1988.

[10] B. P. Agrawal and K. Shenoi, "Design methodology for $\Sigma\Delta$M," *IEEE Trans. Commun.*, vol. COM-31, pp. 360–369, Mar. 1983.

[11] Y. Matsuya *et al.*,, "A 16-bit oversampling a-to-d conversion technology using triple-integration noise shaping," *IEEE J. Solid-State Circuits*, vol. SC-22, no. 6, pp. 921–929, Dec. 1987.

[12] L. R. Carley, "An oversampling analog-to-digital converter topology for high resolution signal aquisition systems," *IEEE Trans. Circuits Syst.*, vol. CAS-34, no. 1, pp. 83–91, Jan. 1987.

[13] J. Kenney and L. R. Carley, "CLANS: A high-level synthesis tool for high resolution data converters," in *Proc. 1988 IEEE Int. Conf. Computer-Aided Des.* (Santa Clara, CA), Nov. 1988, pp. 496–499.

[14] L. R. Carley and J. Kenney, "A 16-bit 4th order noise-shaping D/A converter," in *Proc. 1988 Custom Integrated Circuits Conf.* (Rochester, NY), May 1988.

[15] J. H. Lindholm, "An analysis of the pseudo-randomness properties of subsequences of long *m*-sequences," *IEEE Trans. Inform. Theory*, vol. IT-14, pp. 569–567, July 1968.

[16] J. L. Manos, "Some techniques for testing pseudo-random number sequences," Lincoln Labs., Lexington, MA, Tech. Note 1974-44, 1974.

[17] A. C. Davies, "Properties of waveforms obtained by nonrecursive digital filtering of pseudo-random binary sequences," *IEEE Trans. Computers*, vol. C-20, pp. 270–281, Mar. 1971.

# A Dual-Channel Voice-Band PCM Codec Using ΣΔ Modulation Technique

VLADIMIR FRIEDMAN, MEMBER, IEEE, DOUGLAS M. BRINTHAUPT, DE-PING CHEN, MEMBER, IEEE,
TIMOTHY W. DEPPA, JOHN P. ELWARD, JR., EVELYN M. FIELDS, JEFFREY W. SCOTT, AND
T. R. VISWANATHAN, SENIOR MEMBER, IEEE

*Abstract* —A dual-channel sigma–delta (ΣΔ) voice-band codec meeting AT&T/CCITT specifications is described. Its digital signal processing section has a bit-slice architecture which can be expanded to accommodate higher bit resolution. The active area per channel is 13 mm² in a 1.5-μm CMOS process. It has one power supply only and the maximum power dissipation is 90 mW per channel. The crosstalk between channels is less than −71 dB.

## I. INTRODUCTION

IN THE present 1.5-μm CMOS technology, codecs which use the sigma–delta (ΣΔ) modulation conversion technique become competitive with charge redistribution codecs. The active area necessary for implementing a single codec is around 13 mm². A multicodec device presents the advantage of a reduced component density for line cards in telephone systems. Because it is a 5-V only part, it is also ideal for Integrated Services Digital Network (ISDN) basic access terminals. Other applications include multichannel speech recognition and synthesis, and cellular communication systems. The high level of integration also makes it closer to the optimum manufacturing die size for fine-line technologies. The ΣΔ oversample techniques rely heavily on digital signal processing, thereby reducing the amount of analog circuitry. This helps reduce the crosstalk and PSRR. Lastly, the ΣΔ design technique is well suited to submicrometer technologies.

## II. BLOCK DIAGRAM OF THE DEVICE

The device contains two identical sections. A block diagram of a single channel is shown in Fig. 1. The encoder uses a double-loop ΣΔ modulator at 1.024-MHz oversampling frequency. Its 1-bit output stream contains, besides the signal information, the quantization noise, shaped by the modulator such that most of its energy lies outside the signal band (Fig. 2(a)). PCM samples can be obtained from the output of the modulator by filtering the noise and undersampling the signal down to the Nyquist sampling rate. This is done in two stages. The first stage which decimates the signal to 16 kHz uses an FIR filter. The multiplication operation is reduced to an addition, since the input to the filter is only 1 bit wide. The FIR filter has zeros at multiples of the frequency 16 kHz, because the noise around these frequencies is aliased into the signal band. In order for the contribution of the high-frequency noise components aliased into the signal band to be negligible, the order of multiplicity of these transmission zeros must be three for double-loop ΣΔ modulators [1]. This corresponds to a sinc cubic function. The frequency characteristic of such a filter is shown in Fig. 2(b) and the signal and the noise spectrum after the decimation in Fig. 2(c). The second decimation stage, from 16- to 8-kHz sampling frequency, uses the low-pass section of a bandpass IIR filter. This filter removes the remaining noise (Fig. 2(c)) and compensates for the attenuation at high frequencies introduced by the first decimation stage. The high-pass section is attenuating the frequencies in the 0–200-Hz band according to CCITT specifications. It is followed by the output section which converts the signal to a compressed PCM format.

The decoder also uses the ΣΔ modulation technique for conversion, with the roles of the digital and analog sections interchanged. The 1-MHz bit stream is generated by a digital ΣΔ modulator, while the filtering of the noise is done by the analog section. Because of this change of roles some significant differences exist. In the process of raising the sampling frequency, images of the input spectrum are generated. First the sampling frequency is raised to 32 kHz by repeating each input sample after the expansion to the linear format four times. This is equivalent with an attenuation of the images by a sinc function (Fig. 3(a)) given by the equation

$$S(f) = \frac{1}{N} \frac{\sin\left(N\pi \dfrac{f}{f_s}\right)}{\sin\left(\pi \dfrac{f}{f_s}\right)} \tag{1}$$

Manuscript received August 22, 1988; revised December 2, 1988.
V. Friedman, D. M. Brinthaupt, D.-P. Chen, T. W. Deppa, J. P. Elward, Jr., and E. M. Fields are with AT&T Bell Laboratories, Murray Hill, NJ 07974.
J. W. Scott and T. R. Viswanathan are with AT&T Bell Laboratories, Reading, PA 19604.
IEEE Log Number 8826148.

Reprinted from *IEEE J. Solid-State Circuits*, vol. SC-24, pp. 274–280, April 1989.

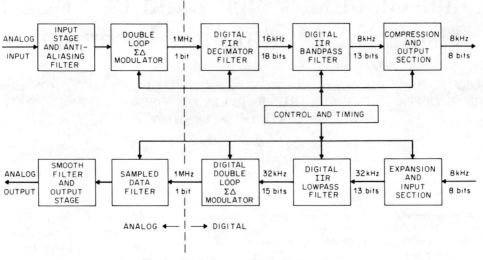

Fig. 1. Block diagram of a single channel.

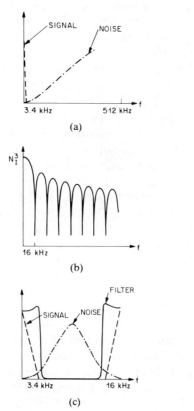

Fig. 2. Encoder signal processing. (a) Signal and noise at the output of the modulator. (b) Frequency characteristic of the sinc cube filter. (c) Signal and noise at the output of the sinc cube decimator.

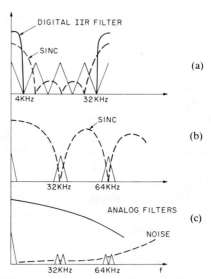

Fig. 3. Decoder signal processing. (a) Input spectrum images at the input of the IIR filter. (b) Input spectrum images at the input of the $\Sigma\Delta$ modulator. (c) Signal and noise spectrum at the output of the modulator.

ated by the digital $\Sigma\Delta$ modulator outside the signal band (Fig. 3(c)).

## III. ANALOG SECTION

A block diagram of the encoder analog section for a single channel is shown in Fig. 4. The input interface is an uncommitted op amp with I/O terminals available off chip for resistor-ratio gain setting (voice-band gain $=1+ R_F/R_{IN}$, adjustable over a 20-dB range). External capacitor $C_C$ is required to ac couple the input signal across the 200-k$\Omega$ resistor. Note that the network is operated by a single $+5$-V power supply and the 200-k$\Omega$ resistor is returned to an on-chip midsupply reference. The signal is subsequently low-pass filtered by an integrated second-order Sallen–Key network (3-dB frequency set at approximately 30 kHz), which provides anti-aliasing for the fol-

with $N=4$ and $f_s=32$ kHz. The residues of the images of the input signal spectrum in the 4–28-kHz band are removed by a low-pass IIR filter. The output signal is applied to a digital $\Sigma\Delta$ modulator with a 1-MHz sampling rate, each input sample being repeated 32 times. Images of the input signal spectrum are created at multiples of 32-kHz frequencies. They are attenuated by a sinc function as in (1) with $N=32$ and $f_s=1.024$ MHz (Fig. 3(b)). It is the task of the analog section to provide enough filtering to attenuate the residues of these images and the noise gener-

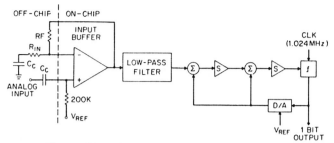

Fig. 4. Block diagram of the encoder analog section.

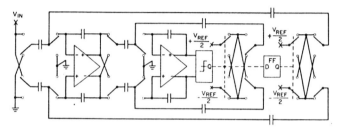

Fig. 5. Circuit diagram of the ΣΔ modulator.

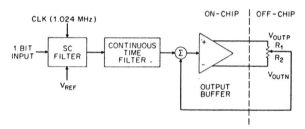

Fig. 6. Block diagram of the decoder analog section.

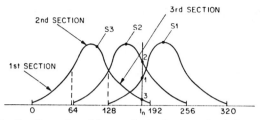

Fig. 7. Processing of the samples by the sinc cube decimator.

The filtered signal is then buffered by an output stage, which consists of a differential class AB power amplifier capable of driving 600-Ω loads and a gain control network. Decoder gain is adjustable by an off-chip resistor ratio over a 12-dB range; voice-band gain = $(1 + R_1/R_2)/(4 + R_1/R_2)$. Maximum output level is 6.0-$V_{p\text{-}p}$ differential with a common-mode voltage of 2.2 V dc.

All required reference voltages and bias currents for both the encoder and decoder are generated from an on-chip bandgap voltage/current reference. Link blowing at the wafer probe is used to trim (separately) encoder and decoder voice-band signal gain.

## IV. DIGITAL SECTION

The digital section uses a bit-slice architecture [4]. Its main blocks are described below.

### A. Sinc Cube Decimator

The $z$ transform of the sinc cube FIR filter is

$$H(z) = \left[ \frac{1 - z^{-64}}{1 - z^{-1}} \right]^3. \qquad (2)$$

The length of the sinc cubic filter is 192. Its coefficients are divided into three sections as shown in Fig. 7 and are given by

$$h_{1,n} = \frac{n(n+1)}{2}, \qquad \text{for } 0 \leqslant n < 64$$

$$h_{2,n} = 2080 + (n-64)(127-n), \qquad \text{for } 64 \leqslant n < 128$$

$$h_{3,n} = \frac{(191-n)(192-n)}{2}, \qquad \text{for } 128 \leqslant n < 192. \quad (3)$$

Because the impulse response of the filter is wider than the decimation factor, the decimator has to keep track of three output samples at the same time (Fig. 7, $t_n$). Depending on the value of the output bit coming from ΣΔ modulator the three corresponding coefficients are added to the respective samples or not. It must be observed that these three coefficients come from the three different sections defined by (3).

The sinc cubic FIR filter consists of two sections. The first computes the values of the coefficients, while the second accumulates these values to the output samples. The increments between two successive coefficients $h_{i,n}$

---

lowing double-loop ΣΔ modulator. The maximum input signal level is 3.0 $V_{p\text{-}p}$.

The encoder modulator, shown in Fig. 5, is a fully differential switched-capacitor implementation of the double-loop oversampled ΣΔ architecture described by Candy [2]. Differential capacitor switching performs single-ended-to-differential conversion at the modulator input. Feedback reference polarity is controlled by the comparator in the inner loop and the half-cycle clock-delayed flip-flop in the outer loop. Nonoverlapping clock phases for switch control are generated from the 1.024-MHz master sampling clock. The operational amplifiers in the switched-capacitor integrators are standard differential folded-cascode topologies [3], with open-loop gain of 60 dB and unity-gain frequency of 20 MHz. The comparator is a positive feedback, latch/reset type with a response time of 20 ns.

A block diagram of the decoder analog section for a single channel is shown in Fig. 6. The 1.024-Mbit/s serial bit stream generated by the digital ΣΔ modulator is filtered in two fully differential stages to attenuate the out-of-band quantization noise. The first stage is a first-order switched-capacitor low-pass filter with a 3-dB cutoff frequency of 64 kHz. The second stage is a 30-kHz second-order active smoothing filter.

and $h_{i,n-1}$ for the three sections $i = 1, 2, 3$ are

$$\Delta_1(n) = n = n_c, \quad \text{for } 0 \leqslant n_c < 64$$

$$\Delta_2(n) = 64 - 2(n - 64) = 64 - 2n_c, \quad \text{for } 0 \leqslant n_c < 64$$

$$\Delta_3(n) = -64 + (n - 128) = -64 + n_c, \quad \text{for } 0 \leqslant n_c < 64.$$
$$(4)$$

The coefficients of the three sections can be computed by adding the corresponding increments, which are linear functions of the pointer $n_c$, to their previous values. The initial coefficient values for the three sections are 0, 2080, and 2016, respectively. The increment $\Delta_1(n)$ is equal to the pointer $n_c$. $\Delta_2(n)$ can be generated directly from the pointer $n_c$. The least significant bit is a ONE; the next five bits are the complement value of the least significant bits of the pointer $n_c$ and the sign is given by the most significant bit of $n_c$. The input carry is forced to ONE when the addition to the previous coefficient value takes place. In a similar way, the increment $\Delta_3(n)$ is obtained by inserting a negative sign before the binary representation of $n_c$.

A block diagram of the coefficient generator is shown in Fig. 8(a). The values of the three coefficients $h_{1,n}$, $h_{2,n}$, and $h_{3,n}$ and of the pointer $n_c$ are stored in dynamic registers as shown in the figure. The operating frequency of the parallel adder is four times the sampling frequency of the $\Sigma\Delta$ modulator. The four phases of the adder operation are the incrementing of the pointer $n_c$ and the generation of the three coefficients. A coefficient is generated by adding its previous value to the value of the appropriate increment, which is generated through the manipulation of the bits of the binary representation of the pointer by the multiplexer as described above. The register is updated with the new computed coefficient value, which is also sent to the output section of the decimator.

The arithmetic unit of the output section (Fig. 8(b)) is synchronized with that of the coefficient section. It is inactive during one of the four phases. During the other three phases, the coefficients of the three sections multiplied with the value of the input coming from the modulator (this is a simple AND operation) are added to the respective samples stored in the registers. The samples are rotated from one register to another. In Fig. 8(b) the status of the registers during one of the phases is shown in parentheses. The output of the last register is returned to one of the inputs to the adder.

### B. IIR Filters

The IIR filter uses a parallel arithmetic unit, which is time-shared between several second-order IIR filter sections, and which is running at a much higher frequency (Fig. 9). There are two such blocks: one is for the encoder low-pass and high-pass filters, and the other is for the decoder low-pass filter. The arithmetic unit consists of a parallel adder and two accumulator registers, $A_y$ for stor-

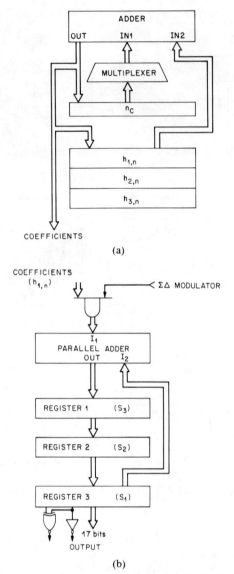

Fig. 8. Architecture of the sinc cube decimator. (a) Coefficient section. (b) Output section.

ing the new state variable $Y$, and $A_0$ for the output $Y_0$. The filter coefficients are stored in a ROM with one row location of the ROM containing all five coefficients $a_s$, $a_2$, $a_1$, $b_2$, and $b_1$ of a second-order IIR section. The selection of a particular coefficient is performed by the multiplexer MUX2. During the multiplication interval, the arithmetic unit is controlled by logic which implements Booth's algorithm. The coefficients are loaded into a register and shifted serially controlling the parallel addition/subtraction of the signal words. The state variables corresponding to a second-order section are stored in the same row of the RAM, and the state variable used for computation is selected by the multiplexer MUX1.

Usually the control unit of such a parallel arithmetic unit consists of an instruction ROM, an instruction sequencer, and instruction decoding circuitry. For the implementation of IIR filters this unit can be simplified significantly. The differences in the implementation of two independent second-order IIR filters consist only of the

314

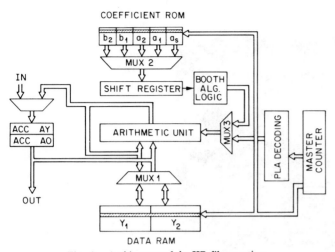

Fig. 9. Architecture of the IIR filter section.

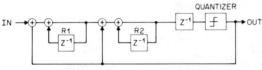

Fig. 10. Block diagram of the digital $\Sigma\Delta$ modulator.

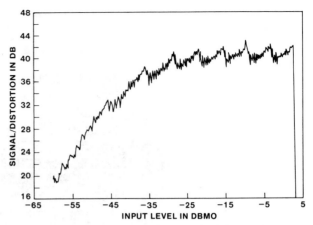

Fig. 11. Signal-to-noise characteristic of the encoder.

different set of coefficients and the different state variables. In other words, the succession of arithmetic operations is the same in both cases, and only the ROM and RAM memory locations with which the arithmetic operations are executed are different. Therefore, signals to control the arithmetic unit can be generated by a PLA which decodes the time slots of a master counter. This counter is incremented at the same frequency as the time-shared arithmetic unit described above. The memory is organized in such a way that all the data necessary for a second-order filter computation is stored in a row of the ROM (the coefficients) and in a row of the RAM (the state variables). Therefore, both the ROM and the RAM rows can be enabled using the same address. This address is generated by the most significant bits of the master counter. For instance, if each coefficient is 11 bits wide, 64 time slots are necessary to execute the operations of a second-order IIR section. (There are five multiplications; 11 time slots are necessary for each for a radix-2 Booth's algorithm implementation. Additional time slots are required for the additions and the updating of the RAM state variables.) Therefore, the six least significant bits of the master counter are decoded by the PLA into control signals to the arithmetic unit. The most significant bits of the master counter (starting from bit seven) address the ROM and the RAM, and their number depends on the number of second-order IIR sections necessary for the implementation of the digital filter function required. If this number is small, then the master counter could directly generate the row enable signals, so the RAM decoder and the ROM decoder are eliminated. In the case of multiple arithmetic units implementing digital filter functions in parallel, the control signals coming from the PLA and the most significant bits of the master counter can be shared between these units.

This parallel implementation of the digital filter has several advantages. It is programmable; the frequency characteristic can be modified for different applications by reprogramming the coefficient ROM. This requires a change of only one processing (photomask) level. Also for higher bit resolution, the width of the signal words can be increased by adding more bit slices to the arithmetic unit, without modifying the control logic.

### C. Digital $\Sigma\Delta$ Modulator

The digital $\Sigma\Delta$ modulator uses the same double-loop configuration as the analog modulator. The integrators are implemented by add and store structures (Fig. 10). The modulator uses a time-shared arithmetic unit, which has two phases of operation corresponding to the two integration operations. During the first phase the input signal is added to the content of the register $R1$. The result is used to update the register $R1$ and it is added to the content of the register $R2$. The quantization is achieved by taking the sign bit from the register $R2$. The 1-bit subtraction of the quantizer bit is done by bit manipulation of the most significant bit of the input signal for the first phase and of the input coming from the register $R1$ during phase two. Because of the finite register length, the values of the input signal and of the contents of the accumulators, normalized with respect to the quantizer step, are rational numbers. Because of this, when the input is a dc value (the most important case is idle code) limit cycles are generated [5]. These limit cycles are perceived as tones and are noticeable at levels much lower than the white noise or the noise characteristic of $\Sigma\Delta$ modulators. In order to prevent such limit cycles, a dither signal is added to the input of the modulator. This dither signal, which is the 1-bit output of a random pattern generator running at 1 MHz, is applied to the carry-in of the adder, during the first computation phase.

### V. Results

The device contains two codecs and the layout of one is the mirrored image of the other. The clocks which control the analog sections are delayed with respect of each other.

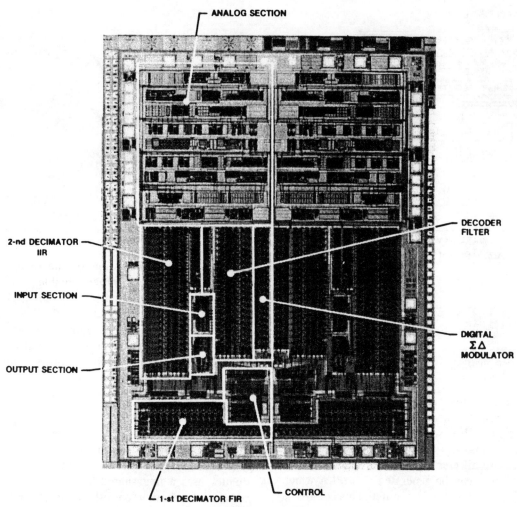

Fig. 12.   Photomicrograph of the device.

The average values for the crosstalk are:

encoder $A$ to decoder $A$     $-90$ dB
encoder $A$ to decoder $B$     $-95$ dB
encoder $A$ to encoder $B$     $-82$ dB
decoder $A$ to encoder $A$     $-76$ dB
decoder $A$ to encoder $B$     $-74$ dB
decoder $A$ to decoder $B$     $-79$ dB.

The values of the crosstalk from the decoder side are much higher because of the output buffer. The measurements were made for the worst case with 3 $V_{pp}$ in a single-ended configuration on a 300-$\Omega$ output load. These results become better if the current in the output load is reduced; for a small load current they become comparable with the crosstalk from the encoder.

The signal-to-noise characteristic of the encoder is shown in Fig. 11. For low input levels the resolution is limited by the rounding operation before the companding. For high input levels the resolution is limited by the companding itself. The idle channel noise measurements are 12 dBrnC0 for the encoder and 7 dBrnC0 for the decoder. A photomicrograph of the device is shown in Fig. 12. The device was fabricated in 1.5-$\mu$m technology and it requires only one 5-V power supply. The total area is 32 mm². It has a maximum power dissipation of 180 mW.

## VI. CONCLUSION

A dual-channel $\Sigma\Delta$ codec has been described in the paper. Its digital signal processing section has a bit-slice architecture which can be easily expanded to accommodate higher bit resolution. The crosstalk performance was better than $-71$ dB. If no large currents are required from the output drivers the crosstalk is better than $-80$ dB. This shows that the conversion techniques using $\Sigma\Delta$ modulation can be used successfully in the implementation of devices which process signals coming from different sources.

## REFERENCES

[1]  J. C. Candy, "Decimation for sigma delta modulation," *IEEE Trans. Commun.*, vol. COM-34, pp. 72–76, Jan. 1986.
[2]  J. C. Candy, "A use of double integration in sigma delta modulation," *IEEE Trans. Commun.*, vol. COM-33, pp. 249–258, Mar. 1985.
[3]  P. R. Gray and R. C. Meyer, "MOS operational amplifier design—A tutorial overview," *IEEE J. Solid-State Circuits*, vol. SC-17, pp. 969–982, Dec. 1982.
[4]  V. Friedman *et al.*, "A bit-slice architecture for sigma-delta analog-to-digital converters," *IEEE J. Selected Areas Commun.*, vol. 6, no. 3, pp. 520–526, Apr. 1988.
[5]  V. Friedman, "The structure of the limit cycles in sigma delta modulation," *IEEE Trans. Commun.*, vol. 36, no. 8, pp. 972–979, Aug. 1988.

## MOS ADC-Filter Combination That Does Not Require Precision Analog Components*

*Max W. Hauser, Paul J. Hurst\*\*, Robert W. Brodersen*

*University of California*

*Berkeley, CA*

AN OVERSAMPLE-AND-DECIMATE ARCHITECTURE yielding a analog-to-digital interface system with a built-in mask-programmable digital antialias filter in standard 5V digital MOS technology will be reported. Unlike other MOS A/D converter techniques, this approach requires no precision component ratios or precision comparator, nor a double-polysilicon or other special capacitor structure. Prototype devices display approximately 12b converter linearity at an 8kHz output rate with a single 5V power supply and a total die area of 4 8mm$^2$. The low supply voltage and simplified process requirements are attractive for scaled MOS fabrication technologies. This is intended as a linear (non-companding) analog-interface block for constant-sampling-rate applications such as speech processing.

The system is based on 1b A/D conversion, downsampled 256 times by a 1024-point, finite-impulse-response (FIR) digital lowpass filter. Such a configuration can display a net resolution exceeding the performance of the internal analog components, and since the bulk of the circuitry is digital, it benefits from continued technology scaling. It differs also from *interpolative* A/D circuits[1], in that it requires no internal D/A converter, nor does it produce a coarser quantization with larger inputs.

A delta-sigma-modulator front end, used elsewhere in discrete[2], bipolar[3] and passive grounded-capacitor MOS[4] circuitry, has been realized with a parasitic-insensitive switched-capacitor integrator; Figure 1. The parasitic insensitivity allows

————
*Research supported by DARPA contract N00034-K-0251, and by Fannie and John Hertz Foundation.

\*\*Current address: Silicon Systems, Inc., Nevada City, CA.

[1] Candy, J.C., Ninke, W.H., and Wooley, B.A., "A Per-Channel A/D Converter Having 15-Segment $\mu$-255 Companding", *IEEE Trans. Communications*, Vol. COM-24, No. 1, p. 33-42; Jan. 1976.

[2] Everard, J.D., "A Single-Channel PCM Codec", *IEEE Journal of Solid-State Circuits*, Vol. SC-14, No. 1, p. 25-37; Feb., 1979.

[3] van de Plassche, R.J., "A Sigma-Delta Modulator as an A/D Converter", *IEEE Trans. Circuits and Systems*, Vol. CAS-25, No. 7, p. 510-514; July, 1978.

[4] Misawa, T., Iwersen, J.E., Loporcaro, L. J., and Ruch, J. G., "Single-Chip per Channel Codec with Filters Utilizing $\Delta$-$\Sigma$ Modulation", *IEEE Journal of Solid-State Circuits*, Vol. SC-16, No. 4, p. 333-341; Aug., 1981.

[5] Candy, J.C., Ching, Y.C., and Alexander, D.S., "Using Triangularly Weighted Interpolation to Get 13-Bit PCM from a Sigma-Delta Modulator", *IEEE Trans. Communications*, Vol. COM-24, No. 11, p. 1268-1275; Nov., 1976.

[6] Ruetz, P.A., Pope, S.P., Solberg, B., and Brodersen, R.W., "Computer Generation of Digital Filter Banks", *ISSCC DIGEST OF TECHNICAL PAPERS*, p. 20-21; Feb., 1984.

good performance from metal/polysilicon capacitors, and the basic circuit requires two capacitors, of similar size, which need not be accurately ratioed.

This front end operates at a sampling rate much higher than the input signal frequencies. It produces a 1b signal that can be expressed as a linear representation of the input (between +$V_{REF}$ and -$V_{REF}$) along with a noise component having a highpass spectrum. A special-purpose digital lowpass filter operates on this 1b signal to remove frequencies above the desired signal bandwidth, and hence most of the quantization noise. Its output is a multibit digital representation at a lower (decimated) sampling rate.

Basic delta-sigma modulator systems generate severe noise components for certain input values; previous designs circumvented this by biasing the input to a low-noise region[4,5] or introducing a squarewave dither signal at a frequency that aliased to dc upon decimation[2]. These methods are unsatisfactory for a general-purpose system that must accommodate dc inputs. Here, a squarewave dither frequency within the decimating filter's stopband randomizes the delta-sigma noise as desired, but does not propagate to the output nor preclude arbitrary dc inputs.

Figure 2 shows the circuit of the switched-capacitor front end in 3$\mu$ P-well CMOS technology. Its active area is 0.3mm$^2$ and its power consumption approximately 2mW. A folded-cascode integrator gain stage drives a single-ended regenerative comparator. In the $\phi_1$ clock phase, the integrator output settles and precharges the comparator input. In the $\phi_2$ phase, the comparator's positive feedback loop is closed and it latches according to the polarity of the integrator output with respect to a threshold. Regenerative comparators, although fast, are rarely used in analog circuits since their offset voltage is difficult to cancel. In this application, the offset voltage is unimportant, since it does not affect the converter's performance, but merely changes the average voltage at the integrator output. Figure 3 is a die photograph of the front-end circuit.

We have fabricated both CMOS and NMOS front-end designs and concluded that CMOS is far preferable, because of the need for speed and gain in the integrator op amp. Nevertheless, the digital filter is realized as a separate NMOS chip in our prototype circuits, to exploit computer-assisted layout tools[6] now being extended to CMOS.

The architecture of the digital filter (Figure 4) exploits the 256:1 decimation factor and 1b digital input. A 1024-point impulse response, of which a symmetric half is stored in a read-only memory, is distributed to four accumulators. No explicit multiplications are necessary because of the one-bit input signal.

The impulse-response coefficients were developed by computer optimization to satisfy the dual objectives of quantization-noise removal and antialias filtering. The prototypes yield a minimum alias rejection of 46dB, consistent with com-

Reprinted from *ISSCC Dig. Tech. Pap.*, pp. 80-81, 313, Feb. 1985.

puter simulation, and only a simple single-pole prefilter is required before the 2MHz analog input sampling. Because FIR lowpass filters are relatively insensitive to coefficient roundoff, six-bit coefficients are adequate to define the impulse response. The die photograph of Figure 5 shows an area of $4.5mm^2$ in $4\mu$ silicon-gate NMOS technology. We expect a $3\mu$ CMOS version, currently under development, to occupy similar area.

The raw 1b delta-sigma output spectrum from dc to 200kHz appears in Figure 6. Desired signal frequencies extend from dc to 4kHz; the rest is unwanted quantization noise, to be removed by the digital filter. A 500Hz input sinusoid produces a 0dB peak very near the lefthand axis. Smaller peaks on the left are

the fundamental and third harmonic of a 16kHz squarewave dither signal; these too will be removed by the filter. Note the highpass trend in the quantization-noise spectrum.

Figure 7 shows the measured signal-to-noise ration performance. Input is a 500Mz sinusoid; input sampling rate is 2MHz and the output rate is 8kHz. The A/D system maintains a 30dB SNR over a 40dB amplitude range and approaches within 2dB the performance of a trimmed conventional 12b converter tested under the same conditions. Since the oversampled system exhibits its errors as ac noise, rather than dc nonlinearity, an SNR/amplitude plot is a more meaningful measure, for the intended signal-processing applications, than the static linearity specifications associated with standard (dc) A/D converters.

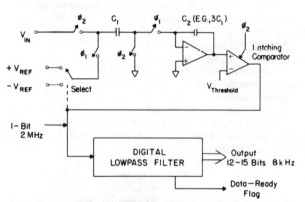

FIGURE 1—System overview showing switched-capacitor input section.

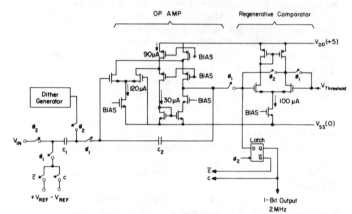

FIGURE 2—Schematic of CMOS switched-capacitor delta-sigma modulator.

*[See page 313 for Figures 5, 6, 7.]*

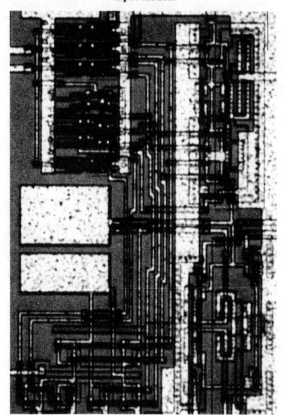

FIGURE 3—Photograph of CMOS switched-capacitor delta-sigma modulator. Dimensions are 0.45mm x 0.65mm.

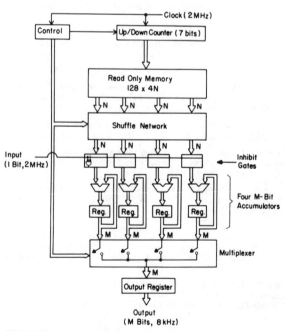

FIGURE 4—Architecture of 1024-point FIR digital filter. Typically N = 6, M = 15.

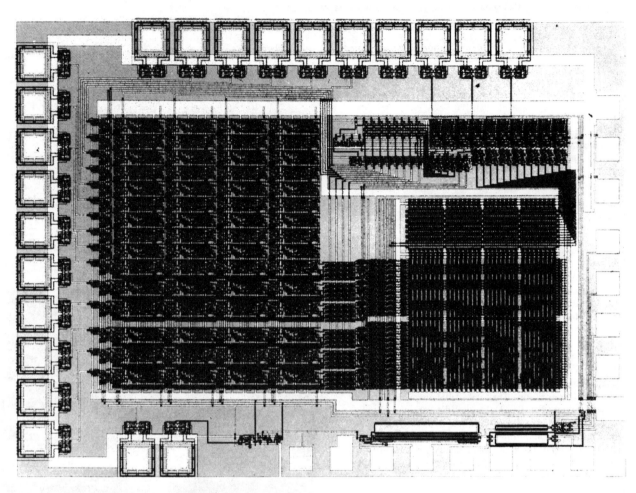

FIGURE 5—Photograph of 1024-point FIR digital filter.
Dimensions are 1.5mm x 3.0mm.

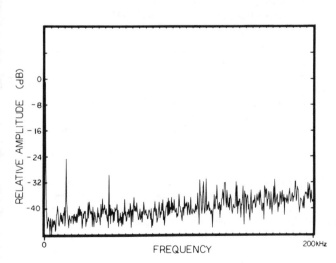

FIGURE 6—Low-frequency detail of 1b modulator output
spectrum.

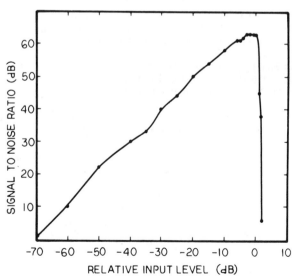

FIGURE 7—SNR performance of modulator-filter combination.
Input is a 500Hz sinusoid.

## A Multistage Delta-Sigma Modulator without Double Integration Loop

*Toshio Hayashi, Yasunobu Inabe, Kuniharu Uchimura, Tadakatsu Kimura*

*NTT Linear Integrated Circuit Section, Electrical Communications Laboratories*

*Kanagawa, Japan*

THE IMPORTANCE OF LOW-NOISE AD/DA CONVERTERS using delta-sigma modulators with smaller chip area has been recognized for such applications as PCM codecs and audio-signal processors. However, the original delta-sigma modulator[1] cannot achieve an adequate signal-to-noise ratio without double integration[2] or interpolation techniques[3]. These techniques require careful stability design or a high precision local DA converter.

This paper will cover the audio-range AD/DA converters with 14b equivalent precision using a multistage delta-sigma modulator. Both converters have been implemented by using 1.5μ silicon gate CMOS technology to integrate with a digital signal processor.

The concept for a multistage delta-sigma modulator for application to an AD converter is shown in Figure 1; DA concept is similar to AD. This converter consists of main and sub delta-sigma modulators. Each of these modulators includes a switched capacitor integrator, a comparator with one threshold level, and a 1b local DA converter. The main delta-sigma modulator converts analog input signals into a digital data sequence, and its integrator output is applied to the sub delta-sigma modulator.

The derivatives of the comparator output in the sub modulator are added to the main digital output. Thus, there is no feedback loop except for each integrator loop. The total digital output $D_{out}$ takes one of four level values (2, 1, 0, −1), because the output of both comparators is 1 or 0 and the derivatives are 1, 0, or −1. Although it seems peculiar that the amplitude of the derivatives is twice as great as the main digital output, the audio frequency range component is small enough to cancel the noise stemming from the main modulator. This is due to the derivation gain designed to be equal to the inverse of the main integrator gain.

The main digital output $D_m$, which is the same as the original delta-sigma, is given by the equation:

$$D_m = \frac{A_1}{A_1 + 1} A_{in} + \frac{1}{A_1 + 1} Q \qquad (1)$$

Here, $A_{in}$ is an input, $A_1$ is the gain of the main integrator given by $1/(1-z^{-1})$, and Q is the quantization noise by the comparator. This equation shows that the maximum (ideal) SNR value is limited to the order of $A_{in} \cdot A_1/Q$.

However, by using the sub delta-sigma modulator, the first-order quantization noise term is canceled at the summing point. As a result, the total output $D_{out}$ is given by the equation:

$$D_{out} = (1 - \frac{1}{A_1 A_2}) A_{in} + \frac{2}{A_1 A_2} Q \qquad (2)$$

where $A_2$ is the gain of the integrator in the sub modulator (similar to $A_1$). Thus, the SNR value ideally improves to the order of

[1] Everard, J., "A Single-Channel PCM Codec", *IEEE J. Solid-State Circuits*, Vol. SC-14, p. 25-37; Feb., 1979.

[2] Candy, J., "A Use of Double Integration in Sigma Delta Modulation", *IEEE Trans. Commun.*, Vol. COM-33, p. 249-258; Mar., 1985.

[3] Eriksson, G., et. al., "Line Circuit Component SLAC for AXE 10", *Ericsson Rev. 60*, p. 192-200; 1983.

[4] Ohno, T., et. al., "A Single-Chip High-Voltage Shallow-Junction BORSHT-LSI", *ISSCC DIGEST OF TECHNICAL PAPERS*, p. 230-231; Feb., 1984.

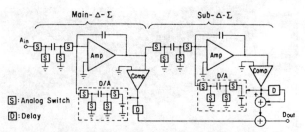

**FIGURE 1—A/D converter using multistage delta-sigma modulators.**

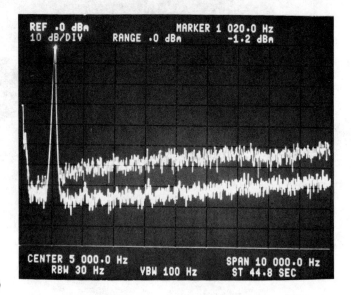

**FIGURE 2—Spectrum of the A/D converter. Input signal level is −22dBmo. Upper spectrum is for main delta-sigma output $D_m$; lower is for total output $D_{out}$.**

$A_{in} \cdot A_1 \cdot A_2/(2Q)$. It is, however, restricted by the circuit noise, the finite gain of the operational amplifiers, and the mismatch of the elements.

This circuit structure, which excludes a double integration loop and multi-bit local DA, is advantageous for circuit stability, analog circuit reduction, and tolerance increases.

The test device consists of AD/DA converters, RC active pre/post-filters, a band-gap reference supplying a 2.4V analog ground, and a PLL generating 2.048MHz clock. The analog signal levels are chosen to be 2.4V peak-to-peak to ensure a single 5V power supply.

The spectra of $D_m$ and $D_{out}$ shown in Figure 2 demonstrate the effectiveness of noise reduction by the multistage structure. Specifically, the noise floor of the $D_{out}$ spectrum is improved by 14dB from that of the $D_m$. In addition, the signal-to-noise ratio and gain tracking of AD and DA converters are plotted in Figures 3 and 4, respectively. Both figures show the 14b resolution required in analog signal processing for telecommunications use. The chip area of the AD and DA converters is only 2.2mm², and all analog sections are 7mm² as shown in Figure 5.

Reprinted from *ISSCC Dig. Tech. Pap.*, pp. 182-183, February 1986.

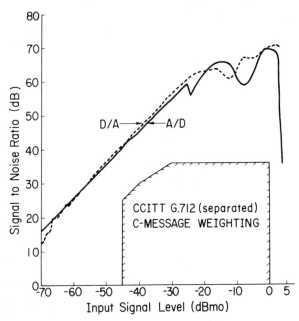

FIGURE 3—Signal-to-noise ratio of A/D and D/A converters.

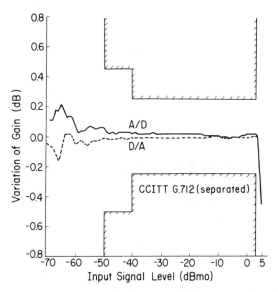

FIGURE 4—Gain tracking of A/D and D/A converters.

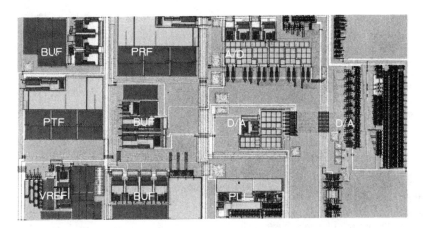

FIGURE 5—Photomicrograph of all analog circuit sections.

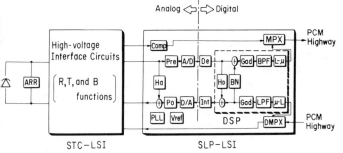

FIGURE 6—Block diagram of BORSCHT circuits. SLP is Subscriber Line Interface Processor; STC is Subscriber Line Termination Circuit.

The AD/DA converters presented show promise for a continuing development of audio signal processor LSI Subscriber Line Interface Processor (SLP) with line termination, two-to-four wire conversion, and channel filters. Moreover, these converters, in addition to the high voltage analog interface LSI Subscriber Termination Circuit (STC), utilizing shallow junction technology[4] implementing ringing, testing, and battery feeding functions, will certainly advance the development of highly-packaged and low-cost subscriber line interface circuits. These dual chip BORSCHT circuits can be installed as shown in Figure 6.

*Acknowledgments*

The authors would like to thank H. Mukai, T. Sudo and A. Iwata for their encouragement.

# AN OVERSAMPLED SIGMA-DELTA A/D CONVERTER CIRCUIT
## USING TWO-STAGE FOURTH ORDER MODULATOR

Teppo Karema,  Tapani Ritoniemi  and  Hannu Tenhunen

Signal Processing Laboratory
Tampere University of Technology
P.O.Box 527, SF-33101 Tampere, Finland

*Abstract - A sigma-delta A/D converter realization using two-stage 4th order modulator architecture and 5th order digital running-sum decimation filter is presented. The analog part of the converter consist of two cascaded second order modulators. Scaling is used between the sections in order to achieve modest requirements for component matching and integrator's gain and phase. A digital running-sum filter is used for the decimation to $4f_S$ or $2f_S$. A dedicated 7 instruction filter processor is designed to perform the final decimation and I/O-communication. The whole system operates on a single 5 V operation voltage.*

## I. INTRODUCTION

Within the past years, the technological improvements in digital VLSI circuits have dramatically increased the need of low cost high resolution A/D and D/A converters. Completely new problems have risen for the analog interfaces in embedded systems because the A/D conversion and the digital signal processing are on the same chip. The use of Nyquist rate A/D converters in such systems limits the resolution below 12 bits because most of the injected noise from the digital part is aliased into the baseband. In addition, their system integration is complex because effective anti-alias filters, high performance sample and hold circuits and jitter-free timing are needed.

Many of the problems mentioned above can be solved using over-sampling technique. Sigma-delta modulation is one of the most promising candidates because it combines oversampling and quantization noise shaping and digital filtering in order to achieve high performance with reduced analog circuit complexity. Sigma-delta converter has excellent linearity because it uses one-bit quantization. The noise coupling from the digital part is small because most of the noise is digitally filtered in the decimation filter. No sample and hold circuit is needed and a simple RC-filter prevents the aliasing of unwanted input signals. Moreover, most of the timing uncertainty becomes noise that spreads equally over the whole band and only a fraction of it lies in the baseband.

This paper presents a high performance sigma-delta A/D converter using 4th order modulator architecture [1], 5th order running-sum decimator and a dedicated filter processor (Figure 1). The filter processor provides complete programmability for the converter's amplitude response. No general multipliers are used which results small silicon area. The converter is designed in such way that it can be integrated using a low priced analog or digital CMOS process with capacitor structures. This makes it suitable for all kinds of high volume digital audio, telecommunication and instrumentation applications. The converter achieves theoretically 18-bit resolution for the audio band at 2.8 MHz sampling frequency as depicted in Figure 2. The measurements with the first prototype indicate that this limit can be achieved in practice by minor improvements in the prototype's analog hardware.

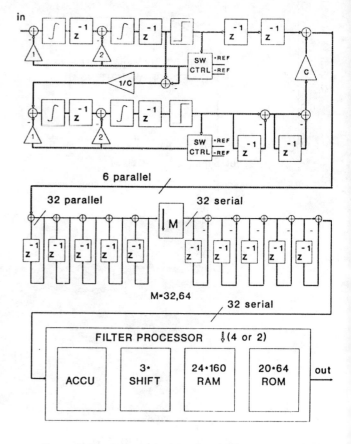

Figure 1. Block diagram of the sigma-delta converter.

## II. MODULATOR

The 4th order noise shaping is produced by cascading two conventional 2nd order modulators [3]. In this way the stability of the whole 4th order system is that of the 2nd order section. No dither signal is required. Scaling is used between the sections in order to achieve modest requirements for the component matching and integrator's gain and phase. Scaling results about 15 dB lower amplifier gain requirement than cascading 1st order sections [4]. In addition, the relative accuracy of the capacitors can be much lower (only 10 % needed). The 4th order modulator was specified to achieve true 16-bit accuracy in 20 kHz baseband with oversampling ratio of 64. Complete documentation of the design principles is provided in [1]. At this moment, 16-bit dynamic range and 14-bit linearity have been measured for the 2nd order subsection of the cascaded structure with single 5 V operation voltage (Figure 3). The prototype's limiting factors were found to be the modulator's 1/f-noise and the clocking of the switches which results signal dependent charge injection.

Reprinted from *IEEE Proc. ISCAS'90*, pp. 3279–3282, May 1990.

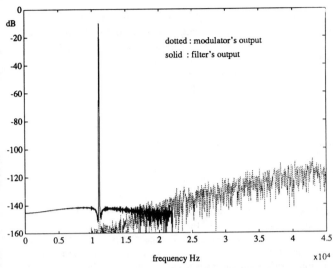

Figure 2. Simulated baseband spectra before (dotted) and after (solid) running-sum decimation filter with sinusoidal -10 dB input and 2.8 MHz sampling frequency. The FFT was computed over $2^{17}$ samples using 4-term Blackman-Harris window.

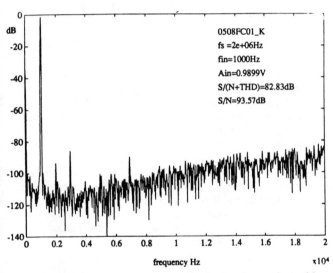

Figure 3. Measured spectrum of the 2nd order section with sinusoidal -1.3 dB input. The FFT was computed over $2^{17}$ samples using 4-term Blackman-Harris window. The signal-to-noise ratio was calculated to 0-4 kHz bandwidth.

## III. RUNNING-SUM FILTER

The decimation filter suppresses the quantization noise generated by the modulator and attenuates those frequencies outside the baseband which have passed the modulator. One of the major obstacles is to implement the decimator in small silicon area. The simplest linear phase FIR decimation filter is constructed by averaging N samples. This operation is very advantageous because it places zero at the final sampling rate and each harmonics of it. However, for the 4th order modulator multiple zeros are required to prevent the aliasing during the decimation [5]. This can be done using an efficient running-sum architecture that performs higher order averaging by separating the recursive and nonrecursive terms of the transfer function [2]. The decimator contains no multipliers and has very few data memory locations, thereby making it easily VLSI realizable.

There are four main modules in the implementation of the running-sum decimator: 1) The IIR2 module containing two feedback loops, 2) the IIR1 module having one feedback loop and parallel to serial register for the decimation, 3) the FIR1 module performing the feedforward loop and 4) the CONTROL module generating the overall control signals. All modules are implemented automatically from the system level specification using GDT module generator software (Silicon Compiler Systems Inc.). The organization of the running-sum decimator is shown in Figure 4. The IIR2 and IIR1 modules use parallel arithmetics and so the one-bit leaf cells are replicated to achieve the specified bit accuracy. The FIR1 differentiator operates in a serial format and only the dynamic flip-flop in the delay loop is replicated for the specified bit accuracy. The CONTROL module consist of several parametrized submodules. The first two submodules on the left bottom of Figure 4 provide the selection of the decimation ratio by choosing whether or not the incoming clock signal is divided by two. The CLK2GEN module generates the nonoverlapping two-phase clock. The COUNT module counts upwards from the parametrized value. When the counter overflows, a one clock period pulse appears in the output and the counter is reloaded to the parametrized value. The DELAY module is a shift register which shifts the pulse from the previous module a parametrized number of the clock periods. In the actual realization, the decimation ratio was specified to 32 or 64. A 32 bit accuracy was chosen for the IIR2, IIR1 and FIR1 modules because in that way the modulo arithmetic range is not exceeded in the filter. In addition, it provides the clocking for the serial data without additional hardware.

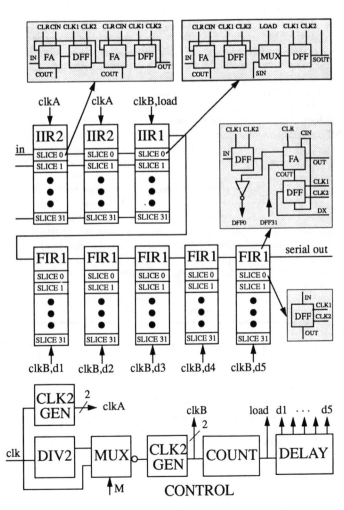

Figure 4. Organization of the running-sum filter.

## IV. FILTER PROCESSOR

When high performance is needed, the running-sum filter can be used only to $2f_S$ or $4f_S$ in order to have small passband attenuation and to have good rejection for input signal frequencies from $f_S$ to $4f_S$. The final decimation must be done using very high order filters. Until now, these filters have been implemented using 16-bit multipliers accompanied with coefficient ROMs and shift registers [6] [7] [8] [9]. Although the realization of those filters is straightforward, it is not optimum when the reusability, chip area and flexibility are considered. To explore the reusability, a dedicated filter processor was designed. The processor's instruction set (Table 1) was optimized for the algorithms reported in [2] and [10]. These have a restricted coefficient memory in such way that the multiplication is done by three adjustable shifters. This results a very efficient VLSI realization for the multiplier. It was found by investigating the algorithms that the processor needs only 5 main instructions for the filtering. However, two instructions were added to implement the subfilters of the algorithm by the subroutines in the program code.

The architecture of the filter processor is shown in Figure 5. The processor has a 24-bit wide RAM as a data storage element, 20-bit wide ROM for the program, and three shifters and one accumulator for the multiplication. The subroutine branch mechanism reduces the original ROM size approximately to 1/3. The selection of the subfilter is controlled by the SEG register, the position in the subfilter by the OF1 register and the position in the ring buffer by the OF2 register. The sum of the OF1 and OF2 registers overlaps 4 bits with the SEG register. In that way, the size of the subfilter can be externally configured to 4, 8, 16, 32, or 64 words. All address registers are 8-bit or less wide. The multiplication is performed by three parallel shift-right registers which are controlled directly by the bits of the program code. The shifted outputs can be added or subtracted from the accumulator. The multiply instruction takes two clock cycles. During the first one the data is transferred to the TMP register and so the second clock cycle is left completely for the shift-add operation. The multiplication is pipelined so that the following instructions are in progress immediately. This means that effectively, all instructions take only one clock cycle. The embedded instructions shown in Table 1 are executed in parallel with each main instruction excluding the "mul" which can have only the "inc(of1)" instruction in parallel. The control of the processor is simple because most of it is derived directly from the program code. The synchronization of the processor to the converter's conversion rate is accomplished by interrupts. The IN and OUT interfaces are 24-bit wide SIPO and PISO registers which have external bit rate clock.

A test chip was implemented using 2.5 μm molybdenum gate CMOS high performance analog process with molybdenum metal capacitors (Micronas Inc. of Finland). The size of the 4th order modulator is about 2 mm² and the 5th order running-sum decimator occupies about 5 mm². The layout of the filter processor is estimated to be less than 10 mm². This means that the area of whole A/D converter circuit is less than 20 mm². The transistor packaging density for the running-sum decimator is 1900 transistors/mm² although automatic layout generator was used. The layout of the modulator and the running-sum filter is shown in Figure 6.

## V. CONCLUSIONS

A sigma-delta A/D converter realization using two-stage 4th order modulator architecture and 5th order digital running-sum decimation filter operating on a single 5 V voltage was presented. Modest requirements were found for the modulator's component matching and integrator's gain and phase. The target technology has no special requirements which implies possibility for mass production using low priced analog or digital CMOS process with capacitor structures. The modulator's output was decimated in the running-sum filter only to $4f_S$ or $2f_S$ in order to have flat passband and good attenuation for the frequency components which are in the beginning of the stopband. The converter operates in a wide frequency range from few kHz to 3 MHz. An 18-bit resolution can be achieved at oversampling ratio of 64. The scaling of the target technology is efficient because most of the converter is digital. A dedicated filter processor was designed for the final filtering. The processor's architecture was tailored for the multiplier-free algorithms reported earlier. The processor explores the reusability because it can be used in various oversampled A/D and D/A converters. The processor has efficient use of the silicon area because it has high density RAM and ROM modules and add-shift multiplier. A simple one clock cycle instruction set provides high operating frequency. This means that in many cases, several modulators can be connected to the same processor. The converter makes the converter suitable for various ASIC applications because it provides a completely programmable amplitude response for the converter. For these reasons, we conclude that the presented system is a very good alternative for all kinds of digital audio, telecommunication and instrumentation applications.

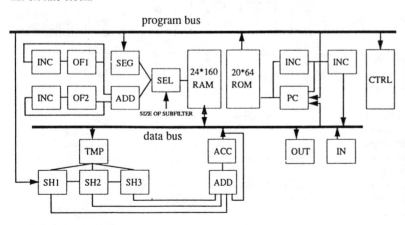

Figure 5. Architecture of the filter processor.

**MAIN INSTRUCTIONS**

| | instruction | operation | plus |
|---|---|---|---|
| 1 | inp | ram ← in | a-f |
| 2 | store | ram ← acc | a-f |
| 3 | mul(sh1,sh2,sh3) | acc ← acc+ram×(sh1,sh2,sh3) | d |
| 4 | out | out ← ram | a-f |
| 5 | jmp(const) | pc ← rom | a-f |
| 6 | savepc | ram ← pc | a-f |
| 7 | ret | pc ← ram | a-f |

**EMBEDDED INSTRUCTIONS**

| | | |
|---|---|---|
| a | clr(of1) | of1 ← 0 |
| b | clr(of2) | of2 ← 0 |
| c | cir(acc) | acc ← 0 |
| d | inc(of1) | of1 ← of1 + 1 |
| e | inc(of2) | of2 ← of2 + 1 |
| f | seg(const) | seg ← rom |

Table 1. Instruction set of the filter processor.

## REFERENCES

[1] T. Karema, T. Ritoniemi H. Tenhunen, "Fourth Order Sigma-Delta Modulator Circuit for Digital Audio and ISDN Applications", in *Proc. IEE European Circuit Theory and Design,* Sep 1989, pp. 223-227.

[2] T. Saramäki, H. Palomäki and H. Tenhunen, "Multiplier-Free Decimators with Efficient VLSI Implementation for Sigma-Delta A/D Converters", in *Proc. 1988 VLSI Signal Processing Workshop,* VLSI Signal Processing III, IEEE Press, 1988.

[3] T. Ritoniemi, T. Karema, H. Tenhunen and M. Lindell, "Fully Differential CMOS Sigma-Delta Modulator for High Performance Analog-to-Digital Conversion with 5 V Operating Voltage", in *Proc. IEEE Int. Symp. on Circuits and Systems,* June 1988, pp. 2321-2326.

[4] M. Rebeschini, N. Bavel, P. Rakers, R. Greene, J. Caldwell and J. Haug, "A High-Resolution CMOS Sigma-Delta A/D Converter with 320 kHz Output Rate", in *Proc. IEEE Int. Symp. on Circuits and Systems,* May 1989, pp. 246-249.

[5] W. Chou, P. Wong and R. Gray, "Multistage Sigma-Delta Modulation", *IEEE Trans. on Information Theory,* vol. 35, pp. 784-796, July 1989.

[6] E. Dijkstra, L. Cardoletti, O. Nys, C. Piguet and M. Degrauwe, "Wave Digital Decimation Filters in Oversampled A/D Converters", in *Proc. IEEE Int. Symp. on Circuits and Systems,* June 1988, pp. 2327-2330.

[7] K. Uchimura, T. Hayashi, T. Kimura and A. Iwata, "Oversampling A-to-D and D-to-A Converters with Multistage Noise Shaping Modulators", *IEEE Trans. on Acoustics, Speech and Signal processing,* vol. 36, pp. 1899-1905, Dec 1988.

[8] "DSP56ADC16 16-Bit Sigma-Delta Analog-to-Digital Converter", *Technical data sheet DSP56ADC16/D,* Motorola Inc.

[9] D. Welland, B. Del Signore, E. Swansson, T. Tanaka K. Hamashita, S. Hara and K. Takasuka, "A Stereo 16-bit Delta-Sigma A/D Converter for Digital Audio", in *Proc. 85th Convention of the Audio Engineering Society,* Nov 1988.

[10] T. Saramäki, T. Karema, J. Isoaho and H. Tenhunen, "VLSI-Realizable Multiplier-Free Interpolators for Sigma-Delta D/A Converters", in *Proc. Int. Conf. on Circuits and Systems,* July 1989, pp. 60-63.

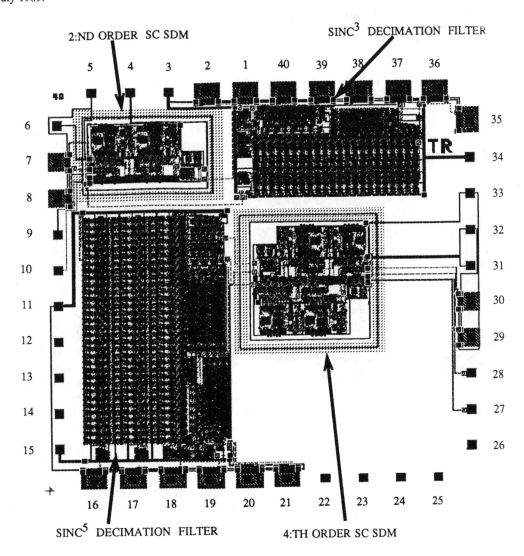

Figure 6. Layout of the fabricated prototype. Uppermost is a 2nd order modulator (1.5 mm²) with 3rd order running-sum decimator (3 mm²), bottom is the presented 4th order modulator (2 mm²) with 5th order running-sum decimator (5 mm²).

# A 12-bit Sigma-Delta Analog-to-Digital Converter with a 15-MHz Clock Rate

RUDOLF KOCH, BERND HEISE, FRANZ ECKBAUER, EDUARD ENGELHARDT,
JOHN A. FISHER, MEMBER, IEEE, AND FRANZ PARZEFALL

*Abstract* —This paper presents a sigma-delta analog-to-digital converter that achieves 12-bit integral and differential linearity and nearly 13-bit resolution without trimming. The baseband width is 120 kHz with a first filter pole at 60 kHz, clock frequency is 15 MHz, and only one 5-V power supply is needed. The circuit was realized in a p-well CMOS technology with 3-μm minimum feature size. Compared with sigma-delta modulators published up to now [1], [2], the input signal frequency and clock rate limit have been increased by one order of magnitude. To achieve this increase, a new integrator concept was developed using bidirectional current sources. The circuit is fully self contained, requiring only a 15-MHz crystal and one blocking capacitor as external elements. This converter was developed as the analog front end of a digital echo cancellation circuit for an integrated services digital network (ISDN).

## I. INTRODUCTION

IN an integrated services digital network (ISDN), full-duplex transmission of digital data with a rate of 160 kbit/s over standard two-wire lines is required for the so-called U interface. Either a ping-pong (burst) or a hybrid balancing method is feasible. Line attenuation increases dramatically for higher signal frequencies and is further degraded at longer lengths. It is therefore important to use the transmission method that needs the smallest bandwidth. The hybrid balancing approach requires by a factor of three less bandwidth than the burst method. The bandwidth needed is further reduced by the use of a ternary block code.

In the hybrid balancing method, both parties transmit data simultaneously. The separation of the transmitted from the received signal is achieved with a hybrid. Due to incomplete matching of the hybrid to the line characteristics, a crosstalk between the transmit and receive path occurs. This crosstalk signal can be 35 dB higher in amplitude than the signal received from the far end. Some additional means is therefore needed to remove this crosstalk from the received sum signal. This is done by means of adaptive echo cancellation. In addition to the crosstalk, signal line echos, e.g., from bridge taps, are also cancelled. Our approach employs a digital, adaptive, linear echo canceller (see Fig. 1). Since nonlinearities of the analog blocks (ADC, DAC, transmitter) cannot be compensated

Manuscript received May 9, 1986; revised July 27, 1986. This work was supported in part by the Federal Department of Research and Development, West Germany.
The authors are with Siemens AG, D-8000 Munich 80, Germany.
IEEE Log Number 8610949.

for, the most stringent demand on the analog parts is a linearity of over 70 dB. This demand, together with the large baseband width, makes the development of the analog parts a very challenging task.

## II. ANALOG-TO-DIGITAL CONVERTER

### A. Functional Principle

A rough specification of the ADC is given in Table I. The extremely high demands on linearity of the ADC, the relatively large baseband width, and the requirement to avoid trimming cannot be fulfilled in a CMOS environment using standard approaches such as simple capacitor or resistor networks. Although one solution might be a self-calibrating ADC, we felt that a sigma-delta modulation converter of second order was a better choice. This choice was proven to meet system specifications and the design parameters were optimized using extensive simulation on the system level.

The functional principles of second-order sigma-delta modulation have been described by Candy [3] and others. A simplified schematic is given in Fig. 2. Two integrators in the forward path, two 1-bit DAC's in the feedback path, and a 1-bit ADC, i.e., a comparator, are used. The dominant noise source of this modulator is the quantization noise of the 1-bit ADC. Oversampling and noise shaping are used to move this noise from the baseband to higher frequencies. Oversampling evenly distributes the quantization noise over the frequency range from dc to half the clock frequency. A noise improvement of 3 dB per octave of oversampling is therefore achieved. The feedback loop contains two integrators, resulting in a high-pass characteristic of second order for the quantization noise. With this filtering, the S/N ratio is improved by an additional 12 dB per octave. With oversampling and filtering, a gain of 15 dB in S/N ratio is theoretically achievable with each octave of oversampling. Our simulations showed that by using ideal elements, a clock rate of 15 MHz resulted in a SNR slightly above 80 dB with a 120-kHz bandwidth. This provides some safety margin in the actual circuit. It is important to notice that the noise shaping is only valid for the quantization noise of the comparator, not for the linearity errors of the DAC's in the feedback path. The performance of the outer DAC is especially crucial for the

Reprinted from *IEEE J. Solid-State Circuits*, vol. SC-21, pp. 1003–1010, December 1986.

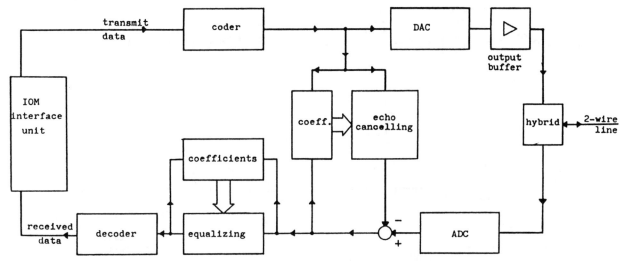

Fig. 1. Simplified block diagram of a digital echo canceller.

TABLE I
ADC SPECIFICATION

| Power supply | 5 V |
|---|---|
| Power consumption | < 20 mW |
| Signal/(noise + distortion) | > 72 dB |
| Baseband width | 3–120 kHz |

performance of the whole modulator. The DAC's linearity must be at least as high as the linearity of the complete circuit reduced by the oversampling ratio. Another problem is the existence of high-frequency noise in the circuit as a natural consequence of the noise shaping principle. Nonlinearities in the circuit give rise to intermodulation products that alias into the baseband.

The output signal of such a sigma-delta modulator is a high-frequency 1-bit signal (in our case the bit rate is 15 MHz). This 1-bit signal contains the analog input signal as its mean value. Digital low-pass filtering is needed to remove the high-frequency noise from the signal. The order of the filter should be higher by one than the order of the noise shaping function. The specified transfer function of that filter is shown in Fig. 3 and can be described by the equation

$$H = (z^0 - 2z^{-64} + 2z^{-192} - z^{-256})/(1 - z^{-1})^3.$$

The filter is realized as a cascade of a 256-stage four-tap FIR part and a third-order IIR part. The complete filter structure is clocked with 15 MHz. At the output, standard multiple bit words with 13 relevant bits are available with this clock rate. Decimation is performed in a separate stage to reduce the data rate to 120 kHz. The low pass and decimation filter were integrated on a separate chip [4]. The performance of a sigma-delta modulator depends heavily on the filter used. All data are only valid for a given filter characteristic.

*B. Measured Results*

Since the digital 1-bit output signal contains the analog input signal as a mean value, one could apply a low-pass

filter between the PDM output and a spectrum analyzer to measure the performance of the modulator. With this approach, however, the output buffer acts as a 1-bit DAC and must have a performance superior to that of the modulator. The design of such a buffer is a rather complex task. Instead, digital samples are taken of the PDM signal. For the measured results given in Figs. 4–7 and in Table II, 4096 samples per measurement were taken with a 15-MHz clock rate and a fast Fourier transform (FFT) was performed to calculate the spectrum. Fig. 4 shows a wideband idle channel output spectrum. No filtering of the signal was performed so the high-frequency noise and the noise shaping behavior can be clearly seen from this measurement. Fig. 5 shows the baseband output spectrum with an 11-kHz sinusoidal input signal of 1-$V_{\text{eff}}$ amplitude. A digital low-pass filter, which was integrated on a separate chip, was used for this measurement. The decrease of noise towards higher frequencies is caused by a first pole of the filter at $f = 60$ kHz. The output word of this filter was fed into a test computer and a FFT was performed to calculate the output spectrum.

There is one line of particular interest close to 7.5 MHz in Fig. 4. With zero input and zero offset voltage to the ADC, the output signal would be a periodic stream of ones and zeros corresponding to exactly half the clock frequency, i.e., 7.5 MHz in our case. With a small dc input voltage there would be a few more ones or zeros in the stream that would show up as discrete lines in the spectrum around dc, 7.5 MHz, 15 MHz, and so on. A line in the baseband would reduce the S/N ratio. This effect is a significant problem with sigma-delta modulators of first order where an additional dither signal is often used to smooth out these discrete lines. Our simulations, as well as our measurements, showed that in a modulator of second order, these discrete lines are so far damped and decorrelated [3] that no dither signal is needed. Some measured results are given in Table II. For these measurements a separate filter chip was again used. The output of the filter was fed into a test computer where a FFT was performed and S/THD and S/N were calculated. An 11-kHz sinusoidal input signal was applied to the ADC as described above.

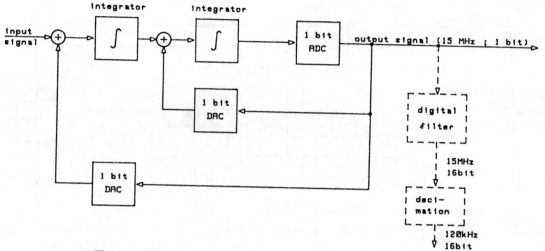

Fig. 2.   Principle circuit schematic of a second-order sigma-delta modulator.

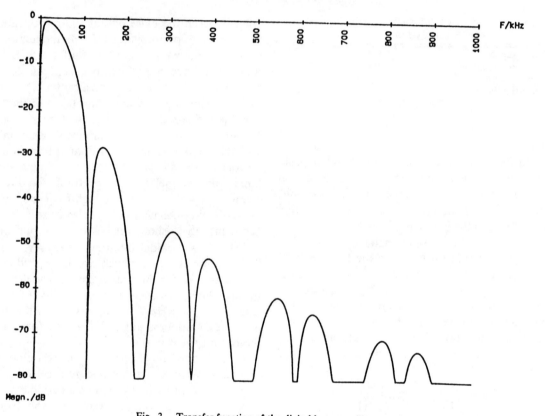

Fig. 3.   Transfer function of the digital low-pass filter.

Figs. 6 and 7 show the S/(N + THD) as function of the input signal amplitude and frequency, respectively. The output signal is evaluated with the low-pass filter and a FFT is performed on the test computer. The measurements for Fig. 6 were taken with an input signal frequency of 11 kHz, as above. With increasing input amplitude, the S/(N + THD) increases until harmonic distortion becomes out of spec. The measurements for Fig. 7 were taken with constant input signal amplitude of 1 $V_{eff}$. The decrease of the S/(N + THD) above 3 kHz is caused by the first pole of the input stage amplifier. The steeper decrease above 60 kHz results from the digital low-pass filter with its first pole at $f = 60$ kHz. For a 100-kHz input signal of 1-$V_{eff}$

amplitude we get S/(N + THD) = 55 dB. Once again it is stressed that these data are only valid together with the filter characteristic specified above.

C. Integrator Concept

Two integrators are needed for a sigma-delta modulator of second order (see Fig. 1). Standard realizations of integrators have traditionally used either a resistor–capacitor network or a switched-capacitor configuration. For the modulator described here, the maximum input frequency to these integrators originates from the feedback path and is in the megahertz range. The clock frequency is 15 MHz.

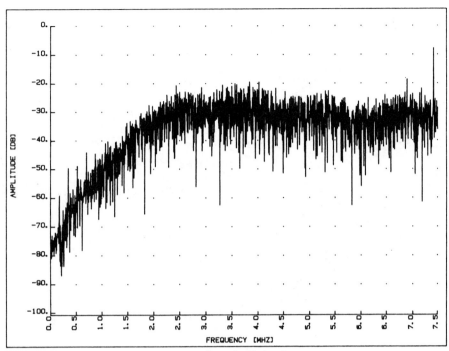

Fig. 4. Measured idle channel wide-band output spectrum.

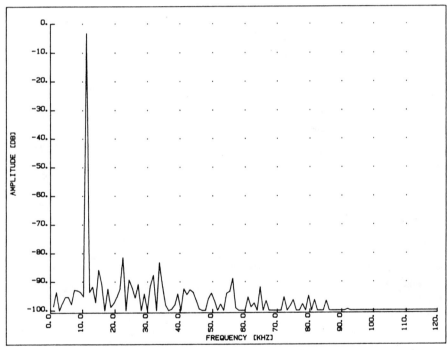

Fig. 5. Measured baseband output spectrum.

In a switched-capacitor integrator with a clock rate of 15 MHz, op amps with about 100-MHz bandwidth and accordingly high power consumption would be needed to ensure proper settling. On the other hand, no output stage is required for these circuits and fast one-stage amplifiers can be employed. Bandwidth requirements would be greatly reduced, compared to the SC approach, with a standard *RC* integrator. To drive the resistive load, a two-stage amplifier would be required. This type of op amp is much slower than a one-stage amplifier and would exhibit prob-

lems of insufficient slew rate and bandwidth. A new circuit configuration was developed that combines the advantages of both approaches. The principle circuit schematic of the sigma-delta modulator employing this new approach is given in Fig. 8. The integrators are built around bidirectional current sources. With this approach a much smaller bandwidth is needed for the op amps as would be needed for an SC integrator. According to system simulation where linear op amps were assumed, 4 MHz of bandwidth would be sufficient for 80 dB of S/N. Measurements using

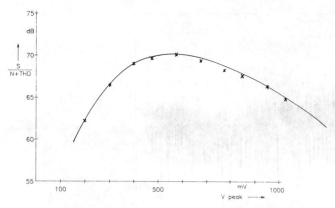

Fig. 6. Measured S/(N + THD) as function of signal amplitude.

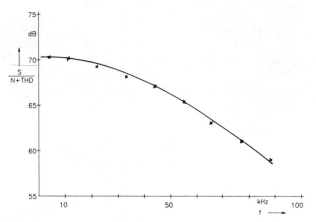

Fig. 7. Measured S/(N + THD) as function of signal frequency.

TABLE II
ADC PERFORMANCE—MEASURED RESULTS

| Clock rate | 15.36 MHz |
|---|---|
| Baseband width | 3–120 kHz |
| Eq. input offset voltage | ±3 mV |
| Signal/noise | 77 dB |
| Signal/total harmonic distortion | 72 dB |
| Power consumption | 15 mW |

variable bandwidth amplifiers showed, however, that 12 MHz is actually needed. We believe the reason for this difference is the presence of frequency-dependent nonlinearities. The op amp has a load of only one transistor gate, hence a one-stage amplifier with high slew rate and large bandwidth can be used. The 1-bit DAC's in the feedback path must be switched current sources in order to be compatible with the bidirectional current sources used in the integrators.

One disadvantage of this approach, compared with SC solutions, is the larger variance of the time constants due to resistor tolerances. The relative variation of the resistors that define the signal current and the feedback current causes different integrator gains for the feedforward and feedback paths. Thus a gain error is introduced into the ADC. However, this gain error is compensated for by a digital AGC in the digital part of the echo canceller. The absolute tolerance of the resistors effects the transfer function of the ADC which, under extreme conditions, can cause stability problems. According to extensive system

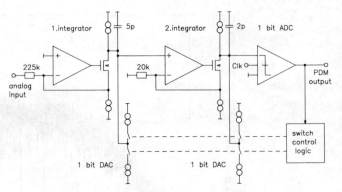

Fig. 8. Sigma-delta modulator realization with current sources.

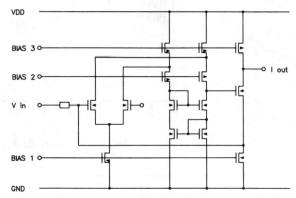

Fig. 9. Circuit schematic of the bidirectional current source.

simulation, a variation in the $RC$ product of $\pm 30$ percent is permissible without stability problems and without degradation of the S/N ratio.

## III. BASIC BUILDING BLOCKS

The basic building blocks of the sigma-delta modulator, the bidirectional current sources, the switched current sources (1-bit DAC's), and the comparator (1-bit ADC) will now be described. In addition to these blocks, a bandgap voltage reference, a biasing network, and an oscillator circuit were also integrated.

### A. Bidirectional Current Source

The linearity and S/N ratio of the integrators, especially of the outer one, are of crucial importance for the performance of the complete modulator. The integrators in this circuit receive signals from the feedback path that can be as high as 7.5 MHz. The linearity of the integrators must be high to avoid intermodulation products and to ensure a S/(N + THD) of over 70 dB over the 120-kHz baseband. Power consumption of the op amp must be low because a limit of 20 mW was set for the whole modulator. A good solution to these requirements is the use of a bidirectional current source as the core of the integrator.

Fig. 9 shows a circuit schematic of this current source that is built around a folded cascode op amp. The additional output stage contains two constant current sources of equal value that force a bias current through the output

330

### TABLE III
#### Op-Amp Performance—Measured Results

| | |
|---|---|
| Open-loop gain at dc | 80 dB |
| Unity-gain bandwidth | 20 MHz |
| Phase margin | 63° |
| Slew rate | 20 V/$\mu$s |
| Power supply rejection at 1 kHz | 55 dB |
| Common-mode rejection at 1 kHz | 80 dB |
| Random offset voltage | $\pm 1.5$ mV |
| Thermal input noise | 25 nV/sqrt (Hz) |

### TABLE IV
#### Bidirectional Current Source—Measured Results

| | |
|---|---|
| Loop gain at dc | 90 dB |
| Unity-gain bandwidth | 18 MHz |
| Random offset voltage | $\pm 3$ mV |
| Maximum signal current | 4.5 $\mu$A resp. 7.5 $\mu$A |
| Power consumption | 1.2 mW |

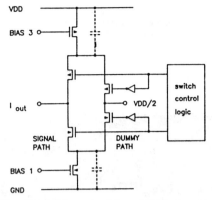

Fig. 10. Circuit schematic of the switched current source.

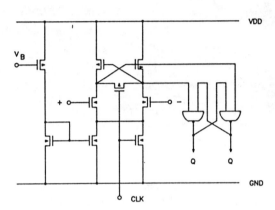

Fig. 11. Circuit schematic of the comparator.

transistor. The input voltage is converted to an input current of positive or negative sign through resistor $R$. This input current flows as additional current through the output transistor. The current is fed in at the source node and flows out of the output stage at the drain node. In this way, the current through the output transistor is modulated as function of the input voltage. The bias current of the output stage has to be sufficiently higher than the maximum signal current to ensure a low harmonic distortion. With a fast high-gain op amp, a current ratio of about two is sufficient.

The output impedance of the current sources forms a leakage path parallel to the integration capacitor. A too low value of leakage resistance reduces the low-frequency gain of the ADC and, consequently, the noise floor at low frequencies is increased. The minimum value permissible for the leakage resistance is 5 M$\Omega$ according to our simulation.

The equivalent offset voltage of the ADC given in Table I is caused in equal parts by the offset voltage of the core amplifier and by mismatch of the upper and lower current sources in the outer integrator. Measured results for the folded cascode core amplifier and the bidirectional current source are given in Tables III and IV.

### B. Switched Current Source

Since bidirectional current sources are used in the feed-forward path, switched current sources are required in the feedback path to realize the 1-bit DAC's. The linearity and S/N ratio of the outer DAC must be superior to the required performance of the complete modulator. The requirements of the DAC, however, are relaxed by the oversampling effect [3]. In our application, a S/(N + THD) of about 55 dB was needed.

One problem inherent to switched current sources is the charging and discharging of parasitic capacitances. The voltage change on the parasitic capacitor prevents fast switching of the current and would cause a loss of charge on the integration capacitor. To achieve fast switching, no loss of charge, and a high internal resistance of the current

sources, a differential configuration was chosen as shown in Fig. 10. The current is never switched off but shifted between the two branches. The switching transistors are operated in the saturation region to guarantee a high internal resistance of the current source, very much like the differential input stage of an op amp. With this configuration, the parasitic capacitance at the common source node of the switching transistors is kept at a constant voltage and thus causes no loss of charge.

No relevant measured results could be obtained for this switched current source since all attempts at probing were unsuccessful because of loading effects. From the performance of the whole modulator, however, one can conclude that the S/(N + THD) of the current sources is better than the 55 dB mentioned above.

### C. Comparator

The third important building block is the 1-bit ADC or comparator. Since it is embedded in two high-gain feedback loops with the switched current sources (see Figs. 2 and 6), its requirements on resolution and offset voltage are moderate (in the range of several tens of millivolts). With a 15-MHz clock, however, the response time is of utmost importance. A latch-type comparator is a good choice for this application. As shown in the circuit schematic in Fig. 11, the comparator consists of a differential stage with a latch as load. In the reset mode, the nodes of the latch are shorted to set it to the astable high-gain

TABLE V
COMPARATOR PERFORMANCE—MEASURED RESULTS

| Resolution | 2 mV |
|---|---|
| Response time | 15 ns |
| Random offset | ±3 mV |
| Averaged power consumption | 1.5 mW |

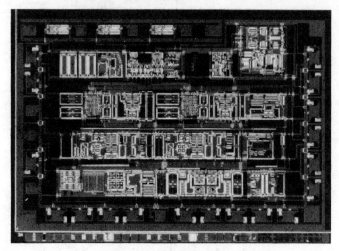

Fig. 12. Chip micrograph.

mode. In the compare mode, these nodes are released and additional current is fed to the differential stage to enable fast switching of the latch. Measured results are given in Table V.

*D. Chip Architecture*

A chip micrograph is shown in Fig. 12. The chip area of this experimental device is about 3 mm$^2$. It was realized in a 3-$\mu$m p-well CMOS technology. A second layer of polysilicon was used for poly1–poly2 capacitors and for high ohmic resistors. A channel length of 6 $\mu$m was frequently used to avoid short-channel effects.

A structured layout was used in which the circuit is organized into several rows. The bottom row contains the static analog parts, i.e., the bandgap voltage reference, two buffer amplifiers for the reference voltages, and a bias current network. The second row from the bottom contains the midfrequency analog parts, i.e., the bidirectional current sources and a biasing network. The next row up contains the high-frequency switched current sources and the associated bias network. The top row consists of an oscillator and small logic block which run at 15 MHz and two comparators. Only one of the comparators was used in the circuit. The other one was for backup purposes.

## IV. SUMMARY

A sigma-delta modulator realized in a 3-$\mu$m CMOS technology has been presented in which 12-bit linearity is achieved without trimming. Baseband width and clock rate are one order of magnitude higher than in modulators described up to now. This high-frequency range was obtained by implementation of a new integrator concept, in which requirements of the op amps were significantly reduced compared with *RC* or SC integrator approaches.

## ACKNOWLEDGMENT

The authors wish to thank Dr. E. Schmid for his suggestions to the basic conept of the modulator and for many stimulating discussions.

## REFERENCES

[1] H. L. Fiedler and B. Hoefflinger, "A CMOS pulse density modulator for high resolution A/D converters," *IEEE J. Solid-State Circuits*, vol. SC-19, pp. 995–996, Dec. 1984.
[2] M. W. Hauser, P. J. Hurst, and R. W. Brodersen "MOS ADC-filter combination that does not require precision analog components," in *ISSCC Dig. Tech. Pap.*, Feb. 1985, pp. 80–81.
[3] J. C. Candy, "A use of double integration in sigma delta modulation," *IEEE Trans. Commun.*, vol. COM-33, pp. 249–258, Mar. 1985.
[4] A. Huber *et al.*, "FIR lowpass filter for signal decimation with 15MHz clock frequency," in *Proc. ICASSP 86*, Apr. 1986.

# Area-Efficient Multichannel Oversampled PCM Voice-Band Coder

BOSCO H. LEUNG, MEMBER, IEEE, ROBERT NEFF, PAUL R. GRAY, FELLOW, IEEE, AND
ROBERT W. BRODERSEN, FELLOW, IEEE

*Abstract* —This paper describes the design of a four-channel oversampled A/D converter with transmit filter for voice-band application. The decimation filter is time-shared between the four channels and the architecture of the sigma–delta coder is selected on the basis of minimizing the chip area. The analog front-end loop is fully differential to minimize the channel-to-channel crosstalk. The key issues in designing multichannel oversampled ADC for area efficiency will be addressed. The proper choice of the coder as well as the filter architecture are discussed. It is concluded that for the multichannel telephony voice-band application implemented in CMOS technology, a first-order loop is most area efficient. The performance of the coder has been evaluated and it has a dynamic range of 79 dB, occupies a total active area of 33 000 mils$^2$ or 8250 mils$^2$ per channel, and meets the D3 specifications for the transmit filter. It runs on a 5-V supply and consumes 50 mW per channel.

## I. INTRODUCTION

BECAUSE of their compatibility with scaled digital processes, oversampled techniques are particularly attractive for the implementation of A/D interfaces for voice-band telecommunication applications. In comparing the oversampled techniques with the conventional analog approach in this application, the key issue is that of minimizing the total chip area. Oversampled coders require a large percentage of the circuit area to implement the digital filter. These digital circuits can easily take advantage of technology scaling and, perhaps more importantly, they can be shared without crosstalk. Considerations of die area create a strong motivation to use a multichannel time-shared filter on a single chip to minimize the per-channel area. Crosstalk between analog circuits from different channels can be reduced by using separate analog front ends, which adds little extra area because the analog circuits make up only a small percentage of the total chip area. In the past, considerations of crosstalk and die area have precluded a multichannel single-chip approach, and previous implementations of both sigma–delta and successive-approximation coders for this application have used a single-channel architecture [1]. In this paper, an experimental four-channel oversampled voice-band coder in a 3-$\mu$m, single-poly, double-metal CMOS technology is described. This prototype achieves a dynamic range of 79 dB, idle channel noise of 12 dBrnC0, channel-to-channel crosstalk of −83 dB, power supply rejection ratio of 44 dB at 1 kHz and more than 60 dB at 20 kHz, and has a frequency response that meets the D3 specifications in a silicon area of 8250 mils$^2$ per channel. The total power consumption is 200 mW, or 50 mW per channel. This power consumption comes mainly from the separate analog loops, and can be powered down if the specific channel is not in use.

This paper consists of three additional parts. In Section II, architectural considerations unique to the design of a multichannel coder are addressed. Section III goes into descriptions of the various circuits used in the prototype. Finally the experimental results will be discussed in Section IV.

## II. CHOICE OF CODER IMPLEMENTATION FOR MULTICHANNEL OVERSAMPLED PCM CODERS

For voice-band telephony applications, first- and second-order sigma–delta coders are of the most practical interest. These coders are shown schematically in Fig. 1. A sigma–delta loop consists of one or two integrators and a 1-bit A/D and 1-bit D/A. The loop generates a 1-bit code whose density is proportional to the input level. A digital low-pass filter averages this signal and generates a digital output. As shown in Fig. 2 both loops have an output quantization noise spectrum that has a high-pass characteristic.

In Fig. 2, $f_s$ is the input sampling frequency, $f_D$ is the downsampled frequency, and $f_o$ is the highest frequency in the passband. The oversampling ratio $D$ is defined as the ratio $f_s/f_D$.

In per-line components for voice-band telephony, per-line component cost is a critically important parameter.

Manuscript received July 11, 1988; revised August 29, 1988. This work was supported by DARPA, the California MICRO Program, Bellcore, Fairchild, AMD, Rockwell International, and Honeywell.

B. H. Leung was with the Electronics Research Laboratory, Department of Electrical Engineering and Computer Sciences, University of California, Berkeley, CA 94720. He is now with the Department of Electrical Engineering, University of Waterloo, Waterloo, Ont., N2L 3G1, Canada.

R. Neff was with the Electronics Research Laboratory, Department of Electrical Engineering and Computer Sciences, University of California, Berkeley, CA 94720. He is now with IBM, San Jose, CA 95193.

P. R. Gray and R. W. Brodersen are with the Electronics Research Laboratory, Department of Electrical Engineering and Computer Sciences, University of California, Berkeley, CA 94720.

IEEE Log Number 8824594.

Reprinted from *IEEE J. Solid-State Circuits*, vol. SC-23, pp. 1351–1357, December 1988.

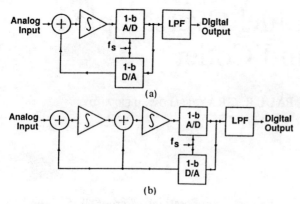

(a)

(b)

Fig. 1.   (a) First-order coder. (b) Second-order coder.

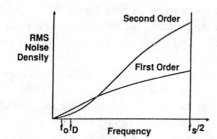

Fig. 2.   Noise spectra of sigma–delta coders.

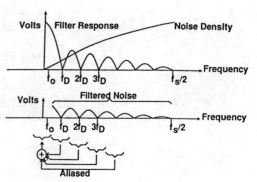

Fig. 3.   Out-of-band quantization noise suppression.

An important objective of the work described here was to reduce the silicon area per channel in the coder function to the lowest possible value. While second-order sigma–delta coders are most frequently used in single-channel applications because a lower sample rate can be used, the realization of multichannel PCM interfaces in a single chip brings about a different set of trade-offs in silicon area minimization which tend to make a first-order implementation more attractive than would be the case for a single-channel coder. As part of this work an area comparison was made to determine the relative area consumptions of first-order and second-order coders in a multichannel implementation.

Since the decimation filter takes up the vast majority of the area of the overall coder, the comparison of the filter area required in the two types of coders is of central importance. The decimation filter has three functions:

1) out-of-band quantization noise suppression;
2) out-of-band signal anti-aliasing; and
3) maintaining the passband ripple to be less than the value dictated by transmission requirements, usually 0.25 dB for typical telephony codecs.

Since the band-edge frequency of the voice channel is only 3 kHz, the digital filtering required to maintain the passband shape can be most efficiently carried out at low sample rate after decimation. As a result this circuitry can be time-shared over the multiple channels and has an area contribution that is small on a per-channel basis. The requirement for signal anti-aliasing can be considerably simplified by incorporating a multistage filter. The second-stage filter can use decimation so that it can then be time multiplexed. Consequently the area of the circuitry

required to do signal anti-aliasing can be made relatively small.

The out-of-band quantization noise suppression must be performed at the input sampling frequency, and as a result the hardware required to achieve this suppression cannot be easily time multiplexed in the technology used for this work. This in turn implies that a critical issue for the resulting area per channel in the final implementation is the means of realizing this initial decimation function. As shown in Fig. 3, the noise density from the sigma–delta loop increases with frequency. This is filtered by the first-stage filter but after decimation the unfiltered out-of-band noise is aliased into the baseband. The filter requirement is to reduce this noise to a level that is small compared to the quantization noise that is in-band before decimation. The area of the filter to achieve this is strongly influenced by the choice of the order of the coder.

A realistic comparison of the area required for the decimation filter required for first- and second-order coders is quite complex and must necessarily be made under a set of assumptions that might be violated in some innovative future design. In the work described here, such a comparison was made in order to compare the areas of such filters as they would be implemented in the prototype described later. These assumptions are as follows.

a) In the telephony codec application, the first-order coder implementation utilizes a $sinc^2$ filter with an $L/D$ of 2, a decimation ratio of 512, and a sampling rate of 4 MHz. The second-order coder utilizes a $sinc^3$ filter with an $L/D$ of 3, a decimation ratio of 128, and a sampling rate of 1 MHz [2]. These filter configurations and sample rates meet the dynamic range and idle channel requirements of PCM telephony with comparable margins. These particular FIR filter configurations have been previously shown to reduce out-of-band aliased noise to a value that is small compared to in-band noise for the two coder cases. Other decimation filters that meet the telephony transmission requirements are certainly possible. For example, use of a second-order loop operating at a relatively high sampling rate with a $sinc^2$ filter gives a coder whose in-band noise is dominated by aliased out-of-band noise which is capable of meeting the noise requirements. Such configurations do not appear to have great promise for having minimum area and were not investigated further.

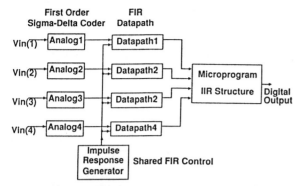

Fig. 4. Block diagram of four-channel oversampled PCM coder.

b) An FIR implementation of the decimation filter is used. Other configurations such as the IIR decimation filter are possible, but since the filter is a decimation filter it is best implemented as an FIR filter, since computation is only needed at the output downsampled frequency $f_D$ instead of the input sampling frequency $f_s$. This does not in itself rule out IIR filters, since there is a class of IIR filter that has a form in which the denominator of the transfer function has only terms that are powers of $D$ (that is, the denominator is a polynomial in $z^{-D}$) where $D = f_s/f_D$ is the decimation ratio. This will also enable the computation of the IIR filter to be performed at the $f_D$ rate. However, in a sigma–delta loop the input signal to the decimation filter is a 1-bit code so that only an AND gate is needed instead of a multiplier to form the product of the input and coefficients. This can be used in an FIR filter to eliminate the need for a multibit multiplier. However, in the IIR filter, because there is feedback there are multiplications in the feedback loop that are between two multibit words. Consequently the IIR filter implementation usually requires significantly more area and FIR filtering was chosen for this work for the first-stage filtering.

c) The overload point of the coder is defined as that point at which the large-signal S/N + D ratio is 40 dB. This impacts the dynamic range and required oversampling ratio of second-order implementations because such coders display significant degradations in SNR due to overload as the signal level approaches full scale.

d) The area of the ROM, PLA, up–down counter, or other circuitry used to develop the impulse response is negligible in the multichannel case because it is shared over many channels. Thus the area of this function has little effect on the comparison of the area of first-order versus second-order filters in the multichannel case.

e) In the sinc$^2$ filter a data-path width of 14 bits is required because the dc gain of a sinc$^2$ filter with an $L$ of 256 is approximately $2^{14}$. Similarly in the sinc$^3$ filter a width of 18 bits is required because the dc gain of a sinc$^3$ filter with an $L$ of 192 is approximately $2^{18}$.

f) The speed capability of the technology used is such that the first stage of decimation required in the digital filter cannot easily be time-shared over multiple channels.

For voice-band telephony interfaces, this corresponds to technologies down to about 2 $\mu$m.

Under the above assumptions the data path dominates the area. The area of the data path is primarily dependent on the number of additions per second that it must perform and the data width it must handle. This depends only on the $L/D$ ratio and the data-path width. Both are smaller for the sinc$^2$ case and hence it is likely to take less area under the assumptions above. Consequently the first-order coder was chosen for use in the particular prototype described here.

## III. PROTOTYPE DISCUSSION

A prototype of a four-channel voice-band oversampled A/D converter with transmit filter realizing a dynamic range of 79 dB and crosstalk of $-83$ dB has been fabricated in a 3-$\mu$m, double-metal, single-poly CMOS technology. The coder uses a first-order analog front end followed by a time-multiplexed digital filter. The coder oversamples at 4 MHz and downsamples to generate the output at 8 kHz. Fig. 4 shows the overall block diagram of a four-channel oversampled A/D converter. There are separate analog front ends for each channel to minimize the channel-to-channel crosstalk. Each front end is working at 4 MHz. The filter is divided into two stages, a high-speed FIR decimation filter followed by an IIR filter. The FIR filter has a shared control since the impulse response is the same for all four channels. The data path is separate for each channel because of the speed limitation imposed by a 3-$\mu$m technology. The FIR filter is working at 8 MHz. The IIR filter is time-shared and has a clock frequency of 4 MHz with an input sampling frequency of 16 kHz.

An additional problem with a first-order loop is that the quantization noise is correlated with the input, giving rise to noise components in the passband. Specifically, at low input level, the quantization noise spectrum exhibits tones that may fall into the passband. To decorrelate the quantization noise at low input level, a dither signal is applied. By adding a square-wave dither signal at the input, the tones in the passband will be phase modulated by the fundamental and harmonics of the square wave [3]. Consequently the tone energy inside the passband is reduced. The reduction of the tone energy depends in a complicated way on the modulation index, which in turn is a function of the dither amplitude. However, the addition of this square-wave dither signal reduces the dynamic range of the coder and so there is a trade-off in the desirable amplitude of the square wave. The dither frequency is chosen to lie at the zeros of the decimation filter so that the dither signal is removed by the filter. Under the above considerations and with the help of system simulations a square-wave dither signal with a frequency of 250 kHz and an amplitude of $-12$ dB from the full scale is used.

The implementation of the decimation filter will now be discussed. The filter needs to suppress the out-of-band quantization noise as discussed above. It also needs to have an anti-alias suppression of more than 33 dB and a

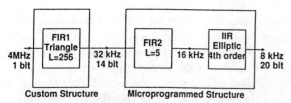

Fig. 5.   Filter structure.

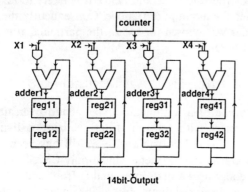

Fig. 6.   Block diagram of FIR1 filter.

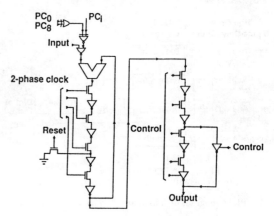

Fig. 7.   One-bit slice of FIR1 filter.

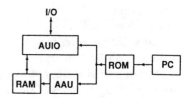

Fig. 8.   FIR2 and IIR filter implementation.

passband ripple of less than 0.25 dB. As discussed in Section II, an $L/D$ of 2 is required to ensure that the out-of-band quantization noise will be sufficiently suppressed. To satisfy the other two requirements the filter would need a higher $L/D$ ratio. Since the FIR filter computes at the high input sampling frequency $f_s$, its complexity should be minimized. By breaking the filter into multiple stages, the requirements can be met while keeping the $L/D$ ratio of the first-stage FIR at 2. The overall filter structure is shown in Fig. 5 where an FIR filter with a triangle impulse response takes a 4-MHz 1-bit input data. The 14-bit output is decimated to 32 kHz and fed into a second-stage FIR filter. The output is further decimated to 16 kHz and fed into the IIR filter where it is decimated to 8 kHz. FIR1 is implemented in a custom structure. FIR2 and IIR are implemented by a time-shared microprogrammed structure.

Next the prototype custom FIR1 design will be discussed. Fig. 6 shows the block diagram of the FIR1 filter. For this four-channel filter one counter provides all the coefficients. Although an FIR decimation filter that does not require an impulse response generator exists [4], [5], this approach does not have a particular advantage in area because the generator is shared over different channels. Furthermore, this approach does not lend itself easily to reduction in area by time multiplexing the data path without introducing extra registers and multiplexers. As shown in the figure, $X1$–$X4$ are the 1-bit output code from the four analog front ends. The distinct feature of this approach is that the adder in each channel is multiplexed. As will be shown later, careful design of the control structure allows this to be implemented in a simple fashion. The detailed implementation of the 1-bit slice of the adder and the registers is shown in Fig. 7. Here $PC_0$ and $PC_8$ are the LSB and MSB from the program counter, which is an up-counter. $PC_i$ is the $i$ bit of the counter. The

$PC_8$ signal is used to invert the counter output after it counts $D$ samples. It converts the up-counter into an up–down counter. As discussed in Section II, since $L/D$ equals 2, two sets of coefficients are needed. If two accumulators are used these two sets of coefficients have to be provided simultaneously. Since only one adder is used, they will have to be provided in consecutive clock cycles. Furthermore, since the impulse response is triangular, the two sets of coefficients are just the inverse of one another, leading to further simplification. In the circuit the $PC_0$ signal is used to invert the counter output between every clock cycle to generate these two sets of coefficients. There are altogether four registers. Two of the registers are used to hold state variables for the addition. The other two registers are used to hold the decimated outputs. These outputs are then latched and transferred to the next stage filter. The control signals are generated by ANDing a latch signal and the two-phase clock. The latch signal happens once every $D$ samples and it latches the new output as well as pushing the old output down like a FIFO. The reset signal is used to reset the FIR filter after the output has been latched.

The IIR filter and the FIR2 filter are both implemented using a microprogrammed architecture. The FIR2 filter is used to provide anti-aliasing around 16 kHz and the IIR filter is used to maintain the passband ripple as well as the stopband attenuation. The IIR filter has to compensate for the droop in the passband due to FIR1 and FIR2 filters while maintaining a ripple of less than 0.25 dB. In the stopband the suppression must be more than 33 dB at 4.6 kHz and beyond. The microprogram structure as shown in Fig. 8 is composed of the program counter (PC), the ROM for the program, the RAM for the state variables, the arithmetic unit and I/O unit (AUIO) for computations,

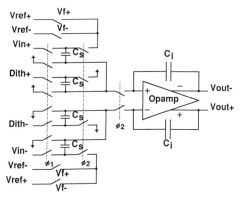

Fig. 9.   Prototype switched-capacitor integrator.

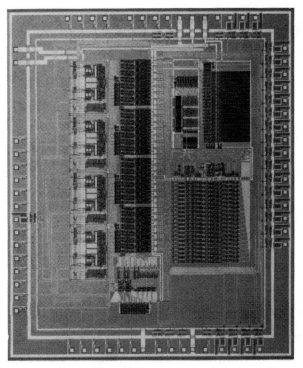

Fig. 10.   Chip photograph.

and the address arithmetic unit (AAU) for calculating addresses for state variables that belong to different channels. The basic hardware is generated by the Lager system [6], a silicon compiler that accepts an assembler program for the filter as the input and generates the layout of the filter. The hardware is heavily pipelined to increase throughput. To save area no hardware multiplier is used, and multiplications are achieved by shifting and adding. In order to reduce the number of shift and add cycles, thus improving the speed of the IIR filter and allowing for more multiplexing, the IIR filter coefficients are optimized by a program Candi [7] that generates the canonical signed digit (CSD) code with the fewest number of bits in the filter coefficients. The program uses a data path of 20 bits which ensures that the idle channel noise is low enough even in the presence of limit cycles. The ROM is 50 words by 26 bits and the RAM is 40 words by 20 bits. The microprogrammed structure runs at 4 MHz. In addition special modifications are introduced to achieve area efficiency for the present application. In particular the AAU is replaced by simple logic to calculate the addresses of the state variables for the different channels.

The switched-capacitor first-order analog front-end integrator is shown in Fig. 9. To minimize the crosstalk, four separate analog front ends are used. With the digital filter being integrated on the chip the power supply rejection is also a primary consideration. A fully differential architecture is used to reduce the crosstalk and improve the power supply rejection. Since the process has only one layer of poly, a metal2–metal1–poly capacitor was used. *Dith* +, *Dith* − are inputs for dithering, and *Vf* +, *Vf* − are outputs from the comparators. Furthermore the output digital pad buffers are switched only at the downsampled frequency to minimize switching transients.

Next the op amp and comparator design will be described. The op-amp gain requirement in a sigma–delta loop is determined by its effect on the quantization noise suppression at low frequency. The effect of the finite gain of the op amp is to modify the frequency response of the integrator from having a pole at dc to having a pole at some finite frequency. Moreover, the dc gain of an integrator is no longer infinite, but equals $A$, the op-amp gain. The effect of changing the frequency response of the

integrator is to modify the noise transfer function. This modification results in more quantization noise in the passband. From simulation it is determined that when the op-amp gain is less than the oversampling ratio, the increase in quantization noise becomes significant [8]. For the prototype the oversampling ratio is 512. Based on the above result it is determined that a gain of about 1000 is needed.

A fully differential folded cascode configuration with dynamic common-mode feedback was chosen. The common-mode feedback [9] is connected directly to the output node to improve the speed at the expense of having a lower gain. From simulations the amplifier has a typical gain of 1000 and settles to 0.01 percent in 75 ns with a 1.2-V differential step into a 1.2-pF load. To reduce the harmonic distortion due to signal-dependent charge injection the sampling switches are turned off sequentially, with the switches connecting to ground being turned off first.

Since the comparator offset does not affect the performance of the converter, a simple regenerative latch is used to achieve fast switching. In simulation the comparator latches correctly in 20 ns with an overdrive of 10 mV. A chip photograph of the core of a prototype is shown in Fig. 10.

## IV. EXPERIMENTAL RESULTS

The measured signal-to-noise ratio versus input amplitude curve is shown in Fig. 11. The signal-to-noise ratio of the converter was evaluated by putting a sine wave of 1.024 kHz into the converter, uploading the digital output to a computer, and then running a 1024-point FFT to compute the ratio of the signal power to the noise power.

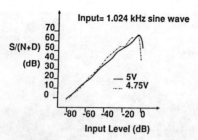

Fig. 11.   Measured signal-to-noise ratio versus input amplitude.

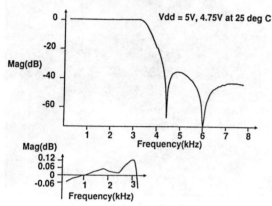

Fig. 12.   Measured frequency response.

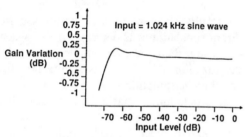

Fig. 13.   Measured gain tracking.

TABLE I
DATA SUMMARY
Typical performance: 5-V power supply and 25°C

| Area | 8250 sq. mils/ch |
|---|---|
| Dynamic range | 79dB |
| Idle channel noise | 12dBrnC0 |
| Crosstalk | -83dB |
| PSRR | 43dB (1 kHz) |
| | 60dB (10 kHz) |
| Harmonic distortion | -76dB (2nd) |
| | -82dB (3rd) |
| Power | 50mW/ch |

ratio between the measured noise and the input signal level at the overload point. The measured value was 12 dBrnC0 C-message weighted. In addition the variation in idle channel noise as a function of dc offset has been evaluated. A dc offset in increments of 5 mV and a range from $-100$ to $+100$ mV was applied to the coder and the measured idle channel noise was found to vary less than 1 dB. The anti-alias requirement was also tested at higher frequencies of 13 and 29 kHz to measure how effective the FIR1 and FIR2 filters are in removing frequency components around 16 and 32 kHz, respectively. At 13 kHz the signal is attenuated by 42 dB and at 29 kHz the signal is attenuated by 39 dB, both attenuations larger than the 33-dB requirement.

## V.   SUMMARY

This paper reports on a prototype four-channel over-sampled CMOS PCM voice-band coder that meets the D3 specifications in an area of 8250 mils$^2$ per channel. In summary, the prototype demonstrates that when compared to the conventional single-channel sigma–delta approach, a multichannel first-order sigma–delta coder in voice-band application can substantially reduce the chip area per channel in a scaled technology.

### ACKNOWLEDGMENT

The authors wish to thank Y.-M. Lin for his valuable discussions, particularly on the implementation of the FIR filter. They would also like to thank MOSIS for fabricating the chip.

### REFERENCES

[1] P. Defraeye et al., "A 3-μm CMOS digital codec with programmable echo cancellation," IEEE J. Solid-State Circuits, vol. SC-20, pp. 679–688, June 1985.
[2] J. C. Candy, "Decimation for sigma delta modulation," IEEE Trans. Commun., vol. COM-34, pp. 72–76, Jan. 1986.
[3] J. E. Iwersen, "Calculated quantizing noise of single-integration delta-modulation coders," Bell Syst. Tech. J., vol. 48, pp. 2359–2389, Sept. 1969.

The measured result meets the D3 specifications. In Fig. 12 the frequency response of the overall coder is shown for power supplies of 5 and 4.75 V, which show no appreciable difference. The response has more than 33-dB suppression at 4.6 kHz and beyond. The passband ripple is less than 0.25 dB. In Fig. 13 the gain tracking characteristic of the coder is shown. The measured result meets the D3 specifications. The observed performance of the prototype is summarized in Table I. The crosstalk was measured by putting a full-scale sine wave at 1 kHz at the input of channel one, a zero input to channel two, and then measuring the 1-kHz component at the output of channel two. The crosstalk was measured to be $-83$ dB. Power supply rejection ratio at 1 kHz was measured by applying a 20-mV peak-to-peak sine wave at 1 kHz at the $V_{dd}$ of the chip, zero input to the chip, and measuring the 1-kHz component at the output. Measurements were repeated for different frequencies from 500 Hz to 20 kHz. The PSRR at 10 kHz is better than that at 1 kHz because the power noise is suppressed by the digital filter. The idle channel noise was measured by putting zero input at the channel input, summing up the noise at the output, and taking the

[4] E. B. Hogenauer, "An economical class of digital filters for decimation and interpolation," *IEEE Trans. Acoust., Speech, Signal Processing*, vol. ASSP-29, pp. 155–162, Apr. 1981.

[5] A. N. Netravali, "Optimum digital filters for interpolative A/D converters," *Bell Syst. Tech. J.*, vol. 56, pp. 1629–1641, Nov. 1977.

[6] J. Raebaey, S. P. Pope, and R. W. Brodersen, "An integrated automated layout generation system for DSP circuits," *IEEE Trans. Computer-Aided Des.*, vol. CAD-4, pp. 285–296, July 1985.

[7] R. Jain *et al.*, "Custom design of a VLSI PCM-FDM transmultiplier," *IEEE J. Solid-State Circuits*, vol. SC-21, pp. 73–86, Feb. 1986.

[8] M. W. Hauser and R. W. Brodersen, "Circuit and technology considerations for MOS delta-sigma A/D converters," in *ISCAS Conf. Proc.*, May 1986, pp. 1310–1315.

[9] R. Castello and P. R. Gray, "A high-performance micropower switched-capacitor filter," *IEEE J. Solid-State Circuits*, vol. SC-20, pp. 1122–1132, Dec. 1985.

# An 18b Oversampling A/D Converter for Digital Audio

Kouichi Matsumoto, Eiichi Ishii

Kazuki Yoshitate, Kaori Amano

NEC Corp.

Kawasaki, Japan

Robert W. Adams

dbx/Division of BSR

Newton, MA

IN RECENT YEARS, there has been significant progress in the field of consumer digital audio. Recently, DAT players employ 16b A-to-D and D-to-A conversion. In the professional audio market, there is a need for higher resolution conversion so that 16b quality can be maintained when many tracks are mixed together. This paper will present a monolithic 16b A/D converter that achieves 105dB S/N and 0.003% THD* by using oversampling and decimation techniques. The converter consists of a front end LSI and two decimator/filter ICs to reduce the sampling rate. The front end contains a 4b quantizer and runs at 6.144MHz, which is 128 times faster than the final sampling rate of 48KHz. When combined with the two decimator/filter ICs, a complete 18b A/D converter is realized. Figure 1 shows the overall block diagram of the converter.

The front end LSI is a noise-shaping coder. It contains three operational amplifiers, a 4b flash converter and circuitry to drive an external resistor network which forms the loop D/A converter. The loop filter consists of a 4th-order filter composed of three active integrators with real zeros and a passive first-order section.

The three active integrators require a slew rate of 40V/μs. To achieve this performance level requires fast transistors. The front end has been implemented in a 2.4μm BiCMOS process to achieve the required performance. Figure 2 shows the schematic diagram of the operational amplifier. The internal compensation capacitor is 10pF for the first operational amplifier (which must be unity-gain stable) and 5pF for the second and third operational amplifiers (which must be stable for gains greater than 2). A slew rate of 46V/μs was measured for the first operational amplifier.

For stability, the slope of the filter response is −6dB/octave at the point where the loop gain is unity. Figure 3 shows the frequency response of the loop.

The loop D/A converter is composed of 15 CMOS buffers and 15 equal value resistors. The matching of these resistors determines the linearity of the converter. To obtain 0.003% harmonic distortion, the resistors are required to match within ±0.01%. For accuracy resistors must be added externally.

Another important specification of the loop D/A converter is waveform symmetry[1]. It was reported in 1986, that the difference between rise and falltimes should be less than 500ps (assuming equal low-to-high and high-to-low propagation delays). Waveform asymmetry is caused by unequal rise and falltimes at the ECL-to-CMOS interface circuitry, as well as in the CMOS output buffers. We attempted to minimize asymmetry in the CMOS buffers by scaling the relative sizes of the N− and P−type devices. It was found that for optimum performance the P-channel FET should be larger than the N-channel FET by a factor of 2.57:1.

The ECL-to-CMOS interface uses a non-saturating design to minimize asymmetry errors and maximize operating speed. This design employs only NPN transistors and is shown in Figure 4. The output voltage swing is centered on 0.5 VCC

---

[1] Adams, R.W., "Design and Implementation of an Audio 18b Analog-to-Digital Converter, Using Oversampling Techniques", J. Audio Eng. Soc., Vol. 34, 3; March, 1986.

*Total Harmonic Distortion.

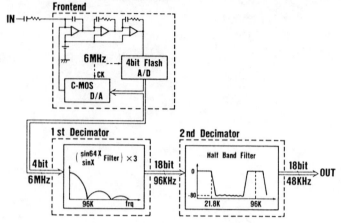

FIGURE 1—Overall block diagram.

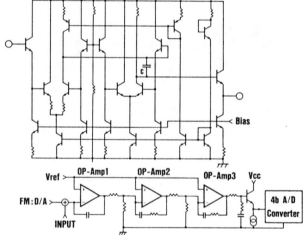

FIGURE 2—Circuit diagram of the operational amplifier.

### Front-end

| | |
|---|---|
| Fabrication technology | Si-gate Bi-CMOS |
| | Dual-layer metal |
| | Twin-well |
| Gate length | 2.4 μm for N-ch |
| | 2.8 μm for P-ch |
| NPN Tr. | $f_T$ = 4GHz |
| | $h_{FE}$ = 120 |
| Power supply | 9v (Analog cell) |
| | 5v (Digital Cell) |
| Chip size | 4.64 × 7.08 mm² |

TABLE 1

Reprinted from ISSCC Dig. Tech. Pap., pp. 202–203, February 1988.

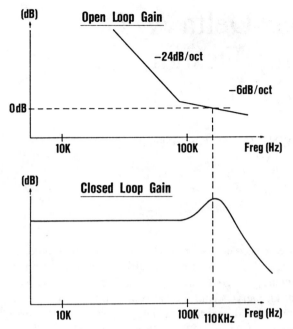

FIGURE 3—Frequency responses.

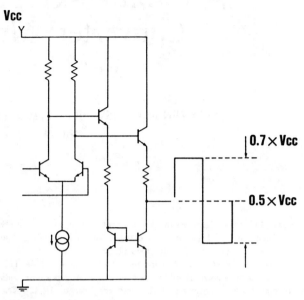

FIGURE 4—Interface circuit.

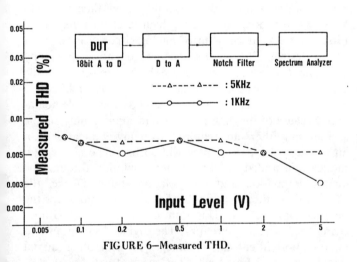

FIGURE 6—Measured THD.

and does not swing from rail-to-rail. The effect of the output voltage swing on the D/A output symmetry has been simulated using SPICE. The results of the simulation indicated that when the peak-to-peak output swing was 0.55 VCC to 0.70 VCC, the rise and falltime matching was very good. An output swing of 0.62 VCC has been used for the interface circuit. Table 1 shows the device characteristics of the front end.

The 4b output of the front end feeds the first decimator/filter IC which consists of three cascaded moving-average filters. The sample rate is reduced to the intermediate sample rate of 96KHz at the output of these filters.

The second decimator/filter is a 105-tap half-band FIR low-pass. It removes noise and signal components above 24KHz and then reduces the sampling rate to 48KHz. This filter also acts as the system anti-aliasing filter, eliminating the need for analog pre-filtering.

The measured THD of a complete conversion system is shown in Figure 6. The low-level measurements were obtained by left-shifting the digital output before D/A conversion. This allowed the D/A converter to operate at close to full level, which minimized the contribution of the D/A converter to the measured distortion.

# A 14-bit 80-kHz Sigma–Delta A/D Converter: Modeling, Design, and Performance Evaluation

STEVEN R. NORSWORTHY, MEMBER, IEEE, IRVING G. POST, MEMBER, IEEE,
AND H. SCOTT FETTERMAN

*Abstract* —This paper describes the development of a sigma–delta A/D converter. Included is a brief overview of sigma–delta conversion. The A/D converter achieves an 88.5-dB dynamic range and a maximum signal-to-noise ratio of 81.5 dB. The harmonic distortion is negligible. This level of performance is about 10 dB higher than previously reported results for oversampled A/D converters in this frequency range [1], [2]. The analog modulator employs a double-integration switched-capacitor architecture with an oversampling rate of 10.24 MHz. Transconductance amplifiers having a 160-MHz $f_t$ were developed for the integrators. The circuit is implemented in a 1.75-µm 5-V CMOS process. The analog circuitry occupies 2 mm$^2$ of silicon area and consumes 75 mW of power. Some of the difficult problems associated with evaluating the performance of sigma–delta converters are described. The design of a sigma–delta development and performance evaluation system is presented. This system includes a custom interface board linking the chip to a Sun workstation, and extensive digital signal processing and analysis software.

## I. INTRODUCTION

SIGMA–DELTA converters have been receiving much attention recently in the VLSI industry. This interest has been fueled by advances in fine-line digital integrated circuit technology, making it possible to integrate complex digital signal processing systems on a single chip. Sigma–delta converters benefit from these technology advances because they typically require several thousand gates to implement digital decimation or interpolation filters. Before the advent of fine-line CMOS VLSI, sigma–delta converters were not economically competitive with other types of data converters, given the same basic conversion specifications. Now, that is no longer the case.

The early use of sigma–delta conversion was limited primarily to voice-band codecs. There is now a growing potential for sigma–delta converters in such areas as ISDN U-interface tranceivers, modems, digital audio tape (DAT), compact disc (CD) players, and sonar signal processing.

Sigma–delta converters are well-suited for most *data* or *signal conversion* applications requiring high resolution in the frequency range from a few kilohertz up to several hundred kilohertz. There is also a growing interest in single-chip signal processors, i.e., general-purpose programmable and ASIC DSP's with on-chip A/D and D/A converters. There are continual market pressures to produce economical digital signal processing *solutions* utilizing high levels of hardware integration.

Sigma–delta converters possess many characteristics which naturally and advantageously lend themselves to VLSI signal processing utilizing high levels of integration. First, only a small amount of analog circuitry is required in the design. The circuits generally do not require any component trimming to achieve high resolution in the conversion process. In addition, the analog architecture lends itself very well to a switched-capacitor implementation. Because of the high rate of oversampling, only simple anti-aliasing filters are required. In many applications, an integrated passive $RC$ filter will suffice. The conversion rate can be traded off directly for resolution through the digital decimation and interpolation filters. Hence, the converter specifications can be digitally customized for the required application. This can often be done without changing the analog circuitry. The digital decimation and interpolation filters can be integrated with other digital signal processing functions. They can be treated as part of the overall signal processing requirements for the system. All of these unique characteristics result in both cost and performance advantages for the end user.

In this paper, we begin with an overview of sigma–delta conversion, familiarizing the reader with some of the basic concepts. Then, the circuit architecture and implementation are discussed. This is followed by a section describing a performance evaluation and development system for the converter. Lastly, we show the performance results.

## II. SIGMA–DELTA CONVERSION

Sigma–delta conversion may be employed for either A/D or D/A conversion. A sigma–delta A/D converter consists of two basic blocks, as illustrated in Fig. 1: an

Manuscript received September 2, 1988; revised December 5, 1988.
S. R. Norsworthy is with AT&T Bell Laboratories, Allentown, PA 18103.
I. G. Post and H. S. Fetterman are with AT&T Bell Laboratories, Reading, PA 19604.
IEEE Log Number 8826149.

Reprinted from *IEEE J. Solid-State Circuits*, vol. SC-24, pp. 256–266, April 1989.

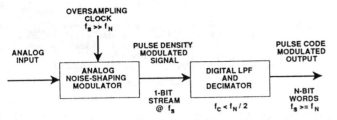

Fig. 1.   Sigma–delta A/D converter.

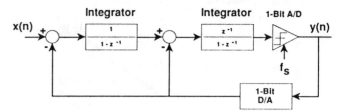

Fig. 2.   Double-integration loop sigma–delta modulator.

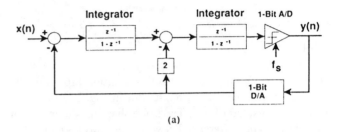

(a)

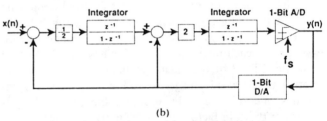

(b)

Fig. 3.   (a), (b) Practical sigma–delta modulator architectures.

analog noise-shaping modulator and a digital decimation filter.

An analog input enters the modulator, where it is sampled at a very high rate—many times the Nyquist rate. It is called a modulator because the analog signal is pulse density modulated, that is, the density of the pulses at the output over a given period is approximately equal to the mean value of the analog input over the same period. In this design, the oversampling rate $f_s$ is 10.24 MHz. The modulator generates a 1-bit output stream, which is then digitally filtered to remove the out-of-band quantization noise that is produced by the modulator. The cutoff frequency $f_c$ of the digital filter is 40 kHz, and the resulting resolution in baseband is ideally 15 bits. The sampling rate may then be lowered to any rate greater than the Nyquist rate $f_N$, which would be 80 kwords/s. This digital filter may also be viewed as a *pulse-density-modulation to pulse-code-modulation converter*.

Fig. 2 shows a block diagram of the modulator, which consists primarily of two integrators, a 1-bit A/D, and a 1-bit D/A. In the sampled-data or discrete-time domain, the previous output of the 1-bit D/A is subtracted from the analog input signal $x(n)$. This results in an error signal, which is integrated. The D/A output is subtracted from this integrator output, and the resulting signal is integrated and quantized. The resulting 1-bit digital output $y(n)$ is then converted into one of two analog levels by the 1-bit D/A converter. The 1-bit quantization that goes on in the modulator generates a high level of quantization noise. This quantization noise is spectrally shaped by the modulator such that most of the energy lies at high frequencies outside the baseband.

### A. Quantization Noise

Candy and Benjamin [3] have given a simplified analysis of sigma–delta quantization noise, whereby the 1-bit A/D is approximated as a white-noise source. The rms value of the quantization noise $e_o$ is given by

$$e_o = \frac{\Delta}{\sqrt{12}}$$

where $\Delta$ is the quantizer step size. With this model, the input–output relationship of the modulator is readily derived, and given by

$$Y(z) = X(z)z^{-1} + e_o(1 - z^{-1})^2.$$

Hence, the input $X(z)$ simply goes straight through the modulator with only a delay, but the quantization noise $e_o$ is shaped by the high-pass function $(1 - z^{-1})^2$. The spectral density of the quantization noise can easily be shown to be

$$N_Q(f) = \frac{4\Delta}{\sqrt{12}} \sin^2\left(\frac{\pi f}{f_s}\right).$$

Therefore, $N_Q(f)$ is zero at $f = 0$ and reaches a maximum at one-half the sampling frequency. The rms quantization noise in the baseband $f_b$ is then found by integration as

$$N_{QB}(\text{rms}) = 20\log_{10}\left(\frac{\pi^2\Delta}{\sqrt{60}}\right) + 50\log_{10}\left(\frac{2f_b}{f_s}\right).$$

Hence, a twofold decrease in the ratio $f_b/f_s$ will result in a 15-dB decrease in the baseband quantization noise, or 2.5 bits of additional resolution.

### B. Architectures for Practical Realization

Fig. 3(a) and (b) shows two architectures which are alternatives to the one given in Fig. 2. The input–output relationships of each are the same (using the linear model for the quantizer), and are given by

$$Y(z) = X(z)z^{-2} + e_o(1 - z^{-1})^2.$$

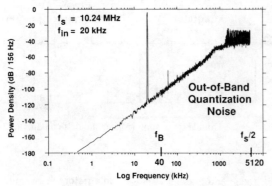

Fig. 4. Broad-band power spectrum of a double-integration loop sigma–delta modulator from a behavioral model simulation. Input frequency is 20 kHz.

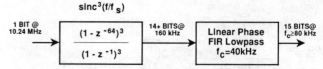

Fig. 5. Two-stage digital decimation filter.

The only difference between this relationship and the previous one given for the modulator in Fig. 2 is the extra delay, $z^{-2}$ instead of $z^{-1}$. The noise shaping is identical in both cases. The difference between Fig. 3(a) and Fig. 2 is that in Fig. 3(a) both integrators have a $z^{-1}$ delay, and there is a gain factor of 2 in the inner feedback loop. The delay in both integrators amounts to pipelining between the two stages, and makes a practical implementation easier, since each integrating amplifier has one clock cycle to settle. Fig. 3(b) represents an implementation having another practical advantage. As in Fig. 3(a), each integrator has one full delay. In addition, the gain scaling is different: the gain factor of 2 in the inner feedback loop is replaced with a gain of $1/2$ before the first integrator and a gain of 2 before the second integrator. The advantage here is that the signal swing at the output of the first integrator is exactly one-half that of Fig. 3(a). This potentially improves the dynamic range and linearity of the converter with respect to amplifier saturation. In addition, the gain of 2 associated with the second integrator can actually be replaced by any arbitrary amount of gain, since the signal is quantized by a 1-bit A/D converter (a comparator). There are, of course, practical lower limits to this gain. The second integrator signal swing can be controlled by choosing the correct gain such that the second integrator has the same maximum signal swing as the first integrator. The value of this gain was determined through time-domain behavioral-level modeling and simulation to be approximately 0.7.

### C. Nonlinearities in Sigma–Delta Converters

Fig. 4 shows the output power spectrum of a sigma–delta modulator from a behavioral model computer simulation, using ideal integrators and an ideal comparator. The input is a 20-kHz sinusoid. The sampling rate is 10.24 MHz. The plot scale is power density (dB/156 Hz) versus log frequency (kHz). The frequency scale goes up to 5.12 MHz, or one-half the sampling rate. One can see that the quantization noise is rising at approximately 15 dB/octave as predicted by the linear model. If the A/D converter were used as a receiver front end for an ISDN U-interface basic access rate transceiver, then the baseband of interest

would be 40 kHz. (This is assuming that the ISDN transceiver is for a 2B1Q North American ANSI Standard system.) The noise integrand from 0 to 40 kHz would yield A/D converter resolution of 15 bits ideally. The noise beyond 40 kHz would be digitally filtered out. For digital audio applications, the bandwidth of interest is 20 kHz. By moving the digital filter cutoff frequency down one octave, from 40 to 20 kHz, the converter would ideally produce 17.5 bits of resolution.

In Fig. 4, a small third-harmonic distortion component can be seen at 60 kHz. Even ideal sigma–delta modulators exhibit certain nonlinear behavior. One form of nonlinear behavior is known as *limit-cycle oscillations*, which occur for certain dc inputs [3]. Limit-cycle behavior is much more severe in first-order modulators than in second-order modulators [4]. Gray [5] has provided an analysis of limit-cycle behavior in single-integration loop sigma–delta modulators. Another type of nonlinear behavior, namely odd harmonic distortion, occurs for an ac input whose signal level is within approximately the top 10 dB of the dynamic range (0 dB is usually defined, by convention, as being a sinusoidal input whose peak-to-peak amplitude equals the spacing between the two outer quantization levels). This type of nonlinearity was discovered by simulation and confirmed later in the actual hardware. The third harmonic is the predominant component. For a well-designed sigma–delta modulator which is carefully balanced to minimize even harmonics, this third harmonic will actually dominate the signal-to-distortion ratio (S/D) performance. This will be seen later in Section VI.

### D. Digital Filtering and Decimation

For the purposes of performance evaluation, the output of the modulator is filtered and decimated in two stages, as shown in Fig. 5. The first stage is a sinc$^3$ digital filter, having a transfer function given by

$$H(z) = \left( \frac{1 - z^{-64}}{1 - z^{-1}} \right)^3.$$

The output word rate is reduced by a factor of 64 from 10.24 MHz to 160 kHz. This is followed by a linear-phase FIR low-pass filter with a cutoff frequency of 40 kHz. A sinc$^3$ filter was chosen for two reasons. First, for simplicity of implementation. It can be implemented with an architecture that requires no hardware multipliers and no ROM coefficient storage. It can be fabricated in a very small chip area [6], [7]. In addition, the sinc$^3$ filter function is nearly ideal for removing out-of-band quantization noise for a double-integration sigma–delta modulator [8]. The high-level simulation model of the digital decimation filter was designed to model a chosen hardware implementation. A digital decimation filter test chip was then successfully designed and fabricated [9] as a parallel effort with the modulator test chip.

*E. Higher-Order ( > 2) Noise-Shaping Modulators*

There has been much interest recently in higher-order oversampled noise-shaping coders [2], [10]–[13]. The basic idea is to *push* even more of the quantization noise out of the baseband. This is accomplished by using a pulse density modulator having a quantization noise-shaping transfer function of an order higher than two, i.e., more than two integrators. Investigation of several such architectures revealed the following.

The advantage is that a lower oversampling ratio can be used to achieve the same baseband resolution as would be achieved by a double-integration loop sigma–delta modulator. Hence, one could use a lower-speed technology if needed, or relax the circuit speed and power requirements for the same technology. The disadvantage, however, is that these architectures are in general not as robust. One must be concerned with either precise circuit-matching problems, or with modulator loop stability problems, or both, depending upon the architecture. Overall, there is increased analog circuit complexity and precision. Also, steeper anti-aliasing filters are required, because the sampling rate is lower with respect to the baseband. In addition, higher-order digital low-pass filtering is required. For example, if a third-order noise-shaping modulator is employed, the quantization noise would rise at 21 dB/octave instead of 15 dB/octave for a second-order modulator. If the sinc class of filtering were used, a third-order modulator would require a $sinc^4$ filter, instead of the $sinc^3$ filter for the second-order modulator. If the decimation factor were 64, then the impulse response of the $sinc^4$ filter would be 256 taps in length, as opposed to 192 taps for the $sinc^3$ filter. Overall, all of the added analog and digital complexity will yield a more expensive design. There may, however, be certain applications where the specifications cannot be met with lower-order modulators. The main conclusion is that if one has a fast-enough technology, trading off speed for circuit complexity is certainly advantageous. Thus, the double-integration sigma–delta architecture represents a good compromise.

## III. Modeling and Simulation

One of the most challenging aspects of designing the converter is determining the relationship between the system parameters (i.e., sampling rate, resolution, linearity, etc.) and the circuit parameters (i.e., amplifier bandwidth and slew rate, comparator speed, switch and capacitor sizes, etc.) which affect the performance. In order to make these determinations, a C-language behavioral model of the sigma–delta converter was created. This involved designing a series of subprograms for the modulator, the

digital decimation filter, and the power spectral estimation. The overall algorithm flow is described in Section V. The programs were run on a Sun workstation having an interface to capture the data from the actual chip. The results obtained could be directly compared with those of the software behavioral model to determine the quality of the design. This performance evaluation system is discussed in more detail in Section V. Some nonideal circuit parameters such as those encountered in the actual chip design were also modeled. This information was obtained from analog circuit designers, so that the design trade-offs could be considered up-front.

Transistor-level simulators such as SPICE and ADVICE [14] are indispensable tools for a circuit designer. But there is no known practical way to simulate enough sample points, using device-level models, in order to verify the overall performance of a sigma–delta modulator. This is primarily because the converter oversamples the incoming signal at such a high rate compared to the input frequency. Each individual sample requires hundreds or even thousands of time steps, primarily because it is a switched-capacitor circuit (reconvergence must occur at every clock edge). One would need at least 32 768 of these individual samples (one new sample would be obtained per clock cycle) in order to determine the baseband performance in the frequency domain. Hence, it would require many millions of time steps. It would literally take days or even weeks to simulate on a super-computer, such as Cray XMP. Even if such a simulation were performed, the accuracy of the outcome would be doubtful, since one is typically looking for performance of 80–90 dB in terms of noise and distortion. The internal algorithms and device models within such a program probably do not have this kind of accuracy.

## IV. Circuit Implementation

Fig. 6 illustrates the overall circuit of the modulator, which was implemented using a 1.75-$\mu$m twin-tub double-poly 5-V CMOS process. Fully differential folded-cascode transconductance amplifiers are employed [15], with the common-mode voltage set to midsupply. The amplifier $g_m$ is flat to approximately 80 MHz for worst-case temperature and processing. Approximately 600 $\mu$A of current is made available for output slewing. Fully differential circuit arrangements are used starting from the inputs all the way to the digital output and latch. The delay from comparator strobe to output data latch is less than 10 ns (worst-case) for a 1-mV input.

Care has been taken to achieve approximately the same delay per stage in the combinatorial logic between the 1-bit A/D output and (including) the transmission gates of the 1-bit D/A. Thus there is a steady progression in device size from the very small gates at the comparator output for fast detection, to the much larger switching devices required for rapid capacitor charging. Total delay, from comparator strobe to the switching of the feedback capacitors, is less than 15 ns.

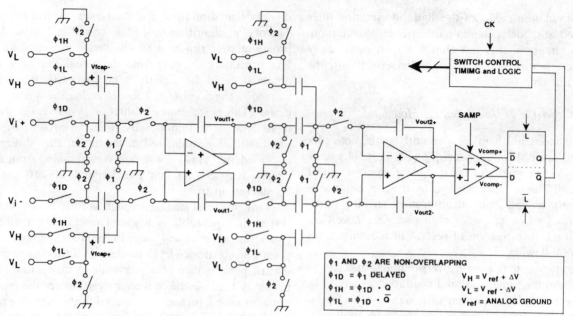

Fig. 6. Fully differential switched-capacitor sigma–delta modulator architecture.

Every pair of components in the balanced signal path, including transmission gates, capacitors, and transistors, is arranged in a common-centroid fashion. This technique tends to reduce balancing errors due to processing and temperature variations across the chip as well as noise induced from transient substrate currents. Wiring runs are also matched carefully.

Analog circuits, transmission gates, and logic are powered from their respective on-chip power and ground wiring to reduce common-mode induced crosstalk.

### A. Limitations to Performance

There are several factors which limit the maximum clock rate at which the circuit can operate. These are the amplifier slew rate, the $RC$ time constants associated with the switch resistance and charging capacitors, the amplifier frequency response, the comparator delay, the switching logic delays, and the nonoverlapping clock duty cycle. All of these delays add up to reduce the time allowed for capacitors to fully charge and amplifiers to settle, which is particularly important for the first stage. The circuit designer seeks to minimize these delays but is ultimately limited by the available technology. In the 1.75-$\mu$m CMOS technology, a noticeable and rapid decrease in the performance of the converter occurs for clock frequencies above 15 MHz. Other factors that can limit performance even at low clock rates are amplifier gain and linearity and $kT/C$ noise.

The feedback voltage levels have been set to prevent output excursions from clipping, which would lead to a rise in the noise floor and distortion. The use of a differential-amplifier scheme with sufficient gain results in negligible second harmonic distortion. Nonlinearity can also result from charge feedthrough and imbalance due to large parasitic capacitances. Distortion resulting from capacitor

nonlinearity is negligible for two reasons: the capacitor dielectric is a polysilicon oxide (not a poly/nitride sandwich), and the fully differential structure minimizes whatever even-order nonlinearity is present. Delayed clocking is used in the switching network to create an equivalent four-phase clock system. A dummy switch is also placed at each amplifier input to affect at least partial cancellation of charge feedthrough.

The main sources of noise (other than quantization noise) considered in the design include the $kT/C$ noise from the switched capacitors [16], and the thermal noise from the amplifiers and the bandgap reference circuit. Only a portion of the total broad-band thermal noise will affect the performance of the converter. This portion is the amount in the baseband $f_b$ plus the high-frequency portions which are aliased around multiples of the sampling rate, $Nf_s \pm f_b$. The remaining portions are digitally filtered out. Proper design trade-offs have been made to ensure that the converter performance is not significantly limited by this noise.

### B. Operational Details

Fig. 7 shows a simulated 250-ns interval plot of some of the signals in the circuit. The simulation has been performed using models for 25°C and nominal processing. The clock frequency is 10.24 MHz. During $\phi_1$ the comparator output is sampled and the integrator input switches are set to charge the input capacitors with the proper polarity voltage. Fig. 7(a) shows the charging of one of the feedback capacitors. Fig. 7(b) shows the differential charging across the feedback capacitors, in order to illustrate the cancellation achieved. During $\phi_2$ the comparator is switched to the sense mode and the charge on the input capacitors is transferred to the integrator feedback capacitors. Fig. 7(c) shows the differential output voltage of the

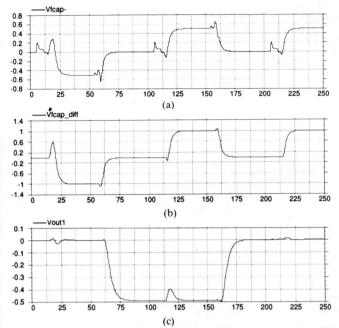

Fig. 7. Transient simulation output of the sigma–delta modulator. (a) Voltage across one of the feedback capacitors. (b) Differential voltage across one pair of the feedback capacitors. (c) Differential voltage at the first integrator output.

first integrator. This illustrates the charging of these capacitors. The large transient at 120 ns in the stage 1 amplifier output is a measure of the recovery time from the closure of the switches connecting the output to the input of stage 2.

The delay between the start of $\phi_1$ and the onset of input capacitor charging is about 11 ns. Of this, 3.5 ns is due to the nonoverlap requirement, another 3.5 ns is for the delay for charge feedthrough control, and the remaining 4 ns is for combinatorial logic delay. In this particular example the comparator delay is only 3.2 ns, so that it does not influence the response. Once the switches are closed, the first-stage $RC$ product is about 2 ns. During $\phi_2$ the initial amplifier slewing is difficult to distinguish from the linear charging of the integrating capacitors, but the time constant associated with charging the integrating capacitors during $\phi_2$ is about 4 ns. It is obvious that any significant reduction of the length of a period will quickly effect circuit performance. For instance, if the clock frequency is doubled, only about 14 ns of each phase remains for settling (after deducting 11 ns for delays). Therefore the first-stage output only has about 3.5 time constants in which to settle.

### C. Subcircuit Characteristics

The amplifier schematic is shown in Fig. 8. The transconductance of the amplifier is almost equal to the transconductance of the source-coupled input pair $M4$ and $M5$. The differential input stage current passes through a common-gate output pair $M8$ and $M9$, which serves to increase the output impedance of the amplifier. The common-mode voltage at the output is maintained by transis-

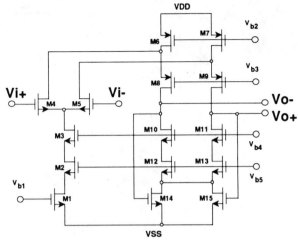

Fig. 8. Fully differential transconductance amplifier.

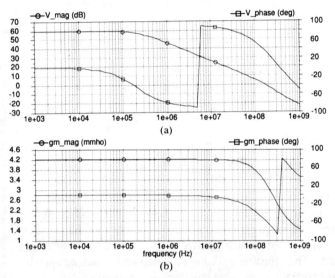

Fig. 9. Simulated amplifier performance for 25°C and nominal processing. (a) Open-loop voltage gain (decibels) and phase (degrees). (b) Transconductance magnitude (mmhos) and phase (degrees).

tors $M14$ and $M15$. The performance of the amplifier is shown in Fig. 9. The 3-dB corner frequency for $g_m$ is approximately 100 MHz at 25°C for nominal processing. The voltage gain (no-load) is about 60 dB at lower frequencies. Capacitors have been added from each output to ground to provide about 40° of phase margin at 160 MHz.

A schematic of the strobed comparator is shown in Fig. 10. The input stage, $M1–M7$, is always active. When the sampling control input $V_{samp}$ is high, the current mirrors $M8$, $M10$, $M12$ and $M9$, $M11$, $M13$ are inoperative and the outputs are pulled high by $M14–M17$. When the sampling control is low the mirrors are active, with $M10$ and $M11$ providing positive feedback. The result is that one of the outputs will go to a low state. As shown in Fig. 11, the dual outputs are fed into a NAND gate whose output is used to temporarily allow the data latch to accept new data. Transistors $M16$ and $M17$ are used to speed the pull-up of the outputs when $V_{samp}$ goes high. Note that the only nodes with a large voltage swing are the digital outputs.

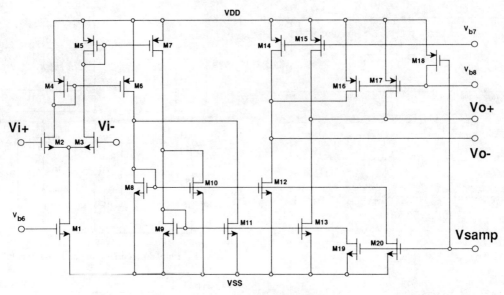

Fig. 10.   Fully differential strobed comparator.

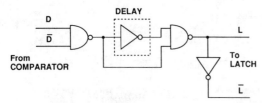

Fig. 11.   Delay latch logic.

The OFF period between phases is nominally about 3.5 ns. In order to have the circuit work with such a short interval, gates associated with clock generation and switching are scaled in size relative to the capacitances of their loads. Thus the amount of nonoverlap and switching speeds tend to track over processing and temperature.

The choice of switch size in the integrator input networks is based on the trade-off between the desire to reduce the series $RC$ time constant and the requirement to keep parasitics and charge feedthrough effects negligible.

A photomicrograph of the test chip is shown in Fig. 12. No attempt was made to optimize the overall area at the time of this design. Support circuitry, which includes current and voltage references, occupies the upper two rows, well away from the digital section which occupies the bottom row. The integrators and comparator occupy the middle row separated from the digital circuitry by the switch networks. The feedback capacitors are located in the middle row while input capacitors are located in the next row below, along with the switches. The circuit for a crystal oscillator was also provided, and it occupies a small row by itself in the lower right portion.

## V.   PERFORMANCE EVALUATION

When measuring the performance of an A/D converter, there are two basic approaches. One is to follow the A/D with a better D/A, then measure the resulting analog output. The second approach is to capture the digital

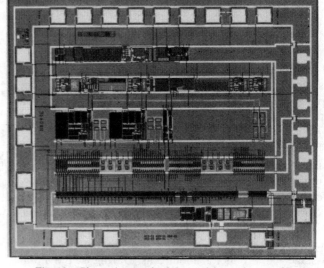

Fig. 12.   Photomicrograph of sigma–delta analog test chip.

output words into a computer and then analyze the power spectrum of the signal in software [18]. Sometimes, a combination of these techniques is used.

### A.   Alternatives for Measuring Sigma–Delta Performance

One way to measure the performance of a sigma–delta A/D converter is to feed the 1-bit output of the modulator into a precise 1-bit D/A converter and then feed that analog output into a spectrum analyzer to measure the noise and distortion in the baseband. This approach is referred to as *analog loopback*, and it is often used when a complete sigma–delta codec (both A/D and D/A) is designed, since the 1-bit D/A converter already exists. The disadvantage is that it is difficult to isolate any design nonideality to the A/D converter alone, since the signal is influenced by the performance of the D/A converter as well.

An alternative is to digitally filter and decimate the modulator output *in hardware* to the Nyquist rate, capture that data through a computer interface, and then spectrally analyze the data in software by using a FFT (or some other spectral estimation method). This approach is advantageous when the desired digital decimation filter is known ahead of time and already exists in hardware. The disadvantage is that one may wish to examine how the analog modulator would perform with a different digital decimation filter, maybe for some other application. If that decimation filter does not exist in hardware, then one would have to rely only on simulation.

A third approach is to capture the 1-bit modulator output samples directly into the computer interface. From here, there are two possibilities.

1) Spectrally analyze the baseband portion of the oversampled pulse-density-modulated signal: This requires very large FFT's (minimum of 32 768 and preferably greater), because the baseband is typically a factor of 64 to 256 times less than the sampling rate. The advantages of this method is that the noise and distortion in baseband can be determined without any effects from the decimation filter.

2) Digitally filter and decimate the modulator data in software, and then examine the power spectrum of that data: The advantage here is that if a different digital decimation filter is desired, then one can simply write the software corresponding to that architecture. The resulting baseband performance can then be compared with the above result (examining the baseband spectrum without digital filtering and decimation), to examine the aliasing performance of the decimator.

The last approach was adopted, with both parts 1) and 2). This method is by far the most thorough and accurate.

*B. Performance Evaluation System*

A sigma–delta performance evaluation system is depicted in Fig. 13. This block diagram shows the flow of data from the actual chip into the computer hardware, and then through the signal processing algorithms. Every function in the program is written entirely in C, with approximately 4000 lines of source code in total. The computer is a Sun 3/160 workstation.

When the program is executed, the user is prompted as to which type of sigma–delta modulator he/she wants: the actual hardware chip or a behavioral model. If the hardware modulator is selected, the program automatically captures 1 Mbit of output samples from the chip and transfers them into the cache memory of a custom data acquisition and interface board attached to the Sun. The data are then transferred from the interface board into the Sun memory, making them available to the signal processing functions which follow. If the behavioral model is chosen, the user is prompted for many parameters, such as input frequency, sampling frequency, input amplitude, etc. Some additional nonideal parameters may also be selected, such as amplifier gain and bandwidth, integrator gains, dc offsets, comparator hysteresis, etc.

The resulting samples, whether they are obtained from the actual chip or from the behavioral model, may then be digitally filtered and decimated in software. The user chooses between one of several different digital decimation filter algorithms that have been included in the program. The decimation factor of the digital filter may also be chosen. The Welch periodogram method [17] is used to compute the power spectrum, yielding a more accurate and less biased result than that obtained using only a single FFT. The size of the desired periodogram may be selected, as well as the number of overlapping window segments. Dynamic memory allocation is used so that very large periodograms may be accommodated efficiently. The user also has the option of skipping the digital filtering and decimation step and going directly to the periodogram. Following the periodogram function, the program computes the signal-to-noise and signal-to-total-harmonic-distortion in the frequency domain. The power spectrum is displayed in a Sun window, showing the results.

In the periodogram, the user may also select the type of window to use on the data. Typically, either a Blackman or a Blackman–Harris [19] window is used. These windows are selected for two reasons. First, the Fourier transform of these windows contains spectral nulls such that when the window function is convolved with a sinusoidal having a frequency equal to one of the FFT bin frequencies, there is no *frequency smearing* of that sinusoid into any other frequency bins except those under the main lobe of the window. The second reason for choosing the Blackman window is that it has a wide main lobe and high sidelobe attenuation. This is desirable since it is difficult to align the sinusoidal input frequency precisely to one of the FFT bin frequencies. Nevertheless, care still must be taken to make this alignment as precise as possible. Two methods are used for synchronizing the sinusoidal input frequency to an integer submultiple of the sampling clock frequency: manual phase-locking and automatic phase-locking using a phase-lock loop. Manual phase locking is accomplished by first dividing the digital sampling clock down to the frequency of the sinusoidal input. Then the sinusoidal input is chopped into a square wave, and the two signals are synchronized together visually using an oscilloscope. A phase-lock loop is implemented as an alternative in order to eliminate the need for manual intervention. This approach, however, results in a small but measurable amount of jitter in the output power spectrum. Therefore, the manual approach is used whenever possible.

*C. Custom Data Acquisition Board*

A custom high-speed data acquisition board, designed to capture the data from the sigma–delta modulator, is shown in Fig. 14. The board has a 1-Mbit cache memory, and contains the control logic necessary to interface with a DR11-W protocol interface. The board is connected to a IKON Model 10084 DR11-W parallel interface board, which resides in the VME-bus backplane of the Sun work-

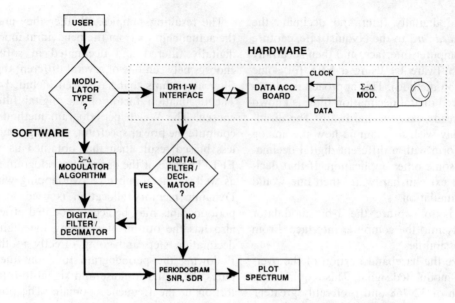

Fig. 13.   Sigma–delta performance evaluation system.

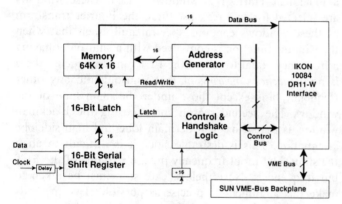

Fig. 14.   Custom data acquisition board to interface the chip to a Sun workstation.

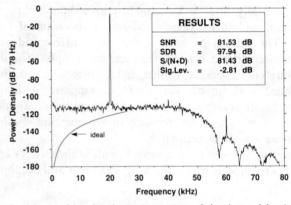

Fig. 15.   Measured baseband power spectrum of the sigma–delta A/D converter with a 20-kHz sinusoidal input.

station. Since the DR11-W is set up to transfer 16-bit words in parallel, the 1 Mbit of memory was organized as 64 kbit × 16. A high-speed shift register and parallel latch precede the memory, converting the 1-bit words from the sigma–delta into 16-bit words at one-sixteenth the input rate.

There are three modes of operation of the data acquisition board: *idle*, *acquire* into memory, and *transfer* into the Sun workstation. The operation of the board is controlled by the DR11-W parallel interface to which it is connected. A function, written in C, controls the operation of the DR11-W board, which in turn controls the operation of the data acquisition board. The DR11-W is treated like a file in the C function: it is assigned a file descriptor, an *open* is issued, and READ and WRITE system calls are issued to transfer data. Various *ioctl* (input–output control) calls within the C function are used to access the control bus of the DR11-W. This provides a straightforward and flexible method of transferring data into the Sun.

The data are transferred to the program memory in the form of a character array. The program then reconverts the data back into 1-bit samples by using the bit-wise *shift* and *mask* operators that are available in the C language.

## VI. RESULTS

Fig. 15 shows the power spectral output of the program after obtaining the data from chip. The input is a 20-kHz sinusoidal whose signal level is −2.81 dB. (We stated earlier that 0 dB is defined as being a sinusoid whose peak-to-peak amplitude equals the spacing between the two outer quantization levels.) The sampling rate is 10.24 MHz. The plot scale is power density (dB/78 Hz) versus frequency (kHz). The cutoff frequency of the digital filters is 40 kHz. The second harmonic distortion is down at −100 dB. The signal-to-noise ratio (SNR) is 81.53 dB, the signal-to-total-harmonic distortion ratio (SDR) is 97.94 dB, combining for a total of 81.43-dB signal-to-noise-

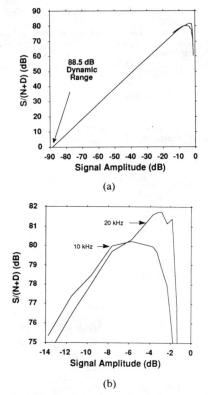

Fig. 16. S/(N+D) versus signal amplitude for 10- and 20-kHz sinusoidal inputs: (a) over the full dynamic range, and (b) over the top 14 dB of dynamic range.

plus-total-harmonic-distortion (S/(N+D)). Ideally, the S/(N+D) performance would be 85 dB, which was determined by using the software modulator described previously. The ideal quantization noise curve is also shown on the same plot. The two curves intersect at approximately 25 kHz. The thermal noise floor encountered is the major difference between the ideal modulator and the actual chip. The four filter ZERO's that appear in the spectrum between 50 and 80 kHz are the result of the second-stage FIR filter which follows the sinc³ decimation filter. The overall performance of the sigma–delta converter. Te converter is 5 dB better with this second stage of digital filtering, as opposed to what is obtained using the sinc³ filter by itself.

Fig. 16(a) and (b) shows plots of S/(N+D) versus signal amplitude. The plot in Fig. 16(a) is scaled over the full dynamic range of the converter, which is 88.5 dB. The performance is linear, i.e., the plot has 1 dB per dB slope, from −88.5 to −10 dB. The plot in Fig. 16(b) is a portion of the previous data from −14 to 0 dB signal amplitude. In this signal amplitude range, nonlinearities begin to arise, mostly due to the odd harmonic distortion phenomenon described earlier. Two curves are shown in this figure, corresponding to two different input frequencies: one at 10 kHz and the other at 20 kHz. The reason that the performance is slightly better for the 20-kHz input is because its third harmonic distortion component lies outside of the baseband and is being attenuated by the digital filter. The peak occurs in both cases for input amplitudes between −3 and −5 dB.

When the converter is evaluated with a digital filter cutoff frequency of 20 kHz instead of 40 kHz, the dynamic range improves by only 4 dB instead of the ideal 15 dB. This is attributed to the fact that the baseband thermal noise plus the aliased thermal noise folding down into the baseband begins to exceed the quantization noise. This conclusion is verified experimentally by scaling the sampling rate and baseband frequency down by a factor of 10. Using a 1.024-MHz clock, the resolution in a 4-kHz baseband is the same as that obtained using a 10.24-MHz clock with a 40-kHz baseband.

## VII. SUMMARY

We have presented the modeling, design, and performance evaluation of a 14-bit 80-kHz sigma–delta A/D converter. The converter is to be used as a receiver input of an ISDN U-interface 2B1Q basic access rate tranceiver. It achieves 88.5 dB of dynamic range and a maximum S/(N+D) of 81.5 dB. The harmonic distortion due to circuit imperfections is negligible. The analog modulator employs a double-integration switched-capacitor architecture. The oversampling rate is 10.24 MHz. The architecture and design are both inherently robust and economical. The technology is 1.75-μm 5-V double-poly CMOS. We also described the development of a performance evaluation system. This system includes a custom interface board which links the chip to a Sun workstation, and extensive digital signal processing and analysis software.

## ACKNOWLEDGMENT

The authors wish to thank the following people for their contributions: W. C. Ballamy, R. B. Blake, J. C. Candy, M. R. Dwarakanath, J. H. Fischer, G. F. Foxhall, V. L. Hein, H. Khorramabadi, S. Kumar, R. M. Montalvo, H. S. Moscovitz, G. L. Mowery, D. H. Nelson, R. A. Pedersen, J. M. Pichardo, D. L. Price, J. W. Scott, J. L. Sonntag, and T. R. Viswanathan.

## REFERENCES

[1] R. Koch *et al.*, "A 12-bit sigma-delta analog-to-digital converter with a 15-MHz clock rate," *IEEE J. Solid-State Circuits*, vol. SC-21, no. 6, pp. 1003–1010, Dec. 1986.
[2] L. Longo, "A 13-bit ISDN-band oversampled ADC using two-stage third order noise shaping," in *Proc. IEEE Custom Int. Circuits Conf.*, May 1988, pp. 21.2.1–21.2.4.
[3] J. C. Candy and O. J. Benjamin, "The structure of quantization noise from sigma–delta modulation," *IEEE Trans. Commun.*, vol. COM-29, pp. 1316–1323, Sept. 1981.
[4] J. C. Candy, "A use of double-integration in sigma–delta modulation," *IEEE Trans. Commun.*, vol. COM-33, pp. 249–258, Mar. 1985.
[5] R. M. Gray, "Oversampled sigma–delta modulation," *IEEE Trans. Commun.*, vol. COM-35, pp. 481–489, May 1987.
[6] H. Meleis and P. LeFur, "A novel architecture design for VLSI implementation of an FIR decimation filter," in *Proc. IEEE Int. Conf. Acoust., Speech, Signal Processing*, Mar. 1985, pp. 1380–1383.
[7] J. C. Candy, AT&T Bell Labs., private communication, 1985.
[8] J. C. Candy, "Decimation for sigma–delta modulation," *IEEE Trans. Commun.*, vol. COM-34, pp. 72–76, Jan. 1986.
[9] AT&T Bell Labs., unpublished development, 1986.

[10] R. W. Adams, "Design and implementation of an audio 18-bit analog-to-digital converter using oversampling techniques," *J. Audio Eng. Soc.*, vol. 34, pp. 153–166, Mar. 1986.

[11] L. R. Carley, "A 16-bit fourth order noise shaping D/A converter," in *Proc. IEEE Custom Int. Circuits Conf.*, May 1988, pp. 21.7.1–21.7.4.

[12] W. L. Lee and C. G. Sodini, "A topology for higher order interpolative coders," in *Proc. IEEE Int. Conf. Circuits and Systems*, May 1987, pp. 459–466.

[13] Y. Matsuya *et al.*, "A 16b oversampled A/D conversion technology using triple integration noise shaping," in *ISSCC Dig. Tech. Papers*, Feb. 1987, pp. 48–49.

[14] L. W. Nagel, "ADVICE for circuit simulation," presented at the IEEE Int. Conf. Circuits and Systems, Apr. 1980.

[15] R. Chio *et al.*, "High-frequency CMOS switched capacitor filters for communications applications," in *ISSCC Dig. Tech. Papers*, Feb. 1983, pp. 246–247.

[16] J. H. Fischer, "Noise sources and calculation techniques for switched capacitor filters," *IEEE J. Solid-State Circuits*, vol. SC-17, pp. 742–752, Aug. 1982.

[17] P. D. Welch, "The use of fast fourier transform for the estimation of power spectra," *IEEE Trans. Audio Electroacoust.*, vol. AU-15, pp. 70–73, June 1970.

[18] B. E. Boser *et al.*, "Simulating and testing oversampled analog-to-digital converters," *IEEE Trans. Computer-Aided Des.*, vol. 7, pp. 668–673, June 1988.

[19] F. J. Harris, "On the use of windows for harmonic analysis with the discrete Fourier transform," *Proc. IEEE*, vol. 66, pp. 51–84, Jan. 1978.

# FULLY DIFFERENTIAL CMOS SIGMA-DELTA MODULATOR FOR HIGH PERFORMANCE ANALOG-TO-DIGITAL CONVERSION WITH 5 V OPERATING VOLTAGE

Tapani Ritoniemi, Teppo Karema, Hannu Tenhunen
and Markku Lindell

Dept. of Electrical Engineering, Tampere University of Technology
P. O. Box 527, SF-33101 Tampere, Finland, Phone: +358-931-162618

## Abstract

This paper presents a high performance second order sigma-delta modulator for modem and ISDN analog to digital conversion applications. The major performance design limiting factors are demonstrated. It will be shown that a true 16-bit analog to digital (A/D) converter with single 5 V power supply for voice band can be realized with oversampling ratio of 512 and 16-bit dynamic range is achieved with oversampling ratio of 256. The die size of the proposed modulator using 2.5 micron CMOS technology is only 0.56 $mm^2$.

## I. Introduction

High resolution low cost A/D-converters find many applications in the areas of instrumentation, measurement, telecommunication, digital signal processing (DSP) and consumer electronics. Recent trend solving this demanding technological challenge has been DSP by means of digital filtering and decimation. The basic idea has been to combine a simple and digital IC-process compatible analog hardware with an accurate and reliable digital postfiltering on the same chip. For this reason delta modulation based sigma-delta A/D-converters have been widely published during last years and summarized in table 1.

The operation of the sigma-delta A/D-converter relies on oversampling and noise shaping. The sigma-delta modulator modulates the analog input signal with high rate using a coarse one-bit quantizer in such way that the large quantization noise is distributed out of the baseband. The quantized data is digitally postfiltered and decimated thus attenuating the quantization noise power under the desired level. The order of the modulator and the oversampling ratio put absolute limits for the signal to noise (S/N) ratio of the whole converter.

Because the oversampling ratio is very high no prefiltering or sample and hold circuit is needed.

In this work we will concentrate on the double integration sigma-delta modulator structure which provides the potential of 15 dB baseband S/N improvement for each doubling of the sampling frequency. This means that theoretically for voiceband the peak S/N of 96 dB could be achieved with oversampling ratio of 256. In practice the transfer function of that kind is not realizable and the corresponding figure with optimum realizable transfer function gives about 12 dB/octave S/N improvement and 84 dB peak S/N. The main obstacle realizing the high performance modulator is implementing the specified dynamic range on silicon without harmonic distortion (THD) and noise. Thus we will concentrate on evaluating the major SNR (S/(N+THD)) limiting factors.

## II. Charge based discrete time simulator for switched capacitor sigma-delta modulator

Modulator will pose strict limits to achievable A/D-converter SNR. The evaluation of the nonidealities with simulation is necessary for optimal implementation. However, we found that the simulations with circuit level simulators like SPICE or SPLICE [17] are too slow due to their heavy calculations. The modulator design by IDAC [8] was also tested but IDAC restricts to analytical idealized models. The combined table look-up and difference equation simulator like, e.g. ZSIM [13], was thought to be unsuitable because it has not accurate way to adjust one parameter in time thus complicating the circuit level optimization.

The system design work was started by optimising the two feedback coefficients a and b from ideal sigma-delta modulator model (fig. 1) using the transfer function design methodology [14]. With

Reprinted from *IEEE Proc. ISCAS'88*, pp. 2321–2326, June 1988.

optimised coefficients the statistical node voltage analyses over $10^6$ samples were performed to evaluate the absolute voltage values in order to avoid distortion caused by the amplifier overload in integrators. The subcircuits (switches, integrators, comparator) were specified according the earlier experiments [2] [5] and the system parameters of the modulator simulator were extracted using SPICE circuit simulator.

The simulator evaluates the net steady state charge for each network node at the end of each clock cycle using Kirchoff's charge conservation law. Thus clock frequency depended charge sharing is included. The modulator model is shown in figure 2. Following performance limiting effects were added to the ideal model:

1. The white noise generated by the operational amplifiers. The total noise voltage in the whole bandwidth of the amplifier was calculated and the equivalent random charge corresponding the noise voltage was added to the integrator charge including the aliased noise from high frequencies to baseband.
2. The operational amplifier gain. Gain model was extracted from SPICE simulations of the used amplifiers.
3. The noise and hysteresis of the comparator.
4. The capacitance voltage nonidealities up to quadratic term.
5. The on-resistances of the transmission gates.
6. The different capacitor ratios of the integrator.
7. The different sampling clock frequencies corresponding to the decimation values of 16, 32, 64, 128, 256.

The nonidealities in the modulator generate noise, distortion and affect to the transfer function of the whole modulator. The sampling and decimation causes noise alising. The sampling band noise which is not properly attenuated will alias into the baseband giving poor SNR. To see the effects of the nonidealities in a proper way the spectral analyses were done for the decimated output. Thus a new high performance digital filter structure with multiple decimating ratios was included [11]. The decimation filter amplitude response with decimation value of 128 (oversampling ratio of 256) is shown in figure 3. Extra effort was taken to keep the numerical inaccuracies under -300 dB level throughout the calculations.

## III. Optimised modulator structure for 5 V operating range using switched capacitor integrators

### 1. Modulator noise

The largest noise contribution is caused by the 1/f-noise and the thermal noise of the amplifiers. The 1/f-noise is added directly to the input signal while the thermal noise is aliased into the sampling band. The noise of the first integrator is most critical because it is amplified the most. The aliasing of the wideband thermal noise can be reduced by the circuit switch resistances and capacitances [15]. The noise of the comparator is very small compared to the 1-bit quantization noise and can be neglected.

The total noise voltage is calculated using amplifier band limited white noise which is convoluted with the switch RC-function. When the amplifier noise band is 60 MHz and the clock frequency is 5 MHz this aliased noise voltage is in our case about 0.2 times the noise voltage using low resistance switches. If the 16-bit SNR is needed the baseband SNR of the amplifier must be better than 96 dB. In 20 kHz baseband, 60 MHz amplifier band and oversampling ratio of 256, the acceptable white noise of the first amplifier becomes about 100 nV·$Hz^{1/2}$. Because the capacitance $C_{s2}$ has always zero charge before sampling, it is possible to lower the settling time requirement of the $C_{s2}$ from 0.1 % to 10 % (fig. 2). To minimize noise aliasing the RC-function should be as large as possible and the amplifier bandwidth should be as small as possible. Thus the subcircuit performance requirements should be balanced or otherwise the total modulator performance is reduced. Figure 4 demonstrates the modulator SNR as a function of the $R_{s2}C_{s2}$ time constant.

### 2. Amplifier gain

The simulations show that integrators must operate in very linear operating range because the nonlinearity in the amplifiers causes direct loss in SNR. The modulator with nonlinear high gain amplifiers have usually much worse SNR than the one with smaller but linear gain amplifiers. Thus the operating range of the integrators should be scaled to the linear region in spite of increased noise level and the size of the capacitors. Figure 5 shows the modulator filtered spectra with nonlinear amplifiers.

The gain required to keep the quantization noise level low is calculated assuming that the input signal and the feedback quantization noise are uncorrelated. For the 96 dB SNR the amplification is estimated from:

$$A = 1/2 \cdot [96 \text{ dB} + 10 \cdot \log_{10}((Q^2/12) \cdot B)] \qquad (1)$$

where B is base band width
$Q^2/12$ is quantization noise power

This gives only 67 dB amplifier open loop gain (A) requirement and a total of 140 dB open loop gain for the input signal. Usually more gain than this is needed for fast settling in integrators and for good amplifier closed loop linearity. When small gain amplifiers are used the linearity is more difficult to achieve than the short settling time.

The gain of the amplifiers influences also to the quantization noise transfer function (NTF). Instead of the ideal z-plane integrator transfer function we get a nonideal integrator [15]. When feedback coefficients a and b are 0.5 the noise transfer function of the modulator is:

$$NTF_0(z) = \frac{(z-1)^2}{(z^2-z+0.5)} \qquad (2)$$

With nonideal integrators and open loop gain of 69 dB this becomes:

$$NTF(z) = \frac{(z-0.9994)^2}{(z^2-0.9998 \cdot z+0.4994)} \qquad (3)$$

It can be seen that the double zero in transfer function is no more exactly one and correspondingly the second order quantization noise attenuation at baseband is decreased. So, the gain and bandwidth of the amplifiers must be kept high enough to prevent loss in SNR which is caused by the zero shift in transfer function.

### 3. Comparator hysteresis and speed

The hysteresis of the comparator does not effect directly to the modulator SNR but internal node voltages of the integrator may increase depending on the feedback coefficients of the modulator. Usually the largest increase is in the second integrator and the capacitor ratios were adjusted correspondingly. These effect will be discussed elsewhere [12].

Fast comparator structure is needed, because the comparator and the nondelaying integrator are working at the same clock phase. With slow comparator there is not enough settling time for integrator.

### 4. Capacitor nonidealities

The capacitor nonidealities in double metal or double poly capacitors are low enough for high performance A/D-converter. This is illustrated in figure 6.

### IV. IC-implementation

The circuit structure was developed based on the extensive simulation. The modulator was designed to be a part of general purpose A/D-converter for 16-bit modem and 12-bit ISDN applications. In modem applications the oversampling ratio of 256 is needed with the sampling frequency of 2.5 MHz. In ISDN applications the baseband Nyquist frequency is 160 kHz and the sampling frequency of 5.12 MHz is needed with oversampling ratio of 64. The different baseband requirements are achieved by external programming of the multirate decimation filter and thus the same A/D-converter chip can be used for both applications.

### 1. Differential circuit topology

The differential circuit architecture for the modulator was chosen in order to have better dynamics and larger linear range of the amplifiers with a single 5 V power supply. In addition the differential structure cancels out the even harmonic distortion components, decreases the consequences of the charge injection from switches and has a better power supply rejection ratio up to high frequencies than the single ended one. The layout of the differential circuit was done such that the components have only random offset errors. The overall modulator circuit topology is shown in figure 7 and the floorplan in figure 8.

### 2. Integrators

The first integrator is a nondelaying and inverting model. The second one is a delaying model because at least one delay is needed for clocked feedback loop system. In the first clock phase the comparison and the first integration are done. Second integration is done during the second clock. The circuit was designed to have maximum clock frequency of 5 MHz. This is limited by the integrator settling time of 100 ns which is achieved by using high slew rate AB-class amplifier structure and unity gain bandwidth of 60 MHz.

### 3. Amplifiers

The input stage of the operational transconductance amplifier (OTA) is shown in figure 9. It has a normal differential pair connected as current gain stage and a push-pull output cascode stage. The input transistors are quite large (1400 um$^2$) to reduce the 1/f-noise generated by the input stage. The cascode output stage is needed to boost the amplifier gain. With used bias current it increases the amplifier gain by 20 dB.

Because the signal is represented by the current, excluding the output stage, the first pole of the amplifier (caused by input stage) is at 120 MHz. This guarantees that the amplifier is stable for the capacitive loading of integrator circuit topology. Because the phase margin of class AB OTA is not slew rate limited [17] large phase margins are not required for the integrators. The large current driving capability of class AB OTA is needed to charge the integrator capacitances with high speed.

Additional transistors has been connected as current sources parallel to the output stages. Their purpose is to keep the amplifier outputs in linear region with the common mode feedback circuit. The SC-type common mode circuit was chosen because of small loading of the OTAs [16].

### 4. Switches

The switches connected to integrator input must be low resistance such that they can quickly drive the charge to the Miller capacitance of the integrator. Other switches are almost minimum size. All switch structures are CMOS transmission gates with same aspect ratios for pmos and nmos devices to prevent the charge injection in spite of the fact that fully differential circuit cancels most of them.

### 5. Comparator

The comparator was designed to be a positive feedback clocked type with delay of 15 ns and resolution of 1 mV. This performance keeps the integrators in linear region. It is important to get a proper digital value from the comparator with small delay to ensure adequate response time for the nondelaying integrator. The comparator has two amplification stages. The first one is a normal differential pair connected to another one which has a positive feedback and is clocked. The output of the second stage is connected to a RS-flip-flop which insures with positive feedback that comparator always has the proper digital output level.

The most hazardous situation for the modulator is a comparator which stays sometimes in a metastable state between two digital levels. In that case the SNR after the decimation filter will be bad because the comparator output level may have been interpreted differently by the modulator and the decimation filter. The loss of the SNR depends on the used filter structure and it will affect the converter output during the timeperiod corresponding the filter latency.

### 6. Area

The modulator was implemented using 2.5 micron molybdenum gate CMOS high performance analog process with molybdenum metal capacitors (Micronas ltd). The differential circuit takes about 1.5 times the area compared to a single ended. The modulator die area is about 0.8·0.7 mm$^2$. The modulator has been designed to be a part of the general purpose A/D-converter VLSI macrocell. The whole A/D-converter including the decimation filter is about 5 mm$^2$.

### V. Conclusions

It is demonstrated that 16-bit SNR A/D-converter for voice band using second order noise shaping sigma-delta modulation with oversampling ratio of 512 and a 14-bit SNR and 16-bit dynamic range converter with 256 oversampling can be designed. The implemented converter uses single 5 V power supply and can be placed on a chip containing digital circuits without loosing its accuracy. With the novel decimation filter structure it is the first high performance macrocell A/D-converter that consumes only 5 mm$^2$.

### Acknowledgement

We like to thank G. Holm and A. Yläjääski of Nokia Corp. for many fruitful discussions. This work is partially supported by National Microelectronics Program of Finland and Micronas ltd.

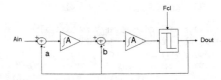

Figure 1. Ideal sigma-delta modulator model.

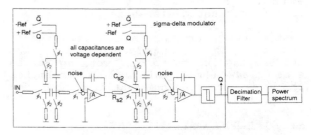

Figure 2. Sigma-delta modulator model for simulations.

| Ref | Year | Type | Bandwidth kHz | S/(N+THD) dB | Clock MHz | Technology | Size mm | Voltage V | Power mW | Decimation Filter | Special |
|---|---|---|---|---|---|---|---|---|---|---|---|
| 1 | 81 | S.I. | 4 | 52 note 1,3,5 | 2.048 | ? um NMOS | 5 note 5 | ? | 21 | 127 tap FIR + 4 tap FIR | exponental modulator |
| 2 | 82 | S.I. | 4 | 60 note 1,3,4,5 | 2.048 | 5 um NMOS | 0.4 note 9 | ? | 25 | 1 order FIR + 256 tap FIR | dicher |
| 3 | 86 | D.I. analog filter | 20 | 84 note 1 | 12 | 3 um CMOS | 3.3 note 7 | +5,-5 | 60 | ? | dicher RC implementation 2-order filter |
| 4 | 86 | D.I. cascoded | 3.4 | 68 note 1,3,4,5 | 2.048 | 1.5 um CMOS | 1.5 | +5 | ? | FIR | two cascoded S.I. modulators |
| 5 | 86 | D.I. | 120 | 70 note 1 | 15.36 | 3 um CMOS | 3 note 6 | +5 | 15 | 4 tap FIR + 3 order IIR | bidirectional current source integrators |
| 6 | 87 | S.I. | very small | 96 note 2 | ? | 4 um CMOS | 1 note 10 | +2.5 -2.5 | 0.15 | counter | simple and slow |
| 7 | 87 | Q.I. analog filter | 0.4 | 112 note 2,3,4 | 2.048 | ? um | one chip | +15,-15 +5 | 9 | 448 tap FIR + 172 tap FIR | modulator has one integrator and 3-order filter |
| 8 | 87 | D.I. note 8 | 4.988 note 8 | 70 note 8 | 1.336 note 8 | note 8 | note 8 | note 8 | note 8 | 1002 order FIR note 8 | interactive design tool for analog CMOS circuits |
| 9 | 87 | T.I. cascoded | 24 | 91 note 2 | 3 | 2 um CMOS | 7.3 | +5 | 35 | 4 order FIR + 256 tap FIR | dicher |
| 10 | 87 | D.I. | 120 | 62 note 2,4 | 15.36 | 2 um CMOS | ? | +5 |  | 3 order IIR |  |
| this work | 87 | D.I. | 4 | 85 note 11 | 2.048 | 2.5 um CMOS | 0.56 | +5 | 10 | 2600 order adjustable 15 tap FIR |  |

Table 1.

Notes:
1: Baseband S/(N+THD) from modulator output spectrum (S/((N+THD) with superior filter).
2: Baseband S/(N+THD) from decimation filter output.
3: THD has not been properly specified.
4: Input test signal has not been specified.
5: Estimated from the photograph.
6: The figure includes filter.
7: Signal substraction and loop are implemented with a passive RC-network outside the chip.
8: The user has to specify the tehnology, the desired building blocks and options. Available are S.I. and D.I. types. The figures are an example from the reference.
9: Comparator only.
10: Control logic outside the chip.
11: Actual circuit has not been measusured yet. The figures represent the results of the simulations.

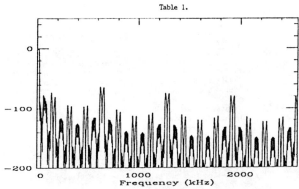

Figure 3. Amplitude response of digital filter with decimation ratio of 128 for full frequency range.

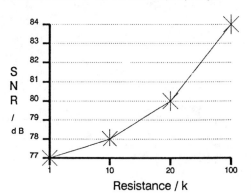

Figure 4. Modulator SNR dependence on the Rs2Cs2 time constant with noisy amplifier, Cs2 of 1 pF and oversampling ratio of 256.

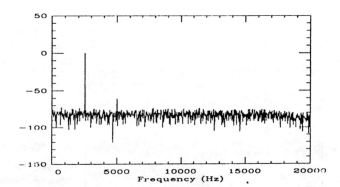

Figure 5. Modulator filtered spectra with nonlinear amplifier and decimation ratio of 128. Gain depends on output voltage (A=66dB when Vout>0.5V, A=60dB when -1V<Vout<0.5V, A=40dB when Vout<-1V). Internal node voltages scaled to be between -1.5 and 1.5V. Sampling frequency is 5.12MHz. Input signal is -10dB 2.5kHz sinusoidal voltage.

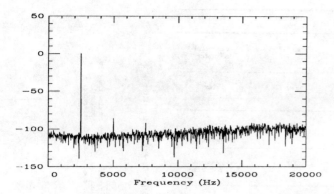

Figure 6. Modulator filtered spectra with decimation ratio of 128. The capacitors have voltage nonidealities of 1000 ppm. Sampling frequency is 5.12MHz. Input signal is -10dB 2.5kHz sinusoidal voltage.

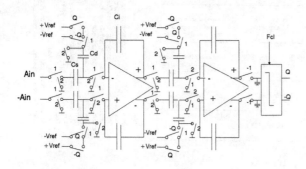

Figure 7. Modulator overall topology.

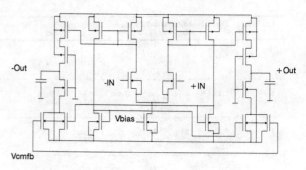

Figure 9. Circuit schematic of the OTA.

References

[1] T. Misawa, J. Iwersen, L. Loporcaro and J. Ruch, "Single-chip per channel codec with filters utilizing sigma-delta modulation", IEEE J. Solid-State Circuits, Vol SC-16, pp. 333-340, Aug 1981.

[2] J. R. Fox and G. J. Garrison, "Analog to digital conversion using sigma-delta modulation and digital signal processing", Conf. Dig. Tech. Papers, 1982 conference on advanced research in VLSI, M.I.T, pp. 101-112, Jan 1987.

[3] U. Roettcher, H. Fiedler and G. Zimmer, "A compatible CMOS-JFET pulse density modulator for interpolative high-resolution A/D conversion", IEEE J. Solid-State Circuits, Vol SC-21, pp.446-452, Jun 1986.

[4] K. Uchimura, T. Hayashi, T. Kimura and A. Iwata, "VLSI-A to D and D to A converters with multi-stage noise shaping modulators", Conf. Dig. Tech. Papers, ICASSP 86, pp. 1545-1548, Japan, Tokyo, 1986.

[5] R Koch, B. Heise, F. Eckbauer, E. Engelhardt, J. Fisher and F. Parzefall, "A 12-bit sigma-delta analog-to-digital converter with a 15-MHz clock rate", IEEE J. Solid-State Circuits, Vol SC-21, pp.1003-1009, Dec 1986.

[6] J. Robert, G. Temes, V. Valencic, R. Dessoulavy and P. Deval, "A 16-bit low-voltage CMOS A/D-converter", IEEE J. Solid-State Circuits, Vol SC-22, pp. 157-163 Apr 1987.

[7] F. Goodenough, "Fast 24-bit ADC converter handles dc-to-410 Hz input signals", Electronic Design, pp. 59-62, Oct 1987.

[8] M. Degrauwe, O. Nys, E. Dijkstra, J. Rijmenants, S. Bitz, B. Goffart, E. Vittoz, S. Cserveny, C. Meixenberger, G. Stappen and H. Oguey, "IDAC: an interactive design tool for analog CMOS circuits", IEEE J. Solid-State Circuits, Vol SC-22, pp. 1106-1115, Dec 1987.

[9] Y. Matsuya, K. Uchimura, A. Iwata, T. Kobayashi, M. Ishikawa, T. Yoshitome, "A 16-bit oversampling A-to-D conversion technology using triple-integration noise shaping", IEEE J. Solid-State Circuits, Vol SC-22, pp. 921-929, Dec 1987.

[10] D. Sallaerts, D. Rabaey, R. Dierckx, J. Sevenhans, D. Haspeslagh and B. Ceulaer, "A single-chip U-interface transceiver for ISDN", IEEE J. Solid-State Circuits, Vol SC-22, pp. 1011-1021, Dec 1987.

[11] T. Saram ki and H. Tenhunen, "Efficient VLSI-realizable decimators for sigma-delta analog-to-digital converters", ISCAS'88, this conference.

[12] T. Karema, T. Ritoniemi and H. Tenhunen, "Statistical properties of advanced sigma-delta modulator with enhanced noise shaping characteristics", to be published in EUROCON 88, Sweden, Stockholm, Jun 1988.

[13] G. Brauns, M. Steer, S. Ardalan and J. Paulos, "ZSIM: a nonlinear z-domain simulator for delta-sigma modulators", product sheet.

[14] B. Agrawal and K. Shenoi, "Design methodology for sigma-delta modulator", IEEE Trans. on Communications, Vol COM-31, pp. 360-369, Mar 1983.

[15] R. Gregorian and G. Temes, Analog MOS integrated circuits for signal processing, pp. 462-524, John Wiley & Sons 1986.

[16] R. Castello, "Low-voltage low-power MOS switched-capacitor signal-processing techniques", Ph. D. Dissertation, University of California, Berkeley, pp. 118-127, Dec 1984.

[17] M. Copeland, G. Bell and T. Kwasniewski, " A Mixed-Mode Sampled-data Simulation program", IEEE J. Solid-State Circuits, Vol. SC-22, pp. 1098-1105, Dec 1987.

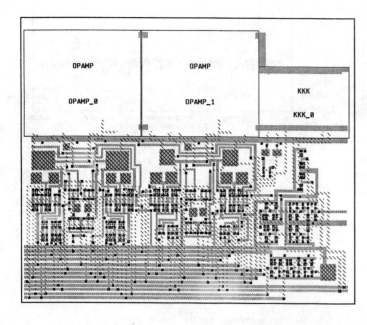

Figure 8. Modulator floorplan.

# A HIGH-RESOLUTION CMOS SIGMA-DELTA A/D CONVERTER WITH 320 KHZ OUTPUT RATE

Mike Rebeschini, Nicholas van Bavel*, Pat Rakers,
Robert Greene, Jim Caldwell, John Haug


Corporate Research Design Labs, Motorola Inc., Schaumburg, IL USA
*Semiconductor Products Sector, Motorola Inc., Austin, TX USA

## ABSTRACT

A third-order switched-capacitor sigma-delta A/D converter is presented. The converter consists of three cascaded first-order sigma-delta modulators, and achieves 88 dB pk/rms S/(N+D) in a 160 kHz passband with a 10.24 MHz sampling rate. The high signal-to-noise ratio is maintained over this wide bandwidth by attenuating the quantization noise with a third-order noise-shaping function and a fourth-order comb decimation filter. A special auto-zeroed integrator having low pole error is required to simultaneously achieve the 10.24 MHz sampling rate and high-order noise-shaping.

## INTRODUCTION

In recent years, oversampled A/D converters using sigma-delta modulation have become increasingly popular for wide bandwidth applications[1]-[3] and high-resolution audio-band applications[4]-[6]. As the reduction in linewidths for CMOS integrated circuit processes yields higher packing densities, digital signal processing (DSP) becomes feasible for more applications. Integrating the A/D and D/A conversion circuitry on to the same die creates a complete and cost-effective DSP system. Sigma-delta modulation is an excellent candidate for the A/D converter in such a DSP chip, since it can achieve excellent linearity and noise performance with reduced analog circuit complexity by using a high sampling rate and digital filtering. In addition, sigma-delta A/Ds are particularly well suited to implementation in a digital CMOS process[7].

Unlike conventional successive approximation or flash A/D converters, sigma-delta A/Ds do not require precisely matched components in order to achieve a high signal-to-noise (S/N) ratio[8]. Instead, a high S/N is obtained by oversampling and digital filtering. The analog input is converted to a low resolution digital signal by the sigma-delta modulator at a high sampling rate. By virtue of the analog low-pass filter inside the modulator, the quantization noise is attenuated at low frequencies in the passband, and amplified at high frequencies. The order of a sigma-delta modulator is defined to be the order of this analog low-pass filter. The subsequent digital low-pass filter then rejects the out-of-band noise, creating a high S/N in a reduced bandwidth. High S/N can be obtained using even a two-level quantizer and feedback[8]. A two-level quantizer and feedback is desirable because it results in simplified modulator implementation and is inherently linear, since level mismatch results in only small gain and offset errors, and not harmonic distortion.

As a result of the digital and analog filtering, the noise at the output of a sigma-delta A/D is close to Gaussian in distribution. Since the Gaussian distribution is unbounded, the A/D performance cannot be measured in terms of resolution or differential and integral linearity. Instead, we use S/N ratio with sinusoidal inputs. For the optimal loop filter, the S/N ratio increases by 6N+3 dB for every doubling of the oversampling ratio, where N is the order of modulator[9]. Thus for a given passband, the S/N can be greatly improved by increasing the sampling rate or the order of the modulator. The sampling rate is limited by the settling time of the fastest op-amp that can be obtained in the chosen process. The order of the modulator can be increased by using higher-order loop filters, although stability becomes a problem for orders greater than two when using one-bit quantizers [6][8].

Another approach to achieve higher-order noise-shaping is to cascade first-order modulators[5]. The first modulator converts the analog input signal, and subsequent modulators convert the quantization noise generated by the previous modulator. The quantization errors of all but the last modulator are digitally cancelled, yielding a noise-shaping function of order equal to the number of first-order modulators. Advantages of the cascaded sigma-delta approach include guaranteed stability to any order, limited signal swing at the output of the integrator, and pipelining of the switched-capacitor integrators[5]. The main disadvantage of the cascaded technique is that the S/N is more sensitive to analog component accuracy[5].

In particular, the transfer function of the switched-capacitor integrator used in the modulator must be close to the ideal, $1/(1-z^{-1})$. Deviations from this function can be classified by gain and pole error. Capacitor mismatch causes gain error, and finite op-amp open-loop gain causes gain and pole error. For reasonable capacitor matching, the dominant noise source is caused by pole error due to finite op-amp gain. It is difficult to maintain sufficiently high open-loop gains for higher sampling rates, because the gain of a CMOS op-amp tends to be inversely proportional to its bandwidth. For this reason, it was previously believed that the maximum sampling rate of the cascaded sigma-delta technique is 4 MHz[5].

The purpose of this paper is to examine the effects of the various error sources on the performance of the converter, and to present the design and measured performance of a third-order, cascaded sigma-delta A/D converter using a novel autozeroed modulator architecture. This autozeroed modulator maintains close to the ideal noise-shaping function even with relatively low op-amp gains, and thereby greatly extends the maximum sampling rate and hence the bandwidth of the converter.

Reprinted from *IEEE Proc. ISCAS'89*, pp. 246–249, May 1989.

## THEORETICAL ANALYSIS

A system block diagram of the third-order cascaded sigma-delta A/D converter described in this paper is shown in Figure 1 with $d=z^{-1}$ and $D=z^{-32}$. The A/D converter consists of three first-order sigma-delta modulators providing three inputs to a fourth-order digital decimation filter. (Note that the digital differentiators used in [5] are eliminated by absorbing them into the first stages of the comb decimation filter.) The quantizer is modeled as an additive noise source labeled $e_i$. The inputs to the second and third modulators are the quantization errors from the previous modulators, obtained by subtracting the output of the modulator from the integrator output.

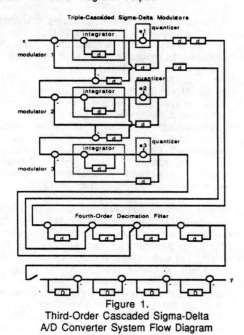

Figure 1.
Third-Order Cascaded Sigma-Delta
A/D Converter System Flow Diagram

The overall system transfer function is given by:

$$y = \left[x\, z^{-2} + e_3(1-z^{-1})^3\right]\frac{(1-z^{-32})^4}{(1-z^{-1})^4} \quad (1)$$

Note that for this ideal system model of the A/D converter, the quantization noises, $e_1$ and $e_2$, have been cancelled in the decimation filter and do not appear in the output. This ideal model assumes that the analog integrators have a transfer function of exactly $1/(1-z^{-1})$. Due to capacitor mismatch and finite op-amp gain, however, the actual transfer function is $\dfrac{1\ (1-\alpha_i)}{1-z^{-1}(1-\beta_i)}$, where $\alpha_i$ is the resultant gain error and $\beta_i$ is the pole error in the $i$th modulator. Assuming that the transfer function across each modulator is unchanged, ignoring the small filtering effects on the input x, and dropping the lower-order error terms, the actual transfer function is closely approximated by:

$$y = \left[x\, z^{-2} + e_3(1-z^{-1})^3 + \alpha_1 e_1(1-z^{-1})z^{-3}\right.$$
$$\left. + \beta_1 e_1 z^{-4}\right]\frac{(1-z^{-32})^4}{(1-z^{-1})^4} \quad (2)$$

Notice that the dominant leakage terms come from the gain and pole error in the first modulator, and therefore only the design of the first modulator requires special attention. A nonzero pole error, $\beta_1$, causes a leakage of unshaped quantization noise, $e_1$. Assuming that the quantization noise is white, $e_1$ is attenuated 18.2 dB by the decimation filter with an oversampling ratio of 32, and decreases only 3 dB per doubling of the oversampling ratio. A nonzero gain error, $\alpha_1$, causes a leakage of first-order shaped quantization noise $e_1(1-z^{-1})$, which is attenuated 46.9 dB by the combined effects of the noise-shaping and decimation filter, and decreases 9 dB per doubling of the oversampling ratio. From this point on we will assume for convenience that the peak input and peak feedback signal are unity, the additive error source $e_i$ is flatly distributed white noise, and a one-bit quantizer and feedback are used. Under this assumption the peak of the error $e_1$ is unity and its power is -4.77 dBV, where 1 V rms=0 dBV. The gain error and the pole error noise will be:

$$N_{ge} = -51.7 + 20\log(\alpha)\ \text{dBV} \quad (3)$$
$$N_{pe} = -23.0 + 20\log(\beta)\ \text{dBV} \quad (4)$$

## MODULATOR CIRCUIT IMPLEMENTATION

To examine the effects of finite op-amp gain on the noise, let us examine a straightforward, single-ended implementation of the first-order modulator as shown in Figure 2.

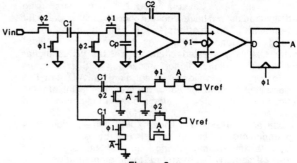

Figure 2.
First-Order Sigma-Delta Modulator Schematic

Due to finite op-amp gain A, the gain and pole errors will be:

$$\alpha = \frac{1}{A}\left(1 + \frac{3C_1}{C_2}\right)\ ,\ \beta = \frac{1}{A}\frac{3C_1}{C_2} \quad (5)$$

Because of noise-shaping, the pole error noise dominates the noise due to gain error. If $C_2=4C_1$ and the noise due to pole error, $N_{pe} = -100.77$ dBV, then by (4) and (5), the open-loop gain of the op-amp needs to be 75.3 dB. To achieve the 10.24 MHz sampling rate, a wide-bandwidth CMOS op-amp is required with narrow gate lengths, operating at high current density. These design criteria make it very difficult to meet the 75.3 dB gain requirement over process and temperature variations.

To alleviate this problem, the autozeroed sigma-delta modulator shown in Figure 3 is used. The circuit is shown single-ended for simplicity, but is implemented fully differential on the actual chip. This modulator incorporates an autozeroed integrator similar to that described in [10]. This integrator has more accurate pole placement for the same op-amp gain, and hence the modulator has less pole-error noise.

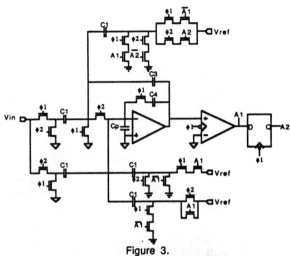

**Figure 3.**
Autozeroed Sigma-Delta Modulator Schematic

The basic idea of the new modulator design is to autozero the op-amp with the output not returning to zero, but held at the previous state. This cancels the D.C. offset,1/f noise, and the error voltage at the input terminals of the op-amp. This error voltage is caused by the finite open-loop gain of the op-amp, and cancelling it has the effect of boosting the effective gain of the op-amp. A mathematical analysis of the circuit yields a gain and pole error of:

$$\alpha=\frac{1}{A}(1+\frac{3C_1+C_p}{C_4})\ ,\ \beta=\frac{3C_1+C_p}{C_4A^2}(1+\frac{2C_1}{C_3}) \qquad (6)$$

Comparing the pole error of the autozeroed modulator expressed in (6) to the previous implementation in (5), the pole error has almost decreased by a factor of A, the op-amp gain. As an example of the benefits of using this modulator architecture, if $C_4=4C_1$, $C_p=C_3=2C_1$, and the pole error noise, $N_{pe}$ = -100.77 dBV, then by (4) and (6) the open-loop gain of the op-amp now needs to be 42.9 dB, instead of 75.3 dB for the previous modulator design.

The schematic of the sigma-delta loop used for the second and third modulators is shown in Figure 4 where A is the output of the previous modulator.

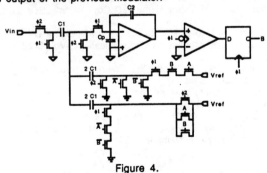

**Figure 4.**
Schematic of Second and Third Modulator

The analog input to these modulators is the output of the integrator in the previous modulator. To get the quantization noise only as the input to this modulator, the output of the previous modulator must be subtracted as shown in the system diagram (Figure 1). This subtraction is incorporated in the feedback circuitry. Thus, instead of feeding back +Vref or -Vref into the integrator depending on the output of this modulator in the previous clock period, either +2Vref,0, or -2Vref is fed back, depending on the output of this modulator minus the previous modulator. Thus, the subtraction is done digitally, and a three-level D/A feeds back the correct signal. As mentioned before, autozeroing is required only in the first modulator, so this modulator is the same as the straightforward one shown in Figure 2, except for the three-level feedback.

## OTHER CIRCUIT CONSIDERATIONS

Gain error may also be caused by capacitor mismatch, which was minimized by using careful layout techniques. Since all capacitor values were in integer ratios, capacitors were made up of integer numbers of an identical unit plate. A common-centroid layout for these unit capacitors was used so that linear gradients in oxide depth were cancelled. Sharp corners were not used on capacitor plates to minimize variations in etching. Using these techniques, a capacitor matching of better than .5% was achieved.

With the low oversampling ratio of 32, the effect of thermal noise had to be considered in the channel resistance of the input switches. The switched input and feedback capacitors in the first stage are 1 pF. The resultant thermal noise is -99.1 dBV.

In the first-modulator, 1/f noise was cancelled by the autozeroing integrator. Because of noise shaping, the thermal and 1/f noise in the op-amps in the second and third modulators was negligible.

To obtain low distortion, voltage-variable switched-charge-injection is eliminated by proper clock phasing. The switches coupled to the summing junction are turned off slightly before the switches on the opposite side of the capacitors. Thus the charge from the switches that are driven by the input, Vref, or op-amp output are not integrated on to the integration capacitor. The summing junction switches still inject charge, but the voltage on the source and drain of these switches prior to turn-off is equal to analog ground. This causes the charge-injection to be input-signal independent and therefore causes only an offset voltage but not harmonic distortion.

As speed was the major concern in the design of the op-amp, a fully-differential folded-cascode structure was chosen. In the comparator design, speed was also the main design constraint, while offset voltage and hysteresis were of little concern, because these errors are attenuated by the noise-shaping function. Therefore, a regenerative type latch similar to a RAM sense amp was used.

## EXPERIMENTAL RESULTS

The third-order cascaded sigma-delta modulator was implemented in a 1.5 µm double-metal double-poly CMOS process. Figure 5 shows an FFT of the output of a simulated fourth-order comb filter with a 10 kHz sinusoidal input.

Figure 6 is a plot of S/(N+D) as a function of input amplitude with a 10 kHz sinusoidal input for an oversampling ratio of 32 and a clock frequency of 10.24 MHz.

Figure 6 shows a degradation in the noise floor with low input amplitudes. This is caused by the quantization noise generated in the first modulator becoming correlated with the signal and itself. Under this condition, the assumption that the quantization noise, $e_1$, is white is no longer valid and, in actuality, has higher baseband energy. Thus, the leakage of this noise from gain error causes a higher baseband noise. This problem can be eliminated by adding a dither signal, at the expense of lowering the peak S/N and dynamic range.

According to the ideal system model described in (1), the S/N ratio should be 97.1 dB pk/rms. The measured S/N of 88 dB is believed to be dominated by a leakage of first-order shaped noise due to capacitor mismatch.

The active area of the modulator was 3 mm$^2$. The measured peak signal-to-noise-plus-distortion ratio was 88 dB pk/rms, and the peak signal-to-distortion ratio was 100 dB pk/rms.

## CONCLUSION

A novel cascaded sigma-delta architecture was presented with improved bandwidth and S/N. The degradation in S/N caused by various analog circuit imperfections were analyzed for the previous and new modulator architectures. The measured performance correlated well with the theoretical analysis.

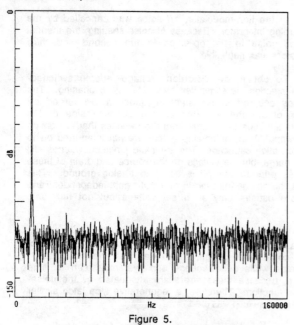

Figure 5.
Converter Output Spectrum with Sinusoidal Input

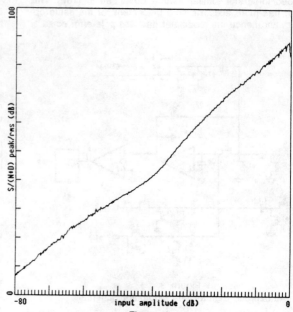

Figure 6.
Plot of S/(N+D) versus Input Amplitude

REFERENCES

[1]    S. Norsworthy, I. Post, "A 13-bit, 160 kHz Sigma-Delta A/D Converter For ISDN", Proc. of the IEEE 1988 CIC Conf., p.21.3.1-4: May 1988.
[2]    L. Longo, M. Copeland, "A 13 bit ISDN-band Oversampled ADC using Two-Stage Third Order Noise Shaping", Proc. of the IEEE 1988 CIC Conf., p21.2.14: May 1988.
[3]    R. Koch, et al., "A 12-bit Sigma-Delta Analog-to-Digital Converter with a 15-MHz Clock Rate", IEEE Journal of Solid-State Circuits, Vol. SC-21, No. 6: December 1986.
[4]    K. Matsumoto, et al., "An 18b Oversampling A/D Converter for Digital Audio", IEEE Solid-State Circuits Conf. Digest of Technical Papers, p. 202-203: February, 1988.
[5]    Y. Matsuya, et al., "A 16-bit Oversampling A-to-D Conversion Technology Using Triple Integration Noise Shaping", IEEE Journal of Solid-State Circuits, Vol. SC-22, No. 6: December 1987.
[6]    D. Welland, et al., "A Stereo 16-bit Delta-Sigma A/D Converter for Digital Audio", 85th Convention of the Audio Engineering Society: November 1988.
[7]    M. Hauser, R. Brodersen, "Circuit and Technology Considerations for MOS Delta-Sigma A/D Converters", ISCAS Proc.: 1986.
[8]    J. Candy, "A Use of Double Integration in Sigma Delta Modulation", IEEE Transactions on Communications, Vol. COM-33, No. 3: March 1985.
[9]    S. Tewksbury, R. Hallock, "Oversampled, Linear Predictive and Noise-Shaping Coders of Order N>1", IEEE Transactions on Circuits and Systems, Vol. CAS-25, No. 7: July 1978.
[10]   K. Haug, F. Maloberti, G. Temes, "Switched-Capacitor Integrators with Low Finite-Gain Sensitivity", Electronic Letters, Vol. 21, No. 24: November 1985.

# A CMOS Slope Adaptive Delta Modulator

*Jeffrey W. Scott\*, Wai Lee, Charles Giancario\*\*, Charles G. Sodini*

MIT

Cambridge, MA

Chairman:  George Erdi

Linear Technology Co.

Milpitas, CA

RECENT INNOVATIONS in audio signal processing have placed high demands on associated data conversion circuitry. Dynamic range requirements of 96dB or more have prevented inexpensive CMOS processing from claiming its share of the audio band A/D market. This paper will demonstrate, theoretically and experimentally, that high resolution CMOS A/D conversion of audio signals can be achieved by oversampling a feedback loop which attenuates quantization noise power in a low-frequency (20kHz) baseband. The technique has been used to perform high resolution A/D conversion over a 4kHz voice-band[1].

A block diagram of the Immediately Adaptive Delta Modulator (IADM) is shown in Figure 1($a$). The presence of the integrating filter H and the digital accumulator ensure sufficient low-frequency loopgain so that $x(n)$ effectively tracks the audio band input $x(t)$. The digital signal $d(n)$ represents the slope of the input, since accumulating $d(n)$ yields the output $x(n)$ (the digital estimate of the input). Hence, the system is slope adaptive (large changes in $x(n)$ during periods of large input slope), as can be seen in the tracking pattern simulated in ($b$) of Figure 1.

Digitizing the slope ($d(t)$) of the input signal provides two system advantages: first, the required dynamic range of the flash A/D converter is reduced below that of the D/A converter, since the slope of an oversampled input is converted to $d(n)$. In fact, the number of comparators in the flash A/D converter is reduced further by asking the digital accumulator to change its contents by only one bit position on any clock cycle, the particular bit position being determined by the magnitude of $d(t)$. This approach, which yields a noisier conversion for input signals with large maximum slope, results in a companded SNR characteristic (to be displayed shortly). The second advantage to slope digitization is a more efficient encoding (less number of bits per sample) of $x(t)$ in the form of $d(n)$.

Slope overload of the converter can be seen in the measured step response shown in Figure 2. Since a step transition in the input contains spectral components outside the audio band, $d(t)$ exceeds the flash A/D converter dynamic range, causing $x(n)$ to take limited step size transitions until it *catches up* to the input. Stability and the slope adaptive nature of the modulator are demonstrated by the flat output response following the initial transient.

The split capacitor D/A converter and the integrating filter H are shown in Figure 3. Capacitor $C_2$ integrates $V_{IN}$ through $C_F$, while capacitors $C_{UNIT}$, $C_X$ and $C_{LSB}$ integrate $V_{D/A}$ through $C_F$. This integrating action is required to provide sufficient low frequency gain

to attenuate quantization noise introduced by the A/D converter. Capacitors $C_1$ and $C_{MSB}$ introduce a zero into the loop transmission required for closed loop stability.

A block diagram of the comparator used in the IADM flash A/D converter is shown in Figure 4($a$). A differential gain stage, shown in ($b$) of Figure 4, is offset-cancelled during phase $\phi_1$ and then ac-coupled to the input voltages to be compared during phase $\phi_2$. Devices MCM1 through MCM4 provide common mode feedback biasing. The amplification provides sufficient overdrive for the following CMOS latch (latched on phase $\phi_4$), shown in ($c$) of Figure 4, so that the comparator can realize rail-to-rail output voltage voltages for a differential input signal $<$1mV in less than 20ns.

The IADM was fabricated in a 3.5$\mu$m twin-tub, 2 level poly CMOS process. Autorouting techniques were used throughout the chip; in fact, only the power supplies were routed by hand. No attempt was made to optimize the area of the 35mm$^2$ test chip.

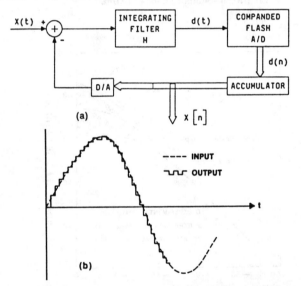

FIGURE 1—($a$) IADM block diagram; ($b$) simulated tracking pattern.

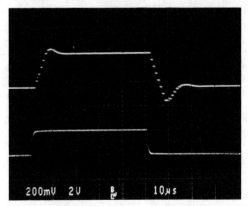

FIGURE 2—Demonstration of slope adaption of IADM. Input: 20kHz square wave.

\*Current address: AT&T Bell Laboratories, Reading, PA.

\*\*Current Address: O'Dowd Communications, Montclair, NJ.

[1] Hauser, M.W., Hurst, P.J., Brodersen, R.W., "MOS ADC-Filter Combination That Does Not Require Precision Analog Components", ISSCC *DIGEST OF TECHNICAL PAPERS*, p. 80-81; Feb., 1985.

[2] Candy, J.C., Wooley, B.A., Benjamin, O.J., "A Voiceband Codec With Digital Filtering", *IEEE Trans. Comm.*, Vol. COM-29, No. 6, p. 815-830; June, 1981.

[3] Giancarlo, C.H., Sodini, C.G., "A Slope Adaptive Delta Modulator For VLSI Signal Processing Systems", to be published in *IEEE Trans. of Circuits and Systems*.

Reprinted from *ISSCC Dig. Tech. Pap.*, pp. 130–131, February 1986.

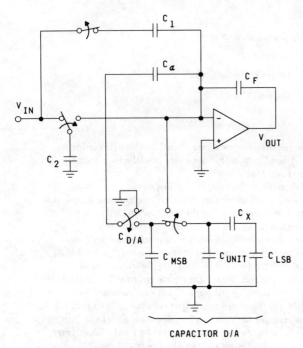

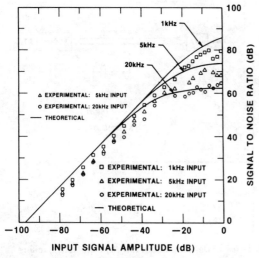

FIGURE 3—Summing node of IADM.

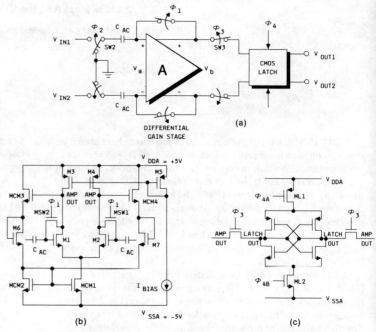

FIGURE 4—Differential ac-coupled comparator: (a) block diagram, (b) differential gain stage, (c) CMOS latch.

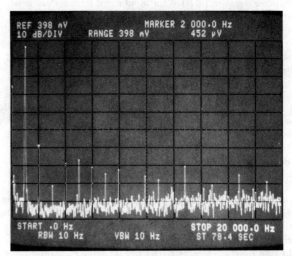

FIGURE 5—IADM signal-to-noise ratio. Sampling frequency = 1MHz.

FIGURE 6—IADM output spectrum for a 1kHz sinusoidal input: THD = 0.17%.

Both the theoretical SNR derived in references[2,3] and experimental measurements made on the IADM are shown in Figure 5. The measured dynamic range of 91dB for a sampling frequency of 1MHz and a 10b accumulator and D/A converter is in close agreement with the theoretical value of 97dB. Total harmonic distortion of the IADM A/D converter is limited by the nonlinearity of the feedback D/A converter.

The output spectrum of the converter for a 1kHz sinusoid input, shown in Figure 6, yields a measured THD of 0.17%.

In summary, experimental results have shown that the IADM architecture can perform high resolution (>15 bits) audio band A/D conversion with standard, untrimmed CMOS technology.

# A Stereo 16-Bit Delta–Sigma A/D Converter for Digital Audio*

D. R. WELLAND, B. P. DEL SIGNORE, AND E. J. SWANSON

*Crystal Semiconductor Corporation, Austin, TX 78744, USA*

**AND**

T. TANAKA, K. HAMASHITA, S. HARA, AND K. TAKASUKA

*Asahi Kasei Microsystems, Inc., Tokyo, Japan*

A two-channel 16-bit analog-to-digital (A/D) converter employing oversampling techniques has been developed. The device contains two fourth-order delta–sigma modulators with 1-bit outputs, each followed by a digital finite impulse response filter/ decimator. The analog inputs are sampled at 3.072 MHz and the digital words are output at 48 kHz.

## 0 INTRODUCTION

The emergence of digital audio has increased the demand for high-performance analog-to-digital (A/D) converters. Delta–sigma conversion has been gaining recognition as having advantages over more classical audio band conversion techniques. In particular, it obviates the need for sample-and-hold amplifiers, eases the design of anti-aliasing filters, and is free of differential nonlinearity errors that distort low-level signals.

This engineering report discusses the development of a two-channel 16-bit audio band delta–sigma converter featuring a high degree of integration, which is suitable for use in stereo digital audio applications. Sec. 1 gives an overview of the concepts pertaining to delta–sigma conversion. The device architecture and a functional partitioning strategy are presented in Sec. 2. Secs. 3 and 4 discuss the design of the two major functional blocks, and measured results are presented in Sec. 5.

## 1 DELTA–SIGMA CONVERSION

### 1.1 Quantization Noise

A/D conversion is a process that necessarily introduces errors into a signal due to quantization. The difference between the output of an otherwise perfect

converter (or "quantizer") and that which might be expected of a converter with unlimited resolution can be modeled as an additive noise signal. The level of this "quantization noise" is reduced in converters of higher resolution with finer quantization levels, but nonetheless must remain nonzero. If the analog input is sufficiently large or if the input is sufficiently random, the spectrum of the quantization noise can be approximated as white [1], with its energy equally distributed between direct current and $f_s/2$, where $f_s$ is the sampling and conversion rate. (It is assumed that the signal is sampled before it is converted.) The effective resolution of a converter can be increased by filtering the output and thereby reducing the level of the quantization noise. A commensurate reduction in the available signal bandwith must be accepted as a consequence of the filtering, but may prove acceptable if the conversion rate is high. The sampling and conversion of a signal at a rate much higher than the signal frequency is a technique termed *oversampling*. The oversampling ratio is the ratio of the actual sampling rate to the Nyquist rate (that is, twice the highest signal frequency of interest).

This process is illustrated in Fig. 1, where an analog sinusoid of frequency $f_0$ is converted to a digital, quantized sinusoid at a rate $f_s$ by a linear pulse-code-modulation (PCM) converter [Fig. 1(a)]. The input spectrum is given in Fig. 1(b), and Fig. 1(c) shows the effects of the quantization process with the addition of white noise. (This and the following spectra repeat at multiples of $f_s$, of course, due to the discrete-time nature of the sampled signal.) Processing by a digital filter

* Presented at the 85th Convention of the Audio Engineering Society, Los Angeles, 1988 November 3–6; revised 1989 March 13.

with a baseband cutoff frequency of $f_b$, as illustrated in Fig. 1(d), yields the output spectrum in Fig. 1(e). The baseband cutoff frequency $f_b$ is presumed to be equal to the highest signal frequency of interest. The remaining quantization noise voltage level will be lower than the original level by a factor of $\sqrt{f_s/2f_b}$. (The quantization noise *energy* is lowered directly by the oversampling ratio $f_s/2f_b$.)

The utility of this technique when applied to audio signals with a baseband frequency of $f_b = 20$ kHz is questionable. To obtain the equivalent of 16-bit performance from, say, a 12-bit converter, the 12-bit quantization noise would have to be lowered by $2^4 = 16$. This would in turn require an oversampling ratio of $f_s/2f_b = 256$, or $f_s \cong 10$ MHz.

Delta–sigma conversion is a technique that employs oversampling to obtain high-resolution (low quantization noise) digital signals from low-resolution (high quantization noise) quantizers. As will be shown below, the delta–sigma converter contrasts with the previous example in that the quantization noise *at the output* of the low-resolution quantizer is not white, but rather

frequency "shaped," such that noise in the baseband ($f < f_b$) is suppressed at the expense of slightly higher out-of-band ($f > f_b$) noise. The power of digital filtering can then be applied to the resultant spectrum to pass the signal (and the residual baseband quantization noise) while rejecting the out-of-band quantization noise.

The converter consists of a modulator and a digital filter. As can be seen in Fig. 2, embedded in the modulator is the low-resolution $N$-bit quantizer, around which frequency-dependent feedback is applied by means of an $N$-bit D/A converter and an analog filter. The analog filter has a frequency response of $H(f)$. The analog input is summed into the modulator loop at a point where, for frequencies with high loop gain, the output of the D/A converter will be substantially equal to the input. To the extent that the D/A converter faithfully reproduces the quantizer digital output, this signal, too, must be substantially representative of the analog input—again, for frequencies with high loop gain.

## 1.2 Loop Analysis

The presence of the nonlinear quantizer renders exact analysis of the loop difficult. A typical and useful approach is to linearize the quantizer by replacing it with a gain stage and a quantization noise source [2]. Again, the latter is only a contrivance to account for the difference between the quantized output and the amplified input. This is illustrated in Fig. 3. The D/A converter has been replaced by a unity-gain stage, as its function is irrelevant for analyzing the effects of quantization noise. Note, though, that nonidealities in the D/A converter can be the chief limitation to the modulator's performance [3].

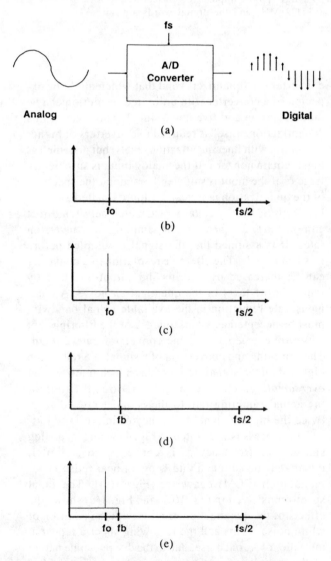

Fig. 1. Increasing resolution with filtering. (a) PCM converter. (b) Input spectrum. (c) Converter output spectrum. (d) Filter response. (e) Filter output spectrum.

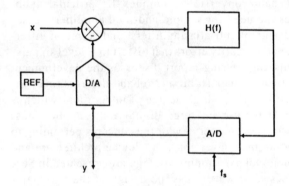

Fig. 2. Delta–sigma modulator.

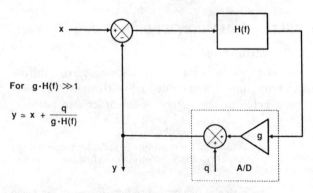

For $g \cdot H(f) \gg 1$

$$y \simeq x + \frac{q}{g \cdot H(f)}$$

Fig. 3. Linearized delta–sigma modulator.

The actual value of the gain $g$ and the rms value of the quantization noise signal $q$ need not be known to obtain an understanding of how the modulator shapes the quantization noise. It must be assumed, however, that successive values of the quantization error are uncorrelated, that is, the quantization noise spectrum is white. The validity of this assumption is borne out empirically.

The modulator output signal $y$ is a function of the analog input $x$ and the quantization noise $q$,

$$y = (x - y)H(f)g + q$$

$$y[1 + H(f)g] = xH(f)g + q$$

$$y = \frac{H(f)gx}{1 + H(f)g} + \frac{q}{1 + H(f)g} .$$

If the loop gain $H(f)g \gg 1$, then

$$y \cong x + \frac{1}{H(f)g} q .$$

That is, the output will be the sum of the input and the quantization noise spectrally shaped by the inverse of the analog filter frequency response.

A glance at the approximate expression for $y$ may lend hope to the prospect of reducing quantization noise by increasing the value of $H$ at all frequencies. However, the effective value of $g$ would change to compensate such a maneuver. Reducing $q$ by introducing a higher resolution quantizer would indeed offer improved performance. But the most effective method of achieving lower baseband quantization noise (for a given oversampling rate) is the selection of a filter function $H(f)$ that possesses high in-band gain and high out-of-band attenuation, thereby shaping the quantization noise spectrum advantageously. Note that the poles of the filter are zeros of the noise transfer function.

## 1.3 Loop Filters

A simple integrator has the desired spectral qualities for a filter. A cascade of two integrators would appear more attractive, and indeed such a "second-order" filter more effectively shifts quantization noise to out-of-band frequencies than the "first-order" filter. Extension to higher order filters is problematic, though, due to stability considerations. A second-order design requires the placement of a single zero in the filter response to obtain a well-behaved modulator. Higher order filters can also have zeros included as an aid to stability, but even so they are conditionally stable due to the high phase shift at baseband frequencies. Conditionally stable loops can become unstable if a frequency-independent gain parameter in the loop is reduced. The effective quantizer gain $g$ is such a parameter and is subject to change under varying operating conditions. As such, the stability of modulators with third and higher order filters is at risk. Nevertheless, the attractiveness of higher order filters has led to a number of solutions to

the stability problem [4], [5]. The device discussed in this report utilizes a fourth-order filter. Stability is discussed below.

Another approach that leads to performance similar to higher order modulators without the attendant stability question has been reported [6]. This architecture has cascaded lower order modulators, with successive modulators measuring the residual in-band quantization noise of previous modulators. The various modulator outputs are processed digitally to lower overall in-band noise. However, accurately matched components and high-gain integrators are necessary to achieve the desired performance.

## 1.4 Filtering and Decimation

Once the quantization noise has been appropriately shaped, it remains the task of the digital filter to remove the out-of-band quantization noise. A straightforward approach of synthesizing a low-pass filter with sufficient stopband attenuation and acceptable passband response would prove inefficient. Instead, a strategy of staged filtering and decimation can be adopted to ease the computational burden [7]. Decimation is the process of sampling a discrete-time signal at a rate lower than its own. The advantage of decimation is that signal processing after decimation can proceed at the lower rate. As a sampling process, though, decimation is subject to the ill effects of aliasing.

In this application, each stage of filtering need only reject signals that will be aliased into the baseband by the immediately subsequent decimation process, since later filter stages will reject signals aliased elsewhere. Fig. 4 illustrates this approach. A signal with a spectrum characteristic of a modulator output signal is input to a filter at a rate $f_s$. The output of this filter is to be decimated to a new rate $f_s'$, where $f_s = Nf_s'$, before being processed further. Aliasing will occur throughout the spectrum, but only components within $\pm f_b$ of integer multiples of $f_s'$ will get aliased into the baseband. A filter designed to have rejection only in these frequency "pockets" requires much less computation than one with rejection across the entire stopband. As the sampling rate gets lower, of course, the pockets become proportionately wider and the filters become more complex. However, they can proceed with their computations at a more leisurely pace.

The final filter stage, operating at the slowest rate, can be a true low-pass filter, eliminating the accumulated out-of-band quantization noise. In addition, it can "tweak" the frequency response of the passband if previous filter stages, the modulator, or even analog processing prior to the modulator have warped the response.

The need to reject the out-of-band quantization noise also represents a benefit, namely, signals outside the baseband up to half the modulator sampling frequency do not get aliased by the modulator and are rejected by the digital filter. Indeed, spurious input signals approaching the sampling frequency do not get aliased into the baseband unless they are within $\pm f_b$ of $f_s$ and are likewise rejected. This characteristic can greatly relax analog anti-aliasing requirements and, for some applications, stands as one of the leading benefits of delta–sigma conversion.

## 2 ARCHITECTURE

### 2.1 Overall Architecture

The device contains two delta–sigma converters suitable for stereo digital audio applications. It is packaged in a 28-pin dual-in-line package with a standard 0.600-in-wide footprint. The cavity of the package is occupied by two silicon dice. One die contains the two modulators, a voltage reference (the value of which determines full-scale signal level), clocking circuitry, and a small amount of digital housekeeping circuitry. The second die contains the digital filter/decimators.

### 2.2 Reasons for Two Dice

Partitioning of the system in such a fashion was motivated by the following considerations.

1) The complexity of the digital filter necessarily creates a large amount of electrical noise during normal operation. Placing this circuitry on a silicon substrate separate from the modulators eases the task of preventing this noise from interfering with the modulators' analog signal processing.

2) The great majority of the silicon area is occupied by the digital filter/decimators, and the manufacturing

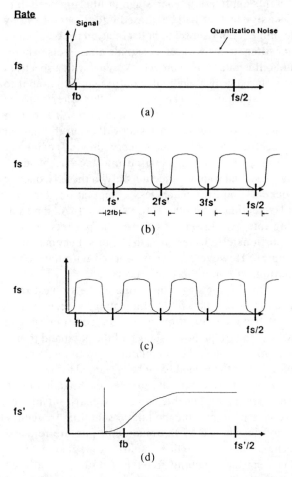

Fig. 4. Filter strategy. (a) Modulator output. (b) Filter response. (c) Filter output. (d) Decimated filter output.

cost is dominated by this circuitry. Shrinking of the geometries comprising this circuitry as the product matures can lead to cost reductions without affecting performance. Shrinking of analog circuitry is risky and difficult, hence the advantage of removing this circuitry from the more cost-sensitive digital die.

3) With separate dice, different processes can be used to fabricate the analog and digital portions on the converters.

Regarding item 3), the digital die is manufactured using a standard 5-V, 2-$\mu$m double-metal digital CMOS process. A 10-V process was chosen for the analog die to allow more headroom for the analog signals. CMOS was chosen to support the switched-capacitor design discussed in the next section. The selected process has 3-$\mu$m line widths, double polysilicon layers for capacitors, and a single metal layer.

### 2.3 Shared Functions

To a large extent, the two channels function independently. However, some circuit blocks are shared. On the analog die, all clocking is common between the two channels to facilitate simultaneous sampling of the left and right channel signals. In addition, a single-voltage reference circuit is utilized by both channels. The voltage reference employs both lateral and vertical bipolar *npn* transistors (both of which are useful parasitics in this process) in a bandgap configuration. Its output is connected to a pin so that it can be capacitively bypassed to reduce crosstalk between the two channels. This capacitor and two simple RC anti-alias filters are all the external elements required of the device other than standard power supply decoupling elements.

The two digital filter/decimators also have common clocking. The various filter coefficients are stored in a single ROM, which is accessed by both right and left channels.

## 3 MODULATOR

### 3.1 Major Characteristics

The major characteristics that need to be determined in the design of a delta–sigma modulator are filter technology, oversampling ratio, quantizer resolution, and filter order.

### 3.2 Discrete-Time Implementation

Although continuous-time filters can be employed in the implementation of a delta–sigma modulator (sampling occurs at the quantizer only), three considerations dictated the choice of sampled-data filters for use in the product. First is the ease with which sampled-data filters can be integrated in comparison with continuous-time filters. Second is that continuous-time filters are sensitive to timing errors in the feedback of the modulator's D/A converter signal [3], whereas sampled-data filters are not. Third, with proper design care sampled-data circuits can provide greater isolation between channels in a stereo application since signal currents are transient. They can be made quite small at the sampling instances and are of no consequence at other times. (Both technologies are subject to capacitive crosstalk.) Thus sampled-data switched-capacitor technology was chosen for the design of the modulator.

### 3.3 Oversampling Ratio

The oversampling ratio is limited by the achievable settling time of analog components as well as the maximum computation rate of the digital filter. It is also preferable that the oversampling ratio be a factor of $2^N$ to ease applications. With a standard baseband of 24 kHz, the oversampling ratio of 64 was chosen for a sampling rate of 3.072 MHz. In a switched-capacitor network each cycle is divided into two phases. With design margin, all modulator circuit blocks were designed to settle in 100 ns to 0.1%.

### 3.4 1-Bit Quantizer

As was mentioned in Sec. 1, higher resolution quantizers embedded in the modulator loop yield lower levels of in-band quantization noise. However, a 1-bit quantizer (that is, a comparator) is simple to implement and minimizes the number of connections between the modulators and their digital filters. More important, a very attractive attribute of the use of a 1-bit quantizer is that errors in the 1-bit feedback D/A converter are not sources of distortion or excess noise, but only gain and offset errors [8]. Therefore, no precision components are necessary. Further, a 1-bit output simplifies the design of the first (and highest speed) digital filter stage.

### 3.5 Modulator Filter Order

The selection of a 1-bit quantizer operating at an oversampling rate of 64 requires at least a third-order modulator filter to obtain 16-bit performance at the digital filter output. The addition of other noise sources (such as, quantization effects in the digital filter) eliminated the candidacy of a third-order filter. A fourth-order modulator filter composed of four cascaded integrators would provide sufficient rejection of baseband quantization noise. However, the modulator's baseband quantization noise can be rendered insignificant by optimization of a fourth-order filter, as follows.

In Sec. 1 it was noted that the poles of the modulator filter are the zeros of the quantization noise transfer function. A filter with four cascaded integrators results in a noise transfer function with four zeros at direct current. Lee and Sodini [9] found that spreading these zeros by application of local feedback around the integrators was effective in lowering the total baseband quantization noise output by the modulator. Optimal placement of all four zeros (two conjugate pairs) results in an 11-dB improvement in baseband quantization noise rejection. Optimal placement of a single conjugate pair with two zeros left at direct current results in a 10-dB improvement.

Implementation of the two-conjugate-pair filter requires feedback to the input summing junction. This requirement has associated undesirable consequences (PSSR degradation, for example), and the two-pair configuration offers little additional noise-shaping improvement above the single pair. So the single-conjugate-pair configuration was adopted.

### 3.6 Modulator Design

Fig. 5 is a block diagram of the modulator. Coefficient $b$ is fed back around the third and fourth integrators to form the conjugate pair of poles in the filter transfer function. The analog input is represented by $x$, and the single-bit digital output is $y$ (which is inverted and summed with $x$ in analog form). The feedforward coefficients $a_1$ through $a_4$ are necessary (although not sufficient) for stable operation. The value of one coefficient is arbitrary. The values of the other three coefficients determine the location of filter zeros, but the effect of these on modulator operation is not easily predictable. Higher ratios of $a_1$ to $a_4$ lead to more stable, noisier operation.

### 3.7 Stability Considerations

As was mentioned in Sec. 1, low values of the "effective gain" of the quantizer (in this case comparator) $g$ can lead to instability. Since the output levels of the comparator are fixed, and since in an unstable mode the integrator output levels can be expected to grow, any linearization criterion for evaluating $g$ should lead to an ever-decreasing value. Similarly, it would not be unreasonable to expect that large values of the input

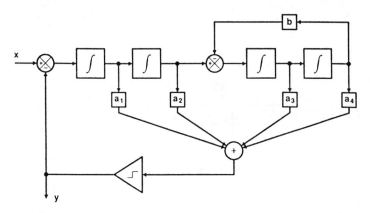

Fig. 5. Modulator block diagram.

would lead to small effective values of $g$, initiating instability.

Simulation and laboratory experience has shown that the modulators do indeed exhibit this behavior. Fortunately, stable regions of operation also exist. The strategy adopted in the design of these modulators is to allow normal operation only well within the stable state space. Circuitry is provided to detect excessively high integrator levels as an indication of unstable operation. If such levels are detected, the integrators are reset to a stable condition. In practice, the reset circuitry is never utilized except at power-up (whereupon the modulator filter may or may not be in a stable state) or during periods when the input is excessively high. "Excessively high" means much higher than full scale, in which case the converter's digital output would be clipped and occasional modulator resets would be of no consequence. As the input returns to a level near full scale, the latest reset event leaves the modulator in a stable state.

### 3.8 Simulated Performance

Fig. 6 shows a typical spectrum of a simulated modulator, including a sinusoid input signal (very close to direct current on the linear frequency scale) plus the quantization noise from direct current to $f_s/2$. An expansion of the low-frequency portion of this figure is shown in Fig. 7. Here the effect of the conjugate pair of quantization noise zeros is evident, as well as that of the pair at direct current.

## 4 FILTER/DECIMATOR

### 4.1 Overall Architecture

Linear-phase finite-impulse-response (FIR) filters are used for decimation. The 1-bit 3.072-MHz outputs of the modulators are decimated in steps of 8, 4, and 2 to yield 16-bit 48-kHz results. A functional block diagram of the digital die appears in Fig. 8. Timing, control, and coefficient ROMs (FIR2 and FIR3) are shared by the two channels. Left- and right-channel data paths operate independently. FIR2 and FIR3 use a per-channel multiplier/accumulator.

### 4.2 Decimation

The decimation strategy includes two stages, FIR1 and FIR2, whose primary responsibility is attenuation of quantization noise prior to decimation and aliasing. As Fig. 6 shows, modulator out-of-band quantization noise spectral density is very high. FIR1 and FIR2 use 17- and 18-bit coefficients to attenuate this noise—and out-of-band input signals—into the converter noise floor. Filter orders are 27 and 30, respectively. Data are processed with 18-bit fixed-point arithmetic.

### 4.3 Passband Shaping

FIR3 performs passband shaping and out-of-band signal attenuation. Passband frequency response errors introduced by the modulator, FIR1, and FIR2 are corrected by FIR3. Overall filter passband ripple is thus reduced to $\pm 0.001$ dB from direct current to 22 kHz.

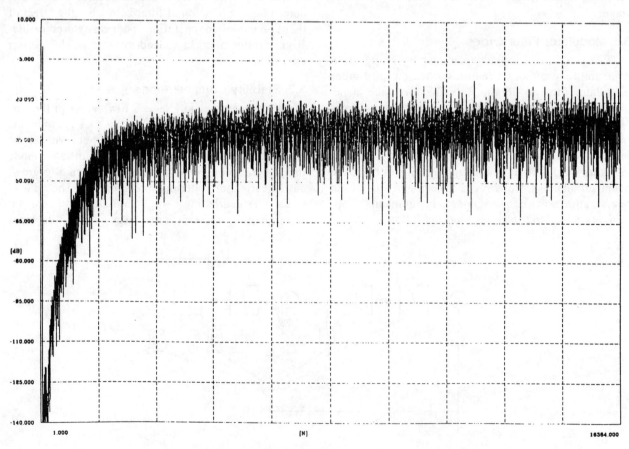

Fig. 6. Simulated modulator output spectrum.

The passband compensation function prevents the use of a half-band filter for FIR3. The filter has 124 nonzero, 18-bit coefficients. Again, data are processed with 18-bit fixed-point arithmetic. Data are truncated to 16 bits at the output, and this operation is the major noise contributor in the system.

### 4.4 Anti-Aliasing Filtering

As indicated in Sec. 1, FIR1, FIR2, and FIR3 also combine to provide anti-aliasing filtering. All analog input frequencies from 26 kHz to 3046 kHz are attenuated by at least 86 dB. The magnitude response of the complete modulator/decimator from direct current to 48 kHz is shown in Figs. 9 and 10. Phase response is precisely linear.

## 5 RESULTS

### 5.1 Modulator Output

Testing of the key specification parameters was performed with the aid of fast Fourier transform (FFT) routines. Fig. 11, for instance, shows the low-frequency portion of an FFT performed on a modulator's single-bit output. The quantization noise shaping is evident in this plot. Absent, however, is the null due to the conjugate-pair noise-shaping zeros, which is visible in Fig. 6. The quantization noise of the modulator is masked by the device's more classical noise mechanisms. Quantization noise can only be seen rising out of the thermal noise floor at frequencies above the audio band.

### 5.2 Digital Filter Output

Fig. 12 shows a plot of the result of an average of 1000 FFTs performed on the 16-bit words output by the digital filter. Averaging of numerous FFTs serves only to smooth the noise floor cosmetically and does not change the ratio of the signal-to-noise level. The width of the fundamental is due to the application of a low side-lobe window to the data stream [10].

In addition to the noise floor and the fundamental, dc and harmonic distortion components are visible. Close inspection reveals a $1/f$ noise corner in the area of 300 Hz and a bump in the noise floor in the area of

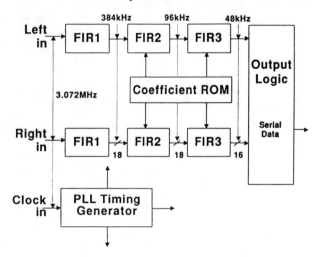

Fig. 8. Filter/decimator block diagram.

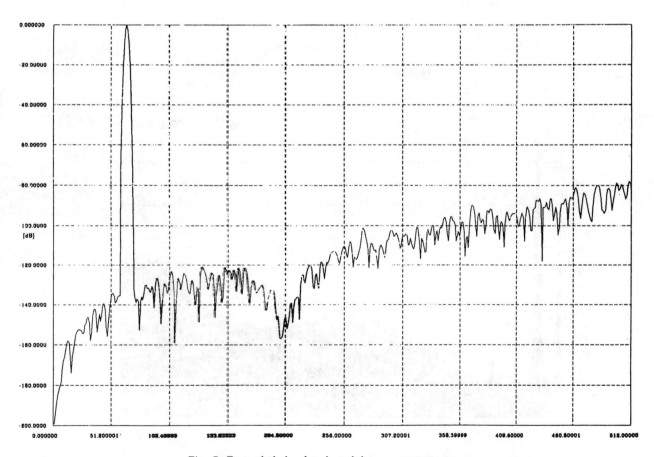

Fig. 7. Expanded simulated modulator output spectrum.

371

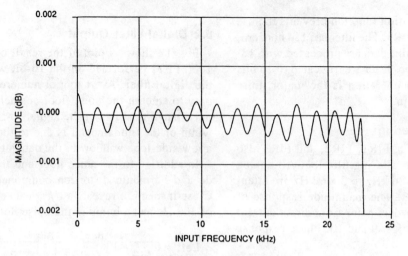

Fig. 9. Filter passband ripple.

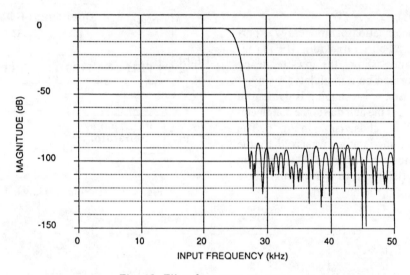

Fig. 10. Filter frequency response.

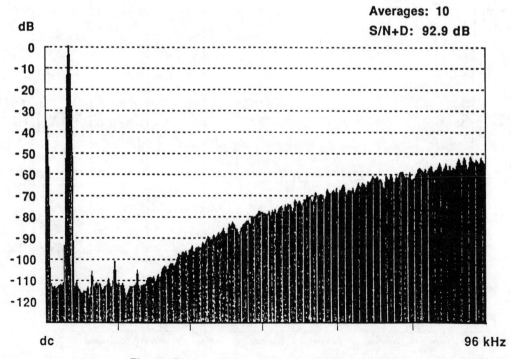

Fig. 11. Partial 16K point FFT of modulator output.

23 kHz. The latter is caused by the rising modulator quantization noise in concert with the falling digital filter characteristic.

Fig. 13 is a plot of measured signal-to-noise plus distortion ratio versus signal level for 1-kHz and 10-kHz input frequencies. High-level performance appears slightly better for the 10-kHz signal because all but the second-harmonic distortion components fall outside the baseband.

## 5.3 Specifications

The key specifications and their measured values are as follows:

| | |
|---|---|
| Oversampling ratio | 64× |
| Signal-to-noise plus distortion | 92 dB |
| Dynamic range | 94 dB |
| Filter passband ripple | <0.001 dB |
| Filter stop-band rejection | >86 dB |
| Calibration error | 5 LSB |
| Channel-to-channel crosstalk | −103 dB at 20 kHz |
| Channel-to-channel gain mismatch | 0.04 dB |
| Gain temperature coefficient | 80 ppm/°C |
| PSRR | 50 dB |
| Power dissipation | 450 mW |

## 6 REFERENCES

[1] A. Gersho, "Quantization," *IEEE Commun. Soc. Mag.*, vol. 15, pp. 20–29 (1977 Sept.).

[2] S. H. Ardalan and J. J. Paulos, "An Analysis of Non-Linear Behavior in Delta–Sigma Modulators," *IEEE Trans. Circuits and Systems*, vol. CAS-34, pp 593–603 (1987 June).

[3] R. W. Adams, "Design and Implementation of an Audio 18-Bit Analog-to-Digital Converter Using Oversampling Techniques," *J. Audio Eng. Soc.*, vol. 34, pp. 153–166 (1986 Mar.).

[4] R. W. Adams, "Companded Predictive Delta Modulation: A Low-Cost Conversion Technique for Digital Recording," *J. Audio Eng. Soc.*, vol. 32, pp. 659–672 (1984 Sept.).

[5] Gould Electronics, Tech. Notes 0141A0860.

[6] Y. Matsuya et al., "A 16-bit Oversampling A/D Conversion Technology Using Triple-Integration Noise Shaping," *IEEE J. Solid-State Circuits*, vol. SC-

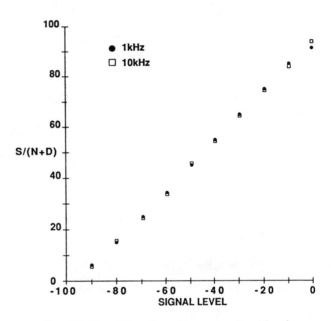

Fig. 13. Signal-to-noise ratio versus signal level.

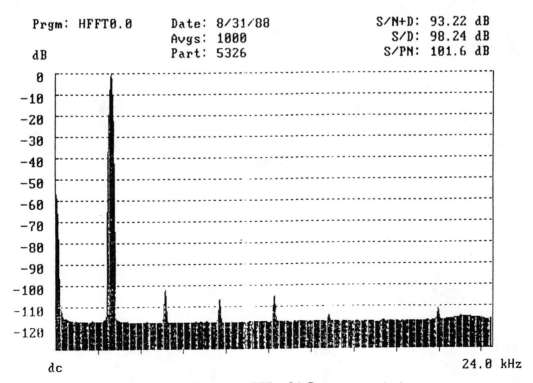

Fig. 12. 1000 averaged FFTs of A/D converter output.

22, pp. 921–929 (1987 Dec.).

[7] R. E. Crochiere and L. R. Rabiner, "Interpolation and Decimation of Digital Signals—A Tutorial Review," *Proc. IEEE*, vol. 69, pp. 300–331 (1981 Mar.).

[8] M. W. Hauser and R. W. Brodersen, "Circuit and Technology Considerations for MOS Delta–Sigma A/D Converter," *ISCAS Dig*. (1986).

[9] W. L. Lee and C. G. Sodini, "A Topology for Higher Order Interpolative Coders," *ISCAS Dig*. (1987), pp. 459–462.

[10] A. H. Nuttal, "Some Windows with Very Good Sidelobe Behavior," *IEEE Trans. Acoust., Speech, Signal Processing*, vol. ASSP-29, pp. 84–91 (1981 Feb.).

# Part 4
# Digital Filters for Oversampling
# A/D Converters

# Using Triangularly Weighted Interpolation to Get 13-Bit PCM from a Sigma–Delta Modulator

JAMES C. CANDY, FELLOW, IEEE,
Y. C. CHING, MEMBER, IEEE, AND
D. S. ALEXANDER

*Abstract*—We present and analyze a method of interpolation that improves the amplitude resolution of an analog-to-digital converter. The technique requires feedback around a quantizer that operates at high speed and digital accumulation of its quantized values to provide a PCM output. We show that use of appropriate weights in the accumulation has important advantages for providing finer resoution, less spectral distortion, and white quantization noise.

The theoretical discussion is supplemented by the report of a practical converter designed especially to show up the strengths and weaknesses of the technique. This converter comprises a sigma–delta modulator operating at 8 MHz and an accumulation of the 1-bit code with triangularly distributed weights. 13-bit resolution at 8 kwords/s is realized by periodically dumping the accumulation to the output. We present a practical method for overcoming a thresholding action that distorts low-amplitude input signals.

## I. INTRODUCTION

A well-known method for PCM encoding, proposed by Goodman [1], [2], uses for the primary encoding an ordinary delta modulator operating at a clock rate that is many times faster than the Nyquist rate of the input signal. The code from the modulator is accumulated, digitally low-pass filtered, and resampled to give a PCM encoding at the Nyquist rate. The resolution obtained with this scheme depends not only on the cycling rate of the delta modulator, but also on characteristics of the digital filter because, in resampling, imperfectly rejected out-of-band quantization noise and signal components are reflected into the signal band and appear as a component of the PCM signal. Thus, Goodman had tradeoffs that exchanged both delta modulater clock speed and complexity of the digital filter for encoding resolution. Several practical encoders that draw upon Goodman's proposal have been reported [3]–[6].

A related method, described in [7], uses as its primary converter a direct feedback encoder; it provides tradeoffs that are different from those of delta modulation. Indeed, a simple accumulate-and-dump is appropriate filtering for generating the PCM words; this technique is akin to ordinary linear interpolation of amplitudes [7], [8]. We now describe an extension of the interpolation process that achieves improved resolution by constructing a weighted accumulation of the quantized values. The method is illustrated by a circuit that processes the output from a sigma–delta modulator [9] operating at $2^{10}$ times the Nyquist rate to provide 13-bit uniform PCM.

Paper approved by the Editor for Data Communication Systems of the IEEE Communications Society for publication without oral presentation. Manuscript received April 26, 1976; revised July 12, 1976.

J. C. Candy and Y. C. Ching are with Bell Laboratories, Holmdel, NJ 07733.

D. S. Alexander was with Bell Laboratories, Holmdel, NJ 07733. She is now at the Stevens Institute of Technology, Hoboken, NJ.

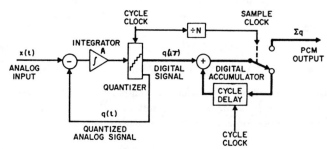

Fig. 1. Outline of an interpolative A/D converter, showing digital accumulation of the output of a direct feedback encoder to generate PCM words.

## II. INTERPOLATION

Fig. 1 shows a circuit that provides interpolative quantization. Here the input $x(t)$ is coarsely quantized at high speed to a sequence of values $q(\cdot)$, which is available in both analog and digital forms. The analog values feed back and subtract from the input to give an error signal, which accumulates in integrating amplifier $A$. At decision times, the input to the quantizer is the signal value plus the accumulated modulation noise from all past cycles. The circuit tries to minimize this noise by having the quantized values oscillate between levels in such a way that the average of $q(t)$ approximates the average of $x(t)$. To obtain PCM words, digital representations of the sum of $N$ consecutive quantizations $\Sigma_N \, q(i\tau)$ are dumped to the output every $N\tau$ seconds.

Analysis of this interpolation process, given in the next section and in [7], demonstrates that, for a given output rate, the resolution of the PCM signal increases in proportion to the number of quantizations $N$ that are averaged for each output word. Thus, increased resolution is the reward for cycling the quantizer at a fast rate, and there is a direct tradeoff between the number of levels in the local quantizing circuit and the speed at which it operates. This concise paper presents a method that provides a more favorable tradeoff of resolution for speed; it constructs a nonuniformly weighted accumulation of the quantized values. We call the process double interpolation. The reason for this name can be seen by studying the following example, which gives an insight to the process.

Consider a converter in which $N$ coarse quantizations are available for interpolating each output sample. The resolution obtained by averaging any $M$ consecutive values from this set is expected to be $M$ times better than selecting any one value at random. But there are $(N + 1 - M)$ different ways of selecting $M$ consecutive values from a string of $N$, so we can construct $(N + 1 - M)$ such interpolations for each sample, Now, if the noises in these interpolations are uncorrelated with one another, an averaging of their values will reduce the noise by a further factor, equal to $\sqrt{N + 1 - M}$, giving an overall improvement of $M\sqrt{N + 1 - M}$ in the resolution. This can exceed the

Reprinted from *IEEE Trans. Commun.*, vol. COM-24, pp. 1268–1275, November 1976.

improvement $N$ obtained from simple interpolation, especially when $N$ is large. The most favorable value for $M$ is $2(N + 1)/3$; then the process is equivalent to an accumulation of the quantized values with weights that have a trapezoidal envelope.

The above description of double interpolation is an oversimplification of the factors involved. More detailed analysis will show that, besides improving the resolution, use of double interpolation results in a flatter spectrum for the net quantization noise and in a narrower aperture in the encoder's sampling process.

## III. ANALYSIS OF A GENERALIZED INTERPOLATION

Before describing a practical implementation of an interpolation encoder, some analytical results will be derived. We will use the notation and a method similar to that in [7].

| | |
|---|---|
| $\tau$ | Cycling period of the quantizer. |
| $N$ | Number of cycles contributing to each output word. |
| $X(t)$ | Input signal. |
| $q(t)$ | Output of the quantizer is constant for each cycle. |
| $e(t)$ | Quantization error. |
| $w_i$ | A set of weights. |
| $W$ | Sum of the weights $\Sigma w_i$. |
| $\Delta$ | A differential operator $\Delta w_i = w_i - w_{i-1}$. |
| $F$ | Reduction in noise power. |
| $\tilde{X}(i\tau)$ | Average input over a cycle; |

$$\tilde{X}(i\tau) = \frac{1}{\tau} \int_{i\tau - \tau}^{i\tau} X(t)\, dt. \tag{1}$$

The action of the feedback loop in Fig. 1 can be described by

$$\frac{1}{\tau} \int_{-\infty}^{i\tau} (X(t) - q(t))\, dt = q(i\tau) + e(i\tau), \tag{2}$$

which can be conveniently rewritten in the form

$$q(i\tau) = \tilde{X}(i\tau) - e(i\tau) + e(i\tau - \tau) = \tilde{X}(i\tau) - \Delta e(i\tau). \tag{3}$$

Now, a generalized accumulation of $N$ quantized values is given by

$$\frac{1}{W} \sum_{1}^{N} w_i q(i\tau) = \frac{1}{W} \sum_{1}^{N} w_i \tilde{X}(i\tau) - \frac{1}{W} \sum_{1}^{N} w_i \Delta e(i\tau). \tag{4}$$

In this expression, the term

$$Y(N\tau) = \frac{1}{W} \sum_{1}^{N} w_i \tilde{X}(i\tau) \tag{5}$$

is an apertured sample of the input signal. We stipulate that it be the desired output value; then the noise is given by the term

$$E(NT) = -\frac{1}{W} \sum_{1}^{N} w_i \Delta e(i\tau). \tag{6}$$

This, rewritten in terms of $e(\cdot)$, is

$$E(N\tau) = \frac{1}{W} \sum_{1}^{N+1} \Delta w_i e(i\tau - \tau). \tag{7}$$

Equation (7) tacitly includes two extra weights, $w_0$ and $w_{N+1}$, both of which are set to zero.

Now consider a situation where the errors $e(i\tau)$ are uncorrelated with one another, but when averaged over many output samples $Y$, have the same rms value $e_{rms}$. Then the mean power in the noise samples $E_{rms}$ can be written in the form

$$E_{rms}{}^2 = \frac{e_{rms}{}^2}{W^2} \sum_{1}^{N+1} [\Delta w_i]^2. \tag{8}$$

This shows that the noise reduction gained by interpolation can be defined by the factor

$$F = W^2 \left[ \sum_{1}^{N+1} [\Delta w_i]^2 \right]^{-1}. \tag{9}$$

Differentiating $F$ with respect to $w_i$ for $0 < i < N$ and equating to zero enables the $N$ unknown weights to be evaluated at extreme values of $F$. By this means, it can be shown that the condition for stationary $F$ with respect to variation in $w_i$ is

$$\Delta^2 w_{i+1} = -\frac{W}{F}, \qquad 0 \leqslant i \leqslant (N + 1). \tag{10}$$

This is the finite difference equation for a set of weights that lie on a parabola. A set that gives minimum noise and is normalized so that $w_1 = 1$ is the *parabolic weights*

$$w_i = \frac{i(N + 1 - i)}{N}, \qquad 0 \leqslant i \leqslant (N + 1) \tag{11}$$

for which

$$F = F_0 = \frac{N(N + 1)(N + 2)}{12}. \tag{12}$$

Our optimization is based on the constraint of constant $N$. Maintaining constant $N$ has practical relevance, and is a simple means for placing a bound of the spectral shaping that can occur during the encoding. Other related sets of weights are interesting because they are easy to implement.

*Uniform Weights*

$$w_i = 1, \quad 0 < i < (N + 1)$$

$$= 0, \quad \text{otherwise}$$

for which

$$F = F_u = \frac{N^2}{2}. \tag{13}$$

*Triangular Weights*

$$w_i = \min \{i, (N + 1 - i)\}, \qquad 0 \leqslant i \leqslant (N + 1)$$

for which

$$F = F_T = \frac{(N+1)^3}{16}.$$ (14)

*Trapezoidal Weights*

$$w_i = \min \{i, (N+1-M), (N+1-i)\}$$

and

$$2M \geqslant (N+1), \qquad 0 \leqslant i \leqslant (N+1)$$

for which

$$F = F_D = \frac{M^2(N+1-M)}{2};$$ (15)

the value of $M$ resulting in maximum $F$ is

$$M_0 = \frac{2(N+1)}{3}$$

for which

$$F = \frac{(N+1)^3}{13 \cdot 5}.$$ (16)

We see that using triangular weighted accumulation for large $N$ results in a very little, 1.25 dB, increase in coding noise over that obtained with optimum weighting. The effects of the weights on the spectrum of the noise and the transmitted signal will be discussed in Section VIII.

The following sections describe the design and performance of an encoder that was chosen to show up the advantages and disadvantages of double interpolation. It is an example of an extreme case where the whole quantization range is interpolated from the two outer levels. $N$ is 1023 and the weights $w_i$ are triangularly distributed, increasing linearly from 0 to 512 and then decreasing linearly back to 0. The cycle rate for the quantizer is 8.2 MHz and the interpolation of its output provides 13-bit resolution at 8 kwords/s for busy inputs. We will see that, for certain steady inputs, the resolution is spoiled by a threshold phenomenon that is a consequence of then having a correlation in the quantization noise. Means for removing it will be described.

## IV. SIGMA–DELTA WITH TRIANGULARLY WEIGHTED INTERPOLATION

Fig. 2 is a schematic of the circuit that was built and tested. The input signal connects to an ordinary sigma–delta modulator whose output, a ONE or a ZERO, controls the clocking of a digital accumulator that takes its input from an up–down counter. The contents of the counter step progressively from 0 to 512 and then back down to 0; it generates the weighting coefficients $w_i$ for the summation in (4). A ONE from the sigma–delta modulator allows the extant weight to increment the accumulation; a ZERO inhibits change of the accumulation.

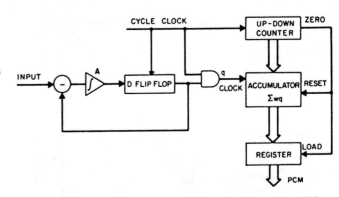

Fig. 2. Outline of the system used to explore double interpolation showing triangularly weighted accumulation of sigma–delta code.

Assigning the values 0 and 1 rather than $-1$ and $+1$ to bits of the sigma–delta code simplifies the circuit and merely introduces a dc shift into the significance of the accumulated code words. When the number in the up–down counter returns to zero, the accumulating register clears while its content transfers to the output. The process repeats for each cycle.

In order to test the encoder, its PCM output was fed to a conventional 16-bit D/A converter. Results of testing the hardware and a computer simulation are reported.

## V. SIGNAL-TO-NOISE RATIOS

The measured signal-to-noise ratio obtained from the complete codec is plotted in Fig. 3 against the amplitude of a 1.02 kHz input sinewave. Dashed diagonal lines on this graph define the responses expected from ideal PCM encoding with bandwidth defined by characteristics of the 4 kHz low-pass filter used in all the experiments. The circled points represent results obtained by computer simulation. Also shown is the measured response for single interpolation ($w_i = 1$).

We see that, for large inputs, the resolution obtained with double interpolation is approximately that of 13-bit PCM, as was expected. But the resolution falls off for low-amplitude inputs. This degradation is caused by a thresholding phenomenon that is explained in the following section. At low amplitude, the real circuit performs better than the simulation. This unexpected result is a consequence of noise and other perturbations dithering over the threshold in the real circuit. These graphs were obtained with 16-bit signals transmitted from the encoder to the decoder; disconnecting the least significant three bits had little effect on the measurements.

## VI. THRESHOLDING IN THE SIGMA–DELTA MODULATOR

Our explanation of the benefits of weighted interpolation assumes that the noise in quantized approximations of the signal is random. This cannot be true in general because certain dc input signals result in repetitive patterns in the code from the sigma–delta modulator and, consequently, in the noise. Three examples of this are given in Fig. 4. It can be demonstrated that periodicities occur in the code for steady input biases that divide the dynamic range of the encoder into two parts whose width has a common integer factor. Fig. 5 illustrates how the noise increases near such input values. The abscissa is the amplitude of a constant input bias that was

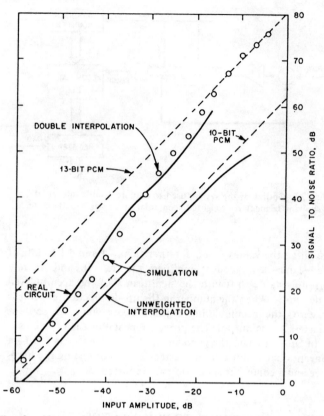

Fig. 3. Measured signal-to-noise ratios of various A/D converters for a 1.02 kHz input, all using the same low-pass filter. A threshold phenomenon at midrange spoils the response of double interpolation to small inputs.

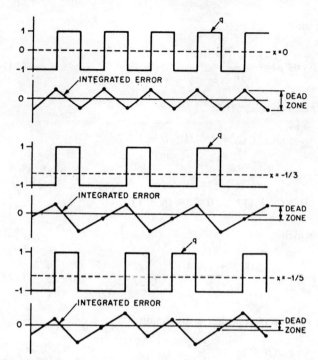

Fig. 4. Three graphs of the quantized signal and the integrated error for constant inputs equal to 0, −1/3, and −1/5. Displacement of the integrated error within a dead zone has no effect on the code.

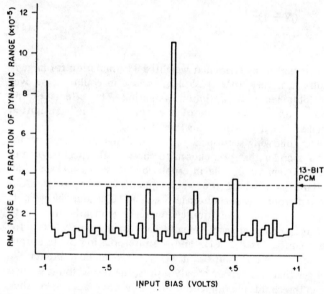

Fig. 5. The rms noise introduced into a small signal that is superimposed on various constant bias levels. The noise is largest when the bias divides the range by a small integer fraction.

applied to a simulation of an encoder having dynamic range ±1. The ordinate is the quantization noise in the output of the codec for small sinusoidal perturbations, $1/64 \sin (2\pi 10^3 t)$, about the bias. For comparison, the rms noise from 13-bit PCM is shown. We expect the resolution of the codec to be the same as 13-bit PCM when the noise in the quantized approximations are uncorrelated; correlated noise can result in performance other than 13-bit, but always better than 10-bit PCM (0.00028 of the dynamic range).

The results plotted in Fig. 3 were obtained using a circuit that was biased to midrange by means of feedback that is described in the Appendix. Zero input signal corresponds to a code of alternating ONES and ZEROS which is a noisy condition of the codec. The nature of the noise can be seen in Fig. 6(c), which shows the unfiltered output of the entire codec for an input that spans only 1/300 of the dynamic range. Fig. 6(a) and (b) show, for comparison, the response of ordinary and interpolating 10-bit PCM. There apparently is a thresholding effect at the center because the encoder does not respond well to small changes of input. This phenomenon is well known for delta modulation; it sometimes fails to respond to amplitude increments that are less than a step size [9]. The sigma–delta modulator can fail to respond to small impulses of input; the size of the dead zones at the output of the integrator is indicated in Fig. 4 for three different bias settings. For a constant input, small offsets of the integrator output that lie within the dead zone will remain uncorrected by the feedback and can introduce a correlation into the quantization noise.

The signal-to-noise ratio is much improved by biasing the modulator to a low-noise state. But then it is necessary to subtract the bias value from the output PCM code. An easy way

of implementing this idea by means of feedback is described in the Appendix. Fig. 7 is a plot of signal-to-noise ratio obtained from a codec biased to a value 1/64 of its dynamic range from the center. Fig. 6(d) and (e) show corresponding waveforms. The choice of this offset was a compromise between preserving dynamic range and placing the threshold where it distorts only large amplitude signals. The threshold noise is apparent in Fig. 7; it distorts signals that are in the 20–30 dB range. The useful dynamic range of the codec is curtailed for symmetric inputs by having this offset, but the loss, 0.27 dB, is small.

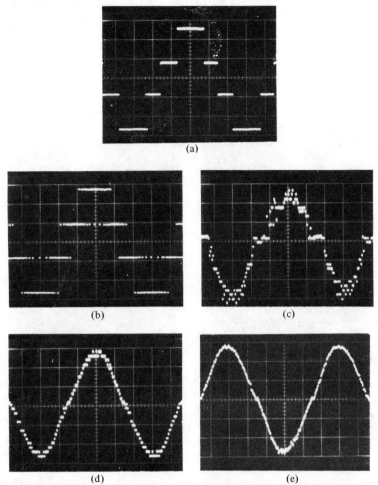

Fig. 6. Pictures of the output of the D/A for an input 60 Hz sinewave of amplitude equal to 1/300 of the encoder's range. (a) 10-bit PCM encoding and decoding. (b) Unweighted interpolative encoding. (c) Decoding 13 bits with the double interpolative encoder biased to midrange. (d) Decoding 13 bits with the double interpolative encoder biased 1/64 from midrange. (e) Decoding 16 bits with the double interpolative encoder biased 1/64 from midrange.

Graphs are plotted in Fig. 7 for the two conditions where 13- and 16-bit words connect the encoder to the D/A. Notice that there is less than 3 dB difference between them. In general, such a rounding off at the output of the accumulator is necessary because the accumulator must be large enough to hold $\log_2 W$ bits, but only the most significant $\log_2 \sqrt{F}$ of these contribute fully to the overall resolution of the encoder.

## VII. HARMONIC DISTORTION AND GAIN TRACKING

Besides degrading signal-to-noise ratios, the threshold phenomenon will deform the overall gain of the codec and introduce harmonic distortion. Figs. 8 and 9 show the deviation of the gain from linearity and the harmonic content of the output, both plotted against input amplitude, for the centered and the offset bias conditions. The offset has a significant advantage in reducing the distortion of small input signals.

## VIII. SPECTRA

Equation (5) defines the aperture of the sampling process because it describes the contribution of the signal to a sample value. This finite aperture will modify the spectrum of the signal in the same way as a prefilter having an impulse response equal to the aperture.

We have obtained expressions for the frequency responses of the encoder by deriving Fourier transforms of the aperture. Table I lists these responses for the four weighting schemes. Also listed are the interpolation advantages and z-transforms of the weighting functions. Fig. 10 is a graph of the frequency responses. The abscissa of the graph is a frequency $\alpha$, the half period of which is normalized to the width of the sampling aperture.

Much of the discussion so far has inferred that each quantized value $q(i\tau)$ contributes to no more than one output sample. Such a restriction is unnecessary because storage of the digital signals permits their contributing to more than one sample. Then the sampling apertures overlap one another. For a given cycling rate, overlapping apertures increase the resolution at the cost of increased spectral shaping of the signal. The expressions in Table I and the graphs in Fig. 10 apply to overlapped apertures.

Another effect of overlapping the apertures concerns the spectrum of the noise. Equation (7) describes the noise. Its

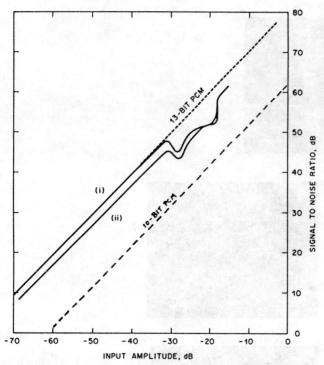

Fig. 7. Signal-to-noise ratios when the encoder is biased 1/64 from midrange. (i) Decoding 16 bits. (ii) Decoding only the most significant 13 bits.

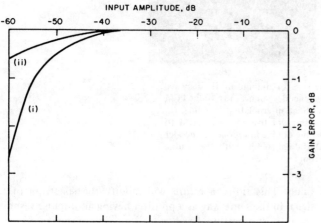

Fig. 8. Variation of the net gain with input amplitude for a 1.02 kHz input. (i) Biased to midrange. (ii) Offset bias.

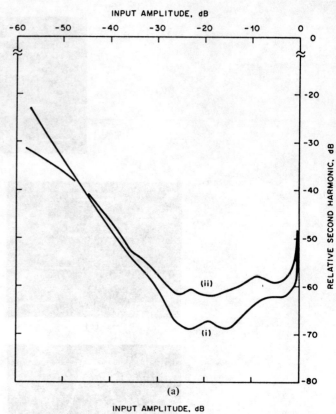

(a)

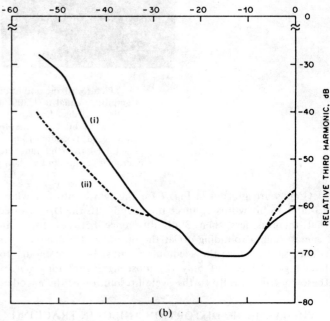

(b)

Fig. 9. Harmonic distortion of a 1.02 kHz sinusoid. (i) Biased to midrange. (ii) Offset bias.

spectral density depends on the manner in which $N$ cycles that affect each sample are selected. For cases where nonoverlapping sets of $(N + 1)$ cycles are assigned to each output word, the noise will be independent from sample to sample, provided the roundoff errors $e(i\tau)$ are independent of one another. Moreover, the output noise will have almost Gaussian amplitude distribution provided $N$ is not very small. Overlapped apertures introduce a correlation in the noise from sample to sample, but this usually is modified by a rounding off that occurs when less than all the bits in the accumulator are taken for output.

Measurements of the frequency response of the entire PCM codec confirm that the falloff of gain with frequency up to 4 kHz is about 4 dB for unweighted accumulation and is less than 2 dB for triangular weighting. The net quantization noise

is white and Gaussian, provided that all 16 bits are connected to the decoder.

## IX. CONCLUDING REMARKS

We have demonstrated that triangularly weighted interpolation has very significant advantages compared to single interpolation: higher resolution, narrower input aperture, and white noise. Our demonstration of a quantizer having only two

TABLE I

| WEIGHTING METHOD | WEIGHTS, $w_i$ $\begin{pmatrix} w_i = 0 \\ N < i < 1 \end{pmatrix}$ | SUM $\sum_N w_i = W$ | INTERPOLATION ADVANTAGE F | z-TRANSFORM OF THE WEIGHT SEQUENCE $\frac{1}{W}\sum_N z^{-i} w_i$ | FREQUENCY RESPONSE OF THE ENCODER $y(\omega)/x(\omega)$ |
|---|---|---|---|---|---|
| UNIFORM | 1 | $N$ | $\dfrac{N^2}{2}$ | $\dfrac{(1-z^{-N})}{N(1-z^{-1})}$ | $\mathrm{SINC}\left(\dfrac{N\omega\tau}{2}\right)$ |
| TRAPEZOIDAL | $\mathrm{MIN}\{i, N+1-M, N+1-i\}$ $2M > N$ | $M(1+N-M)$ | $\dfrac{M^2(N+1-M)}{2}$ | $\dfrac{(1-z^{-M})(1-z^{-(N+1-M)})}{W(1-z^{-1})^2}$ | $\dfrac{\mathrm{SINC}\left(\frac{M\omega\tau}{2}\right)\mathrm{SINC}\left(\frac{(N+1-M)\omega\tau}{2}\right)}{\mathrm{SINC}\left(\frac{\omega\tau}{2}\right)}$ |
| TRIANGULAR | $\mathrm{MIN}\{i, N+1-i\}$ $N = 2M-1$ | $\dfrac{(N+1)^2}{4}$ | $\dfrac{(N+1)^3}{16}$ | $\dfrac{1}{W}\left[\dfrac{1-z^{-\frac{N+1}{2}}}{(1-z^{-1})}\right]^2$ | $\dfrac{\mathrm{SINC}^2\left(\frac{(N+1)\omega\tau}{2}\right)}{\mathrm{SINC}\left(\frac{\omega\tau}{2}\right)}$ |
| PARABOLIC | $i(N+1-i)$ | $\dfrac{N(N+1)(N+2)}{6}$ | $\dfrac{N(N+1)(N+2)}{12}$ | $\dfrac{N\left(1-z^{-(N+2)}\right)-z^{-1}(N+2)(1-z^{-N})}{W(1-z^{-1})^3}$ | $\dfrac{3}{2}\dfrac{\mathrm{SINC}\left(\frac{N\omega\tau}{2}\right)-\mathrm{SINC}\left(\frac{(N+2)\omega\tau}{2}\right)}{(N+1)\mathrm{SIN}^2\left(\frac{\omega\tau}{2}\right)}$ |
| IDEAL LOW PASS FILTER | $\delta(t-i\tau)\mathrm{SINC}\left(\dfrac{\pi t}{N\tau}\right)$ | $N$ | $\dfrac{N^3}{\pi^2}$ | | $\mathrm{SINC}\left(\dfrac{\omega\tau}{2}\right), \quad |\omega| < \dfrac{\pi}{N\tau}$ $0 \qquad , \quad |\omega| > \dfrac{\pi}{N\tau}$ |

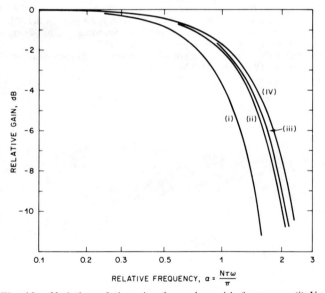

Fig. 10. Variation of the gain of encoders with frequency. (i) Unweighted accumulation. (ii) Parabolic weights. (iii) Trapezoidal weights $M = 2/3(N + 1)$. (iv) Triangular weights $M = 1/2(N + 1)$.

levels draws attention to threshold distortion, but we have mitigated its effect by offsetting the input bias from center. This offset will be unnecessary when multilevel quantizers are employed because the scale of the distortion decreases with the level spacing, enabling small perturbations of the signal to smooth over the threshold. The use of double interpolation for applications in group band encoding [11] seems to be very

attractive; 14-bit resolution at 112 kHz could be obtained from a 16-level quantizer operating at 28.7 MHz.

It is interesting to compare the interpolation sigma–delta encoder with other well-known techniques. Thus, we expect 13-bit resolution at 8 kwords/s using an 8 MHz cycle rate and nonoverlapping triangular weights. Under similar conditions, unweighted interpolation gives 9.5-bit resolution; nonoverlapping parabolic weighting should give 13.2-bit resolution. Ideal low-pass filtering of a sigma–delta or a delta modulator would give 14-bit resolution. In all cases, we have assumed no thresholding fault and no overloading of a full amplitude 4 kHz input sinewave. We know of no easy way for removing the threshold imperfection from small signals in delta modulation. The threshold is not apparent in unweighted interpolation and its effect can easily be mitigated for our technique of weighted interpolation.

Triangularly weighted interpolation requires relatively simple circuits and gives good resolution compared with ideal low-pass filters, yet retains the robustness [7] of ordinary interpolation, especially when the quantizer has only two levels.

## APPENDIX

### CIRCUIT ARRANGEMENT

Fig. 11 shows an outline of the circuit used to test double interpolation. The input signal couples, through a capacitor, to a sigma–delta modulator whose input is biased by a feedback from the most significant bit of the PCM output. This feedback, smoothed by large capacitors, tries to bias the

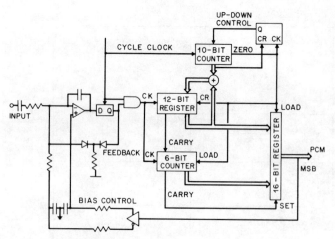

Fig. 11. Schematic of the circuit used to accumulate triangular weights showing the feedback that controls the bias setting.

encoder to a state where the output word is positive for one half of any long time interval. A design for this kind of dc control is given in an Appendix of [10].

Code from the sigma–delta modulator controls the clocking of an 18-bit digital accumulator that comprises a combination of registers and counters. Input to this accumulator is a 10-bit word that represents a triangularly distributed set of weights; they are generated in an up–down counter. When the content of this counter reaches zero, a clock pulse loads the output register with the accumulated data and rests the accumulator. Only the most significant 16 bits in the accumulator are fed to the output, and these are protected from overloading by setting them all to ONE when a carry emerges from the top of accumulator.

In Section VI, we showed that introducing a small offset bias can remove a threshold imperfection from small signals. Resetting the accumulator to 1/64 instead of zero at the beginning of each sample is a simple method for introducing a suitable bias and yet retaining zero output code for zero input.

## ACKNOWLEDGMENT

We are especially grateful to W. H. Ninke for his helpful advice, and to many other of our colleagues at Bell Laboratories for their interest and encouragement.

## REFERENCES

[1] D. J. Goodman, "The application of delta modulation to analog-to-digital PCM encoding," *Bell Syst. Tech. J.*, vol. 48, pp. 321–343, Feb. 1969.

[2] D. J. Goodman and L. J. Greenstein, "Quantization noise of DM/PCM encoders," *Bell Syst. Tech. J.*, vol. 52, pp. 183–204, Feb. 1973.

[3] S. L. Freeney, R. B. Kieburtz, K. V. Mina, and S. K. Tewksbury, "System analysis of a TDM-FDM translator/digital *A*-type channel bank," *IEEE Trans. Commun.*, vol. COM-19, pp. 1050–1069, Dec. 1971.

[4] T. Ishiguro, T. Oshima, and H. Kaneko, "Digital DPCM codec for TV signals based on DM/DPCM digital conversions," *IEEE Trans. Commun.*, vol. COM-22, pp. 970–976, July 1974.

[5] L. D. J. Eggermont, "A single-channel PCM coder with companded DM and bandwidth-restricting filtering," in *Conf. Rec., IEEE Int. Conf. Commun.*, vol. III, June 1975, pp. (40-2)-(40-6).

[6] J. H. Condon and H. T. Breece, III, "Low cost analog-digital interface for telephone switching," in *Conf. Rec., IEEE Int. Conf. Commun.*, June 1974, pp. 13b1–13b4.

[7] J. C. Candy, "A use of limit cycle oscillations to obtain robust analog-to-digital converters," *IEEE Trans. Commun.*, vol. COM-22, pp. 298–305, Mar. 1974.

[8] G. R. Ritchie, J. C. Candy, and W. H. Ninke, "Interpolative digital-to-analog converters," *IEEE Trans. Commun.*, vol. COM-22, pp. 1797–1806, Nov. 1974.

[9] R. Steele, *Delta Modulation Systems.* New York: Wiley, 1974.

[10] J. C. Candy, W. H. Ninke, and B. A. Wooley, "A perchannel A/D converter having 15-segment $\mu$-255 companding," *IEEE Trans. Commun.*, vol. COM-24, pp. 33–42, Jan. 1976.

[11] Bell Laboratories, *Transmission Systems for Communications*, 4th ed., 1970, pp. 131–134.

# A Voiceband Codec with Digital Filtering

JAMES C. CANDY, FELLOW, IEEE, BRUCE A. WOOLEY, SENIOR MEMBER, IEEE, AND OCONNELL J. BENJAMIN

*Abstract*—Oversampling and digital filtering have been used to design a per-channel voiceband codec with resolution that exceeds the typical transmission system requirement by more than 15 dB. This extended dynamic range will allow for the use of digital processing in the management of signal levels and system characteristics in many telecommunication applications. Digital filtering contained in the codec provides rejection of out-of-band inputs and smoothing of the analog output that is sufficient to eliminate the need for analog filtering in most telephone applications. Some analog filtering may be required only to maintain the expanded dynamic range in cases where there is a danger of large amounts of out-of-band energy on the analog input impairing the dynamic range of the modulator.

The encoder portion of the oversampled codec comprises an interpolating modulator that samples at 256 kHz followed by digital filtering that produces a 16-bit PCM code at a sample rate of 8 kHz. In the decoder, digital processing is used to raise the sampling rate to 1 MHz prior to demodulation in a 17-level interpolating demodulator. The circuits in the codec are designed to be suitable for large-scale integration. Component matching tolerances required in the analog circuits are of the order of only ±1 percent, while the digital circuits can be implemented with fewer than 5000 gates with delays on the order of 0.1 μs.

In this paper the response of the codec is described mathematically and the results are confirmed by measurements of experimental breadboard models.

## I. INTRODUCTION

OVERSAMPLING as a means of converting between analog and digital signals is especially well suited to per-channel voiceband telephone applications. Oversampled codecs inherently incorporate digital low-pass filtering which, when properly designed, can perform the antialiasing and reconstruction functions required in the telephone network. Furthermore, the analog circuits associated with these codecs can generally be designed under constraints that are less severe than those encountered in more conventional approaches, such as successive approximation, where sample values are operated on individually.

The oversampling method [1]–[12] is illustrated in Fig. 1 for both analog-to-digital (A/D) and digital-to-analog (D/A) conversion. In the A/D converter, an analog input signal is first encoded into a digital format by a modulator operating at high speed; for example, a 4 kHz voiceband signal might be sampled at 256 kHz. The resulting digital signal is then digitally processed (filtered) to permit lowering the sampling to the Nyquist rate (8 kHz) while preserving the quality of the baseband signal. For D/A conversion, digital processing is used to increase the sampling rate so that demodulation is done at high speed. The high-speed modulation and demodulation in

Paper approved by the Editor for Data Communication Systems of the IEEE Communications Society for publication without oral presentation. Manuscript received August 8, 1980; revised October 20, 1980.

The authors are with Bell Laboratories, Holmdel, NJ 07733.

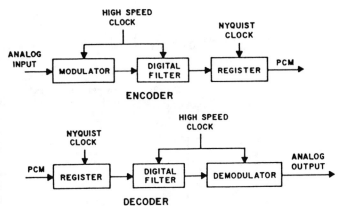

Fig. 1.   Outline of oversamples analog-to-digital and digital-to-analog conversion.

an oversampled codec permits the relatively coarse quantization of analog amplitudes and eliminates the need for precision analog antialiasing and reconstruction filters. The in-band response of the resulting system is particularly insensitive to component variations, and the preponderance of digital circuitry simplifies testing.

This paper describes the design of an oversampled per-channel codec suitable for efficient large-scale integration. This codec, the structure of which is shown in Fig. 2, converts between analog signals and a uniform PCM code with signal-to-noise ratios more than 15 dB above typical digital channel bank objectives. The resolution for small signals is greater than 15 bits. Such performance allows for the presence of out-of-band energy at the modulator input and will permit more extensive use of digital processing for purposes such as gain management, equalization, conferencing, and echo cancellation.

There follows a description of the various sections of the oversampled codec. Much of the discussion concerns the design of these sections so as to satisfy the requirements for the rejection of out-of-band signal components. The effect of each section on quantization noise is treated in Appendixes I and II. The management of signal levels, overloading, and gain are discussed in Appendix IV.

## II. ENCODER

In the A/D converter of Fig. 2, a modulator sampling at 256 kHz digitally encodes the analog input signal. The digital output of the modulator is processed so as to lower its sampling rate to 8 kHz while increasing the word length in such a way as to preserve its baseband (0–4 kHz) quality. A single low-pass digital filter cutting off between 3 and 4 kHz could serve this purpose. However, the speed and power required

Reprinted from *IEEE Trans. Commun.*, vol. COM-29, pp. 815–830, June 1981.

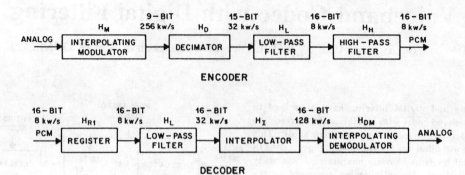

Fig. 2. The oversampled per-channel codec. Data rates are shown in kilowords/s (kw/s).

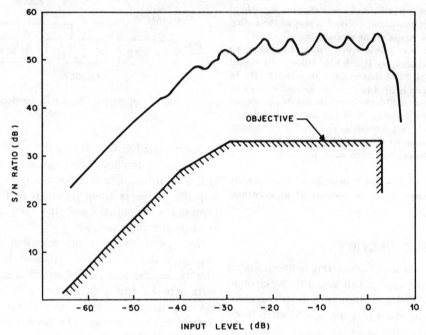

Fig. 3. Signal-to-noise ratio for an integrated modulator with a 1.02 kHz input.

for such a filter is excessive. Considerable advantage is to be gained by lowering the sampling rate in stages [13], and making use of a 32 kHz intermediate rate as shown in Fig. 2.

*Modulator*

The first stage of the encoder is an interpolating modulator that employs 17 logarithmically spaced analog quantization levels [11]. This modulator generates 9-bit words at 256 kwords/s. Fig. 3 shows the 0–4 kHz baseband signal-to-noise ratio measured for an integrated circuit realization of the modulator [14]. The idle channel noise is comparable to that expected for 15-bit encoding.

The frequency response of the modulator can be expressed as [10]

$$|H_M(f/f_0)| = \left| \frac{\sin (\pi f/f_0)}{\pi (f/f_0)} \right| = |\text{sinc} (f/f_0)| \qquad (1)$$

where $f_0$ is the output word (sampling) rate. Input noise in the frequency bands $nf_0 \pm 4$ kHz will be aliased into the 4 kHz baseband by the modulation, the distortion being most trou-

blesome when $n = 1$. For $f_0 = 256$ kHz, the filtering inherent in the modulator provides adequate antialiasing protection. Namely, from (1)

$$|H_M(252 \text{ kHz}/256 \text{ kHz})| = -36.0 \text{ dB}$$

and

$$|H_M(260 \text{ kHz}/256 \text{ kHz})| = -36.3 \text{ dB}$$

whereas the typical rejection required for telephone applications is only −32 dB. The filtering incorporated in the modulator would be insufficient if the sampling rate were reduced to 128 kHz since $|H_M(124 \text{ kHz}/128 \text{ kHz})| = -29.8$ dB.

*Decimator*

The term decimation has become associated, somewhat inappropriately, with digital processes for lowering sampling rates [13]. In Fig. 2, a decimation filter (decimator) is used to lower the sampling rate of the modulator output from 256 kHz to the intermediate frequency of 32 kHz. The use of a

CANDY *et al.*: VOICEBAND CODEC WITH DIGITAL FILTERING

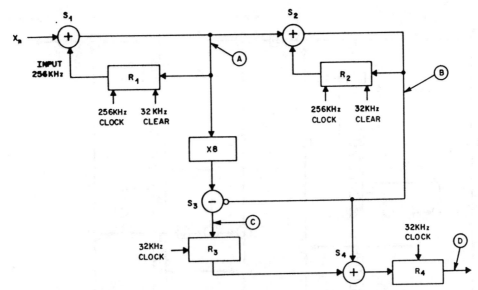

Fig. 4.   Decimator that lowers the word rate from 256 to 32 kwords/s by generating a triangularly weighted sum of 16 high frequency words.

decimator greatly simplifies the design of the subsequent low-pass filter, the complexity of which increases significantly as its sampling rate is increased. The 32 kHz intermediate sampling rate is high enough relative to baseband to allow for the use of a fairly simple decimation circuit. At the same time, this frequency is low enough so that subsequent processing can be performed in a serial mode without clock rates exceeding 1 MHz.

An adequate filtering function for the decimator is that having, for a sample rate of 256 kHz, the *z*-transform

$$H_D(z) = \frac{1}{64}\left[\frac{1-z^{-8}}{1-z^{-1}}\right]^2. \qquad (2)$$

The corresponding frequency response is

$$|H_D(f/f_0)| = \left|\frac{\mathrm{sinc}\,(8f/f_0)}{\mathrm{sinc}\,(f/f_0)}\right|^2 \qquad (3)$$

where $f_0$ is the input sampling rate (256 kHz). The net filtering that results when the decimator is cascaded with the modulator is thus,

$$|H_M(f/f_0)H_D(f/f_0)| = \left|\frac{\mathrm{sinc}^2\,(8f/f_0)}{\mathrm{sinc}\,(f/f_0)}\right|. \qquad (4)$$

This filtering adequately protects against aliasing into baseband when the output of the decimator is resampled at 32 kHz. In particular,

$$|H_M(f/f_0)H_D(f/f_0)|_{f=28\,\mathrm{kHz}} = -34.1\ \mathrm{dB}$$

and

$$|H_M(f/f_0)H_D(f/f_0)|_{f=36\,\mathrm{kHz}} = -38.3\ \mathrm{dB}.$$

As demonstrated in Appendix I, this decimator is nearly

ideal for reducing the quantization noise generated by the interpolating modulator.

The digital filter represented by (2) is relatively easy to implement. It has an impulse response that is triangularly shaped and spans 16 consecutive 256 kHz samples. If $x_m$ represents an input word (output of the modulator) and $y_m$ is the output of the decimator (neglecting the gain constant), then

$$y_0 = x_1 + 2x_2 + 3x_3 + 4x_4 + 5x_5 + 6x_6 + 7x_7 + 8x_8$$
$$+ 7x_9 + 6x_{10} + 5x_{11} + 4x_{12} + 3x_{13} + 2x_{14} + x_{15}$$

$$(5)$$

and, in general,

$$y_m = \sum_{k=8m}^{8m+7} (k-8m)x_k + \sum_{k=8(m+1)}^{8m+15} [8(m+2)-k]x_k. \quad (6)$$

An efficient means of realizing this decimator, and at the same time performing the output resampling, is shown in Fig. 4. The circuit uses parallel arithmetic and comprises four adders ($S_1 - S_4$), four registers ($R_1 - R_4$) each holding one word, and a multiply-by-8 circuit that is simply a 3-bit shift. Adder $S_1$ together with register $R_1$ and adder $S_2$ with register $R_2$ each perform an accumulate and dump function; the registers $R_1$ and $R_2$ are clocked at 256 kHz and cleared at 32 kHz. Registers $R_3$ and $R_4$ are clocked at 32 kHz at the same time $R_1$ and $R_2$ are cleared. $R_3$ provides a (1/32) ms delay, and $R_4$ is simply an output holding register.

The operation of the decimator is easily understood from the expressions for signals at the points *A*, *B*, *C*, and *D* in Fig. 4 at the time the 32 kHz clock is applied. Namely,

$$A_m = \sum_{k=8m}^{8m+7} x_k \qquad (7)$$

387

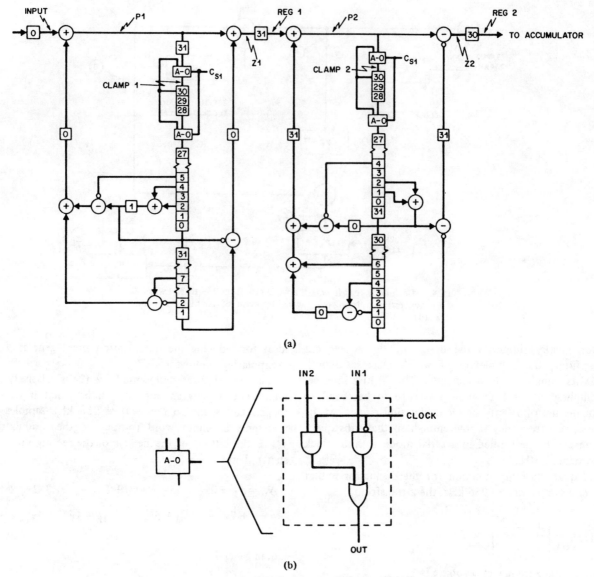

Fig. 5. (a) Fourth-order low pass filter, registers labeled with the same number contain bits having the same significance in the 32-bit words. (b) The AND-OR gate.

$$B_m = \sum_{k=8m}^{8m+7} [8(m+1) - k] x_k \qquad (8)$$

$$C_m = 8A_m - B_m = \sum_{k=8m}^{8m+7} (k - 8m) x_k \qquad (9)$$

and

$$D_m = C_m + B_{m+1}$$

$$= \sum_{k=8m}^{8m+7} (k - 8m) x_k + \sum_{k=8(m+1)}^{8m+15} [8(m+2) - k] x_k. \qquad (10)$$

$D_m$ is the desired output.

### Low-Pass Filter

Reduction of the sampling rate from 32 kHz to 8 kHz

requires a low-pass filter that cuts off sharply between 3.5 and 4.5 kHz and provides at least 32 dB out-of-band rejection. Precise requirements for this filter are derived from typical published specifications for the frequency response of the analog–digital interface [15], [16]. In the design of the filter, these specifications are modified to account for the filtering already provided by the modulator and decimator. Preliminary estimates, and a consideration of analog anti-aliasing filters [17], suggest that at least a fifth-order filter is required.

The design of digital low-pass filters of the type we require is a well-documented science. Procedures have been established for "optimizing" a filter structure while maintaining its spectral response within specified bounds. The "optimum" is generally taken to mean a structure with a minimum number of certain types of operations, such as multiplication or addition. However, such a structure may be far from optimum for a single-chip integrated circuit imple-

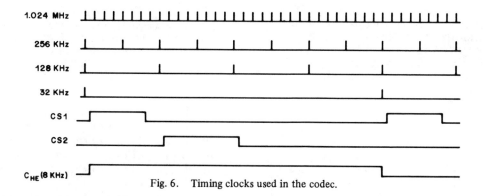

Fig. 6.   Timing clocks used in the codec.

mentation, particularly in dedicated applications where the filter coefficients are fixed. Integrated circuit technology has progressed to the point where the area required for elements such as gates, registers, and simple arithmetic operators can be very small. Unless great care is taken, a large fraction of the chip area may be devoted simply to interconnections. Clock generation and distribution are particularly troublesome in this regard. Seemingly simple circuits may require the distribution of a large number of clocks to many points; the area needed to generate these clocks, keep them aligned with the data, and distribute them can substantially exceed the area consumed by the signal path. Moreover, such circuits can be especially difficult to diagnose and test.

In view of the above concerns, the basic low-pass filter has been designed using only two clocks: a single 1.024 MHz clock that activates all shift register delay elements and word-framing clock that is distributed to only one clamp circuit in each second-order filter section. Furthermore, the filter is constructed using only three simple circuit elements: a 1-bit delay register, a 1-bit adder, and an AND-OR gating circuit.

The structure of the low-pass filter comprises a fourth-order filter operating at a word rate of 32 kHz, followed by an accumulate and dump circuit that lowers the sampling rate to 8 kHz. The accumulate and dump circuit is the simplest means of augmenting the response of the fourth-order filter to meet telephone system requirements.

Our goal for the low-pass filter design was to achieve precision equivalent to a 16-bit word at 8 kHz while defining the in-band spectral response to within ±0.08 dB. A design that meets this goal and requires coefficients with only six or fewer bits is represented, for sampling at 32 kHz, by the $z$-transform response

$$H_L(z) = H_{L4}(z)H_A(z) \qquad (11)$$

where

$$H_{L4}(z) = \frac{1}{8}\left[\frac{1 - \frac{5}{4}z^{-1} + z^{-2}}{1 - \frac{19}{16}z^{-1} + \frac{31}{64}z^{-2}}\right]$$

$$\cdot \left[\frac{1 - \frac{3}{4}z^{-1} + z^{-2}}{1 - \frac{23}{16}z^{-1} + \frac{55}{64}z^{-2}}\right] \qquad (12)$$

and

$$H_A(z) = \frac{1}{4}\left[\frac{1 - z^{-4}}{1 - z^{-1}}\right]. \qquad (13)$$

$H_A(z)$ is the response of the accumulate and dump circuit at the output of the filter. The design of this circuit is treated in the next section along with the high-pass filter because its registers are used to align the signal so as to simplify timing in the high-pass circuit.

The transfer function $H_{L4}(z)$ is implemented as a cascade of two second-order sections, as shown in Fig. 5. In this circuit all register cells, the numbered elements in the figure, are clocked synchronously at 1.024 MHz. The input word enters the filter serially, least significant bit first, in the form of 32 bits at 32 kwords/s. The essential 16 bits of the input are preceded by three ZERO bits and followed by 13 extensions of the sign bit. The 26th bit of the word assumes a special significance; it is regarded as a protected sign bit. When this bit appears in the register cells numbered 30, the framing clock $C_{S1}$ switches the AND-OR gates for six clock beats. This effectively overwrites bits 24–32 with the protected sign bit. The resulting string of sign bits separates consecutive words, helps to preserve the protected sign bit, and prevents limit cycle oscillations [18]. The main 1.024 MHz clock is the only timing signal fed to the delay elements of the filter. The framing clock $C_{S1}$ is distributed only to the AND-OR gates. The CARRY registers (not shown in Fig. 5) associated with the 1-bit adders are neither preset nor cleared between words.

In the circuit of Fig. 5 no more than two operations, two additions, or an addition and a gating, occur between retiming operations. Consequently, delays as long as 300 ns/operation can be safely accommodated. The clock waveforms for the circuit are included in Fig. 6, which illustrates all of the clocks for the encoder of Fig. 2. The framing clock $C_{S1}$ is derived from the trailing edge of the 1.024 MHz clock, $C_1$, while all data transfer in the filter occurs on the leading edge.

Table I is a chart that illustrates the form of the data word at various points in the filter as identified in Fig. 5. This information clearly demonstrates that the use of a long (32-bit) word, although requiring extra stages of delay, greatly simplifies the clocking and management of data in filters of this type. This tradeoff is particularly beneficial where integration of the filter is concerned since delay circuits are regular in structure and can be packed very densely in a circuit layout, whereas clocking and data management functions are generally of an irregular nature and lay out very inefficiently. Crucial to this method of exchanging additional delay for clock simplicity is the ability to express the filter

TABLE I
IDENTIFICATION OF BITS FOR ONE FRAME OF THE BIT STREAM AT VARIOUS LOCATIONS IN THE FOURTH-ORDER LOW-PASS FILTER OF FIG. 5

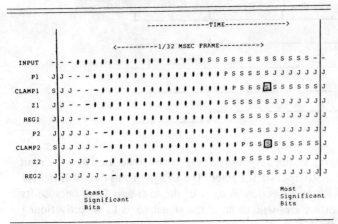

TABLE II
IDENTIFICATION OF BITS AT SEVERAL POINTS IN THE ACCUMULATE AND DUMP CIRCUIT AND HIGH-PASS FILTER OF FIG. 7. BITS ARE IDENTIFIED AS IN TABLE I

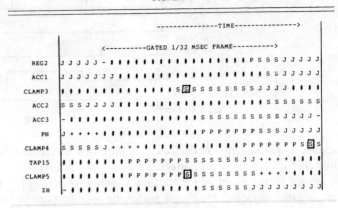

Bits are Identified as Follows:
# = Data bit.
P = Data bit due to peaking in the frequency response.
− = Valid least significant bit.
S = Extended sign bit.
☐S☐ = Clamped sign bit.
+ = Valid bit, possibly data, temporarily overwritten by noise.
J = Meaningless bit.

coefficients as short words (in our case 6 bits or less). It is the length of these coefficients that determines the number of sign bits to be forced by the framing clock.

### High-Pass Filter

High-pass filtering is needed in the encoder in order to attenuate 60 Hz, or 50 Hz, interference. Typical specifications require that such interference be attenuated by at least 20 dB to prevent its interacting with the signal in a way that increases the quantization noise during $\mu$-law encoding.

The high-pass filter used is a single second-order section with a complex zero at 56.3 Hz and a complex pole at 114.4 Hz. These frequencies correspond to filter coefficients that can be expressed as binary numbers that are short and contain few ONES. The $z$-transform response of the filter, for 8 kHz sampling, is

$$H_H(z) = \frac{1 - 2[1 - 2^{-10}]z^{-1} + z^{-2}}{1 - 2[1 - \frac{35}{512}]z^{-1} + [1 - \frac{1}{8}]z^{-2}} \quad (14)$$

The choice of the zero location for the high-pass section is tightly constrained, but there is considerable latitude in choosing the pole. A relatively high frequency pole was used because the dynamic range of signals within the filter tends to decrease as the pole frequency is increased. In addition, a higher pole frequency provides more attenuation at dc.

The high-pass filter is implemented as shown in Fig. 7, following the style of the low-pass sections. Included in Fig. 7 is the accumulate and dump circuit that provides part of the low-pass filtering and resampling of the low-pass filter output at 8 kHz. The registers and clamp in the accumulate and dump circuit are configured to appropriately align the input word for the high-pass section.

Delay elements in the accumulate and dump circuit and in the high-pass filter are clocked with the same 1.024 MHz clock $C_1$ used in the low-pass filter. However, in the high-pass section this clock is gated by an 8 kHz, 31.25 $\mu$s wide enable pulse, $C_{HE}$ in Fig. 6, that is also used to clear the accumulate and dump register.

The high-pass section makes use of the same 32 kHz framing clock $C_{S1}$ used in the low-pass filter to inject a string of sign bits into each word. In addition, a second 32 kHz clock $C_{S2}$ is used to inject sign bits into the least significant segment of the signal that determines the pole frequency. This is necessary because of the length (10 bits) of the coefficients in the high-pass filter. $C_{S2}$ is also used in the accumulate and dump circuit to inject a string of sign bits into the input of the high-pass section.

The coefficients of the high-pass filter are such that special precautions are needed to prevent overflow oscillations. The required protection is provided by the clamp circuit of Fig. 7(b). In this circuit exclusive-OR gate EO detects the occurrence of a small overflow and, in response, injects the complement of the sign bit into the code word. The effect of this action is discussed in Appendix IV.

Table II illustrates the form of data word at various points in the circuit of Fig. 7, following the notation of Table I. Note that at point PH in the high-pass filter the noise encroaches on the four least significant bits of the signal. This is not of concern because, as explained in Appendix III, noise at this point of the circuit has little effect on the output.

### Simulated Encoder Response

Stage-by-stage simulations have been used to establish the overall properties of the encoder under ideal conditions. The frequency response is shown in Fig. 8, along with typical telephone system objectives. Fig. 9 shows the deviation of the encoder phase from linearity with 1850 Hz as the reference frequency. Simulations also indicate that noise introduced by the digital processing due to the restricted word length (roundoff noise) is more than 100 dB below the maximum signal level.

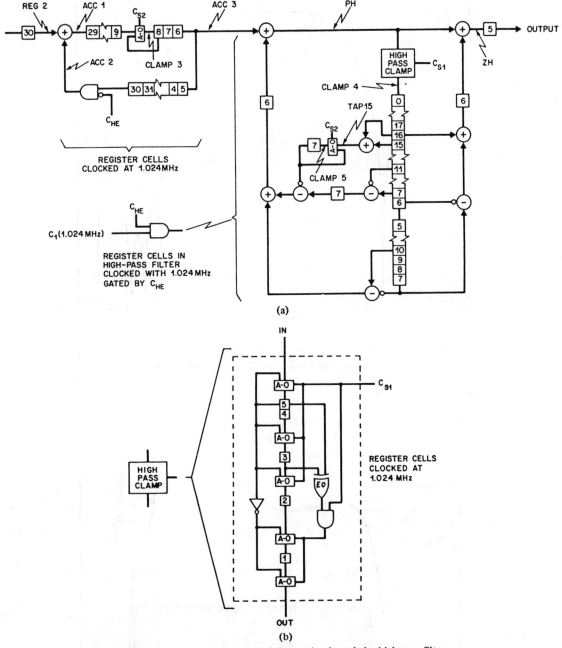

Fig. 7.   (a) The accumulate-and-dump circuit and the high-pass filter.
(b) Expanded view of the clamp circuit.

### III. DECODER

The input to the decoder of Fig. 2 is 16-bit, 8 kword/s PCM. If this signal were demodulated directly, the resulting spectrum would contain images of the base signal centered on 8 kHz and all its harmonics. In an oversampled decoder, the sampling rate is raised prior to demodulation, and digital processing is used to remove all of the baseband images except for those adjacent to the new sampling rate and its harmonics. As in the encoder, this processing is best accomplished in stages rather than with a single low-pass filter.

*Low-Pass Filter*

The first stage of the decoder is a low-pass filter identical to the fourth-order low-pass filter used in the encoder. The

32 kword/s rate of this filter is accommodated simply by repeating each of the decoder input samples four times. The repetition of the input samples is essentially a filtering function with the 32 kHz sampling rate $z$-transform response,

$$H_{R1}(z) = \frac{1}{4}\left[\frac{1 - z^{-4}}{1 - z^{-1}}\right] \tag{15}$$

which is identical to the response of the accumulate and dump circuit used as part of the low-pass filter in the encoder (13).

The output of the low-pass filter approximates a PCM representation of the baseband signal sampled at 32 kHz, with images of the signal adjacent to 8, 16, and 24 kHz atten-

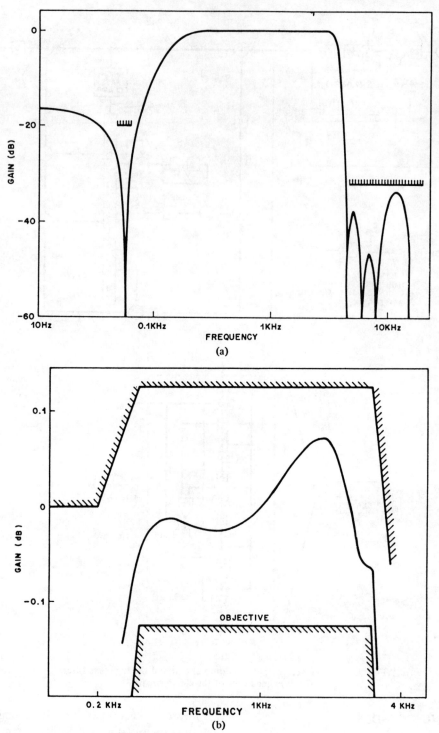

Fig. 8. (a) Frequency response of the simulated encoder. (b) In-band response.

uated by more than 33 dB. The required rejection of these images for telephone applications is only 28 dB.

*Interpolator*

The second stage of the decoder is an interpolator that raises the data rate to 128 kword/s by linearly interpolating three new words between each pair of consecutive words appearing at the low-pass filter output. For example, if $x_{4m}$ and $x_{4(m+1)}$ denote the $m$th and $(m+1)$th samples at the interpolator input, and $y_{4m+n}$, $0 \leq n \leq 3$, represents the output samples interpolated from these inputs, then

$$y_{4m+n} = \left[\frac{4-n}{4}\right] x_{4m} + \frac{n}{4} x_{4(m+1)}. \tag{16}$$

This equation indicates that the impulse response of the interpolating filter has a triangular shape with the $z$-transform (for 128 kHz sampling)

$$H_I(z) = \frac{1}{16}\left[\frac{1-z^{-4}}{1-z^{-1}}\right]^2. \tag{17}$$

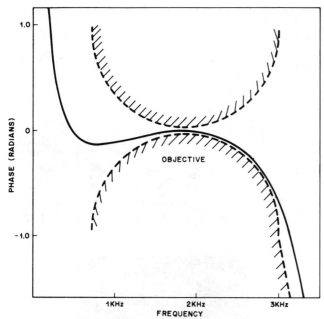

Fig. 9.   Phase dispersion of the encoder.

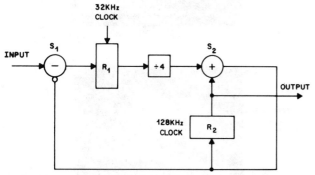

Fig. 10.   Linear interpolator that raises the word rate from 32 to 128 kwords/s.

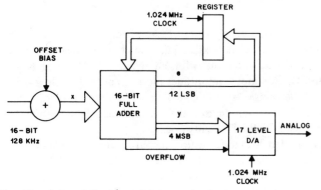

Fig. 11.   Interpolating demodulator; $x$ is the input, $y$ the output, and $e$ the quantization error.

The frequency response of the filter is thus

$$|H_I(f/f_1)| = \left| \frac{\text{sinc}\,(4f/f_1)}{\text{sinc}\,(f/f_1)} \right|^2 \tag{18}$$

where $f_1 = 128$ kHz is the sampling rate of the interpolator output. Since $|H_I(28\text{ kHz}/128\text{ kHz})| = -33.6$ dB and $|H_I(36\text{ kHz}/128\text{ kHz})| = -37.5$ dB, the interpolator clearly provides

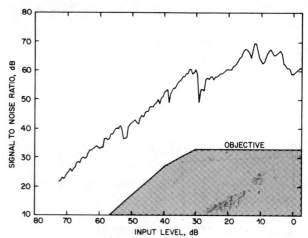

Fig. 12.   Measured signal-to noise ratio of the decoder for an ideal PCM input corresponding to a 1.02 kHz sinusoid.

adequate rejection of baseband images at 32, 64, and 96 kHz.

A straightforward means of realizing the interpolator is suggested by rewriting (16) in a recursive form,

$$y_{4m+n+1} = y_{4m+n} + \tfrac{1}{4}[x_{4(m+1)} - x_{4m}] \tag{19}$$

where $y_0 = x_0$. It is evident from (19) that the interpolated output samples $y_{4m+n}$ can be generated simply by accumulating the input sample differences $(x_{4(m+1)} - x_{4m})$ at the increased (128 kHz) sampling rate. However, a direct implementation of this process suffers from the disadvantage that spurious offsets appearing in the accumulator will remain there indefinitely. This impediment is easily overcome by noting that, for proper operation, $y_{4m} = x_{4m}$ and $y_{4m}$ can be substituted for $x_{4m}$ in (19) to obtain

$$y_{4m+n+1} = y_{4m+n} + \tfrac{1}{4}[x_{4(m+1)} - y_{4m}]. \tag{20}$$

In an implementation of this expression, an offset $\delta$ appearing in the accumulator at the $4m$th sample time,

$$y_{4m} = x_{4m} + \delta \tag{21}$$

is removed on the next 32 kHz clock. That is,

$$y_{4(m+1)} = x_{4(m+1)} \tag{22}$$

independent of the error $\delta$ in $y_{4m}$.

A circuit that performs the function of equation (20) is shown in Fig. 10. Register $R_2$ and adder $S_2$ comprise an accumulator, while subtractor $S_1$ generates the difference $(x_{4(m+1)} - y_{4m})$ and register $R_2$ holds this difference for four 128 kHz clock times. Since any offsets appearing in the accumulator are removed on the following 32 kHz clock, there is no need to initialize the circuit.

*Demodulator*

The final stage of the decoder is a demodulator clocked at 1.024 MHz. This sampling rate is accommodated at the demodulator input simply by repeating the 128 kHz samples

393

eight times. The filtering action introduced by this repetition is

$$|H_{R2}(f/8f_1)| = \left| \frac{\text{sinc } (f/f_1)}{\text{sinc } (f/8f_1)} \right|. \qquad (23)$$

The analog output of the demodulator is a sequence of $(1/1.024)$ $\mu$s amplitude modulated pulses, and the filtering due to the aperture of this signal is

$$|H_{AP}(f/8f_1)| = |\text{sinc } (f/8f_1)| \qquad (24)$$

where $8f_1 = 1.024$ MHz. The net frequency response of the demodulator is thus

$$|H_{DM}(f)| = |H_{R2}(f/8f_1) \cdot H_{AP}(f/8f_1)|$$

$$= |\text{sinc } (f/f_1)|. \qquad (25)$$

$|H_{DM} (124 \text{ kHz})| = -29.8$ dB and $|H_{DM} (132 \text{ kHz})| = -30.4$ dB, indicating adequate rejection of images at all harmonics of 128 kHz (including 1.024 MHz).

The demodulator itself is a digital-to-analog converter of the type described in [19]. Its output is generated by interpolating at high speed (1 MHz) between 17 uniformly spaced quantization levels in such a way that the average of the output has the desired value.

The detailed structure of the demodulator is illustrated in Fig. 11. The input $x$ is a 16-bit PCM code that is fed to the input of a digital full adder. The 16-bit parallel output of the adder is separated into two paths. The least significant 12 bits, $e$, are delayed in an error accumulating register and then added to the next input. The most significant 4 bits, $y$, together with the overflow carry bit, provide the input to a digital-to-analog (D/A) conversion network that generates the analog output. In effect, the D/A network is controlled by an overflowing accumulation of the PCM input.

A concern of this approach to demodulation is that, for certain constant inputs, the interpolating oscillations may contain strong components within baseband. For telephone applications the only constant input encountered is the idle channel. Therefore, the demodulator of Fig. 11 is biased to a quiet state in the idle channel by digitally adding a small constant offset to the signal prior to demodulation. This bias is chosen to place the idle channel 3/16ths of a quantization step away from the central level of the D/A converter.

An analysis of the demodulator resolution is included in Appendix II.

## IV. EXPERIMENTAL CODEC RESPONSE

An experimental model of the oversampled codec was implemented using the integrated modulator described in [14]. The decimator, low-pass filter, high-pass filter, interpolator, and demodulator were constructed using standard commercial logic elements, along with discrete resistors in the D/A network of the demodulator.

The $C$-message weighted signal-to-noise ratio measured for the experimental decoder alone is plotted in Fig. 12 as a function of signal level. The input to the decoder was a digi-

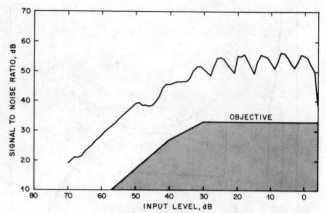

Fig. 13. Measured signal-to-noise ratio of the encoder for a 1.02 kHz input.

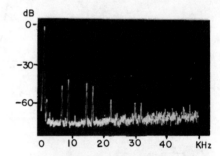

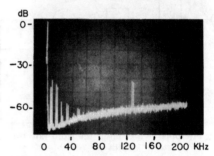

Fig. 14. Spectrum of the decoder output for a full amplitude 1.02 kHz signal.

tally generated ideal 16-bit PCM encoding of a 1.02 kHz sinusoid sampled at 8 kHz. The idle channel noise measured at the decoder output with an inactive PCM input is $-3$ dBrnCO. For input amplitudes more than 21 dB below overload the noise is dominated by roundoff (quantization) noise. For larger inputs the performance is governed by imprecision in the D/A network.

The signal-to-noise ratio measured for the encoder is plotted in Fig. 13. In this case, the response is determined largely by the performance of the modulator. The measured idle channel noise for the encoder is $-1$ dBrnCO.

Fig. 14 shows the spectrum of the decoder output when the input is an ideally encoded full amplitude 1.02 kHz sinusoid. The small amounts of discernible second and third harmonic distortion result from imprecision in the analog circuits of the modulator and demodulator. The sampling images of the input signal are all more than 40 dB below the signal level. The interpolation noise is seen to increase with frequency but remain more than 55 dB below the signal level.

The measured frequency response of the encoder when coupled to a conventional digital-to-analog converter is shown

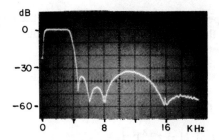

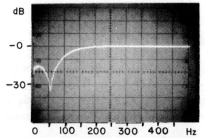

Fig. 15.  Spectral response of the encoder.

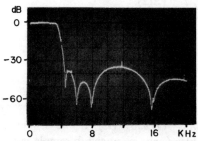

Fig. 16.  Spectral response of the decoder.

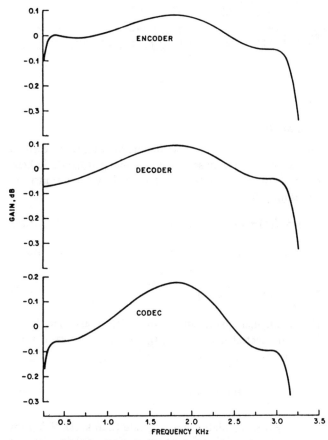

Fig. 17.  Measured in-band spectral responses.

Fig. 15. The frequency response of the decoder driven by a conventional analog-to-digital converter is shown in Fig. 16. The baseband frequency responses measured for the encoder, the decoder and the full codec are shown on an expanded scale in Fig. 17.

## V. CONCLUDING REMARKS

In voiceband codecs oversampling offers a means of obtaining resolution significantly higher than can be achieved with more conventional approaches. In the encoder, the digital modulation of the analog input at a relatively high sampling rate permits the use of digital filters to prevent the aliasing of out-of-band energy into baseband when the signal is resampled at the Nyquist rate. Such digital filtering removes quantization noise, noise generated in the analog circuits and power-supply noise as well as noise on the input. In contrast, analog antialiasing filters remove noise only from the input. In the decoder of an oversampled codec, demodulation at a high sampling rate eliminates the need for reconstruction filtering and prevents spurious noise from aliasing into baseband.

Oversampled codecs have typically made use of simple two-level modulators [7]-[9] together with finite impulse response filters. However, a multilevel companded quantizer offers an important advantage by provide high resolution at a moderate sampling rate (256 kHz). Such quantizers can be implemented with circuits that are very tolerant of component

imperfections and can be easily integrated [14]. The apparent disadvantage of having the modulator generate a 9-bit code has been overcome through the design of recursive digital filters that do not require the use of conventional multipliers.

Since a practical per-channel codec must be viewed in terms of a single-chip integrated circuit implementation, the major contraints on the design are circuit area and power dissipation. Such an approach leads to a design that differs markedly from that which might be expected if the intent were simply to minimize either gate count or the number of arithmetic operations. In particular, considerable effort is given to simplifying the clock circuits and implementing the filters through the repeated use of a few simple circuit structures that dissipate little power and can be packed very densely. Preliminary design of an integrated codec indicates that the digital filters for the oversampled codec can be implemented with fewer than 5000 gates, a number well within the capability of current LSI technology.

## APPENDIX I

### QUANTIZATION NOISE IN THE ENCODER

Shown in Fig. 18 is a model suitable for analysis of quantization noise in the encoder of Fig. 2. Represented in this model are the modulator, the decimator and the low-pass filter, as well as the resampling operations following the decimator and low-pass filter. The effect of the high-pass filter on the encoder quantization noise is negligible and can be disregarded.

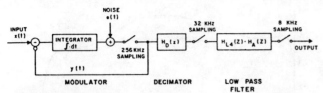

Fig. 18.   Model of the encoder for quantization noise analysis.

## The Modulator

Quantization in the interpolating modulation is represented in Fig. 18 by additive noise $e(t)$ and sampling at $f_0 = 256$ kHz. The action of the modulator is described by the $z$-transform relationship [10]

$$Y(z) = X(z) + (1 - z^{-1})E(z).\tag{26}$$

The noise added to the input signal $Y(z)$ by the modulator is thus

$$N_M(z) = (1 - z^{-1})E(z)\tag{27}$$

If the original quantization noise $e(t)$ is assumed to be white with spectral density $e_0/\sqrt{f_0}$ then the spectral density of the noise at the modulator output is

$$N_M(f) = \frac{2e_0}{\sqrt{f_0}} \sin(\pi f/f_0)\tag{28}$$

where $f_0$ is the modulator sampling rate (256 kHz). If $N_{M0}$ is defined to be the rms baseband component of this noise and $f_B = 4$ kHz is the baseband frequency, then

$$N_{M0}^2 = \int_0^{f_B} N_M{}^2(f)\, df$$

$$= 2e_0{}^2 \frac{f_B}{f_0} \left[ 1 - \left[ \frac{f_0}{2\pi f_B} \right] \sin(2\pi f_B/f_0) \right]\tag{29}$$

when $f_B \ll f_0$,

$$N_{M0} = \frac{2}{\sqrt{3}}\, \pi e_0 \left( \frac{f_B}{f_0} \right)^{3/2}.\tag{30}$$

In the nonlinear interpolating modulator [11] the effective spacing of the finest (inner) quantization levels is $A/512$, where $A$ is the peak-to-peak signal range of the modulator; the sampling rate is $f_0 = 256$ kHz. Therefore,

$$e_0 = \frac{1}{\sqrt{6}}\, (A/512)\tag{31}$$

and, since $f_B/f_0 = 4/256 = 1/64$

$$N_{M0} \cong \frac{\pi}{32\sqrt{3}} \left[ \frac{A}{512\sqrt{12}} \right] \sqrt{\frac{2f_B}{f_0}}$$

$$= \frac{A}{\sqrt{12}} \left[ \frac{\pi}{2^{16}\sqrt{6}} \right]$$

$$< \frac{1}{2^{15}}\, (A/\sqrt{12}).\tag{32}$$

Thus, the expected idle channel resolution of the modulator is better than that provided by 15-bit PCM encoding.

## Decimation

The $z$-transform response of the decimator is given in (2). This response modifies the noise at the modulator output so as to produce the noise

$$N_D{}'(z) = \frac{1}{64} \left[ \frac{(1 - z^{-8})^2}{1 - z^{-1}} \right] E(z)$$

$$= \frac{1}{64} (1 - z^{-8}) \left[ \sum_{i=0}^{7} z^{-i} \right] E(z)\tag{33}$$

at the output of the decimator, prior to resampling at 32 kHz.

The spectral density of the noise following resampling is most readily established by reverting to analysis in the time, rather than frequency, domain. Let $E'(z)$ represent the accumulated noise

$$E'(z) = \left[ \sum_{i=0}^{7} z^{-i} \right] E(z).\tag{34}$$

In the time domain (34) is equivalent to

$$e'(nT) = \sum_{i=0}^{7} e(nT - iT)\tag{35}$$

where $T = 1/f_0 = (1/256$ kHz). If the sequence $e'(nT)$ is sampled at 32 kHz (only every eighth sample is retained) we obtain the sequence

$$e''(8mT) = \sum_{i=0}^{7} e(8mT - iT).\tag{36}$$

Since there is no correlation between the samples $e''(8mT)$, the spectrum of the noise following the 32 kHz sampling will be white, with spectral density $8e_0/\sqrt{f_0}$, if the original noise $e(t)$ is white. It follows that, when the noise at the decimator output, given by (33), is resampled at 32 kHz, the result is described by the $z$-transform

$$N_D(Z) = \tfrac{1}{64} (1 - Z^{-1})E''(Z)\tag{37}$$

where the transform variable $Z$ corresponds to sampling at $f_1 = 32$ kHz ($Z = z^8$). In the frequency domain

$$N_D(f) = \frac{e_0}{4\sqrt{f_0}} \sin(\pi f/f_1).\tag{38}$$

The rms baseband component of this noise is

$$N_{D0} = \left[ \frac{e_0{}^2 f_B}{32 f_0} \left[ 1 - \frac{4}{\pi} \sin(\pi/4) \right] \right]^{1/2}\tag{39}$$

where $f_B = 4$ kHz.

A comparison of (29) and (39) indicates that the decimation reduces the net baseband noise by 0.14 dB. Approxi-

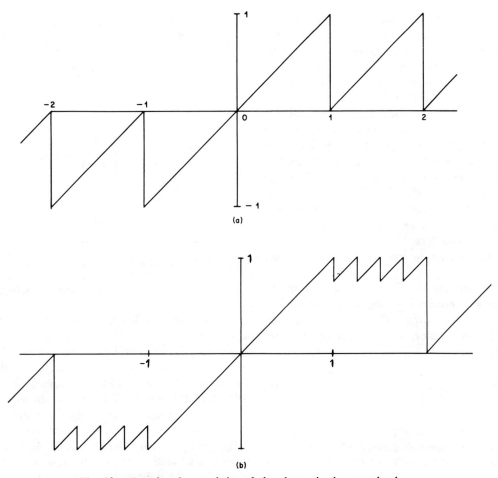

Fig. 19. Transfer characteristics of the clamps in the second-order recursive filter sections. (a) Low-pass filter. (b) High-pass filter.

mately 0.26 dB noise reduction is expected due to the baseband attenuation of the decimator. Therefore, the penalty for lowering the sampling rate to 32 kHz following decimation is only an insignificant 0.12 dB increase in the baseband quantization noise.

*Low-Pass Filtering*

Ideal low-pass filtering of the 32 kHz decimator output, followed by resampling at 8 kHz, would not affect the baseband quantization noise. Since the low-pass filter of Fig. 18 is not ideal, some high frequency noise will be aliased into baseband by the resampling. However, the out-of-band attenuation of the filter is large enough (>33 dB) to ensure a less than 0.05 dB increase in the baseband noise.

## APPENDIX II

### ANALYSIS OF THE DEMODULATOR

The operation of the demodulator in Fig. 11 is described by the expression

$$x(nT) + e[(n-1)T] = y(nT) + e(nT) \tag{40}$$

or as a $z$-transform relationship

$$X(z) - Y(z) = (1 - z^{-1})E(z) \tag{41}$$

where the sampling rate is $f_D = 1/T = 1024$ kHz. Equation (41) has the same form as the expression (26) that describes

the operation of the modulator. When the error, or noise, in (40) is assumed to be white with spectral density $e_0/\sqrt{f_D}$ the spectral density of the noise in the demodulator output is

$$N_{DM}(f) = \frac{2e_0}{\sqrt{f_D}} \sin(\pi f/f_D). \tag{42}$$

Since the peak-to-peak signal range $A$ is quantized into 16 uniform intervals in the demodulator, the spectral density of the noise is

$$e_0 = \frac{1}{\sqrt{6}} (A/16). \tag{43}$$

By an analysis similar to that leading to (32), it follows that the rms baseband noise is approximated by

$$N_{DM_0} \cong \frac{A}{\sqrt{12}} \left[ \frac{\pi}{2^{14}\sqrt{6}} \right] = \frac{A}{\sqrt{12}} \left[ \frac{1.28}{2^{14}} \right]. \tag{44}$$

It is apparent from (44) that the noise introduced by the demodulator corresponds to slightly worse than 14-bit PCM encoding. This performance is more than adequate to meet typical digital channel bank specifications, which require only about 12-bit resolution in the idle channel. However, our goal is to provide better than 15-bit resolution.

The idle channel resolution of the demodulator can be improved by exploiting the fact that the noise $e(t)$ is not white for steady input signals. This approach is especially suitable for digital implementation of interpolating circuits, in contrast to analog realizations wherein circuit imperfections often tend to modify the spectrum of the noise.

The spectrum of $e(t)$ for a steady input comprises sets of discrete lines [20] whose frequencies can be shifted by means of a bias added to the input signal. In the demodulator, the only expected steady input is the idle channel. If the idle channel output lies close to a quantization level, half way between levels, or on a small whole number division of the quantization interval, the noise in baseband tends to be large. At other biases this noise will be, relatively, much smaller. The demodulator has been biased so that the idle channel lies 3/16ths of a quantization interval away from the center level.

In addition to achieving the desired resolution in the demodulator, it is also necessary to insure adequately low out-of-band energy in the output. The interpolation noise generated by the demodulator, (42), is maximum when $f = f_D/2 = 512$ kHz; then

$$N_{DM}(512 \text{ kHz}) = \frac{2e_0}{\sqrt{f_D}} = \frac{A}{\sqrt{12}} \left[ \frac{\sqrt{2}}{256} \right]. \tag{45}$$

This is reduced 4 dB by the rectangular aperture of the decoding switch (24). The spectral density of the baseband signal for large inputs is of the order of $A/(4\sqrt{2})$. Thus, the decoding noise density is more than 40 dB below the signal density, whereas only 28 dB is called for.

## APPENDIX III

### OVERFLOW OSCILLATIONS AND ROUNDOFF NOISE IN THE SECOND-ORDER FILTER SECTIONS

Tables I and II illustrate that, in normal operation, the second-order filter sections used for both low-pass and high-pass filtering will not overload. However, when switched on the initial state of the registers may be such that overflow occurs, and this overflow may initiate oscillations that do not decay [18]. Both simulations and experiments show that, in the low-pass filter sections, these oscillations are prevented simply by preserving the sign of the word during overflow; this is accomplished by the clamp circuits in Fig. 5 that are clocked by $C_{S1}$. Fig. 19(a) shows the form of the transfer characteristic for each second-order section that results from such clamping.

The coefficients of the second-order high-pass filter are such that more elaborate protection against overflow oscillations is required. Specifically, not only is the sign of the word preserved, but the magnitude of the word is set to 0.75 for small overflows. This is accomplished, as shown in Fig. 7, by injecting the complement of the sign bit into two bits of the word when a small overflow is detected. The resulting transfer characteristic for the section is shown in Fig. 19 (b). Simulations and tests on breadboard models confirm that the

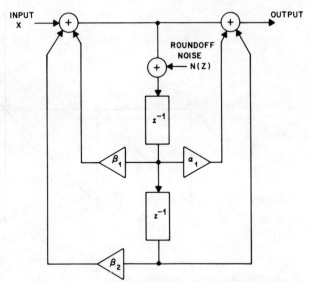

Fig. 20. Model of second-order filter section for noise analysis.

circuit of Fig. 7 will come out of saturation within a few clock beats.

In the second-order filter sections of Figs. 5 and 7, roundoff noise is added to the signal when the framing pulse injects the string of successive sign bits to separate words. This action is illustrated in Fig. 20. If the gain of the second-order section for an input signal, is $G(z)$, then the equivalent gain for the roundoff noise is $G(z) - 1$. For frequencies where $G(z)$ is near unity, the noise at the output is, thus, highly attenuated. This noise cancellation is of crucial importance in the high-pass filter where the noise at the framing clamp is allowed to overwrite the lower four bits of the signal, as shown in Table II.

## APPENDIX IV

### GAIN MANAGEMENT

A major consideration in the design of the codec in Fig. 2 is the gain in each of the various stages. These gains establish the signal levels throughout the codec and must be managed so as to avoid both the saturation of large signals and excessive roundoff noise in small signals. Arbitrary digital gain adjustments can be very expensive because multiplication is required. However, changes in gain by integer powers of two are simply shifts of bit position. Each stage of the codec is assumed to incorporate a shift that brings its gain close to unity. Under this assumption we now consider the gain of each stage of the circuit in Fig. 2.

The modulator converts the analog input to a 9-bit code and saturates gracefully by clipping the signal amplitude.

The decimator (2) requires no adjustment of gain; its gain is unity at low frequencies and falls to −0.027 dB at 1 kHz and −0.44 dB at 4 kHz. This small droop across baseband is compensated for in the low-pass filter. The decimator is designed so as not to saturate in response to any signal that can be generated by the modulator.

At 1 kHz the gain of the fourth-order low-pass filter (12) is −0.578 dB and that of the accumulate and dump circuit is −0.014 dB. The gain of the high-pass filter (14) at 1 kHz is +0.603 dB. All of these circuits transmit 32-bit words and, as illustrated in Tables I and II, are designed not to saturate.

The net gain of the digital circuits in the encoder is −0.016 dB at 1 kHz and varies ±0.07 dB throughout baseband. The output buffer circuit that follows the high-pass filter truncates the 32-bit word to 16-bits and is designed to saturate gracefully by clipping amplitudes.

The decoder comprises a low-pass filter identical to that used in the encoder followed by an interpolator (17) and the demodulator. The net gain of the low-pass filter and interpolator is −0.62 dB at 1 kHz, a loss that is easily compensated for in the analog circuits of the demodulator.

## ACKNOWLEDGMENT

We are grateful for the helpful advice offered by many of our associates; in particular, J. E. Iwersen, T. Misawa, J. Condon, and W. H. Ninke. We thank M. T. Doland and J. F. Kaiser for their indispensable assistance in obtaining coefficients for the low-pass filter, and we are indebted to J. R. Perucca for performing experiments on the decoder.

## REFERENCES

[1] H. S. McDonald, "Impact of large scale integrated circuits on communication equipment," in *Proc. Nat. Elec. Conf.*, Dec. 1968, pp. 569–572.

[2] D. J. Goodman, "The application of delta modulation to analog-to-digital PCM encoding," *Bell Syst. Tech. J.*, vol. 48, pp. 321–343, Feb. 1969.

[3] S. L. Freeny, R. B. Kieburtz, K. V. Mina, and S. K. Tewksbury, "System analysis of a TDM-FDM translator/digital a-type channel bank," *IEEE Trans. Commun.*, vol. COM-19, pp. 1050–1069, Dec. 1971.

[4] D. J. Goodman and L. J. Greenstein, "Quantizing noise of DM/PCM encoders," *Bell Syst. Tech. J.*, vol. 52, pp. 183–204, Feb. 1973.

[5] J. H. Condon and H. T. Breece, "Low cost analog-digital interface for telephone switching," in *Conf. Rec., IEEE Int. Conf. Commun.*, June 1974, pp. 13B1–13B4.

[6] T. Ishiguro, T. Oshima, and H. Kaneko, "Digital DPCM codec for TV signals based on DM/DPCM digital conversions," *IEEE Trans. Commun.*, vol. COM-22, pp. 970–976, July 1974.

[7] L. D. J. Eggermont, "A single-channel PCM coder with companded DM and bandwidth-restricting filtering," in *Conf. Rec., IEEE Int. Conf. Commun.*, vol. III, June 1975, pp. (40-2)–(40-6).

[8] L. D. J. Eggermont, M. H. H. Hofelt, and R. H. W. Salters, "A delta-modulation to PCM converter," *Phillips Tech. Rev.*, vol. 37, pp. 313–329, Dec. 1977.

[9] J. D. Everhard, "A single-channel PCM codec," *IEEE J. Solid-State Circuits*, vol. SC-11, pp. 25–38, Feb. 1979.

[10] J. C. Candy, "A use of limit cycle oscillations to obtain robust analog-to-digital converters," *IEEE Trans. Commun.*, vol. COM-22, pp. 298–305, Mar. 1974.

[11] J. C. Candy, W. H. Ninke, and B. A. Wooley, "A per-channel A/D converter having 15-segment $\mu$-225 companding," *IEEE Trans. Commun.*, vol. COM-24, pp. 33–42, Jan. 1976.

[12] J. C. Candy, Y. C. Ching, and D. S. Alexander, "Using triangularly weighted interpolation to get 13-bit PCM from a sigma-delta modulator," *IEEE Trans. Commun.*, vol. COM-24, pp. 1268–1275, Nov. 1976.

[13] D. J. Goodman and M. J. Carey, "Nine digital filters for decimation and interpolation," *IEEE Trans. Acoust., Speech, Signal Processing.*, vol. ASSP-25, vol. 2, pp. 121–126, Apr. 1977.

[14] B. A. Wooley and J. L. Henry, "An integrated per-channel PCM encoder based on interpolation," *IEEE J. Solid-State Circuits*, vol. SC-14, pp. 14–20, Feb. 1979.

[15] *The D3 Channel Bank Compatibility Specification, Issue 2*, Eng. Dep., Amer. Telephone and Telegraph Co., Oct. 14, 1974.

[16] CCITT, *Pulse Code Modulation of Voice Frequencies*, Rec. G.711, Geneva, 1972; amended at Geneva, 1976.

[17] R. A. Friedenson, R. W. Daniels, R. J. Dow, and P. H. McDonald, "RC active filters for the D3 channel bank," *Bell Syst. Tech. J.*, vol. 54, pp. 507–529, Mar. 1975.

[18] P. M. Ebert, J. E. Mazo, and M. G. Taylor, "Overflow oscillations in digital filters," *Bell Syst. Tech. J.*, vol. 48, pp. 2999–3020, Nov. 1969.

[19] G. R. Ritchie, J. C. Candy, and W. H. Ninke, "Interpolative digital-to-analog converters," *IEEE Trans. Commun.*, vol. COM-22, pp. 1797–1806, Nov. 1974.

[20] J. E. Iwersen, "Calculated quantizing noise of single-integration delta modulation coders," *Bell Syst. Tech. J.*, vol. 48, pp. 2359–2389, Sept. 1969.

# Decimation for Sigma Delta Modulation

JAMES C. CANDY, FELLOW, IEEE

*Abstract*—Decimation is an important component of oversampled analog-to-digital conversion. It transforms the digitally modulated signal from short words occurring at high sampling rate to longer words at the Nyquist rate. Here we are concerned with the initial stage of decimation, where the word rate decreases to about four times the Nyquist rate. We show that digital filters comprising cascades of integrate-and-dump functions can match the structure of the noise from sigma delta modulation to provide decimation with negligible loss of signal-to-noise ratio. Explicit formulas evaluate particular tradeoffs between modulation rate, signal-to-noise ratio, length of digital words, and complexity of the modulating and decimating functions.

## I. INTRODUCTION

DECIMATION is the name given to processes that lower the word rate of digitally encoded signals which are sampled above their Nyquist rate [1]–[3]. Important examples of decimation increase the length of the words in order to conserve resolution; examples of this occur in oversampled codecs [4].

The trading of word length for word rate usually results in a reduction in the net bit rate, thus improving the efficiency of the encoding. The value of this trading depends on the type of encoding involved. For example, the noise from ordinary PCM quantization decreases by 6 dB for every significant bit extended onto the word, but decreases only 3 dB for every doubling of the sampling rate. Therefore, decimating oversampled pulse code modulation can result in dramatic decrease in the bit rate, because the encoding is so inefficient. A more efficient means of oversampling is sigma delta modulation, wherein quantization noise can be reduced by as much as 15 dB for a doubling of the sampling rate [6]. However, it is not only the efficiency of sigma delta modulation that makes it attractive but also the ease with which it can be implemented. Appropriate decimation is needed to lower the bit rate of sigma delta modulation and convert it to a form that is more suitable for processing and transmission.

We will show that filters with response $\text{sinc}^k(f)$ are appropriate for decimating sigma delta modulation down to four times the Nyquist rate. Further decimation usually requires filters that cut off more sharply at the edge of baseband. We will evaluate filters by determining the amount of noise that aliases into baseband when their outputs are resampled at the reduced rate. This noise originates from two sources, the quantization noise of the encoding and spurious noise from analog circuits.

The next section summarizes the main features of sigma delta modulators and calculates their signal and noise properties. Section III determines the effect that decimation has on the noise, Section IV describes how the length of the digital words change, Section V describes the range of signal amplitudes that can be accommodated, and Section VI discusses the rejection of spurious circuit noise. The main

Paper approved by the Editor for Signal Processing and Communication Electronics of the IEEE Communications Society. Manuscript received July 18, 1984; revised July 2, 1985.

The author is with AT&T Bell Laboratories, Holmdel, NJ 07733.

IEEE Log Number 8906425.

results of this study are summarized in design charts which are described in Section VII.

## II. SIGMA DELTA MODULATION

Fig. 1 shows a diagram of a basic sigma delta modulator. This circuit represents quantization by an additive noise $E$, and we shall assume that the signals are sufficiently busy to make this noise random. The rms spectra density of the added noise is then given by

$$E(f) = \sigma \sqrt{\frac{\tau}{6}} = \bar{e}\sqrt{2\tau} \qquad (1)$$

where $\sigma$ is the spacing of the quantization levels, $\tau$ is the spacing of the sample times, and $\bar{e}$ is the rms noise, $\sigma/\sqrt{12}$. The assumption that this noise is random and uncorrelated with the input does not hold for slowly changing input amplitudes [5]. However, measurements and simulations have shown [6] that analysis based on (1) gives good descriptions of average noise values even when the quantization has only two levels.

The $z$-transform of the output from the modulator is

$$Y(z) = z^{-1}X(z) + (1 - z^{-1})E(z). \qquad (2)$$

The spectral density of the noise in this signal can be expressed as

$$N(f) = 2\bar{e}\sqrt{2\tau}\,\sin\,(\pi f\tau) \qquad (3)$$

and the net rms noise in baseband $0 < f < f_0$ is given by

$$N_0 = 2\bar{e}\sqrt{f_0\tau(1 - \text{sinc}\,(2f_0\tau))} \approx \frac{\bar{e}\pi}{\sqrt{3}}(2f_0\tau)^{3/2}. \qquad (4)$$

We use the notation

$$\text{sinc}\,(f) \equiv \frac{\sin\,(\pi f)}{\pi f}. \qquad (5)$$

An important development [6] of this basic modulator is the double feedback circuit of Fig. 2. Its output is given by

$$Y(z) = z^{-1}X(z) + (1 - z^{-1})^2 E(z) \qquad (6)$$

which contains noise having spectral density

$$N(f) = 2\bar{e}\sqrt{2\tau}\,(1 - \cos\,(2\pi f\tau)), \qquad (7)$$

the net baseband component of which is

$$N_0 = 2\bar{e}(f_0\tau(3 - 4\,\text{sinc}\,(2f_0\tau) + \text{sinc}\,(4f_0\tau)))^{1/2} \qquad (8)$$

and

$$N_0 \approx \frac{\bar{e}\pi^2}{\sqrt{5}}(2f_0\tau)^{5/2}, \qquad f_0\tau \ll 1. \qquad (9)$$

Results (3) and (7) show that the integrating feedback loop

Reprinted from *IEEE Trans. Commun.*, vol. COM-34, pp. 72–76, January 1986.

400

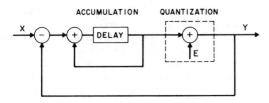

Fig. 1. A sigma delta modulator in which quantization is represented by added noise.

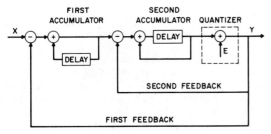

Fig. 2. A second-order sigma delta modulator, showing the two feedback loops.

placed around quantizers moves the noise to high frequencies, and when the sampling rate is high, most of it lies above baseband.

Data encoded at high sampling rates usually are unsuitable for processing or transmission, so that decimation to a low word rate is often necessary. If the modulated signal is subsampled directly with period $N\tau$, that is, every $N$th word is selected, the noise in baseband increases by an amount equal to the sum of the noises in all the intervals

$$\left(\frac{n}{N\tau}-f_0\right)<f<\left(\frac{n}{N\tau}+f_0\right) \qquad (10)$$

for which $f_0 < f < 1/2\tau$ and $n$ is an integer.

In order to preserve the resolution, we need to filter away the high-frequency noise before subsampling it. Providing a conventional low-pass filter for this purpose is costly, especially when the sampling rates are high. Fortunately, when $N\tau f_0 \ll 1$, much simpler filters will attenuate the noise in the critical regions defined by (10). Particularly attractive are comb filters made up of the accumulate-and-dump function

$$\frac{1}{N}\sum_{i=0}^{N-1}z^{-i}\equiv\frac{1}{N}\left(\frac{1-z^{-N}}{1-z^{-1}}\right)\equiv\frac{\text{sinc }(fN\tau)}{\text{sinc }(f\tau)}\text{ (delay).} \quad (11)$$

These filters are easy to build and their action is easy to predict.

## III. DECIMATING WITH $\text{sinc}^k$ $(Nf)$

Although practical considerations restrict the usefulness of all but the simplest of these circuits, it is valuable to now analyze a general case. Consider a signal $x$, modulated by a sigma delta modulator that comprises $l$ feedback loops [6]. Figs. 1 and 2 show examples where $l$ is 1 and 2, respectively. Let the output $y$ be decimated from a sampling period $\tau$ to $N\tau$ by means of a filter

$$D(z)=\left(\frac{1}{N}\frac{(1-z^{-N})}{(1-z^{-1})}\right)^k \qquad (12)$$

and let

$$2N\tau f_0<1. \qquad (13)$$

By iteration from results (2) and (6), it can be shown that the output from the modulator is given by

$$Y(z)=X(z)+(1-z^{-1})^l E(z) \qquad (14)$$

and the spectral density of its noise can be expressed as

$$N(f)=\bar{e}\sqrt{2\tau}\ (2\ \sin\ (\pi f\tau))^l. \qquad (15)$$

Let us express the decimated signal as

$$Y'(Z)=X'(Z)+F(Z)E'(Z) \qquad (16)$$

where $Z$ is the $z$-transform corresponding to the reduced sampling rate $Z = z^N$. $X'(Z)$ represents $N\tau$ sampling of a filtered form of the input signal. Its baseband spectral density is given by

$$X'(f)=X(f)D(z)=X(f)\left(\frac{\text{sinc }(N\tau f)}{\text{sinc }(\tau f)}\right)^k. \qquad (17)$$

This filtering of the signal is a primary limitation on the lowest word rate that can be achieved with this form of decimation.

Expressions for the white noise $E'$ and the filter $F$ depend on the relative values of $l$ and $k$, and the following sections examine some particular cases.

### A. Decimation with $k = l$

When the order of the modulating and decimating filters are the same, $k = l$, the noise can be expressed as

$$F(Z)E'(Z)=\left(\frac{1-Z^{-1}}{N}\right)^l E(Z) \qquad (18)$$

where $E(Z)$ represents the original noise $E$ sampled with period $N\tau$. Comparing this noise to that in (14), we see that it is identical to the noise introduced by a sigma delta modulator having $l$ feedback loops, cycling with period $N\tau$, and with its quantization steps spaced by

$$\sigma'=\frac{\sigma}{N^k}. \qquad (19)$$

The spectral density of the noise (18) can be written as

$$N'(f)=\bar{e}\sqrt{2N\tau}\left(\frac{2\ \sin\ (\pi fN\tau)}{N}\right)^l. \qquad (20)$$

In order to determine the effect this noise has on the baseband resolution, we need to equalize it for the filtering described by (17). The effective noise in baseband is then given by

$$N_0'(f)=\bar{e}\sqrt{2N\tau}\ (2\ \sin\ (\pi f\tau))^l, \qquad 0<f<f_0. \quad (21)$$

Comparison with the noise before decimation (15) shows an increase of $\sqrt{N}$ and no change of the spectral shape.

### B. Decimation with $k = l + 1$

When the order of the decimating filter exceeds that of the modulation filter, $k = l + 1$, the resampled noise can be expressed in the form

$$F(Z)E'(Z)=\left(\frac{1-Z^{-1}}{N}\right)^l E'(Z) \qquad (22)$$

where $E'(Z)$ represents $N\tau$ sampling of the signal

$$\frac{1}{N}\left(\frac{1-z^{-N}}{1-z^{-1}}\right)E(z).$$

Each sample of $E'$ is made up of the average of $N$ samples of $E$:

$$e_n' = \frac{1}{N}\sum_{i=0}^{N-1} e_{nN-i}. \tag{23}$$

Since we are assuming no correlation between the samples $e_n$, there will be no correlation of the samples $e_n'$. The rms spectral density of $E'(Z)$ will therefore be given by

$$E'(f) = \frac{\bar{e}}{\sqrt{N}}\sqrt{2N\tau} = \bar{e}\sqrt{2\tau} \tag{24}$$

and the spectral density of the decimated noise will be

$$N'(f) = \bar{e}\sqrt{2\tau}\left(\frac{2\sin(\pi fN\tau)}{N}\right)^l. \tag{25}$$

When this is equalized to compensate for the signal filtering (17), it becomes

$$N_0'(f) = \bar{e}\sqrt{2\tau}\,(2\sin(\pi f\tau))^l\left(\frac{\mathrm{sinc}(\tau f)}{\mathrm{sinc}(N\tau f)}\right), \qquad 0<f<f_0. \tag{26}$$

In cases of interest to us, $N\tau f_0 < 1/8$, this is within 0.25 dB of the baseband component of the noise before decimation (15). The spectral density (25) is similar to that generated by a sigma delta modulator having $l$ feedback loops, sampling with period $N\tau$, and having quantization levels spaced by

$$\sigma' = \frac{\sigma\sqrt{N}}{N^k}. \tag{27}$$

### C. Decimating with $k = l - 1$

For illustration we include one useless case, $k = l - 1$, for which the resampled noise can be expressed as

$$F(Z)E'(Z) = \left(\frac{1-Z^{-N}}{N}\right)^{l-1}E'(Z) \tag{28}$$

where $E'(Z)$ now represents a resampling of the signal $(1 - z^{-1})E(z)$,

$$e_n' = e_{nN} - e_{nN-1}. \tag{29}$$

The rms spectral density of the decimated noise is

$$E'(f) = 2\bar{e}\sqrt{N\tau}\left(\frac{2\sin(\pi fN\tau)}{N}\right)^{l-1}. \tag{30}$$

Comparison with (20) shows that the noise could be reduced by removing one feedback loop from the modulator to make $l = k$. The design has no value.

### IV. The Length of Digital Words

We have seen that decimation with the filters (12) can preserve the rising spectral shape of the noise from sigma delta (15), keeping most of its power at high frequencies. We now show that this decimation can be designed to provide relatively short code words compared to PCM. This is important, because rounding off surplus bits introduces white noise unless special circuits are provided to shape its spectral density.

During the analysis of the decimation it was convenient to include the term $N^{-k}$ in the filter response (12) in order to set the dc gain at unity. In concept the decimation ratio $N$ may be any integer that satisfies (13), but it is advantageous to make it an integer power of 2. Then the adjustment of gain is a simple change of significance, with no change in the form of the words. We shall ignore this gain adjustment during the following discussion, merely noting that, in general, it must be implemented with care to avoid intolerable lengthening of the words or introduction of roundoff noise.

When the term $N^{-k}$ is removed, the filtering (12) becomes

$$N^k D(z) = \left(\sum_{i=0}^{N-1} z^{-i}\right)^k. \tag{31}$$

It has a finite impulse response spanning $(k(N-1)+1)$ samples, with integer coefficients including unity, and dc gain $N^k$. When the quantizer has $Q = 2^b$ levels, the modulated signal comprises $b$-bit words having integer values 0 through $(Q-1)$. The code generated by the decimator then takes on integer values in the range 0 through $(Q-1)N^k$ it can be described by $(b + k\log_2(N))$ bit words, but the $(N^k - 1)$ largest states of this code will not be utilized. An explanation for this fact, given in Section V, concerns the overloading properties of sigma delta modulation. A case of practical importance is two-level quantization, $Q = 2$. It sometimes pays to suppress the largest state of its decimated code; then the remaining $N^k$ states can be signaled with only $k\log_2(N)$ bits.

We have seen in (19) that when $k = l$, the decimating filters reduce the effective spacing of the quantization step size by a factor $N^k$. Increasing in word length by $k\log_2(N)$ bits is appropriate for this decimation. In Section III-B where $k = l + 1$, decimation reduces the effective step size by a factor $N^k/\sqrt{N}$. This reduction should require the word length to increase by only $(k - 1/2)\log_2(N)$ bits. Multilevel sigma delta modulation would require $0.5\log_2(N)$ fewer bits to provide the same resolution as this decimated signal. But for most applications in telecommunications, the difference is small, usually less than 4 bits.

### V. Overloading Characteristics

In (1) we have expressed the quantization noise as a multiple of $\sigma$, the spacing of quantization levels. We can also express the maximum range of signal amplitudes as a multiple of $\sigma$. Therefore, the maximum signal-to-noise ratio is independent of the level spacing, but it depends on the number of levels $Q$ in the quantizer.

For ordinary quantization without feedback, $l = 0$, the range of signal amplitudes for which (1) holds is given by $\sigma Q$. Then, for sinusoidal signals the maximum signal-to-noise ratio is given by

$$\Gamma_0 = \sqrt{1.5}\,Q(2f_0\tau)^{-1/2}, \qquad l = 0. \tag{32}$$

Since sigma delta modulation overloads when input amplitudes lie outside the outermost quantization levels, its range of useful amplitudes is $\sigma(Q-1)$. The maximum signal-to-noise ratio for sinusoidal signals is given by

$$\Gamma_1 = \frac{\sqrt{4.5}\,(Q-1)}{\pi}(2f_0\tau)^{-3/2}, \qquad l = 1 \tag{33}$$

and

$$\Gamma_2 = \frac{\sqrt{7.5}\,(Q-1)}{\pi^2}(2f_0\tau)^{-5/2}, \qquad l = 2 \tag{34}$$

for one and two feedback loops, respectively [5].

The reduction of signal range from $\sigma Q$ to $\sigma(Q-1)$ when feedback is included accounts for the fact that some states of the code generated by the decimator are not utilized. This reduction is most significant when $Q$ is small, especially for the important case when only two quantization levels are used. Then the 6 dB loss of range is the penalty paid for the important practical advantage of having only one decision threshold [6].

## VI. DECIMATION OF SPURIOUS NOISES

Besides attenuating quantization noise, decimating filters must also attenuate any high frequency noise that accompanies the signal, so that when it aliases into baseband, the loss of resolution will be tolerably small. For a particular stage of decimation, only the noise in bands (10) need be considered. Decimating filter (12) provides least attenuation for this noise at frequency $((1/N\tau) - f_0)$ and it attenuates the signal most at frequency $f_0$. The ratio of the filter's attenuations at these two frequencies is a figure of merit for the filter. It gauges the amount that noise is rejected relative to the signal and is given by

$$\Phi = \left( \frac{\sin \left( \pi \left( \dfrac{1}{N} - \tau f_0 \right) \right)}{\sin (\pi \tau f_0)} \right)^k . \quad (35)$$

When the decimating filter is the sole protection against this out-of-band noise, the specification of toll network codecs can be interpreted [4] as requiring that $\Phi$ exceed 34 dB. This can be met by having $N\tau f_0 < 1/8$, $N > 4$, and $k = 2$, which is the decimation proposed in [4].

Fig. 3 presents graphs of $\Phi$ plotted against $1/2N\tau f_0$ for various values of $N$ and $k$. $\Phi$ often is the factor that limits the lowest sampling for which the decimation is useful. The decimation described in this study is often satisfactory for reducing word rates to about four times the Nyquist rate. Further reduction calls for filters that cut off more sharply at the edge of baseband.

## VII. GRAPHICAL PRESENTATIONS

Figs. 3–6 present the main results of this study as charts that are useful for designing oversampled codecs. Fig. 3 was described in the preceding section. Fig. 4 plots baseband quantization noise against the oversampling ratio for various kinds of modulation. The oversampling ratio is the sampling rate $1/\tau$ expressed as a multiple of the Nyquist rate, $2f_0$. The noise is expressed in decibels relative to $\bar{e}$, which is the noise from PCM sampled at the Nyquist rate and having the same spacing of quantization levels. Graphs are presented for three modulations, oversampled PCM and sigma delta modulation with one and two feedback loops. Notice that it is permissible to include the PCM as the case of sigma delta modulation with no feedback, $l = 0$. Cases for $l > 2$ are not included because they have practical difficulties [6]. The grid lines on this graph show the effects of decimation. When $k = l + 1$, decimation traces approximately horizontal lines, lowering the frequency without significantly changing the noise level. When $k = l$, decimation traces the sloping lines, lowering the frequency raises the noise 3 dB per octave. For example, suppose we want to encode a sinewave at four times the Nyquist rate with a signal-to-noise ratio $\Gamma$ dB. First we choose a value for $Q$, the number of levels of the sigma delta modulation. An ordinary PCM encoding with this number of levels at the Nyquist rate would provide a signal-to-noise ratio

$$\bar{e} = \frac{Q\sigma}{\sqrt{8}} \bigg/ \frac{\sigma}{\sqrt{12}} = 1.2 Q.$$

Next, we locate the point $(4, 20 \log (1.2 \, Q) - \Gamma)$ on Fig. 4,

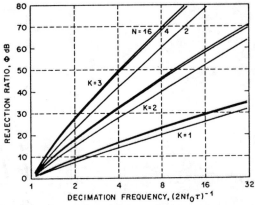

Fig. 3. A graph of the attenuation of spurious noise, relative to the attenuation of signal, plotted against the word rate at the output of the decimator.

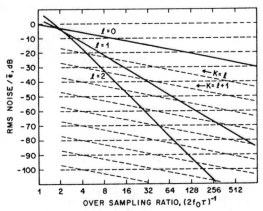

Fig. 4. Quantization noise plotted against the sampling rate for 0, 1, and 2 feedback loops. Grid lines trace the level of the noise with word rate for two orders of decimation, $k = l$ and $k = l + 1$.

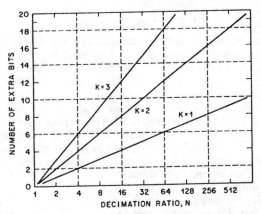

Fig. 5. A graph showing the increase in word length due to decimation for various values of $N$ and $k$. In the case of two-level initial modulation, the decimated words can be shortened by one bit if the largest state of the decimated code is suppressed.

and trace either the horizontal or the sloping grids to a point on one of the curves. The abscissa of this point gives the required modulation rate for the conditions selected. Table I lists conditions which satisfy the requirement that $\Gamma > 60$ dB. Fig. 5 shows a graph of the increase in word length caused by decimation under various conditions.

The filter used for decimation introduces spectral distortion into the signal path which must be equalized. Fig. 6 provides a measure of the amount of equalization that will be necessary.

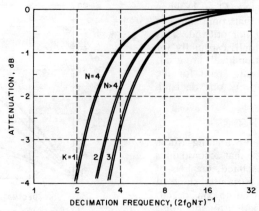

Fig. 6. The attenuation at the edge of baseband caused by decimation for various frequencies and orders of decimation.

TABLE I
ANALOG-TO-DIGITAL CONVERSION OF 4 KHz SIGNALS INTO 32 000 WORDS/s

| | MODULATION | | | DECIMATION to 32KHz. | | | |
|---|---|---|---|---|---|---|---|
| TYPE | FREQUENCY 1/T KHz | ♦LEVELS Q | ♦LOOPS $\ell$ | RATIO N | ORDER k | ♦BITS | S/N dB. |
| PCM | 32 | 1024 | 0 | 1 | 0 | 10 | 68 |
| $\Sigma\Delta$ | 32 | 512 | 1 | 1 | 0 | 9 | 69 |
| $(\Sigma\Delta)^2$ | 32 | 256 | 2 | 1 | 0 | 8 | 67 |
| $\Sigma\Delta$ | 8096 | 2 | 1 | 256 | 1 | 8 | 63 |
| $\Sigma\Delta$ | 2048 | 2 | 1 | 64 | 2 | 12 | 69 |
| $(\Sigma\Delta)^2$ | 512 | 2 | 2 | 16 | 2 | 8 | 67 |
| $(\Sigma\Delta)^2$ | 256 | 2 | 2 | 8 | 3 | 9 | 64 |

It plots the attenuation at the edge of baseband $f_0$ against the word rate $1/N\tau$ for various values of $k$ and $N$.

## VIII. CONCLUDING REMARKS

We have demonstrated that decimation with simple filter functions can lower the word rate of sigma delta modulation while preserving both the spectral shape of its noise and the advantage of having short code words. The techniques are useful for decimation down to about four times the Nyquist rate. Attempts to use it for lower frequencies results in significant spectral distortion in baseband and poor rejection of spurious noise. There is no incentive to retain the form of sigma delta modulation for the lower sampling rates. Fig. 4 shows that at twice the Nyquist rate, sigma delta modulation has approximately the same signal-to-noise ratio as PCM, and at the Nyquist rate it is substantially worse.

Only low-order modulation has been considered in detail because sigma delta circuits that use more than two integrators, $l > 2$, have practical dangers [6]. As a result, decimators with $k \leq 3$ are sufficient and have easy implementations [4].

The techniques that have been described for sigma delta modulation also apply to delta modulation, but with the following proviso. The quantity $x(t)$ which represented the signal must now represent the derivative of the signal, and the noise introduced into the output of the delta modulator will be the integral of the noise introduced by sigma delta modulation. Our decimation does not change the shape of the spectral density of the baseband noise with either type of modulation, and it changes the net noise power by the same factors. Note that the quantization levels of the delta modulator will lie at $\pm\sigma/2$: the step size is half of the level spacing.

A number of codecs have been built and tested both as breadboards and as integrated circuits [7]; their performance confirms the results predicted in this work.

## REFERENCES

[1] D. J. Goodman and M. J. Carey, "Nine digital filters for decimation and interpolation," IEEE Trans. Acoust., Speech, Signal Processing, vol. ASSP-25, pp. 121-126, Apr. 1977.

[2] R. E. Crochiere and L. R. Rabiner, "Interpolation and decimation of digital signals—A tutorial review," Proc. IEEE, vol. 69, pp. 300-331, Mar. 1981.

[3] ———, Multirate Digital Signal Processing. Englewood Cliffs, NJ: Prentice-Hall, 1983.

[4] J. C. Candy, B. A. Wooley, and O. J. Benjamin, "A voiceband codec with digital filtering," IEEE Trans. Commun., vol. COM-29, pp. 815-830, June 1981.

[5] J. C. Candy and O. J. Benjamin, "The structure of quantization noise from sigma-delta modulation," IEEE Trans. Commun., vol. COM-29, pp. 1316-1323, Sept. 1981.

[6] J. C. Candy, "A use of double integration in sigma delta modulation," IEEE Trans. Commun., vol. COM-33, pp. 249-258, Mar. 1985.

[7] J. H. Fischer and T. Misawa, unpublished report on integrated circuits.

# Multirate Filter Designs Using Comb Filters

SHUNI CHU, MEMBER, IEEE, AND C. SIDNEY BURRUS, FELLOW, IEEE

*Abstract* —Results on multistage multirate digital filter design indicate most of the stages can be designed to control aliasing with only slight regard for the passband which is controlled by a single stage compensator. Because of this, the aliasing controlling stages can be made very simple. This paper considers comb filter structures for decimators and interpolators in multistage structures. Design procedures are developed and examples shown that have a very low multiplication rate, very few filter coefficients, low storage requirements, and a simple structure.

## I. INTRODUCTION

**M**ULTIRATE filters are members of a class which has different sampling rates in various stages of the filtering operation. This class of filters includes decimators, interpolators, and narrow-band low-pass filters implemented with decimation, low-pass filtering, and interpolation. A multistage implementation of these filters has the sample rate changed in several steps where each step is a combined filtering and sample rate change operation. Crochiere and Rabiner [1]–[4] gave the standard multistage design method for these filters which has each stage as a low-pass filter where one optimally chooses the decimation (or interpolation) ratio at each stage. A design method was presented in [5] which uses a different design criterion for each stage. It only requires that each stage have enough aliasing attenuation but has no passband specifications.

Using the design described in [5] with no passband specifications for each stage allows simple filters to be employed and gives a satisfactory frequency response. Let $H(z)$ and $D$ be the transfer function and decimation ratio of one stage of a multistage decimator. We propose to design $H(z)$ such that $H(z) = f(z)g(z^D)$. In the implementation, by the commutative rule [5], the transfer function $g(z^D)$ can be implemented at the lower rate (after decimation) as $g(z)$. This implementation reduces the filter order, storage requirement, and the arithmetic.

In this paper, to simplify arithmetic, further requirements are put on $H(z)$ to allow only simple integer coefficients. This is feasible because there are no passband specifications on the frequency response. A cascade of comb filters is a particular case of these filters where the coefficients are only 1 or $-1$ and, therefore, no multiplications are needed. Hogenauer [6] had also used a cascade of comb filters as a one-stage decimator or interpolator but with a limited frequency-response characteristic. Here the cascade of comb filters is used as one stage of a multistage multirate filter with just the right frequency response. More comb filter structures are easily derived using the commutative rule.

The FIR filter optimizing procedure used in this paper minimizes the Chebyshev norm of the approximation error and this is done using the Remez exchange algorithm. The IIR filter optimizing procedure used minimizes the $l_p$ error norm which approaches the Chebyshev norm when $p$ is large.

## II. THE NEW MULTISTAGE MULTIRATE DIGITAL FILTER DESIGN METHOD

In a paper for limited range DFT computation using decimation [7], Cooley and Winograd pointed out that the passband response of a decimator can be neglected and be taken care of after decimation. A multistage multirate digital filter design method which has no passband specification but using passband and stopband gain difference as an aliasing attenuation criterion for each stage is described in [5]. The design method and equations used in that paper which are needed for the comb filter structure are outlined in this section.

The commutative rule introduced in [5] states that the filter structures in Fig. 1(a) and (b) are equivalent. It means that a filter can commute with a rate changing switch provided that the filter has its transfer function changed from $H(z)$ to $H(z^D)$ or vice versa. Fig. 1 illustrates the case for decimation, and it is also true for interpolation. This rule is very useful in finding equivalent multirate filter structures and in deriving the transfer function of a multistage multirate filter.

For example, Fig. 2(a) shows the filter structure of a multistage decimator where $f_{rk}$, $k = 0, 1, \cdots, K$, is the sampling rate at each stage, and a one-stage equivalent decimator shown in Fig. 2(b) is found by repeatedly applying the commutative rule to move the latter stages forward. From the one-stage equivalent, it is clear that the transfer function and frequency response of the multistage decimator are

$$H(z) = H_1(z)H_2(z^{D_1})H_3(z^{D_1 D_2}) \cdots H_c(z^D) \quad (1)$$

and[1]

$$H(\omega) = H_1(\omega)H_2(D_1\omega)H_3(D_1 D_2\omega) \cdots H_c(D\omega) \quad (2)$$

where $D = D_1 D_2 \cdots D_k$. The filtering function of $H_c(z)$ does not involve a sampling rate change. It is used to

Manuscript received November 24, 1981; revised November 2, 1982 and August 22, 1983. This work was supported by the National Science Foundation under Grant ECS 81-00453.

S. Chu was with the Rice University, Houston, TX. He is now with Bell Laboratories, Holmdel, NJ 07733.

C. S. Burrus is with the Department of Electrical Engineering, Rice University, Houston, TX 77251-1892.

[1]$H(\omega)$ is used for $H(e^{j\omega})$ and this notation will be used hereafter.

Reprinted from *IEEE Trans. Circuits and Sys.*, vol. CAS-31, pp. 913–924, November 1984.

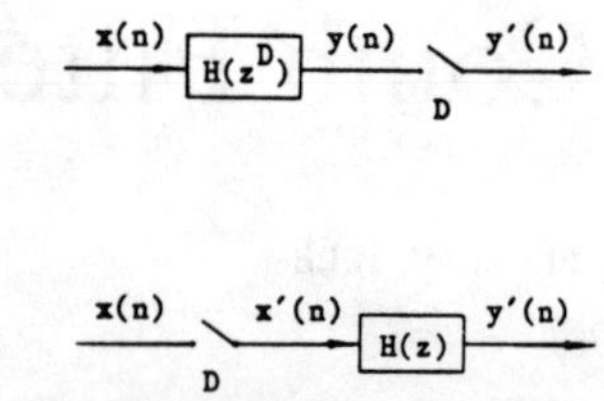

Fig. 1. Illustration of the commutative rule.

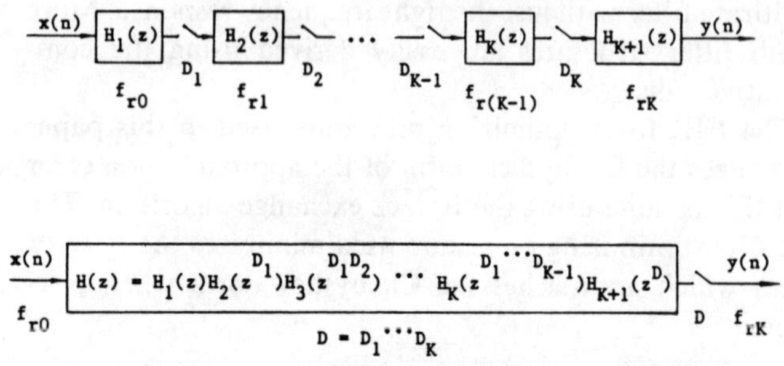

Fig. 2. Illustration of the equivalence of multistage decimators.

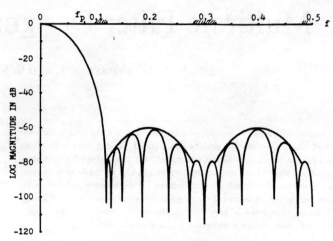

//// Area that will alias to the transition region.

Fig. 3. An example of frequency response of $H_{D_1}(\omega)$.

compensate the passband frequency responses of previous stages, and hence, is called the *compensator*.

Each decimation stage is designed successively. At the time of designing the $i$th stage filter, all the previous $i-1$ stages have already been designed and the transfer functions known. The requirement on $H_i(z)$ is that the composite frequency response $H_{D_i}(\omega)$ of the first stage to the $i$th stage have enough aliasing attenuation where

$$H_{D_i}(\omega) \equiv H_1\left(\frac{\omega}{D_1 \cdots D_{i-1}}\right) H_2\left(\frac{\omega}{D_2 \cdots D_{i-1}}\right)$$

$$\cdots H_{i-1}\left(\frac{\omega}{D_{i-1}}\right) H_i(\omega) \quad (3)$$

referenced to $f_{r(i-1)} = 1$. Enough aliasing attenuation means that those frequency components which will alias into the passband at the current decimation process will have adequate attenuation with respect to the corresponding passband components. Fig. 3 shows an example frequency response of $H_{D_i}(\omega)$ which has an aliasing attenuation exceeding 60 dB. In Fig. 3, the passband response is repeated in the stopbands but has been moved down by 60 dB. They are used as the attenuation bounds for the stopbands. If the stopband response is below these bounds, it will have enough aliasing attenuation.

The overall filter frequency response is $H_c(\omega) H_{D_K}(\omega/D_K)$ referenced to $f_{rK} = 1$. The design of the compensator transfer function is to make the overall frequency response approximate one in the passband. The frequency-response error in the passband is

$$E(\omega) = 1 - H_c(\omega) H_{D_K}\left(\frac{\omega}{D_K}\right)$$

$$= H_{D_K}\left(\frac{\omega}{D_K}\right) \left[ \frac{1}{H_{D_K}\left(\frac{\omega}{D_k}\right)} - H_c(\omega) \right] \quad (4)$$

for $\omega \in [0, \omega_p]$. To give attenuation to the first band that

will alias to the transition band, it is required that $|H_c(\omega) H_{D_K}(\omega/D_K)| \leqslant \delta_s$ for $\omega \in [\pi, 2\pi - \omega_p]$, or equivalently, $|H_c(\omega) H_{D_K}((2\pi - \omega)/D_K)| \leqslant \delta_s$ for $\omega \in [\omega_p, \pi]$. The frequency band $(\omega_p, \pi]$ can be considered as the stopband of the compensator and the frequency-response error is

$$E(\omega) = 0 - H_c(\omega) H_{D_K}\left(\frac{2\pi - \omega}{D_K}\right)$$

$$= H_{D_K}\left(\frac{2\pi - \omega}{D_K}\right) [0 - H_c(\omega)] \quad (5)$$

for $\omega \in (\omega_p, \pi]$. Equations (4) and (5) can be combined to give an error function of

$$E(\omega) = W(\omega)[H_{Des}(\omega) - H_c(\omega)] \quad (6)$$

for $\omega \in [0, \pi]$, where

$$W(\omega) = \begin{cases} H_{D_K}\left(\dfrac{\omega}{D_K}\right), & \text{for } \omega \in [0, \omega_p] \\[2mm] W_r H_{D_K}\left(\dfrac{2\pi - \omega}{D_K}\right), & \text{for } \omega \in (\omega_p, \pi] \end{cases}$$

$$(7)$$

$$H_{Des}(\omega) = \begin{cases} \dfrac{1}{H_{D_K}\left(\dfrac{\omega}{D_K}\right)}, & \text{for } \omega \in [0, \omega_p] \\[2mm] 0, & \text{for } \omega \in (\omega_p, \pi] \end{cases} \quad (8)$$

and $W_r = \delta_p/\delta_s$ which is the error weighting of the stopband with respect to the passband. The optimal $H_c(z)$ is obtained by minimizing the error norm $\|E\|$ of (6). The solution depends on the definition of the norm.

The multistage interpolator design is the same as the multistage decimator design but with the filter structure reversed.

The multirate low-pass filter structure is a multistage decimator followed by a multistage interpolator and, in between, there is a compensator operated at the lowest sampling rate with no rate change. If the aliasing attenuation requirement for the decimator is the same as the imaging attenuation requirement for the interpolator, the design of the multistage decimator part and that of the interpolator part can be the same. The overall frequency response is

$$H(\omega) = G(\omega) H_c(D\omega) G\left(\omega \bmod \frac{2\pi}{D}\right) \quad (9)$$

where

$$G(\omega) \equiv H_1(\omega) H_2(D_1\omega) \cdots H_K(D_1 \cdots D_{K-1}\omega). \quad (10)$$

$H_i(\omega)$ is the frequency response of each decimator (or interpolator) stage and "mod" means a modulo operation. The frequency response of (9) is the output baseband response due to the whole input in terms of the input frequency as in the case of decimator. It is also the output response due to the baseband input in terms of the output frequency as in the case of interpolator.

In the multirate low-pass filter design, each decimation or interpolation stage design is the same as that in a multistage decimator design. The compensator is to give the desired frequency response in the baseband where the baseband is the frequency band that never aliases. Its design is to minimize $\|E\|$ of (6) with the weighting and desired functions given by

$$W(\omega) = \begin{cases} G^2\left(\dfrac{\omega}{D}\right), & \text{for } \omega \in [0, \omega_p] \\ W_r G\left(\dfrac{\omega}{D}\right) G\left(\dfrac{2\pi - \omega}{D}\right), & \text{for } \omega \in (\omega_p, \omega_s) \\ W_r G^2\left(\dfrac{\omega}{D}\right), & \text{for } \omega \in [\omega_s, \pi] \end{cases}$$

$$\quad (11)$$

and

$$H_{Des}(\omega) = \begin{cases} \dfrac{1}{G^2\left(\dfrac{\omega}{D}\right)}, & \text{for } \omega \in [0, \omega_p] \\ 0, & \text{for } \omega \in (\omega_p, \pi]. \end{cases} \quad (12)$$

In the case where there is not a full decimation, i.e., $\omega_s < \pi$ referenced to $f_{rK} = 1$, there is a stopband for the compensator design. The transition region can also be viewed as the stopband of the compensator with requirement to limit the transition region aliasing.

## III. COMB FILTER STRUCTURES AS DECIMATORS OR INTERPOLATORS

This section exploits some simple efficient filter structures which can be used in the decimation or interpolation stages of the multistage multirate filter. The requirement on these filters is that they have enough aliasing attenuation such as shown in the example frequency response of Fig. 3. Since the operation and structure of an interpolator are the duals of a decimator, most explanation in this section will be for the decimator case only. Extension to the interpolator case is simple and straighforward.

Let $H(z)$ and $D$ be the transfer function and decimation ratio for one stage of a multistage decimator. The filter

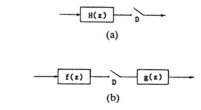

Fig. 4. Decimator implementation with a special transfer function. (a) A decimator structure. (b) The decimator structure with $H(z) = f(z)g(z^D)$.

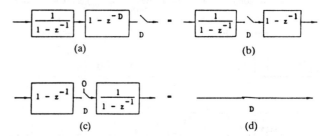

Fig. 5. (a), (b) Illustrations of equivalence of comb decimators. (c), (d) Illustration of equivalence of comb interpolators.

structure is shown in Fig. 4(a). One method to make the filter efficient is to design $H(z)$ such that it has the form

$$H(z) = f(z)g(z^D) \quad (13)$$

and the factor $g(z^D)$ can be implemented at the lower rate as $g(z)$ as shown in Fig. 4(b). By this implementation, a high-order $H(z)$ can be implemented at the low rate as a low-order filter. The arithmetic rate, number of filter coefficients, and number of registers used are, therefore, reduced. Further improvement in arithmetic rate can be achieved by simplifying the filter coefficients of $f(z)$ and $g(z)$ in (13) to be simple integers and using additions instead of multiplications.

One example of this kind of filter is a cascade of comb filters. We will show some filter structures first and discuss the filter operations in the next section.

A comb filter of length $D$ is an FIR filter with all $D$ coefficients equal to one. The transfer function of this comb filter is

$$H(z) = \sum_{n=0}^{D-1} z^{-n} = \frac{1 - z^{-D}}{1 - z^{-1}}. \quad (14)$$

A comb filter with length $D$ followed by decimation with a ratio $D$ is shown in Fig. 5(a). The commutative rule can be applied to the numerator to get the structure of Fig. 5(b). The new comb decimator structure needs two registers, one addition at the high rate, and one addition at the low rate regardless of the decimation ratio $D$, i.e., the filter length.

The comb interpolator structure is shown in Fig. 5(c). It is the reverse of the decimator structure with the sampler replaced by a zero padder. The realization of the transfer function $1/(1 - z^{-1})$ is an accumulator. Since the accumulator has $D - 1$ out of every $D$ inputs as zero, it can take advantage of this to accumulate only once for every $D$ inputs. This is equivalent to operating the accumulator at the lower rate and each output is used $D$ times at the higher rate. When the accumulator is moved to the lower

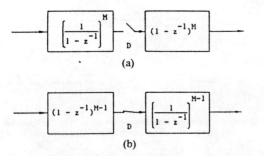

(a)

(b)

Fig. 6. (a) Cascaded comb filters as decimator. (b) Cascaded comb filters as inteprolator.

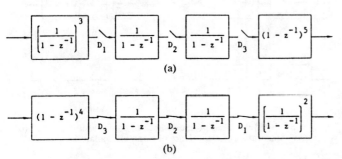

(a)

(b)

Fig. 7. (a) An example of multistage comb decimator. (b) An example of multistage comb interpolator.

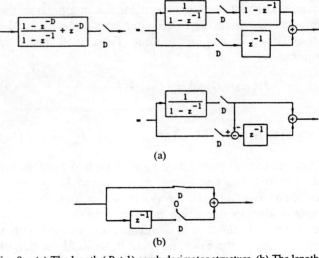

(a)

(b)

Fig. 8. (a) The length-$(D+1)$ comb decimator structure. (b) The length-$(D+1)$ comb interpolator structure.

$$\underset{x(n)}{\longrightarrow} \boxed{1+z^{-1}} \overset{}{\underset{2}{\searrow}} \boxed{1+z^{-1}} \overset{}{\underset{2}{\searrow}} \boxed{1+z^{-1}} \overset{}{\underset{2}{\searrow}} \boxed{1+z^{-1}} \overset{}{\underset{2}{\searrow}} \underset{y(n)}{\longrightarrow}$$

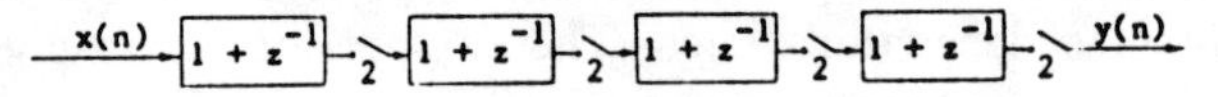

Fig. 9. A length-16 comb decimator structure.

rate stage, it cancels the $(1-z^{-1})$ section and leaves a sample and hold switch alone as a comb interpolator, as shown in Fig. 5(d). To distinguish the sample and hold switch from the sampling switch of the decimator and to indicate the sampling rate increase after a sample and hold switch, the sample and hold switch is represented by a normally closed switch. The commutative rule can be applied across a sample and hold switch since it applies when there is a rate change.

A single comb filter generally will not give enough stopband attenuation, however, cascaded comb filters can often meet requirements. Cascading $M$ length-$D$ comb filters will have a transfer function

$$H(z) = \left[ \frac{1 - z^{-D}}{1 - z^{-1}} \right]^M. \tag{15}$$

Fig. 6(a) shows a comb decimator with $M$ length-$D$ comb filters in cascade where all the accumulators are cascaded before the sampler and all the $(1-z^{-1})$ sections are cascaded after the sampler. When the reverse of the structure of Fig. 6(a) is used as an interpolator, one of the comb filters can be realized as a sample and hold switch. This interpolator structure is shown in Fig. 6(b).

In a multistage decimator design, a latter stage usually needs more comb filters in cascade to give adequate stopband attenuation because of the relatively wider stopband(s) and narrower transition region. Fig. 7(a) shows an equivalent three-stage comb decimator structure. The first, second, and the third stages have three, four, and five length-$D_1$, length-$D_2$, and length-$D_3$ comb filters in cascade, respectively. Fig. 7(b) shows the corresponding equivalent comb interpolator structure using sample and hold switches. These equivalent structures are obtained by applying the commutative rule. Because of the propagation of the $(1-z^{-1})$ section, some $(1/(1-z^{-1}))$ sections and $(1-z^{-1})$

sections have canceled each other. This multistage comb filter structure is called a *merged structure*.

A length-$D$ comb filter has $D-1$ zeros. In the frequency range between 0 and 0.5, it has $[D/2]$ zeros located at $f = i/D$, $i = 1, 2, \cdots, [D/2]$, where $[X]$ represents the maximum integer not greater than $X$. The other zeros are the images of these zeros at negative frequencies. There is one zero at the center of each stopband when used as a decimator with decimation ratio $D$. Cascading length-$D$ comb filters often does not give enough attenuation at the first stopband edge. The length-$(D+1)$ or length-$(D+2)$ comb filter which has the first zero at $f = 1/(D+1)$ or $f = 1/(D+2)$ can be used to give a better result. Fig. 8(a) shows a length-$(D+1)$ comb decimator structure obtained by superposition and application of the commutative rule. Fig. 8(b) shows the corresponding comb interpolator structure. The length-$(D+2)$ comb decimator and interpolator structures can be obtained similarly. It is better to use only one non-$D$ length comb filter in each decimator stage for simplicity of the filter structure.

Nonrecursive comb filter structures can also be derived by using the commutative rule. For example, the transfer function of a length-16 comb filter can be factored as

$$H(z) = \sum_{n=0}^{15} z^{-n}$$

$$= (1+z^{-1})(1+z^{-2})(1+z^{-4})(1+z^{-8}). \tag{16}$$

If this filter is used as a decimator with a decimation ratio 16, by applying the commutative rule, the filter structure is derived, as shown in Fig. 9. Cascading the length-16 comb filters is equivalent to raising the power of $(1+z^{-1})$ term in each stage of Fig. 9. This kind of structure has the advantage of not having the special modulo arithmetic and stability problem of the recursive ones as will be ad-

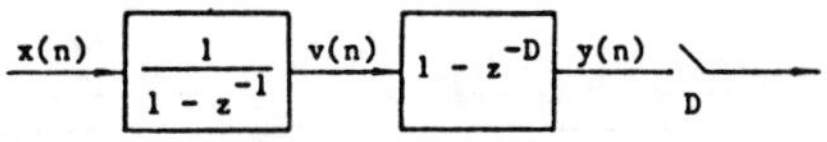

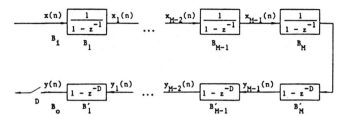

Fig. 10.   The equivalent cascaded comb decimator structure.

Fig. 11.   The equivalence cascaded comb decimator arithmetic.

dressed in the next section. In general, the transfer function of a length-$D^K$ comb filter can be factored as

$$H(z) = \sum_{n=0}^{D^K-1} z^{-n} = \prod_{k=0}^{K-1} \left[ \sum_{i=0}^{D-1} z^{-iD^k} \right]. \qquad (17)$$

When this filter is used as a decimator with a decimation ratio $D^K$, it can be implemented as a $K$-stage decimator which has each stage as a length-$D$ comb decimator with a decimation ratio $D$.

## IV.   THE STABILITY AND ARITHMETIC OF COMB FILTERS

This section discusses the special modulo arithmetic needed for the simple filter structures of last section, and the stability and overflow problems.

The filter structure in Fig. 5(a) and (b) has a recursive stage with a pole at $z = 1$, which is not asymptotically stable, and therefore may overflow. The filter operation depends on a "wrap around" number system similar to the two's compliment number system. With this kind of number system, overflow is a modulo operation, i.e., what is actually stored in the register is the residue of the true value. This filter will work if the arithmetic is a modulo arithmetic, and the output does not overflow. This will be explained as follows.

A comb decimator realized as in Fig. 5(b) is equivalent to the structure in Fig. 5(a). The latter structure is easier to analyze and its structure is repeated in Fig. 10. In Fig. 10, let the number range of the accumulator $1/(1-z^{-1})$ be the discrete points in $[-R/2, R/2]$. With the notation $\langle x \rangle_R$ as the residue of $x$ modulo $R$ with range in $[-R/2, R_2)$, then

$$v(n) = \left\langle \sum_{i=0}^{\infty} x(n-i) \right\rangle_R \qquad (18)$$

and

$$y(n) = \langle v(n) - v(n-D) \rangle_{R_0}$$
$$= \left\langle \left\langle \sum_{i=0}^{\infty} x(n-i) \right\rangle_R - \left\langle \sum_{i=0}^{\infty} x(n-D-i) \right\rangle_R \right\rangle_{R_0} \qquad (19)$$

where $R_0$ represents the number range in the $(1-z^{-D})$ section or, equivalently, the $(1-z^{-1})$ section in the actual realization. In order for $y(n)$ to recover information about $(x(n)$, (19) should have only one modulus, i.e., $R_0 = R$. Under this condition (19) becomes

$$y(n) = \left\langle \sum_{i=0}^{\infty} x(n-i) - \sum_{i=0}^{\infty} x(n-D-i) \right\rangle_R$$
$$= \left\langle \sum_{i=0}^{D-1} x(n-i) \right\rangle_R. \qquad (20)$$

This filter structure works if

$$\left\langle \sum_{i=0}^{D-1} x(n-i) \right\rangle_R = \sum_{i=0}^{D-1} x(n-i). \qquad (21)$$

Let $R_i$ denote the number range for the input $x(n)$, then

$$\left| \sum_{i=0}^{D-1} x(n-i) \right| \leqslant \sum_{i=0}^{D-1} |x(n-i)| \leqslant D\frac{R_i}{2}. \qquad (22)$$

A necessary and sufficient condition for the filter to work for any input is $R \geqslant DR_i$ and $R_0 = R$.

The condition $R_0 = R$ does not mean that they must be represented by the same wordlength, $v(n)$ may be rounded before being fed to the $(1-z^{-D})$ section. Rounding means dropping the least significant bits below a lower limit; modulo arithmetic means dropping the most significant bits above some upper limit. The lower limit is set by the quantization step chosen; the upper limit is set by the modulus $R$. These two operations are independent, and the upper limit can be fixed and the lower limit varied.

Fig. 11 shows the equivalent decimator structure of $M$ length-$D$ comb filters in cascade. Each $(1-z^{-D})$ section corresponds to a $(1-z^{-1})$ section at the same position after the sampling switch in the actual realization. $R_m$ and $R'_m$ represent the number range at each stage. The necessary condition for $y(n)$ to recover information about $x(n)$ is $R_m = R'_m$, $m = 1, 2, \cdots, M$. The most economic scheme is that all moduli be equal, and $R_m = R'_m = R$, $m = 1, 2, \cdots, M$. With this condition, it can be shown that

$$y_{M-m}(n)$$
$$= \left\langle \sum_{i_1=0}^{D-1} \sum_{i_2=0}^{D-1} \cdots \sum_{i_m=0}^{D-1} x_{M-m}(n-i_1-i_2\cdots-i_m) \right\rangle_R.$$
$$\qquad (23)$$

The condition on $R$ for $y(n)$ not to overflow for any input is

$$R \geqslant D^M R_i. \qquad (24)$$

$D^M$ is the factor of wordlength growth. In the case where different comb filter lengths are used in the decimator, the wordlength growth factor is the product of all the comb filter lengths. In the case of the multistage decimator with each stage implemented as a cascaded comb filter, this factor is the product of the comb filter lengths of all the stages.

The condition of (24) or the equivalent is often overly pessimistic for the signal growth in many practical applications. Define the term *cascaded comb decimator realization*

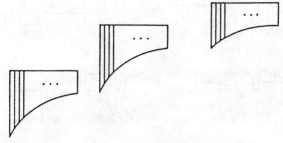

Fig. 12. The wordlength distribution in a comb decimator.

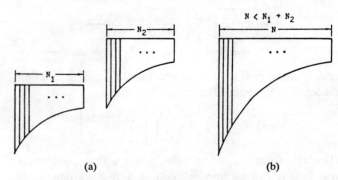

(a)                                    (b)

Fig. 13. (a) The wordlength distribution of a two-stage cascaded comb decimator. (b) The wordlength distribution of merged realization.

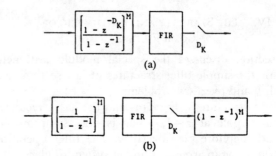

(a)

(b)

Fig. 14. Illustration of comb filters mixed with FIR Filters.

as a merged multistage cascaded comb decimator, i.e., in one cascaded comb decimator realization, there is only one modulus. A general comb decimator is a cascade of several stages of cascaded comb decimator realizations defined above with each stage having a decimation ratio $D_i$. Let $F_i(z)$ be the transfer function of the equivalent one-stage decimator from the input $x(n)$ to the $i$th stage cascaded comb decimator realization output $v_i(n)$, then

$$|v_i(n)| \leqslant \|F_i\|_p \|X\|_q \qquad (25)$$

with

$$\frac{1}{p} + \frac{1}{q} = 1$$

where the $l_p$ norm of a function $A(\omega)$ is defined as

$$\|A\|_p = \left[ \frac{1}{2\pi} \int_{-\pi}^{\pi} |A(\omega)|^p \, d\omega \right]^{1/p}. \qquad (26)$$

If the input satisfies the condition $\|X\|_q \leqslant 1$ for some $q \geqslant 1$, then

$$|v_i(n)| < \|F_i\|_p, \qquad \text{for } p = \frac{q}{q-1}. \qquad (27)$$

$\|F_i\|_p$ is the wordlength growth factor and it is used to determine the modulus in the $i$th stage of the cascaded comb decimator realization.

Fig. 12 shows the wordlength distribution in a general comb decimator. Each vertical length denotes the wordlength of each register in a cascaded comb decimator realization, the vertical position denotes the relative weighting of the bits in the register. The upper end of the wordlength is determined by the modulus in each cascaded comb decimator realization; the lower end variation is due to rounding.

The merging of two cascaded comb decimator stages (or realizations) is explained in Fig. 13. The two-stage realization has a total of $N_1 + N_2$ registers. A merged realization has fewer registers because it lets the $1 - z^{-1}$ sections of the first stage cancel some $1/(1 - z^{-1})$ sections of the second stage, but each register has a longer wordlength since the merged realization has only one bigger modulus. The choice depends on the total number of bits used and the complexity of the structure.

In the realization of a comb decimator, the last stage decimator often can not be realized with comb filters alone because of its sharp transition frequency response, and a short length FIR filter is required in cascade to meet the specifications. In this case, the commutative rule can be applied to the numerator of the comb filters across the FIR filter. An example is shown in Fig. 14. However, since the comb filter depends on the modulo arithmetic, the FIR filter must have its coefficients represented by integers and use the same modulo arithmetic as the comb filters. This kind of realization will increase the overall wordlength growth, hence, the most significant bits for every register, so it is desirable just to use the realization of Fig. 14(a) and implement the comb filter by direct summation.

The operation of the comb interpolator does not depend on modulo arithmetic, and there is no error if overflows do not occur. No rounding can be made inside a comb interpolator or the roundoff error variance will grow without bound. Each register inside a comb interpolator must have enough wordlength so that it will not overflow, otherwise the overflow error will also grow without bound. Each $(1 - z^{-D})$ section in a comb interpolator results in a wordlength growth factor of 2; each $1/(1 - z^{-1})$ section combined with a $(1 - z^{-D})$ section results in a wordlength growth factor of $D$. In the case of a multipath structure, the wordlength growth after summing these paths is the sum of the wordlength growth in each path.

Consider the term *cascaded comb interpolator realization* as a merged multistage cascaded comb interpolator, i.e., the least significant bit of each register in a cascaded comb interpolator realization is the same, and there is no rounding inside a realization. The output of a cascaded comb interpolator realization may have some of its most significant bits dropped according to the actual signal growth as indicated by (27).

A general comb interpolator structure is a cascade of several cascaded comb interpolator realizations. Fig. 15 shows the wordlength distribution of a comb interpolator.

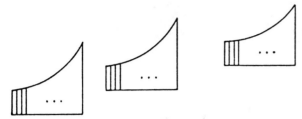

Fig. 15.    The wordlength distribution of a comb interpolator.

The lower limit of the wordlength is fixed for each cascaded comb interpolator realization because of no rounding; the upper limit grows because of the signal grows. Again, there is a choice of merging these realizations to cancel some $1/(1-z^{-1})$ and $(1-z^{-1})$ sections but with more wordlength growth, or implementing each realization separately without cancellation but with a smaller wordlength growth.

## V.    Multistage Multirate Filter Design Using Comb Filters

In this section, a multistage multirate filter design is presented which uses comb filters as decimators and/or interpolators and uses either an IIR or an FIR filter as the compensator.

A cascade of comb filters may not be the optimal choice for a decimator or interpolator. However, it has a frequency response with zeros at the center of each stopband which is generally desired, and it has a readily known explicit frequency-response expression which simplifies the design problem. Also, the comb filter realization is exact because there is no coefficient quantization error.

The design of a comb decimator with a decimation ratio $D$ is done by cascading length-$D$ comb filters until enough aliasing attenuation is obtained. A length-$(D+1)$ or length-$(D+2)$ comb filter is also used in cascade when necessary. The last stage decimator is a cascade of comb filters and a short length FIR filter since a comb filter alone usually can not give enough aliasing attenuation for this stage. With all the comb filters chosen, the design problem is of this FIR filter and the compensator.

### A.    The IIR Design

In the case where phase response is not important, an IIR compensator can be used to give high efficiency. For the multistage decimator design, define

$$H_p(\omega) \equiv H_{D_K}(\omega) H_c(D_K \omega) \qquad (28)$$

whose response over $[0, \pi]$ is the

$$\left[0, \frac{\pi}{D_1 \cdots D_{K-1}}\right]$$

part of the overall response of (2), and this is the part of response which remains to be designed when designing the

last stage decimator and compensator. Equation (28) is the same as (2) except that it is referenced to $f_{r(k-1)} = 1$. The design with $H_K(z)$ and $H_c(z)$ is such that $H_p(\omega)$ has a passband ripple less than $\delta_p$ for $\omega \in [0, \omega_p]$ and a stopband ripple less than $\delta_s$ for $\omega \in [\omega_s, \pi]$ referenced to $f_{r(K-1)} = 1$.

If, in the $k$th decimation stage, $k = 1, 2, \cdots, K$, there are $J_k$ comb filters with lengths $N_{kj}$, $j = 0, 1, \cdots, J_k - 1$, then

$$H_k(\omega) = \prod_{j=0}^{J_k-1} \frac{\sin \frac{\omega N_{kj}}{2}}{N_{kj} \sin \frac{\omega}{2}}, \qquad k = 1, 2, \cdots, K-1 \quad (29)$$

and

$$H_K(\omega) = \left[ \prod_{j=0}^{J_k-1} \frac{\sin \frac{\omega N_{Kj}}{2}}{N_{Kj} \sin \frac{\omega}{2}} \right] H_F(\omega) \qquad (30)$$

where $H_F(\omega)$ is the frequency response of the short length FIR filter in the $K$th decimation stage. Equation (3) can be written as

$$H_{D_K}(\omega) = H_{\text{comb}}(\omega) H_F(\omega) \qquad (31)$$

where $H_{\text{comb}}(\omega)$ is the response of $H_{D_K}(\omega)$ due to the comb filters in all stages,

$$H_{\text{comb}}(\omega) \equiv H_1\left(\frac{\omega}{D_1 \cdots D_{K-1}}\right)$$

$$\cdots H_{K-1}\left(\frac{\omega}{D_{K-1}}\right) \prod_{j=0}^{J_K-1} \frac{\sin \frac{\omega N_{Kj}}{2}}{N_{Kj} \sin \frac{\omega}{2}}. \qquad (32)$$

Let $H_F(z)$ have $N_{Kz}$ zeros on the unit circle with polar coordinates $(1, \theta_i)$, $i = 0, 1, \cdots, N_{Kz} - 1$. $H_F(z)$ and $H_F(\omega)$ can be written as

$$H_F(z) = \prod_{i=0}^{N_{Kz}-1} \left[1 - 2\cos\theta_i z^{-1} + z^{-2}\right] \qquad (33)$$

and

$$H_F(\omega) = \prod_{i=1}^{N_{Kz}-1} 2(\cos\omega - \cos\theta_i). \qquad (34)$$

The compensator is chosen to have $N_p$ complex pole pairs, $N_z$ unit circle zero pairs, and $N_{rz}$ real zeros at $z = -1$. It has a transfer function

$$H_c(z) = A \prod_{i=1}^{N_p} \frac{1}{1 - 2r_i \cos\theta_{pi} z^{-1} + r_i^2 z^{-2}}$$

$$\cdot \prod_{i=1}^{N_z} \left[1 - 2\cos\theta_{0i} z^{-1} + z^{-2}\right] \left[\frac{1 + z^{-1}}{2}\right]^{N_{rz}} \qquad (35)$$

and a frequency response

$$|H_c(\omega)| = A \prod_{i=1}^{N_p} \frac{1}{\left[1 - 2r_{pi}\cos(\omega - \theta_{pi}) + r_{pi}^2\right]^{1/2} \left[1 - 2r_{pi}\cos(\omega + \theta_{pi}) + r_{pi}^2\right]^{1/2}} \prod_{i=1}^{N_z} 2(\cos\omega - \cos\theta_{0i}) \cdot \left[\cos\frac{\omega}{2}\right]^{N_{rz}}. \qquad (36)$$

From (28) and (32), $H_p(\omega)$ can be written as

$$H_p(\omega) = H_{\text{comb}}(\omega)H_F(\omega)H_c(D_K\omega). \quad (37)$$

The design method chosen uses the Fletcher–Powell algorithm [5], [8] to minimize the $l_p$ error norm of

$$E = \sum_{i=1}^{N_g} \left[ W(\omega_i)\left( H_{Des}(\omega_i) - |H_p(\omega_i)| \right) \right]^{2p} \quad (38)$$

where $W(\omega_i)$ is the weighting function for the passband and stopband, $N_g$ is the number of grid points, and

$$H_{Des}(\omega) = \begin{cases} 1, & \omega \in [0, \omega_p] \\ 0, & \omega \in [\omega_s, \pi]. \end{cases} \quad (39)$$

For the case of multistage multirate low-pass filter designs, the design problem is still of the short length FIR filter which is in both the last stage decimator and the first stage interpolator and the compensator. The frequency response $H_p(\omega)$ is now

$$H_p(\omega) = H_{D_K}(\omega)H_c(D_K\omega)H_{D_K}\left(\omega \bmod \frac{2\pi}{D_K}\right)$$

$$= H_{\text{comb}}(\omega)H_{\text{comb}}\left(\omega \bmod \frac{2\pi}{D_K}\right)H_F(\omega)$$

$$\cdot H_F\left(\omega \bmod \frac{2\pi}{D_k}\right)H_c(D_K\omega). \quad (40)$$

For the case without full decimation ($f_s < (0.5/D)$) and $f_s$ is significantly smaller than $0.5/D$, the last stage decimator can be implemented with comb filters alone and the design is solely of the compensator.

### B. The FIR Design

When linear phase is desired, an FIR compensator must be used. The last stage decimator and the compensator are designed separately. Using (31), the design of $H_F(z)$ of the last stage decimator can use the iterative method in [5]. The FIR compensator is designed using the Remez exchange algorithm to minimize the maximum error in (6) with the weighting and desired functions in (7) and (8) or (11) and (12) for the multistage decimator design, or the multistage multirate low-pass filter design, respectively.

### VI. Examples and Comparisons

In this section, several practical examples are presented to explain the design procedure and results compared to those of other designs. The comparisons are made on multiplication rate, addition rate, number of filter coefficients and number of storage registers needed. Scaling in each stage is considered as a shift and not counted as a multiplication. In the new designs, filter coefficients do not include those of comb filters which are incorporated into the filter structure and do not require coefficient storage.

### A. Example 1

This example is from [1] and [9]. The specifications are:

$$f_p = 0.0225 \qquad \delta_p = 0.050$$
$$f_s = 0.0250 \qquad \delta_s = 0.005$$
$$D = 20.$$

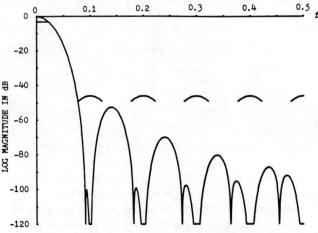

Fig. 16. Frequency response and aliasing attenuation bound of the first stage decimator. Comb $= 11, 10, 10, 10$.

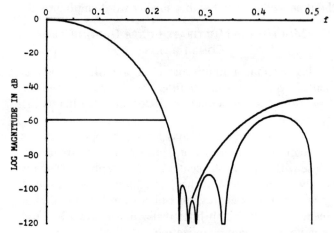

Fig. 17. Composite frequency response and aliasing attenuation bound of the second stage decimator. $N_{Kz} = 2$, comb $= 4, 3, 3, 3$.

The new filter uses a two-stage design with $D_1 = 10$ and $D_2 = 2$. The first stage decimator is a cascade of four comb filters with lengths 10, 10, 10, and 11. The frequency response and the bounds for aliasing attenuation are shown in Fig. 16. For an IIR design, the second stage decimator has three comb filters with lengths 4, 3, and 3 in cascade with a length-5 FIR filter (two pairs of zeros on the unit circle). The compensator is an all-pole IIR filter with two pairs of poles. The resulting responses of $H_{D_2}(\omega)$ and $H_c(\omega)$ are shown in Figs. 17 and 18. These two responses combined to give $H_p(\omega) = H_{D_2}(\omega)H_c(2\omega)$ with which we optimized to get $H_F(z)$ and $H_c(z)$. The response of $H_p(\omega)$ is shown in Fig. 19 which is the [0, 0.05] part of the overall response shown in Fig. 20 according to (2).

For the FIR design, the second stage decimator is still the same but with the zero positions of the short length FIR filter slightly changed. A length-45 FIR compensator is designed with a frequency response shown in Fig. 21.

The multistage decimator structure is shown in Fig. 22. This structure employs the comb decimator structures in Section III. The only difference between an IIR design and an FIR design is the compensator.

Table I shows the comparison of the new IIR and FIR designs to other designs. The new designs are shown to simultaneously require fewer coefficients, less data storage,

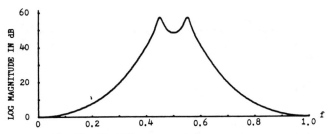

Fig. 18. The IIR compensator frequency responses.

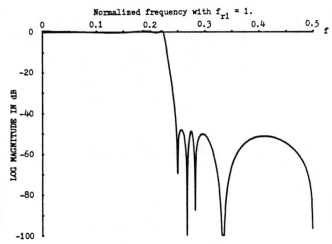

Fig. 19. Low frequency part of the overall frequency response.

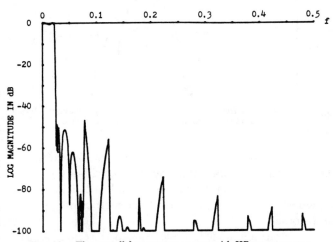

Fig. 20. The overall frequency response with IIR compensator.

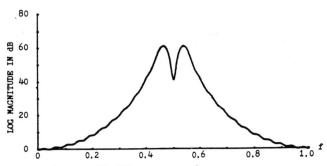

Fig. 21. The FIR compensator frequency response.

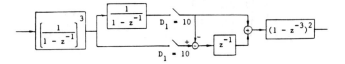

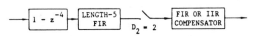

Fig. 22. Multistage decimator structure for Example 1.

TABLE I
COMPARISONS FOR EXAMPLE 1

| Type of filter | Decimation ratios | Filter order | No. of coeff's. | Mult. rate | Add. rate | Data storage |
|---|---|---|---|---|---|---|
| One-stage FIR | $D = 20$ | $N = 653$ | 327 | 16.35 | 32.6 | 653 |
| One-stage elliptic | $D = 20$ | $N = 7$ | 12 | 1. 7.4<br>2. 10.1 | 7.35<br>12.1 | 7 |
| Two-stage FIR | $D_1 = 10$<br>$D_2 = 2$ | $N_1 = 36$<br>$N_2 = 73$ | 55 | 3.65 | 7.1 | 109 |
| Two-stage elliptic | $D_1 = 10$<br>$D_2 = 2$ | $N_1 = 3$<br>$N_2 = 7$ | 17 | 1. 4.4<br>2. 5.2 | 4.35<br>5.5 | 10 |
| Two-stage Martinez's | $D_1 = 10$<br>$D_2 = 2$ | $N_1 = 27, M_1 = 1$<br>$N_2 = 11, M_2 = 6$ | 27 | 2.1 | 3.5 | 44 |
| Two-stage new IIR | $D_1 = 10$<br>$D_2 = 2$ | $N_2 = 5$<br>$M = 4$ | 7 | 0.35 | 4.85 | 23 |
| Two-stage new FIR | $D_1 = 10$<br>$D_2 = 2$ | $N_2 = 5$<br>$N_3 = 45$ | 26 | 1.3 | 6.85 | 64 |

1. Implemented in Direct Form II.
2. Implemented in cascaded form.

and lower multiplication rates. In addition, the new filters have a stopband response better than that of an equal-ripple filter.

For the case of the regular low-pass filter, the new filters realized as in Fig. 2(b) without decimation will require a multication rate of 26 and 7 for the case of FIR and IIR, respectively. Compared to 327 and 12 of direct FIR and IIR implementations, they are considerably more efficient.

Since the compensator frequency response is very peaking at the passband edge, it may have a serious roundoff noise problem. There are two ways to solve this problem. One is to merge the compensator with the second stage decimator. Instead of compensating the passband drop of Fig. 17, it compensates the passband of Fig. 16 which has a very small drop. The filter efficiency is a little lower, but it drastically reduces the compensator peakedness. Example 3

presents this method. The second way is to increase the data wordlength of the compensator stage. The extreme case is to use double wordlength which means no output rounding for the second stage decimator. In this case, one multiplication in the compensator can be considered as two single word multiplications. But from Table I, the new filter is still more efficient. Reference [10] gives a detailed roundoff noise analysis for multirate filters.

B. Example 2

This is a narrow-band low-pass filter example taken from [2]. The specifications are:

$$f_p = 0.025 \qquad \delta_p = 0.010$$
$$f_s = 0.050 \qquad \delta_s = 0.001.$$

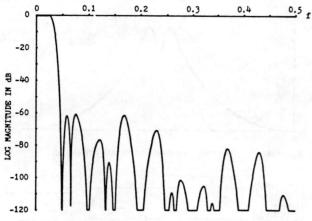

Fig. 23.  The overall frequency response with FIR compensator.

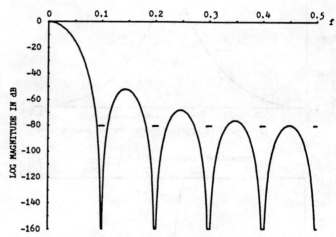

Fig. 24.  Frequency response and aliasing attenuation bound of the first stage decimator. Comb = 10, 10, 10, 10.

TABLE II
COMPARISONS FOR EXAMPLE 2

| Type of filter | Decimation ratios | Filter order | No. of coeff's. | Mult. rate | Add. rate | Data storage |
|---|---|---|---|---|---|---|
| Straight FIR | – | N = 110 | 55 | 55 | 109 | 110 |
| Straight elliptic | – | N = 5 | 9 | 9 | 10 | 5 |
| 1.5-stage FIR | D = 5 | $N_1 = 22$ $N_2 = 25$ | 24 | 1. 7.0 2. 9.2 | 13.2 12.4 | 69 52 |
| Two-stage new IIR | $D_1 = 5$ $D_2 = 2$ | M = 4 | 5 | 0.5 | 8.8 | 28 |
| Two-stage new FIR | $D_1 = 5$ $D_2 = 2$ | $N_3 = 10$ | 5 | 0.5 | 9.3 | 36 |

1. Interpolator in transposed form.
2. Interpolator in direct form.

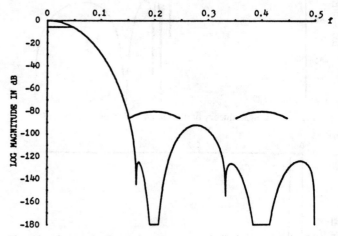

Fig. 25.  Composite frequency response and aliasing attenuation bound of the second stage decimator. Comb = 6, 5, 5, 5, 5, 5, 5.

The new two-stage design has $D_1 = 5$ and $D_2 = 2$; this is the case with full decimation. The interpolation stages are the same as the decimation stages. The first stage decimator is a cascade of four length-5 comb filters. For this wide transition case, the second stage decimator specification can be satisfied by a cascade of five comb filters with lengths 3, 3, 2, 2, and 2. For the IIR case, an IIR compensator with two pairs of poles is designed to satisfy the requirement.

For the FIR design, the second stage has five comb filters with lengths 4, 3, 3, 2, and 2. This will give more transition region aliasing attenuation and reduce the order of the FIR compensator. The compensator is a length-10 FIR filter. The overall frequency response is shown in Fig. 23.

The comparisons of the new designs to other designs are summarized in Table II. The 1.5-stage FIR filter is the conventional design with one stage decimation and interpolation and one stage of low-pass filtering in between. The name *1.5-stage* is used to differentiate it from the conventional one-stage design of one-stage decimation and interpolation. This is a better design than the conventional two-stage FIR. FIR interpolators can take advantage of symmetric coefficients and save multiplications by a half if they are implemented in the transposed form. For direct

form implementation, it is generally impossible (except for $D = 2$) to save multiplications by symmetry, but the implementation in [2] and [3], which is equivalent to a polyphase implementation, with scrambled filter coefficients can save data storage.

For this example of a wide transition region, an IIR compensator is not particularly more efficient than an FIR compensator. But in the case of a narrow transition region, the IIR compensator will be much more efficient as can be seen from Example 1 and the next example. Also, for this wide transition example, the compensator frequency response is very little peaked, so that the roundoff noise is not a problem here.

### C. Example 3

As a final example, this is a very narrow-band low-pass filter example from [2] with very tight specifications:

$$f_p = 0.00475 \qquad \delta_p = 0.001$$
$$f_s = 0.005 \qquad \delta_s = 0.0001.$$

Two new filter designs are presented. One is a two-stage design without full decimation; the other is a three-stage design with full decimation.

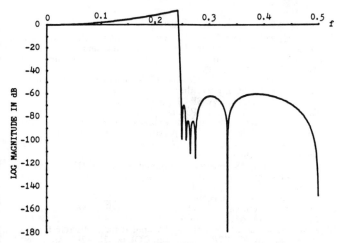

Fig. 26. Frequency response of the IIR compensator for the two-stage design. $N_p = 7$, $N_z = 3$, comb = 4, 4, 3.

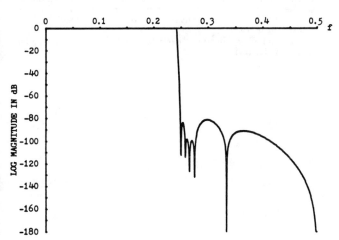

Fig. 27. Low-frequency part of the overall low-pass filter frequency response. (Normalized frequency with $f_{r2} = 1$.)

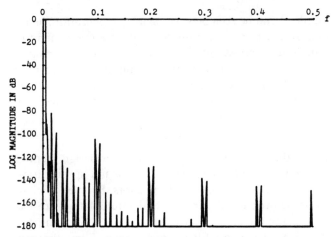

Fig. 28. The overall frequency response of the two-stage design.

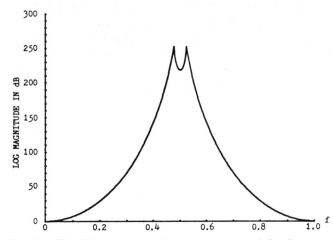

Fig. 29. The IIR compensator frequency response of the three-stage design. $N_p = 8$.

The two-stage design uses two comb decimators with $D_1 = 10$ and $D_2 = 5$. The same filters are used for interpolation. The first stage decimator is a cascade of four length-10 comb filters, its frequency response is shown in Fig. 24, together with the aliasing attenuation bounds. The second stage decimator is a cascade of six length-5 comb filters and a length-6 comb filter, its frequency response $H_{D_2}(\omega)$ and the aliasing attenuation bounds are shown in Fig. 25. The low-pass compensator is an IIR filter of seven pole pairs and three zero pairs in cascade with three comb filters of lengths 4, 4, and 3. Fig. 26 shows the frequency response of the compensator while Fig. 27 shows the optimized frequency response of $G^2(\omega/D)H_c(\omega)$, which is the $[0, 0.01]$ part of the overall frequency response shown in Fig. 28 according to (9).

The three-stage new design has a third stage $D_3 = 2$ which consists of five comb filters with lengths 4, 3, 3, 3, 3, and four pairs of zeros on the unit circle. The compensator is an IIR filter with eight pairs of poles, its frequency response is shown in Fig. 29.

Table III summarizes the results of comparisons of these two new designs to other designs. The 2.5-stage FIR is the conventional optimum FIR design with two stages of deci-

mation and interpolation and one stage of low-pass filtering in between.

The two-stage new filter is less efficient than the three-stage design in terms of multiplication rate. But, as its structure is simpler, it needs less data storage and has a much better roundoff error characteristic [10]. From this

## TABLE III
### COMPARISONS FOR EXAMPLE 3

| Type of filter | Decimation ratios | Filter order | No. of coeff's. | Mult. rate | Add. rate | Data storage |
|---|---|---|---|---|---|---|
| Straight FIR | — | N = 15590* | 7795 | 7795 | 15589 | 15590 |
| Straight IIR | — | N = 14 | 22 | 22 | 28 | 14 |
| 2.5-stage FIR | $D_1 = 10$ $D_2 = 5$ | $N_1 = 33$ $N_2 = 40$ $N_3 = 353$* | 214 | 1. 7.74 2. 9.74 | 15.00 14.02 | 499 438 |
| Two-stage new IIR | $D_1 = 10$ $D_2 = 5$ | $N_3 = 6$ $M = 14$ | 18 | 0.36 | 8.30 | 42 |
| Three-stage new IIR | $D_1 = 10$ $D_2 = 5$ $D_3 = 2$ | $N_2 = 9$ $M = 16$ | 21 | 0.25 | 8.21 | 77 |

\* Estimated.

1. Interpolators in transposed form.
2. Interpolators in direct form.

example, it can be seen how passband compensation, hence roundoff noise, can be greatly reduced by going from a full decimation (three-stage) design to a partial decimation (two-stage) design, particularly for this narrow-transition example. In an actual application, there are many considerations and a two-stage design may be much more favorable than a three-stage design.

## VII. Conclusions

Using the new multistage multirate filter design method in [5] allows simple filters without passband specification to be used as decimators or interpolators. Simple, efficient filter structures for decimators and interpolators are exploited using the commutative rule to save both arithmetic and storage. The comb filter is one of these filters that has the simplest structure. A particularly efficient multistage multirate filter structure is proposed to use comb decimators and/or interpolators and an IIR or FIR compensator. Examples show that the new filter structures can have, besides a simple structure, impressive simultaneous improvements in multiplication rates, addition rates, number of filter coefficients, and amount of data storage.

## References

[1] R. E. Crochiere and L. R. Rabiner, "Optimum FIR digital filter implementation for decimation, interpolation, and narrow-band filtering," *IEEE Trans. Acoust. Speech, Signal Processing*, vol. ASSP-23, pp. 444–456, Oct. 1975.

[2] L. R. Rabiner and R. E. Crochiere, "A novel implementation for narrow-band FIR digital filters," *IEEE Trans. Acoust. Speech, Signal Processing*, vol. ASSP-23, pp. 457–464, Oct. 1975.

[3] R. E. Crochiere and L. R. Rabiner, "Further considerations in the design of decimators and interpolators," *IEEE Trans. Acoust. Speech, Signal Processing*, vol. ASSP-24, pp. 296–311, Aug. 1976.

[4] ____, *Multirate Digital Signal Processing*. Englewood Cliffs, NJ: Prentice-Hall, 1983.

[5] S. Chu and C. S. Burrus, "Optimum FIR and IIR multistage multirate filter design," *Circuits, Systems and Signal Processing*, vol. 2, no. 3, pp. 361–386, July 1983.

[6] E. B. Hogenauer, "An economical class of digital filters for decimation and interpolation," *IEEE Trans. Acoust. Speech, Signal Processing*, vol. ASSP-29, pp. 155–162 Apr. 1981.

[7] J. W. Cooley and S. Winograd, "A limited range discrete Fourier transform," in *ICASSP-80*, (Denver, CO), pp. 213–217, Apr. 1980.

[8] R. Fletcher and M. J. D. Powell, "A rapidly convergent decent method for minimization," *Comput. J.*, vol. 6, pp. 163–168, 1963.

[9] H. G. Martinez and T. W. Parks, "A class of infinite-duration impulse response digital filters for sampling rate reduction," *IEEE Trans. Acoust. Speech, Signal Processing*, vol. ASSP-27, pp. 154–162, Apr. 1979.

[10] S. Chu and C. S. Burrus, "Roundoff noise in multirate digital filters," under review.

# Interpolation and Decimation of Digital Signals— A Tutorial Review

RONALD E. CROCHIERE, SENIOR MEMBER, IEEE, AND LAWRENCE R. RABINER, FELLOW, IEEE

*Invited Paper*

*Abstract*—The concepts of digital signal processing are playing an increasingly important role in the area of multirate signal processing, i.e. signal processing algorithms that involve more than one sampling rate. In this paper we present a tutorial overview of multirate digital signal processing as applied to systems for decimation and interpolation. We first discuss a theoretical model for such systems (based on the sampling theorem) and then show how various structures can be derived to provide efficient implementations of these systems. Design techniques for the linear-time-invariant components of these systems (the digital filter) are discussed, and finally the ideas behind multistage implementations for increased efficiency are presented.

## I. INTRODUCTION

ONE OF THE MOST fundamental concepts of digital signal processing is the idea of sampling a continuous process to provide a set of numbers which, in some sense, is representative of the characteristics of the process being sampled. If we denote a continuous function from the process being sampled as $x_C(t)$, $-\infty \leqslant t \leqslant \infty$ where $x_C$ is a continuous function of the continuous variable $t$ ($t$ may represent time, space, or any other continuous physical variable), then we can define the set of samples as $x_D(n)$, $-\infty \leqslant n \leqslant \infty$ where the correspondence between $t$ and $n$ is essentially specified by the sampling process, i.e.,

$$n = q(t) \tag{1a}$$

Many types of sampling have been discussed in the literature [1]–[3] including nonuniform sampling, uniform sampling, and multiple function uniform sampling. The most common form of sampling, and the one which we will refer to throughout this paper is uniform (periodic) sampling in which

$$q(t) = t/T = n \tag{1b}$$

i.e., the samples $x_D(n)$ are uniformly spaced in the dimension $t$, occurring $nT$ apart. For uniform sampling we define the sampling period as $T$ and the sampling rate as

$$F = 1/T \tag{2}$$

It should be clear from the above discussion that $x_C(t)$ can be sampled with any sampling period $T$. However, for a unique correspondence between the continuous function $x_C(t)$ and the discrete sequence $x_D(n)$, it is necessary that the sampling period $T$ be chosen to satisfy the requirements of the Nyquist

Manuscript received June 25, 1980.
The authors are with Bell Laboratories, Murray Hill, NJ 07974.

sampling theorem. This concept of a unique analog waveform corresponding to a digital sequence will often be used in the course of our discussion to provide greater intuitive insights into the nature of the processing algorithms that we will be considering.

The sampling period $T$ is a fundamental consideration in many signal processing techniques and applications. It often determines the convenience, efficiency, and/or accuracy in which the signal processing can be performed. In some cases an input signal may already be sampled at some predetermined sampling period $T$ and the goal is to convert this sampled signal to a new sampled signal at a different sampling period $T'$ such that the resulting signal corresponds to the same analog function. In other cases it may be more efficient or convenient to perform different parts of a processing algorithm at different sampling rates in which case it may be necessary to convert the sampling rates of the signals in the system from one rate to another.

The process of digitally converting the sampling rate of a signal from a given rate $F = 1/T$ to a different rate $F' = 1/T'$ is called *sampling rate conversion*. When the new sampling rate is higher than the original sampling rate, i.e.,

$$F' > F \tag{3a}$$

or

$$T' < T \tag{3b}$$

the process is generally called *interpolation* since we are creating samples of the original physical process from a reduced set of samples. Historically the mathematical process of interpolation, or "reading between the lines," has received widespread attention from mathematicians who were interested in the problem of tabulating useful mathematical functions. The question was how often a given function had to be tabulated (sampled) so that someone could use a simple interpolation rule to obtain accurate values of the function at any higher sampling rate [4]. Not only did this early work lead to an appreciation of the sampling process, but it also led to several interesting classes of "interpolation functions" which could provide almost arbitrarily high accuracy in the interpolated values, provided that sufficient tabulated values of the function were available.

The process of digitally converting the sampling rate of a signal from a given rate $F$ to a lower rate $F'$, i.e.,

$$F' < F \tag{4a}$$

Reprinted from *Proc. IEEE*, vol. 69, pp. 300–331, March 1981.

or

$$T' > T \qquad (4b)$$

is called *decimation*.[1] It will be shown in Section III that decimation and interpolation of signals are dual processes—i.e., a digital system which implements a decimator can be transformed into a dual digital system which implements an interpolator using straightforward transposition techniques.

The techniques to be described in this paper have been applied in a wide variety of areas including:

1) communications systems [5], [6];
2) speech processing systems [7]–[9];
3) antenna systems [10];
4) radar systems [11], [12].

The above list contains only a few representative examples of multirate digital systems.

From a digital signal processing point of view, both the processes of interpolation and decimation can be well formulated in terms of linear filtering operations. This is the basic point of view we have taken in this paper. We begin in Section II with the mathematical (and signal processing) framework of sampling, interpolation, and decimation. In Section III we discuss digital networks (structures) which can be used to implement the conversion from one sampling rate to another. Included in this section is a brief review of signal-flowgraph representations of digital systems, and of structures for implementing the digital filters required for all sampling rate conversion systems. It is then shown how efficient implementations of sampling rate conversion systems can be obtained by simple manipulations of the proposed canonic structures.

In Section IV, we discuss the question of how to design the digital filter used in the systems presented in Sections II and III. It is shown that two general structures can be used to aid in the design of the special filters required in sampling rate conversion systems. Based on these structures, a number of special purpose design algorithms are described.

Finally Section V addresses the question of special structures for handling two special cases of sampling rate conversion, namely: 1) large changes in sampling rates within the system and 2) changes in sampling rate requiring large sampling rate changes internally in the structure—e.g., sampling rate conversion by a factor of 97/151. Each of these cases can be handled most efficiently in a multistage structure in which the sampling rate conversion occurs in a series of 2 or more distinct stages. Questions of computational, storage, and control efficiency are of paramount concern in the discussions in this section.

In this paper, we only consider decimation and interpolation systems based on finite impulse response (FIR) realizations. Another broad class of sampling rate conversion systems that can also be defined is based on infinite impulse response (IIR) realizations. However, they do not conveniently permit linear phase designs and a discussion of these issues was considered to be beyond the scope of this paper.

## II. Basic Concepts of Sampling Rate Conversion

Fig. 1 provides a general description of a sampling rate conversion system. We are given the signal $x(n)$, sampled at the rate $F = 1/T$, and wish to compute the signal $y(m)$ with a new

---

[1] Strictly speaking decimation means a reduction by 10 percent. In signal processing decimation has come to mean a reduction in sampling rate by any factor.

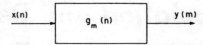

Fig. 1. Basic process of digital sampling rate conversion.

sampling rate $F' = 1/T'$. We will assume throughout this paper that the ratio of sampling periods of $y(m)$ and $x(n)$ can be expressed as a rational fraction, i.e.,

$$T'/T = F/F' = M/L \qquad (5)$$

where $M$ and $L$ are integers.

A close examination of the structure of Fig. 1 shows that the systems we are dealing with for digital-to-digital sampling rate conversion are inherently linear time-varying systems, i.e., $g_m(n)$ is the response of the system at the output sample time $m$ to an input at the input sample time $\lfloor mM/L \rfloor - n$ where $\lfloor u \rfloor$ denotes the integer less than or equal to $u$ (this will become clearer in later discussion).

Since the system is linear, each output sample $y(m)$ can be expressed as a linear combination of input samples. A general form for this expression, which is used extensively in this paper, is [13]

$$y(m) = \sum_{n=-\infty}^{\infty} g_m(n)\, x\left(\left\lfloor \frac{mM}{L} \right\rfloor - n\right). \qquad (6)$$

A derivation of (6) is given shortly where it is also seen that the system response $g_m(n)$ is periodic in $m$ with period $L$, i.e.,

$$g_m(n) = g_{m+rL}(n), \qquad r = 0, \pm 1, \pm 2, \cdots. \qquad (7)$$

Thus the system $g_m(n)$ belongs to the class of linear, *periodically* time-varying systems. Such systems have been extensively studied for a wide range of applications [14], [15].

In the trivial case when $T' = T$, or $L = M = 1$, equation (6) reduces to the simple time-invariant digital convolution equation

$$y(m) = \sum_{n=-\infty}^{\infty} g(n)\, x(m-n) \qquad (8)$$

since the period of $g_m(n)$ in this case is 1, and the integer part of $m - n$ is the same as $m - n$.

In the next few sections we study in some detail the structure and properties of systems that perform two special cases of sampling rate conversion, namely decimation by integer factors, and interpolation by integer factors [16]. We then consider the general case of a sampling rate change by a factor of $L/M$.

### A. Sampling Rate Reduction–Decimation by an Integer Factor M

Consider the process of reducing the sampling rate of $x(n)$ by an integer factor $M$, i.e.,

$$T'/T = M/1. \qquad (9)$$

Then the new sampling rate is $F' = F/M$. Assume that $x(n)$ represents a full band signal, i.e., its spectrum is nonzero for all frequencies $f$ in the range $-F/2 \leqslant f \leqslant F/2$, with $\omega = 2\pi fT$, i.e.,

$$|X(e^{j\omega})| \neq 0, \qquad |\omega| = |2\pi fT| \leqslant \frac{2\pi FT}{2} = \pi \qquad (10)$$

except possibly at an isolated set of points. Based on well known sampling theory, in order to lower the sampling rate and to avoid aliasing at this lower rate, it is necessary to filter the signal $x(n)$ with a digital low-pass filter which approximates the ideal characteristic

$$\tilde{H}(e^{j\omega}) = \begin{cases} 1, & |\omega| \leq 2\pi F'T/2 = \pi/M \\ 0, & \text{otherwise.} \end{cases} \quad (11)$$

The sampling rate reduction is then achieved by forming the sequence $y(m)$ by extracting every $M$th sample of the filtered output. This process is illustrated in Fig. 2(a). If we denote the actual low-pass filter unit sample response as $h(n)$, then we have

$$w(n) = \sum_{n=-\infty}^{\infty} h(k) \, x \, (n-k) \quad (12)$$

where $w(n)$ is the filtered output as seen in Fig. 2(a). The final output $y(m)$ is then obtained as

$$y(m) = w(Mm) \quad (13)$$

as denoted by the operation of the second box in Fig. 2(a). This block diagram symbol, which will be referred to as a *sampling rate compressor*, will be used consistently throughout this paper, and it corresponds to the resampling operation given by (13).

Fig. 2(b) shows typical spectra (magnitude of the Fourier transforms) of the signals $x(n)$, $h(n)$, $w(n)$, and $y(m)$ for an $M$ to 1 reduction in sampling rate.

By combining (12) and (13) the relation between $y(m)$ and $x(n)$ is of the form

$$y(m) = \sum_{k=-\infty}^{\infty} h(k) \, x \, (Mm-k) \quad (14)$$

which is seen to be a special case of (6). Thus for decimation by integer factors of $M$ we have

$$g_m(n) = g(n) = h(n), \quad \text{for all } m \text{ and all } n. \quad (15)$$

Although $g_m(n)$ is *not* a function of $m$ for this case, it can readily be seen that the overall system of (14) and Fig. 2(a) is not time-invariant by considering the output signal obtained when $x(n)$ is shifted by an integer number of samples. For this case, unless the shift is a multiple of $M$, the output is *not* a shifted version of the output for 0 shift, i.e.,

$$x(n) \to y(m) \quad (16a)$$

but

$$x(n - \delta) \not\to y \, (m - \delta/M) \text{ unless } \delta = rM. \quad (16b)$$

It is of value to derive the relationship between the $z$-transforms of $y(m)$ and $x(n)$ so as to be able to study the nature of the errors in $y(m)$ caused by the imperfect low-pass filter. To obtain this relationship we define the signal

$$w'(n) = \begin{cases} w(n), & n = 0, \pm M, \pm 2M, \cdots \\ 0, & \text{otherwise} \end{cases} \quad (17)$$

i.e., $w'(n) = w(n)$ at the sampling instants of $y(m)$, but is zero otherwise. A convenient and useful representation of $w'(n)$ is then

$$w'(n) = w(n) \left\{ \frac{1}{M} \sum_{l=0}^{M-1} e^{j2\pi ln/M} \right\}, \quad -\infty < n < \infty \quad (18)$$

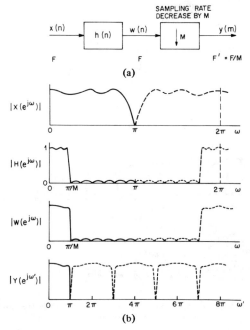

SAMPLING RATE
DECREASE BY M

(a)

(b)

Fig. 2. Block diagram and typical spectra for sampling rate reduction by a factor of $M$

where the term in brackets corresponds to a discrete Fourier series representation of a periodic impulse train with a period of $M$ samples. Thus we have

$$y(m) = w'(Mm) = w(Mm) \quad (19)$$

We now write the $z$-transform of $y(m)$ as

$$Y(z) = \sum_{m=-\infty}^{\infty} y(m) \, z^{-m}$$

$$= \sum_{m=-\infty}^{\infty} w'(Mm) \, z^{-m} \quad (20)$$

and since $w'(m)$ is zero except at integer multiples of $M$, equation (20) becomes

$$Y(z) = \sum_{m=-\infty}^{\infty} w'(m) \, z^{-m/M}$$

$$= \sum_{m=-\infty}^{\infty} w(m) \left[ \frac{1}{M} \sum_{l=0}^{M-1} e^{j2\pi lm/M} \right] z^{-m/M}$$

$$= \frac{1}{M} \sum_{l=0}^{M-1} \left[ \sum_{m=-\infty}^{\infty} w(m) \, e^{j2\pi lm/M} z^{-m/M} \right]$$

$$Y(z) = \frac{1}{M} \sum_{l=0}^{M-1} W(e^{-j2\pi l/M} z^{1/M}). \quad (21)$$

Since

$$W(z) = H(z) \, X(z) \quad (22)$$

we can express $Y(z)$ as

$$Y(z) = \frac{1}{M} \sum_{l=0}^{M-1} H(e^{-j2\pi l/M} z^{1/M}) \, X(e^{-j2\pi l/M} z^{1/M}). \quad (23)$$

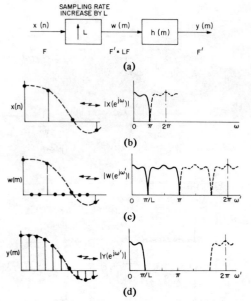

SAMPLING RATE
INCREASE BY L

**(a)**

**(b)**

**(c)**

**(d)**

Fig. 3. Block diagram and typical waveforms and spectra for sampling rate increase by a factor of $L$.

Evaluating $Y(z)$ on the unit circle, $z = e^{j\omega'}$, leads to the result

$$Y(e^{j\omega'}) = \frac{1}{M} \sum_{l=0}^{M-1} H(e^{j(\omega' - 2\pi l)/M}) X(e^{j(\omega' - 2\pi l)/M}) \quad (24a)$$

where

$$\omega' = 2\pi f T' \quad \text{(in radians relative to sampling period } T'). \quad (24b)$$

Equation (24) expresses the Fourier transform of the output signal $y(m)$ in terms of the transforms of the aliased components of the filtered input signal $x(n)$. By writing the individual components of (24) directly we see that

$$Y(e^{j\omega'}) = \frac{1}{M} \left[ H(e^{j\omega'/M}) X(e^{j\omega'/M}) \right.$$

$$\left. + H(e^{j(\omega' - 2\pi)/M}) X(e^{j(\omega' - 2\pi)/M}) + \cdots \right]. \quad (25)$$

The purpose of the low-pass filter $H(e^{j\omega})$ is to sufficiently filter $x(n)$ so that its components above the frequency $\omega = \pi/M$ are negligible. In terms of (24) this implies that all terms for $l \neq 0$ are removed and if the filter $H(e^{j\omega})$ closely approximates the ideal response of (11) then (24) becomes

$$Y(e^{j\omega'}) \cong \frac{1}{M} X(e^{j\omega'/M}), \quad \text{for } |\omega'| \leqslant \pi. \quad (26)$$

*B. Sampling Rate Increase–Interpolation by an Integer Factor L*

If the sampling rate is increased by an integer factor $L$, then the new sampling period $T'$ is

$$\frac{T'}{T} = \frac{1}{L} \quad (27)$$

and the new sampling rate $F'$ is $F' = LF$. This process of increasing the sampling rate of a signal $x(n)$ by $L$ implies that we must interpolate $L - 1$ new sample values between each pair of sample values of $x(n)$.

Fig. 3 illustrates this process of increasing the sampling rate by a factor $L = 3$. The input signal $x(n)$ is "filled-in" with $L - 1$ zero-valued samples between each pair of samples of $x(n)$ giving the signal

$$w(m) = \begin{cases} x(m/L), & m = 0, \pm L, \pm 2L, \cdots \\ 0, & \text{otherwise.} \end{cases} \quad (28)$$

As with the resampling operation, the block diagram symbol of an up-arrow with an integer corresponds to increasing the sampling rate as given by (28) and it will be referred to as a *sampling rate expander*. The resulting signal $w(n)$ has the $z$-transform

$$W(z) = \sum_{m=-\infty}^{\infty} w(m) z^{-m} \quad (29a)$$

$$= \sum_{m=-\infty}^{\infty} x(m) z^{-mL} \quad (29b)$$

$$= X(z^L). \quad (29c)$$

Evaluating $W(z)$ on the unit circle $z = e^{j\omega'}$, gives the result

$$W(e^{j\omega'}) = X(e^{j\omega'L}) \quad (30)$$

which is the Fourier transform of the signal $w(m)$ expressed in terms of the spectrum of the input signal $x(n)$ (where $\omega' = 2\pi f T'$ and $\omega = 2\pi f T$).

As illustrated by the spectral interpretation in Fig. 3(c) the spectrum of $w(n)$ contains not only the baseband frequencies of interest (i.e., $-\pi/L$ to $\pi/L$) but also images of the baseband centered at harmonics of the original sampling frequency $\pm 2\pi/L, \pm 4\pi/L, \cdots$. To recover the baseband signal of interest and eliminate the unwanted higher frequency components it is necessary to filter the signal $w(m)$ with a digital low-pass filter which approximates the ideal characteristic

$$\tilde{H}(e^{j\omega'}) = \begin{cases} G, & |\omega'| \leqslant \dfrac{2\pi F T'}{2} = \pi/L \\ 0, & \text{otherwise.} \end{cases} \quad (31)$$

It will be shown that in order to ensure that the amplitude of $y(m)$ is correct, the gain of the filter $G$ must be $L$ in the passband.

Letting $H(e^{j\omega'})$ denote the frequency response of an actual filter that approximates the characteristic in (31) it is seen that

$$Y(e^{j\omega'}) = H(e^{j\omega'}) X(e^{j\omega'L}) \quad (32)$$

and within the approximation of (31)

$$Y(e^{j\omega'}) \cong \begin{cases} G X(e^{j\omega'L}), & |\omega'| \leqslant \pi/L \\ 0, & \text{otherwise.} \end{cases} \quad (33)$$

It is easy to see why we need a gain of $G$ in $\tilde{H}(e^{j\omega})$, whereas for the decimation filter a gain of 1 is adequate, by examining the zeroth sample of the sequences. From Fig. 2 it is clear that

$$y(0) = w(0)$$

$$= x(0)$$

if we assume that $|H(e^{j\omega})| = 1$ for $|\omega| < \pi/M$ and $|X(e^{j\omega})| = 0$ for $|\omega| > \pi/M$. Alternatively for the interpolator it is seen with the aid of Fig. 3 and (33) that

$$y(0) = \int_{-\pi}^{\pi} Y(e^{j\omega'}) \, d\omega'$$

$$= \int_{-\pi}^{\pi} H(e^{j\omega'}) X(e^{j\omega'L}) \, d\omega'$$

$$= G \int_{-\pi/L}^{\pi/L} X(e^{j\omega'L}) \, d\omega'$$

$$= G \int_{-\pi}^{\pi} X(e^{j\omega}) \, d\omega/L$$

$$= \frac{G}{L} x(0).$$

Therefore, a gain $G = L$ is required to match the amplitudes of the envelopes of the signals $y(m)$ and $x(n)$.

If $h(m)$ denotes the unit sample response of $H(e^{j\omega'})$, then $y(m)$ can be expressed as

$$y(m) = \sum_{k=-\infty}^{\infty} h(m-k) w(k). \tag{34}$$

Combining (28) and (34) leads to

$$y(m) = \sum_{k=-\infty}^{\infty} h(m-k) x(k/L)$$

$$= \sum_{r=-\infty}^{\infty} h(m-rL) x(r). \tag{35}$$

Next we introduce the change of variables

$$r = \left\lfloor \frac{m}{L} \right\rfloor - n \tag{36}$$

and the identity

$$mM - \left\lfloor \frac{mM}{L} \right\rfloor L = (mM) \oplus L \tag{37}$$

where $\lfloor u \rfloor$ denotes the integer less than or equal to $u$ and $(i) \oplus L$ denotes the value of $i$ modulo $L$. Applying (36) and (37) (with $M = 1$) to (35) then gives

$$y(m) = \sum_{n=-\infty}^{\infty} h\left(m - \left\lfloor \frac{m}{L} \right\rfloor L + nL\right) x\left(\left\lfloor \frac{m}{L} \right\rfloor - n\right)$$

$$= \sum_{n=-\infty}^{\infty} h(nL + m \oplus L) x\left(\left\lfloor \frac{m}{L} \right\rfloor - n\right). \tag{38}$$

Equation (38) expresses the output $y(m)$ in terms of the input $x(n)$ and the filter coefficients $h(m)$ and it is again seen to be a special case of (6). Thus for interpolation by integer factors of $L$ we have

$$g_m(n) = h(nL + m \oplus L), \quad \text{for all } m \text{ and all } n \tag{39}$$

and it is seen that $g_m(n)$ is periodic in $m$ with period $L$ as indicated by (7).

*C. Sampling Rate Conversion by a Rational Factor M/L*

In the previous two sections we have considered the cases of sampling rate reduction by an integer factor $M$ and sampling

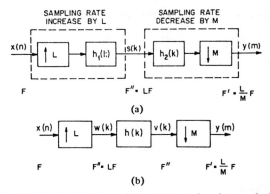

Fig. 4. (a) Cascade of an integer interpolator and an integer decimator for achieving sampling rate conversion by rational fractions. (b) A more efficient implementation of this process.

rate increase by an integer factor $L$. In this section, we consider the general case of conversion by the ratio

$$\frac{T'}{T} = \frac{M}{L} \tag{40}$$

or

$$F' = \frac{L}{M} F. \tag{41}$$

This conversion can be achieved by a cascade of the two above processes of integer conversions by first increasing the sampling rate by $L$ and then decreasing it by $M$. Fig. 4(a) illustrates this process. It is important to recognize that the interpolation by $L$ *must* precede the decimation process by $M$ so that the width of the baseband of the intermediate signal $s(k)$ is greater than or equal to the width of the basebands of $x(n)$ or $y(m)$.

It can be seen from Fig. 4(a) that the two filters $h_1(k)$ and $h_2(k)$ are operating in cascade at the same sampling rate $LF$. Thus a more efficient implementation of the overall process can be achieved if the filters are combined into one composite lowpass filter as shown in Fig. 4(b). Since this digital filter $h(k)$ must serve the purposes of both the decimation and interpolation operations described in the previous two sections it is clear from (11) and (31) that it must approximate the ideal digital low-pass characteristic

$$\tilde{H}(e^{j\omega''}) = \begin{cases} L, & |\omega''| = |2\pi f T''| \leqslant \min \left| \frac{\pi}{L}, \frac{\pi}{M} \right| \\ 0, & \text{otherwise} \end{cases} \tag{42}$$

where

$$\omega'' = 2\pi f T'' = 2\pi f T/L \tag{43}$$

i.e., the ideal cutoff frequency must be the minimum of the two cutoff frequency requirements for the decimator and interpolator and the sampling rate of the filter is $F'' = LF$.

The time domain input-to-output relation for the general conversion circuit of Fig. 4(b) can be derived by considering the integer interpolation and decimation relations derived in Sections II-A and II-B, i.e., from (35) it can be seen that $v(k)$ can be expressed as

$$v(k) = \sum_{r=-\infty}^{\infty} h(k-rL) x(r) \tag{44}$$

and from (13) $y(m)$ can be expressed in terms of $v(k)$ as

$$y(m) = v(Mm).\qquad(45)$$

Combining (44) and (45) gives

$$y(m) = \sum_{r=-\infty}^{\infty} h(Mm - rL)\, x(r)\qquad(46)$$

and making the change of variables

$$r = \left\lfloor \frac{mM}{L} \right\rfloor - n\qquad(47)$$

and applying (37) gives

$$y(m) = \sum_{n=-\infty}^{\infty} h\left(Mm - \left\lfloor \frac{mM}{L} \right\rfloor L + nL\right) x\left(\left\lfloor \frac{mM}{L} \right\rfloor - n\right)$$

$$= \sum_{n=-\infty}^{\infty} h(nL + mM \oplus L)\, x\left(\left\lfloor \frac{mM}{L} \right\rfloor - n\right).\qquad(48)$$

It is seen that (48) corresponds to the general form of the time-varying digital-to-digital conversion system described by (6) and that the time-varying unit sample response $g_m(n)$ can be expressed as

$$g_m(n) = h(nL + mM \oplus L), \qquad \text{for all } m \text{ and all } n\qquad(49)$$

where $h(k)$ is the time-invariant unit sample response of the low-pass digital filter at the sampling rate $LF$ [13].

Similarly, by considering the transform relationships of the individual integer decimation and interpolation systems, the output spectrum $Y(e^{j\omega'})$ can be determined in terms of the input spectrum $X(e^{j\omega})$ and the frequency response of the filter $H(e^{j\omega''})$. From (32) it is seen that $V(e^{j\omega''})$ can be expressed in terms of $X(e^{j\omega})$ and $H(e^{j\omega''})$ as

$$V(e^{j\omega''}) = H(e^{j\omega''})\, X(e^{j\omega''L})\qquad(50)$$

and from (21) $Y(e^{j\omega'})$ can be expressed in terms of $V(e^{j\omega''})$ as

$$Y(e^{j\omega'}) = \frac{1}{M} \sum_{l=0}^{M-1} V(e^{j(\omega'-2\pi l)/M})$$

$$= \frac{1}{M} \sum_{l=0}^{M-1} H(e^{j(\omega'-2\pi l)/M})\, X(e^{j(\omega'L-2\pi l)/M}).\qquad(51)$$

When $H(e^{j\omega'})$ closely approximates the ideal characteristic of (42) it is seen that this expression reduces to

$$Y(e^{j\omega'}) \cong \begin{cases} \dfrac{L}{M} X(e^{j\omega'L/M}), & \text{for } |\omega'| \leqslant \min(\pi, \pi M/L) \\[2mm] 0, & \text{otherwise} \end{cases}\qquad(52)$$

Thus far, we have developed the general system for sampling rate conversion of low-pass signals by arbitrary rational factors $L/M$. It was shown that the process of sampling rate conversion could be modeled as a linear, periodically time-varying system, and that the unit sample response of this system, $g_m(n)$ could be expressed in terms of the unit sample response $h(k)$ of a time-invariant digital filter designed for the highest system sampling rate $LF$.

### D. Sampling Rate Conversion of Bandpass Signals

In the preceding sections it was assumed that the signals that we are dealing with are low-pass signals and therefore the filters

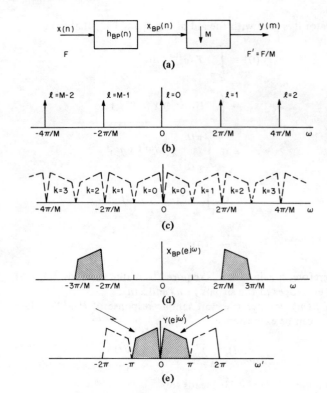

Fig. 5. Decimation of a bandpass signal and a spectral interpretation for the $k = 2$ band.

required for decimation and interpolation are low-pass filters which preserve the baseband signals of interest. In many practical systems it is also necessary to deal with bandpass signals and many of the results discussed in this paper can be logically extended to the bandpass case as well. While it is not our intention to go into detail on these issues in this paper we will briefly give an example in this section of one way in which this can be done. A more involved discussion of these issues can be found in [10].

Perhaps the simplest and most direct approach to decimating or interpolating digital bandpass signals is to take advantage of the inherent frequency translating (i.e., aliasing) properties of decimation and interpolation. This property can be used to advantage when dealing with bandpass signals by associating the bandpass signal with one of these modulated "harmonics" instead of with the baseband. Fig. 5(a) illustrates an example of this process for the case of decimation by the integer factor $M$. The input signal $x(n)$ is first filtered by the bandpass filter $h_{BP}(n)$ to isolate the frequency band of interest. The resulting bandpass signal $x_{BP}(n)$, is then directly reduced in sampling rate by selecting one out of every $M$ samples giving the final output $y(m)$. It is seen that this system is identical to that of the integer low-pass decimator with the exception that the filter is a bandpass filter instead of a low-pass filter. Thus the output signal $Y(e^{j\omega'})$ can be expressed as

$$Y(e^{j\omega'}) = \frac{1}{M} \sum_{l=0}^{M-1} H_{BP}(e^{j(\omega'-2\pi l)/M})\, X(e^{j(\omega'-2\pi l)/M}).\qquad(53)$$

From (53) it is seen that $Y(e^{j\omega'})$ is composed of $M$ aliased components of $X(e^{j\omega'}) H_{BP}(e^{j\omega'})$ modulated by factors of $2\pi l/M$. The function of the filter $H_{BP}(e^{j\omega})$ is to remove (attenuate) all aliasing components except those associated with the desired band of interest. Since the modulation is restricted to values of $2\pi l/M$ it can be seen that only specific frequency bands are allowed by this method. As a consequence the choice of the filter $H_{BP}(e^{j\omega})$ is restricted to approximate one of the $M$ ideal characteristics

$$\tilde{H}_{BP}(e^{j\omega}) = \begin{cases} 1, & k\pi/M < |\omega| < (k+1)\pi/M \\ 0, & \text{otherwise} \end{cases} \quad (54)$$

where $k = 0, 1, 2, \cdots, M - 1$, i.e., $H_{BP}(e^{j\omega})$ is restricted to bands $\omega = k\pi/M$ to $\omega = (k+1)\pi/M$ where $\pi/M$ is the bandwidth.

Figs. 5(b)–(e) illustrate this approach. Fig. 5(b) shows the $M$ possible modulating frequencies which are a consequence of the $M$ to 1 sampling rate reduction, i.e., the digital sampling function (a periodic train of unit samples spaced $M$ samples apart) has spectral components spaced $2\pi l/M$ apart. Fig. 5(c) shows the "sidebands" that are associated with these spectral components which correspond to the $M$ choices of bands as defined by (54). They correspond to the bands that are aliased into the baseband of the output signal $Y(e^{j\omega'})$ according to (54). (As seen by (53) and (54) and Figs. 5(b) and (c), the relationship between $k$ and $l$ is nontrivial).

Fig. 5(d) illustrates an example in which the $k = 2$ band is used, such that $X_{BP}(e^{j\omega})$ is bandlimited to the range $2\pi/M < |\omega| < 3\pi/M$. Since the process of sampling rate reduction by $M$ to 1 corresponds to a convolution of the spectra of $X_{BP}(e^{j\omega})$ (Fig. 5(d)) and the sampling function (Fig. 5(b)) this band is lowpass translated to the baseband of $Y(e^{j\omega'})$ as seen in Fig. 5(e). Thus, the processes of modulation and sampling rate reduction are achieved simultaneously by the $M$ to 1 sampling rate reduction.

The process of bandpass interpolation is the inverse to that of bandpass decimation and it can be accomplished in a similar manner. Referring to Fig. 3(c) it is seen that we can use a bandpass filter with a characteristic similar to that described by (54) (with $M$ replaced by $L$) to remove one of the harmonic images of the baseband signal rather than the baseband signal itself. The net result is that we achieve both an interpolation and a modulation of the input signal to one of its harmonic locations in the spectrum.

## III. SIGNAL PROCESSING STRUCTURES FOR DECIMATORS AND INTERPOLATORS

It is easy to understand the need for studying structures for realizing sampling rate conversion systems by examining the simple block diagram of Fig. 4(b) which can be used to convert the sampling rate of a signal by a factor of $L/M$. As discussed in Section II the theoretical model for this system is increasing the signal sampling rate by a factor of $L$ (by filling in $L - 1$ zero-valued samples between each sample of $x(n)$ to give the signal $w(k)$), filtering $w(k)$ to eliminate the images of $X(e^{j\omega})$ by a standard linear time-invariant low-pass filter, $h(k)$, to give $v(k)$, and sampling rate compressing $v(k)$ by a factor $M$ (by retaining 1 of each $M$ samples of $v(k)$). A direct implementation of the system of Fig. 4(b) is grossly inefficient since the low-pass filter $h(k)$ is operating at the high sampling rate on a signal for which $L - 1$ out of each $L$ input values are zero, and the values of the filtered output are required only once each $M$ samples. For this example, one can directly apply this knowledge in implementing the system of Fig. 4(b) in a more efficient manner as will be discussed in this section. Later in Section V we will extend these concepts to include

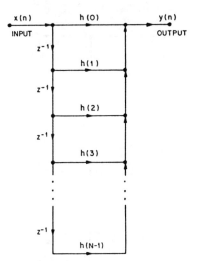

Fig. 6. Direct form structure for an FIR filter.

multistage implementations which can achieve greater efficiencies than single stage designs when the conversion ratios are large.

Before discussing specific classes of structures for sampling rate conversion we will first briefly review in Section III-A a number of fundamental network and signal-flowgraph concepts which will be used in developing these structures. We will then discuss three principle classes of FIR structures for realizing single stage interpolators and decimators and compare their properties.

### A. Signal-Flowgraphs

In order to precisely define the sets of operations necessary to implement these digital systems we will strongly rely on the concepts of signal-flowgraph representation in this section [17]–[19]. Signal-flowgraphs provide a graphical representation of the explicit set of equations that are used to implement such systems. Furthermore, manipulating the flowgraphs in a pictorial way is equivalent to manipulation of the mathematical equations.

Fig. 6 illustrates an example of a signal-flowgraph of a direct-form FIR digital filter. The input branch applies the external signal $x(n)$ to the network and the output of the network $y(n)$ is identified as one of the node values. Branches define the signal operations in the structure such as delays, gains, and sampling rate expanders and compressors. Nodes define the connection points and summing points. The signal entering a branch is taken as the signal associated with the input node value of the branch. The node value of a branch is the sum of all branch signals entering the node.

Therefore from the signal-flowgraph (Fig. 6) we can immediately write down the network equation as

$$y(n) = x(n)h(0) + x(n-1)h(1) + \cdots$$
$$+ x(n-N+1)h(N-1).$$

An important concept in the manipulation of signal-flowgraphs is the principle of commutation of branch operations. Two branch operations commute if the order of their cascade operation can be interchanged without affecting the input-to-output response of the cascaded system. Thus interchanging commutable branches in a network is one way of modifying

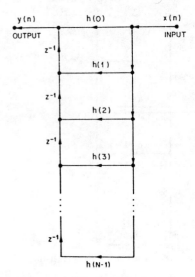

Fig. 7. Transposed direct form structure for an FIR filter.

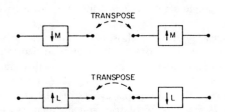

Fig. 8. Transpositions of the sampling rate compressor and expander.

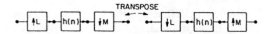

Fig. 9. Transpose of a generalized $L/M$ sampling rate changer.

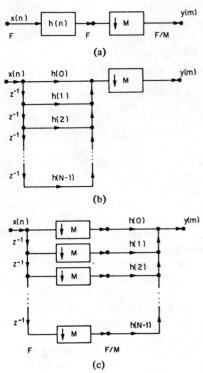

Fig. 10. Generation of an efficient direct form structure of an $M$ to 1 decimator.

the network without affecting the desired input-to-output network response. This operation will be used extensively in constructing efficient structures for decimation and interpolation as we shall see shortly.

Another important network concept that we rely heavily on is that of transposition and duality [17]–[21]. Basically a dual system is one which performs a complementary operation to that of an original system and it can be constructed from the original system through the process of transposition. We have already seen an example of dual systems, namely the integer decimator and interpolator (Fig. 2(a) and Fig. 3(a)) for the case $M = L$.

Basically the transposition operation is one in which the direction of all branches in the network are reversed and the roles of the input and output of the network are interchanged. Furthermore all branch operations are replaced by their transpose operations. In the case of linear time-invariant branch operations, such as gains and delays, these branch operations remain unchanged. Thus, for example, the transpose of the direct form structure of Fig. 6 is the transposed direct form structure shown in Fig. 7. Also it can be shown [17]–[21] that for the case of linear time-invariant systems the input-to-output system response of a system and its dual are identical (e.g., it can be verified that the networks of Fig. 6 and Fig. 7 have identical system functions).

For time-varying systems this is not necessarily the case. For example, the transpose of a sampling rate compressor is a sampling rate expander and the transpose of a sampling rate expander is a sampling rate compressor as shown in Fig. 8. Clearly these systems do not have the same system response.

By extending the concepts of transposition rigorously it can also be shown that the transposition of a network that performs a sampling rate conversion by the factor $L/M$ is a network that performs a sampling rate conversion by the factor $M/L$. This is illustrated in Fig. 9.

### B. Direct Form FIR Structures for Integer Changes in Sampling Rates

Consider the model of an $M$ to 1 decimator as developed in Section II, and as shown in Fig. 10(a). According to this model the filter $h(n)$ operates at the high sampling rate $F$ and $M - 1$ out of every $M$ output samples of the filter are discarded by the $M$ to 1 sampling rate compressor. In particular if we assume that the filter $h(n)$ is an $N$-point FIR filter realized with a direct form structure, the network of Fig. 10(b) results. The multiplications by $h(0)$, $h(1)$, $\cdots$, $h(N-1)$ and the associated summations in this network must be performed at the rate $F$.

A more efficient realization of the above structure can be achieved by noting that the branch operations of sampling rate compression and gain can be commuted. By performing a series of commutative operations on the network, the modified network of Fig. 10(c) results. The multiplications and additions associated with the coefficients $h(0)$ to $h(N-1)$ now occur at the low sampling rate $F/M$ and therefore the total computation rate in the system has been reduced by a factor $M$. For every $M$ samples of $x(n)$ which are shifted into the structure (the cascade of delays) one output sample $y(m)$ is

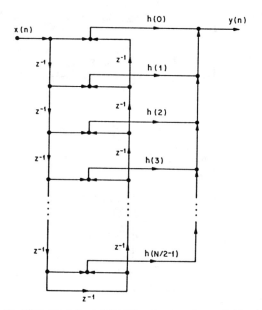

Fig. 11. Modified direct form FIR filter structure for exploiting impulse response symmetry.

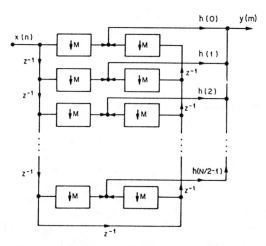

Fig. 12. Direct form realization of an $M$ to 1 decimator that exploits symmetry in $h(n)$ for even values of $N$.

computed. Thus the structure of Fig. 10(c) is seen to be a direct realization of (14) in Section II-A.

An alternate form of this structure which can exploit symmetry in $h(n)$ (for linear phase designs) can be derived by using the modified direct form structure of Fig. 11 (for $N$ even). This leads to the $M$ to 1 decimator structure shown in Fig. 12 and it requires approximately a factor of 2 less multiplications than the structure of Fig. 10(c).

An efficient structure for the 1 to $L$ integer interpolator using an FIR filter can be derived in a similar manner. We begin with the cascade model for the interpolator shown in Fig. 13(a). In this case however, if $h(m)$ is realized with the direct form structure of Fig. 6 we are faced with the problem of commuting the 1 to $L$ sampling rate expander with a series of unit delays. One way around this problem is to realize $h(m)$ with the transposed direct form FIR structure as shown in Fig. 7 [22]. The sampling rate expander can then be commuted into the network as shown by the series of operations in Fig. 13. Since the coefficients $h(0)$, $h(1)$, $\cdots$,

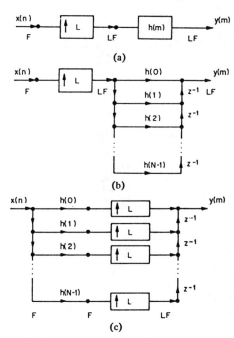

Fig. 13. Steps in the generation of an efficient structure of a 1 to $L$ interpolator.

$h(N-1)$ in Fig. 13(c) are now commuted to the low sampling rate side of the network this structure requires a factor of $L$ times less computation than the structure in Fig. 13(b).

An alternative way of deriving the structure of Fig. 13(c) is by a direct transposition of the network of Fig. 10(c) (letting $L = M$). This is a direct consequence of the fact that decimators and interpolators are duals. Similarly an efficient direct form interpolator structure which can exploit symmetry in $h(n)$ can be obtained by transposing the structure of Fig. 12 and letting $L = M$. A further property of transposition is that for the resulting network, neither the number of multipliers nor the rate at which these multipliers operate will change [21]. Thus if we are given a network that is minimized with respect to its multiplication rate, then its transpose will also be minimized with respect to its multiplication rate.

## C. Polyphase FIR Structures for Integer Decimators and Interpolators

A second general class of structures that are of interest in multirate digital systems are the polyphase networks (sometimes referred to as $N$ path networks) [22], [23]. We will find it convenient to first derive this structure for the $L$ to 1 interpolator and then obtain the structure for the decimator by transposing the interpolator structure.

In Section II it was shown that a general form for the input-to-output time-domain relationship for the 1 to $L$ interpolator is

$$y(m) = \sum_{n=-\infty}^{\infty} g_m(n) x\left(\left\lfloor \frac{m}{L} \right\rfloor - n\right) \qquad (55)$$

where

$$g_m(n) = h(nL + m \oplus L), \qquad \text{for all } m \text{ and } n \qquad (56)$$

is a periodically time-varying filter with period $L$. Thus to generate each output sample $y(m)$, $m = 0, 1, 2, \cdots, L - 1$,

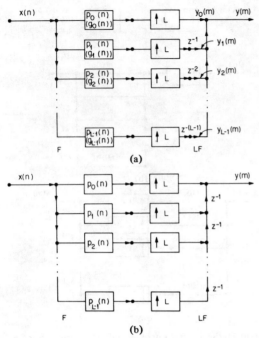

**Fig. 14.** Polyphase structures for a 1 to $L$ interpolator.

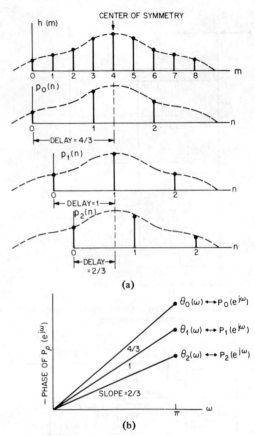

**Fig. 15.** Illustration of the properties of polyphase networks.

a different set of coefficients $g_m(n)$ are used. After $L$ outputs are generated, the coefficient pattern repeats; thus $y(L)$ is generated using the same set of coefficients $g_0(n)$ as $y(0)$, $y(L + 1)$ uses the same set of coefficients $g_1(n)$ as $y(1)$, etc.

Similarly the term $\lfloor m/L \rfloor$ in (55) increases by one for every $L$ samples of $y(m)$. Thus for output samples $y(L)$, $y(L + 1)$, $\cdots$, $y(2L - 1)$ the coefficients $g_m(n)$ are multiplied by samples $x(1 - n)$. In general, for output samples $y(rL)$, $y(rL + 1)$, $\cdots$, $y(rL + L - 1)$ the coefficients $g_m(n)$ are multiplied by samples $x(r - n)$. Thus it is seen that $x(n)$ in (55) is updated at the low sampling rate $F$, whereas $y(m)$ is evaluated at the high sampling rate $LF$.

An implementation of the 1 to $L$ interpolator based on the computation of (55) is shown in Fig. 14(a). The way in which this structure operates is as follows. The partitioned subsets, $g_0(n)$, $g_1(n)$, $\cdots$, $g_{L-1}(n)$, of $h(m)$ can be identified with $L$ separate linear, time invariant filters which operate at the low sampling rate $F$. To make this subtle notational distinction between the time-varying coefficients and the time-invariant filters we will refer to the time-invariant filters respectively as $p_0(n)$, $p_1(n)$, $\cdots$, $p_{L-1}(n)$. Thus

$$p_\rho(n) = g_\rho(n), \quad \text{for } \rho = 0, 1, 2, \cdots, L - 1 \text{ and all } n.$$

$$(57)$$

These filters $p_\rho(n)$ will be referred to as the polyphase filters. Furthermore by combining (56) and (57) it is apparent that

$$p_\rho(n) = h(nL + \rho), \quad \text{for } \rho = 0, 1, 2, \cdots, L - 1 \text{ and all } n$$

$$(58)$$

For each new input sample $x(n)$ there are $L$ output samples (see Fig. 14). The output from the upper path $y_0(m)$ has nonzero values for $m = nL$, $n = 0, \pm 1, \pm 2, \cdots$, which correspond to system outputs $y(nL)$, $n = 0, \pm 1, \cdots$. The output from the next path $y_1(m)$ is nonzero for $m = nL + 1$, $n = 0, \pm 1, \pm 2, \cdots$

because of the delay of one sample at the high sampling rate. Thus $y_1(m)$ corresponds to the interpolation output samples $y(nL + 1)$, $n = 0, \pm 1, \cdots$. In general the output of the $\rho$th path, $y_\rho(m)$ corresponds to the interpolation output samples $y(nL + \rho)$, $n = 0, \pm 1, \cdots$. Thus for each input sample $x(n)$ each of the $L$ branches of the polyphase network contributes one nonzero output which corresponds to one of the $L$ outputs of the network. The polyphase interpolation network of Fig. 14(a) has the property that the filtering is performed at the low sampling rate and thus it is an efficient structure. A simple manipulation of the structure of Fig. 14(a) leads to the equivalent network of Fig. 14(b) in which all the delays are single sample delays.

The individual polyphase filters $p_\rho(n)$, $\rho = 0, 1, 2, \cdots, L - 1$ have a number of interesting properties. This is a consequence of the fact that the impulse responses $p_\rho(n)$, $\rho = 0, 1, 2, \cdots$, $L - 1$, correspond to decimated versions of the impulse response of the prototype filter $h(m)$ (decimated by a factor of $L$ according to (56) or (58)). Fig. 15 illustrates this for the case $L = 3$ and for an FIR filter $h(m)$ with $N = 9$ taps. The upper figure shows the samples of $h(m)$ where it is assumed that $h(m)$ is symmetric about $m = 4$. Thus $h(m)$ has a flat delay of 4 samples [17]. The filter $p_0(n)$ has three samples corresponding to $h(0)$, $h(3)$, $h(6) = h(2)$. Since the point of symmetry of the envelope of $p_0(n)$ is $n = \frac{4}{3}$ it has a flat delay of $\frac{4}{3}$ samples. Similarly $p_1(n)$ has samples $h(1)$, $h(4)$, $h(7) = h(1)$, and because its zero reference ($n = 0$) is offset by $\frac{1}{3}$ sample (with respect to $m = 0$) it has a flat delay of 1 sample. Thus different fractional sample delays and consequently different phase shifts are associated with the different filters

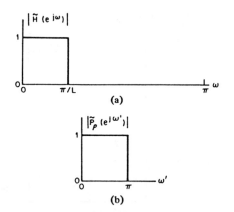

(a)

(b)

Fig. 16. Ideal frequency response of the polyphase networks.

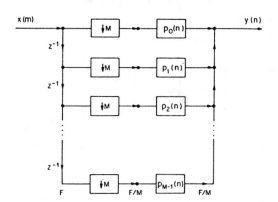

Fig. 17. Polyphase structure for an $M$ to 1 decimator.

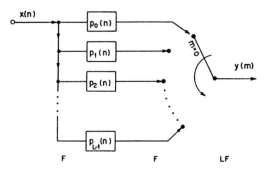

Fig. 18. Commutator model for the 1 to $L$ polyphase interpolator.

$p_\rho(n)$ as seen in Fig. 15(b). These delays are compensated for by the delays which occur at the high sampling rate $LF$ in the network (see Fig. 14). The fact that different phases are associated with different paths of the network is, of course, the reason for the term polyphase network.

A second property of the polyphase filters is shown in Fig. 16. The frequency response of the prototype filter $h(m)$ approximates the ideal low-pass characteristic $\tilde{H}(e^{j\omega})$ shown in Fig. 16(a).[2] Since the polyphase filters $p_\rho(n)$ are decimated versions of $h(m)$ (decimated by $L$) the frequency response $0 \leqslant \omega \leqslant \pi/L$ of $\tilde{H}(e^{j\omega})$ scales to the range $0 \leqslant \omega' \leqslant \pi$ for $\tilde{P}_e(e^{j\omega'})$ as seen in Fig. 16 where $\tilde{P}_\rho(e^{j\omega'})$ is the ideal characteristic that the polyphase filter $p_\rho(n)$ approximates. Thus the polyphase filters approximate all-pass functions and each value of $\rho$, $\rho = 0, 1, 2, \cdots, L - 1$, corresponds to a different phase shift.

The polyphase filters can be realized in a variety of ways. If the prototype filter $h(m)$ is an FIR filter of length $N$ then the filters $p_\rho(n)$ will be FIR filters of length $N/L$. In this case it is often convenient to choose $N$ to be a multiple of $L$ so that all of the polyphase filters are of equal length. These filters may be realized by any of the conventional methods for implementing FIR filters such as the direct form structure or the methods based on fast convolution [17], [24]. If a direct form FIR structure is used for the polyphase filters, the polyphase structure of Fig. 14 will require the same multiplication rate as the direct form interpolator structure of Fig. 13. Exploiting symmetry in $h(m)$ is more difficult in this class of

structures since, at most, only one of the $p_\rho(n)$ subfilters are symmetric.

By transposing the structure of the polyphase 1 to $L$ interpolator of Fig. 14(b), we get the polyphase $M$ to 1 decimator structure of Fig. 17 where $L$ is replaced by $M$. Again the filtering operations of the polyphase filters occur at the low sampling rate side of the network and they can be implemented by any of the conventional structures discussed above.

In the above discussion for the 1 to $L$ interpolator we have identified the coefficients of the polyphase filters $p_\rho(n)$ with the coefficient sets $g_m(n)$ of the time-varying filter model. In the case of the $M$ to 1 decimator, however, this identification cannot be made directly. According to the time-varying filter model, discussed in Section II, the coefficients $g_m(n)$ for the $M$ to 1 decimator are

$$g_m(n) = g(n) = h(n), \quad \text{for all } n \text{ and } m. \quad (59)$$

Alternatively, according to the transpose network of Fig. 17, the coefficients of the $M$ to 1 polyphase decimator are

$$p_\rho(n) = h(nM + \rho), \quad \text{for } \rho = 0, 1, 2, \cdots, M - 1, \text{ and all } n$$

$$(60)$$

where $\rho$ denotes the $\rho$th polyphase filter. Thus the polyphase filters $p_\rho(n)$ for the $M$ to 1 decimator are equal to the time-varying coefficients $g_m(n)$ of the transpose (interpolator) of this decimator.

From a practical point of view it is often convenient to implement the polyphase structures in terms of a commutator model. By careful examination of the interpolator structure of Fig. 14 it can be seen that the outputs of each of the polyphase branches contributes samples of $y(m)$ for different time slots. Thus the 1 to $L$ sampling rate expander and delays can be replaced by a commutator as shown in Fig. 18. The commutator rotates in a counterclockwise direction starting with the zeroth-polyphase branch at time $m = 0$.

A similar commutator model can be developed for the $M$ to 1 polyphase decimator by starting with the structure of Fig. 17 and replacing the delays and $M$ to 1 sampling rate compressors with a commutator. This leads to the structure of Fig. 19. Again the commutator rotates in a counterclockwise direction starting with the zeroth-polyphase branch at time $m = 0$.

At this point the reader should be cautioned that an alternate formulation of these polyphase structures can be developed such that the commutators have clockwise rotations and a *different* but equivalent set of polyphase filters are defined.[3]

---

[2] Recall also that there is an additional gain of $L$ required in the interpolator which we have ignored in this discussion (see Section II-B).

[3] This alternate formulation can be developed by defining a set of polyphase filters such that $\rho$ is replaced by $-\rho$ on the right-hand side of (58).

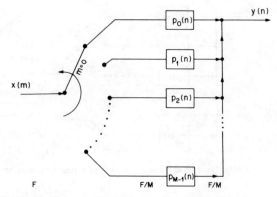

Fig. 19. Commutator model for the $M$ to 1 polyphase decimator.

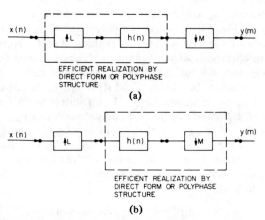

Fig. 20. Possible realization of an $L/M$ sampling rate converter.

Both formulations have been used in the literature and they should not be confused.

### D. FIR Structures with Time-Varying Coefficients for Interpolation/Decimation by a Factor of L/M

In the previous two sections we have considered implementations of decimators and interpolators using the direct form and polyphase structures for the case of integer changes in the sampling rate. Efficient realizations of these structures were obtained by commuting the filtering operations to occur at the low sampling rate. For the case of a network which realizes a change in sampling rate by a factor of $L/M$, it is difficult to achieve such efficiencies. The difficulty is illustrated in Fig. 20. If we realize the 1 to $L$ interpolation part of the structure using the techniques described earlier, then we are faced with the problem of commuting the $M$ to 1 sampling rate compressor into the resulting network (Fig. 20(a)). If we realize the decimator part of the structure first, then the 1 to $L$ sampling rate expander must be commuted into the structure (Fig. 20(b)). In both cases difficulties arise and we are faced with a network which cannot be implemented efficiently simply using the techniques of commutation and transposition.

Efficient structures exist for implementing a sampling rate converter with a ratio in sampling rates of $L/M$, and in this section we discuss one such class of FIR structures with time-varying coefficients [13]. This structure can be derived from the time domain input-to-output relation of the network, as derived in Section II, namely

$$y(m) = \sum_{n=-\infty}^{\infty} g_m(n) x\left(\left\lfloor\frac{mM}{L}\right\rfloor - n\right) \qquad (61)$$

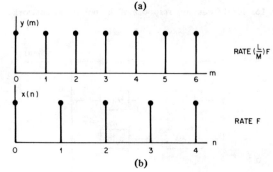

| $y(m)$ | $x\left(\left\lfloor\frac{mM}{L}\right\rfloor\right)$ | $g_{m\oplus L}(0)$ |
|---|---|---|
| $m$ | $\left\lfloor\frac{2m}{3}\right\rfloor$ | $m \oplus L$ |
| 0 | 0 | 0 |
| 1 | 0 | 1 |
| 2 | 1 | 2 |
| 3 | 2 | 0 |
| 4 | 2 | 1 |
| 5 | 3 | 2 |
| 6 | 4 | 0 |

(a)

(b)

Fig. 21. Timing relationships between $y(m)$ and $x(n)$ for the case $M = 2, L = 3$.

where

$$g_m(n) = h(nL + mM \oplus L), \qquad \text{for all } m \text{ and all } n \qquad (62)$$

and $h(k)$ corresponds to the low-pass (or bandpass) FIR prototype filter. It will be convenient for our discussion to assume that the length of the filter $h(k)$ is a multiple of $L$, i.e.,

$$N = QL \qquad (63)$$

where $Q$ is an integer. Then all of the coefficient sets $g_m(n)$, $m = 0, 1, 2, \cdots, L - 1$ contain *exactly* $Q$ coefficients. Furthermore $g_m(n)$ is periodic in $m$ with period $L$, i.e.,

$$g_m(n) = g_{m+rL}(n), \qquad r = 0, \pm 1, \pm 2, \cdots. \qquad (64)$$

Therefore, equation (61) can be expressed as

$$y(m) = \sum_{n=0}^{Q-1} g_{m\oplus L}(n) x\left(\left\lfloor\frac{mM}{L}\right\rfloor - n\right). \qquad (65)$$

Equation (65) shows that the computation of an output sample $y(m)$ is obtained as a weighted sum of $Q$ sequential samples of $x(n)$ starting at the $x(\lfloor mM/L \rfloor)$ sample and going backwards in $n$ sequentially. The weighting coefficients are periodically time varying so that the $m \oplus L$ coefficient set $g_{m\oplus L}(n)$, $n = 0, 1, 2, \cdots, Q - 1$, is used for the $m$th output sample. Fig. 21 illustrates this timing relationship for the $n = 0$ term in (65) and for the case $M = 2$ and $L = 3$. The table in Fig. 21(a) shows the index values of $y(m), x(\lfloor mM/L \rfloor)$ and $g_{m\oplus L}(0)$ for $m = 0$, to $m = 6$. Fig. 21(b) illustrates the relative timing positions of the signals $y(m)$ and $x(n)$ drawn on an absolute time scale. By comparison of the table and the figure it can be seen that the value $x(\lfloor mM/L \rfloor)$ always represents the most recent available sample of $x(n)$, i.e., $y(0)$ and $y(1)$ are computed on the basis of $x(0 - n)$. For $y(2)$ the most recent available value of $x(n)$ is $x(1)$, for $y(3)$ it is $x(2)$, etc.

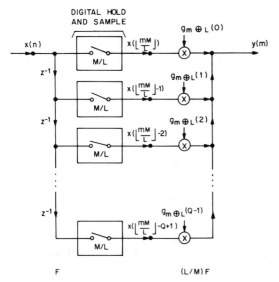

Fig. 22. Efficient structure for realizing an $L/M$ sampling rate converter.

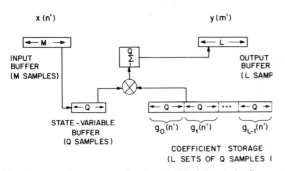

Fig. 23. Diagram of a program structure to implement the flowgraph of Fig. 22 in a block-by-block manner.

Based on (65) and the above description of how the input, output, and coefficients enter into the computation, the structure of Fig. 22 is suggested for realizing an $L/M$ sampling rate converter. The structure consists of:

1) a $Q$ sample "shift register" operating at the input sampling rate $F$ which stores sequential samples of the input signal;

2) a direct form FIR structure with time-varying coefficients $(g_{m \oplus L}(n), n = 0, 1, \cdots, Q - 1)$ which operates at the output sampling rate $(L/M)F$;

3) a series of digital "hold-and-sample" boxes which couple the two sampling rates. The input side of the box "holds" the most recent input value until the next input value comes along; the output side of the box "samples" the input values at times $n = mM/L$. For times when $mM/L$ is an integer (i.e., input and output sampling times are the same), the input changes first and the output samples the changed input.

It should be clear that the structure of Fig. 22 is an efficient one for implementing an $(L/M)$ sampling rate converter since the filtering operations are all performed at the output sampling rate with the minimum required number of coefficients used to generate each output.

Fig. 23 shows a diagram of a program configuration to implement this structure in a block by block manner. The

program takes in a block of $M$ samples of the input signal, denoted as $x(n')$, $n' = 0, 1, 2, \cdots, M - 1$, and computes a block of $L$ output samples $y(m')$ $m' = 0, 1, 2, \cdots, L - 1$. For each output sample time $m'$, $m' = 0, 1, \cdots, L - 1$, the $Q$ samples from the state-variable buffer are multiplied respectively with $Q$ coefficients from one of the coefficient sets $g_{m'}(n')$ and the products are accumulated to give the output $y(m')$. Each time the quantity $\lfloor m'M/L \rfloor$ increases by one, one sample from the input buffer is shifted into the state-variable buffer. (This information can be stored in a control array.) Thus after $L$ output values are computed $M$ input samples have been shifted into the state-variable buffer and the process can be repeated for the next block of data. In the course of processing one block of data ($M$ input samples and $L$ output samples) the state-variable buffer is sequentially addressed $L$ times and the coefficient storage buffer is sequentially addressed once. A program which performs this computation can be found in [25].

### E. Comparisons of Structures

In this section we have discussed three principle classes of FIR structures for decimators and interpolators. In addition, in Section V we discuss multistage cascades of these structures and show how this can lead to additional gains in computational efficiency when conversion ratios are large. A natural question to ask at this point is which of these methods is most efficient? The answer, unfortunately, is nontrivial and is highly dependent on the application being considered. Some insight and direction, however, can be provided by observing some general properties of the above classes of structures.

The direct form structures have the advantage that they can be easily modified to exploit symmetry in the system function to gain an additional reduction in computation by a factor of approximately two. The polyphase structures have the advantage that the filters $p_\rho(n)$ can be easily realized with efficient techniques such as the fast convolution methods based on the FFT [24]. As such this structure has been found useful for filter banks [23]. The structures with time-varying coefficients are particularly useful when considering conversions by factors of $L/M$.

There are many other considerations which determine overall efficiency of these structures. Most of these considerations, however, are filter design ones and hence we must defer further comparisons of single stage structures for decimators and interpolators until we have discussed the filter design issues in some detail.

## IV. DESIGN OF FIR FILTERS FOR DECIMATION AND INTERPOLATION

In the discussion in the previous chapters we have assumed that the filter $h(k)$ approximates some ideal low-pass (or bandpass) characteristic (see Figs 2-4). As such the effectiveness of these systems is directly related to the type and quality of design of this digital filter. The purpose of this section is to review digital filter design techniques, and discuss those methods that are especially applicable to the design of the digital filter in sampling rate changing systems.

The filter design problem is essentially one of determining suitable values of $h(k)$ to meet given performance specifications on the filter. Such performance specifications can be made on the time response $h(k)$, or the frequency response of the filter $H(e^{j\omega})$ defined as

$$H(e^{j\omega}) = \sum_{k=-\infty}^{\infty} h(k) e^{-j\omega k} \qquad (66)$$

$$= H(z)|_{z=e^{j\omega}}. \qquad (67)$$

The frequency response is, in general, a complex function of $\omega$. Thus it is convenient to represent it in terms of its magnitude $|H(e^{j\omega})|$ and phase $\theta(\omega)$ as

$$H(e^{j\omega}) = |H(e^{j\omega})| e^{j\theta(\omega)} \qquad (68)$$

where

$$|H(e^{j\omega})| = \sqrt{\text{Re}^2 [H(e^{j\omega})] + \text{Im}^2 [H(e^{j\omega})]} \qquad (69a)$$

$$\theta(\omega) = \tan^{-1} \left[ \frac{\text{Im}[H(e^{j\omega})]}{\text{Re}[H(e^{j\omega})]} \right]. \qquad (69b)$$

An important filter parameter is the group delay $\tau(\omega)$ defined as

$$\tau(\omega) = \frac{-d\theta(\omega)}{d\omega}. \qquad (70)$$

The group delay is a measure of time delay as a function of frequency of a signal as it passes through the filter. Nondispersive filters have the property that $\tau(\omega)$ is a constant (i.e., a fixed delay) over the frequency range of interest.

Before proceeding to a discussion of filter design techniques for decimators and interpolators, it is important to consider the ideal frequency domain and time domain criteria that specify such designs. It is also important to consider, in more detail, the representation of such filters in terms of a single prototype filter or as a set of polyphase filters. Although both representations are equivalent, it is sometimes easier to view filter design criteria in terms of one representation or the other. Also, some filter design techniques are directed at the design of a single prototype filter such as in the classical filter design methods, whereas other filter design techniques are directed at the design of the polyphase filters. Thus we will consider both representations in this section.

A. *Relationship Between the Prototype Filter and its Polyphase Representation*

As discussed in Section III, the coefficients, or impulse responses, of the polyphase filters correspond to sampled (and delayed) versions of the impulse response of the prototype filter. For a 1 to $L$ interpolator there are $L$ polyphase filters and they are defined as (see Fig. 15).

$$p_\rho(n) = h(\rho + nL), \qquad \rho = 0, 1, 2, \cdots, L-1, \text{ and all } n. \qquad (71)$$

Similarly for an $M$ to 1 decimator there are $M$ polyphase filters in the polyphase structure and they are defined as

$$p_\rho(n) = h(\rho + nM), \qquad \rho = 0, 1, 2, \cdots, M-1, \text{ and all } n. \qquad (72)$$

Taken as a set, the samples $p_\rho(n)$ ($\rho = 0, 1, \cdots, L-1$ for an interpolator, or $\rho = 0, 1, \cdots, M-1$ for a decimator) represent all of the samples of $h(k)$. Since the development of the filter specifications is identical for both cases (1 to $L$ interpolators and $M$ to 1 decimators) we will only consider the case of inter-

polators. The results for decimators can then simply be obtained by replacing $L$ by $M$ in the appropriate equations.

The samples $h(k)$ can be recovered from $p_\rho(n)$ by sampling rate expanding the sequences $p_\rho(n)$ by a factor $L$. Each expanded set, is then delayed by $\rho$ samples and the $L$ sets are then summed to give $h(k)$ (the reverse operation to that of Fig. 15). If we let $\hat{p}_\rho(k)$ represent the sampling rate expanded set

$$\hat{p}_\rho(k) = \begin{cases} p_\rho(k/L), & k = 0, \pm L, \pm 2L, \cdots \\ 0, & \text{otherwise} \end{cases} \qquad (73)$$

then $h(k)$ can be reconstructed from $\hat{p}_\rho(k)$ via the summation

$$h(k) = \sum_{\rho=0}^{L-1} \hat{p}_\rho(k - \rho). \qquad (74)$$

The $z$-transform $H(z)$ of the prototype filter can similarly be expressed in terms of the $z$-transforms of the polyphase filters $P_\rho(z)$. It can be shown that

$$H(z) = \sum_{\rho=0}^{L-1} z^{-\rho} P_\rho(z^L). \qquad (75)$$

Finally, the $z$-transform $P_\rho(z)$, can be expressed in terms of $H(z)$ according to the following derivation. If we define a sampling function $\delta_\rho(k)$, such that

$$\delta_\rho(k) = \begin{cases} 1, & k = \rho, \rho \pm L, \rho \pm 2L, \cdots \\ 0, & \text{otherwise} \end{cases} \qquad (76a)$$

$$= \frac{1}{L} \sum_{l=0}^{L-1} e^{j2\pi l(k-\rho)/L} \qquad (76b)$$

then the sampling rate expanded sequences $\hat{p}_\rho(k)$ in (73) can be expressed as

$$\hat{p}_\rho(k) = \delta_\rho(k) h(k) = h(k) \frac{1}{L} \sum_{l=0}^{L-1} e^{j2\pi l(k-\rho)/L}. \qquad (77)$$

The $z$-transform $P_\rho(z)$ can then be expressed in the form

$$P_\rho(z) = \sum_{n=-\infty}^{\infty} p_\rho(n) z^{-n} = \sum_{n=-\infty}^{\infty} \hat{p}_\rho(\rho + nL) z^{-n} \qquad (78)$$

and by the substitution of variables $k = \rho + nL$,

$$P_\rho(z) = \sum_{k=-\infty}^{\infty} \hat{p}_\rho(k) z^{-(k-\rho)/L}. \qquad (79)$$

By substituting (77) into (79) we get

$$P_\rho(z) = \frac{1}{L} \sum_{k=-\infty}^{\infty} \sum_{l=0}^{L-1} h(k) e^{j2\pi l(k-\rho)/L} z^{-(k-\rho)/L}. \qquad (80)$$

Letting $z = e^{j\omega}$ and rearranging terms gives

$$P_\rho(e^{j\omega}) = \frac{1}{L} \sum_{l=0}^{L-1} e^{j(\omega-2\pi l)\rho/L} \sum_{k=-\infty}^{\infty} h(k) e^{-j(\omega-2\pi l)k/L}$$

$$= \frac{1}{L} \sum_{l=0}^{L-1} e^{j(\omega-2\pi l)\rho/L} H(e^{j(\omega-2\pi l)/L}),$$

$$\rho = 0, 1, 2, \cdots, L-1. \qquad (81)$$

Equation (81) shows the relationships of the Fourier transforms of the polyphase filters to the Fourier transform of the prototype filter.

Equations (71)–(74), therefore, illustrate the time-domain relationships between $h(k)$ and $p_\rho(k)$ and (75) and (81) show their frequency-domain relationships.

### B. Ideal Frequency Domain Characteristics for Interpolation and Decimation Filters

In the previous sections we have assumed that the filter $h(k)$ approximates some ideal low-pass (or bandpass) characteristic. We will elaborate on these "ideal" characteristics in somewhat more detail in the next two sections. In practice it is also necessary to specify a performance criterion to measure (in a consistent manner) how closely an actual filter design approximates this ideal characteristic. Since different design techniques are often based on different criteria, we will consider these criteria as they arise.

Recall from the discussion in Section II-B that the interpolator filter $h(k)$ must approximate the ideal[4] low-pass characteristic

$$\tilde{H}(e^{j\omega'}) = \begin{cases} L, & |\omega'| < \pi/L \\ 0, & \text{otherwise} \end{cases} \qquad (82)$$

as illustrated in Fig. 3.

By applying (82) to (81) it is possible to derive the equivalent ideal characteristics, $\tilde{P}_\rho(e^{j\omega})$, that are implied in the polyphase filters.[5] Because of the constraint imposed by (82), only the $l = 0$ term in (81) is nonzero and thus it simplifies to the form

$$\tilde{P}_\rho(e^{j\omega}) = \frac{1}{L} e^{j\omega\rho/L} \tilde{H}(e^{j\omega/L})$$

$$= e^{j\omega\rho/L}, \qquad \rho = 0, 1, 2, \cdots, L - 1. \qquad (83)$$

Equation (83) shows that the "ideal" polyphase filters $\tilde{p}_\rho(n)$ should approximate all-pass filters with linear phase shifts corresponding to fractional advances of $\rho/L$ samples ($\rho = 0, 1, 2, \cdots, L - 1$) (ignoring any fixed delays that must be introduced in practical implementations of such filters). A further interpretation of the reason for this phase advance can be found in Section III-C on the discussion of polyphase structures.

In some cases it is known that the spectrum of $x(n)$ does not occupy its full bandwidth. This property can be used to advantage in the filter design and we will see examples of this in the next section on cascaded (multistage) implementations of sampling rate changing systems. If we define $\omega_c$ as the highest frequency of interest in $X(e^{j\omega})$, i.e.,

$$|X(e^{j\omega})| < \epsilon, \qquad \text{for} \quad \pi > |\omega| > \omega_c \qquad (84)$$

where $\epsilon$ is a small quantity (relative to the peak of $|X(e^{j\omega})|$), then $W(e^{j\omega'})$ is an $L$-fold periodic repetition of $X(e^{j\omega})$ as shown in Fig. 24 (for $L = 5$). In this case, the ideal interpolator filter only has to remove the $(L - 1)$ repetitions of the band of $X(e^{j\omega})$ where $|X(e^{j\omega})| > \epsilon$. Thus in the frequency

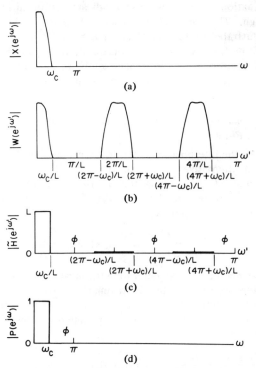

Fig. 24. Illustrations of $\phi$ bands in the specification of an interpolator filter ($L = 5$).

domain, the ideal interpolator filter satisfies the constraints

$$\tilde{H}(e^{j\omega'}) = \begin{cases} L, & 0 \leqslant |\omega'| \leqslant \omega_c/L \\ 0, & (2\pi r - \omega_c)/L \leqslant |\omega'| \leqslant (2\pi r + \omega_c)/L, \\ & \qquad r = 1, 2, \cdots, L - 1 \end{cases} \qquad (85)$$

as illustrated in Fig. 24(c). The bands from $(2\pi r + \omega_c)/L$ to $(2\pi(r + 1) - \omega_c)/L$, $(r = 0, 1, \cdots)$ are "don't care" ($\phi$) bands in which the filter frequency response is essentially unconstrained. (In practice, however, $|H(e^{j\omega'})|$ should not be very large in these $\phi$ bands, e.g., not larger than $L$, to avoid amplification of any noise (or tails of $X(e^{j\omega})$) that may exist in these bands). We will see later how these $\phi$ bands can have a significant effect on the filter design problem. Fig. 24(d) shows the response of the ideal polyphase filter which is converted from an allpass to a low-pass filter with cutoff frequency $\omega_c$. Of course, the phase response of each polyphase filter is unaltered by the don't care bands.

As discussed in Section II-A for a decimator, the filter $H(e^{j\omega'})$ should approximate the ideal low-pass characteristic

$$\tilde{H}(e^{j\omega'}) = \begin{cases} 1, & 0 \leqslant \omega' \leqslant \pi/M \\ 0, & \text{otherwise.} \end{cases} \qquad (86)$$

Alternatively, the polyphase filters should approximate the ideal allpass characteristics

$$\tilde{P}_\rho(e^{j\omega}) = \frac{1}{M} e^{j\omega\rho/M}, \qquad \rho = 0, 1, 2, \cdots, M - 1. \qquad (87)$$

If we are only interested in preventing aliasing in a band from 0 to $\omega_c$, where $\omega_c < \pi/M$, and we are willing to tolerate aliased components for frequencies above $\omega_c$, then we again

---

[4] The "ideal" characteristic for a filter is denoted by a tilde over the variable throughout this section.

[5] Since the conventional filter $h(m)$ is implemented at the higher sampling rate, the frequency variable (in (82) for example) is $\omega'$, whereas since the polyphase filters are implemented at the lower sampling rate, the frequency variable is $\omega = \omega' L$.

have a situation where don't care bands are permitted in the filter design. The don't care regions are the same as those in (85) as illustrated in Fig. 24(c) (with $L$ replaced by $M$). In fact all of the frequency-domain constraints that apply to the design of interpolation filters also apply to the design of decimation filters, a consequence of the property that they are transpose systems.

### C. Time-Domain Properties of Ideal Interpolation and Decimation Filters

If we view the interpolation filter design problem in the time domain, an alternative picture of the "ideal" interpolation filter is obtained. By taking the inverse transform of the ideal filter characteristic defined by (82) we get the well-known $\sin(x)/x$ characteristic

$$\tilde{h}(k) = \frac{\sin(\pi k/L)}{(\pi k/L)}, \qquad k = 0, \pm 1, \pm 2, \cdots. \qquad (88)$$

In a similar manner we can determine the ideal time responses of the polyphase filters, either by taking the inverse transform of (87), or by sampling the above time response $\tilde{h}(k)$ according to (71). The net result is that the ideal time responses of the polyphase filters are

$$\tilde{p}_\rho(n) = \frac{\sin[\pi(n + \rho/L)]}{\pi(n + \rho/L)}, \qquad \rho = 0, 1, 2, \cdots, L - 1, \text{ and all } n.$$

$$(89)$$

A number of interesting observations can be made about the above ideal time responses. First we see that they constrain every $L$th value of $\tilde{h}(k)$ such that

$$\tilde{h}(k) = \begin{cases} 1, & k = 0 \\ 0, & k = rL, r = \pm 1, \pm 2, \cdots. \end{cases} \qquad (90)$$

Alternatively, this implies the constraint that the zeroth-polyphase filter have an impulse response that is a unit pulse, i.e.,

$$\tilde{p}_0(n) = \delta(n), \qquad \text{for all } n. \qquad (91)$$

In terms of the polyphase structure of Fig. 14 and its signal processing interpretation of Fig. 15, the above constraint is easy to visualize. It simply implies that the output $y_0(m)$, of the zeroth polyphase branch is identical to the input $x(n)$ filled in with $L - 1$ zeros, i.e., these sample values are already known. The remaining $L - 1$ samples in between these values must be interpolated by the polyphase filters $p_\rho(m)$, $\rho = 1$, $2, \cdots, L - 1$. Since these filters are theoretically infinite in duration, they must be approximated, in practice, with finite duration filters. Thus the interpolation "error" between the outputs of a practical system and an ideal system can be zero for $m = 0, \pm L, \pm 2L, \cdots$. However "in-between" these samples, the error will always be nonzero.

By choosing a design that does not specifically satisfy the constraint of (90) or (91) a tradeoff can be made between errors that occur at sample times $m = 0, \pm L, \pm 2L, \cdots$ and errors that occur between these samples.

Another "time-domain" property that can be observed is that the ideal filter $\tilde{h}(k)$ is symmetric about zero, i.e.,

$$\tilde{h}(k) = \tilde{h}(-k). \qquad (92)$$

(Alternatively, for practical systems it may be symmetrical

about some fixed nonzero delay.) This symmetry does not necessarily extend directly to the polyphase filters since they correspond to sampled values of $\tilde{h}(k)$ offset by some fraction of a sample. Their envelopes, however, are symmetrical (see Fig. 15).

The above symmetry property does, however, imply a form of mirror image symmetry between pairs of polyphase filters $\tilde{p}_\rho(n)$ and $\tilde{p}_{L-\rho}(n)$. Applying (92) to (71) gives

$$\tilde{p}_\rho(n) = \tilde{h}(-\rho - nL)$$
$$= \tilde{h}(L - \rho - (n + 1)L). \qquad (93)$$

Also noting that

$$\tilde{p}_{L-\rho}(n) = \tilde{h}(L - \rho + nL) \qquad (94)$$

it can be seen that this symmetry is of the form

$$\tilde{p}_\rho(n) = \tilde{p}_{L-\rho}(-n - 1). \qquad (95)$$

In the case of decimators it is not possible to identify the outputs of specific polyphase branches with specific output samples of the network. All branches contribute to each output. Thus it is not as convenient to give meaningful time domain interpretations to the operation of the filter in a decimator.

The ideal time responses for $\tilde{h}(k)$ and $\tilde{p}_\rho(n)$ for decimators, however, are the same as those of (88) and (89), respectively, with $L$ replaced by $M$.

### D. Filter Design Procedures

In the remainder of this section we will discuss a number of filter design procedures which apply to the design of multirate systems. Since the filter design problem for such systems generally is a low-pass (or bandpass) design problem, nearly all of the work in digital signal processing filter theory can be brought to bear on this problem. We will not attempt to discuss all of these methods in detail, since they are well documented elsewhere [11], [17], but rather we will try to point to the relevant issues involved in these designs that particularly apply to multirate systems.

We will discuss five main categories of filter design procedures, namely:

1) window designs [11], [17], [26];
2) optimal, equiripple linear phase designs [27]–[30];
3) half-band designs [31]–[33];
4) special FIR interpolator designs based on time domain filter specifications [34]–[38], or stochastic properties of the signal [39];
5) classical interpolation designs, namely linear and Lagrangian interpolators [4], [16].

### E. FIR Filters Based on Window Designs

One straightforward approach to designing FIR filters for decimators and interpolators is by the well-known method of windowing or truncating the ideal prototype response $\tilde{h}(k)$. A direct truncation (i.e., a rectangular windowing of $\tilde{h}(k)$), however, leads to the Gibbs phenomenon which manifests itself as a large (9 percent) ripple in the frequency behavior of the filter in the vicinity of filter magnitude discontinues (i.e., near the edges of the passband and stopband). Furthermore, the amplitude of this ripple does not decrease with increasing duration of the filter (it only becomes narrower in width).

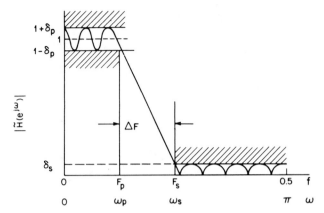

Fig. 25. Illustration of a tolerance scheme for a practical low-pass filter.

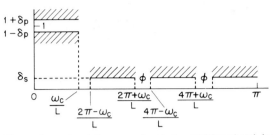

Fig. 26. Illustration of a tolerance scheme for multistopband inter-polator filters.

Thus a direct truncation, or a rectangular windowing, of $\tilde{h}(k)$ is rarely used in practice.

A more successful way of windowing the ideal characteristic $\tilde{h}(k)$ is by more gradually tapering its amplitude to zero near the ends of the filter with a weighting sequence $w(k)$ known as a window. The resulting filter design $h(k)$ is thus the product of the window $w(k)$ with the ideal response $h(k)$, i.e.,

$$h(k) = \tilde{h}(k)\, w(k), \quad -(N-1)/2 \leqslant k \leqslant (N-1)/2 \quad (96)$$

where we assume that $w(k)$ is a symmetric $N$-point ($N$ odd) window. A number of windows have been proposed in the literature for controlling the effects of the Gibbs phenomenon, and their properties are well understood [11], [17], [26]. Two commonly used types of windows are the "generalized" Hamming windows and the Kaiser windows.

The window designs have the property that they preserve the zero crossing pattern of $\tilde{h}(k)$ in the actual filter design $h(k)$. Thus if $\tilde{h}(k)$ is obtained from (88), then the time-domain properties discussed in Section IV-C apply to this class of filters.

Among the advantages of the window design approach is that it is simple, easy to use, and can readily be implemented in a direct manner (i.e., closed form expressions are available for the window coefficients, hence the filter responses can be obtained simply from the ideal filter response). Among the disadvantages are that there is only limited control in choosing cutoff frequencies and passband and stopband errors for most cases. The resulting filter designs are also suboptimal in that a smaller value of $N$ can be found (using other design methods) such that all specifications on the filter characteristics are met or exceeded.

### F. Equiripple (Optimal) FIR Designs

The windowing technique of the previous section represented a simple straightforward approach to the design of the digital filter required in all sampling rate conversion systems. However considerably more sophisticated design techniques have been developed for FIR digital filters [11], [27]–[30]. One such technique is the method of equiripple design based on Chebyshev approximation methods. The filters designed by this technique are optimal in the sense that the peak (weighted) approximation error in the frequency domain over the frequency range of interest is minimized.

To apply the method of equiripple design to the digital filter required for sampling rate conversion, a tolerance scheme on

the filter must be defined. Fig. 25 shows an example of a tolerance scheme for a low-pass filter where

$\delta_p$  ripple (deviation) in the passband from the ideal response

$\delta_s$  ripple (deviation) in the stopband from the ideal response

$F_p$  passband edge frequency $= \omega_p/2\pi$

$F_s$  stopband edge frequency $= \omega_s/2\pi$

$N$  number of taps in the FIR filter.

Alternatively, Fig. 26 shows a practical tolerance scheme for a multistopband design of a 1 to $L$ interpolator when it is known that the input spectrum does not occupy its full bandwidth (see Section IV-B and Fig. 24).

Given the tolerance schemes of Figs. 25 or 26, it is a simple matter to set up a filter approximation problem based on Chebyshev approximation methods. Several highly developed techniques have been presented in the literature for solving the Chebyshev approximation problem [27]–[29], including a well documented, widely used computer program [30]. The solutions are based on either a multiple exchange Remez algorithm, or a single exchange linear programming solution. We will not be concerned with details of the various solution methods.

For the case of the low-pass characteristic of Fig. 25 an empirical formula has been derived that relates the filter parameters. It can be expressed in the form

$$N \cong \frac{D_\infty(\delta_p, \delta_s)}{(\omega_s - \omega_p)/(2\pi)} - f(\delta_p, \delta_s)\frac{(\omega_s - \omega_p)}{2\pi} + 1 \quad (97a)$$

$$\approx \frac{D_\infty(\delta_p, \delta_s)}{(\omega_s - \omega_p)/(2\pi)} \quad (\text{when } \omega_s - \omega_p \text{ is small}) \quad (97b)$$

$$D_\infty(\delta_p, \delta_s) = D + f(\delta_p, \delta_s)(\Delta F)^2 \quad (98)$$

$$D = (N-1)\Delta F = (N-1)(\omega_s - \omega_p)/2\pi \quad (99)$$

$$D_\infty(\delta_p, \delta_s) = \log_{10}\delta_s[a_1(\log_{10}\delta_p)^2 + a_2\log_{10}\delta_p + a_3]$$
$$+ [a_4(\log_{10}\delta_p)^2 + a_5\log_{10}\delta_p + a_6] \quad (100)$$

$$f(\delta_p, \delta_s) = 11.012 + 0.512(\log_{10}\delta_p - \log_{10}\delta_s),$$
$$\text{for } |\delta_s| \leqslant |\delta_p| \quad (101)$$

and where $a_1 = 0.00539$, $a_2 = 0.07114$, $a_3 = -0.4761$, $a_4 = -0.00266$, $a_5 = -0.5941$, and $a_6 = -0.4278$.

Although the design relationships of (97)–(101) appear complex, they are fairly simple to apply. By way of example, Fig. 27 shows a series of plots of the quantity $D_\infty(\delta_p, \delta_s)$ as a function of $\delta_s$ for several values of $\delta_p$. Through the use of

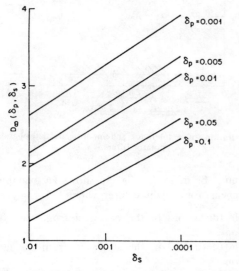

Fig. 27. Plot of $D_\infty(\delta_p, \delta_s)$ for practical values of $\delta_p$ and $\delta_s$.

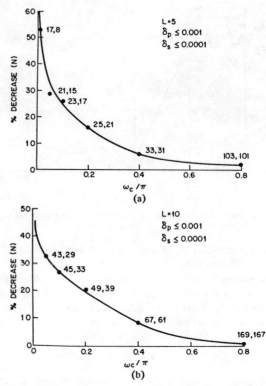

Fig. 28. Percentage decrease in required filter order by using a multi-band design instead of a low-pass design, as a function of $\omega_c/\pi$ for (a) $L = 5$ and (b) $L = 10$, respectively.

either the analytic form (e.g., (97)) or from widely available design charts and tables, it is a simple matter to determine the value of $N$ needed for an FIR low-pass filter to meet any set of design specifications [11], [38].

The above design relationships are for low-pass filters. However, as shown in Fig. 24 and 26, digital filters for interpolation and decimation need not be strictly low-pass filters, but instead can also include $\phi$, or don't care bands, which can influence the filter design problem. This is especially true when the total width of the $\phi$ bands is a significant portion of the total frequency band. For such cases no simple design formula such as (97) exists which relates the relevant filter parameters. Thus the simplest way of illustrating the effects of the $\phi$ bands on the filter design problem is by way of example.

Consider the design of a set of interpolators with specifications:

$$L = 5$$

$$\delta_p \leq 0.001$$

$$\delta_s \leq 0.0001$$

$$0 \leq \omega_c \leq \pi$$

i.e., the parameter that we allow to vary is $\omega_c$ (see Fig. 26). First we design a series of low-pass filters with passband cutoff frequencies $\omega_p = \omega_c/5$, and stopband cutoff frequencies $\omega_s = (2\pi - \omega_c)/5$. Next we design a series of multiband filters with a single passband (identical to that of the low-pass design), and a series of stopbands, separated by don't care bands (see Fig. 26). If we compare the required impulse response duration of the lowpass filter to the required impulse duration of the multiband filter we get a result of the type shown in Fig. 28(a) which gives the percentage decrease in $N$ as a function of $\omega_c/\pi$. The heavy dots shown in this figure are measured values (i.e., they are not theoretical computations), and the smooth curve shows the trend in the data. The numbers next to each heavy dot are the minimum required impulse response durations (in samples) for the low-pass, and multiband designs, respectively.

The trends in the curve are quite clear. For $\omega_c/\pi$ close to 1.0, there is essentially no gain in designing and using a multi-band filter instead of a standard low-pass filter since the total width of the $\phi$ bands is small. However, as $\omega_c/\pi$ tends to 0, significant reductions in $N$ are possible; e.g., for $\omega_c/\pi \approx 0.05$, a 50 percent reduction in impulse response duration is possible. Fig. 28(b) shows similar trends for a series of interpolators with a value of $L = 10$.

Figs. 29 and 30 show typical frequency responses for low-pass and multiband interpolation filters. For the example of Fig. 29(a) the specifications were $L = 5$, $\delta_p = 0.001$, $\delta_s = 0.0001$, and $\omega_c = 0.5\pi$. The required values of $N$ were 41 for the low-pass filter, and 39 for the multiband design. Thus for this case the reduction in filter order was insignificant. However, the change in filter frequency response (as seen in Fig. 29(b)) was highly significant. Fig. 30 shows the same comparisons for designs with $L = 10$, $\delta_p = 0.001$, $\delta_s = 0.0001$, and $\omega_c = 0.5\pi$. In this case a 28-percent reduction in filter order (from 45 for the low-pass filter to 33 for the multiband filter) was obtained since the frequency span of the stopbands was small compared to the frequency span of the don't care bands. In both these examples we see that, in the $\phi$ bands, the frequency response of the resulting filter is truly unconstrained, and can, in fact, become extremely large (as compared to the passband response). The practical implication is that some care must be taken to ensure that the amplitude response in the $\phi$ bands stays below some well specified level to guarantee that the noise in the input signal, in the $\phi$ bands, is not amplified so much that it becomes excessive in the output signal.

The above examples have shown that when $\omega_c/L$, the cutoff frequency of the passband, is relatively small (compared to $\pi/L$), significant reductions in computation can be obtained by exploiting the don't care bands in the design of the interpolation (or decimation) digital filter. In practice such conditions

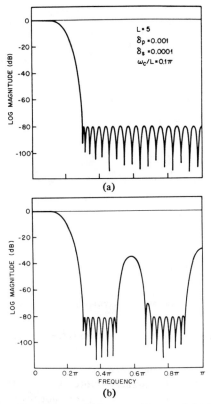

Fig. 29. Comparison between (a) low-pass filter, and (b) its equivalent multiband design for $L = 5$.

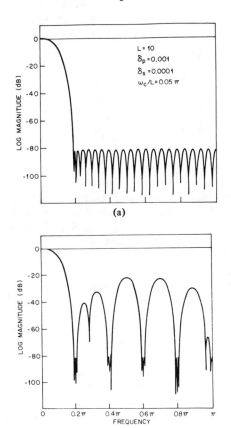

Fig. 30. Comparison between (a) a low-pass filter, and (b) its equivalent multiband design for $L = 10$.

rarely occur, i.e., we are usually dealing with signals where $\omega_c/L$ is relatively large. As such the question arises as to whether the multiband filter approach is of practical utility in the implementation of digital sampling rate changing systems. We will see in Section V that the techniques discussed here are of value in multistage implementation of decimators and interpolators involving large changes in sampling rates.

### G. Half-Band FIR Filters—A Special Case of FIR Designs for Conversion by Factors of Two

Let us again consider the tolerance specifications of the ideal low-pass filter of Fig. 25. If we consider the special case

$$\delta_s = \delta_p = \delta \tag{102}$$

$$\omega_s = \pi - \omega_p \tag{103}$$

then the resulting equiripple optimal solution to the approximation problem has the property that

$$H(e^{j\omega}) = 1 - H(e^{j(\pi-\omega)}) \tag{104}$$

i.e., the frequency response of the optimal filter is symmetric around $\omega = \pi/2$, and at $\omega = \pi/2$, $H(e^{j\omega}) = 0.5$. It can also be readily shown that any symmetric FIR filter satisfying (104), also satisfies the ideal time-domain constraints discussed in Section IV-C, equation (90), i.e., every other impulse response coefficient (except for $k = 0$) is *exactly* 0.

Filters designed using the constraints of (102) and (103) have been called "half-band" filters [31]–[33], and their properties have been intensively studied. They can be designed in a variety of ways including the window designs and equiripple designs discussed previously.

### H. Minimum Mean-Square-Error Design of FIR Interpolators—Deterministic Signals

Thus far we have considered the design of filters for interpolators and decimators from the point of view of designing the prototype filter $h(m)$ such that it satisfies a prescribed set of frequency-domain specifications. In the remainder of Section IV we consider an alternative point of view in designing filters (particularly for integer interpolators). In this approach the error criterion to be minimized is a function of the difference between the actual interpolated *signal* and its ideal value rather than a direct specification on the filter itself. We see in this section and in following sections that such an approach leads to a number of filter design techniques [34]–[37], [39] which are capable of accounting directly for the spectrum of the signal being interpolated.

Fig. 31(a) depicts the basic theoretical framework used for defining the above interpolator error criterion. We wish to design the FIR filter $h(m)$ such that it can be used to interpolate the signal $x(n)$ by a factor of $L$ with minimum interpolation error. To define this error we need to compare the output of this actual interpolator with that of an ideal (infinite duration) interpolator $\tilde{h}(m)$ whose characteristics were derived in Sections IV-B and IV-C. This signal error is defined as

$$\Delta y(m) = y(m) - \tilde{y}(m) \tag{105}$$

where $y(m)$ is the output of the actual interpolator and $\tilde{y}(m)$ is the ideal output.

In this section we will consider interpolator designs which

minimize the mean-square value of $\Delta y(m)$, defined as

$$E^2 = \|\Delta y(m)\|^2 = \lim_{K \to \infty} \frac{1}{2K+1} \sum_{m=-K}^{K} \Delta y(m)^2$$

(106a)

$$= \frac{1}{2\pi} \int_{-\pi}^{\pi} |\Delta Y(e^{j\omega'})|^2 \, d\omega'.$$

(106b)

Later in Section IV-I we will consider designs which minimize the maximum value of $|\Delta Y(e^{j\omega'})|$ over a prescribed frequency range, and in Section IV-J we will refer to designs which minimize the maximum value of $|\Delta y(m)|$ in the time domain.

The above design problems are greatly simplified by considering them in the framework of the polyphase structures as illustrated in Fig. 14. Here it is seen that the signal $y(m)$ is actually composed of interleaved samples of the signals $u_\rho(n)$, $\rho = 0, 1, 2, \cdots, L - 1$, as shown by Fig. 31(b), where $u_\rho(n)$ is the output of the $\rho$th polyphase filter. Thus the errors introduced by each polyphase branch are orthogonal to each other (since they do not coincide in time) and we can define the error in the $\rho$th branch as the error between the actual output and the output of the $\rho$th branch of an ideal polyphase interpolator as shown in Fig. 31(b), i.e.,

$$\Delta u_\rho(n) = u_\rho(n) - \tilde{u}_\rho(n).$$

(107)

Because of this orthogonality property we can *separately* and *independently* design each of the polyphase filters for minimum error and arrive at an overall interpolator design which minimizes the error $\|\Delta y(m)\|$. Thus a large (multirate) filter design problem can be broken down into $L$ smaller (time-invariant) filter design problems.

In the case of the mean-square-error criterion it can be seen that

$$E^2 = \|\Delta y(m)\|^2 = \frac{1}{L} \sum_{\rho=0}^{L-1} E_\rho^2$$

(108a)

where

$$E_\rho^2 = \|\Delta u_\rho(n)\|^2.$$

(108b)

To minimize $E^2$ we then need to design $L$ independent polyphase filters $p_\rho(n), \rho = 0, 1, 2, \cdots, L - 1$, which independently minimize the respective mean-square errors $E_\rho^2$.

In order to analytically set up the filter design problem, it can be noted that the ideal polyphase filter response is

$$\tilde{P}_\rho(e^{j\omega}) = e^{j\omega\rho/L}$$

(109)

which then leads to the form

$$E_\rho^2 = \|\Delta u_\rho(n)\|^2$$

$$= \frac{1}{2\pi} \int_{-\pi}^{\pi} |P_\rho(e^{j\omega}) - e^{j\omega\rho/L}|^2 |X(e^{j\omega})|^2 \, d\omega.$$

(110)

Equation (110) reveals that in the minimum mean-square-error design, we are in fact attempting to design a polyphase filter such that the integral of the squared difference between its frequency response $P_\rho(e^{j\omega})$ and a linear (fractional sample)

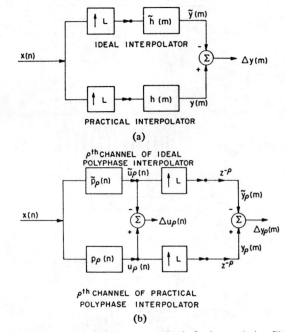

Fig. 31. Framework for defining error criteria for interpolation filters.

phase delay $e^{j\omega\rho/L}$, weighted by the spectrum of the input signal $|X(e^{j\omega})|^2$, is minimized. Note also that the integral from $-\pi$ to $\pi$ in (110) is taken over the frequency range of the input signal of the interpolator, not the output signal.

In practice this error criterion is often modified slightly [34]–[37], [39] by specifying that $X(e^{j\omega})$ is bandlimited to the range $0 \leqslant \omega \leqslant \alpha\pi$ where $0 < \alpha < 1$, i.e.,

$$|X(e^{j\omega})| = 0, \quad \text{for} \quad |\omega| \geqslant \alpha\pi$$

(111)

Then (110) can be expressed as

$$E_{\rho,\alpha}^2 = \frac{1}{2\pi} \int_{-\alpha\pi}^{\alpha\pi} |P_\rho(e^{j\omega}) - e^{j\omega\rho/L}|^2 |X(e^{j\omega})|^2 \, d\omega$$

(112)

where the subscript $\alpha$ will be used to distinguish this norm from the one in (110). Alternatively we can consider the above modification as a means of specifying that we want the design of $P_\rho(e^{j\omega})$ to be minimized only over the frequency range $0 \leqslant \omega \leqslant \alpha\pi$, and that the range $\alpha\pi \leqslant \omega \leqslant \pi$ is allowed to be a transition region. Then $\alpha$ can be used as a parameter in the filter design procedure.

The solution to the minimization problem of (112) involves expressing the norm $E_{\rho,\alpha}^2$ directly in terms of the filter coefficients $p_\rho(n)$. Then, since the problem is formulated in a classical mean-square sense, it can be seen that $E_{\rho,\alpha}^2$ is a quadratic function of the coefficients $p_\rho(n)$ and thus it has a single, unique minimum for some optimum choice of coefficients. At this minimum point, the derivative of $E_{\rho,\alpha}^2$ with respect to all of the coefficients $p_\rho(n)$ is zero. Thus the second step in the solution is to take the derivative of $E_{\rho,\alpha}^2$ with respect to the coefficients $p_\rho(n)$ and set it equal to zero. This leads to a set of linear equations in terms of the coefficients $p_\rho(n)$ and the solution to this set of equations gives the optimum choice of coefficients which minimize $E_{\rho,\alpha}^2$. This minimization problem is solved for each value of $\rho, \rho = 0, 1, 2, \cdots, L - 1$ and each solution provides the optimum solution for one of the polyphase filters. Finally, these optimum polyphase filters can be combined, as in (73)–(75), to obtain the optimum prototype filter $h(m)$ which minimizes the overall norm. The details for this approach can be found in [35]. Also, this same reference, or [40] by Oetken, Parks, and Schuessler, contains a computer program which designs minimum mean-square interpolators according to the above techniques and it greatly simplifies the task of designing these filters.

The minimum mean-square-error interpolators designed using the procedure described have a number of interesting properties [36].

1) The resulting filters have the same symmetry properties as the ideal filters (92) and (95)).

2) The minimum error min $E_{\rho,\alpha}^2$ for the polyphase filters also satisfies the symmetry condition

$$\min E_{\rho,\alpha}^2 = \min E_{L-\rho,\alpha}^2. \qquad (113a)$$

This error increases monotonically as $\rho$ increases (starting with $E_{0,\alpha}^2 = 0$) until $\rho = L/2$ at which point it decreases monotonically according to (113a). Thus the greatest error occurs in interpolating sample values which are halfway between two given samples. This normalized error is closely approximated by the sine-squared function [36], i.e.,

$$\frac{\min E_{\rho,\alpha}^2}{\min E_{L/2,\alpha}^2} \approx \sin^2\left(\frac{\rho\pi}{L}\right) \qquad (113b)$$

3) If an interpolator is designed for a given signal with a large value of $L$, all interpolators whose lengths are fractions of $L$ are obtained by simply sampling the original filter, i.e. if we design an interpolator for $L = 100$, then for the same parameters $\alpha$ and $R$ we can derive from this filter the optimum mean-square error interpolators for $L = 50, 25, 20, 10, 5$, and 2 by taking appropriate samples (or appropriate polyphase filters).

Fig. 32 shows an example of the impulse response and frequency response for a minimum mean-square-error interpolation filter with parameter values $\alpha = 0.5$, $N = 49$, $R = (N-1)/2L = 3$, $L = 8$, and assuming that $|X(e^{j\omega})| = 1$.

*I. Design of FIR Interpolators with Minimax Error in the Frequency Domain*

In the previous section we considered the design of FIR interpolators based on minimizing the mean-square error norm. In this section we will consider another class of designs of interpolation filters in which the maximum of the error $|\Delta Y(e^{j\omega'})|$ over the frequency range of interest is minimized. This type of design gives a greater degree of control over the errors at specific frequencies [36]. In minimax designs of this type, the error $\Delta Y(e^{j\omega'})$ oscillates (ripples) in the frequency domain between positive and negative values of this maximum error. If the number of taps in the overall filter is

$$N = 2RL + 1 \qquad (114)$$

then each polyphase filter has $2R$ taps and therefore $2R$ degrees of freedom. These $2R$ degrees of freedom allow the error $|\Delta Y(e^{j\omega'})|$ to be exactly zero at $2R$ frequencies or $R$ conjugate pairs of frequencies. Thus $\Delta Y(e^{j\omega'})$ will have $R+1$ extremal values in the range of positive frequencies $0 \leqslant \omega' \leqslant \alpha\pi/L$. The frequencies at which $|\Delta Y(e^{j\omega'})|$ is zero will be

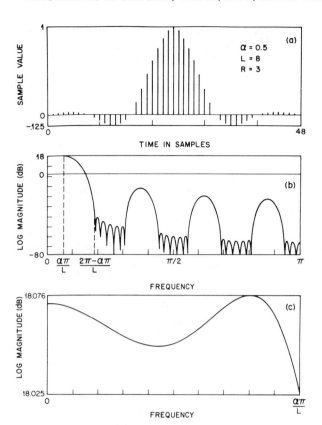

Fig. 32. The impulse and frequency responses of a minimum mean-squared error interpolation filter with $\alpha = 0.5$, $R = 3$, and $L = 8$.

denoted as

$$0 \leqslant \omega_\lambda = \omega_\lambda' L \leqslant \alpha\pi, \quad \lambda = 1, 2, \cdots, R \qquad (115)$$

where the reader may recall that $\omega$ refers to frequencies specified with reference to the low sampling rate and $\omega'$ refers to frequencies specified with reference to the high sampling rate. From (110) it then follows that

$$P_\rho(e^{j\omega_\lambda}) = e^{j\omega_\lambda\rho/L}, \quad \text{for} \quad \lambda = 1, 2, \cdots, R,$$
$$\text{and for} \quad \rho = 0, 1, 2, \cdots, L-1. \qquad (116)$$

Thus at these frequencies the polyphase filter responses $P_\rho(e^{j\omega_\lambda})$ are all equal to the ideal filter responses $\tilde{P}_\rho(e^{j\omega_\lambda})$, $\lambda = 1, 2, \cdots, R$.

The above problem has now been converted to a problem of finding the set of $R$ frequencies $\omega_\lambda$, $\lambda = 1, 2, \cdots, R$. The assumption that these frequencies be distinct is not essential to the validity of the result. Indeed if we assume that $\omega_\lambda = 0$ for all $\lambda$ and $X(e^{j\omega}) = $ constant for $|\omega| \leqslant \alpha\pi$, then all of the zeros of $\|\Delta Y(e^{j\omega})\|^2$ occur at $\omega' = 0$. In this case we get a maximally flat behavior of $\|\Delta Y(e^{j\omega})\|^2$ such that its first $4R$ derivatives with respect to $\omega'$ are zero. The solution then leads to the well-known class of Lagrange interpolators [36].

For equiripple designs a procedure for finding the frequencies $\omega_\lambda$ has been proposed by Oetken [36]. First he showed that the error $\|\Delta Y(e^{j\omega'})\|^2$ could be expressed as the sum of the errors due to each polyphase filter. He then showed that the individual errors $\|\Delta Y_\rho(e^{j\omega'})\|^2$ are almost exactly proportional to each other over the whole frequency range. This proportionality has the form (similar to that of (113b))

$$\|\Delta Y_\rho(e^{j\omega'})\| \approx \|\Delta Y_{L/2}(e^{j\omega'})\| \sin(\rho\pi/L) \qquad (117)$$

where $\|\Delta Y_{L/2}(e^{j\omega'})\|^2$ denotes the error for the polyphase filter $\rho = L/2$, i.e. it is the filter which interpolates samples exactly half way between the input samples. This form is a stronger condition than that of (113b) in that it applies to individual frequencies as opposed to the mean-square error integrated over the entire frequency range. The deviations to the approximation have been found to be smaller than 1 percent [36].

Because of the above condition, the design problem may be converted to a simpler problem of designing one of the polyphase filters such that it has the desired minimax error. The zeros of this filter can be calculated to obtain the frequencies $\omega_\lambda$, $\lambda = 1, 2, \cdots, R$ and they can be applied to obtain the minimax solutions to the other polyphase filters. It is convenient to choose the polyphase filter $\rho = L/2$ which interpolates sample values half way between input samples. Also, as in the case of mean-square-error designs, the design of the polyphase filters is independent of $L$ (assuming that there are always $2R$ taps for each polyphase filter). Therefore, it is convenient to choose $L = 2$ so that the above design requirements are those of a minimax half-band filter design. This design can be obtained using the techniques described in Section IV-G on half-band filters, with the appropriate weighting factor $|X(e^{j\omega})|$.

Fig. 33 shows an example of a minimax intepolator design for $\alpha = 0.5$, $R = 3$, $L = 8$, and assuming that $|X(e^{j\omega})| = 1$. Fig. 33(a) shows the individual errors $\|\Delta Y_\rho(e^{j\omega'})\|$ (note that $\|\Delta Y_\rho(e^{j\omega'})\| = \|\Delta Y_{L-\rho}(e^{j\omega'})\|$) and Fig. 33(b) shows the total error $\|\Delta Y(e^{j\omega'})\|$. Figs. 33(c)–(e) shows the impulse response and frequency response of the final prototype filter $H(e^{j\omega'})$. Note that although the error signal $\|\Delta Y(e^{j\omega'})\|$ exhibits an equal-ripple behavior (as specified by the design criterion) $H(e^{j\omega'})$ does not exhibit equal-ripple behavior.

### J. Other Designs

In addition to the above design procedures a number of other methods have been proposed in the literature. In this section, we briefly refer to these methods.

Parks and Kolba [37] proposed a design technique for interpolators that minimizes the error $|\Delta y(m)|$ in a minimax sense in the time domain. Under the assumption that $|X(e^{j\omega})|$ is bandlimited and spectrally flat within this band this procedure leads to the same designs as the minimum mean-square designs of Section IV-H. Matrinex and Parks also investigated the use of IIR filters for interpolators. The interested reader is referred to [41] and [42] for details.

Other classical interpolation techniques are those based on linear and Lagrange methods [4], [16]. Linear interpolation has obvious limitations since it can be interpreted as only a two point filter. Lagrange interpolators are of historical importance and they can be shown to be related to the case of maximally flat designs at $\omega = 0$ as mentioned in Section IV-I.

This concludes our discussion of FIR design techniques for the filters for decimators and interpolators. As we have shown, there are many techniques for designing such filters, and they are all based on slightly different criteria. Some techniques are convenient because of their simplicity, some because they optimize a specific error criterion, and others are of interest strictly from a historical point of view. As in most filter design problems, it is up to the user to decide which of a set of alternative solutions to the filter design problem is most applicable to the problem at hand.

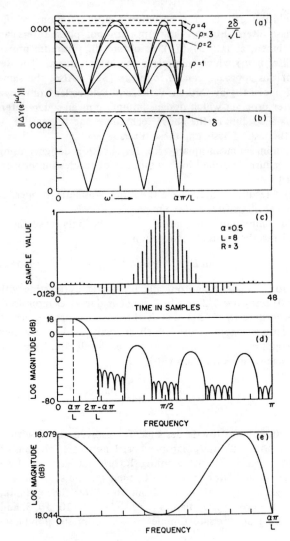

Fig. 33. The impulse and frequency responses of a minimax design interpolation filter with $\alpha = 0.5$, $R = 3$, and $L = 8$.

## V. Multistage Implementations of Sampling Rate Conversion

The concept of using a series of stages to implement a sampling rate conversion system can be extended to the case of simple interpolators and decimators [13], [31]–[33], [43]–[45] as shown in Figs. 34 and 35. Consider first a system for interpolating a signal by a factor of $L$ as shown in Fig. 34(a). We denote the original sampling frequency of the input signal $x(n)$ as $F_0$ and the interpolated signal $y(m)$ has a sampling rate of $LF_0$. If the interpolation rate $L$ can be factored into the product

$$L = \prod_{i=1}^{I} L_i \qquad (118)$$

where each $L_i$ is an integer, then we can express this network in the form shown in Fig. 34(b). This structure, by itself, does not provide any inherent advantage over the structure of Fig. 34(a). However, if we modify the structure by introducing a lowpass filter between *each* of the sample rate increasing boxes, we produce the structure of Fig. 34(c). This structure has the property that the sampling rate increase occurs in a series of $I$

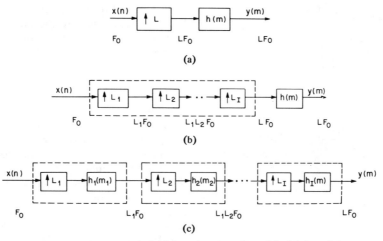

Fig. 34. Single stage and multistage structures for a 1 to $L$ interpolator.

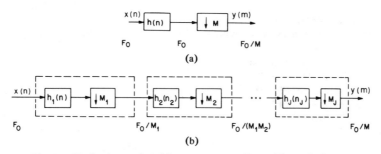

Fig. 35. Single stage and multistage structures for an $M$ to 1 decimator.

stages, where each stage (shown within dashed boxes) is an *independent* interpolation stage.

Similarly, for an $M$ to 1 decimator, if the overall decimation rate $M$ can be factored into the product

$$M = \prod_{j=1}^{J} M_j \qquad (119)$$

Then the general single-stage decimator structure of Fig. 35(a) can be converted into the multistage structure of Fig. 35(b). Again each of the stages within the structure of Fig. 35(b) is an *independent* decimation stage.

Perhaps the most obious question that arises from the above discussion is why consider such multistage structures. At first glance it would appear as if we are greatly increasing the over-all computation (since we have inserted filters between each stage) of the structure. This, however, is precisely the opposite of what occurs in practice. The reasons for considering multistage structures, of the types shown in Figs. 34(c) and 35(b), are:

1) significantly reduced computation to implement the system;
2) reduced storage in the system;
3) simplified filter design problem;
4) reduced finite word length effects, i.e., roundoff noise, coefficient sensitivity, in the implementations of the digital filters.

These structures however are not without some drawbacks.

These include:

1) increased control structure required to implement a multistage process;
2) difficulty in choosing the appropriate values of $I$ (or $J$) of (118) and the best factors $L_i$ (or $M_j$).

It is the purpose of this section to show why and how a multistage implementation of a sampling rate conversion system can be (and generally is) more efficient than the standard single stage structure for the following cases:

1) $L \gg 1$        ($M = 1$)        Case 1

2) $M \gg 1$        ($L = 1$)        Case 2

3) $L/M \approx 1$    but    $L \gg 1$, $M \gg 1$    Case 3

Cases 1 and 2 are high-order interpolation and decimation systems, and Case 3 is when a slight change in sampling rate is required (e.g., $L/M = 80/69$).

### A. Computational Efficiency of a Two-Stage Structure—A Design Example

Since the motivation for considering multistage implementations of sampling rate conversion systems is the potential reduction in computation, it is worthwhile presenting a simple deisign example which illustrates the manner in which the computational efficiency is achieved.

The design example is one in which a signal $x(n)$ with a sampling rate of 10 000 Hz, is to be decimated by a factor of

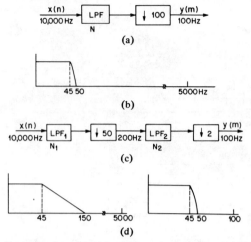

Fig. 36. Example of a one-stage and two-stage network for decimation by a factor of 100.

$M = 100$ to give the signal $y(m)$ at a 100-Hz rate. Fig. 36(a) shows the standard, single stage, decimation network which implements the desired process. It is assumed that the passband of the signal is from 0 to 45 Hz, and that the band from 45 to 50 Hz is a transition band. Hence the specifications of the required low-pass filter are as shown in Fig. 36(b). We assume, for simplicity, that the design formula (97b)

$$N \approx \frac{D(\delta_p, \delta_s)}{(\Delta F/F)} \qquad (120)$$

can be used to give the order $N$ of a symmetric FIR filter with maximum passband ripple $\delta_p$, maximum stopband ripple $\delta_s$, transition width $\Delta F$ and sampling frequency $F$. For the low-pass filter of Fig. 36(b) we have

$$\Delta F = 50 - 45 = 5 \text{ Hz}$$
$$F = 10,000 \text{ Hz}$$
$$\delta_p = 0.01$$
$$\delta_s = 0.001$$
$$D(\delta_p, \delta_s) = 2.54$$

giving, from (120), $N \approx 5080$ TAPS. The overall computation in multiplications per second necessary to implement this system is

$$R = \frac{NF}{2M} = \frac{(5080)10\,000}{2(100)} = 250\,000 \text{ multiplications/sample}$$

i.e., a total of 250 000 multiplications per sample at the 10 000-Hz rate is required to implement the system of Fig. 36(a) (assuming the use of symmetry of $h(n)$).

Consider now the 2-stage implementation shown in Fig. 36(c). The first stage decimates the signal by a factor of 50,[6] and the second stage decimates the (already decimated) signal by a factor of 2, giving a total decimation factor of 100. The resulting filter specifications are illustrated in Fig. 36(d). For the first stage the passband is from 0 to 45 Hz, but the transition band extends from 45 to 150 Hz. Since the sampling rate at the output of the first stage is 200 Hz the residual signal energy

[6] We will explain later in this section how the individual decimation factors of the stages are obtained.

from 100 to 150 Hz gets aliased back into the range 50 to 100 Hz after decimation by the factor of 50. This aliased signal then gets removed in the second stage. For the second stage the passband extends from 0 to 45 Hz and the transition band extends from 45 to 50 Hz with a sampling rate of 200 Hz. One other change in the filter specifications occurs because we are using a two-stage filtering operation. The passband ripple specification of the two-stage structure is reduced to $\delta_p/2$ (since each stage can theoretically add passband ripple to each preceding stage). The stopband ripple specification does not change since the cascade of two low-pass filters only reduces the stopband ripples. Hence the $D(\delta_p, \delta_s)$ function in the filter design equation becomes $D(\delta_p/2, \delta_s)$ for the filters in the two-stage implementation. Since $D(\delta_p, \delta_s)$ is relatively insensitive to factors of two, only slight changes occur (from 2.54 to 2.76) due to this factor. For the specific example of Fig. 36(c) we get (for the first stage)

$$N_1 = \frac{2.76}{((150 - 45)/10\,000)} = 263 \text{ TAPS}$$

$$R_1 = \frac{N_1 F}{2(M_1)} = \frac{263(10\,000)}{(2)(50)}$$

$$= 26\,300 \text{ multiplications per second.}$$

For the second stage we get

$$N_2 = \frac{2.76}{(5/200)} = 110.4 \text{ TAPS}$$

$$R_2 = \frac{110.4(200)}{(2)(2)} = 5500 \text{ multiplications per second.}$$

The total computation for the two-stage implementation is

$$R_1 + R_2 = 26\,300 + 5500 = 31\,800 \text{ multiplications per second.}$$

Thus a reduction in computation of almost 8 to 1 is achieved in the two-stage decimator over a single-stage decimation for this design example.

It is easy to see where the reduction in computation comes from for the multistage decimator structure by examining (120). We see that the required filter orders are directly proportional to $D(\delta_p, \delta_s)$ and $F$, and inversely proportional to $\Delta F$, the filter transition width. For the early stages of a multistage decimator, although the sampling rates are large, equivalently the transition widths are very large; thereby leading to relatively small values of filter length $N$. For the last stages of a multistage decimator, the transition width becomes small but so does the sampling rate and the combination again leads to relatively small values of required filter lengths. We see from the above analysis that computation is kept low in each stage of the overall multistage structure.

The simple example presented above is by no means a complete picture of the capabilities and sophistication that can be found in multistage structures for sampling rate conversion. It is merely intended to show why such structures are of fundamental importance for many practical systems in which sampling rate conversion is required. In the next section we set up a formal structure for dealing with multistage sampling rate conversion networks, and show how it can be used in a variety of implementations.

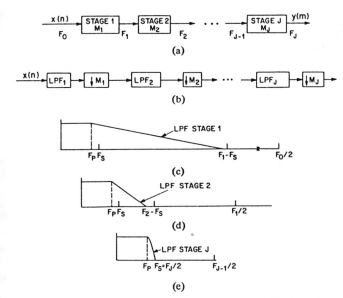

(a)

(b)

(c)

(d)

(e)

Fig. 37. Signal processing operations and filter specifications for a J-stage decimator.

*B. Parameter Specifications for Multistage Implementations*

Consider the J-stage decimator of Fig. 35(b) where the total decimation rate of the system is $M$. The sampling rate (frequency) at the output of the jth stage ($j = 1, 2, \cdots, J$) is

$$F_j = \frac{F_{j-1}}{M_j}, \qquad j = 1, 2, \cdots, J \qquad (121)$$

with initial input sampling frequency $F_0$, and final output sampling frequency $F_J$, where

$$F_J = \frac{F_0}{\displaystyle\prod_{j=1}^{J} M_j} = \frac{F_0}{M}. \qquad (122)$$

Any of the structures discussed in Section III can be used for each stage of the network. Using the ideas developed above, we define the frequency range of the output signal $y(m)$ as

$$0 \leqslant f \leqslant F_p \qquad \text{Passband} \qquad (123a)$$

$$F_p \leqslant f \leqslant F_s = \frac{F_J}{2} \qquad \text{Transition Band.} \qquad (123b)$$

In each stage of the processing, the baseband from 0 to $F_s$ must be protected from aliasing. Any other frequency band can be aliased in an early stage since subsequent processing will remove any signal components from the band as seen in Fig. 37. For the first stage the passband is defined from

$$0 \leqslant f \leqslant F_p \qquad \text{Stage 1 Passband} \qquad (124a)$$

but the transition band is from

$$F_p \leqslant f \leqslant F_1 - F_s \qquad \text{Stage 1 Transition Band.} \qquad (124b)$$

The transition band will alias back upon itself (after the decimation by $M_1$) only from $f = F_s$ up to $f = F_1/2$; hence the baseband of (123) is protected against aliasing. The stopband of the Stage 1 low-pass filter is from

$$F_1 - F_s \leqslant f \leqslant F_0/2 \qquad \text{Stage 1 Stopband.} \qquad (124c)$$

For the second stage of the system, the specifications on the low-pass filter are defined for

$$0 \leqslant f \leqslant F_p \qquad \text{Stage 2 Passband} \qquad (125a)$$

$$F_2 - F_s \leqslant f \leqslant F_1/2 \qquad \text{Stage 2 Stopband} \qquad (125b)$$

as shown in Fig. 37(d). Again the band from $0 \leqslant f \leqslant F_s$ is protected from aliasing in the decimation stage.

For the rth stage of the system, the low-pass filter band specifications are

$$0 \leqslant f \leqslant F_p \qquad \text{Stage } r \text{ Passband} \qquad (126a)$$

$$F_r - F_s \leqslant f \leqslant F_{r-1}/2 \qquad \text{Stage } r \text{ Stopband.} \qquad (126b)$$

Alternatively if it is permissible to allow aliasing into the transition region $F_p$ to $F_s$, $F_s$ in (123)–(126) can be replaced by $F_p$.

It is readily seen from (122)–(126) and Fig. 37 that, for the last stage, the transition band of the low-pass filter is the same as the transition band of the one-stage implementation filter. However, the sampling rate of the system is substantially reduced in most cases.

Up to now we have been concerned solely with the regions of definition of the individual low-pass filter frequency bands. Another consideration in the design equations is the magnitude specifications on the filter response in each of the frequency bands. If it is desired that the overall passband response for the cascade of $J$ stages be maintained within $1 \pm \delta_p$, it is necessary to require more severe constraints on the passband ripple of the individual filters in the cascade. A convenient choice which will satisfy this requirement is to specify the passband ripple constraints for each stage $j$ to be within $1 \pm \delta_{pj}$ where $\delta_{pj} = \delta_p/J$. In the stopband the ripple constraint for the composite filter must be $\delta_s$ and this constraint must be imposed on each of the individual low-pass filters as well in order to suppress the effects of aliasing.

Fig. 38 illustrates the signal processing operations and low-pass filter specifications for an I-stage implementation of an interpolator with an overall change in sampling rates of 1 to $L$. One small change in notation is used in defining the I-stage interpolator—namely the stages are numbered backwards from $I$ to 1. The reason we do this is to clearly show that the I-stage interpolator with interpolation factors $L_i$ is a dual of the I-stage decimator with decimation factors $L_i$. This may be trivially seen by taking the transpose of the network of Fig. 38(a). The result is the network of Fig. 37(a) (with $M_i = L_i$ and $J = I$). Hence, in order to understand the behavior and properties of multistage structures for interpolators, we need only study the multistage decimator.

Given the network structure of Fig. 37 for the J-stage decimator, we will be interested in the specification of the following:

1) the number of stages $J$ to realize an overall decimation factor of $M$;
2) the choice of decimation factors $M_j, j = 1, 2, \cdots, J$, that are appropriate for the chosen implementation;
3) the types of digital filters used in each stage of the structure—e.g., FIR versus IIR designs, equiripple versus half-band versus specialized designs;
4) the structure used to implement the filter chosen for each stage;
5) the required filter order (impluse response duration) required in each stage of the structure;

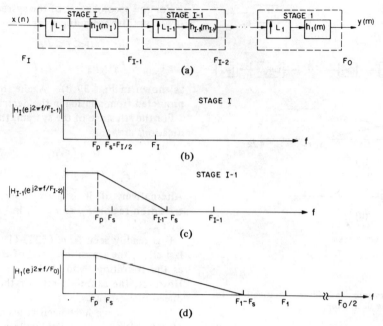

Fig. 38. Signal processing operations and filter specifications for an
*I*-stage interpolator.

6) the resulting amount of computation and storage required for each stage, and for the overall structure.

As in most signal processing design problems, there are a number of factors that influence each of the above choices, and it is not a simple matter to select any one choice over all others.

Three distinct approaches to the design of multistage structures have been proposed by various authors. In one approach, suggested by Bellanger *et al.* [32], [33] and Rorabacher [31], a choice of factors of $M_j = 2$ for all stages is suggested in order to take advantage of the properties of half-band filters. Shively [44] and Crochiere and Rabiner [13], [49] have suggested an approach in which general equiripple filters are used and the choice of the factors $M_j$ are chosen through an optimization procedure which minimizes the overall computation rate (or storage) and the number of stages. Finally, Goodman and Carey [45] have suggested a family of specific filter designs which can be applied to early stages of multistage decimators or final stages of multistage interpolators.

In practice any of the above approaches or any combination of them can be applied to obtain a multistage implementation. The tradeoffs are highly applications dependent. Furthermore any of the structures discussed in Section III can be be used to implement individual stages. In the following discussion we will briefly discuss the above approaches.

*C. Half-Band Designs*

The half-band filter structure is based on the symmetrical half-band FIR filter discussed in Section IV-G. Such filters naturally have the property that approximately half of the filter coefficients are exactly zero. Hence the number of multiplications in implementing such filters is half of that needed for a linear phase design, and a quarter of that needed for an arbitrary phase FIR filter. The half-band filter is appropriate only for sampling rate changes of 2 to 1. Hence the half-band multistage structure consists of a cascade of *J*-stages with 2 to 1 reductions in the sampling rate, and, if necessary, a final

stage with a reduction of $M_\epsilon/L_\epsilon$ in sampling rate [32], [33] where

$$M_\epsilon/L_\epsilon = M/2^J \qquad (127)$$

This final stage can be realized by any of the structures for the general $L/M$ sampling rate conversion discussed in Section III.

Several constraints must be observed with the half-band filters, as discussed in Section IV-G. First, the filter tolerances in the passband and stopband must be identical as seen from (102). Therefore, the smallest of the two required tolerances $\delta_p$ or $\delta_s$, for the passband or stopband, respectively, must be used in the design for both bands. A second constraint is that the filter response for the half-band filter is symmetrical about one quarter of the sampling rate, as seen by (104) and the attenuation at this frequency is only 6 dB. Therefore, if the last stage of a decimator (or the first stage of an interpolator) is a 2 to 1 stage, a filter other than a half-band design may still be needed if this attenuation is not sufficient. We now consider a simple design to illustrate this method.

*Design Example 1*—Consider the design of a six stage structure to decimate a signal by a factor of $M = 2^6 = 64$, with filter specifications

$$\delta_p = 0.01, \quad \delta_s = 0.001, \quad F_p = 0.45 \text{ Hz}, \quad F_s = 0.5 \text{ Hz}$$

The input sampling frequency is $F_0 = 64$ Hz, and the output sampling frequency is $F_6 = F_0/64 = 1.0$ Hz. For a single-stage implementation we get

$$N = \frac{D_\infty(0.01, 0.001)}{(0.05)/64} = (2.54)(20)(64) = 3251 \text{ TAPS}$$

$$R = \frac{NF_0}{2M} = \frac{3251(64)}{2(64)} = 1625 \text{ multiplications per second}$$

where the factor of two in the above expression for $R$ is due to the fact that the filter is symmetrical.

For a six stage half-band implementation we require:

$$\delta = \min\left(\frac{\delta_p}{6}, \delta_s\right) = 0.001$$

therefore,

$$D_\infty(0.001, 0.001) = 3.25.$$

In the implementation we will permit aliasing into the transition region $F_p$ to $F_s$ to permit the use of a half-band filter in the final stage. The filter order for the $j$th stage of the design can then be expressed as

$$N_j \cong \frac{D_\infty(0.001, 0.001)}{\Delta F_j / F_{j-1}}$$

where

$$\Delta F_j = F_j - 2F_p.$$

Therefore, the filter orders are approximately (based on the next largest odd order filter from the above estimate)

$$
\begin{aligned}
N_1 &\simeq 7 \quad \text{TAPS} \\
N_2 &\simeq 7 \quad \text{TAPS} \\
N_3 &\simeq 9 \quad \text{TAPS} \\
N_4 &\simeq 9 \quad \text{TAPS} \\
N_5 &\simeq 13 \quad \text{TAPS} \\
N_6 &\simeq 65 \quad \text{TAPS.}
\end{aligned}
$$

The number of multiplications per second necessary to implement each of these filters is

$$R_j \simeq \tfrac{1}{4} N_j F_j$$

where the factor of $\frac{1}{4}$ is due to the fact that approximately one-half of the coefficients in the half-band filters are zero and the impulse response is symmetric. Then the multiplication rates for the stages are

$$R_1 \simeq \frac{(7)(32)}{4} = 56 \quad \text{multiplications per second}$$

$$R_2 \simeq \frac{(7)(16)}{4} = 28 \quad \text{multiplications per second}$$

$$R_3 \simeq 18 \quad \text{multiplications per second}$$

$$R_4 \simeq 9 \quad \text{multiplications per second}$$

$$R_5 \simeq 6.5 \quad \text{multiplications per second}$$

$$R_6 \simeq 16.25 \quad \text{mulitplications per second}$$

and the total rate is

$$R_T = \sum_{j=1}^{6} R_j \simeq 134 \text{ multiplications per second}$$

If we do not allow aliasing into the transition band from $F_p$ to $F_s$, then all of the transition bands of the above filters become smaller by $2(F_s - F_p) = 0.1$ Hz. This does not significantly affect the orders of the filters $N_1$ to $N_5$. The transition band of the last filter $N_6$, however, is effectively reduced by a factor of two and it can no longer be a half-band design. Thus $N_6$ and $R_6$ become

$$N_6 \simeq 130 \text{ TAPS}$$

$$R_6 \simeq \tfrac{1}{2} N_6 F_6 = 65 \text{ multiplications per second.}$$

This increases the total rate to

$$R_T \simeq 183 \text{ multiplications per second}$$

Thus it is seen that, for the above example, the multistage structure is more efficient than the single-stage design by a factor $1625/183 = 8.9$ in multiplications per second. If aliasing is permitted in the band from 0.45 to 0.5 Hz then the filter order for the single-stage design can be reduced by two and the multistage implementation is more efficient by a factor $(1625/2)/134 = 6.1$ in multiplications per second.

### D. Designs Based on a Multistage Optimization Procedure

A second approach to the selection of parameters $M_j$, $j = 1, 2, \cdots, J$ and the number of stages $J$ is based on setting up the problem in terms of an optimization procedure [13], [44], [49]. The order of the filters for each stage can be expressed as

$$N_j = \frac{D_\infty[(\delta_p/J), \delta_s] F_{j-1}}{(F_j - F_p - F_s)}, \qquad j = 1, 2, \cdots, J \quad (128)$$

and the computation rate (multiplications per second) can be expressed as

$$R_j = \tfrac{1}{2} N_j F_j \quad (129)$$

where the factor of $\frac{1}{2}$ is due to the assumption that the symmetry of the filter is used to reduce multiplications. (This optimization approach cannot, however, account for special features of specific filter designs such as half-band filters or those discussed in the next section.)

From (119)–(126), (128), (129) it can be shown [13] that the total multiplication rate $R_T$ for the design can be expressed as the product of three terms

$$R_T = \frac{1}{2} \cdot D_\infty\left(\frac{\delta_p}{J}, \delta_s\right) \cdot S \cdot F_0 \text{ (multiplications per second)}$$

$$(130)$$

where $S$ is a function of $J, \Delta f, M, M_1, M_2, \cdots, M_{J-1}$, i.e.,

$$S = S(J, \Delta f, M, M_1, M_2, \cdots, M_{J-1}) \quad (131)$$

and where

$$\Delta f = \frac{F_s - F_p}{F_s}. \quad (132)$$

This function can be minimized as a function of the parameters $M_j$, $j = 1, 2, \cdots, J$ for each choice of $J$ by minimizing $S(\cdot)$. For $J = 2$ this can be achieved by taking the derivative of $S(\cdot)$ with respect to $M_1$ and setting it to zero. This leads to the ideal choice of $M_1$ and $M_2$ as

$$M_{1_{\text{opt}}} = \frac{2M(1 - \sqrt{M\Delta f/(2 - \Delta f)})}{2 - \Delta f(M+1)} \quad (133a)$$

$$M_{2_{\text{opt}}} = \frac{M}{M_{1_{\text{opt}}}}. \quad (133b)$$

In practice the nearest integer values for $M_1$ and $M_2$ must be used.

For $J > 2$ an analytical approach is not possible. However, it is possible to find the values of $M_j$, $j = 1, 2, \cdots, J$ with the aid of a computer-aided optimization procedure. It was found in [13] that the Hooke and Jeaves [46], [47] procedure worked

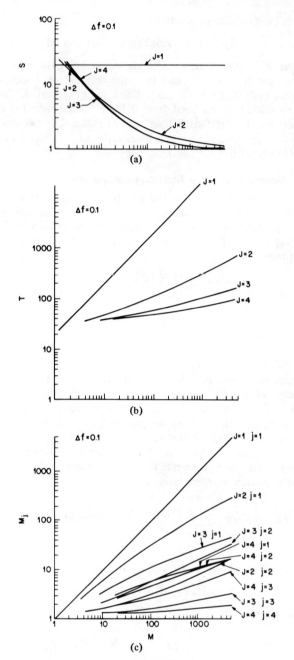

Fig. 39. Minimized values of $S$ and $T$ for ideal values of $M_j$, $j = 1, 2, \cdots$, $J$, and $\Delta f = 0.1$.

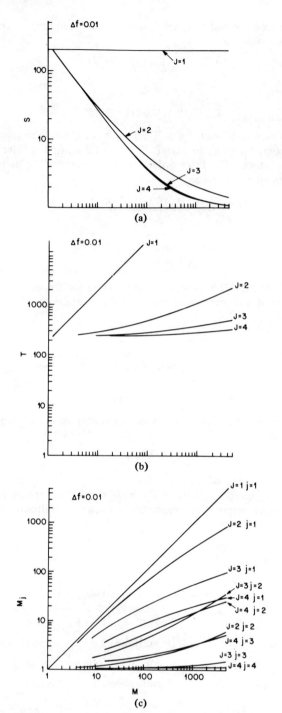

Fig. 40. Minimized values of $S$ and $T$ for ideal values of $M_j$, $j = 1, 2, \cdots$, $J$, and $\Delta f = 0.01$.

well on this problem and did not require the evaluation of derivatives.

A set of design curves was generated based on the above procedure and they can be found in [13]. Alternatively, a similar approach can be used for designs which minimize total storage, i.e.,

$$N_T = \sum_{j=1}^{J} N_j \qquad (134a)$$

$$= D_\infty \left( \frac{\delta_p}{J}, \delta_s \right) \cdot T \qquad (134b)$$

where $T$ is another function of $J$, $\Delta f$, $M$, $M_1$, $M_2$, $\cdots$, $M_{J-1}$,

i.e.,

$$T = T(J, \Delta f, M, M_1, M_2, \cdots, M_{J-1}). \qquad (134c)$$

These design curves can be found in [49]. It turns out that designs based on minimizing $N_T$ also result in designs that are essentially minimized for $R_T$ as well since the minima for $R_T$ are relatively broad whereas minima for $N_T$ are slightly narrower. Therefore, the design curves in [49] are preferred.

Figs. 39 and 40 illustrate two example of these design curves for the cases $\Delta f = 0.1$ and $\Delta f = 0.01$, i.e., a ten percent and a one percent transition band. Figs. 39(a) and 40(a) show plots of minimized values of $S$ in (130) and (131) as a function of $M$ for each value of $J$. Figs. 39(b) and 40(b) show similar plots of minimized values of $T$ in (134a)–(134c). Finally Figs. 39(c) and 40(c) show ideal values of $M_j$, $j = 1, 2, \cdots, J$ as a function of $M$ for each $J$ which result in the minimized values of $S$ and $T$. Several important properties of this design procedure can be seen from these figures. They include the following.

1) For optimized computation, most of the gain in efficiency is achieved in a two stage structure ($J = 2$), with only small gains being achieved with three or four stage designs.

2) For optimized storage, substantial reductions in storage can still be achieved in going from two to three or four stages, although the largest decrease is obtained in going from one to two stages.

3) For optimized storage designs, the actual computation is essentially identical to that of optimized computation designs [49]. Thus a design that is minimized for storage is also minimized for computation. This result is a consequence of the fact that the function $S$ has a broad minimum, whereas $T$ has a somewhat narrower minimum.

4) The gain in efficiency (either computation or storage) in going from 1 to $J$ stages increases dramatically as $\Delta f$ gets smaller, and as $M$ gets larger. For example a computational reduction of about 200 to 1 can be achieved with a three stage structure implementing a 1000 to 1 decimation for a normalized transition width of $\Delta f = 0.01$.

5) The decimation ratios of a $J$-stage optimized design follow the relation

$$M_j > M_{j+1} > \cdots > M_J. \tag{135}$$

6) The required computation (or storage) of an optimized $J$-stage design is relatively insensitive to small changes in the $M_j$ for each stage. Thus nearest integer values of $M_j$ can be chosen in practical designs with little loss in efficiency.

The curves of Figs. 39 and 40 provide a set of design guidelines for the optimized $J$-stage decimator. To illustrate how these curves can be applied we give a design example.

*Design Example 2*—Consider the design of a $J$-stage structure to decimate a signal by a factor of $M = 64$, with filter specifications $\delta_p = 0.01$, $\delta_s = 0.001$, $F_p = 0.45$ Hz, $F_s = 0.5$ Hz (this is identical to *Design Example 1*). The input sampling frequency is $F_0 = 64$ Hz, and the output sampling frequency is $F_{J6} = F_0/64 = 1.0$ Hz.

For a $J = 1$ stage design we get (see Example 1)

$$N = 3251 \quad \text{TAPS}$$
$$R = 1625 \quad \text{multiplications per second.}$$

For a $J = 2$ stage we can determine the ideal choices of $M_1$ and $M_2$ from (133). Alternatively, noting that

$$\Delta f = \frac{(0.5 - 0.45)}{0.5} = 0.1,$$

we may use the appropriate design curves for $\Delta f = 0.1$ in Fig. 39. Using either method we get ideal values $\tilde{M}_1 \cong 23$ and $\tilde{M}_1 \cong 2.7$. The nearest choice of integer values such that $M_1 M_2 = 64$ is either ($M_1 = 32$, $M_2 = 2$) or ($M_1 = 16$, $M_2 = 4$). We will use the second choice in this example. This leads to the design parameters:

$$N_1 = 58 \quad \text{TAPS}$$
$$N_2 = 221 \quad \text{TAPS}$$

$$R_1 = 116 \quad \text{multiplications per second}$$
$$R_2 = 111 \quad \text{multiplications per second}$$
$$R_T = R_1 + R_2 = 227 \quad \text{multiplications per second.}$$

From the curves of $S$ in Fig. 39(a) it is seen that a $J = 3$ stage design can be expected to result in a slightly lower overall computation rate. Thus following the same procedure we get

$$M_1 = 8, \quad M_2 = 4, \quad M_3 = 2$$
$$N_1 = 26 \quad R_1 = 104$$
$$N_2 = 22 \quad R_2 = 22$$
$$N_3 = 115 \quad R_3 = 57$$
$$R_T = R_1 + R_2 + R_3 = 183$$

Thus we see that a three stage design achieves a reduction in computation (multiplications/second) of a factor of 8.9 over a one-stage design or essentially the same as that of the 6-stage halfband approach.

By using the multistopband designs discussed in Section IV-F for early stages in the above two- or three-stage design it is possible to reduce the order of the filters in those stages. For example from Fig. 28 it can be estimated that for the three-stage design, $N_1$ can be reduced by a factor of about 22 percent or $N_1 \cong 20$ and $N_2$ can be reduced by a factor of about 10 percent or $N_2 \cong 20$. This results in an overall computation rate of $R_T = 157$ (multiplications per second).

### E. Multistage Designs Based on a Specific Family of Filter Designs

The third approach [45] to multistage design is to use a specialized family of filter designs shown in Table I for early stages of the design and possibly a general FIR filter or halfband filter designed according to one of the methods in Section IV for the final stage. $F1$, the first filter in Table I is a filter with $M_1$ one's for coefficients and it can be used for arbitrary decimation factors $M_1$. Filters $F2$ through $F9$ are half-band filters for decimation by 2 to 1 with impulse response durations from 3 to 9 samples and short coefficient wordlengths. Filters $F2$, $F3$, and $F5$ have monotone passband frequency responses, and filters $F4$, $F6$–$F9$ have 2 or more ripples in the passband [45].

The way in which specific filters from Table I are selected for each stage of a $J$-stage design is as follows. The filter specifications $F_p$, $F_s$, $\delta_p$, $\delta_s$ are defined as discussed previously. A level ripple factor $D$ is then defined as

$$D = -20 \log(\delta_s) \quad \text{for filters} \quad F1, F2, F3, F5 \tag{136a}$$

or

$$D = -20 \log[\min(\delta_p/J, \delta_s)] \quad \text{for filters} \quad F4, F6\text{–}F9 \tag{136b}$$

and an "oversampling" frequency ratio $O_j$ for each stage $j$ is defined as

$$O_j = \frac{F_{j-1}}{F_J}. \tag{137}$$

The procedure at each stage is to determine the range of $D$ and $O_j$ such that each of the filters of Table I is applicable. This range is plotted in the chart of Fig. 41 which shows boundaries, in the plane, for which each of the filters meets specifications. The way the chart is used is as follows. For stage 1, the grid point ($O_1, D$) is located. If the boundary line of filter $F1$ lies

TABLE I
COEFFICIENTS OF SPECIALIZED FIR FILTERS

| Filter | $N$ | $h(0)$ | $h(1)$ | $h(3)$ | $h(5)$ | $h(7)$ | $h(9)$ |
|---|---|---|---|---|---|---|---|
| $F1^*$ | 3 | 1 | 1 | | | | |
| $F2$ | 3 | 2 | 1 | | | | |
| $F3$ | 7 | 16 | 9 | -1 | | | |
| $F4$ | 7 | 32 | 19 | -3 | | | |
| $F5$ | 11 | 256 | 150 | -25 | 3 | | |
| $F6$ | 11 | 346 | 208 | -44 | 9 | | |
| $F7$ | 11 | 512 | 302 | -53 | 7 | | |
| $F8$ | 15 | 802 | 490 | -116 | 33 | -6 | |
| $F9$ | 19 | 8192 | 5042 | -1277 | 429 | -116 | 18 |

*$F1$ can be of any length with $h(i) = 1$ all $i$

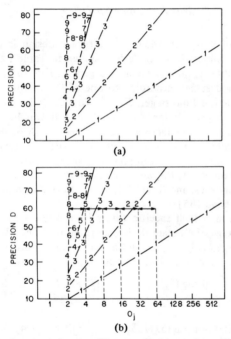

Fig. 41. Design chart showing filters required to meet specifications on $D$ and $O_j$ and a typical example of the use of the chart.

to the left of the grid point, filter $F1$ is used in the first stage, and the decimation rate of this stage $M_1$, is the largest integer such that the grid point $(O_1/M_1, D)$ also lies to the right of the $F1$ boundary line. For the second, and subsequent stages, decimation ratios of 2 are used, and the filter required is the filter whose boundary line lies to the left of the grid point $(O_j/2, D)$ for stage $j$. This procedure can be used until the next to last stage at which a general FIR filter is required to meet overall filter specifications.

We now illustrate this design method with the following example.

*Design Example 3*—To illustrate the above procedure, consider again the $M = 64$ decimator with $\delta_p = 0.01$, $\delta_s = 0.001$, $F_p = 0.45$, $F_s = 0.5$, $F_0 = 64$ Hz, and $F_J = 1$ Hz. We calculate first the ripple factor $D$ as

$$D = -20 \log (0.001) = -60 \text{ dB } (F1, F2, F3, F5)$$

$$D = -20 \log (\min (0.01/6, 0.001)) = -60 \text{ dB } (F5, F5, F6-F9)$$

where we have used 6 as the largest possible number of stages in the design. The oversampling index for the first stage goes from 64 to 32, showing that an $F2$ filter is required, as shown in Fig. 41(b). For stages 2 and 3, filter $F3$ can be used as $O_j$ goes from 32 to 16 (stage 2), and 16 to 8 (stage 3). For stage 4 (as $O_j$ goes from 8 to 4) an $F5$ filter is required, whereas for stage 5 (as $O_j$ goes from 4 to 2) an $F8$ filter is required. For the final stage, a 115 point FIR filter (the same used in Design Example 2, stage 3 of the three-stage design) is required.

The total computation (using symmetry and excluding multiplication by 1, -1, or 2) of the resulting six-stage design is:

| Stage | $M_j$ | Filter | Multiplications per second |
|---|---|---|---|
| $j = 1$ | 2 | $F2$ | $R_1 = 0$ |
| 2 | 2 | $F3$ | $R_2 = 32$ |
| 3 | 2 | $F3$ | $R_3 = 16$ |
| 4 | 2 | $F5$ | $R_4 = 16$ |
| 5 | 2 | $F8$ | $R_5 = 10$ |
| 6 | 2 | $N_6 = 115$ | $R_6 = 57$ |
| | | | $R_T = 131$ |

From this example it can be seen that some savings in computation (multiplications per second) can be achieved over previous designs discussed. However, this is achieved at the expense of extra manipulations of the data to avoid coefficients of 2, 1, and -1.

### F. Other Considerations in Multistage Designs

In the preceding sections we have considered three approaches to multistage designs of decimators or, equivalently, interpolators by use of transposition. In practice there are many factors besides computation rate in multiplications/sec that determine the tradeoffs in these designs.

One consideration is that of the cost of control of data flow in the structure when many stages are involved. Fig. 42(a) illustrates an example of a block diagram of a three-stage decimation structure and Fig. 42(b) shows the corresponding control sequence that must be used to implement this structure [49]. In practice a block of $M$ samples of $x(n)$ are entered into the main input buffer and, after processing, one sample $y(m)$ is obtained at the output. Three state-variable buffers $S1$, $S2$, and $S3$ are used to hold internal data for the filters. The control structure consists of three loops, one within the other, to maintain the proper timing and flow of data in the structure. The actual operations are self-explanatory. In [50] a program is available to implement this structure.

Many other factors must also be considered in the design of multistage decimators and interpolators. As in most real world situations there is no simple or universal answer as to what approach is best and in practice a combination of the above techniques often yields the most appropriate tradeoffs.

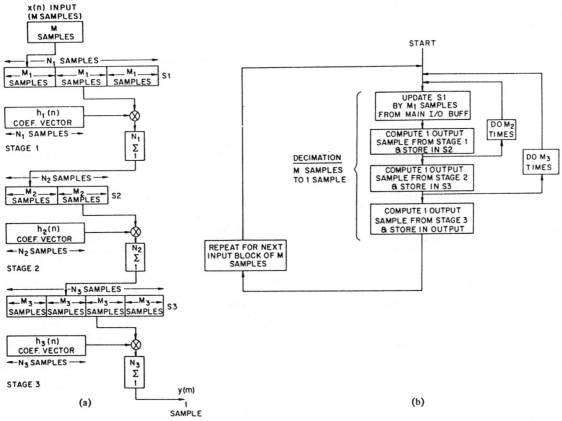

Fig. 42. Computational operations and control structure for a three-stage decimator.

## VI. Discussion

In this paper, we have attempted to present a tutorial coverage of the topic of decimation and interpolation of digital signals. We began in Section II by introducing the basic concepts of sampling rate conversion and its relationship to periodically time-varying linear systems. In Section III we discussed three categories of structures, direct-form, polyphase, and time-varying structures for implementing single stage designs of decimators or interpolators. In Section IV we examined properties of ideal filters for sampling rate conversion systems and then briefly discussed a variety of filter design techniques for designing practical filters for these systems. Finally in Section V we considered multistage structures and showed that under certain circumstances they could be considerably more efficient than single stage designs. Three approaches were then presented for designing such multistage structures.

While we have covered a large scope of material in this paper there were numerous important topics that had to be left out. For example a large body of literature is developing on the use of IIR designs for sampling rate conversion, particularly where linear phase response is not of concern. References [23], [41], [42], [49] illustrate examples of some of this work.

Another area of work that has not been covered in this paper is the use of decimation and interpolation concepts for the efficient implementation or design of other digital signal processing operations. For example, the polyphase concept is very useful in designing phase shifters with fractional sample delays since each polyphase filter component by itself approxi-

mates an allpass system [51]. The cascade of a decimator followed by an interpolator also leads to an efficient approach to the implementation of narrow-band low-pass or bandpass filters [13], [32], [52], [53] and these concepts have not been covered in this paper.

## References

[1] C. E. Shannon, "Communications in the presence of noise," *Proc. IRE*, vol. 37, pp. 10–21, Jan. 1949.
[2] D. A. Linden, "A discussion of sampling theorems," *Proc. IRE*, vol. 47, pp. 1219–1226, July 1959.
[3] H. Nyquist, "Certain topics in telegraph transmission theory," *Trans. AIEE*, vol. 47, pp. 617–664, Feb. 1928.
[4] R. W. Hamming, *Numerical Methods for Scientists and Engineers*. New York: McGraw-Hill, 1962.
[5] S. L. Freeny, R. B. Kieburtz, K. V. Mina, and S. K. Tewksbury, "Design of digital filters for an all digital frequency division multiplex–time division multiplex translator," *IEEE Trans. Circuit Theory*, vol. CT-18, pp. 702–711, Nov. 1971.
[6] S. L. Freeny, J. F. Kaiser, and H. S. McDonald, "Some applications of digital signal processing in telecommunications," in *Applications of Digital Signal Processing*, A. V. Oppenheim, Ed. Englewood Cliffs, NJ: Prentice-Hall, 1978, ch. 1, pp. 1–28.
[7] L. R. Rabiner and R. W. Schafer, *Digital Processing of Speech Signals*. Englewood Cliffs, NJ: Prentice-Hall, 1978.
[8] R. E. Crochiere, S. A. Webber, and J. L. Flanagan, "Digital coding of speech in subbands," *Bell Syst. Tech. J.*, vol. 55, no. 8, pp. 1069–1085, Oct. 1976.
[9] D. J. Goodman and J. L. Flanagan, "Direct digital conversion between linear and adaptive delta modulation formats," *Proc. IEEE Int. Communications Conf.*, (Montreal, P.Q., Canada), June 1971.
[10] R. G. Pridham and R. A. Mucci, "Digital interpolation beam forming for lowpass and bandpass signals," *Proc. IEEE*, vol. 67, no. 6, pp. 904–919, June 1979.
[11] L. R. Rabiner and B. Gold, *Theory and Application of Digital Signal Processing*. Englewood Cliffs, NJ: Prentice-Hall, 1975, ch. 13.

[12] J. H. McClellan and R. J. Purdy, "Applications of digital signal processing to radar," in *Applications of Digital Signal Processing*, A. V. Oppenheim, Ed., Englewood Cliffs, NJ: Prentice-Hall, 1978, pp. 239–330.

[13] R. E. Crochiere and L. R. Rabiner, "Optimum FIR digital filter implementations for decimation, interpolation, and narrow-band filtering," *IEEE Trans. Acoust., Speech, Signal Proc.*, vol. ASSP-23, no. 5, pp. 444–456, Oct. 1975.

[14] R. A. Meyer and C. S. Burrus, "A unified analysis of multirate and periodically time-varying digital filters," *IEEE Trans. Circuits Syst.*, vol. CAS-22, pp. 162–168, Mar. 1975.

[15] B. Liu and P. A. Franaszek, "A class of time-varying digital filters," *IEEE Trans. Circuit Theory*, vol. CT-16, pp. 467–471, Nov. 1969.

[16] R. W. Schafer and L. R. Rabiner, "A digital signal processing approach to interpolation," *Proc. IEEE*, vol. 61, pp. 692–702, June 1973.

[17] A. V. Oppenheim and R. W. Schafer, *Digital Signal Processing*. Englewood Cliffs, NJ: Prentice-Hall, 1975.

[18] R. E. Crochiere and A. V. Oppenheim, "Analysis of linear digital networks," *Proc. IEEE*, vol. 63, pp. 581–595, Apr. 1975.

[19] A. Fettweis, "A general theorem for signal-flow networks, and applications," *Arch. Elek. Übertragung*, vol. 25, pp. 557–561, Dec. 1971. (Also in *Digital Signal Processing*, L. R. Rabiner and C. M. Rader, Eds. New York: IEEE Press, pp. 126–130, 1972.)

[20] L. B. Jackson, "On the interaction of roundoff noise and dynamic range in digital filters," *Bell Syst. Tech. J.*, vol. 49, no. 2, pp. 159–184, Feb. 1970.

[21] T. A. C. M. Claasen and W. F. G. Mecklenbrauker, "On the transposition of linear time-varying discrete-time networks and its application to multirate digital systems," *Philips J. Res.*, vol. 23, pp. 78–102, 1978.

[22] M. G. Bellanger and G. Bonnerot, "Premultiplication scheme for digital FIR filters with application to multirate filtering," *IEEE Trans. Acoust., Speech, Signal Proc.*, vol. ASSP-26, pp. 50–55, Feb. 1978.

[23] M. G. Bellanger, G. Bonnerot, and M. Coudreuse, "Digital filtering by polyphase network: Application to sample rate alteration and filter banks," *IEEE Trans. Acous., Speech, Signal Proc.*, vol. ASSP-24, pp. 109–114, Apr. 1976.

[24] R. A. Meyer and C. S. Burrus, "Design and implementation of multirate digital filters," *IEEE Trans. Acoust., Speech, Signal Proc.*, vol. ASSP-24, pp. 53–58, Feb. 1976.

[25] R. E. Crochiere, "A general program to perform sampling rate conversion of data by rational ratios," in *Programs for Digital Signal Processing*. New York: IEEE Press, 1979, pp. 8.2-1 to 8.2-7.

[26] J. F. Kaiser, "Nonrecursive digital filter design using the $I_0$−Sinh window function," in *Proc. 1974, IEEE Int. Symp. Circuits and Systems*, pp. 20–23, Apr. 1974. (Also reprinted in *Digital Signal Processing, II*. New York: IEEE Press, 1975, pp. 123–126.)

[27] T. W. Parks and J. H. McClellan, "Chebyshev approximation for nonrecursive digital filters with linear phase," *IEEE Trans. Circuit Theory*, vol. CT-19, pp. 189–194, Mar. 1972.

[28] E. Hofstetter, A. V. Oppenheim, and J. Siegel, "A new technique for the design of nonrecursive digital filters," in *Proc. 5th Annu. Princeton Conf. Information Sciences Systems*, pp. 64–72, 1971.

[29] L. R. Rabiner, "The design of finite impulse response digital filters using linear programming techniques," *Bell Syst. Tech. J.*, vol. 51, no. 6, pp. 1177–1198, July–Aug. 1972.

[30] J. H. McClellan, T. W. Parks, and L. R. Rabiner, "A computer program for designing optimum FIR linear phase digital filters," *IEEE Trans. Audio Electroacoust.*, vol. AU-21, pp. 506–526, Dec. 1973. (See also *Programs for Digital Signal Processing*. New York: IEEE Press, 1979, pp. 5.1-1 to 5.1-13.)

[31] D. W. Rorabacher, "Efficient FIR filter design for sample rate reduction or interpolation," in *Proc. 1975 Int. Symp. Circuits and Systems*, Apr. 1975.

[32] M. G. Bellanger, J. L. Daguet and G. P. Lepagnol, "Interpolation, extrapolation, and reduction of computation speed in digital filters," *IEEE Trans. Acoust., Speech, Signal Proc.*, vol. ASSP-22, pp. 231–235, Aug. 1974.

[33] M. G. Bellanger, "Computation rate and storage estimation in multirate digital filtering with halfband filters," *IEEE Trans. Acoust., Speech, Signal Proc.*, vol. ASSP-25, pp. 344–346, Aug. 1977.

[34] G. Oetken and H. W. Schuessler, "On the design of digital filters for interpolation," *Arch. Elek. Übertragung*, vol. 27, pp. 471–476, 1973.

[35] G. Oetken, T. W. Parks and H. W. Schuessler, "New results in the design of digital interpolators," *IEEE Trans. Acoust., Speech, Signal Proc.*, vol. ASSP-23, pp. 301–309, June 1975.

[36] G. Oetken, "New approaches for the design of digital interpolating filters," *IEEE Trans. Acoust., Speech, Signal Proc.*, vol. ASSP-27, pp. 637–643, Dec. 1979.

[37] T. W. Parks and D. P. Kolba, "Interpolation minimizing maximum normalized error for band-limited signals," *IEEE Trans. Acoust., Speech, Signal Proc.*, vol. ASSP-26, pp. 381–384, Aug. 1978.

[38] O. Herrmann, L. R. Rabiner, and D. S. Chan, "Practical design rules for optimum finite impulse response lowpass digital filters," *Bell Syst. Tech. J.*, vol. 52, no. 6, pp. 769–799, July–Aug. 1973.

[39] A. Polydoros and E. N. Protonotarios, "Digital interpolation of stochastic signals," in *Proc. Zurich Seminar on Communications*, pp. 349–367, Mar. 1978.

[40] G. Oetken, T. W. Parks, and H. W. Schuessler, "A computer program for digital interpolator design," in *Programs for Digital Signal Processing*. New York: IEEE Press, 1979, pp. 8.1-1 to 8.1-6.

[41] H. G. Martinez and T. W. Parks, "A class of infinite-duration response digital filters for sampling rate reduction," *IEEE Trans. Acoust., Speech, Signal Proc.*, vol. ASSP-27, pp. 154–162, Apr. 1979.

[42] ——, "Design of recursive digital filters with optimum magnitude attenuation poles on the unit circle," *IEEE Trans. Acoust., Speech, Signal Proc.*, vol. ASSP-26, pp. 150–156, Apr. 1978.

[43] G. A. Nelson, L. L. Pfeifer, and R. C. Wood, "High-speed octave band digital filtering," *IEEE Trans. Audio Electroacoust.*, vol. AU-20, pp. 58–65, Mar. 1972.

[44] R. R. Shively, "On multistage FIR filters with decimation," *IEEE Trans. Acoust., Speech, Signal Proc.*, vol. ASSP-23, pp. 353–357, Aug. 1975.

[45] D. J. Goodman and M. J. Carey, "Nine digital filters for decimation and interpolation," *IEEE Trans. Acoust., Speech, Signal Proc.*, vol. ASSP-25, pp. 121–126, Apr. 1977.

[46] R. Hooke and T. A. Jeaves, "Direct search solution of numerical and statistical problems," *J. Assoc. Comput. Mach.*, vol. 8, no. 4, pp. 212–229, Apr. 1961.

[47] J. L. Kuester and J. H. Mize, *Optimization Techniques with Fortran*. New York: McGraw-Hill, 1973.

[48] L. R. Rabiner, J. F. Kaiser, O. Herrmann, and M. T. Dolan, "Some comparisons between FIR and IIR digital filters," *Bell Syst. Tech. J.*, vol. 53, no. 2, pp. 305–331, Feb. 1974.

[49] R. E. Crochiere and L. R. Rabiner, "Further considerations in the design of deimators and interpolators," *IEEE Trans. Acoust., Speech, Signal Proc.*, vol. ASSP-24, pp. 296–311, Aug. 1976.

[50] ——, "A program for multistage decimation, interpolation, and narrow band filtering," in *Programs for Digital Signal Processing*. New York: IEEE Press, 1979, pp. 8.3-1 to 8.3-14.

[51] R. E. Crochiere, L. R. Rabiner, and R. R. Shively, "A novel implementation of digital phase shifters," *Bell Syst. Tech. J.*, vol. 54, pp. 1497–1502, Oct. 1975.

[52] L. R. Rabiner and R. E. Crochiere, "A novel implementation for narrow-band FIR digital filters," *IEEE Trans. Acoust., Speech, Signal Proc.*, vol. ASSP-23, pp. 457–464, Oct. 1975.

[53] F. Mintzer and B. Liu, "The design of optimal multirate bandpass and bandstop filters," *IEEE Trans. Acoust., Speech, Signal Proc.*, vol. ASSP-26, pp. 534–543, Dec. 1978.

# WAVE DIGITAL DECIMATION FILTERS IN OVERSAMPLED A/D CONVERTERS

E. Dijkstra, L. Cardoletti, O. Nys,
C. Piguet and M. Degrauwe

CSEM, Centre Suisse d'Electronique et de Microtechnique S.A.
Maladiere 71, 2007 Neuchatel, Switzerland

## ABSTRACT

In this paper a digital decimation filter system for oversampled A/D converters is addressed. In order to achieve very low passband ripples for data acquisition converters a cascade of a comb-filter and some bireciprocal equiripple Wave Digital Decimation filters is applied. The complexity of the filters is minimized by taking into account the noise shaping of the modulator.
A design example has been given.

## I Introduction

Since the advent of VLSI technology, the demand for performant A/D converters is constantly growing. Oversampled converters are very well suited for VLSI technologies since a high resolution can be achieved without the need of highly matched components in the analog part of the converter[1]. The large digital part of the converter is a digital low-pass decimation filter whose function is to filter out the noise shaped quantization noise and to compress sample rate again to a more convenient frequency.

The digital decimation filter can be realized in several ways. In previous work[2] we discussed a design methodology for very high order FIR decimation filters in Sigma-Delta converters. Such filter realizations become however unrealistic for very high performance A/D converters. In this paper we will, therefore, focus on a cascade of a simple comb-filter and some wave digital decimation filters. The principle motivation of this cascade is to be able to realize efficiently b-bits data acquistion converters with pass-band ripples smaller than $2^{-(b+1)}$.

Section II will give an overview of the system to be considered. Thereafter, in section III we will discuss the filter transferfunctions which are needed. The complexity of the filters has been minimized by taking into account the noise shaping of the modulator. Finally, in section IV we will give a typical design example for a 16 bits data acquisition Sigma-Delta converter, followed by the conclusions.

## II System overview

As stated in the introduction several filter algorithms could be used to filter out the quantization noise. In this paper we will focus on the data acquisition converter system of fig.(1). In this scheme, a first filtering and decimation is performed by a comb-decimation filter. Their transferfunction can be expressed as:

$$\left( \sum_{i=0}^{N-1} z^{-i} \right)^n = \left( \frac{1}{N} \frac{1 - z^{-N}}{1 - z^{-1}} \right)^n$$

where N=decimation factor of the comb filter
n=order of the comb filter (usually 2 for a first order modulator and 3 for a second order modulator)

As Candy showed[4] such filters are attractive candidates for a first decimation stage since they can be efficiently implemented. One such an implementation using modulo arithmetic has been described in [5].
The comb filter introduces a $\sin(x)/x$ attenuation in the frequency domain. This behaviour is the main reason to sample down only to about 4 times the Nyquist rate with the combfilter.

The $\sin(x)/x$ corrector in fig(1) corrects the $\sin(x)/x$ attenuation of the combfilter. Fig(2) shows the flowgraph of the corrector we used which is basically a lossy Wave Digital Filter (WDF) [6]. The performance of the corrector depends on the position in the cascade chain of filters. The best correction is obtained at the highest sample rate, thus directly after the combfilter. On the other hand, from an implementation point of view it would be better to operate this filter at the lowest possible rate. Therefore, the definite position of the corrector will depend on the application and the desired correction performance.
The coefficients of the corrector are determined by applying a Newton-Raphson iteration algorithm which minimizes the remaining passband ripple after correction. In this way remaining passband ripples can be achieved which are smaller than $10^{-5}$ dB.

Reprinted from *IEEE Proc. ISCAS'88*, pp. 2327–2330, June 1988.

After the sin(x)/x corrector the chain of fig(1) continues with a cascade of bireciprocal equiripple Wave Digital lattice Filters[7]. Between each WDF a decimation by a factor of two is performed. The advantages or this cascade are in this context threefold:

a). The passband behaviour and the coeffients sensibility in the passband are excellent for WDF structures. As explained before, for data acquisition converters this is of prime importance. The "poorer" stopband properties are less important because we do not need very severe filter characteristics due to the first prefiltering by the comb filter.

b). For bireciprocal WDF's complexity is reduced by the possibility of an interchangeability between sample rate compression and filtering function. Concretely, this means that decimation (by a factor of two) can be performed before filtering and therefore each WDF will operate at only half of its sample frequency. Furthermore, bireciprocal WDF strucures do use only $(N-1)/2$ multiplications for a filter order $N[7]$.

c). Bireciprocal WDF's are inherently scaled in an optimum way for sinusoidal signals[7].

In the next section we will discuss the filter characteristics of the WDF's which are needed and the different design trade-off's which have to be made.

## III The required filter complexity for the WDF's

The main system parameters of interest are the imposed passband ripple and the tolerated loss in SNR due to the non-ideal filtering. With these parameters one should try to fix the filter specifications leading to a minimum overall filter complexity. In this section we will review successively the loss in SNR, the WDF filter specifications, the finite wordlength effects and some architectural issues.

### III.1 Loss of SNR due to non-ideal filtering

Following the same reasoning as in [2], the converter will loose due to non-ideal filtering

$$SNR-loss = 10 \log [1+P2/P1]$$

where P1=power of the quantization noise in the baseband
P2=power of the quantization noise after filtering outside the baseband

Assuming that the quantization noise injected into the comparator of the modulator was white, Candy[4] derived formula's for the noise spectral density after the combfilter. Knowing this spectal density we can calculate for different filter parameters the resulting loss in SNR.

Obviously, as discussed in [2,3] the loss in SNR should be compensated by an "overdimensioning" of the oversampling rate pushing more quantization noise outside the baseband. Similar performance trade-off's between the analog loop and the complexity of the decimation filter as proposed for FIR filters [3] are valid for the in this paper presented system.

### III.2 The filter specifications

Starting from the tolerated SNR loss and the passband ripple to be achieved we will determine in this subsection the optimum filter specifications leading to a minimum complexity. We will demonstrate this for a cascade of two bireciprocal WDF's but the results can be easily generalized. Fig.(3). defines the parameters which should be fixed.

Due to the noise shaping most of the noise is rejected into higher frequencies. This means that in order to conserve a reasonnable SNR we should require

$$as1 > as2 \qquad [dB] \qquad (1)$$

For bireciprocal WDF this inherently means:

$$ap1 < ap2 \qquad [dB] \qquad (2)$$

Furthermore in order to satisfy the passband ripple we require:

$$ap1 + ap2 = ap \qquad [dB] \qquad (3)$$

The first WDF should normally avoid the aliasing of the [Fc1/2-Fmax, Fc1/2] band into the baseband, but due to the fact that only bireciprocal filters are considered, the passband edge continues until Fc1/8 and therefore the second filter can not filter out any noise in the [Fmax, Fc1/8] range. Hence, our Fs1 choice is fixed as:

$$Fs1 = 3/4 * Fc1/2 \qquad (4)$$

Due to the bireciprocal WDF option the parameters as1 and as2 are related to respectively ap1 and ap2. This together with (3) and (4) implies that Fs2 and ap2 remain the only real degrees of freedom.

The algorithm of fig.(4) scans the different possible values of Fs2 and ap2. Each time, the orders n1 and n2 of both WDF's and the resulting SNR loss are outputted. For a given maximum loss of SNR we can sort out the satisfying couples (n1,n2). It turns out that for a large majority of applications the couple which minimizes the chip complexity can easily be found by minimizing (n1+n2).

### III.3. Finite wordlength effects

The finite wordlength of the coefficients affect obviously the transferfunctions of the sin(x)/x corrector and the WDF's. Due to the excellent sensitivity properties of

WDF's the number of bits on which the coefficients should be coded can be kept surprisingly low while still satisfying the transferfunction.

The inevitable wordlength truncation after each multiplication introduces extra noise sources, which degrade again the SNR.

### III.4.  Architectural issues

The comb filter in the chain of fig (1) can be efficiently implemented by using the algorithmic decomposition proposed in [5], whereas the sin(x)/x corrector and the WDF's can efficiently be implemented by the microprogram controlled datapath of fig(5). This datapath essentially consists of a programmable barrel shifter cascaded with an addition unit. Compared to e.g.[8] the architecture has been slightly changed (i.e. two slave-accumulators + some multiplexers) in order to be able to handle more efficiently the rather complicated flowgraph of WDF's.

### IV A design example

In this design example we will consider the design of a digital decimation filter for a 16 bits Sigma-Delta A/D converter. Fig.(6) resumes the main characteristics which have been imposed.

In this example we will first sample down the oversampling frequency by the combfilter with a factor 78. This means that the output of the combfilter operates at 32 kHz. Since the imposed pass-band ripple is severe, the sin(x)/x corrector should be placed directly after the combfilter thus operating also at 32 kHz. In this case all coefficients of the sin(x)/x corrector can be coded on 11 bits. The resulting passband ripple after correction is 3.7E-6 dB.

After this correction a cascade of two WDF's has been dimensionned. Fig.(7) gives the best solution for the tolerated SNR loss. Thus, by taking N1=5 and N2=9 (orders should always be impair) the filter characteristics can be realized. Figs. (8) and (9) give a graphical representation[9] of the imposed filter specifications and the transferfunctions for coefficients coded on 8 bits.

### V Conclusions

The proposed cascade of filters allows to design efficiently decimation filters for oversampled Sigma-Delta data acquisition converters. The needed filter specifications are derived from the noise spectral density function at the output of the modulator in view of minimizing the complexity of the filters.

References:

[1] M.W. Hauser, P. Hurst, R. Brodersen: "MOS ADC filter combination that does not require precision analog components". Proc. of ISSCC 85, pp 80-81.

[2] E. Dijkstra, M. Degrauwe, J. Rijmenants, O. Nys: "A design methodology for decimation filters in Sigma Delta A/D converters". ISCAS 87 pp.479-482.

[3] O. Nys, M. Degrauwe, E.Dijkstra. "A CAD tool for oversampled CMOS A/D converters", Proc. of 1987 Symp. on VLSI circuits, May 1987, Karuizawa, pp 113-114

[4] J.C. Candy "Decimation for Sigma-Delta modulation" IEEE Trans. on comm. Vol COM-34, No.1, Jan 1986, pp 72-76.

[5] E. Dijkstra, O. Nys, C. Piguet, M. Degrauwe "On the use of modulo arithmetic combfilters in Sigma Delta converters" to be published at ICASSP 88, New York.

[6] G. Lucioni, "Alternative method to magnitude truncation in WDF", IEEE Trans. on circuits and systems, Vol CAS-33

[7] L. Gazsi, "Explicit formulas for lattice Wave Digital Filters", IEEE Trans. on circuits and systems, Vol CAS-32, No. 1, Jan 1985, pp. 68-88.

[8] A. Rainer, "Adder based digital signal processor architecture for 80 ns cycle time". Proc. ICASSP 84, pp.16-9.1-4, San Diego, March 1984

[9] L. Gazsi, "Reference manual Falcon", Ruhr University, Bochum, West-Germany.

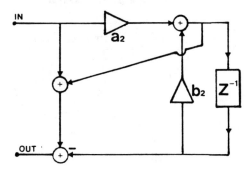

fig.2  Flow-graph of the sinX/X corrector.

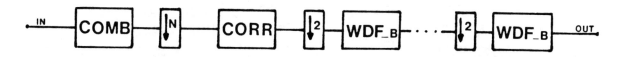

fig.1  Decimation filter structure to be considered.

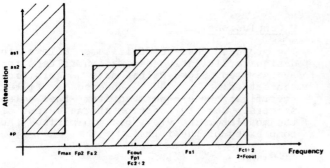

Fmax = maximum signal input frequency.
Fcout = output rate of the converter.
Fci = sample frequency of i-th WDF.
Fsi = stopband-edge of i-th WDF.
Fpi = pass band-edge of i-th WDF.
    This corresponds for bireciprocal WDF's to Fci/4.
    The attenuation at this frequency is 3 dB.
api = passband ripple of i-th WDF in dB.
ap = sum of api.
asi = stopband attenuation of i-th WDF in dB.

fig.3  parameter definition for the filterspecifications.

User's specifications :

    Fmax=?          Fcout=?
    SNR_loss=?      ap=?

Algorithm :

```
Fp2:=Fcout/2;  Fp1:=Fcout;
Fs1:=3/2*Fcout; Fs2:=Fp2+epsilon;
while Fs2<(Fcout-Fmax) do
    begin
        ap2:=ap/2;
        while ap2<ap do
            begin
                ap1:=ap-ap2;
                as1:=fct(ap1);
                as2:=fct(ap2);
                order_n1:=fct(ap1,as1,Fs1);
                order_n1:=fct(ap2,as2,Fs2);
                SNR_loss:=fct(ap1,ap2,as1,as2,Fs1,Fs2);
                ap2:=ap2+epsilon2;
            end;
        Fs2:=Fs2+epsilon1;
    end;
```

fig.4

Algorithm determining the possible couples (n1,n2).

| | | |
|---|---|---|
| nbr_of significant bits | 16 | bits |
| passband ripple | 10E-4 | dB |
| tolerated SNR-loss | 5 | dB |
| max. freq. of input signal | 3.4 | kHz |
| output rate | 8 | kHz |
| order of analog loop | 2 | |
| oversamping frequency calculated with program described in [3] | 2496 | kHz |

fig 6. characteristics of the 16 bits data
        acquisition A/D converter

| | | |
|---|---|---|
| Fs1 | 12 | kHz |
| Fs2 | 4.6 | kHz |
| ap1 | 3.4E-5 | dB |
| ap2 | 6.6E-5 | dB |
| as1 | 51 | dB |
| as2 | 48 | dB |
| n1 | 4.18 | |
| n2 | 7.15 | |
| SNR_loss | 4.6 | dB |

fig. 7. optimum filter specifications leading to a
        minimum complexity

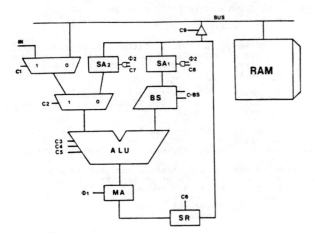

Fig.5. Architecture of the datapath

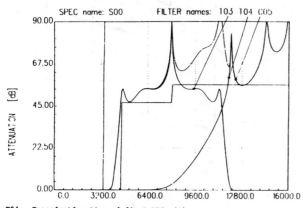

T04 = Transfertfunction of first WDF with
      coefficients coded on 8 bits.

T03 = Transfertfunction of second WDF with
      coefficients coded on 8 bits.

C05 = Cascade of T03 and T04.

    fig.8  Transfertfunction of the WDF's cascade for
           a 16 bits A/D converter.

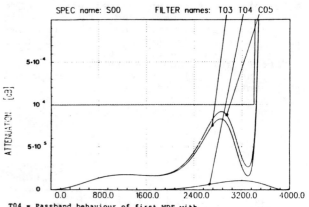

T04 = Passband behaviour of first WDF with
      coefficients coded on 8 bits.

T03 = Passband behaviour of second WDF with
      coefficients coded on 8 bits.

C05 = Cascade of T03 and T04.

    fig.9  Passband behaviour of the WDF's cascade for
           a 16 bits A/D converter.

452

# A DESIGN METHODOLOGY FOR DECIMATION FILTERS IN SIGMA DELTA A/D CONVERTERS

E. Dijkstra, M. Degrauwe, J. Rijmenants, O. Nys

CENTRE SUISSE D'ELECTRONIQUE ET DE MICROTECHNIQUE S.A.
Maladière 71, 2007 Neuchâtel, Switzerland

## ABSTRACT

A design methodology is presented for FIR decimation filters in sigma delta A/D converters. The realized filter transfer function takes into account the noise shaping in order to minimize filter complexity. A simulated annealing optimization is performed to minimize the wordlength of the filter coefficients. This reduces typically the wordlength with two bits compared to classical approaches.

The methodology has been integrated in a CAD-tool for the design of high performance analog circuits. This permits to make rapidly trade-off's between the performance requirements of the analog loop and the complexity of the decimation filter. A design example has been given.

## I. INTRODUCTION

Decimation is the process of sample rate reduction in digital filters. Usually this is done by FIR filters because of the interchangeability between sample rate compression and filtering functions [1] A typical and popular application is that of a FIR decimation filter for oversampled A/D converters. In this case, due to the high performance requirements (narrow base-band and transition range, high stop-band attenuation), high filter orders are required. Often, this is still realizable in an efficient way, since the multiplications of the innerproducts are reduced to simple additions/substractions. Fig. 1 shows the basic architecture of this solution [2]. The ROM which contains half of the coefficients of a symmetric FIR filter is scanned with an up-down counter, whereas the convolution sum between coefficient and data is accumulated.

The accumulator has L-coef full adder (FA) slices. On the ROM-side, L+1 bits has to be connected to the logical "0". The NXOR function between the sign-bit of the coefficients and the output of the analog loop determines whether an addition or a substraction is performed.

Time multiplexing on M adders can be used in order to handle oversampling rates of 1/M times the filter order N. In this case, either the ROM should be enlarged with new tables (fig. 1) or a shuffle network should be applied [2]. If a new table is added to the ROM, the new up-down counter should start

operation N/M clock cycles later than its predecessor.

Other interesting realizations which are recently published are based on dump- and accumulate filters [3] or using an IIR configuration [4]. As such approaches are only appropriate for decimation down to 4 times the Nyquist rate, a second decimation stage should be used to arrive at the Nyquist rate.

In this paper we will, therefore, only focus on the architecture of fig. 1. First, we will discuss the filter transfer function which is needed. The complexity of the filter is minimized by taking into account the noise shaping. Furthermore, the most important trade-off's between the performance requirements of the analog loop and the complexity of the decimation filter are discussed.

In sections III and IV, a simulated annealing optimization is presented for the minimization of the wordlength of the filter coefficients. Finally, the software environment into which the methodology has been integrated is discussed and a design example is given.

## II. THE REQUIRED FILTER COMPLEXITY

The SNR of the A/D converter can be calculated if the spectral density of the quantization noise at the output of the $\Sigma$-$\delta$ modulator is known. This spectral density can be easily derived by recognizing that the sum of the signal and the noise power is equal to $V_{DD}^2$. As the power of the input signal is

$$V_{DD}^2 \left( \frac{V_{in}}{V_{in,max}} \right)^2,$$

where $V_{in,max}^2$ is the full scale input signal, the power of the quantization noise is given by:

$$P_N = V_{DD}^2 \left[ 1 - \left( \frac{V_{in}}{V_{in,max}} \right)^2 \right] \tag{1}$$

Assuming that the quantization noise injected into the comparator is white, it can be deduced that the noise spectral density at the output of the modulator is given by:

$$S_{nn}(f) = V_{DD}^2 \left[ 1 - \left( \frac{V_{in}}{V_{in,max}} \right)^2 \right] \cdot \frac{\left| T^2(f) \right|}{\int_0^{f_H/2} \left| T^2(f) \right| df} \tag{2}$$

Reprinted from *IEEE Proc. ISCAS'87*, pp. 479–482, May 1987.

where: $T(f)$ = noise transfer function

$f_H$ = sample frequency.

The decimation filter function can be splitted up into three parts (fig. 2) and the noise contributions of each part can be calculated.

In the base-band, the power of the noise is given by:

$$P1 = V_{DD}^2 \left[1 - \left(\frac{V_{in}}{V_{in,max}}\right)^2\right] \cdot \frac{\int_0^{f_s} |T^2(f)| \, df}{\int_0^{f_H/2} |T^2(f)| \, df} \quad (3)$$

In the transition band, it is given by:

$$P_2 = V_{DD}^2 \left[1 - \left(\frac{V_{in}}{V_{in,max}}\right)^2\right] \cdot$$
$$\cdot \frac{\int_{f_s}^{f_s + \delta f} |T^2(f)| \cdot \left|1 - \frac{(f-f_s)^2}{\delta f}\right| }{\int_0^{f_H/2} |T^2(f)| \, df} \quad (4)$$

Finally, for the stop-band, the noise power is given by:

$$P_3 = V_{DD}^2 \left[1 - \left(\frac{V_{in}}{V_{in,max}}\right)^2\right] \cdot$$
$$\cdot \frac{\int_{f_s + \delta f}^{f_H/2} \delta s^2 \, df}{\int_0^{f_H/2} |T^2(f)| \, df} \leq \quad (5)$$
$$\leq \frac{V_{DD}^2 \left[1 - \left(\frac{V_{in}}{V_{in,max}}\right)^2\right] \delta s^2 \cdot \frac{f_H}{2}}{\int_0^{f_H/2} |T^2(f)| \, df}$$

By supposing a sinusoïdal input signal with amplitude A, the SNR is given by

$$SNR = 10.\text{Log} \frac{\frac{1}{2}\left(\frac{A}{V_{in,max}}\right)^2 V_{DD}^2}{P_1 + P_2 + P_3} \quad (6)$$

For an ideal low pass filter, the SNR would be:

$$SNR = 10.\text{Log} \frac{\frac{1}{2}\left(\frac{A}{V_{in,max}}\right)^2 V_{DD}^2}{P_1} \quad (7)$$

Thus, due to the non-ideal filtering of the noise, the converter will loose:

$$SNR = 10.\text{Log}\left[1 + \frac{P_2}{P_1} + \frac{P_3}{P_1}\right] \quad (8)$$

of its signal to noise ratio.

On the other hand, for full scale input signals ($A = V_{in,max}$), a n-bits converter requires a SNR of:

$$SNR > 20.\text{Log } 2^{n+1} \quad (9)$$

Thus, in order to satisfy (6) and (9), the following equation should hold:

$$\frac{P_1 + P_2 + P_3}{V_{DD}^2} \leq \left(\frac{1}{2}\right)^{2n+3} \quad (10)$$

Likewise, an ideal decimation filter requires that:

$$\frac{P_1}{V_{DD}^2} < \left(\frac{1}{2}\right)^{2n+3} \quad (11)$$

where $P_1/V_{DD}^2$ is proportional to the third or fifth power of $f_s/f_H$ for respectively a first and second order sigma-delta converter.

Equation (10) shows that a trade-off between the performance requirements of the analog part (over-sampling rate) and the digital part should be made. Either $P_1$ is chosen much smaller than (11) and the filter performances are relaxed or it is chosen close to (11) and the filter performances are enhanced.

From (8), it is clear that if a X dB loss of SNR in the decimation filter is accepted, then the following equations should be satisfied:

$$\frac{\int_{f_s}^{f_s + \delta f} |T^2(f)| \cdot \left|1 - \frac{f-f_s}{f}\right|^2 df + \delta_s^2 \cdot f_H/2}{\int_0^{f_s} |T^2(f)| \, df} = 10^{X/10} - 1 \quad (12)$$

and

$$\frac{P_1}{V_{DD}^2} = \frac{10^{-X/10}}{2^{2n+3}} \quad (13)$$

Thus, we should calculate $\delta f$ and $\delta s$ in such a way that a minimal filter order Nmin is obtained for a given loss of X dB SNR. As the order of a FIR can be approximated as [5] :

$$N = \frac{-10 \, \log(\delta p * \delta s) - 15}{14. \delta f} \, fh + 1 \quad (14)$$

we should minimize N under the constraint of (12). However, as N can only take multiple values of $f_h/f_n$ ($f_n$ = output rate of the digital filter), the minimizing of the filter's complexity can still be a little refined. The margin between M.fh/fn and Nmin can be exploited by taking $\delta f$ as small as possible. Consequently, for the same SNR loss, the stop-band attenuation requirement $\delta s$ can be relaxed and can, therefore, reduce the wordlength of the coefficients [8].

## III. FINITE WORDLENGTH OF THE COEFFICIENTS

Obviously, while satisfying the specifications, the coefficients of the filter should be calculated with a minimal finite wordlength. Traditionally this problem is tackled by applying the Remez-algorithm [6,7] for the design of an infinite precision solution. A b-bit finite wordlength solution is then found by rounding or truncation of the coefficients to b-bit.

In [8] it has been pointed out that a substantial (5-7 dB) improvement in the stop-band can be obtained if the discrete constraint on the coefficients is included in the optimization process. Mixed Integer Programming can, at the expense of much CPU time and for low filter orders be used for this optimization [8].

As our filters will have a very high order (200--4000 taps), we can rule out this optimization method. Therefore, we applied a simulated annealing optimization algorithm [9].

## IV. SIMULATED ANNEALING OPTIMIZATION OF THE WORDLENGTH

The optimization problem can be stated as follows:

Given a FIR filter of order N, find coefficients which can all be coded with a minimum number of L-coef bits and a maximum number of L-leading zero bits (see fig. 1, definition representation of coefficients) and which respect the imposed transfer function H(jw).

An initial solution can be found by synthesizing the infinite precision filter combined with coefficients rounding. Such a solution can be optimized with simulated annealing. The basic ideas behind this optimization are:

1) Choose at random a coefficient h(k).

2) Choose at random a small coefficient change $\delta h(k)$.

3) Establish a cost (=energy) function. This function should be a function of l(h(k)), L-coef(h(k)) and the difference between the target transfer function and the actual transfer function.

4) The cost-function can easily be updated in an incremental way for small changes of h(k). It is easy to derive that:

$$\frac{\delta H(jw)}{\delta h(k)} = \begin{cases} 1 & k = 0 \\ 2\cos(2\pi k) & k <> 0 \end{cases}$$

5) In order to avoid a prohibitive CPU-time, an adaptive number of iterations per temperature step and an adaptive temperature scheme has been implemented as in [10]. The number of iterations per temperature step is determined by building up a Markov chain. After reaching a steady state behaviour, the annealing is continued at a lower temperature. This new temperature is also chosen as a function of the length of the Markov chain. In this way, the optimiza-

tion takes a CPU-time which is in the same order of magnitude as the synthesis of the infinite coefficients, i.e. a few minutes on a VAX 8600 computer.

## V. SOFTWARE ENVIRONMENT

The synthesis of decimation filters and its optimization for sigma-delta converters have been incorporated in a CAD-tool for the design of high-performance analog circuits [11]. As this tool is, amongst others, also able to synthetize the analog loop of the A/D converter (1st and 2nd order), a trade-off between the required performances of both parts of the converter can easily be made. This performance trade-off depends strongly on the application and the available VLSI technology. Therefore, such a tool is absolutely necessary for a fast "customization" of a state of the art A/D converter.

## VI. DESIGN EXAMPLE

We will illustrate the design of the sigma-delta decimation filter and its optimization by means of a very simple example. Consider as example the design of a 8-bit first order sigma-delta A/D converter. Following the methodology proposed above, we find:

| | | | |
|---|---|---|---|
| SNR loss in filter | = | 6 dB | (imposed) |
| $\delta p$ | = | 11 dB | ( " ) |
| $f_n$ | = | 8 kHz | ( " ) |
| $f_s$ | = | 4 kHz | ( " ) |
| $f_H$ | = | 800 kHz | |
| M | = | 4 adders | |
| N | = | 400 | |
| $\delta_f$ | = | 5,09 kHz | |
| $\delta_s$ | = | 62,7 dB | |

An initial synthesis of the coefficients in infinite precision with rounding of the coefficients [7] reveals that we should take L-coef = 18, L = 6 in order to satisfy the requirements. Thus a 12 bits wordlength of the coefficients should be implemented in the ROM-memory. A simulated annealing optimization reduced L-coef to 17 bits and increased 1 to 7 bits. Thus after optimization only wordlengths of 10 bits are necessary for the ROM-implementation. Fig. 3 shows the transfer function before and after the simulated annealing optimization.

## VII. CONCLUSION

The design of decimation filters for sigma-delta converters can efficiently be realized by a high order FIR filter. A simulated annealing optimization has been proposed to reduce the wordlength of the coefficients with several bits.

The synthesis and optimization of the decimation filter has, together with the synthesizing of the analog loop, been integrated in a CAD-tool, enabling us to make rapidly suitable trade-off's between the required performances of both parts. Some further examples of this trade-off are given [12].

### REFERENCES

[1] R.E. Crochiere, L.R. Rabiner:"Interpolation and decimation of Digital signals-A tutorial review", Proc. IEEE, Vol. 69, No 3, March 1981.

[2] M.W. Hauser, P.J. Hurst, R.W. Brodersen:"MOS ADC Filter combination that does not require precision analog components, ISSCC 1985.

[3] J.C. Candy:"Decimation for Sigma-Delta Modulation", IEEE Trans. on Comm., Vol. COM-34, No 1, January 1986.

[4] A. Huber et al:"FIR lowpass filter for signal decimation with 15 MHz clock-frequency", Proc. ICASSP 86, Tokyo, pp. 1533-1537.

[5] L.R. Rabiner et al:"Some comparisons between FIR and IIR digital filters",The Bell System Techn. Journal, Vol. 53, pp. 305-331, Feb.74.

[6] J.H. McClellan:"A computer program for designing optimum FIR linear phase digital filters", IEEE Trans. on Audio and Electroacoustics, Vol. AU-21, No 6, Dec. 1973, pp. 506-525.

[7] U. Heute:"A subroutine for finite wordlength FIR filter design", published in IEEE Press, Programs for digital signal processing.

[8] V.B. Lawrence, A.C. Salazar:"Finite precision design of linear phase FIR filters", The Bell System Techn. Journal, Vol. 59, No 9, Nov. 80.

[9] S. Kirkpatrick:"Optimization by simulated annealing, Science 220 (1983), pp. 671-680.

[10] F. Catthoor:"Characterization of Finite wordlength effects and optimization of coefficients by Simulated annealing", Workshop CAD for DSP, Sept. 9-12, IMEC, Leuven, Belgium.

[11] M. Degrauwe et al:"An analog design expert system", ISSCC 1987.

[12] O. Nys, M. Degrauwe, E. Dijkstra:"A CAD tool for oversampled CMOS A/D converters", submitted for publication at 1987 Symp. on VLSI Technology, May 18-21, Nagano, Japan.

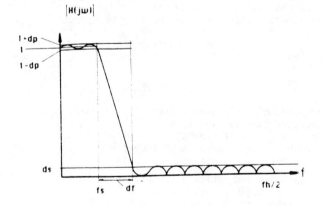

Fig. 2    Filter Characteristic.

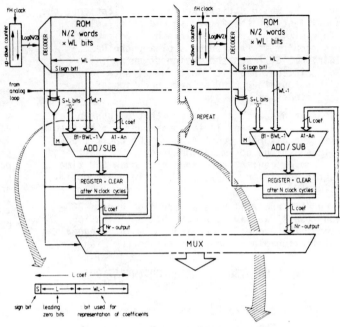

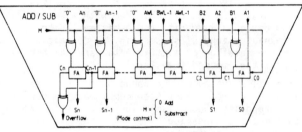

Fig. 1    Architecture of the decimation filter.

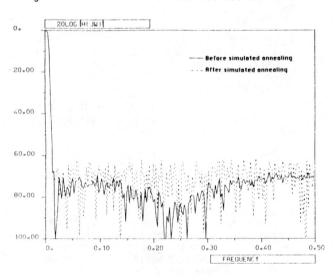

Fig. 3    Transfer functions before and after simulated annealing optimization.

# ON THE USE OF MODULO ARITHMETIC COMB FILTERS IN SIGMA DELTA MODULATORS

E. Dijkstra, O. Nys, C. Piguet and M. Degrauwe

CENTRE SUISSE D'ELECTRONIQUE ET DE MICROTECHNIQUE S.A.
Maladiere 71, 2007 Neuchatel, Switzerland

## ABSTRACT

A novel architecture of one stage Comb decimation filters for Sigma Delta Modulators is described. It performs the decimation of a 1 bit oversampled modulator output to an arbitrary lower output frequency.
The use of modulo arithmetic throughout the filter together with the proposed algorithmic decomposition allows a low power and area efficient implementation. This also avoids the storing of the coefficients in a ROM or the generation of the coefficients with rather complicated up/down counters.
The architecture is applicable to all comb decimation filters with $\text{sinc}^k(f)$ response. A filter with $k=3$ and a programmable decimation factor has been integrated in a 3 $\mu$m CMOS process.

## 1. INTRODUCTION

With the advent of VLSI technology, oversampled Sigma-Delta modulators are becoming more and more popular for A/D conversion. The reason is that a large resolution can be obtained without highly matched components[1]. The converter consists of two parts: a relative small analog modulator and a large digital decimation filter. The oversampled analog modulator shapes the quantization noise in such a way that most of it is pushed into the higher frequencies. There this noise can be filtered out by a digital low pass filter. A frequency decimation is simultaneously performed in order to sample down the oversampled frequency to a more convenient one.

In previous work we presented a design methodology for of very long equiripple FIR decimation filters[2]. Furthermore, in [3] the front-end of a silicon compiler for sigma-delta converters has been discussed which is able to determine the system parameters of the converter (clock-frequency, filter characteristics, voltage reference specifications etc.) starting from the system-specifications (nb of bits, gain error, offset error etc).

In this paper, we will focus on a non-convential architecture for comb decimation filters. As Candy[4] showed, such filters are attractive candidates for replacing the very long equiripple FIR decimation filters[1,2].
Their transfer function can be expressed as:

$$\left( \sum_{i=0}^{N-1} z^{-i} \right)^k = \left( \frac{1}{N} \frac{1 - z^{-N}}{1 - z^{-1}} \right)^k \qquad (1)$$

where:  N=decimation factor
k=order of the comb filter (usually 2 for a first order modulator and 3 for a second order modulator)

In the time domain the behaviour of the decimation filter is to output after each N input samples a sample which is a weighted average of the last $(N-1)*k+1$ input samples. The frequency response of the filter is:

$$H(jw) = (\sin(\omega NT)/(N*\sin(\omega T)))^k \qquad (2)$$

Due to this $\text{sinc}(f)$ attenuation we can with this filter only sample down to about 4 times the Nyquist rate[4]. Then a second decimation stage should be used to arrive at the Nyquist rate. In [5] we describe Wave Digital Decimation filters which are suitable for this task.

## 2. ARCHITECTURAL CONSIDERATIONS

Equation (1) can be realized in several ways.
An obvious attempt is to store the N*k filter coefficients in a ROM and to perform directly the convolution equation. However, because of the fact that one output sample should be delivered after N input samples, time multiplexing on k accumulators is necessary. The architectue will strongly resemble the one described in [1,2].
A more subtle variant is described in [6] for the special case k=3. The coefficients are not stored in a ROM but generated by rather complicated up/down counter combinations. Time-multplexing on k=3 accumulators is of course still required.

Reprinted from *IEEE Proc. ICASSP'88*, pp. 2001-2004, April 1988.

Fig(1) shows an architecture presented in [7] for k=3. The transfer function has been splitted up into a FIR part realizing $(1-z^{-N})^k$ and an IIR part realizing $(z^{-1}/(1-z^{-1}))^k$. The disadvantage of this architecture is that one should implement N*k shift registers for the FIR part. Despite the fact that the input is often coded on one bit only, this necessitates a substantial area (N is usually >100). Moreover, as the decimation can only be done after the IIR part, the whole filter operates at the oversampling rate.

The attractive alternative we focussed on is depicted in fig(2). In this case we interchanged the IIR and FIR parts. This allows a decimation in between the IIR and FIR parts and therefore $z^{-N}$ can be replaced by $z^{-1}$. As a consequence we will use significantly less memory and less power consumption than the previous decomposition.

The recursive stage of the filter structure in fig(2) has a pole at z=1 which is not asymptotically stable and therefore may overflow. It can, however, be proven[8] that overflow can be avoided by using a modulo arithmetic system (or a "wrap around") everywhere in the system.

A sufficient condition for the filter to work correctly is to choose a modulo which is larger than $(N^k+1)*D$, where D is the dynamic range of the input signal. For Sigma-Delta converters the dynamic range D is usually only one bit. Due to this limited dynamic inputrange modulo arithmetic comb filters are very suitable for this kind of VLSI applications. Note that by taking a modulo $2^b \geq (N^k+1)$ all calculations can be performed by ordinary two complement's operators. Carry handling can simply be ignored.

The crucial difference between the decomposition of figs. (1) and (2) is the different need for memory resources in both FIR parts. The FIR part of fig. (1) will need N*k bit locations whereas the structure of fig(2) needs $k^2*\log_2(N)$ bit locations for its FIR part. This means that for k=3 and N>16 the latter will take less memory locations. Besides, in fig(1) all bits should be organized in a large N*k shiftregister, whereas in fig(2) bit locations could be advantageously organized in a small k words of $k*\log_2(N)$ bits RAM.

This difference in memory requirements compensates largely the slightly more complex IIR part of fig(2). (i.e. in fig(2) all integrators use everywhere b-bits, whereas in fig(1) the first and second integrator can eventually be implemented with a few bits less.) The b-bits additions in the FIR part of fig.(2) are performed at the lower frequency and therefore timemultiplexing could be done on serial adders, making the FIR adder complexity comparable to the FIR adder complexity of fig. (1).

## 3. VLSI IMPLEMENTATION

The structure of fig(2) has been integrated in a 3 $\mu$m CMOS technology. For this integration we focussed on a minimum complexity for the control unit and a maximum layout regularity. Moreover, by making the decimation factor programmable, the inherent trade-off between conversion speed and resolution has been exploited. In this way the same chip can be used for several applications.

The integrators of the IIR part are directly "mapped into silicon", i.e. each integrator has his own adder and no multiplexing is performed due to a limited system clock. Because of the significantly lower speed of the FIR part such a direct mapping of this part would spoil much area. Therefore we implemented the calculation of $(1-z^{-1})^3 = 1-3z^{-1}+3z^{-2}-z^{-3}$ with the distributed arithmetic architecture of fig(3). The state variables are written horizontally word by word and read vertically bitslice by bitslice. In order to avoid shifting of state variables a "wrap-around" pointer mechanism provides the new memory location where the data from the IIR part can be stored.

Obviously by using a "wrap-around" pointer mechanism for the state variables we should indicate to the ROM in which order the state variables are classified. This means that 2 extra address lines are needed.

As the ROM output wordlength can be coded on 4 bits, the ROM will contain only $(2^6)*4=256$ bits. With such a small ROM we can even suppress the adder by performing the add functions also in the ROM (fig(4)). The resulting ROM will still be acceptable small (2048 bits) and the layout regularity will be significantly improved.

Fig (5) shows the chip photo of the circuit. The total number of transistors is 7800 on an area of 3.5 $mm^2$ (3 $\mu$m CMOS). The oversampling frequency was 40kHz and the decimation factor N programmable. Power consumption is 5 $\mu$A at 1.5 Volts. First silicon was working.

## 4. CONCLUSIONS

The proposed decomposition together with the use of modulo arithmetic througoutr the filter leads to a very compact, flexible and extensible architecture of Comb decimation filters for $\Sigma-\delta$ converters. It is also very suitable for comb filters in the recently proposed multi-bits "interleaved" modulators[9].

Furthermore, if speed becomes a bottleneck for the structure the throughput can be elegantly enhanced by applying Residue Number System (RNS) arithmetic [10]. In this case, the filter structure (fig(2)) with one large modulo $(m>N^k+1)$ could be splitted up in several units with smaller modulo's $(\Pi_{m_i} \geq N^k+1)$. The RNS encoding is for our applications obviously not necessary (the input is coded on 1 bit), whereas the RNS decoding can be performed after decimation and thus at a significantly lower rate.

References:

[1] M.W. Hauser, P. Hurst, R. Brodersen: "MOS ADC filter combination that does not require precision analog components" Proc. of ISSCC 85, pp 80-81.

[2] E. Dijkstra, M. Degrauwe, J. Rijmenants, O. Nys. "A design methodology for decimation filters in Sigma Delta A/D converters". ISCAS 87 pp.479-482.

[3] O. Nys, M. Degrauwe, E.Dijkstra. "A CAD tool for oversampled CMOS A/D converters" Proc. of 1987 Symp. on VLSI circuits, May 1987, Karuizawa, pp 113-114

[4] J.C. Candy "Decimation for Sigma-Delta modulation" IEEE Trans. on comm. Vol COM-34, No.1, Jan 1986, pp 72-76.

[5] E. Dijkstra et al. "Wave Digital Decimation filters in oversampled A/D converters", subm. for publ. at ISCAS 88

[6] H. Meleis et al. "A novel architecture design for VLSI implementation of an FIR decimation filter" Proc. ICASSP 85, Tampa, pp. 1380-1383

[7] A. Huber et al. "FIR lowpass filter for signal decimation with 15 Mhz clock frequency.", Proc. ICASSP 86, Tokyo, pp 1533-1537.

[8] S. Chu, C. Sidney Burrus "Multirate filter designs using comb filters", IEEE Trans. on CAS, Vol CAS-31, No. 11, Nov 84, pp.913-924.

[9] Y. Matsuya et al. "A 16 bit Oversampling ADC" Proc. ISSCC 87, Feb. 87, New York.

[10] N.S. Szabo, R. Tanaka. "Residue Arithmetic and its applications to computer technology, New-York, McGraw-Hill 1967

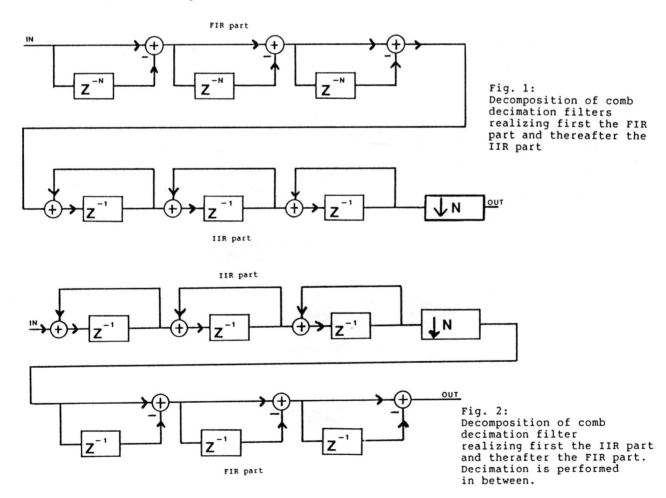

Fig. 1:
Decomposition of comb decimation filters realizing first the FIR part and thereafter the IIR part

Fig. 2:
Decomposition of comb decimation filter realizing first the IIR part and therafter the FIR part. Decimation is performed in between.

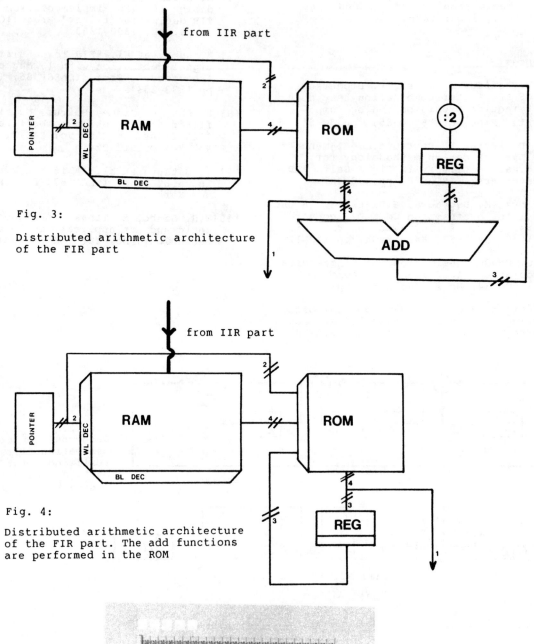

Fig. 3:

Distributed arithmetic architecture
of the FIR part

Fig. 4:

Distributed arithmetic architecture
of the FIR part. The add functions
are performed in the ROM

Fig. 5: Chip photograph

# Nine Digital Filters for Decimation and Interpolation

DAVID J. GOODMAN, MEMBER, IEEE, AND MICHAEL J. CAREY

*Abstract*—Filtering is necessary in decimation (decreasing the sampling rate of) or interpolation (increasing the sampling rate of) a digital signal. If the rate change is substantial, the process is more efficient when the decimation or interpolation occurs in stages rather than in one step. Half-band filters are particularly efficient for effecting octave changes in sampling rate and nine digital filters are presented, eight of them half-band filters, to be used as components of multistage interpolators and decimators. Also presented is a procedure for combining the filters to produce multistage designs that meet a very wide range of accuracy requirements (stopband attenuation to 77 dB, passband ripple as low as 0.00014).

The nine filters admit changes between sampling rates above $4W$, where $W$ is the nominal bandwidth of the signal. Established design techniques may be used to obtain efficient filters for conversion between $4W$ Hz sampling and $2W$ Hz, the "baseband sampling rate." With these multistage filters, the possible interpolation and decimation ratios are all integer multiples of powers of two. To overcome this restriction we present a simple resampling technique that extends the range of designs to conversions between any two rates. The interpolation or decimation ratio need not be an integer or even rational. In fact, it can vary slightly as in a practical situation where the input signal and output signal are under the control of autonomous clocks.

We demonstrate the approach by means of several design examples and compare its results with those obtained from the optimization scheme of Crochiere and Rabiner.

## I. INTRODUCTION

### A. Background

**S**EVERAL RECENT PAPERS [1]–[8] have considered filtering problems that arise in changing the sampling rate of a digital representation of a continuous waveform. It is now recognized that, when the initial rate and final rate are widely separated, it is more efficient to change the sampling rate in stages by means of a sequence of filters than it is to do so all at once with a single filter. The previous papers fall in two categories: some of them offer design methods for choosing decimation (rate reducing) and interpolation (rate increasing) filters from the class of all symmetric finite impulse response (FIR) filters [1], [5]–[8], while others focus on the special class of symmetric FIR filters known as half-band filters [9] in which nearly half of the impulse response coefficients are zero, making them particularly efficient for 2-to-1 decimation and interpolation [2]–[4]. These latter papers describe the efficient use of half-band filters and give examples of practical applications, but none of them provides a design method for finding sequences of half-band filters that meet prespecified fidelity criteria.

Manuscript received December 17, 1975; revised July 12, 1976, and November 10, 1976.

D. J. Goodman is with the Acoustics Research Department, Bell Laboratories, Murray Hill, NJ 07974.

M. J. Carey was with the Post Office Research Center, Martlesham Heath, England. He is now with the University of Keele, Staffordshire, England.

## TABLE I
### FILTER COEFFICIENTS

| Filter | Order | $h(0)$ | $h(1)$ | $h(3)$ | $h(5)$ | $h(7)$ | $h(9)$ |
|--------|-------|--------|--------|--------|--------|--------|--------|
| $F1$ | 3 | 1 | 1 | | | | |
| $F2$ | 3 | 2 | 1 | | | | |
| $F3$ | 7 | 16 | 9 | −1 | | | |
| $F4$ | 7 | 32 | 19 | −3 | | | |
| $F5$ | 11 | 256 | 150 | −25 | 3 | | |
| $F6$ | 11 | 346 | 208 | −44 | 9 | | |
| $F7$ | 11 | 512 | 302 | −53 | 7 | | |
| $F8$ | 15 | 802 | 490 | −116 | 33 | −6 | |
| $F9$ | 19 | 8192 | 5042 | −1277 | 429 | −116 | 18 |

$F1$ can be of any order, with all $h(i) = 1$.
$F2$–$F9$ are half-band filters with $h(-i) = h(i)$ and $h(2) = h(4) = h(6) = 0$.

### B. This Paper

Such a design method is the contribution of the present paper, which offers the set of nine filters, denoted $F1$–$F9$, in Table I, and a simple method for selecting cascade combinations of these filters that satisfy a very wide range of accuracy requirements. The filters have been selected with hardware efficiency in mind; some of them can be realized recursively with accumulators alone and in most of the others, the coefficient word lengths are very modest. An important consequence of this hardware-oriented approach is the discovery that even though the required computation rate grows linearly with the ratio of the two sampling rates, the amount of hardware can remain constant beyond a certain ratio.

Although the half-band filters admit 2:1 sampling rate changes, it is possible, by means of a resampling register, to use them to convert between sampling rates that are not related by powers of two. In fact with the resampling method we present, the input and output rates need not even maintain a fixed ratio: the two clocks can be autonomous and slightly variable.

### C. Filter Requirements

If a signal, sampled at $f_s$ Hz, has essential information in the band 0, $W$ Hz, we refer to $2W$ as the "baseband sampling rate" and introduce the parameter

$$R = f_s/2W \qquad (1)$$

the bandwidth expansion ratio. Fig. 1 shows requirements on filters used in baseband sampling and desampling (conversion between $R = 1$ and $R = \infty$). The passband ripple limit $d_p$ controls linear distortion over the passband, 0 to $aW$ Hz. The stopband attenuation requirement $d_s$ at $f > bW$ controls aliasing in sampling and suppression of spectral images in desampling.

With decimation viewed as an intermediate step in the process of baseband sampling, Fig. 1 is a constraint on the cascade

Reprinted from *IEEE Trans. Acoust., Speech, Signal Proc.*, vol. ASSP-25, pp. 121–126, April 1977.

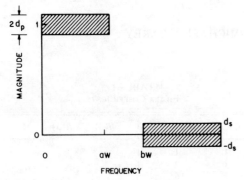

Fig. 1. Filter requirements.

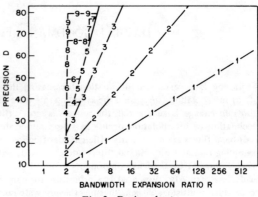

Fig. 2. Design chart.

combination of analog presampling filter and digital decimation filters. Taking the converse view of interpolation, we adopt Fig. 1 as a constraint on the cascade combination of digital interpolating filters and analog desampling filter. Contained in [3] is an example of interpolating and decimating filter sequences that satisfy requirements of telephone transmission systems.

## II. DESIGN METHOD

### A. Range of Performance Levels

Our filter set and the method for finding filter sequences that satisfy decimation and interpolation requirements have been derived from an approach that is opposite to that of many filter design techniques. Rather than start with a set of requirements and search for filters that satisfy them, we have selected certain filters that are easy to implement and calculated the sets of requirements satisfied by each filter. These requirements (ranges of speed and accuracy) appear in Fig. 2 as the regions to the right of and below each curve. The ordinate $D$ is either $-20 \log d_s$ or $-20 \log d_p$ and, for a given $R$, the value of $D$ on a filter locus corresponds to the most stringent accuracy requirement satisfied by that filter. The Appendix provides details of filter characteristics and the derivation of Fig. 2.

The nine filters in our set admit transformations between any pair of bandwidth expansion ratios $R_1$ and $R_2$ with $2 \leqslant R_1 < R_2$ with any set of accuracy requirements in the range $-20 \log d_p$ and $-20 \log d_s \leqslant 77$ dB. The filter set does not admit the transformation between baseband sampling rate and $R = 2$. For this transformation, a specially designed filter is necessary and here our work merges with the optimization studies of Crochiere and Rabiner [6], [7] in which the initial or final filter of every efficient sequence is used for conversion between $R = 2$ and $R = 1$.

### B. Using the Design Chart

We view multistage decimation as a walk in Fig. 2 from right to left at a certain height $D$ determined by $d_p$ and $d_s$. Each step in the walk introduces a new filter to the decimation sequence and with the nomenclature $F1$ to $F9$ placing filters in order of increasing complexity, the design strategy is to use at each step in the walk the filter whose locus is immediately to the left of the end of the step. For interpolation, the walk

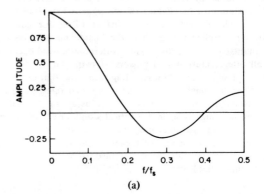

(a)

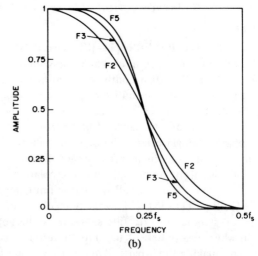

(b)

Fig. 3. (a) Amplitude response of $F1$, $N = 5$. (b) Monotonic responses of half-band filters $F2$, $F3$, $F5$.

is from left to right and the strategy is to use the filter with locus immediately to the left of the beginning of each step.

The filter impulse response coefficients are given in Table I. Except for $F1$, an $N$th order filter with all coefficients unity, all of the filters are half-band filters. The amplitude responses fall into two categories. Those of $F1$, $F2$, $F3$, and $F5$, shown in Fig. 3, are monotonic over the band $0$, $W$. They have no passband ripple and their rolloff may be equalized by a low-speed filter [3], or the equalization may be built into the passband filter. Consequently, only the stopband re-

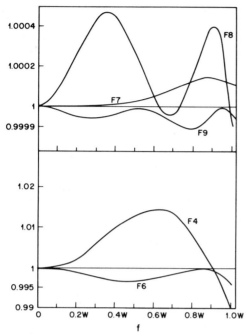

Fig. 4. Passband responses of $F4, F6$–$F9$.

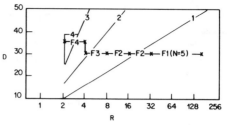

Fig. 5. Use of the design chart to find multistage filters for $R_1 = 1$, $R_2 = 160$.

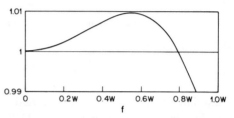

Fig. 6. Passband response of $F1(N = 5), F2, F2, F3, F4$ in cascade.

quirement $d_s$ influences their use. The use of the other filters $F4$ and $F6$–$F9$, with passband responses shown in Fig. 4, is constrained by the passband ripple requirement, as well as by $d_s$.

To obtain a filter sequence, take steps to the left or right in Fig. 2, using at each step the filter with locus just to the left of the entire step. The height $D$ of a step is determined by

$$D = -20 \log d_s,$$ 
provided this admits the choice of $F1, F2, F3,$ or $F5$;

$$D = -20 \log [\min (d_p, d_s)], \quad \text{otherwise.} \quad (2)$$

Each step length is one octave with $F2$–$F9$. With $F1$, the ratio of the endpoints of the step can be any integer.

### C. The Baseband Filter

In the Appendix, it is shown that the filters selected according to the above rules introduce no more than about one-half the allowed passband ripple. It follows that approximate design data for the baseband filter are $d_p/2, d_s, a, b$, with a more precise passband limit obtainable through computation of the overall passband response of the filters chosen from our set. The structure of the baseband filter will depend on the application. Often an infinite impulse response design will be most efficient although there are advantages to the FIR form. In either case many design methods are available [11]. To obtain a half-band design, one may use the optimization program of McClellan et al. [12], placing passbands and stop-bands with equal error weighting coefficients symmetrically about $f = 0.25$.

### D. An Example

Consider a speech signal with nominal bandwidth 4 kHz and design a multistage filter for changing the sampling rate from 1280 kHz ($R = 160$) to 8 kHz ($R = 1$) or vice versa.

Assume that the requirements on an 8 kHz presampling filter are: ripple within $0 \pm 0.13$ dB over 0, 3200 Hz, attenuation at least 30 dB at frequencies above 4800 Hz. In the nomenclature of Fig. 1 these requirements translate to $a = 0.8, b = 1.2$, $d_p = 0.015, d_s = 0.031$. For decimation, the walk in Fig. 2 begins at $R = 160$ and with $F1$ admissible, (2) sets the height of the walk at 30 dB. The $F1$ step ends at $R = 32$ ($N = 5$), allowing octave steps for the remaining filters, which are $F2$, $F2$, and $F3$ for conversion from $R = 32$ to $R = 4$. The step from $R = 4$ to $R = 2$ requires one of the filters with passband ripple so that the second line of (2) determines the height of this step, $D = -20 \log d_p = 36.6$ dB, which dictates the selection of $F4$. The steps in the multistage decimation are shown in Fig. 5. (The same steps in reverse order admit interpolation from $R = 2$ to $R = 160$.)

Fig. 6 shows the overall passband response of the five filters used for conversion from $R = 160$ to $R = 2$. It indicates that the rolloff of the first 4 stages and the ripple of $F4$ have partially cancelled one another, producing an overall ripple of magnitude 0.005, which leaves 0.008, at least, for the ripple allowance of the baseband filter. To complete the design, we have derived a half-band filter for conversion between $R = 2$ and $R = 1$. This one is of order 19, with coefficients listed in Table II, which summarizes the characteristic of the 6-stage conversion between $R = 160$ and $R = 1$. Of the 26 multiplications required per baseband sampling interval, 10 are by powers of two, and only 6 have coefficients with more than 4 bits.

## III. RESAMPLING

### A. General Principle

Although sequences of filters from our set can implement only a limited number of decimation or interpolation ratios, $R_2 : R_1$ (all of them integer multiples of a power of two), an interpretation of the $F1$ locus in Fig. 2 leads to an extension of the design procedure to accommodate any pair of sampling

TABLE II
6-STAGE CONVERSION BETWEEN $f_s = 2W$ AND $f_s = 320W$

| Passband: bandedge = 0.8W | | | ripple = ±0.0149 (±0.13 dB) | | | |
|---|---|---|---|---|---|---|
| Stopband: bandedge = 1.2W | | | attenuation = 0.0169 (35.4 db) | | | |

| | A | B | C | D | E | F | Total |
|---|---|---|---|---|---|---|---|
| Filter | F1 | F2 | F2 | F3 | F4 | b | |
| Order | 5 | 3 | 3 | 7 | 7 | 19 | |
| Storage (interpolation) | 1 | 1 | 1 | 3 | 3 | 9 | 18 |
| Storage (decimation) | 1 | 1 | 1 | 4 | 4 | 13 | 24 |
| Multiplies[a] | 0 | 0 | 0 | 12 | 8 | 6 | 26 |
| Adds[a] (interpolation) | 0 | 64 | 32 | 12 | 6 | 9 | 123 |
| Adds[a] (decimation) | 160 | 64 | 32 | 16 | 8 | 10 | 290 |

[a]Computations per $1/2W$ seconds, assumes F2 realized with only an accumulator [3].

[b]Baseband filter coefficients:

| $i$: | 0 | 1 | 3 | 5 | 7 | 9 |
|---|---|---|---|---|---|---|
| $h(i)$: | 238 | 149 | −46 | 22 | −12 | 6. |

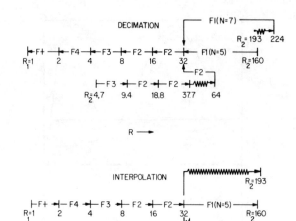

Fig. 7. Examples of interpolation and decimation filters for three values of $R_2$.

rates. In this interpretation, we imagine a zero-order desampling (holding the value of each sample until the arrival of the next sample) of a signal with $R = R_2$. This desampling, equivalent to filtering with $F1, N = \infty$, is followed by sampling at $R = R' > R_2$. The energy aliased to baseband by the resampling originates at frequencies above $(R' - 1) W$ and it follows that if the point $(R_2, -20 \log d_s)$ is in the admissible region of $F1$, then

$$H_\infty(f) = \frac{\sin(\pi f/2RW)}{(\pi f/2RW)} < d_s, \quad f > (R' - 1)W \quad (3)$$

which implies that the hold and resampling together meet the stopband constraint. Thus any signal to the right of the $F1$ curve (with $D = -20 \log d_s$) may be resampled at a higher rate without incurring excessive aliasing.

Augmented by a resampling register, the filter set can transform (with accuracy up to 77 dB) between any pair of sampling rates greater than $4W$. The two rates need not be related by rational numbers; as in practice, the clocks can be autonomous and each slightly variable. To generalize the design rules to incorporate resampling, we introduce a sampling rate $R_2'$ which is an appropriate multiple (integer times a power of 2) of $R_1$ and obtain a filter sequence that transforms between $R_1$ and $R_2'$. We then use a resampling register (and perhaps additional filters) to convert between $R_2'$ and $R_2$.

### B. Examples

In the example of Section II-D, let the initial sampling rate be 1544 kHz ($R_2 = 193$), the bit rate of a T1 transmission line. Because $(R,D) = (193,30)$ is within the admissible region of $F1$, it is possible to resample the signal at $R_2' = 224 = 7 \times 2^5$. Decimation to $R = 1$ can then proceed with $F1, N = 7$ for decimation to $R = 32$ and filters $B - F$ in Table II for decimation to $R = 1$. For interpolation from $R = 1$ to $R = 193$, it suffices to use filters $B - F$ to increase the bandwidth expansion ratio to $R = 32$ and the resampling register to increase the rate to $R = 193$.

A more complicated task is to convert between $f_s = 8$ kHz ($R = 1$) and the 37.7 kHz rate ($R = 193/41$) of the delta modulators in a subscriber loop carrier system [10]. The decimation task requires, initially, interpolation to a bandwidth expansion ratio greater than 30 (the value of $R$ on the $F1$ locus at which $D = 30$). To this end one can interpolate (by means of the three filters $F3, F2, F2$) to $R = 8 \times 193/41 = 37.7$ and then resample at $R_2' = 64$. Now the 64:1 decimation can be accomplished with an $F2$ decimator followed by filters $B - F$. For interpolation from $R_1 = 1$ to $R_2 = 193/41$, use filters $B–F$ to produce $R_2' = 32$. Then resample at $R = 37.7 = 8R_2$ and use $F2, F2, F3$ for 8:1 decimation to $R_2$. The filtering and resampling operations of the examples are summarized in Fig. 7.

### IV. IMPLEMENTATION

The filters in our set have been chosen with hardware efficiency in mind: careful attention has been given to coefficient word lengths. When half-band filters are used as interpolators and decimators, their complexity increases relatively slowly [2] with filter order $N$, as indicated in Table III, which shows the number of multiplications increasing as $N/4$, additions as $N/2$ and storage registers as $N/2$ (interpolation) or $3N/4$ (decimation).

$F1$ and $F2$ can be realized without multipliers. $F1$ as an $N:1$ interpolator is a "hold-for-$N$" circuit, a single register from which $N$ identical output words are read after each input word is written. With $F1$ an $N:1$ decimator, an output is the sum of $N$ inputs, which can be obtained from a single accumulator that is reset after each output is generated. It follows that from any point, $R$, to the right of the $F1$ curve in Fig. 2 it is possible to interpolate to a bandwidth expansion ratio $NR$ with only a single memory element; conversely it is possible to decimate from $NR$ to $R$ with only a single accumulator, regardless of the size of $N$. The amount of hardware is constant to the right of the $F1$ locus; filter complexity does not grow indefinitely with the highest sampling rate to be accommodated.

TABLE III
COMPLEXITY OF HALF-BAND FILTERS

| Order | Multipliers | Decimation | | Interpolation | |
|---|---|---|---|---|---|
| | | Storage | Adders | Storage | Adders |
| 3 | 2 | 1 | 2 | 1 | 1 |
| 7 | 3 | 4 | 4 | 3 | 3 |
| 11 | 4 | 7 | 6 | 5 | 5 |
| 15 | 5 | 10 | 8 | 7 | 7 |
| 19 | 6 | 13 | 10 | 9 | 9 |
| $N$ | $(N+5)/4$ | $(3N-5)/4$ | $(N+1)/2$ | $(N-1)/2$ | $(N-1)/2$ |

Reference [3] shows how $F2$ may be realized with a single accumulator, operating at a rate $2f_s$. As an interpolator, the accumulator output is, alternately, a) the sum of two successive inputs and b) a single input added to itself. As a decimator, the $F2$ accumulator accepts one input sample, adds the next sample twice and a third sample once before generating an output and resetting to zero.

## V. COMPARISON WITH AN OPTIMIZATION METHOD

The optimization procedure of Crochiere and Rabiner [6], [7] considers the class of all symmetric FIR filters and produces sequences that minimize either the number of multiplications or the number of storage registers. Their optimum sequences usually contain three or four stages in which the first (interpolation) or last (decimation) stage converts between $R = 2$ and $R = 1$. Half-band filters do not normally appear in their solutions because their procedure does not distinguish between multiplication by zero (the value of the even numbered half-band filter coefficients) and multiplication by nonzero coefficients.

To compare their procedure with our approach we have studied with Crochiere and Rabiner three decimator design problems. The first is the example in Section III-D of this paper which imposes relatively loose accuracy and transition band constraints. The second design appears in [6] and is quite stringent, while the third example, taken from [2], is intermediate between the other two.

The results are summarized in Table IV. In Example A, our method has a significant overall advantage because the baseband filter accounts for only a fraction of the overall complexity, making the efficiencies of $F1$ and half-band filters in the earlier stages quite salient. In Example B, the 100:1 decimation ratio puts our method at a severe disadvantage. With $d_s = 0.0001$, our resampling principle demands interpolation to an intermediate bandwidth expansion ratio of 6400 followed by resampling at $R = 8192$ and decimation to $R = 128$, a procedure that involves hundreds of additions per input sample. Table IV shows the results of only the subsequent 128:1 decimation in Example B. There, as in Example C, the baseband filter accounts for a substantial fraction of the computational complexity and the overall complexity of the two schemes is similar. The added overheads associated with a large number of stages are a disadvantage of our approach which may be only partially offset by the economies of short coefficient words.

TABLE IV
COMPARISONS OF DESIGN METHODS

| Example A | | | | |
|---|---|---|---|---|
| $R_2:R_1 = 160$, | $d_p = 0.015$, | $d_s = 0.03$, | $a = 0.8$, | $b = 1.2$ |

Crochiere-Rabiner 3-stage design:

| Stage | $D$[a] | $N$ | Multiplies[b] | Adds[b] |
|---|---|---|---|---|
| 1 | 20 | 42.4 | 1.05 | 2.10 |
| 2 | 4 | 13 | 0.0875 | 0.175 |
| 3 | 2 | 19 | 0.0375 | 0.063 |
| | | | 1.1750 | 2.338 |

Goodman-Carey 6-stage design:

| | | | | |
|---|---|---|---|---|
| 1 | 5 | 5 | – | 1.00 |
| 2 | 2 | 3 | – | 0.4 |
| 3 | 2 | 3 | – | 0.2 |
| 4 | 2 | 7 | 0.075 | 0.1 |
| 5 | 2 | 7 | 0.0375 | 0.05 |
| 6 | 2 | 19 | 0.0375 | 0.063 |
| | | | 0.150 | 1.813 |

| Example B | | | | |
|---|---|---|---|---|
| $R_2:R_1 = 100$, | $d_p = 0.001$, | $d_s = 0.0001$, | $a = 0.95$, | $b = 1$ |

Crochiere-Rabiner 3-stage design:

| Stage | $D$ | $N$ | Multiplies | Adds |
|---|---|---|---|---|
| 1 | 10 | 38 | 1.9 | 3.8 |
| 2 | 5 | 38 | 0.38 | 0.76 |
| 3 | 2 | 356 | 1.78 | 3.56 |
| | | | 4.06 | 8.12 |

Goodman-Carey 7-stage design[c]:

| | | | | |
|---|---|---|---|---|
| 1 | 2 | 3 | – | 2.56 |
| 2 | 2 | 7 | 0.96 | 1.28 |
| 3 | 2 | 7 | 0.48 | 0.64 |
| 4 | 2 | 11 | 0.32 | 0.48 |
| 5 | 2 | 19 | 0.24 | 0.4 |
| 6 | 2 | 19 | 0.12 | 0.2 |
| 7 | 2 | 356 | 1.78 | 3.56 |
| | | | 3.90 | 9.12 |

| Example C | | | | |
|---|---|---|---|---|
| $R_2:R_1 = 32$, | $d_p = 0.00316 = d_s$, | | $a = 0.8$, | $b = 1$ |

Crochiere-Rabiner 3-stage design:

| Stage | $D$ | $N$ | Multiplies | Adds |
|---|---|---|---|---|
| 1 | 8 | 27 | 1.75 | 3.5 |
| 2 | 2 | 8 | 0.25 | 0.5 |
| 3 | 2 | 52 | 0.813 | 1.62 |
| | | | 2.813 | 5.62 |

Goodman-Carey 5-stage design:

| | | | | |
|---|---|---|---|---|
| 1 | 2 | 3 | – | 2.0 |
| 2 | 2 | 7 | 0.75 | 1.0 |
| 3 | 2 | 11 | 0.5 | 0.75 |
| 4 | 2 | 15 | 0.313 | 0.5 |
| 5 | 2 | 52 | 0.813 | 1.62 |
| | | | 2.376 | 5.87 |

[a] Ratio of input rate to output rate.
[b] Per input sample.
[c] Assumes initial $R = 128$ (after resampling). Multiplications and additions per input samples are referred to $R = 100$.

## VI. CONCLUSIONS

We have shown how a set of nine specially designed filters can be used to cover a wide range of decimation and interpolation accuracy requirements. The filters are efficient in terms of number of multiplications and coefficient word length. Filter sequences (excluding a baseband filter) can be designed quickly, without a computer. The results often compare favorably with those of a more elaborate optimization procedure.

Although a decimator or interpolator with many half-band filters is efficient computationally, it requires more elaborate timing and control circuitry than a scheme with fewer stages. Consideration of these overheads may in practice lead to a single stage design or one produced by the Crochiere and Rabiner method or perhaps a combination of their method and ours using $F1$ or $F2$ at high speeds and filters with interpolation or decimation ratios greater than 2 at lower speeds.

## APPENDIX

### Filter Characteristics

Except for $F1$, all of the filters are half-band filters: approximations to the ideal low-pass filter with cutoff frequency $f_c = f_s/4$ at the center of the filter operating band. The impulse response coefficients $h_i$ have the property

$$h_{\pm 2} = h_{\pm 4} = \cdots = h_{\pm(N-3)/2} = 0. \quad (A1)$$

They also satisfy the constraint

$$h_0 = 2 \sum_{i=1}^{N-1} h_i \quad (A2)$$

which implies that each amplitude response function has odd symmetry about $f_c$

$$H(f_c + \psi) + H(f_c - \psi) = H(0). \quad (A3)$$

Consequently, $H(2f_c) = H(\frac{1}{2}f_s) = 0$ and each filter has at least a double zero at the bandedge $f = \frac{1}{2}f_s$. The zeros strongly suppress spectral images of narrow-band, low-frequency signals such as power line hum and certain speech sounds.

The curves for $F2$–$F9$ in Fig. 2 show the minimum $R$ for which

$$-20 \log \left| \frac{H(f)}{H(0)} \right| \geqslant D, \quad (2R-1)W \leqslant f \leqslant 2RW. \quad (A4)$$

This frequency range contains the spectral images to be suppressed in interpolation from $f_s = 2RW$ to a higher rate. It is the set of frequencies aliased into the baseband when decimation reduces the rate to $2RW$. It follows that if $D_s = -20 \log d_s$, the intersection of a filter curve and the horizontal line $D = D_s$, indicates the lowest $R$ for which that filter meets the stopband specification. The passband ripple requirement $d_p$ influences the selection of $F4$, $F6$–$F9$ the filters with nonmonotonic responses over 0, $W$ Hz. The symmetry of the response of each filter implies that with the sampling rate in the admissible region shown in Fig. 2,

$$1 - d \leqslant H(f)/H(0) \leqslant 1 + d, \quad f\epsilon(0,W) \quad (A5)$$

where $D = -20 \log d$ is the height of the horizontal line for the filter. Moreover, Fig. 4 shows that for each filter one of the two bounds in (A5) is quite loose, i.e., that the peak-to-peak ripple of each filter is nearer $d$ than $2d$. It follows that a filter

with $D = -20 \log d_p$ uses about half of the peak-to-peak ripple allowance of the entire sequence.

$F1$ is the class of filters in which all impulse response coefficients are unity. The frequency response of the $N$th order $F1$ filter is

$$H_N(f) = \frac{\sin(\pi f/2RW)}{N \sin(\pi f/2NRW)} \quad (A6)$$

in which $2NRW$ is the filter sampling rate. In the band $(0, 2NRW)$ there are $N-1$ zeros located at $2RW, 4RW, \cdots, (N-1)2RW$. For decimation, these are the centers of the frequency bands that are aliased into the signal baseband. In the case of interpolation they are the center frequencies of the spectral images to be suppressed. Over all of these bands, each of width $2W$, the most critical frequency is $2RW - W$. If

$$|H_N(2RW - W)| \leqslant d_s \quad (A7)$$

the filter is sufficiently accurate over its entire range. The $F1$ curve in Fig. 2 corresponds to the equation $H_3(2RW - W) = d_s$. As $N$ increases, precision improves, so that curves for $N > 3$ can be drawn above the $F1$ curve in Fig. 4. This curve does, however, provide a reasonably accurate lower bound on precision. It is within 1.65 dB of the $N = \infty$ curve for all values of $R$.

## REFERENCES

[1] R. W. Schafer and L. R. Rabiner, "A digital signal processing approach to interpolation," *Proc. IEEE*, vol. 61, pp. 692–702, June 1973.
[2] M. G. Bellanger, J. L. Daquet, and G. P. Lepagnol, "Interpolation, extrapolation and reduction of computation speed in digital filters," *IEEE Trans. Acoust., Speech, Signal Processing*, vol. ASSP-22, pp. 231–235, Aug. 1974.
[3] D. J. Goodman, "Digital filters for code format conversion," *Electron. Lett.*, vol. 11, pp. 89–90, Feb. 20, 1975.
[4] D. W. Rorabacher, "Efficient FIR filter design for sample-rate reduction and interpolation," in *Proc. 1975 IEEE Int. Symp. Circuits and Systems*, Apr. 21–23, 1975, pp. 396–399.
[5] R. R. Shively, "On multistage finite impulse response (FIR) filters with decimation," *IEEE Trans. Acoust., Speech, Signal Processing*, vol. ASSP-23, pp. 353–357, Aug. 1975.
[6] R. E. Crochiere and L. R. Rabiner, "Optimum FIR digital filter implementation for decimation, interpolation and narrow band filtering," *IEEE Trans. Acoust., Speech, Signal Processing*, vol. ASSP-23, pp. 444–456, Oct. 1975.
[7] ——, "Further considerations on the design of decimators and interpolators," *IEEE Trans. Acoust., Speech, Signal Processing*, vol. ASSP-24, pp. 296–311 Aug. 1976.
[8] G. Oetken, T. W. Parks, and H. W. Schussler, "New results in the design of digital interpolators," *IEEE Trans. Acoust., Speech, Signal Processing*, vol. ASSP-23, pp. 301–309, June 1975.
[9] D. W. Tufts, D. W. Rorabacher, and W. E. Mosier, "Designing simple, effective digital filters," *IEEE Trans. Audio Electroacoust.*, vol. AU-18, pp. 142–158, June 1970.
[10] R. J. Canniff, "Signal processing in SLC-40, a 40 channel rural subscriber carrier," in *Conf. Rec., 1975 IEEE Int. Conf. Communications*, vol. III, pp. 40.7–40.11, June 1975.
[11] L. R. Rabiner and B. Gold, *Theory and Application of Digital Signal Processing*. Englewood Cliffs, NJ: Prentice-Hall, 1975, chs. 3 and 4.
[12] J. H. McClellan, T. W. Parks, and L. R. Rabiner, "A computer program for designing optimum FIR linear phase digital filters," *IEEE Trans. Audio Electroacoust.*, vol. AU-21, pp. 506–526, Dec. 1973.

# A NOVEL ARCHITECTURE DESIGN FOR VLSI IMPLEMENTATION OF AN FIR DECIMATION FILTER

Hanafy Meleis
Pierre Le Fur


AT&T Bell Laboratories
Murray Hill, NJ 07974

## ABSTRACT

A novel architecture design of a one stage FIR filter for decimation is described. It performs the decimation of a 1-bit code at 1024KHz of double integration Sigma Delta modulation output to PCM at 16KHz. This architecture is designed in such a way that it needs only a simple control structure suitable for VLSI implementation. We devised an algorithm for generating the coefficients of the filter with a minimum of required hardware. It does not require storing the coefficients in a ROM and continuously reading it to calculate the convolution. The accumulators needed to perform the direct convolution are arranged in a way that simplifies and minimizes the hardware required for the filter implementation.

The filter response is $Sinc^3(f)$ which provides sufficient attenuation for modulation generated by means of double integration. The implementation of this filter requires the generation of the coefficients and the performance of the convolution. Three coefficients are needed with every input to obtain the output sequence. The major feature of this architecture is the use of an efficient algorithm to obtain the coefficients thereby reducing the area and power consumption. It is very suitable for VLSI implementation in CMOS technology.

## 1. INTRODUCTION

This paper describes the design of an FIR filter which is used to perform the decimation of a 1-bit code at 1024KHz of a double integration Sigma Delta modulation[1] output to PCM sampled at 16KHz. The filter response is $Sinc^3(f)$ and the duration of the filter impulse response is three resampling periods. Its frequency response has a third order zero at 16KHz resampling rate and all its harmonics.

The problem of implementing the $Sinc^3(f)$ in hardware lies in generating the coefficients of the filter. Previously a $Sinc^2(f)$ filter[2] was used for decimation and conversion of Sigma Delta modulation output to PCM. This filter has a triangularly shaped impulse response and its coefficients can be generated using only a six bit counter. A $Sinc^2(f)$ filter response does not provide sufficient attenuation for modulations generated by means of double integration. A filter with a frequency response equal to $Sinc^3(f)$ is needed. This filter would have incrementally non linear coefficients. In this paper we describe a novel method of generating the coefficients of this filter with a minimum of required hardware. Also, two alternative architectures for performing the convolution will be described.

## 2. FIR FILTER REALIZATION FOR DECIMATION

The process of digitally converting the sampling rate of a signal from a given rate $f_i = 1/T_i$ to a lower rate $f_o = 1/T_o$ is called decimation[3]. Assume x[n] is an input signal with a sampling rate $f_i$ and with full band, i.e., its spectrum is nonzero for all frequency f in the range $-f_i/2 \leqslant f \leqslant f_i/2$. In order to lower the sampling rate and avoid aliasing at the new rate, it is necessary to filter the signal with a digital low-pass filter. An FIR filter can be used to obtain a band-limited input signal[4]. It can be realized by direct application of the convolution equation:

$$y[n] = \sum_{k=0}^{N-1} h[k] x[n-k] \qquad (1)$$

where h[n] is the impulse response of the filter and $N = f_i/f_o$ is the decimation factor. x[n] is the input signal and y[n] is the filter output signal. The sample rate reduction is achieved by extracting y[n] every Nth sample and forming a new sequence as an output of the decimator.

In applications wherein the analog signal is encoded by means of delta modulation, the requirement is that the low pass filter supply an adequate attenuation of the modulation noise[2]. In the case of double integration Sigma Delta modulation, a filter having $Sinc^3(f)$ response is required. The impulse response of this filter is:

$$h[n] = \frac{n(n+1)}{2} \qquad \text{for } 1 \leqslant n < N \qquad (2)$$

$$h[n] = \frac{N(N+1)}{2} + (n-N)(2N-1-n) \qquad \text{for } N \leqslant n < 2N \qquad (3)$$

$$h[n] = \frac{(3N-n-1)}{2} \qquad \text{for } 2N \leqslant n < 3N \qquad (4)$$

where N is the decimation factor.

The filter we designed decimates the 1-bit code at 1024KHz to 16KHz PCM. In this case the filter has 190 coefficients and N=64. The action of the filter as a decimator in the time domain is an averaging of 190 samples with 1μsec being the period of the input sequence and 64 the period of the output sequence. In the frequency domain the filter response is:

$$H(\omega) = \left[ \frac{\sin \omega NT}{\sin \omega T} \right]^3 \qquad (5)$$

This filter has a third-order zero at 16KHz resampling rate and all its harmonics. The realization of this filter in the time domain using the following convolution equation:

$$y[n] = \sum_{k=0}^{189} h[k] x[n-k] \qquad (6)$$

requires the generation of the coefficients and performance of the convolution. The input sequence is a 1-bit sequence, this reduces the complexity of the filter by eliminating the required multi-bit multiplier for calculating the convolution. Only accumulators are needed. Also, the decimation of the input sequence with period T =1μsec to a new sequence with period T =64μsec requires generating three different coefficients every period of 1μsec and performs three accumulations in this period. Figure 1 shows the three simultaneous accumulations needed to filter the data and lower the sampling rate by a factor of 64. The envelope in this figure represents the values of the coefficient. For every 64 samples of the input sequence, a calculation of an output starts by accumulating the corresponding coefficient if the input sample equals 1 for a period of 190 input samples. This output will be extracted after 192 samples of the input signals to form the new sequence.

Reprinted from *IEEE Proc. ICASSP'85*, pp. 1380–1383, March 1985.

An efficient design for VLSI implementation of this decimator required minimum storage to minimize the area required on the chip and a simple control to minimize the power consumption. We designed this decimator with a minimum of required hardware. In the next section we discuss the architecture for this decimator.

## 3. ARCHITECTURE DESIGN

We designed the $Sinc^3(f)$ filter for decimating the delta modulation output sequence as a one stage decimator. The architecture design consists of two modules, one for generating the coefficients and the other for performing the direct convolution.

### 3.1 Generating the coefficients

Equations 2, 3 and 4 represent the coefficients of the filter. One method of implementing this filter can be accomplished by storing the coefficients in a ROM and reading it three times every period of the input sequence to form the output sequence as described in section 2. This implementation does require more hardware than in our method and the rate of changing the control signals in the ROM implementation will lead to a higher power consumption in CMOS implementation.

The coefficients of this filter are incremently nonlinear as shown in equations 2, 3 and 4. We generated these coefficients with a minimum of required hardware by calculating the incremental values in every section of the impulse response. From equation 2, the incremental value between two successive coefficients is:

$$\Delta_1[n] = n+1 = n_c \qquad 1 \leqslant n_c \leqslant 64 \ . \tag{7}$$

Where $n_c$ is the output of a seven bit counter which counts from 1 to 64. From equation 3 the incremental value will be :

$$\Delta_2[n] = N - 2(n - N - 1) \qquad N \leqslant n < 2N \ .$$

The value n-N-1 can be replaced by $n_c$ and $\Delta_2$ can be written as follows:

$$\Delta_2[n] = N - 2n_c \qquad 1 \leqslant n_c \leqslant 64 \ . \tag{8}$$

In the same manner equation 4 can be used to calculate the incremental values in the third portion of the impulse response :

$$\Delta_3[n] = -N + (n - 2N + 1) \qquad 2N \leqslant n < 3N$$

It can be written in relation to $n_c$ as follows·

$$\Delta_3[n] = -N + n_c \qquad 1 \leqslant n_c \leqslant 64 \tag{9}$$

The hardware used to generate the coefficients are one seven bit counter, three twelve bit adders, three twelve bit registers, three half adders and control logic. The counter is driven by a 1024KHz clock and changes its output with the negative edge of this clock. The three registers latch with the positive edge of the 1024KHz clock. The adders used are carry ripple adders. Figure 2 shows the architecture of the first module of the design. CR1, CR2, and CR3 are the coefficient registers. MC is the master counter which counts from 1 to 64. The control logic generates a reset pulse every 64th cycle of 1024KHz. It also generates two signals, DSC0 and DSC1, which count three frames of 64 cycles each. These signals will be used in the second module to generate the output sequence.

The first section of the impulse response contains the coefficients from n=1 to n=64. The value of the first coefficient equals 1 and the 64th coefficient equals 2080 (equation 2). The incremental value between any two successive coefficients equals the counter output nc( equation 7). Figure 3 shows the time diagram for the counter and the coefficient registers. With every negative edge of the 1024KHz clock the counter generates its output. A twelve bit adder is used to accumulate the previous coefficient to the output of the counter. The new coefficient is stored in the first coefficient's register CR1 at the

positive edge of the 1024KHz clock . At the end of the 64th cycle the reset pulse (figure 3) initializes CR1 to a value zero and the coefficient starts again from the first value.

The second section of the impulse response contains the coefficients from n=65 to n=128. The 65th coefficient equals 2142 and the 128th equals 2016 (equation 3). Equation 8 shows the incremental value in this portion of the impulse response. To explain the method of generating the value N-2nc from the counter output we can rewrite this equation as follows:

$$\Delta_2[n] = 64 + 2(-n_c)$$
$$= 64 + \overline{2n_c} + 1 \ . \tag{10}$$

where $\overline{2n_c}$ is the complement of nc after it shifted to the left by 1 bit. The value 1 in equation 10 is added in order to obtain the 2's complement of 2nc. Figure 4 shows the process of obtaining the value of equation 10. Generating $\Delta_2[n]$ will require only three half adders. Adding 1 in the first bit is accomplished by holding the carry input of the adder to a logic level one. Adding this incremental value to the previous coefficient in the CR2 register will generate the new coefficient. After 64 cycles, the reset pulse will initialize the CR2 register to the value 2080. The first coefficient in this section is generated by adding the first incremental value (62) and so on.

The third section of the impulse response represents the 129th coefficient to the 190th. The coefficient register CR3 is initialized with the value of the 128th coefficient and the incremental value is added to generate the corresponding coefficient in this section. From equation 9, the incremental value -N+nc can be constructed from the first six bits of the counter. The value -N (where N=64) in 2's complement is 7700 in octal. Thus the last six bits of the adder are connected to logic level one and the first six bits to the first six bits of the counter MC.

From the above discussion, we can summarize the algorithm of generating the coefficients in two steps. The first step is initializing CR1, CR2 and CR3 every 64 cycles of 1024KHz. CR1 is initialized to zero, CR2 to the 64th coefficient and CR3 to the 128th coefficient. Secondly, the incremental value in every section of the impulse is generated every clock cycle from the output of the counter MC. These values are accumulated to the values in CR1, CR2 and CR3 to generate the new values of the coefficients.

### 3.2 Performing the direct convolution

The second module of the architecture performs the convolution to obtain the output sequence as shown in equation 6.

As discussed in section 2, the input data is a 1-bit sequence at 1024KHz and three accumulations are needed for every cycle (see figure 1). A calculation of an output starts every 64 samples of the input sequence. The corresponding coefficient accumulates if the input sample equals 1 for a period of 190 input samples. The output is extracted every 64th sample of the input sequence. This will achieve the decimation of the data by an integer factor of 64.

There are two ways to implement this module depending on the frequency of the master clock. If the available master clock frequency is 1024KHz, the three accumulations needed during every sample are obtained in a parallel fashion. Figure 5 shows the hardware required in this case. It consists of three 12 bit multiplexers, three 19 bit adders, three 19 bit registers, a 19 bit multiplexer and a control logic. The 19 bits are needed to accommodate the maximum number that can be obtained from the accumulations of all the coefficients. The control logic uses DSC0 and DSC1 signals which generate from the first module. Then it generates the control signals for the multiplexer and a reset signal for the registers DR1, DR2 and DR3 . The control signals CM1, CM2 and CM3 of the multiplexer are generated in such a way that when MUX1 is using the first section of the impulse response, MUX2 is using the third section and MUX3 is using the second section. The output of these multiplexers equals zero if the input sequence is zero. C1, C2 and C3 represents the coefficient values from the registers CR1, CR2 and CR3. In order to understand the function of the second module examine figure 1 and figure 5 simultaneously. Every series combination of multiplexer, adder and register in figure 5 accumulates one output. Figure 1 shows that three outputs are constructed simultaneously. And 190 samples of the input are needed to form an output. The output sequence is extracted every 64th input sequence. The control signals of the multiplexer MUXO are designed in such a way to pass the data from the register which is ready to give an output. A reset signal will clear this register after reading its content and a new cycle of forming the output will start.

The second method of implementing the accumulators needed to perform the convolution required a 4096KHz master clock. Figure 6 shows the hardware used. It consists of a 12 bit multiplexer, a 19 bit adder, three 19 bit registers and a 19 bit output register. The adder and the three 12 bit registers are connected in a series array. The inputs of the adder are a coefficient and the output of the last register DR1 in the array. The value in DR1 is added to a coefficient and stored in DR3, the value of DR3 is shifted to DR2 and DR2 is shifted to DR1. This process takes place three times in a 1μsec period. This requires the generation of a 3072KHz clock to latch the registers DR1, DR2 and DR3. Figure 7 shows the 3072KHz master-slave clock generated from the 4096KHz clock. In a period of 1 μsec there are three slave signals and three master signals. With every master-slave the MUXI supplies the adder with the necessary coefficient to be added to DR1. The output register reads the output of DR1 every 64 cycles of 1024 KHz during a slave clock. It reads the first frame during SL1, the second during SL2 and the third during SL3.

In the previous section we demonstrated two alternative methods of implementing the hardware required to perform the convolution. We demonstrated that less hardware is required if 4096KHz is available. As shown in figures 5 and 6 use of the higher master clock frequency eliminates two 19 bit adders and two 19 bit multiplexers.

## 4. CONCLUSION

A novel architecture design of a one stage FIR filter for decimation with a $Sinc^3(f)$ response has been described. It performs the decimation of a 1-bit code at 1024KHz to PCM at 16KHz. The algorithm used here for generating the coefficients of the filter minimized the hardware required for implementation.

The architecture and logic design presented in this paper has been tested using the standard TTL board level implementation.

## REFERENCES

[1] J. C. Candy, "A Use of Double Integration in Sigma Delta Modulation," IEEE Transactions on Communications Vol. COM-33, March 1985.

[2] J. C. Candy, Bruce A. Wooley and O'Connell J. Benjamin, "A Voiceband Codec with Digital Filtering," IEEE Transactions on Communications, vol. COM-29, June 1981.

[3] R. E. Crochiere and Lawrence R. Rabiner, "Interpolation and Decimation of Digital Signals - A Tutorial Review," Proc. IEEE, vol. 69, pp. 300-331, March 1981.

[4] R. R. Shively, "On Multistage Finite Impulse Response (FIR) Filters with Decimation," IEEE Trans. Acoust., Speech, Signal Processing, vol. ASSP-23, August 1975.

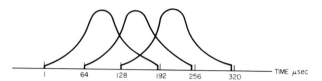

FIGURE 1 EVERY 1μsec THREE ACCUMULATIONS ARE NEEDED. THE ENVELOPE REPRESENTS THE VALUE OF THE COEFFICIENTS

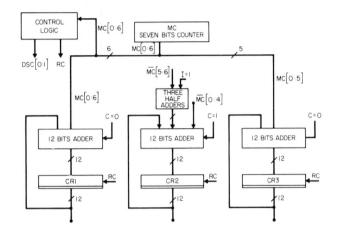

FIGURE 2 THE HARDWARE REQUIRED TO GENERATE THE COEFFICIENTS OF $SINC^3(f)$ FILTER WITH CUTOFF FREQUENCY AT 16 kHz

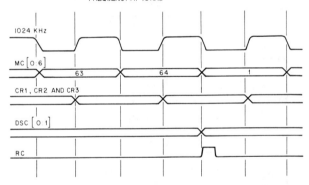

FIGURE 3 TIMING DIAGRAM OF THE ARCHITECTURE SHOWN IN FIGURE 2

| BIT NUMBER | 8 | 7 | 6 | 5 | 4 | 3 | 2 | 1 | 0 |
|---|---|---|---|---|---|---|---|---|---|
| $\overline{n_c}$ | 1 | 1 | X | X | X | X | X | X | X |
| $\overline{2n_c}+1$ | 1 | X | X | X | X | X | X | X | 1 |
| 64 | 0 | 0 | 0 | 1 | 0 | 0 | 0 | 0 | 0 |
| | $B_8$ | $B_7$ | $B_6$ | X | X | X | X | X | 1 |

FIGURE 4 GENERATING $\triangle_2[n] = 64 + \overline{2n_c} + 1$ $B_6, B_7, B_8$ ARE GENERATED USING THREE HALF ADDERS

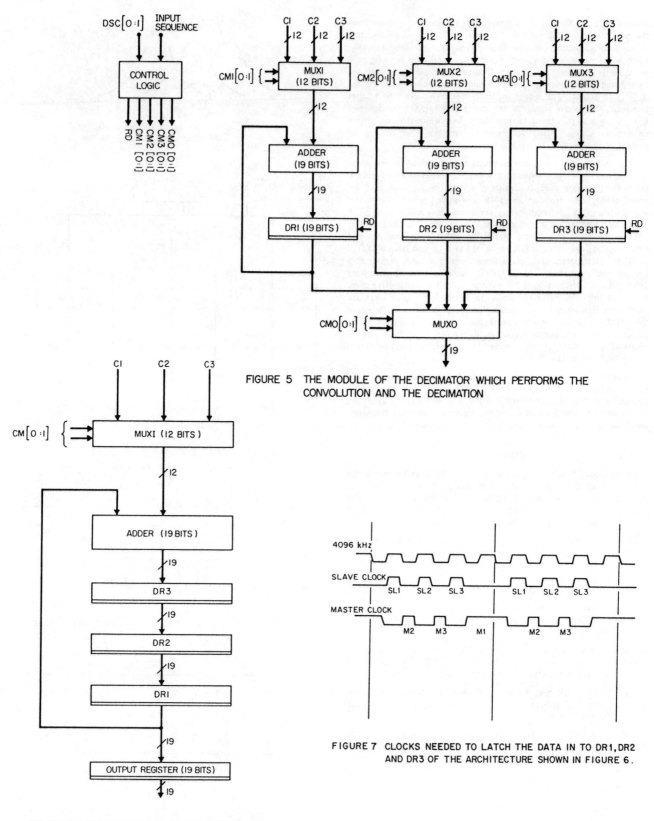

FIGURE 5 THE MODULE OF THE DECIMATOR WHICH PERFORMS THE CONVOLUTION AND THE DECIMATION

FIGURE 6 ALTERNATIVE DESIGN FOR THE MODULE SHOWN IN FIGURE 5

FIGURE 7 CLOCKS NEEDED TO LATCH THE DATA IN TO DR1, DR2 AND DR3 OF THE ARCHITECTURE SHOWN IN FIGURE 6.

# Efficient VLSI-Realizable Decimators for Sigma-Delta Analog-to-Digital Converters

Tapio Saramäki and Hannu Tenhunen

Department of Electrical Engineering
Tampere University of Technology
P. O. Box 527, SF–33101 Tampere, Finland

*Abstract* – This paper introduces a class of efficient linear-phase FIR decimators for attenuating the out-of-band noise generated by a sigma-delta analog-to-digital modulator. These decimators contain no general multipliers and very few data memory locations, thereby making them easily VLSI-realizable. This is achieved by using several decimation stages with each stage containing a small number of delays and arithmetic operations. The output sampling rate of these decimators is the minimum possible one, unlike for most other existing designs and the proposed decimators can be used, with very slight changes, for many ovesampling ratios. Futhermore, these decimators attenuate highly the undesired out-of-band signal components of the input signal, thus significantly relaxing the anti-aliasing prefilter requirements.

## I. INTRODUCTION

Efficient high resolution analog-to-digital conversion is obtained by using oversampled sigma-delta modulation with one-bit quantization [1], [2]. Modulation together with oversampling moves most of the quantization noise out of the baseband [3], [4]. The noise lying out of the baseband can then be reduced by using a decimator [5].

One of the major obstacles in the sigma-delta converter design is the efficient small area integration of the decimator. In this paper, we report a novel high performance linear-phase FIR filter structure which can be easily implemented in small area. In order to optimize both the decimator performance (noise, baseband frequency) and the VLSI realizability (circuit area, power, speed), the proposed decimators are designed to consist of several stages with each stage requiring a small number of arithmetic operations. The optimization is performed in such a way that no general multipliers are required. The overall filter is constructed using a fixed part and an adjustable part. With slight programmable changes in the adjustable filter part, the overall filter can be used for many oversampling ratios. Moreover, the output sampling rate is the minimum possible one, unlike for most other existing designs, and the proposed decimators attenuate highly the undesired input signal components lying out of the baseband, thereby relaxing the anti-aliasing prefilter requirements.

## II. STATEMENT OF THE PROBLEM

The block diagram for the overall system is depicted in Fig 1. The output sampling rate of the sigma-delta modulator is $M$ times the final sampling rate $f_s$. We assume that the sigma-delta modulator is of second order so that the spectral density of the noise at its output is [5]

$$E(f) = \frac{\sigma^2}{12} \frac{2}{Mf_s} [2(1 - \cos(\frac{2\pi f}{Mf_s}))]^2, \quad (1)$$

where $\sigma$ is the spacing of the quantization levels. We state the

following amplitude requirements for the decimator

$$1 + \delta_p \le |H(e^{j2\pi f/(Mf_s)})| \le 1 - \delta_p \text{ for } 0 \le f \le \frac{3}{4}\frac{f_s}{2} \quad (2a)$$

$$|H(e^{j2\pi f/(Mf_s)})| \le \delta_s \text{ for } \frac{5}{4}\frac{f_s}{2} \le f \le M\frac{f_s}{2}, \quad (2b)$$

where $\delta_p = 0.01$ and $\delta_s = 0.001$ (60-dB stopband attenuation). In addition, it is required that that the noise spectral density at the output of the decimator filter before sampling rate reduction

$$\widehat{E}(f) = |H(e^{j2\pi f/(Mf_s)})|^2 E(f) \quad (3)$$

is on the region $[f_s/2, Mf_s/2]$ well below the level on the baseband $[0, f_s/2]$. This guarantees that the contribution of the aliased components to the overall baseband noise becomes very small.

## III. PROPOSED CLASS OF DECIMATORS

To reduce the arithmetic complexity of the decimator, it is preferred to construct it using several low-order stages, instead of one high-order stage. A multistage implementation of the proposed decimator is given in Fig. 2. It consists of an adjustable filter part and a fixed filter part. The adjustable filter part enables us to use the same structure for various values of $M$. The values of $M$ in use are $M = 16, 32, 64, 128, 256, 512$. The transfer function of the equivalent single-stage design can be written in the form

$$H(z) = H_3(z^M)H_2(z^{M/2})H_1(z^{M/4})F_2(z^{M/16})F_1(z), \quad (4a)$$

where

$$F_2(z) = 2^{-8}\left[\frac{1 - z^{-4}}{1 - z^{-1}}\right]^4 \quad (4b)$$

$$F_1(z) = 2^{-P}\left[\frac{1 - z^{-K}}{1 - z^{-1}}\right]^3 \quad (4c)$$

with

$$K = \begin{cases} 32 & \text{for } M = 512 \\ 16 & \text{for } M = 256 \\ 8 & \text{for } M = 128 \\ 4 & \text{for } M = 64 \\ 2 & \text{for } M = 32 \\ 1 & \text{for } M = 16 \end{cases} \quad (4d)$$

and

$$P = \begin{cases} 15 & \text{for } M = 512 \\ 12 & \text{for } M = 256 \\ 9 & \text{for } M = 128 \\ 6 & \text{for } M = 64 \\ 3 & \text{for } M = 32 \\ 0 & \text{for } M = 16. \end{cases} \quad (4e)$$

Here, $F_1(z)$ is the transfer function of the adjustable filter part in the case where the sampling rate reduction is not performed and $F_2(z)$ is the transfer function from the input of the fixed filter

Reprinted from *IEEE Proc. ISCAS'88*, pp. 1525–1528, June 1988.

part to the input of $H_1(z)$. The term in parentheses in Eqn. (4c) can be rewritten in the form

$$\frac{1-z^{-K}}{1-z^{-1}} = \sum_{r=0}^{K-1} z^{-r}.$$

Similarly, the term in the parentheses of Eqn. (4b) can be expressed in the above form. Thus these terms correspond to linear-phase FIR filters. Linear-phase filters with transfer functions consisting of the above recursive terms have been been used for sampling rate alteration in [6] and [7]. Using the techniques proposed in these papers, we can implement $F_1(z)$ and $F_2(z)$ using the substructures shown in Fig. 2. We note that when the feedforward term $1 - z^{-K}$ is transfered after the sampling rate reduction by a factor of $K$, it becomes $1 - z^{-1}$. It should be noted also that if 1's or 2's complement arithmetic (or modulo arithmetic in general) and the worst-case scaling are used, the output of the filters $F_1(z)$ and $F_2(z)$ implemeted as shown in Fig. 2 is correct even though there may occur overflows in the feedback loops realizing the term $1/(1 - z^{-1})$. The proofs of this fact can be found in [7] and [8]. Also, under the above conditions, the effect of temporary miscalculations vanishes from the output in finite time and initial resetting is not necessary needed. The scaling constant $2^{-P}$ and $2^{-8}$ have been selected according the worst-case scaling. Later on, we return to the selection of the number of terms in $F_1(z)$ and $F_2(z)$.

We have designed the remaining transfer functions $H_1(z)$, $H_2(z)$, and $H_3(z)$ using two different approaches to be described later on. Figure 3 gives the amplitude response of the overall design in these two cases for $M = 64$. For the first design, $H_3(z)$ is absent and $H_1(z)$ and $H_2(z)$ are linear-phase FIR filters of orders 5 and 22, respectively. By exploiting the symmetry in the filter coefficients, these filters require 3 and 12 multipliers, respectively. These two filters have been optimized in such a way that the overall filter meets the specifications of Eqn. (2) in the frequency region $[0, 2f_s]$. The optimization has been accomplished using the methods proposed in [9] and [10]. The transfer function $H_1(z)$ has all the zeros on the unit circle and the locations of these zeros have been determined such that the overall filter response exhibits an equiripple behavior on the stopband region $[11f_s/8, 2f_s]$ which alias to the region $[0, 5f_s/8]$ after decimating by a factor of two at the output of $H_1(z)$. $H_2(z)$, in turn, has been designed to provide for the overall amplitude response the desired equiripple nature on the passband $[0, 3f_s/8]$ and on the stopband region $[5f_s/8, f_s]$. In addition to providing the desired attenuation, this filter equalizes the passband distortion caused by the earlier filter stages. We note that when changing $M$, the change in the response of $F_1(z)$ is negligible on $[0, 2f_s]$ and the response of $F_2(z)$ remains the same on this region. This enables us to use the same filters $H_1(z)$ and $H_2(z)$ for all the values of $M$. In the actual implementations of $H_1(z)$ and $H_2(z)$ we have exploited the fact that only every second output needs to be computed. For the multirate FIR filter structures exploiting the coefficient symmetry, see, e.g., [11].

The disadvantage of the first design is that it is not possible to use powers-of-two or sums or differences of two powers-of-two to represent the filter coefficients. If rounding is used, the filter coefficients need approximately 10-bit representations to meet the given amplitude criteria. To overcome this limitation, we have designed, as a second alternative, special tailored subfilters which have been optimized such that the resulting overall filter does not require general multipliers. The transfer functions are

$$H_1(z) = (2^{-5} - 2^{-10})(1 + z^{-5}) + (2^{-3} + 2^{-9})(z^{-1} + z^{-4})$$
$$+ (2^{-2} - 2^{-6})(z^{-2} + z^{-3}), \qquad (5)$$

$$H_2(z) = 2^{-1}z^{-15} + \widehat{H}_2(z^2), \qquad (6a)$$

where

$$\widehat{H}_2(z) = F(z)[(2^0 - 2^{-2})z^{-5} + (-2^{-2} + 2^{-9})[F(z)]^2], \quad (6b)$$

with

$$F(z) = (2^{-4} + 2^{-6})(1 + z^{-5}) + (-2^{-3} - 2^{-5} - 2^{-6})(z^{-1} + z^{-4})$$
$$+ (2^{-1} + 2^{-3})(z^{-2} + z^{-3}), \qquad (6c)$$

and

$$H_3(z) = (2^0 + 2^{-1} - 2^{-6})[(2^{-6} + 2^{-9})(1 + z^{-4})$$
$$+ (-2^{-4} - 2^{-5})(z^{-1} + z^{-3}) + (2^0 + 2^{-6})z^{-2}]. \quad (7)$$

As for the first design, $H_1(z)$ provides the desired attenuation on the region $[11f_s/8, 2f_s]$. $H_2(z)$ is a special half-band filter which can be implemented effectively using a polyphase structure based on the commutative model [11]. The resulting structure is shown in Fig. 4. One of the branches is a pure delay term. The other branch is a tapped cascaded interconnection of three identical fifth-order filters. This filter has been designed to provide the desired attenuation on $[5f_s/8, f_s]$. The actual design of this filter has been accomplished by properly modifying the methods proposed in [12] for optimally designing FIR filters as a tapped cascaded interconnection of identical subfilters. Since $H_2(z)$ is a half-band filter, it cannot be used for compensating the passband distortion caused by the earlier filter stages. For this purpose we use $H_3(z)$.

### IV. FILTER PERFORMANCE

The performance of the overall A/D converter of Fig. 1 is limited by the modulator limitations [13] and by the performance of the decimator due to the aliasing of the noise into the baseband. The decimator filter performance has been examined by assuming that the sigma-delta modulator is ideal with the output noise spectral density as given by (1). For the first decimator design, the overall output noise powers for the used values of $M$ have been evaluated. An illustrative way is to express these values in terms of the number of additional bits defined by

$$\text{number of additional bits} = \log_4\left(\frac{\sigma_0^2}{\sigma_{\text{out}}^2}\right),$$

where $\sigma_{\text{out}}^2$ is the output noise power and $\sigma_0^2 = \sigma^2/12$ is the noise power generated by using direct rounding of data at the final sampling rate of $f_s$. Note the accuracy increases by one bit if the noise power is reduced to be one fourth. The evaluated number of additional bits for various values of $M$ is

| | |
|---|---|
| 8.2 | for $M = 16$ |
| 10.7 | for $M = 32$ |
| 13.2 | for $M = 64$ |
| 15.7 | for $M = 128$ |
| 18.2 | for $M = 256$ |
| 20.7 | for $M = 512$. |

In calculating the above figures, the overall output noise power obtained after decimation has been taken into consideration (not only in the passband). The dashed and solid lines in Fig. 5 give before decimation the noise spectra before and after filtering, respectively, whereas the dashed and solid lines in Fig. 6 show the spectra on the region $[0, f_s/2]$ at the ouputs of the modulator and the overall system, respectively. The spectrum given by the solid line of Fig. 6 contains thus also the contribution of the aliased components. The spectra have been scaled in such a way that 0 dB is the level of the noise spectrum which is obtained by using direct rounding of data at the final sampling rate of $f_s$. For the second tailored design, the number of additional bits is the same. It is interesting to observe from Fig. 6 that in the passband region

$[0, (3/4)f_s/2]$ the contribution of the aliased noise to the overall output noise is negligible. The only exception is the beginning of the passband where the output of the sigma-delta modulator contains very little amount of noise. Another interesting feature is that the noise level of the overall system is lower than that of the sigma-delta modulator in the region $[(3/4)f_s/2, f_s/2]$. This shows that the proposed decimator attenuates very effectively the out-of-band noise. If only the noise in the passband region $[0, (3/4)f_s/2]$ before decimation is taken into consideration, the above values increase only by 0.7 bits.

From Fig. 3, it is seen that except for the very beginning of the stopband, the stopband attenuation of both designs is much higher than 60 dB. This is a desired property since filters of this kind attenuate the stopband noise generated by the sigma-delta modulation well below the noise level in the baseband (see Fig. 5). Therefore, the contribution of the stopband noise to the overall noise obtained after decimation is very small. The fixed filter part $F_1(z)$ controls the stopband peaks around the frequencies $32f_s/2$ and $64f_s/2$. Selecting three terms in $F_1(z)$, as given in Eqn. (4c), guarantees that these peaks in the filtered noise spectrum are well below the level of the first peak around the zero frequency (see Fig. 5). If only two terms had been selected, then these peak would be at the same level. The other peaks in the stopband are controlled by $F_2(z)$ and as seen from Figs. 3 and 5, four terms in $F_2(z)$ guarantee that these peaks are well below the noise level in the baseband and the overall filter satisfies the given amplitude criteria.

To show the efficiency of the proposed structure, we have estimated the minimum lengths of optimal direct-form FIR filters to meet the specifications of Eqn. (2) for the various values of $M$. The minimum lengths are 163, 326, 651, 1302, 2603, and 5206 for $M = 16$, $M = 32$, $M = 64$, $M = 128$, $M = 256$, and $M = 512$, respectively. The number of multipliers required by these designs for large values of $M$ becomes too high for practical implementation. Another disadvantage of these equiripple designs is that the level of the noise lying in the stopband is larger than that of the noise lying in the baseband. In this case, the contribution of the stopband noise to the overall noise obtained after decimation is dominating.

## V. FILTER ARCHITECTURE

The above filter structures will lead directly to a very efficient VLSI implementation. The filter parts $F_1(z)$ and $F_2(z)$, given by Eqns. (4c) and (4b), respectively, have been implemented for programmable decimation ratios 32–128 using combined bit-parallel and bit-serial architectures for maximum layout compactness and speed. The first sections, feeded directly by a one-bit stream from the modulator, have been implemented parallelly because of speed limitations. After decimating by a factor of $K$, the bit-rate is reduced and bit-serial structures can be utilized. Bit-serial and bit-parallel sections have been synchronized such that $K$ parallel sums take as long as 16 serial cycles. The total area for implementing $F_1(z)$ and $F_2(z)$ is 2.2 mm$^2$ using 2.5 micron CMOS technology. The maximum bit-rate can be as high as 20 MHz for the current design.

The remaining fixed filter sections can be implemented either using the first design with some general multipliers or the second multiplier-free design. For the first design, a multiplier-based architecture is used. It is bit serial and follows the structure proposed in [14]. The estimated area is 4.7 mm$^2$ for 12-bit coefficients. An optimal implementation for the multiplier-free solution is a dedicated minimal core processor because of the modularity of the design. This will result in a similar silicon area. The overall sigma-delta A/D converter with 16-bit dynamic range for modem applications can be integrated with 2.5 micron CMOS technology in area less than 8 mm$^2$.

## VI. CONCLUSION

An efficient linear-phase FIR filter structure has been proposed for eliminating the out-of-band noise generated by a sigma-delta analog-to-digital converter. The main advantages of the proposed filter structure are:

1. It can be easily implemented in CMOS VLSI.
2. The quantization noise generated by the sigma-delta modulator is effectively attenuated.
3. The output sampling rate is the minimum possible one, unlike for most other proposed designs.
4. Input signal components, such as possible sinusoidal components, lying in the frequency band $[(5/4)(f_s/2), M(f_s/2)]$ are highly attenuated, thus relaxing the anti-aliasing prefilter requirements.
5. The same structure can be used for many oversampling ratios.
6. The area for implementing the overall A/D converter is less than 8 mm$^2$ with 2.5 micron CMOS technology.

## ACKNOWLEDGEMENT

This work has been supported by the National Microelectronics Program of Finland. The authors also thank the Median-Free Group International for excellent working atmosphere and fruitful discussions during the course of this work.

## REFERENCES

[1] M. W. Hauser, P. J. Hurst, and R. W. Brodersen, "MOS ADC filter combination that does not require precision analog components," in *Proc. IEEE Int. Solid-State Circuit Conf.*, pp. 80–82, Feb. 1985.

[2] M. W. Hauser and R. W. Brodersen, "Circuit and technology considerations for MOS delta-sigma A/D converters, in *Proc. IEEE Int. Symp. Circuits Syst.*, pp. 1310–1315, May 1986.

[3] R. M. Gray, "Oversampled sigma-delta modulation," *IEEE Trans. Comm.*, vol. COM-35, pp. 481–489, May 1987.

[4] S. H. Ardalan, J. J. Paulos, "An analysis of nonlinear behavior in sigma-delta modulators," *IEEE Trans. Circuits Syst.*, vol. CAS-34, pp. 593–603, June 1987.

[5] J. C. Candy, "Decimation for sigma delta modulation," *IEEE Trans. Comm.*, vol. COM-34, pp. 72–76, Jan. 1986.

[6] T. Saramäki, "Efficient recursive digital filters for sampling rate conversion," in *Proc. IEEE Int. Symp. Circuits Syst.* (Newport Beach, CA), pp. 1322-1326, May 1983.

[7] S. Chu and C. S. Burrus, "Multirate filter design using comb filters," *IEEE Trans. Circuits Syst.*, vol. CAS-31, pp. 913–924, Nov. 1984.

[8] T. Saramäki, Y. Neuvo, and S. K. Mitra, "Design of computationally efficient interpolated FIR filters," *IEEE Trans. Circuits Syst.*, pp. 70–88, vol CAS-35, Jan. 1988.

[9] T. Saramäki, "A class of linear-phase FIR filters for decimation, interpolation, and narrow-band filtering," *IEEE Trans. Acoust., Speech, Signal Processing,* vol. ASSP-32, pp. 1023–1036, Oct. 1984.

[10] T. Saramäki, "Design of optimal multistage IIR and FIR filters for sampling rate alteration", in *Proc. IEEE Int. Symp. Circuits Syst.* (San Jose, CA), pp. 227–230, May 1986.

[11] R. E. Crochiere and L. R. Rabiner, *Multirate Digital Signal Processing.* Englewood Cliffs, NJ: Prentice-Hall, 1983.

[12] T. Saramäki, "Design of FIR filters as a tapped cascaded interconnection of identical subfilters," *IEEE Trans. Circuits Syst.*, vol. CAS-34, pp. 1011–1029, Sept. 1987.

[13] T. Ritoniemi, T. Karema, H. Tenhunen, and M. Lindell, "Fully differential CMOS sigma-delta modulator for high performance analog-to-digital conversion with 5 V operating voltage," in *IEEE Int. Conf. Circuits Systs.*, 1988, this conference.

[14] Y. Matsuya, K. Uchimura, A. Iwata, T. Kobayaski, M. Ishikawa, and T. Yoshitome, "A 16-bit oversampling A-to-D conversion technology using triple-integration noise shaping," *IEEE J. Solid-State Circuits*, vol. SC-22, pp. 921-929, Dec. 1987.

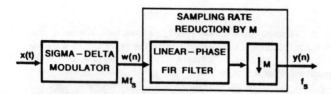

**Fig. 1.** Block diagram for the A/D converter consisting of an oversampled sigma-delta modulator and a decimator filter.

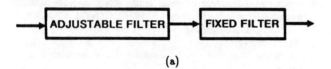

(a)

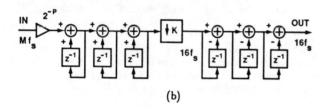

(b)

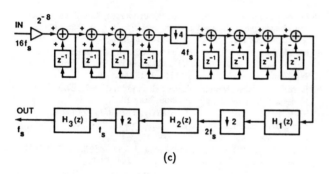

(c)

**Fig. 2.** Implementation of the proposed decimator. (a) Overall filter structure. (b) Adjustable filter part. (c) Fixed filter part.

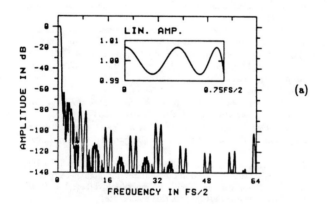

(a)

**Fig. 3.** Amplitude responses for the proposed decimators for $M = 64$. (a) Multiplier-based design.

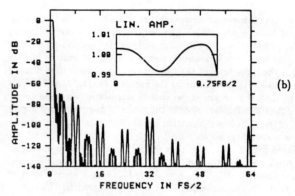

(b)

**Fig. 3.** (Continued.) (b) Multiplier-free design.

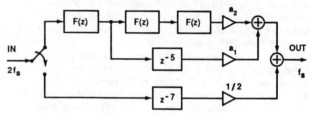

**Fig. 4.** Efficient implementation of the half-band filter $H_2(z)$. $a_1 = 2^0 - 2^{-2}$ and $a_2 = -2^{-2} + 2^{-9}$.

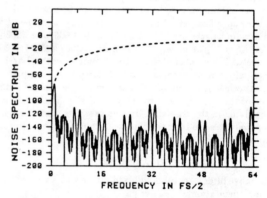

**Fig. 5.** Noise spectra before (dashed line) and after (solid line) filtering. These are the spectra before decimation. 0 dB corresponds to the noise level obtained when using direct rounding of data at the final output sampling rate of $f_s$.

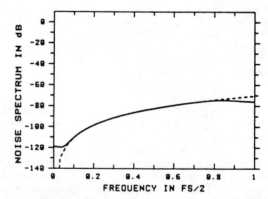

**Fig. 6.** Noise spectra in the baseband $[0, f_s/2]$. The solid line gives the spectrum at the output of the overall system after filtering and decimation and the dashed line gives the spectrum at the output of the modulator.

# Part 5
# Theory and Implementations
# of Oversampling D/A Converters

# Double Interpolation for Digital-to-Analog Conversion

JAMES C. CANDY, FELLOW, IEEE, AND AN-NI HUYNH, STUDENT MEMBER, IEEE

*Abstract*—Interpolative digital-to-analog converters generate an output that has only a few analog levels. They provide fine resolution by oscillating rapidly between these levels in such a manner that the average output represents the value of the applied code. Here we describe an improved method of interpolating that results in reduced noise in the signal band. A theory of the interpolation, confirmed by experiments, demonstrates that switching between only two levels at 1.3 mHz could provide 16 bit resolution for telephone signals.

## I. INTRODUCTION

ORDINARY digital-to-analog converters (D/A's) provide a discrete output level for every value of the digital word that is applied to their input. There is difficulty in implementing these converters for long digital words because of the need to generate a large number of distinct output levels. A method [1] for circumventing the difficulty calls for spanning the signal range with a few widely spaced levels and interpolating values between them. The interpolating mechanism causes the output to oscillate rapidly between the levels, in such a manner that the average output represents the value of the input code. This technique provides a useful tradeoff between the complexity of the analog circuits and the speed at which they operate.

Essential to the technique is an interpolating circuit for truncating the input words to shorter output words. These shorter words change their value at high speed in such a manner that the truncation noise that lies in the bandwidth of the signal is satisfactorily small. The present work describes an improved interpolating circuit that permits significant reduction in the rate at which outputs oscillate, or reduction in the number of levels needed for the output. The technique is suitable for use with pulse code modulated signals that are sampled regularly and quantized uniformly.

The following sections describe and analyze methods of interpolation, while Section V describes the results of measurements on a circuit model.

## II. FIRST-ORDER INTERPOLATION

Fig. 1 shows a circuit of a basic interpolating D/A. The input word, held in register $R_0$, feeds one port of a binary adder, the output of which separates into two paths. The more significant component $y_1$ feeds to the output, while the less significant component $e_1$ feeds back to the second port of the adder via register $R_1$. The least significant bits accumulate until they overflow into the most significant bits and, thus, contribute to the output. The input register loads at the incoming word rate $2f_0$, while register $R_1$ and the output D/A operate $k$ times faster at $2kf_0$. We represent the period of the faster clock by $\tau$ where

$$2kf_0\tau = 1. \tag{1}$$

Paper approved by the Editor for Signal Processing and Communication Electronics of the IEEE Communications Society. Manuscript received April 3, 1985.

The authors are with AT&T Bell Laboratories, Holmdel, NJ 07733.

IEEE Log Number 8406424.

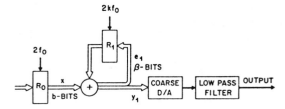

Fig. 1.   An outline of a single interpolative D/A.

The output from this circuit, expressed as its $z$-transform, is

$$Y_1(z) = X(z) - (1 - z^{-1})E_1(z) \tag{2}$$

where

$$z = \exp\ (j\omega\tau).$$

$Y_1$ represents the input contaminated by a truncation noise $E_1$ which is filtered by the high-pass function $(1 - z^{-1})$.

The interpolating converter is implemented most easily, and its action is easily explained when the digital signals are expressed in displaced binary notation rather than in two's-complement or sign-magnitude. We use this notation in all the following discussion. Let the input word comprise $b$ bits and let the error $e_1$ comprise the $\beta$ least significant bits of the sum. Then $y_1$ comprises the $(b - \beta + 1)$ most significant bits, the extra bit being the carry from the top of the adder. Input codes can assume integer values from 0 through $(2^b - 1)$, the error integer values from 0 through $(2^\beta - 1)$, while the output assumes integer multiples of $2^\beta$ in the range 0 through $2^b$. The number of levels needed to represent the output is only

$$l_1 = (2^{b-\beta} + 1) \tag{3}$$

but the switching between values must occur fast enough to suppress the truncation noise that enters the signal band. We will now calculate the frequency ratio $k$ that is needed by this circuit in order to obtain resolutions comparable with $b$-bit PCM.

The truncation error $e$ comprises a constant term $0.5\ (2^\beta - 1)$ and a noise that can fluctuate with uniform probability in the range $\pm 0.5\ (2^\beta - 1)$, its rms value being $(2^\beta - 1)/\sqrt{12}$. If we now assume that the signals applied to the converter are sufficiently busy to make this noise random with white rms spectral density

$$E_1(f) = \frac{(2^\beta - 1)}{\sqrt{12kf_0}}, \tag{4}$$

then the spectral density of the noise in the output is given by

$$N_1(f) = |E_1(z)(1 - z^{-1})| = \frac{(2^\beta - 1)}{\sqrt{3kf_0}}\left|\sin\left(\frac{\pi f}{2kf_0}\right)\right|. \tag{5}$$

Notice that the dc offset filters away. The net noise in the

Reprinted from *IEEE Trans. Commun.*, vol. COM-34, pp. 77–81, January 1986.

signal band $0 < f < f_0$ can be expressed as

$$N_{10} = (2^\beta - 1) \left( \frac{1 - \text{sinc} \left( \frac{1}{k} \right)}{6k} \right)^{1/2}$$

and

$$N_{10} \approx \frac{\pi (2^\beta - 1)}{6k \sqrt{k}} \qquad \text{when } k^2 \gg 0.5. \qquad (6)$$

We now compare this noise to the quantization noise that is inherent in the input, its rms value being $1/\sqrt{12}$. In order for the interpolation noise (6) to be smaller, it is required that

$$k^3 > \frac{1}{3} \pi^2 (2^\beta - 1)^2. \qquad (7)$$

For example, when $b = 16$ and $\beta = 12$, $k$ should exceed 381. This requires an interpolation rate in excess of 3 MHz and 17 levels of output signal for 4 kHz voiceband signals [2]. The case where the output has only two levels is particularly important for practical implementation. For this converter to have 16 bit resolution requires that $\beta = 16$ and $k$ exceed 2418, which corresponds with a 19 MHz interpolation rate for voiceband signals. Such high rates are a handicap that we can avoid by improving the filtering of the truncation noise. One method replaces register $R_1$ with more complex digital processing [3], but a better method uses the multiple interpolation described in the next section.

### III. SECOND-ORDER INTERPOLATION

Fig. 2 shows a converter that uses two accumulations to reduce the amount of truncation noise that enters the signal band. Its output may be expressed in the form

$$Y(z) = Y_1(z) + (1 - z^{-1}) Y_2(z) \qquad (8)$$

$$= X(z) - (1 - z^{-1})^2 E_2(z). \qquad (9)$$

When the error $e_2$ is random, the spectral density of the noise present in the output is given by

$$N_2(f) = |(1 - z^{-1})^2 E_2(f)| = \frac{(2^\beta - 1)}{3kf_0} \left( 1 - \cos \left( \frac{\pi f}{kf_0} \right) \right) \qquad (10)$$

and the net noise in the signal band is

$$N_{20} \approx \frac{\pi^2 (2^\beta - 1)}{2k^2 \sqrt{15k}}; \qquad k^2 > 1. \qquad (11)$$

The number of levels needed in the output is

$$l_2 = (2^{b - \beta} + 3). \qquad (12)$$

In order for the noise (11) to be less than the noise in $b$-bit PCM, it is required that

$$k^5 > \frac{\pi^4 (2^\beta - 1)^2}{5}. \qquad (13)$$

For example with $b = 16$ and $\beta = 12$, $k$ should exceed 51; this corresponds to an interpolation rate of only 404 kHz and 19 level outputs for voiceband signals. When $\beta = b$, four-level output interpolating in excess of 1.25 MHz would provide the resolution of 16 bit PCM. The next section presents a simplified implementation of this converter.

### IV. A DOUBLE INTERPOLATING D/A CONVERTER

The output (8) comprises two components; $y_1$ carries the signal contaminated with noise (2) while $y_2$ provides a first-

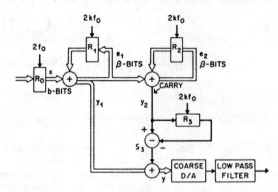

Fig. 2.   An outline of a double interpolative D/A.

order compensation for the noise. Converting these two signals into analog form by separate means significantly improves the tolerance of the circuit to inaccuracies. Two-level conversion of $y_1$ avoids signal distortion caused by misplaced levels. Likewise, two-level conversion for $y_2$ is desirable, but this requires analog differentiation to replace the digital differentiation in register $R_3$ and subtractor $S_3$.

The approximation of digital differentiation by analog differentiation is satisfactory in circuits such as this, where the word rate far exceeds the signal frequency, because

$$(1 - z^{-1}) \equiv 2j \exp \left( -\frac{j\omega\tau}{2} \right) \sin \left( \omega \frac{\tau}{2} \right)$$

and

$$(1 - z^{-1}) \approx j\omega\tau \exp \left( \frac{-j\omega\tau}{2} \right) \qquad \text{when } (\omega\tau)^2 \ll 24 \quad (14)$$

or

$$(1 - z^{-1}) \approx j\omega\tau \qquad \text{when } \omega\tau \ll 2. \qquad (15)$$

The circuit in Fig. 3 uses approximation (14): register $R_3$ provides a half period delay and $C$ differentiates $y_2$. The net output of this circuit may be expressed as

$$Y = G \left( Y_1 + j\omega RC \exp \left( -\frac{j\omega\tau}{2} \right) Y_2 \right) \qquad (16)$$

and this can be equivalent to (8) in the signal band, provided that (14) is valid and that

$$RC \approx \tau = \frac{1}{2kf_0}. \qquad (17)$$

The net gain of the analog circuit to the signal is

$$G = \frac{r_1}{(R + r)} \left( 1 + \frac{j\omega rRC}{r + R} \right). \qquad (18)$$

It cuts off at a frequency that is $k(R + r)/\pi r$ higher than signal frequencies. The purpose of this low-pass filtering, introduced by the presence of resistor $r$, is to stop high-frequency components of the binary signal $y_2$ from hitting the amplifier.

Analysis in the Appendix shows that approximation (14) is good for this application, and that relationship (17) must be satisfied to one part in $k$. It also demonstrates that the least significant $3/5$ $(\beta - 1)$ bits of the signal that feeds from the first to the second accumulator may be truncated. Measurements on a circuit model described in the next section confirm these results. They also show a close resemblance between properties of these interpolating converters and those of sigma delta modulators.

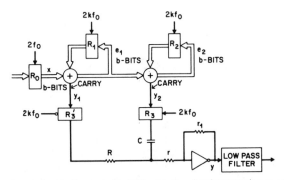

Fig. 3.   A double interpolative D/A using two-level conversion to analog.

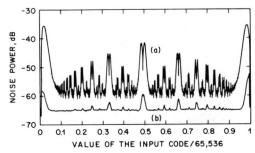

Fig. 4.   The dependence of interpolation noise on the value of the applied digital word. Noise is expressed in decibels with respect to $(2^\beta - 1)$. The interpolation rate $2kf_0 = 128$ kHz. (a) Single interpolation, calculated mean $-45$ dB. (b) Double interpolation, calculated mean $-64$ dB.

## V. Measurements on a Circuit Model

The circuits to be described here use relatively low switching rates in order that the interpolation noise be easily distinguishable from spurious circuit imperfection and from the quantization of the input signal. The input comprised 16 bit words generated by a computer at 8 kHz. It represented dc levels and 870 Hz sinewaves of various amplitudes. The circuits employed two-level D/A's and the low-pass filter at their output approximated $C$-message weighting; its cutoff frequency was about 3 kHz.

Fig. 4 shows graphs of the noise at the output of the converter, plotted in decibels against the value of the binary code as it swept slowly through the entire range 0–65 535. Curve (a) is for single interpolation, $y_2$ disconnected, and curve (b) is for double interpolation. We see that double interpolation lowers and decorrelates the noise in much the same way as it does in sigma delta modulation [4]. The theory developed in [5] applies to curve (a).

Fig. 5 is a graph of signal-to-noise ratio plotted against the amplitude of the input sinewave. Curve (a) is for single accumulation with the sinewave biased at code 32 768, the center of the range. Curve (b) is for the sinewave biased at 31 744, 1/64 of the range from center. Curve (c) is for double integration with the bias at center. Dashed lines show the result derived from (6) and (11).

Fig. 6 shows the signal-to-noise ratio plotted against the amplitude of the input sinewave for double integration at various interpolation rates.

Fig. 7 shows the signal-to-noise ratio plotted against the deviation of the time constant RC from the ideal value $\tau$. Result (26) of the Appendix calls for $\pm 8$ percent precision. Finally, Fig. 8 shows how the signal-to-noise ratio depends on the number of bits in the second accumulation; result (22) requires that at least 6 bits be processed.

The results of these measurements agree very well with the

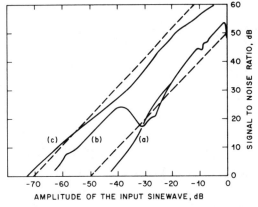

Fig. 5.   Signal-to-noise ratio plotted against the amplitude of the applied sinewave. The interpolation rate is 256 kHz. (a) Single interpolation biased to center. (b) Single interpolation biased 1/64 from center. (c) Double interpolation biased to center. Dashed lines show responses calculated for uncorrelated noise.

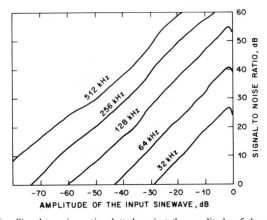

Fig. 6.   Signal-to-noise ratio plotted against the amplitude of the applied sinewave for double interpolation at various rates.

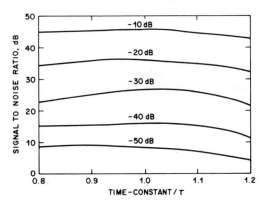

Fig. 7.   The variation of signal-to-noise ratio with the time constant RC for various amplitudes of the applied sinewave.

calculated noise values, especially at the lower cycling rates. At rates above 256 kHz, additional noise is introduced by the limited switching speeds of the commercial TTL components used in our implementation. We anticipate no difficulty in realizing resolutions corresponding to 16 bit PCM in a single chip implementation of the converter using CMOS technology.

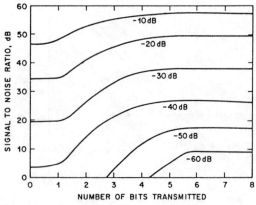

Fig. 8. The dependence of signal-to-noise ratio on the number of bits that are transmitted from the first to the second accumulator for various amplitudes of the applied sinewave.

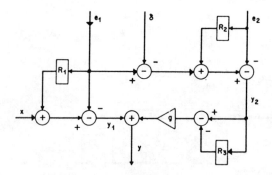

Fig. 9. A model of the double interpolative D/A. $e_1$, $e_2$, and $\delta$ represent truncation noises. $g$ is a gain factor that ideally is unity.

## VI. Concluding Remarks

Use of double accumulation greatly enhances the performance of these interpolating converters, in much the same manner as it improves sigma delta modulation. The technique can clearly be extended to more than two accumulations. Limit cycle oscillations [4] that spoil the performance of high-order sigma delta modulations cannot occur in these circuits because their accumulators are not included in feedback loops. Notice that the accumulations used in these circuits are not cleared at the start of each new word as they are in [1]. Such resetting significantly increases the magnitude of the interpolation noise and its correlation with input code values.

The specification of filters needed to smooth the output signal will depend on the application. Relatively simple filters will usually suffice, because the interpolation noise increases no more than 12 dB per octave and the restriction on out-of-band noise in most applications is usually much less severe than the restriction on in-band noise.

## Appendix

### Precision of the Circuit Parameters

Fig. 9 shows the main functions of the double interpolating D/A. $X$ represents the input. $E_1$ and $E_2$ are the truncation errors at the output of the first and second accumulations, respectively. $\delta$ represents a $d$-bit truncation that may be introduced in the connection between the accumulators. $g$ is a gain factor that may differ from unity because of circuit imperfections.

The output from this circuit may be expressed as

$$Y = X - (1 - z^{-1})(1 - g)E_1 - (1 - z^{-1})g\delta - (1 - z^{-1})^2 gE_2.$$

$$(19)$$

If the noise from the three sources $E_1$, $E_2$, and $\delta$ are uncorrelated, then the net noise power in baseband may be derived using results (6) and (11). It is

$$N_0^2 = \frac{\pi^2(2^\beta - 1)(1 - g)^2}{9k^3} + \frac{\pi^2(2^d - 1)^2 g^2}{9k^3} + \frac{\pi^4(2^\beta - 1)^2 g^2}{60k^5}. \quad (20)$$

Our design should make this noise less than the quantization noise in the input signal; it provides tradeoffs between required values of $k$, $\beta$, $\delta$, and $g$. In order to illustrate a possible design, we will assume that $k$ has been chosen and we determine the conditions that cause the noise introduced by $g$ and $\delta$, to be small compared with the inherent interpolation noise (11).

In order that the noise introduced by the truncation $\delta$ be small, it is required that

$$\frac{\pi^2 g^2 (2^d - 1)^2}{9k^3} < \frac{\pi^4 (2^\beta - 1)^2}{60k^5} \quad (21)$$

which reduces to

$$\left(\frac{2^d - 1}{2^\beta - 1}\right) < \frac{\pi}{k}\sqrt{\frac{3}{20}}; \qquad g \approx 1 \quad (22)$$

when the value of $k$ just satisfies (13).

The number of bits that may be truncated is given approximately by

$$d < \frac{3}{5}(\beta - 1). \quad (23)$$

A significant reduction in the size of the second accumulation is usually permissible.

The condition that the noise introduced by imprecise gain $g$ be small requires that

$$\frac{\pi^2(2^\beta - 1)^2(1 - g)^2}{9k^3} < \frac{\pi^4(2^\beta - 1)^2}{60k^5} \quad (24)$$

which gives

$$\Delta g = (1 - g) < \frac{1.2}{k}. \quad (25)$$

A major reason for the gain $g$ not equaling unity is error in the value of capacitor $C$ in Fig. 4. Result (25) requires that the time constant $RC$ should satisfy

$$\left|\frac{RC}{\tau} - 1\right| < \frac{1.2}{k}. \quad (26)$$

Another reason for inaccurate gain is the approximation (14). This approximation introduces a change in gain that is dependent on frequency. It may be expressed as a per-unit error

$$\Delta g = \frac{(\omega \tau)^2}{24} = \frac{\pi^2}{24k^2}; \qquad f = f_0. \quad (27)$$

This satisfies condition (25) especially when $k$ is large. In comparison, approximation (15) entails a per-unit change of gain

$$\Delta g = \frac{\omega \tau}{2} = \frac{\pi}{2k}; \qquad f = f_0 \quad (28)$$

which does not satisfy (23). Approximation (15) can be

acceptable, however, if sufficient margin is allowed in satisfying (13).

## REFERENCES

[1] G. R. Ritchie, J. C. Candy, and W. H. Ninke, "Interpolative digital-to-analog converters," *IEEE Trans. Commun.*, vol. COM-22, pp. 1797–1806, Nov. 1974.

[2] J. C. Candy, B. A. Wooley, and O. J. Benjamin, "A voiceband codec with digital filtering," *IEEE Trans. Commun.*, vol. COM-29, pp. 815–830, June 1981.

[3] H. G. Musmann and W. W. Korte, "Generalized interpolative method for digital/analog conversion of PCM signals," US Patent 4 467 316, Aug. 1984.

[4] J. C. Candy, "A use of double integration in sigma delta modulation," *IEEE Trans. Commun.*, vol. COM-33, pp. 249–258, Mar. 1985.

[5] J. C. Candy and O. J. Benjamin, "The structure of quantization noise from sigma-delta modulation," *IEEE Trans. Commun.*, vol. COM-29, pp. 1316–1323, Sept. 1981.

# A 16-BIT 4'TH ORDER NOISE-SHAPING D/A CONVERTER

L. Richard Carley and John Kenney

Department of Electrical and Computer Engineering
Carnegie Mellon University
Pittsburgh PA 15213

A 16-bit oversampling D/A converter has been designed using a 4'th order all-digital noise-shaping loop followed by a 3-bit D/A converter. The 3-bit D/A converter, which employs a novel form of dynamic element matching, achieves high accuracy and long-term stability without requiring precision matching of components. The harmonic distortion of the untrimmed monolithic CMOS prototype D/A converter is less than −90dB. This multi-bit noise-shaping D/A converter achieves performance comparable to that of a 1-bit noise-shaping D/A that operates at nearly 4 times its' clock rate.

## I. Introduction

Delta−Sigma Modulation (DSM) converters have recently achieved popularity for use in integrated circuit data converters, both A/D[1] and D/A[2] converters. Their attractiveness for IC systems, in part, is due to the fact that they employ a 1-bit D/A converter which does not require precision component matching. However, experience has revealed that high integral linearity is very difficult to achieve, perhaps because the 1-bit loop quantizer is frequently overloaded. The all-digital noise-shaping loop presented in this paper avoids quantizer overload by employing a quantizer which truncates the signal to 3 bits rather than 1 bit as in DSM-type converters. Eliminating quantizer overload, in addition to improving integral linearity, also allows higher order (> 2) noise-shaping loops to be employed, since the low frequency oscillations observed in higher order delta−sigma modulation loops[1] are a result of quantizer overload.[7] For these reasons, a multi-bit noise-shaping D/A converter system can achieve a signal−to−noise (SNR) ratio comparable to that of a second order 1-bit noise-shaping (DSM) D/A converter operating at a much higher sampling rate. For example, the prototype system, which operates at 3.2MHz, achieves performance comparable to that of the DSM-type converter which operates at 11.3MHz described by Naus, et. al..[2]

As stated above, many advantages result from using a multi-bit noise-shaping converter instead of a DSM converter. However, there is one major disadvantage; a multi-bit D/A converter typically *requires* precision component matching.[3,4]

This paper presents a D/A converter architecture that employs a novel form of "dynamic element matching"[5,6] particularly suited to oversampled data systems, to achieve excellent integral and differential linearity, while requiring only modest component matching. For example, the prototype system has a peak component mismatch of approximately 0.3%, and a measured peak integral linearity error of 0.0022% of full scale. This new topology allows multi-bit noise-shaping coders to employ D/A converters which achieve high integral linearity without requiring component trimming. Therefore, this new D/A converter topology is easily implemented in IC technology and does not require on-chip component trimming.

## System Topology

Figure 1 shows the topology of the 16-bit D/A converter system. Interpolation, the process of increasing the sampling rate, is performed in two stages. The input signal is interpotated by a factor of 64 (L in figure 1) in two stages. First, the input is interpolated by a factor of two and filtered by a standard 64-tap FIR filter topology. Then, the FIR filter's output is interpolated by a factor of 32 and filtered using a comb filter.[8] The combined filter attenuates frequencies above 30KHz by more than 80dB. Extensive simulations indicate that the SNR at the output of the noise-shaping loop with this interpolator architecture is less than 3dB below the SNR with an ideal interpolating filter.

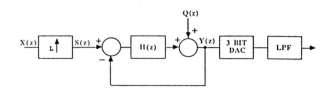

Figure 1 − Topology of the 16-Bit D/A converter system.

Reprinted from *IEEE Proc. CICC*, pp. 21.7.1–21.7.4, 1988.

The noise–shaping loop filter (H(z) in figure 1) employs a 4'th order loop filter with 4 poles at z=1 and 3 zeros positioned to maintain system stability and to minimize the "noise gain" of the loop.[9,10] The worst–case noise gain is determined by assuming that the quantization error will take on a sequence of worst case values, +1/2 LSB or −1/2 LSB.[12] The noise gain can be computed by summing the magnitude of the impulse response of the transfer function from the quantizer's output back to its' input. Noise gain is a constraint used in choosing the location of the poles of the loop transmission of the noise–shaping loop. A maximum noise gain of 4 was chosen as the constraint when placing the closed–loop poles for the noise–shaping loop of the prototype system. Therefore, the quantization error at the noise–shaping loop's output, which is in the range +1/2 LSB to −1/2 LSB, is amplified in the worst case by a factor of 4. Hence, the noise signal at the quantizer's input will be in the range +2 LSB to −2 LSB. In order to prevent the noise–shaping quantizer from being overloaded, the 16–bit input signal is scaled so that its' full–scale range is between +2 LSB and −2 LSB.

### 3–Bit D/A Converter Topology

The 3–bit D/A converter topology achieves high integral linearity without requiring precisely matched components through the application of a modified form of dynamic element matching. The D/A converter's n'th output level is generated by charging all capacitors to a +5 volt reference level and then switching n of them into the summing junction of an operational amplifier during each clock cycle. Note, the capacitors which are not switched into the summing junction are switched into ground in order to maintain a constant current in the +5 volt reference line. A resistor version of the 3–bit D/A converter was also fabricated. As with the capacitor version, the n'th output level was generated by switching n resistors into the summing junction of an operation amplifier. The other end of the resistors was tied to a reference voltage.

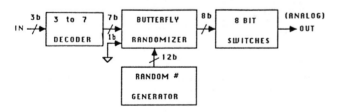

Figure 2 – Topology of the 3–Bit D/A converter.

Dynamic element matching is implemented by choosing different elements to represent the n'th level as a function of

time. The "randomizer" selects each element on n out of 8 clock cycles (see figure 2). The goal of this approach is to decorrelate the error on successive samples. Ideally, a mismatch between the capacitors is converted into a white noise signal; and, since the D/A converter is being used in an oversampling application, a large portion of the error power is filtered out. One simple approach to decorrelating the error on successive samples would be to have a barrel shifter which rotated one increment after each clock. This approach would completely decorrelate successive output errors only if the mismatch between capacitors were independent of the capacitor's position on the die. Unfortunately, just the opposite is true. Adjacent capacitors are much more likely to match than distant capacitors due to gradients in oxide thickness across the wafer. An ideal randomizer, one which connects each of the 8 inputs to all 8 possible outputs in a random fashion, would have to include 40,320 possible permutations. In order to conserve die area, a selected set of random combinations was allowed. The randomizer circuit consists of a series of 3 "butterfly" networks coupling the inputs to the output (see figure 3). This randomizer implements 4096 different combinations. A pseudo–random sequence generator is used to generate the random 12–bit control sequences for the butterfly switches. Simulations verified that this partial randomizer resulted in good decorrelation of successive converter errors in the face of both linear and quadratic gradients in oxide thickness across the wafer. Figure 4 shows a photomicrograph of the 3–bit D/A converter, the digital randomizer circuitry, and the pseudo–random sequence generator.

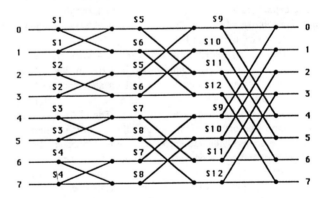

Figure 3 – Topology of the "Butterfly" type randomizer.

For a DC input code of n, each capacitor is discharged into the output on average n out of every 8 clock cycles. Therefore, the D/A converter acts as a duty–cycle modulator and the integral linearity is limited only by the product of the fractional element mismatch ($\Delta E/E$) and the fractional clock jitter ($\Delta T/T$).[5,6] Extremely high DC integral linearity can be achieved, even if the elements match very poorly, as long as a

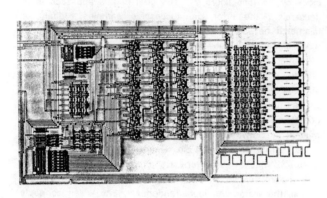

Figure 4 — Photomicrograph of randomizer and 3—Bit D/A converter.

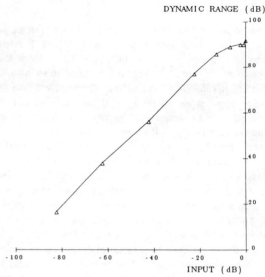

Figure 5 — Plot of SNR as a function of input amplitude. In this case distortion is *included* with the noise. Input signal is a 1KHz sinewave.

precise clock signal is used. However, the element mismatch will appear as an AC noise signal added to the D/A converter's output.

The maximum noise signal, as a fraction of the 3-bit D/A converter's full scale (FS), is $\frac{RMS[\Delta E/E]}{\sqrt{2M}}$. In an oversampling converter, only noise power in the signal pass-band is important. Assuming that the randomizer's pattern is not correlated with the element variations, the noise signal will be white and the RMS fraction of FS for the noise signal in the passband will be $\frac{RMS[\Delta E/E]}{\sqrt{2M \times OV}}$ where OV is the oversampling ratio. For example, if a 3-bit D/A converter was constructed using capacitors with an RMS variation of 0.1% and the system oversamples by a factor of 64, the in-band RMS noise signal is approximately 0.003% of FS.

The analog output from the 3-bit D/A converter (see figure 1) was filtered by a fifth-order Butterworth filter. The real axis pole was provided by the first op amp and the two complex pole pairs were implemented as a cascade of two Sallen-and-Key second order sections. The SNR and the open-loop gain of the operational amplifiers limit the system's SNR. The analog filter was implemented off-chip using ultra-low noise operational amplifiers. Although the prototype CMOS IC did not include the analog filter, low-noise monolithic CMOS operational amplifiers with performance suitable for this task have been designed and fabricated.[2,11]

### Results

Figure 5 shows the SNR at the analog filter's output as a function of the input amplitude for a 1 KHz sinusoidal input. The system achieves a total dynamic range greater than 92dB.

Figure 6 shows the power spectral density (PSD) of the 3-bit D/A converter's output without the analog filter for a 0dB 1KHz sinusoidal input. The flattening of the PSD of the noise at

approximately 1 MHz is a result of the poles incorporated into the noise-shaping loop's transfer function to control the noise gain. Note, this signal contains noise components from two sources: quantization noise from the noise-shaping loop and the approximately "white" noise resulting from the randomized capacitor mismatch.

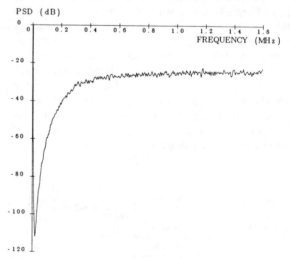

Figure 6 — Power Spectral Density of the signal at D/A converter output with a 0dB 1KHz input signal.

Figure 7 shows the power spectral density at the output of the analog filter. Simulations with an ideal 3-bit D/A converter indicate that the PSD of the in-band noise is dominated, in the audio frequency range, by the component mismatch noise rather than quantization noise from the noise-shaping loop, The harmonic distortion components are each more than 96dB below the maximum input level.

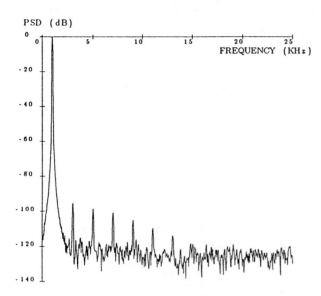

Figure 7 – Power Spectral Density of the signal at output of the analog filter with a 0dB 1KHZ input signal. Note, the amplitude of the input signal has been artificially reduced by passing it though a 1 KHz notch filter in order to decrease the dynamic range to allow accurate measurement of the PSD.

## Conclusions

A 16–bit oversampling D/A converter has been designed using an all–digital noise–shaping loop followed by a 3–bit D/A converter that employs a modified form of dynamic element matching to achieve high accuracy without the need for precision matching of components. Using a version of the 3–bit D/A converter which was fabricated in 3$\mu$m CMOS, the 16–bit D/A conversion system achieved a dynamic range of 90dB and a THD < −94dB at a sampling rate of only 3.2MHZ. It is the topology of the 3–bit D/A converter which enables us to design a noise–shaping loop which can attain performance equivalent to that of DSM converters operating at much higher clock rates, without incurring the penalty of high integral nonlinearity.

Although all of the system was not fabricated in IC form, other researchers have fabricated digital interpolating filters, digital noise–shaping loops, and analog output filters equivalent in performance to those required in this system. For example, a higher order (1:256) interpolating filter and a 2'nd order noise–shaping loop were implemented by Naus et. al.[2] in approximately 18mm[2]. Therefore, we conclude that by testing the novel portion of this system in IC form we have proved that integration of the entire system is viable.

## Acknowledgements

This work was supported in part by a grant from the Analog Devices Corporation and by the NSF under grant ENG–8451496.

## References

[1] J. C. Candy, "A Use of Double Integration in Sigma–Delta Modulation", *IEEE Trans. Commun.*, vol. COM–33, pp. 249–258, March 1985.

[2] P. J. A. Naus, E. C. Dijkmans, E. F. Stikvoort, A. J. McKnight, D. J. Holland, and W. Brandinal, "A CMOS Stereo 16–bit D/A Converter for Digital Audio", *JSSC* Vol. SC–22, No. 3, pp. 390–395, June 1987.

[3] J. W. Scott. W. Lee, C. Giancarlo, and C. G. Sodini, "A CMOS Slope Adaptive Delta Modulator", Proceedings of the 1986 ISSCC, pp. 130–131, Feb. 1986.

[4] R. W. Adams, "Design and Implementation of an Audio 18–Bit A/D Converter using Oversampling Techniques." *J. Audio Eng. Soc.* Vol. 34, No. 3, pp. 153–166, March 1986.

[5] R. J. Van De Plassche, "A monolithic 14–bit D/A converter", *JSSC* Vol. SC–14, No. 3, pp. 552–556, June 1979.

[6] K. B. Klaassen, "Digitally controlled absolute voltage division", *IEEE Trans. on Instrumentation and Measurement* Vol. 24, No. 3, pp. 106–112, June 1975.

[7] C. Wolff and L. R. Carley, "Modeling the Quantizer in Higher–Order Delta–Sigma Modulators", *International Symposium on Circuits and Systems*, Helsinki Finland, June 1988.

[8] S. Chu and C. S. Burrus, "Multirate Filter Designs using Comb Filters", *IEEE Trans. on Circuits and Systems*, vol. CAS–31, pp. 913–924, Nov. 1984.

[9] J. Kenney, "Design Methodology for N–th Order Noise–Shaping Coders", Masters Project Report, Department of ECE, CMU, Pittsburgh PA, Jan. 1988.

[10] B. P. Agrawal and K. Shenol, "Design Methodology for $\Sigma\Delta M$", *IEEE Trans. Commun.* COM–31:360–369, March 1983.

[11] Y. Matsuya, K. Uchimura, A. Iwata, T. Kobayashi, M. Ishikawa, and T. Yoshitome, "A 16–bit Oversampling A–to–D Conversion Technology Using Triple–Integration Noise Shaping", *JSSC* Vol. SC–22, No. 6, pp. 921–929, December 1987.

[12] L. R. Carley, "An Oversampling Analog–to–Digital Converter Topology for High Resolution Signal Acquisition Systems", *IEEE Trans. Circuits and Systems* CAS–34, #1, 83–91, January 1987.

# A CMOS Stereo 16-bit D/A Converter for Digital Audio

PETER J. A. NAUS, EISE CAREL DIJKMANS, EDUARD F. STIKVOORT,
ANDREW J. McKNIGHT, DAVID J. HOLLAND, AND WERNER BRADINAL

*Abstract* — A complete monolithic stereo 16-bit D/A converter primarily intended for use in compact-disc players and digital audio tape recorders is described in this paper. The D/A converter achieves 16-bit resolution by using a code-conversion technique based upon oversampling and noise shaping. The band-limiting filters required for waveform smoothing and out-of-band noise reduction are included. Owing to the oversampling principle most applications will require only a few components for an analog post-filter. The converter has a linear characteristic and linear phase response. The chip is processed in a 2-μm CMOS process and the die size is 44 mm$^2$. Only a single 5-V supply is needed.

## I. Introduction

IN CONVENTIONAL linear PCM the quantizing noise is assumed to be white noise having a power $q^2/12$, where $q$ is the quantizing step [1], [2]. If the signal bandwidth is less than half the sample frequency $f_s$, the noise spectrum can be reshaped in order to decrease the in-band noise [3]. For equal in-band noise, this noise shaping results in a lower number of bits per sample, together with a higher sample rate than is used in conventional PCM. In the system presented here (see Fig. 1), a performance, equivalent to 16-bit D/A conversion at a sample rate of 44.1 kHz, has been obtained by using a 1-bit D/A converter at a sample rate of 11.2896 MHz ($256f_s$).

In the system, the stereo 16-bit PCM 44.1-kHz input signal is filtered and upsampled to $4f_s$ in the first oversampling section. After the first oversampling section the signal is demultiplexed and from this point on both channels (left and right) are treated separately. The sample rate is increased to $256f_s$ in the second oversampling section. The word length is reduced to 1 bit, by a noise-shaping code conversion, and the 1-bit code passes through a 1-bit D/A converter. A small analog post-filter completes the D/A conversion of the original 16-bit PCM signal.

Manuscript received October 6, 1986; revised December 19, 1986.
P. J. A. Naus, E. C. Dijkmans, and E. F. Stikvoort are with Philips Research Laboratory, Eindhoven 5600 JA, The Netherlands.
A. J. McKnight and D. J. Holland are with Mullard, Southampton, England.
W. Bradinal is with Valvo, Hamburg, Germany.
IEEE Log Number 8714201.

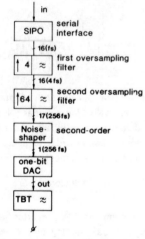

Fig. 1. Block diagram of one channel of the D/A converter.

## II. Oversampling and Filtering

The 44.1-kHz sample rate of the PCM input signal is increased by a factor of 4 in the first oversampling section. It contains a finite-impulse-response (FIR) low-pass filter, which has a 20-kHz bandwidth, a passband ripple of $\pm 0.02$ dB, and a stopband rejection of 60 dB for frequencies above 24 kHz. Its hardware contains a ROM, a RAM, and an array multiplier. Although the upsampling filter structure is multiplexed between the two channels, there is no phase shift between the two analog outputs.

The second oversampling section is an interpolating filter based on an adder structure. The sample rate is increased by a factor of 64, by using a linear interpolator ($\uparrow 128f_s$) and a sample-and-hold circuit ($\uparrow 256f_s$). An internally generated out-of-band dither signal is used to prevent audible idling patterns of the noise shaper at low-input signal levels. The amplitude increases because of the dither signal and therefore an extra bit is needed. The word length at the output of the linear interpolator has to be 17 bits.

All signal processing steps involved in generating the 1-bit code use FIR filters, ensuring a linear phase characteristic. Owing to the digital oversampling filters only a simple analog post-filter is required, which is realized by a

Reprinted from *IEEE J. Solid-State Circuits*, vol. SC-22, pp. 390–395, June 1987.

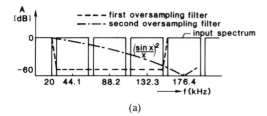

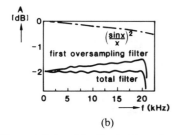

Fig. 2. Characteristics of the filters: (a) stopband, and (b) passband.

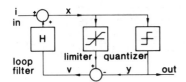

Fig. 3. Block diagram of the noise shaper.

third-order Butterworth filter with a $-3$-dB cutoff frequency at 60 kHz. The frequency responses of the digital filter and the analog output filter are shown in Fig. 2. Frequencies at multiples of the sample frequency $f_s$ are attenuated by the first oversampling filter. The linear interpolator has a transfer function of $(\sin(x)/x)^2$, in which $x = \pi f/4f_s$, and attenuates the frequencies at multiples of $4f_s$. The first oversampling filter contains compensation for the $(\sin(x)/x)^2$ roll-off due to the interpolator plus the roll-off of the Butterworth filter. The combined filter curve will be flat over the audio band with a ripple of 0.02 dB.

## III. NOISE SHAPER

The 17-bit oversampled PCM signal is supplied to a second-order noise shaper. Fig. 3 shows the block diagram of the noise shaper. The noise shaper is a feedback loop consisting of a quantizer which reduces the word length to 1 bit, a limiter which prevents overflow, and a loop filter which shapes the spectral distribution of the quantization noise generated by the quantizer.

Fig. 4(a) and (b) shows the characteristics of the quantizer and the limiter as a function of the input $x$, respectively. The operation of the noise shaper can easily be explained, if one crudely models the quantizer by

$$y = cx + r \tag{1}$$

where $r$ is the quantizing error of the quantizer and $c$ is the gain attributed to the quantizer in the loop. Together with

$$X = I + HV$$
$$V = X - Y$$

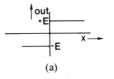

(a)

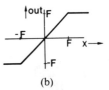

(b)

Fig. 4. (a) Output of the quantizer as a function of the input. (b) Output of the limiter as a function of the input.

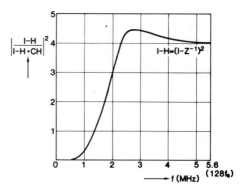

Fig. 5. Noise density at the output of the noise shaper.

it follows that

$$Y = \frac{cI}{1 - H + cH} + \frac{R(1 - H)}{1 - H + cH}. \tag{2}$$

$I$, $X$, $V$, and $Y$ are the frequency-domain representations of $i$, $x$, $v$, and $y$. $R$ is the spectral density of the quantizing error. The spectral components of the quantizing error in the signal band $(0, \Theta_b)$ must be minimized, hence $|1 - H(e^{j\Theta})|$ must be minimized for $\Theta \in [0, \Theta_b]$. If $|1 - H(e^{j\Theta})|$ is small relative to one, (2) may be approximated by

$$Y = I + R(1 - H)/c. \tag{3}$$

The coder output $Y$ contains the undistorted spectrum of the input signal $I$, added to the spectrally shaped quantizing error.

In our design we have chosen for a rather uncomplicated type of loop filter $H(z) = z^{-1}(2 - z^{-1})$. The resulting noise transfer function, in the signal band, of the 1-bit quantization noise has the shape

$$1 - H(z) = (1 - z^{-1})^2. \tag{4}$$

This noise shaping changes the quantizing noise spectrum (see Fig. 5). In the signal band the noise density increases 12 dB/octave with frequency. The power in the quantizer output in the loop can be computed from the noise density for zero input signal to the coder:

$$\bar{y}^2 = \int_{-\pi}^{\pi} \frac{E^2}{6\Pi} \left| \frac{1 - H(\theta)}{1 - H(\theta) + cH(\theta)} \right|^2 d\theta. \tag{5}$$

The signal $Y$ only takes the values $\pm E$, hence $c$ follows from

$$\int_{-\pi}^{\pi} \left| \frac{1-H(\theta)}{1+(c-1)H(\theta)} \right|^2 d\theta = 6\Pi. \qquad (6)$$

The idle channel in-band noise power $N_i$ is approximated for $\Theta_b \ll 1$ by

$$N_i = \frac{E^2 \Theta_b^5}{15\pi c^2} \qquad (7)$$

where $E$ is the amplitude of the output of the quantizer. In our design the value of $c$ is 0.667. For an oversampling factor of 256 and an audio bandwidth of 20 kHz, the in-band noise power $N_i = -107.8$ dB$_n$, where 0 dB$_n$ corresponds to the maximum signal power $E^2/2$. It should be noted that the maximum signal power which the coder can handle is less than $E^2/2$. For increasing input signal level, the quantizer input level increases sharply in the proximity of overload of the coder, the signal-to-noise ratio in the output decreases, and the coder tends towards instability. Hence, in a practical device, a limiter in the loop is necessary for reasons of stability and overflow protection.

## IV. THE 1-BIT D/A CONVERTER

The D/A converter circuit, as shown in Fig. 6(a), has a very simple structure. Its function is to modulate a dc voltage with the 1-bit data stream. However, as the data contain a wide-band frequency spectrum, the modulator needs a high linearity, otherwise distortion will cause inter-modulation with the out-of-band quantization noise, folding it back into the audio band.

The circuit is implemented as a switched-capacitor circuit. The timing diagram is shown in Fig. 6(b). During the first half of the sample period, either $C_1$ is charged by drawing a unity charge out of the summing node of the op amp or capacitor $C_2$ is discharged by pushing a unity charge into the op amp. During the second half, $C_1$ and $C_2$ are discharged and charged, respectively. The current flowing through the feedback network of the op amp, disregarding the quantization noise, can be approximated as follows:

$$I^+ = \frac{C_1 f_s}{2}(V_1 - V_2)(1 + m\sin(pt) + \cdots)$$

$$I^- = \frac{C_2 f_s}{2}(-V_1 - V_2)(1 - m\sin(pt) + \cdots)$$

$$I^+ + I^- = \frac{f_s}{2}\{ V_1(C_1 - C_2) - V_2(C_1 + C_2)$$

$$+ mV_1(C_1 + C_2)(\sin(pt) + \cdots)$$

$$- mV_2(C_1 - C_2)(\sin(pt) + \cdots) \}. \qquad (8)$$

Clearly, the resulting current contains three components—a

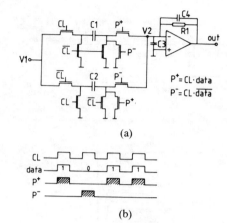

(a)

(b)

Fig. 6. One-bit D/A converter: (a) circuit diagram, and (b) timing diagram.

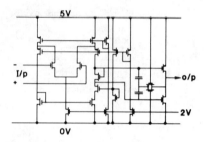

Fig. 7. Circuit diagram of the op amp.

dc term, the wanted signal, and a term which contains the signal $V_2$—that can be described as

$$V_2 = V_o - \frac{V_{\text{out}}}{A} \qquad (9)$$

where $V_o$ is the dc offset voltage of the op amp and $A$ is the open-loop gain of the amplifier. From (8) and (9) it follows that distortion products will be generated where the second harmonic distortion will dominate. From

$$\frac{V_{\text{dist}}}{V_{\text{signal}}} = \frac{mR_1 f_s}{4}\left( \frac{C_1 - C_2}{A} \right) \qquad (10)$$

it follows that this second-harmonic distortion is very dependent on the gain of the op amp and the matching between $C_1$ and $C_2$.

As the input signal is the sum of a low-frequency audio signal and high-frequency noise, any nonlinearity will result not only in harmonic distortion of the audio signal, but also in the folding back of intermodulation products of the high-frequency noise components into the audio band.

The voltage step on the input of the summing node of the op amp generates slew-rate distortion of the op amp. The amplitude of this voltage step is dependent on the ratio of the capacitors $C_4$ and $C_1$, $C_2$ and the ratio of the ON resistance of the switches and the high-frequency output impedance of the op amp. The high-frequency output impedance of the op amp is roughly determined by the $1/g_m$ of the output stage. Capacitor $C_3$ is added to the

TABLE I
OP-AMP SPECIFICATIONS

| | |
|---|---|
| DC-Gain | 90 dB |
| Phase margin (0 dB) $C_{load} = 20\,pF$ | 45 deg. at 45 MHz |
| Phase margin (20 dB)$C_{load} = 20\,pF$ | 85 deg. at 4.5 MHz |
| Output impedance | 80 ohm |
| Power consumption | 18 mW |
| Input offset voltage | 2 mV |
| Slew-rate | 30 V/$\mu$sec |
| Size | 610 $\mu$m x 350 $\mu$m |

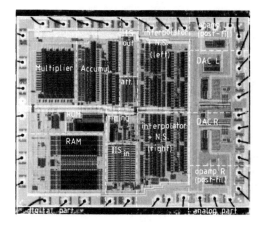

Fig. 8.   Photograph of the stereo 16-bit D/A converter.

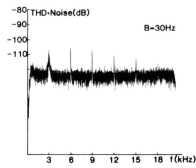

Fig. 9.   Measured total harmonic distortion plus noise in the audio band.

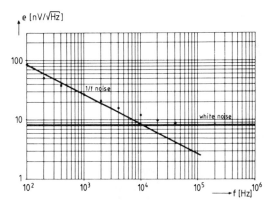

Fig. 10.   Measured $1/f$ plus thermal noise of the op amp.

TABLE II
NOISE CALCULATIONS

| | |
|---|---|
| Input quantization noise (16 bit) | - 98.4 dB |
| Quantization noise first filter section | -104.4 dB |
| Noise-shaper | -105.8 dB |
| Opamp 1/f noise relative to 0.7Vrms | -104.1 dB |
| Opamp thermal noise relative to 0.7 Vrms (folded noise included) | -103.0 dB |
| Theoretical total noise | - 95.6 dB |

summing node of the op amp to decrease the high-frequency components in $V_2$ sufficiently. The op amp with the feedback network $R_1$ and $C_4$ serves as a first-order low-pass filter reducing the high-frequency noise. Fig. 7 shows the circuit diagram of the op amp. It is a classical design [4] with a low-gain differential input stage driving a very high-gain cascode stage from both ends. This drives a Miller gain stage to give good gain, bandwidth, and output impedance with medium power consumption. The transistors are all large to give a low $1/f$ noise figure. The specifications of the op amp are shown in Table I.

## V.   PRACTICAL D/A CONVERTER CHIP

Fig. 8 shows a photograph of the stereo 16-bit converter. The circuit is processed in a 2-$\mu$m CMOS technology. The stereo D/A converter needs a chip area of 44 mm$^2$ and is

mounted in a 40-pin dual-in-line plastic package or in a 44-pin quad flat pack.

Two inter IC signal (I$^2$S) [5] ports have been incorporated, one to receive data from the compact-disc decoder IC (or any 16-bit 44.1-kHz I$^2$S source), and the other to transmit the four-times oversampled data to an external device such as a dual 16-bit D/A converter [6]. The I$^2$S standard, in which the two channels are multiplexed, provides easy interfacing between digital signal processing devices operating at various word lengths. For compact-disc applications an input has been provided which will allow an attenuation of the analog output by $-12$ dB during track search. A de-emphasis input can control the roll-off of the integrator by switching in an extra (external) feedback network.

## VI.   MEASUREMENTS

Fig. 9 shows the noise and the total harmonic distortion as a function of the frequency measured with an HP 339A distortion-measurement set and a spectrum analyzer with a resolution bandwidth of 30 Hz. The input signal has a frequency of 3 kHz. At full signal level ($-2$ dB$_n$) the harmonic distortion is less than $-100$ dB$_n$.

Fig. 10 shows the thermal noise and the $1/f$ noise measured at the op amps. It can be seen that the $1/f$ noise plays a dominant role.

The minimum noise level generated by a D/A converter system is given by the resolution of the input signal. Extra noise will be added owing to imperfections of the converter and noise generated by the low-pass filter. In the case of a partially digital and partially analog filter, round-

TABLE III
MEASURED DATA OF THE STEREO 16-BIT D/A CONVERTER

| Dynamic range | > 94 dB |
|---|---|
| Distortion | < -90 dB |
| Dissipation | < 250 mW |
| Pass-band ripple | < 0.02 dB |
| Stop-band attenuation | > 60 dB |
| Power supply | single 5V ±10% |
| Package | 44 pin QFP or 40 pin DIL |
| De-emphasis switch | internal |
| Output low-pass filter | internal |
| Serial input interface | I²S |
| Serial output interface | I²S |
| X-tal oscillator | internal |
| Die size | 44 mm² |
| Process | 2 μm CMOS |

ing noise is generated in the digital filter and $1/f$ and thermal noise in the analog filter. The noise contributions, over a 20-kHz band, of the several components used in our system are given in Table II (0 dB corresponds to a full-level sine wave). Measurements show a noise level of $-94$ dB which is fairly close to the calculated noise level. The main noise contribution is given by the noise of the op amp.

Table III gives some additional data for the stereo 16-bit D/A converter.

## VII. CONCLUSIONS

A stereo 16-bit D/A converter for use in digital audio equipment has been realized in a 2-μm CMOS process by using a code-conversion technique based upon oversampling and noise shaping. The oversampling filters, the noise shaper, the 1-bit D/A converter, and the analog post-filter are integrated on one chip. For most applications only a few external components are required. The power consumption is < 300 mW and only a single 5-V supply is needed.

## ACKNOWLEDGMENT

The authors wish to acknowledge the valuable contributions made by D. Goedhart of the CD-lab, Eindhoven, The Netherlands, and A. Durham and D. Braithwaite of Mullard, Southampton, United Kingdom.

## REFERENCES

[1] L. R. Rabiner and B. Gold, *Theory and Application of Digital Signal Processing.* Englewood Cliffs, NJ: Prentice-Hall, 1975.
[2] A. Oppenheim and R. Schafer, *Digital Signal Processing.* Englewood Cliffs, NJ: Prentice-Hall, 1975.
[3] H. A. Spang and P. M. Schultheiss, "Reduction of quantizing noise by use of feedback," *IRE Trans. Commun. Syst.*, pp. 373–380, Dec. 1962.
[4] P. R. Gray and R. G. Meyer, "MOS operational amplifier design—A tutorial overview," *IEEE J. Solid-State Circuits*, vol. SC-17, no. 6, pp. 969–982, Dec. 1982.
[5] Philips publication, "I²S bus specification," Feb. 1986.
[6] H. J. Schouwenaars, E. C. Dijkmans, B. M. J. Kup, and E. J. M. van Tuijl, "A monolithic dual 16-bit D/A converter," *IEEE J. Solid-State Circuits*, vol. SC-21, pp. 424–429, June 1986.

# Author Index

## A

Adams, R. W., 206, 279, 340
Agrawal, B. P., 153
Alexander, D. S., 377
Amano, K., 340
Ardalan, S. H., 33, 163, 245

## B

Benjamin, O. J., 52, 385
Bishop, R. J., 163
Boser, B. E., 168, 293
Bradinal, W., 486
Brauns, G. T., 245
Brinthaupt, D. M., 311
Brodersen, R. W., 213, 317, 333
Burrus, C. S., 405
Buzo, A., 106

## C

Caldwell, J., 359
Candy, J. C., 1, 44, 52, 174, 377, 385, 400, 477
Cardoletti, L., 449
Carey, M. J., 461
Carley, L. R., 184, 304, 482
Cataltepe, T., 192, 227, 266
Chao, K. C.-H., 196
Chen, D.-P., 311
Ching, Y. C., 377
Chou, W., 60
Chu, S., 405
Copeland, M., 233
Crochiere, R. E., 417

## D

Degrauwe, M., 449, 453, 457
Del Signore, B. P., 365
Deppa, T. W., 311
Deval, P., 249
Dijkmans, E. C., 486
Dijkstra, E., 449, 453, 457

## E

Eckbauer, F., 326
Elward, Jr., J. P., 311
Engelhardt, E., 326

## F

Ferguson Jr., P. F., 206
Fetterman, H. S., 342
Fields, E. M., 311
Fisher, J. A., 326
Friedman, V., 311

## G

Ganesan, A., 206
Giancario, C., 363
Goodman, D. J., 461
Gosslau, A., 209
Gottwald, A., 209
Gray, P. R., 333
Gray, R. M., 60, 73, 81
Greene, R., 359

## H

Hallock, R. W., 139
Hamashita, K., 365
Hara, S., 365
Haug, J., 359
Hauser, M. W., 213, 219, 317
Hayashi, T., 259, 320
Heise, B., 326
He, N., 106
Holland, D., 486
Hurst, P. J., 223, 317
Huynh, A.-N., 477

## I

Inabe, Y., 320
Inose, H., 115
Ishii, E., 340
Ishikawa, M., 237
Iwata, A., 237, 259

## K

Karema, T., 127, 322, 353
Karmann, K. P., 168
Kenney, J., 482
Kimura, T., 259
Kobayashi, T., 237
Koch, R., 326
Kramer, A. R., 192
Kuhlmann, F., 106

491

# L

Larson, L. E., 192, 227
Le Fur, P., 467
Lee, W. L., 196, 363
Leslie, T. C., 229
Leung, B. H., 333
Levinson, R. A., 223
Lindell, M., 353
Longo, L., 233

# M

Martin, H., 168
Matsumoto, K., 340
Matsuya, Y., 237
McKnight, A. J., 486
Meleis, H., 467

# N

Nadeem, S., 196
Naus, P. J. A., 486
Neff, R., 333
Norsworthy, S. R., 342
Nys, O., 60, 449, 453, 457

# P

Parzefall, F., 326
Paulos, J. J., 33, 163, 245
Piguet, C., 449, 457
Post, I. G., 342

# R

Rabiner, L. R., 417
Rakers, P., 359
Rebeschini, M., 359
Rijmenants, J., 453
Ritoniemi, T., 127, 322, 353
Robert, J., 249

# S

Saramäki, T., 471
Schultheiss, P. M., 131
Scott, J. W., 311, 363
Shenoi, K., 153
Shoji, Y., 255
Singh, B., 229
Sodini, C. G., 196
Spang III, H. A., 131
Steer, M. B., 163, 245
Stikvourt, E. F., 486
Suzuki, T., 255
Swanson, E. J., 365

# T

Takasuka, K., 365
Tanaka, T., 365
Temes, G. C., 1, 192, 227, 266
Tenhunen, H., 127, 322, 353, 471
Tewksbury, S. K., 139

# U

Uchimura, K., 237, 259, 320

# V

van Bavel, N., 359
Viswanathan, T. R., 311

# W

Walden, R. H., 192, 266
Welland, D. R., 365
Wong, P.-W., 60
Wooley, B. A., 168, 293, 385

# Y

Yasuda, Y., 115
Yoshitate, K., 340
Yoshitome, T., 237
Yukawa, A., 270

# Subject Index

**Abstract dynamical system, 76**
Analog section, dual-channel voice-band PCM codec, 312-13
Arbitrary loop gain, analysis of system with, 49-51
Audio A/D converter:
  D/A conversion, 290-91
  decimator design, 286-90
    second stage of decimator, 288-89
    truncation, effect on output spectrum, 289-90
  design/implementation, 279-92
  extension to higher resolution, 291
  front-end A/D design, 280-82
  front-end topology:
    linearity considerations, 283
    selection of, 282-83
    waveform symmetry, 283-86
    integrator considerations, 291
    performance of prototype, 291

**Baseband filter, 463**
Baseband noise, 55-57
Binary quantizer error, moments of, 65-68
Block diagram, dual-channel voice-band PCM codec, 311-12

**Cascaded demodulators, 23-24**
Cascaded modulators, 11-12
Circuit parameters:
  effect on signal, 46-48
  loop gain/circuit stability, 46-48
  threshold settings, 48
CMOS sigma-delta A/D converter:
  with 5 V operating voltage, 353-58
    design, 353-54
  amplifier gain, 354-55
    amplifiers, 355-56
    area, 356
    capacitor nonidealities, 355
    comparator, 356
    comparator hysteresis and speed, 355
    differential circuit topology, 355
    integrators, 355
    switches, 356
  modulator noise, 354
  with 350-kHz output rate, 359-62
    capacitor mismatch, 361
    experimental results, 361-62
    folded cascade structure, 361
    modulator circuit implementation, 360-61
    theoretical analysis, 360
    thermal noise, 361
    voltage-variable switched-charge-injection, 361
CMOS slope adaptive delta modulator, 363-64
CMOS stereo 16-bit D/A converter for digital audio, 486-90
  measurements, 489-90

noise shaper, 487-88
one-bit D/A converter, 488-89
oversampling and filtering, 486-87
practical D/A converter chip, 489

**D/A quantization levels, positioning, 6**
DC idle-channel noise, interpolative modulators, 202-3
Dead zones, 5
Decimating the modulated signal, 18-21
  first-stage decimator design, 19-20
  low-pass filter, 21
  multistage decimation, 18-19
  sinc decimators, implementing, 20-21
Decimation, 400-404
  decimating with $sinc^K$, 401-2
  defined, 400
  of digital signals, 417-48
  digital words, length of, 402
  first-stage decimator design, 19-20
  graphical presentations, 403-4
  low-pass filter, 21
  modulator, 400-401
  multistage, 18-19
  overloading characteristics, 402-3
  sinc decimators, implementing, 20-21
  of spurious noises, 403
Decimation filter design, 453-56
  design example, 455
  finite wordlength of coefficients, 455
  required filter complexity, 453-54
  simulated annealing optimization of wordlength, 455
  software environment, 455
Delta modulation, 12, 115
  noise from, 58-59
Delta-sigma A/D converters:
  analog constraints, 219-21
  area and power, 221-22
  basic building blocks, 330-32
    bidirectional current source, 330-31
    chip architecture, 332
    comparator, 331-32
    switched current source, 331
  circuit implementation, 345-48
    operational details, 346-47
    performance limitations, 346
    subcircuit characteristics, 347-48
  core circuitry, 219
  decimation filter design, 453-56
  with 15-MHz clock rate, 326-32
  functional principle, 326-27
  integrator concept, 328-30
  measured results, 327-28
  mixed A/D constraints, 221
  modeling/simulation, 345
  MOS converters, 213-18
  performance evaluation, 348-50
    alternative measures, 348-49
    custom data acquisition board, 349-50

Delta-sigma converters (*cont.*)
    performance evaluation system, 349
  with reduced sensitivity to op-amp noise and gain, 223-26
  sigma-delta conversion, 342-45
    architectures for practical realization, 343-44
    digital filtering/decimation, 344-45
    higher-order noise-shaping modulators, 345
    nonlinearities in, 344
    quantization noise, 343
  VLSI-realizable decimators for, 471-74
Delta-sigma demodulation, circuit design for, 24
Delta-sigma interpolative modulator, 196-97
Delta-sigma modulation, 3-7, 34-35, 60-72, 175-76
  alternative modulator structures, 10-12
    cascaded modulators, 11-12
    delta modulation, 12
    error feedback, 10-11
  circuit parameters, influence of, 5-7
  with DC inputs, pattern noise from, 4-5
  dead zones in, 5
  decimation for, 400-404
  design methodology, 153-62
    circuit-level design, 157-58
    D/A or digital design, 158-59
    discrete-time model, 154-55
    parameter definitions, 155-56
    systems-level design, 156-57
    uniform quantizer, analogy with, 156
  double integration in, 174-83, 255-57
    simulation results, 257
  first-order feedback circuit, 3
  gain compensation, 224-25
  loop stability, 259
  low-frequency noise compensation, 224
  model, 53-54
    analysis of, 54-57
  modulation noise in busy signals, 3-4
  multistage, 60-72
  multilevel quantization, 58
  noise from, 52-53
  noise in, 7-8
  nonlinear behavior of, 159-61
  performance limits due to circuit non-idealities, 223-24
  quantization noise, 75-76
  random noise, analysis based on, 58
  stability, 159-61
  tri-level, signal-to-noise ratio using, 245-48
  *See also* Multistage sigma-delta modulation
Delta-sigma modulators:
  basic theory of operation, 229-30
  coder topologies, 230
  DC input, 35-37
  improved architecture, 229-31
    simulation program, 231
    simulation results, 231-32
  modulo arithmetic comb filters in, 457-60
  MOS ADC-filter combination, 317-19
  noise spectra and signal-to-noise ratio, 38-41
  nonlinear behavior in, 33-43
  nonlinear quantizer, modeling of, 35-41
  optimization by use of a slow ADC, 209-12
  sinusoid input and nonlinearity modeling, 37-38
  stability analysis, 41-42
  table-based simulation of, 163-67
  thresholding in, 379-81
  triangularly weighted interpolation, 377-84
  using multibit quantizers, 16-18

*See also* Oversampled A/D converters; Oversampled D/A converters
Digital filters, 461-66
  baseband filter, 463
  characteristics, 466
  design chart, using, 462-63
  design method, 462-63
    performance levels, range of, 462
  example, 463
  implementation, 464-65
  optimization method, comparison with, 465
  requirements, 461-62
  resampling, 463-64
    examples, 464
    general principle, 463-64
Digitally corrected multi-bit sigma-delta data converters, 192-95
  A/D converter, 193
  D/A converter, 192-93
  experimental results, 193-94
  self-calibration of, 193
Digital modulation, 2-12
  delta-sigma modulation, 3-7
  high-order modulators, 7-10
  quantization, 2-3
Digital section, dual-channel voice-band PCM codec, 313-15
Digital signals:
  FIR filter design, 429-38
  interpolation/decimation of, 417-48
  quantizing, 23
  sample rate conversion, multistage implementations of, 438-46
  sampling rate conversion, basic concepts of, 418-23
  signal processing structures for decimators/interpolators, 423-29
  comparisons of structures, 429
  direct form FIR structures for integer changes in sampling rates, 424-25
  FIR structures with time-varying coefficients, 428-29
  polyphase FIR structures for integer decimators and interpolators, 425-28
  signal-flowgraphs, 423-24
Dither, 91-93, 185-87
  A/D converter system, modeling, 186
  dither distribution vs. quantizer error, 185-86
  and output truncation, 187
  quantization error:
    modeling, 186
    PSD of, 186-87
Double integration in sigma-delta modulation, 174-83
  decimator, design of, 181-82
  digital processor, design of, 179-80
  feedback, penalties for using, 176-77
  limit cycles that overload the quantizer, 178-79
  method, 175-76
  modulation noise, 180-81
  quantization with feedback, 174-75
  toll network use requirement, 175
  two integrators, use of, 179
  two-level quantization, 177-78
Double integrator, 118
Double interpolation for D/A conversion, 477-81
  circuit parameters, precision of, 480-81
  first-order interpolation, 477-78
  measurements on a circuit model, 479
  second-order interpolation, 478-79
Double-loop sigma-delta modulation, 106-14
  discrete-time model, 107-8
    difference equation solution, 108
    difference equations, 107
    stability problems, 107-8
  long-term time average properties, 108-11
    fundamental result, 109-10

preliminaries, 108-9
properties of various processes, 110-11
system performance, 111-12
comparisons, 112
optimum FIR filter, 111
sigma-delta quantization noise, 111
sinc filters, 111-12
Dual-channel voice-band PCM codec, 311-16
analog section, 312-13
block diagram, 311-12
digital section, 313-15
digital sigma-delta modulator, 315
IIR filters, 314-15
sinc cube decimator, 313-14
experimental results, 315-16
Dynamical system, 76-77

**Elliptical filter design techniques, 204**
Equiripple (optimal) FIR designs, 433-35
Error feedback, 10-11
quantization with, 23
Esaki diode, 122-24

**Feedback:**
penalties for using, 176-77
quantization with, 174-75
reducing quantization noise by use of, 131-38
Feedback loop, net gain in, 5-6
Finite op-amp gain, interpolative modulators, 201
FIR filter design, 429-38
architectural design, 468-69
generating the coefficients, 468
performing the direct convolution, 468-69
decimation, realization for, 467-68
equiripple (optimal) FIR designs, 433-35
filters based on window designs, 432-33
half-band FIR filters, 435
ideal frequency domain characteristics, 431-32
minimum error in frequency domain, 437-38
minimum mean-square-error design, 435-36
optimum FIR filter, 111
procedures, 432
prototype filter/polyphase representation, relationship between, 430-31
time-domain properties of ideal interpolation and decimation filters, 432
First-order delta-sigma modulators, 13-14, 249
First-order feedback circuit, 3
First-order incremental A/D converter, 250-51
First-stage decimator, design, 19-20

**Half-band FIR filters, 435, 461**
Higher-order cascade modulators, 16
Higher-order single-stage modulators, 15-16
High-order modulators, 7-10, 196-205
architectures for, 265-69
multistage sigma-delta modulators, 266-67
dynamic range of, 8-9
in-band values of quantization error, predicting in, 7

noise in, 7-8
noise shaping, 10
second-order modulators, influence of circuit parameters on, 9
third-order modulators, limit cycles in, 9-10
High-order one-bit sigma-delta A/D converters, 206-8
experimental results, 208
theory, 206-8
High-order one-bit sigma-delta modulators:
design of, 127-30
examples, 128-29
modulator analysis, 127-28
modulator structures, 128
optimized NTF, 128

**Integration:**
oversampled A/D converters, 191
range of, 9
Integrators, leakage in, 6-7, 9
Interpolation, 377-79
analysis of, 378-79
triangularly weighted, 377-84
Interpolative modulators, 196-205
elliptical filter implementation, 204
experimental results, 203-4
Nth-order topology, 198-200
loop coefficients, design of, 199-200
stability, 199
system function, 198-99
quantization noise, 200
simulation results, 200-203
coefficient errors, 202
DC idle-channel noise, 202-3
finite op-amp gain, 201
integrator settling time, 207-8
limited op-amp swing, 202

**Limit cycle oscillations, 44-51**
arbitrary loop gain, analysis of system with, 49-51
circuit parameters, effect on signal, 46-48
operating modes, 45
pulse-code modulation (PCM), circuit for, 44-45
resolution, 45-46
time-shared operation, 48-49
timing circuit operations, 48
Limited op-amp swing, interpolative modulators, 202
Low-pass filter, 388-92
and decimation of modulated signal, 21
ideal low-pass filter decoder, 69

**MASH modulator circuit configuration, 14**
for oversampled A/D converters, 262-63
for oversampled D/A converters, 263-64
Modulation noise, 3, 7-8, 180-81
Modulator architectures, 13-18
criteria for choosing the architecture, 13
oversampling modulator architectures, 13-18
scaling signal amplitudes in modulators, 18

Modulo arithmetic comb filters, 457-60
  architectural considerations, 457-58
  VLSI implementation, 458
MOS A/D converter filter combination, 317-19
MOS delta-sigma A/D converters, 213-18
  capacitor linearity, 216
  op amp requirements, 216-17
  oversampling converters in MOS, 213-14
  performance limits due to noise, 215-16
  ratio- and parasitic-insensitivity, 214-15
  simulation, 216
$M$th-order incremental A/D converter, 252
Multi-bit quantizers, delta-sigma modulators using, 16-18
Multi-bit sigma-delta data converters, digitally corrected, 192, 227-28
Multichannel oversampled PCM voice-band coder, 333-39
  coder implementation, choice of, 333-35
  experimental results, 337-38
  prototype, 335-37
Multilevel quantization, 58
Multirate filter designs, 405-16
  multistage, 405-7
    examples/comparisons, 412-16
    FIR design, 412
    IIR design, 411-12
  using comb filters:
    stability/arithmetic of, 409-11
    structures, 407-9
Multistage decimation, 18-19
Multistage delta-sigma modulator, without double integration loop, 320-21
Multistage sigma-delta modulation, 60-72, 266-67
  average quantization noise power, 71-72
  binary quantizer error, 64-69
    moments of, 65-69
  decoding filters, 69-70
    ideal low-pass filter decoder, 69
    $sinc^K$ filter decoder, 69-70
  difference equation and linear network, 61-64
  digital correction, 267
  multiple sinusoids, inputs with, 71
  simulation results, 70

Noise shaping, 174
  high-order modulators, 10, 23
Noise-shaping coders of order $N > 1$, 139-49
  design, 148-49
  optimum oversampled higher order linear noise-shaping coder structures,
    43-44
  optimum oversampled higher order linear predictive coder structures,
    141-43
  oversampled linear feedback coder structures, 139-41
  predictive coder design, 146-48
  third-order predictive coder, 144-46
Noise-shaping coder topology, 304-5
  design example, 307-10
    results, 309-10
    system topology, 307
    three-bit internal D/A converter, 307-9
  for 15+ bit converters, 304-10
  internal D/A converter topology, 305-7
    dynamic element matching approach, 306-7
    randomizer, 307
Nonlinear behavior, in delta-sigma modulators, 159-61, 297-98, 344
Nonlinear quantizer, modeling of, 35-41
$N$th-order topology, 198-200

loop coefficients, design of, 199-200
quantization noise, 200
stability, 199
system function, 198-99

**Output truncation, and dither, 187**
Oversampled A/D converters:
  advantage of, 196
  constraints analysis, 270-75
    integrator settling and quantizer uncertainty effects, 274-75
    with lossy integrator, 273-74
  design, 293-303
    electronic noise, 295
    implementation, 299-301
    integrator and comparator design, 296-99
    sampling jitter, 295-96
    second-order sigma-delta modulator, 294
    signal range, 294-95
  for digital audio, 340-41
  with digital error correction, 227-28
  dither, 185-87
DPCM loop, 187-89
  random errors, 188
  start-up problem, 188-89
  tracking error, 188
  experimental results, 189-90, 301-2
  implementation, 299-301
    circuit topology, 299-300
    comparator design, 301
    integrator design, 300
  implementations/applications, 277-374
  integration, potential for, 191
  integrator and comparator:
    bandwidth, 297
    comparator hysteresis, 298-99
    design, 300-301
    gain variations, 296
    leak, 296-97
    nonlinearity, 297-98
    slew rate, 297
  interpolative modulators for, 196-205
  MASH modulator circuit configuration for, 262-63
  multistage noise shaping modulator performance, 261-62
  multistage noise shaping modulator quantization noise, 259-61
  performance of, 168-69
  simulating/testing, 168-73
    comparison with other techniques, 171-72
    practical considerations, 170-71
  sinusoidal minimum error method, 169-70
  smoothing signal, 189
  structures, 270-72
  switched capacitor integrator models, 272-74
    integrator without offset cancelling, 273
    offset cancelling integrator, 272-73
  thirteen-bit ISDN-band oversampled A/D converter, 233-36
  topology, 184-85
  wave digital decimation filters in, 449-52
Oversampled D/A converters, 259-65
  MASH modulator circuit configuration for, 263-64
  multistage noise shaping modulator performance, 261-62
  multistage noise shaping modulator quantization noise, 259-61
Oversampled linear feedback coder structures, 139-41
Oversampled sigma-delta modulation, 73-80, 106

496

quantization noise, 75-76
quantizer performance, 78-79
using two-stage fourth-order modulator, 322-25
Oversampling converters, theory of, 279-80
Oversampling methods, 1-25
    D/A converters, 21-24
        demodulator stage, 23-24
        demodulating signals at elevated word rates, 21-22
        interpolating with $sinc^K$-shaped filter functions, 22-23
    decimating the modulated signal, 18-21
    digital modulation, 2-12
    modulator architectures, 13-18
    popularity of, 1
    resolution, 24-25
Oversampling modulator architectures, 13-18
    delta-sigma modulators using multibit quantizers, 16-18
    first-order delta-sigma modulators, 13-14
    higher-order cascade modulators, 16
    higher-order single-stage modulators, 15-16
    second-order cascaded modulators, 14-15
    second-order delta-sigma modulators, 14

**Pattern noise, 4-5**
PCM quantization noise, 87-91
Performance evaluation, sigma-delta A/D converter, 348-50
Periodic noise, 119-20
Predictive coder, design, 146-48
Pulse-code modulation (PCM), circuit for, 44-45

**Quantization, 2-3**
    two-level, 177-78
Quantization error, 84, 132-34
    modeling, 186
    predicting i-band values of, 7
    PSD of, 186-87
    vs. dither distribution, 185-86
Quantization levels, alignment of, 48
Quantization noise, 75-76, 87-91, 116-18, 200
    interpolative modulators, 200
    oversampled sigma-delta modulation, 75-76
    reducing by use of feedback, 131-38
        error spectrum, 132-34, 137-38
        optimization of system, 134-37
    structure of, 52-59
    in uniform quantizers, 75
Quantization noise spectra, 81-105
    dithering, 91-93
    extensions, 103-4
    PCM quantization noise, 87-91
    second-order sigma-delta modulation, 102-3
    single-loop sigma-delta modulator, 93-100
    two-stage sigma-delta modulation, 100-102
    uniform quantization, 81, 84-87
Quantization thresholds, positioning, 6

**Random noise, 58**
Reduction, sample rate, 418-20
Running-sum filter, 323-24
    filter processor, 324

**Sampling rate conversion, 418-23**
    by bandpass signals, 422-23
    by rational factor M/L, 421-22
    increase, 420-21
    multistage implementations of, 438-46
        designs based on specific family of filter designs, 445-46
        half-band designs, 442-43
        multistage optimization procedures, designs based on, 443-45
        parameter specifications, 441-42
        two-stage structure, computational efficiency of, 439-40
    reduction, 418-20
Scaling signal amplitudes, in modulators, 18
Second-order cascaded modulators, 14-15
Second-order delta-sigma modulators, 14, 102-3
    circuit tolerances, 9
Second-order incremental A/D converter, 249-54
    compared to sigma-delta converters, 253
    experimental results, 253-54
    offset and charge injection compensation, 252-53
Second-order interpolative modulator, 197-98
Second-order modulators, influence of circuit parameters on, 9
Settling time:
    interpolative modulators, 207-8
    oversampled A/D converters, 274-75
        16-bit oversampling A/D converters, 239-40
Signal-to-noise ratio, 38-41, 379
    numerical calculation of, 39-40
    optimization of, 212
    using tri-level delta-sigma modulation, 245-48
Signal-to-quantization-noise ratio, 82, 120
$Sinc^K$ filter decoder, 69-70
Sinc decimators, implementing, 20-21
Sinc filters, 111
Single-loop sigma-delta modulation, 93-100, 106
    DC input, 94-96
    sinusoidal inputs, 96-100
Sinusoidal minimum error method, 169-70
16-bit fourth order noise-shaping D/A converter, 482-85
    experimental results, 484-85
    system topology, 482-83
    3-bit D/A converter topology, 483-84
16-bit oversampling A/D converters, 237-44
    circuit considerations, 241-42
        analog circuit configuration and operation, 241-42
        block diagram, 241
        digital filter configuration and operation, 242
    measurement characteristics, 242-44
    practical accuracy limiting factors, 239-40
        capacitance-matching tolerance, 240
        integrator gain and settling speed, 239-40
        reducing noise from digital circuits, 240
    theoretical accuracy limiting factors, 238
    three-stage MASH operation, 238-39
Stereo 16-bit delta-sigma A/D converter, 365-74
    architecture, 368
    delta-sigma conversion, 365-68
        filtering and decimation, 367-68
        loop analysis, 366-67
        loop filters, 367
        quantization noise, 365-66
    design, 369
    digital filter output, 371-73
    discrete-time implementation, 368-69
    filter/decimator, 370-71
        anti-aliasing filtering, 371

decimation, 370
  passband shaping, 370-71
filter order, 369
major characteristics, 368
modulator output, 371
one-bit quantizer, 369
oversampling ratio, 369
simulated performance, 370
specifications, 373
stability, 369-70

input-output characteristics, 120
principle, 115-16
signal-to-noise characteristics, 116-20
  periodic noise, 119-20
  quantizing noise, 116-18
  SNR and signal amplitude, 119
signal-to-quantizing-noise ratio, 120
video signal coding, 122-24

**Table-based simulation of delta-sigma modulators, 163-67**
circuit description, 164
generating tables, 164-65
ZSIM, 163-64
ZSIM simulation results, 165-67
  enlarged switches, 166-67
  increased clock rate, 166
  nominal case, 165-66
Third-order modulators, limit cycles in, 9-10
13-bit ISDN-band oversampled A/D converter, 233-36
architecture, 233-34
implementation, 235
performance, 235-36
Triangularly weighted interpolation:
delta-sigma modulators, 377-84
  circuit arrangement, 383-84
  harmonic distortion and gain tracking, 381
  signal-to-noise ratios, 379
  spectra, 381-82
  thresholding, 379-81
Tri-level delta-sigma modulation:
signal-to-noise ratio using, 245-48
simulation, 246
tri-level coding, 245-46
  implementation of, 246-47
Two-level quantization, 6
Two-stage fourth-order modulator:
oversampled sigma-delta A/D converter circuit using, 322-25
  running-sum filter, 323-24
Two-stage sigma-delta modulation, 100-102

**VLSI-realizable decimators, 471-74**
filter architecture, 473
filter performance, 472-73
problem statement, 471
proposed class of decimators, 471-72
Voiceband codec with digital filtering, 385-99
decoder, 391-94
  demodulator, 393-94
  interpolator, 392-93
  low-pass filter, 391-92
encoder, 385-90
  decimator, 386-88
  high-pass filter, 390
  low-pass filter, 388-90
  modulator, 386
  quantization noise in, 395-98
  simulated encoder response, 390
experimental codec response, 394-95
gain management, 398-99
second-order filter sections, overflow oscillations and roundoff noise in, 398

**Wave digital decimation filters, 449-52**
architectural issues, 451
design example, 451
filter specifications, 450
finite wordlength effects, 450-51
system overview, 449-50
Wordlength, decimation filters, 455

**Uniform quantization, 81, 84-87**
analyzing, 83
quantization error, 84
quantization noise, 75-76, 116-18
Unity bit coding method by negative feedback, 115-26
frequency characteristics, 120

**ZSIM, 163-67**
simulation results, 165-67
  enlarged switches, 166-67
  increased clock rate, 166
  nominal case, 165-66

# Editors' Biographies

**James C. Candy** was born in Crickhowell, South Wales in 1929. He received the B.Sc. and Ph.D. degrees in engineering from the University of North Wales, Bangor in 1951 and 1954 respectively.

From 1954 to 1956 he was with S. Smith and Sons, Guided Weapons Department, Cheltenham, and for the next three years he worked on nuclear instrumentation at the British Atomic Energy Research Establishment, Harwell. In 1959 he came to the United States to take up an appointment as Research Associate at the University of Minnesota. A year later he joined AT & T Bell Laboratories, New Jersey where he has been engaged in research on digital signal processing, including efficient encoding of video and speech signals. He has also been interested in methods for converting signals between digital and analog formats.

Dr. Candy is a fellow of the IEEE. He has published 30 technical papers and holds twenty-nine U.S. patents.

**Gabor C. Temes** received his Dipl.Ing. from the Technical University of Budapest in 1952, his Dipl. Phys. from Eotvos University, Budapest in 1954, and the Ph.D. in Electrical Engineering from the University of Ottawa in 1961.

Dr. Temes was a member of the faculty of the Technical University of Budapest from 1952 to 1956. He was employed by Measurement Engineering Ltd., Arnprior, Ontario, Canada, from 1957 to 1959. From 1959 to 1964 he was with Northern Electric R&D Laboratories, Ottawa, Ontario, Canada. From 1964 to 1966 he was a research group leader at Stanford University, Stanford, California, and from 1966–1969, a Corporate Consultant at Ampex Corporation, Redwood City, California. He is now on the faculty of the University of California, Los Angeles, as a Professor in the Department of Electrical Engineering. Between 1975 and 1979, he was also Chairman of the Department.

Dr. Temes is a former Associate Editor of the *Journal of the Franklin Institute*, a former Editor of the *IEEE Transactions on Circuit Theory*, and a former Vice-President of the IEEE Circuits and Systems Society. He is a Fellow of the IEEE.

In 1968 and 1981, Dr. Temes was a cowinner of the Darlington Award of the IEEE Circuits and Systems Society. In 1981, he received the Outstanding Engineer Merit Award of the Institute for the Advancement of Engineering. In 1982, he was awarded the Western Electric Fund Award of the American Society for Engineering Education and in 1984, he was awarded the Centennial Medal of the IEEE. He received the Andrew Chi Prize Award of the IEEE Instrumentation and Measurement Society in 1985, the Education Award of the IEEE Circuits and Systems Society in 1987, and the Technical Achievement Award of the same Society in 1989. He is coeditor (with S. K. Mitra) and coauthor of *Modern Filter Theory and Design*, Wiley, 1973; coauthor of *Introduction to Circuit Synthesis and Design*, McGraw-Hill, New York, 1977; coauthor of *Analog MOS Integrated Circuits for Signal Processing*, Wiley, 1986, and a contributor to several other edited volumes. He has published approximately 150 papers in engineering journals and conference proceedings.